A-LEVEL CHEMISTRY

E. N. Ramsden

B. Sc., Ph. D., D. Phil.
Wolfreton School, Hull

Stanley Thornes (Publishers) Ltd.

First published 1985 by
Stanley Thornes (Publishers) Ltd,
Old Station Drive,
Leckhampton Road,
CHELTENHAM GL53 0DN

Front cover: Photomicrograph of a smectic liquid crystal.
Courtesy Standard Telecommunication Laboratories, Harlow, England.

British Library Cataloguing in Publication Data
Ramsden, E.N.
A-level chemistry.
1. Chemistry
I. Title
540   QD33

ISBN 0–85950–154–X

Typeset by Grafikon, Brugge, Belgium
Printed and bound in Great Britain at The Bath Press, Avon

CONTENTS

PART 1: THE FOUNDATION

PART 2: PHYSICAL CHEMISTRY

PART 3: INORGANIC CHEMISTRY

CHAPTER 15: PATTERNS OF CHANGE IN THE PERIODIC TABLE

CHAPTER 16: THE NOBLE GASES

CHAPTER 17: HYDROGEN

CHAPTER 18: THE S BLOCK METALS; GROUPS 1A AND 2A

CHAPTER 19: GROUP 3B

PART 4: ORGANIC CHEMISTRY

CHAPTER 25: ORGANIC CHEMISTRY

CHAPTER 26: THE ALKANES

CHAPTER 27: ALKENES AND ALKYNES

CHAPTER 28: BENZENE AND OTHER ARENES

CHAPTER 29: HALOGENOALKANES AND HALOGENOARENES

CHAPTER 30: ALCOHOLS, PHENOLS AND ETHERS

CHAPTER 31: ALDEHYDES AND KETONES

Note Checkpoint question numbers marked with an asterisk are difficult questions. A dagger denotes a question which should be tackled on re-reading, as it involves material which is covered in later chapters.

PREFACE

This text will prepare students for A-level and S-level examinations in chemistry. It includes all the topics covered by the Examination Boards. Some topics are not common to all the Boards and students will need to refer to their particular syllabus for guidance. Each topic is treated from the beginning, without assuming that O-level work has been remembered. The more advanced S-level topics are marked with an asterisk.

I do not wish to prescribe an order in which teachers take classes through their chemistry syllabus. Nevertheless, every book has to have its contents arranged in a linear manner and I have had to order mine. I have presented atomic structure, chemical bonding, equations and equilibria early in the text. Since studies of systems at equilibrium pervade all chemical topics, I have found it necessary to put some qualitative work on equilibria into the introductory section. Once this foundation has been laid, I envisage students being taken through the organic, inorganic and physical sections of the subject simultaneously. Since thermodynamic considerations throw light on much inorganic and organic chemistry, students will find it an advantage to take this chapter early in their course.

At intervals in each chapter, 'checkpoints' are included so that students can pause and test their understanding and, if necessary, revise a section before they pass on to new material. The A-level syllabus covers so much ground that many teachers find it difficult to take their classes through all the material, while still leaving time for practical work. I hope that teachers will be able to allow students to cover parts of the syllabus on their own from this text, assisted by the checkpoints, thus reducing the amount of note-making which needs to be done in class and releasing time for discussion, reinforcement and practical work. Each chapter ends with a searching set of questions, including some from examination papers. At the end of each section is a collection of questions which span the chapters in that section.

The margin carries a summary of the text. On reaching the end of a chapter, a reader can glance back through the summary to see whether he or she has assimilated all the material. If the reader notes any points which need further study, he or she has only to glance at the text alongside the summary to find the relevant passage.

I hope that students will like the technique I have devised for integrating descriptive material with diagrams, so that the reader's eyes do not have to travel constantly to and fro between the diagram and the text. I have used this technique largely in the physical chemistry section of the book. The annotated diagrams have been consumer-tested and approved by sixth formers in my own school.

Much of the detailed inorganic chemistry is summarised in the form of tables and reaction schemes. Students find these helpful for revision. My preference is to take the s block metals first, follow them with the halogens, and then work through the non-metallic elements of Groups 6 and 5 to arrive at Group 4, with its interesting gradation from non-metallic to metallic behaviour, and end with the transition metals. Teachers who prefer a different order will find no difficulty in taking the inorganic chapters in a different sequence.

In my experience, even students with a fair knowledge of the various series of organic compounds find difficulty in tackling problems which require a knowledge of several series of compounds. The method of converting, say, **A** into **D** via the route

$$\mathbf{A} \rightarrow \mathbf{B} \rightarrow \mathbf{C} \rightarrow \mathbf{D}$$

may well be difficult to formulate. I have tackled this problem in stages by summarising at the end of each chapter the relationships between the series of compounds studied in that chapter and those covered in previous chapters. At the end of the section on organic chemistry, all the synthetic routes are summarised in a few reaction schemes. A number of threads which run through the separate chapters of organic chemistry are also drawn together at the end of this section.

There is more to chemistry than the content of any A-level syllabus. I assume that all readers will be following a course of practical work, but I have not found room in the text for instructions for experiments. I would also like to think that A-level students are reading outside the confines of their syllabus. At A-level, their understanding is sufficiently advanced to open the door to the study of many fascinating topics. I would have liked to find more room for topics of general interest to whet students' appetites for finding out more about science in the world outside the school laboratory. I have added postscripts to some chapters to show students a few examples of the importance of chemistry in society (e.g., DDT and aspirin), the balance of economic factors and environmental factors (e.g., the salt-based industries), the history of science (e.g., the 'new gas') and the applications of chemistry (e.g., synthetic fibres). Since I wrote the postscript on liquid crystals, I have read that liquid crystals have now moved into pocket-sized colour television sets. To keep up to date with developments in chemistry, students will have to read newspapers and periodicals. Their advanced chemistry course equips them to understand many of the scientific issues which they will see reported. I hope they will continue to take an interest in scientific topics long after their examinations are behind them.

E. N. Ramsden,
Hull, 1985

Acknowledgements

My task has been made possible through my being able to draw on the counsel of the staff of the Chemistry Department of the University of Hull. I am indebted to Professor R R Baldwin, Professor N B Chapman, Dr P J Francis, Professor G W Gray, FRS, Professor W C E Higginson and Dr J R Shorter for excellent advice. I have been fortunate in receiving once again the guidance of my former supervisor, Professor R P Bell, FRS. My work has benefited from the advice on content and presentation which I received from Dr G H Davies, Dr J J Guy and Dr G H Pratt.

The numerical values in the text have been taken largely from *Chemistry Data* by J G Stark and H G Wallace (John Murray, 1982).

I thank the following examination boards for permission to reprint questions from their papers.
The Associated Examining Board   (AEB)
The Joint Matriculation Board   (JMB)
The Northern Ireland Schools Examination Council   (NI)
The Oxford and Cambridge Schools Examination Board   (O & C)
The Oxford Delegacy of Local Examinations   (O)
The Southern Universities Joint Board   (SUJB)
The University of Cambridge Schools Local Examinations Syndicate   (C)
The University of London School Examinations Council   (L)
The Welsh Joint Education Committee   (WJEC)

The following people and organisations have kindly supplied me with photographs and given permission for their inclusion.

Bristol Uniforms Ltd	Figure 19.3
British Aerospace plc	Figure 19.4
British Petroleum plc	Figures 6.15, 26.1, 26.2 and back cover (left and right)
British Steel Corporation	Figures 24, 19(b), 24.21(b) & (c), 24.33(b) and back cover (centre)
British Telecom plc	Figures 19.5 and 24.4(b)
Capper Pass	Figure 24.28
Chubb Security Services Ltd	Figure 23.5
De Beers Consolidated Mines Ltd	Figure 6.16
ICI plc Agricultural Division	Figures 22.5(b) and 23.6
ICI plc Mond Division	Figures 12.2, 18.5(b) and 18.7(b)
IMI Refiners Ltd	Figure 12.3
Ind Coope Burton Brewery	Figures 19.1 and 19.2
J Allan Cash Photolibrary	Figure 24.23
Perkin Elmer	Figure 34.6
Picturepoint	Figure 33.10
Pilkington Brothers plc	p. 518
RTZ Services Ltd	Figures 24.22 and 24.30
STEAM ICI	Figure 34.2
United Kingdom Atomic Energy Authority	Figures 1.20 and 1.21
Vidocq Photo Library	Figure 24.20

Illustration Acknowledgements

Figures 4.2, 4.3 and 4.13 after H Witte and E Wolfel, *Reviews of Modern Physics*, 30, 51–5, used by permission of the American Physical Society.
Figure 4.8, C A Coulson, *Proc. Cam. Phil. Soc.* 34, 210 (1938) used by permission of the Cambridge Philosophical Society.

Figures 4.12, 15.6, Linus Pauling, *The Nature of the Chemical Bond*, Second Edition (1939), used by permission of Cornell University Press.
Figure 4.25 adapted from Pauling, Corey and Branson *Proc. Natl. Acad. Sci.*, US37, 205 (1951).
Figure 6.24 after G W Gray.

I thank Stanley Thornes (Publishers) for the commitment which they have shown to the production of this volume and my family for the encouragement which has sustained me during its preparation.

E N Ramsden,
Wolfreton School,
Hull.
1985

Part 1

THE FOUNDATION

1

THE ATOM

1.1 THE ATOMIC THEORY

The idea that matter consists of atoms is an ancient one

One of the oldest ideas in science is that matter can be divided and further divided until the smallest possible particles of matter are obtained. This idea was put forward by the Greek philosopher, Democritus, in 400 BC. He called the particles **atoms**. (Greek: *atomos*, indivisible.)

His idea was not accepted. It was not until the eighteenth century that chemists began to explain their results in terms of atoms. They needed to account for the relationships between the masses of reacting substances which they observed in their experiments. All the calculations on reacting masses and volumes of chemicals which you see in Chapter 3 are based on the idea that each chemical element has a characteristic atomic mass.

Dalton advanced the theory, postulating that atoms cannot be created, or destroyed, or divided

In 1808, the British chemist John Dalton, formulated his **Atomic Theory**. He postulated that <u>all matter consists of atoms, minute particles which cannot be created, or destroyed, or split</u> [but see p. 12]. He theorised that all the atoms of an element are identical in every respect, for example their mass [but see p. 8]. Nineteenth-century chemists were quick to accept Dalton's ideas.

1.2 THE SIZE OF THE ATOM

Twentieth-century X ray work [p. 111] has shown that the diameters of atoms are of the order of 2×10^{-10} m, which is 0.2 nm (1 nm = 1 nanometre = 10^{-9} m).

The masses of atoms [p. 8] range from 10^{-27} to 10^{-25} kg. They are often stated in atomic mass units, m_u (where $1 m_u = 1.661 \times 10^{-27}$ kg).

1.3 THE NATURE OF THE ATOM

Around the year 1900, physicists began to find evidence that atoms are made up of smaller particles.

1.3.1 CATHODE RAYS

Sir William Crookes was experimenting in 1895 on the discharge of electricity through gases at low pressure. Using a tube now called a 'Crookes tube' [see Figure 1.1], he observed that at very low pressures (10^{-4} atm), the tube glowed. If the glass at the end of the tube opposite to the **cathode** (the negative electrode) was treated with phosphorescent material, a bright luminescence was seen on the glass. Using his 'Maltese cross tube', he found that if an obstruction was placed in the tube, its shadow appeared in the luminescence. He deduced that a beam of rays, which like light rays travelled in straight lines, was given off by the cathode. Unlike light rays, however, these rays were deflected by a magnetic field and by an electric field. The direction of the deflection, as shown by the movement of the shadow, suggested that the rays were negatively charged. Crookes called the rays **cathode rays**.

Crookes detected cathode rays by means of the luminescence they produce

FIGURE 1.1 Crookes' Maltese Cross Tube

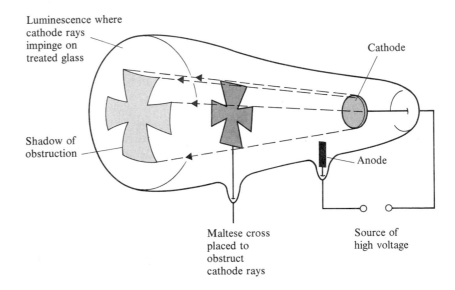

The rays also have the properties of particles

Crookes showed that cathode rays also behave like particles. He made them drive a paddle wheel in the discharge tube. Like particles, cathode rays are able to impart kinetic energy to an object in their path. Cathode rays are negatively charged particles.

1.3.2 ELECTRONS

Sir J J Thomson studied the deflection of cathode rays in electric and magnetic fields; Figure 1.2 shows the kind of apparatus he used. From his measurements he

FIGURE 1.2 Thomson's Apparatus for measuring *e/m*

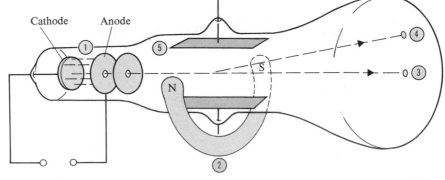

The ratio e/m for cathode ray particles

calculated that the ratio of charge/mass, e/m, was $-1.76 \times 10^{11}\,\text{C}\,\text{kg}^{-1}$ (C = coulomb, the SI unit of charge). Since he obtained the same value, regardless of what gas was used or what kind of electrodes were used, he deduced that these negatively charged particles are present in all matter. They were named **electrons**, and were recognised as the particles of which an electric current is composed.

Thomson's apparatus for finding e/m led to the development of the mass spectrometer, which separates ions according to the value of e/m [see Figures 1.8, 1.9, pp. 9, 10].

THE CHARGE ON AN ELECTRON AND THE MASS OF AN ELECTRON

The charge on an electron and its mass

R A Millikan found the value of the electric charge carried by an electron. His 'oil drop' experiments, carried out from 1909 to 1917, are described in many physics books*. From his experiments, he obtained the value of $-1.60 \times 10^{-19}\,\text{C}$. This amount of charge is called 1 elementary charge unit. Combining this value of charge with Thomson's value of charge/mass gave a value of $9.11 \times 10^{-31}\,\text{kg}$ for the mass of an electron. This is 1/1840 times the mass of a hydrogen atom.

1.3.3 X RAYS

X rays arise when electrons hit a metal

In 1895, W R Röntgen was experimenting with gas discharge tubes. He noticed that when **cathode rays**, that is electrons, impinged on the glass surface of the tube, they gave rise to rays of a different kind. These rays could penetrate material. If the electrons were made to hit a metal target placed in the discharge tube, the emission of penetrating radiation was much stronger [see Figure 1.3]. The penetrating rays were named **X rays** by Röntgen. X rays are now known to be a form of electromagnetic radiation with a wavelength of about $10^{-10}\,\text{m}$. Visible light, radio waves, ultraviolet rays and X rays are electromagnetic radiation, as distinct from particle radiation. Visible light has a wavelength of around $10^{-6}\,\text{m}$. [See Figure 34.4, p. 728.]

FIGURE 1.3 Röntgen's X Ray Apparatus

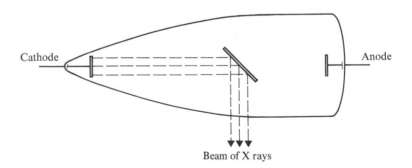

Cathode   Anode

Beam of X rays

1.4 THE ATOMIC NUCLEUS

The Thomson model of the atom

In 1898, Thomson surveyed all the evidence that atoms consist of charged particles. He described an atom as a sphere of positive electricity, in which negative electrons are embedded. Other people described this as the 'plum pudding' picture of the atom!

If this model of the atom is correct, then a metal foil is a film of positive electricity containing electrons. A beam of α particles fired at it should pass straight through. In 1909, Lord Rutherford's colleagues, Geiger and Marsden, tested this prediction [see Figure 1.4].

Geiger and Marsden tested the model...

*See, e.g., R Muncaster, *A-Level Physics* (Stanley Thornes)

FIGURE 1.4 Illustration of Geiger–Marsden Experiment

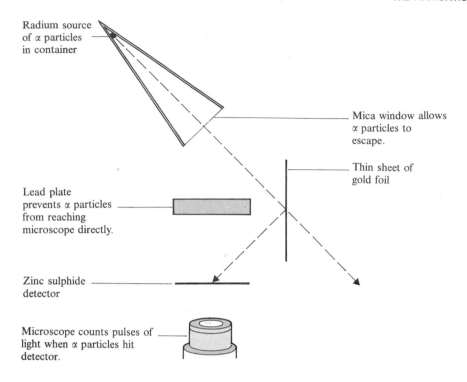

Radium source of α particles in container

Mica window allows α particles to escape.

Thin sheet of gold foil

Lead plate prevents α particles from reaching microscope directly.

Zinc sulphide detector

Microscope counts pulses of light when α particles hit detector.

...their results... They found, as they expected, that α particles penetrated the gold foil. They also found, to their amazement, that a small fraction (about 1 in 8000) of the α particles were deflected through large angles and even turned back on their tracks. Rutherford described this as 'about as incredible as if you fired a 15-inch shell at a piece of tissue paper and it came back and hit you'.

...Rutherford's explanation Rutherford deduced that the mass and the positive charge must be concentrated in a tiny fraction of the atom, called the **nucleus**. Figure 1.5 shows his interpretation of the results.

FIGURE 1.5 Scattering of α particles by the Nuclei of Metal Atoms

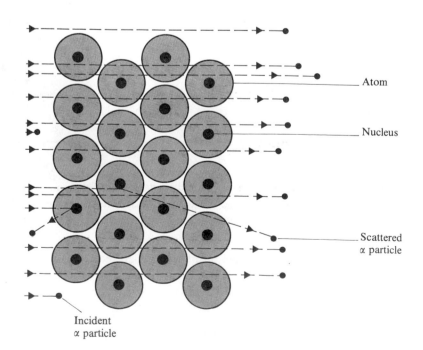

Atom

Nucleus

Scattered α particle

Incident α particle

Only the α particles which collide with the nuclei are deflected; the vast majority pass through the spaces in between the nuclei. The figure is not drawn to scale: a nucleus of the size shown here would belong to an atom the size of your classroom. An atom of diameter 10^{-10} m has a nucleus of diameter 10^{-15} m.

The mass of an atom is concentrated in its nucleus

The model of the atom which Rutherford put forward in 1911 was like the solar system. At the centre was the positively charged nucleus. Around the nucleus moved electrons in circular paths called **orbits**. The size of the orbits determined the size of the atom.

Rutherford's model of the atom

1.5 ATOMIC NUMBER OR PROTON NUMBER

Moseley related wavelengths of X rays to the numerical order of an element in the Periodic Table

H G Moseley in 1913 was investigating the wavelengths of the X rays produced by different metals in a cathode ray tube such as that shown in Figure 1.3. He plotted the reciprocal of the square root of the wavelength $(1/\sqrt{\lambda})$ for the X rays emitted against the order of the element in the Periodic Table. His plot [see Figure 1.6] shows that the numerical order of the element in the Periodic Table represents an important property of the element. Moseley supposed that this was the charge on the nucleus, and that the charge increases by one unit from each element to the next. He called the order of the element in the Periodic Table the **atomic number**, and gave it the symbol Z. It is now called the **proton number**, as explained on p. 8.

He called the numerical order the atomic number Z...

...It is now called the proton number

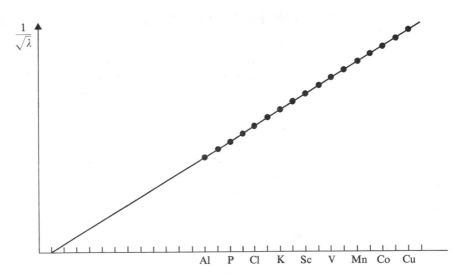

FIGURE 1.6 Moseley's Plot of $1/\sqrt{\lambda}$ against Atomic (Proton) Number

Elements in order of periodic table (now known to be in order of proton number)

1.6 THE NEUTRON

Moseley had postulated that the atomic number of an element is equal to the nuclear charge. He further suggested that the multiple charge on the nucleus arose from the presence of **protons**, which contribute the charge. Since atoms are neutral, the number

The atomic number is the number of protons in the nucleus which equals the number of electrons in the atom

of electrons must be the same as the number of protons. Atomic masses are greater than the mass of the protons in the atom. To make up the extra mass, the existence of **neutrons** was postulated. These particles should have the same mass as a proton and zero charge. The search for the neutron began.

It was a member of Rutherford's team, J Chadwick, who established the existence of the neutron in 1934. He found that beryllium emitted uncharged particles when it was bombarded with α particles (which are helium nuclei). His experiment is illustrated in Figure 1.7.

Neutrons have mass but no charge

FIGURE 1.7 Chadwick's experiment

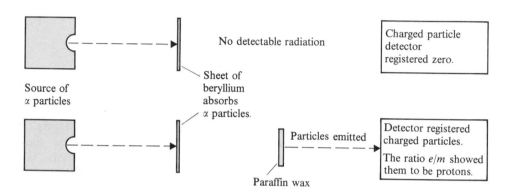

The uncharged radiation that affected paraffin wax was a stream of neutrons. The equation for the reaction between α particles and beryllium is shown below.

1.7 THE FUNDAMENTAL PARTICLES

Proton number or atomic number. Nucleon number or mass number

The nucleus was thus shown to consist of protons and neutrons. The number of protons has been called the **atomic number** [p. 7] but is now known as the **proton number**. Protons and neutrons are both **nucleons**. The number of protons and neutrons is called the **nucleon number**, or, alternatively, the **mass number**.

TABLE 1.1 The Mass and Charge of Sub-atomic Particles

Particle	Charge/C	Relative charge	Mass/kg	Mass/m_u
Proton	$+1.6022 \times 10^{-19}$	$+1$	1.6726×10^{-27}	1.0073
Neutron	0	0	1.6750×10^{-27}	1.0087
Electron	-1.6022×10^{-19}	-1	9.1095×10^{-31}	5.4858×10^{-4}

1.8 NUCLIDES AND ISOTOPES

Notation for nuclides

The word **nuclide** is used to describe any atomic species of which the proton number and the nucleon number are specified. Nuclides are written as $^{\text{nucleon number}}_{\text{proton number}}$Symbol (i.e. $^{\text{mass number}}_{\text{atomic number}}$Symbol). The species $^{12}_{6}$C and $^{9}_{4}$Be are nuclides. Protons are represented as $^{1}_{1}$H, neutrons as $^{1}_{0}$n, α particles as $^{4}_{2}$He and electrons as $_{-1}^{0}$e. Using this notation, the equation for Chadwick's reaction is

Equation for the production of neutrons

$$^{4}_{2}\text{He} + ^{9}_{4}\text{Be} \rightarrow ^{1}_{0}\text{n} + ^{12}_{6}\text{C}$$

Isotopes contain the same number of protons and different numbers of neutrons

When an element has a relative atomic mass [p. 47] which is not a whole number, it is because it consists of a mixture of **isotopes**. Isotopes are nuclides of the same element. They have the same proton number but different nucleon numbers, i.e. they differ in the number of neutrons in the nucleus. Since chemical properties depend upon the nuclear charge and electronic structure of an atom, with mass having little effect, isotopes show the same chemical behaviour. The isotopes of chlorine, $^{35}_{17}Cl$ and $^{37}_{17}Cl$, have the same proton number, 17. The difference between the nucleon numbers shows that one isotope has 18 neutrons and the other has 20 neutrons. The chemical reactions of the two isotopes are identical. Their names can be written as chlorine-35 and chlorine-37. The isotopes of hydrogen differ more than isotopes of other elements [p. 346].

CHECKPOINT 1A: ATOMS

1. 'Dalton was incorrect in saying that all the atoms of a particular element are identical.' Discuss this statement.

2. Why did Chadwick look for the neutron? Why was it hard to find?

3. Point out the importance of Moseley's discovery.

1.9 MASS SPECTROMETRY

In mass spectrometry...

...ions are deflected in a magnetic field

Atomic masses are determined by **mass spectrometry**. The mass spectrometer was developed by F W Aston from J J Thomson's apparatus for measuring the ratio e/m for a particle. In the mass spectrometer, atoms and molecules are converted into ions. The ions are separated as a result of the deflection which occurs in a magnetic field. Figure 1.9 shows how a mass spectrometer operates, and Figure 1.8 is a photograph of an instrument. Figure 1.10 shows a mass spectrometer trace for copper(II) nitrate.

FIGURE 1.8 A Mass Spectrometer

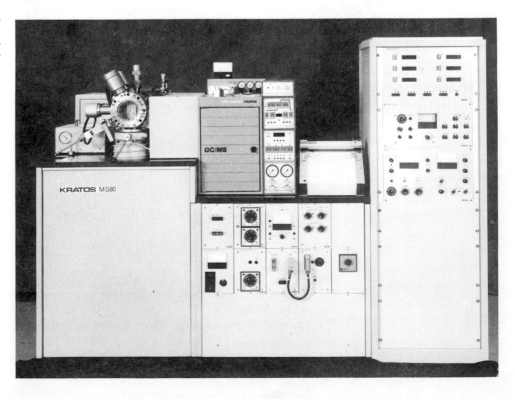

FIGURE 1.9 A Mass Spectrometer: How It Works

4 Electrons pass through a hole in the plate into the ionisation chamber.

5 Trap. Electrons are directed towards the trap by an electric field.

6 The sample is injected as a gas into the ionisation chamber. Electrons collide with molecules of the sample and remove electrons to give positive ions. Some molecules break into fragments. The largest ion is the molecular ion.

3 A voltage is applied to accelerate the electrons towards the plate.

2 Heated filament gives electrons.

1 Pump evacuates instrument.

7 A slit collimates the beam of positive ions.

8 To this plate, a negative potential is applied (about 8000 V). The electric field accelerates the positive ions.

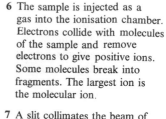
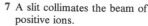
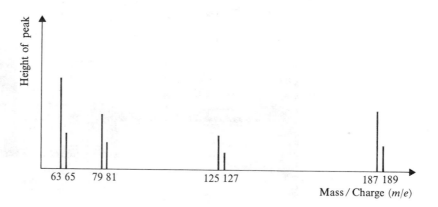

14 Either the accelerating voltage or the magnetic field can be varied. If the magnetic field is kept constant while the accelerating voltage is continuously varied, one species after another is deflected into the ion collector. A trace such as that in Figure 1.10 is obtained.

9 An electromagnet produces a magnetic field. The direction of the field is perpendicular to and coming out of the plane of the paper. The field deflects the beam of ions into circular paths. Ions with a high ratio of mass / charge are deflected less than those with a low ratio of mass / charge.

10 These ions have the correct ratio of mass / charge to pass through the slit and arrive at the collector.

11 Collector. The ions give up their charge.

13 Recorder. The electric current operates a pen which traces a peak on a recording.

12 Amplifier. Here the charge received by the collector is turned into a sizeable electric current.

FIGURE 1.10
The Mass Spectrum of Copper(II) Nitrate

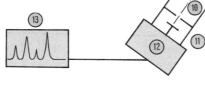

63 65   79 81   125 127   187 189

Mass / Charge (*m/e*)

Notes

(*1*) The height of each peak measures the relative abundance of the ion which gives rise to that peak.

(*2*) The ratio of mass/charge for each species is found from the value of the accelerating voltage associated with a particular peak. Many ions have a charge of +1 **elementary charge unit**, and the ratio *m/e* is numerically equal to *m*, the mass of the ion (1 elementary charge unit = 1.60×10^{-19} C).

...deflection of ion depends on ratio m/e

(3) The peaks on this trace correspond to the ions

$$63 = {}^{63}Cu^+, \ 65 = {}^{65}Cu^+, \ 79 = {}^{63}CuO^+, \ 81 = {}^{65}CuO^+,$$

$$125 = {}^{63}CuNO_3{}^+, \ 127 = {}^{65}CuNO_3{}^+, \ 187 = {}^{63}Cu(NO_3)_2{}^+,$$

$$189 = {}^{65}Cu(NO_3)_2{}^+$$

1.9.1 USES OF MASS SPECTROMETRY

DETERMINATION OF THE RELATIVE ATOMIC MASS OF AN ELEMENT

Mass spectrometry is used for the determination of relative atomic mass

Relative atomic mass, A_r, is defined on p. 47. Figure 1.11 shows the mass spectrum of neon.

FIGURE 1.11
The Mass Spectrum of Neon

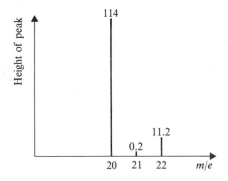

The average atomic mass of neon is calculated as follows.

The calculation of the average atomic mass of neon . . .

Multiply the relative abundance (the height of the peak) by the mass number to find the total mass of each isotope present:

$$\text{Mass of } {}^{22}Ne = 11.2 \times 22.0 = 246.4\,m_u$$

$$\text{Mass of } {}^{21}Ne = 0.2 \times 21.0 = 4.2\,m_u$$

$$\text{Mass of } {}^{20}Ne = 114 \times 20.0 = 2280.0\,m_u$$

$$\text{Totals} = 125.4 = 2530.6\,m_u$$

$$\text{Average mass of Ne} = 2530.6/125.4\,m_u$$

$$= 20.18\,m_u$$

. . . and the relative atomic mass

The average atomic mass of neon is $20.2\,m_u$, and the relative atomic mass is 20.2.

DETERMINATION OF THE RELATIVE MOLECULAR MASS OF A COMPOUND

Mass spectrometry also gives the relative molecular masses of compounds . . .

The ion with the highest value of m/e is the molecular ion, and its mass gives the molecular mass of the compound. If isotopes are present, the average molecular mass and the relative molecular mass [p. 48] are found as in the neon example. Some large molecules (e.g. polymers) are fragmented, and do not give molecular ions [p. 732].

IDENTIFICATION OF COMPOUNDS

. . . and can be used for the identification of compounds . . .

A mass spectrum is obtained, and information about the peak heights and m/e values is fed into a computer. The computer compares the spectrum of the unknown compound with those in its data bank, and thus identifies the compound.

FORENSIC SCIENCE

...and in forensic science where it is useful as a small sample is enough to give results

The sensitivity of the mass spectrometer makes it an admirable tool for forensic scientists. The size of sample which they receive for analysis is often very small. A mass spectrum can be obtained on as little as 10^{-12} g. Small amounts of drugs can be identified by mass spectrometry. A fibre left at the scene of a crime can be compared by mass spectrometry with a fibre from a suspect's clothing.

CHECKPOINT 1B: MASS SPECTROMETRY

1. Describe how, in a mass spectrometer, ions are (*a*) formed, (*b*) accelerated, (*c*) separated and (*d*) detected.

2. Define the terms nucleon number, isotope, relative atomic mass.

Chlorine has two isotopes of relative atomic masses 34.97 and 36.96 and relative abundance 75.77% and 24.23% respectively.

Calculate the mean relative atomic mass of naturally occurring chlorine.

3. The mass spectrum of dichloromethane shows peaks at 84, 86 and 88. The intensities of the lines at 84, 86 and 88 m_u are in the ratio 9 : 6 : 1. What species give rise to these lines? How do you account for the relative intensities of the lines?

*4. Carbon consists of 99% of ^{12}C and 1% of ^{13}C. In the mass spectrum of a hydrocarbon, the peak with the highest mass number occurs at mass $M + 1$. It corresponds to the molecular ion containing one ^{13}C atom. The peak at mass M corresponds to the molecular ion containing only ^{12}C. The peak at mass M is 16.5 times as intense as that at mass $M + 1$. How many C atoms are there in a molecule of the hydrocarbon?

5. Figure 1.12 shows the mass spectrum of zirconium, and Figure 1.13 shows that of lead. The heights of the peaks and the nucleon numbers of the isotopes are shown on the figures. Calculate the average atomic masses of (*a*) zirconium and (*b*) lead.

*Note An asterisk denotes a difficult question.

FIGURE 1.12
The Mass Spectrum of Zirconium

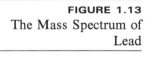

FIGURE 1.13
The Mass Spectrum of Lead

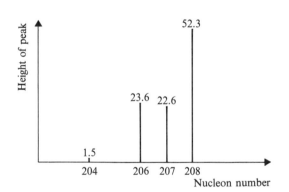

1.10 NUCLEAR REACTIONS

In nuclear reactions new elements are formed

A nuclear reaction is different from a chemical reaction. In a chemical reaction, the atoms which make up the reactants enter into different combinations to form the products, but the nuclei of the atoms are unchanged. In a nuclear reaction, a rearrangement of the protons and neutrons in the nucleus of the atom takes place, and new elements are formed.

1.10.1 RADIOACTIVITY

Some nuclides are unstable and split up to form smaller atoms

A number of elements have atoms which are unstable and split up to form smaller atoms. The nucleus splits, and the protons and neutrons in it form two new nuclei. The electrons divide themselves between the two. Sometimes, protons, neutrons and electrons fly out when the original nucleus divides. The process is called **radioactive decay,** and the element is said to be **radioactive**. The particles and energy are called **radioactivity**. Radioactive isotopes have unstable nuclei.

Radioactive substances give three types of radiation... Three types of radiation are given off by radioactive substances. They all cause certain substances, such as zinc sulphide, to luminesce, and they all ionise gases through which they pass. They differ in their response to an electric field, in the manner shown in Figure 1.14. The uncharged rays, γ (**gamma**) rays, are similar to X rays. They have high penetrating power, being able to pass through 0.1 m of metal. Measurements of e/m identified α (**alpha**) rays as the nuclei of helium atoms and β (**beta**) rays as electrons. β rays can pass through 0.01 m of metal, and α rays can penetrate no more than 0.01 mm of metal.

...α, β and γ rays

FIGURE 1.14 Effect of an Electric Field on Radiation

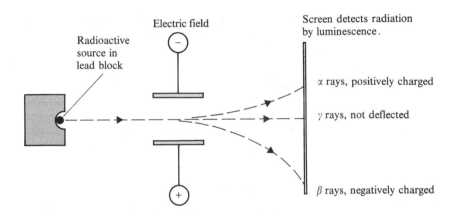

DISCOVERY

In 1896, during the course of his experiments on the fluorescence of uranium salts, A H Becquerel left a wrapped photographic plate in a drawer. On developing the plate, he found that it had been exposed. Since no light could penetrate the wrapping, Becquerel concluded that the plate had been fogged by some rays coming from a uranium salt which was stored in the same drawer. There was no known type of radiation that had this effect.

Becquerel discovered radioactivity. It fogs photographic plates

Marie Curie discovered radium and polonium Marie and Pierre Curie started to investigate this new type of radiation. Marie Curie succeeded in isolating compounds of two new elements from the uranium ore, **pitchblende**. These new elements gave off the sort of radiation which Becquerel had discovered and which we call radioactivity. Marie Curie called her elements **polonium** and **radium**.

1.10.2 NATURAL RADIOACTIVITY

The uranium series There are three naturally occurring series of radioactive elements. The **uranium series** starts with $^{238}_{92}U$ and decays through a series of unstable isotopes to $^{206}_{82}Pb$. The first two steps in the decay are:

$$^{238}_{92}U \rightarrow {}^{234}_{90}Th + {}^{4}_{2}He \qquad (\alpha \text{ decay})$$

$$^{234}_{90}Th \rightarrow {}^{234}_{91}Pa + {}^{0}_{-1}e \qquad (\beta \text{ decay})$$

α decay and β decay   When an isotope undergoes α decay (with the emission of an α particle), its proton number decreases by 2, and its nucleon number decreases by 4. The isotope produced is two groups to the left in the Periodic Table. When an isotope undergoes β decay (with the emission of an electron), its proton number increases by 1, and its nucleon number is unchanged. The isotope produced is one group to the right in the Periodic Table.

$$\underline{\text{Group 4}} \qquad \underline{\text{Group 5}} \qquad \underline{\text{Group 6}}$$

$$^{234}_{90}\text{Th} \xleftarrow{\quad -\alpha \quad} \qquad\qquad ^{238}_{92}\text{U}$$

$$\searrow {-\beta}$$

$$^{234}_{91}\text{Pa} \xrightarrow{\quad -\beta \quad} ^{234}_{92}\text{U}$$

The actinium and thorium series   The other naturally occurring series is the **actinium series**, which decays from $^{235}_{92}\text{U}$ to $^{207}_{82}\text{Pb}$, and the **thorium series**, which starts with $^{232}_{90}\text{Th}$ and ends with $^{208}_{82}\text{Pb}$.

There is a fourth series of radioisotopes, which do not occur in nature but have been made by nuclear reactions.

1.10.3 BALANCING NUCLEAR EQUATIONS

Balancing nucleon (mass) numbers and proton (atomic) numbers   In the equation for a nuclear reaction, the sum of the nucleon numbers (mass numbers) is the same on both sides, and the sum of the proton numbers (atomic numbers) is the same on both sides of the equation. For example, when nitrogen-16 undergoes β decay

$$^{16}_{7}\text{N} \rightarrow {}^{a}_{b}\text{O} + {}^{0}_{-1}\text{e}$$

Considering nucleon numbers gives $16 = a + 0 \therefore a = 16$
Considering proton numbers gives $7 = b + (-1) \therefore b = 8$
The isotope produced is $^{16}_{8}\text{O}$
The equation is $^{16}_{7}\text{N} \rightarrow {}^{16}_{8}\text{O} + {}^{0}_{-1}\text{e}$

1.10.4 ARTIFICIAL RADIOACTIVITY

Radioactivity can be induced   Some nuclear reactions are not spontaneous. They occur when stable isotopes are bombarded with particles such as α particles or neutrons. Rutherford was the first person to bring about a nuclear reaction. He was experimenting on the bombardment of nitrogen with α particles in a cloud chamber of the type invented by C T R Wilson. Figure 1.15 shows the kind of photograph he obtained.

FIGURE 1.15 Drawing of a Cloud Chamber Photograph of a Nuclear Reaction

1 The cloud chamber is filled with air which is supersaturated with water vapour. If any ions are produced, they cause condensation.

2 Radioactive source of α particles

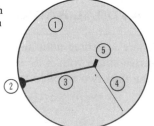

5 Short, thick track of $^{17}_{8}\text{O}$

4 Long, thin track of $^{1}_{1}\text{H}$

3 This track is a trail of condensation produced by an α particle.

Rutherford used bombardment by α particles

Figure 1.15 shows the track of an α particle coming to an end and being replaced by a short, thick track and a long, thin track. Rutherford realised that two particles had been formed in a nuclear reaction. He attributed the short, thick track to $^{17}_{8}O$ and the long, thin track to $^{1}_{1}H$. He proposed that they had been formed by the nuclear reaction

Nuclear reactions

$$^{14}_{7}N + ^{4}_{2}He \rightarrow ^{17}_{8}O + ^{1}_{1}H \qquad (\alpha, p)$$

... (α, p)

This is classified as an (α, p) reaction since the projectile is an α particle and a proton (p) is produced in the reaction.

Other bombarding particles were used, and more nuclear reactions were observed. Examples are

... (α, n)

$$^{9}_{4}Be + ^{4}_{2}He \rightarrow ^{12}_{6}C + ^{1}_{0}n \qquad (\alpha, n)$$

... (p, α)

$$^{7}_{3}Li + ^{1}_{1}H \rightarrow 2^{4}_{2}He \qquad (p, \alpha)$$

... (n, α)

$$^{16}_{8}O + ^{1}_{0}n \rightarrow ^{13}_{6}C + ^{4}_{2}He \qquad (n, \alpha)$$

Elements with Z greater than 92 are artificially made

Neutrons (n) have the advantage over α particles and protons in that, being uncharged, they are not repelled by the positive nuclei of the bombarded atoms. Since 1940, a set of new elements with proton numbers greater than 92, the proton number of the heaviest naturally occurring element, uranium, have been made. They are called the **transuranium elements**. The element neptunium is made by neutron bombardment of $^{238}_{92}U$, followed by radioactive decay of the isotope formed:

$$^{238}_{92}U + ^{1}_{0}n \rightarrow ^{239}_{92}U$$

$$^{239}_{92}U \rightarrow ^{239}_{93}Np + ^{0}_{-1}e$$

Another new element, plutonium, was formed by the decay of neptunium-239:

$$^{239}_{93}Np \rightarrow ^{239}_{94}Pu + ^{0}_{-1}e$$

It is unstable and decays to form the stable isotope, uranium-235:

$$^{239}_{94}Pu \rightarrow ^{235}_{92}U + ^{4}_{2}He$$

The first artificially made radioisotope

The first artificially produced **radioisotope** was made by Irene Curie and J Joliot in 1934 by an (α, n) reaction. An isotope of boron was converted into a radioactive isotope of nitrogen. This decayed by emitting **positrons** (positive electrons):

$$^{10}_{5}B + ^{4}_{2}He \rightarrow ^{13}_{7}N + ^{1}_{0}n \quad (\alpha, n)$$

$$^{13}_{7}N \rightarrow ^{13}_{6}C + ^{0}_{+1}e \qquad \text{(positron emission)}$$

1.10.5 RATE OF RADIOACTIVE DECAY

The rate of radioactive decay is proportional to the number of radioactive atoms present

The half-life is the time taken to decay to half the number of radioactive atoms

The rate at which a radioactive isotope decays cannot be speeded up or slowed down. It depends only on the identity of the isotope and the amount of isotope present. The nature of **nuclear decay** is illustrated in Figure 1.16. The time taken for a number N_0 of radioactive atoms to decay to $N_0/2$ atoms is called the **half-life**, $t_{1/2}$, of the radioactive isotope. The times taken for $N_0/2$ atoms to decay to $N_0/4$ and for $N_0/4$ atoms to decay to $N_0/8$ atoms are the same and have the same value as $t_{1/2}$. The rate of decay is thus proportional to the number of atoms present. Such reactions are described as **first-order** reactions [p. 296].

FIGURE 1.16
Radioactive Decay

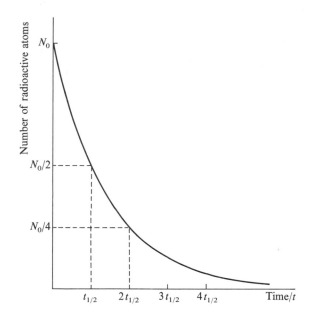

1.10.6 METHODS OF DETECTING AND MEASURING RADIOACTIVITY

WILSON CLOUD CHAMBER

Radioactivity is detected by means of a cloud chamber...

The tracks of condensation produced by ionising radiations in the Wilson cloud chamber can be photographed [see Figure 1.15, p. 14].

THE GEIGER–MÜLLER TUBE

...and measured in a Geiger–Müller counter...

H Geiger and F M Müller invented the device shown in Figure 1.17. It enables the particles emitted in radioactive decay to be counted as pulses of electric current.

FIGURE 1.17
A Geiger–Müller Tube

1 Liquid under investigation sends β particle or γ ray into the Geiger-Müller tube through a thin-walled glass tube.

2 Geiger-Müller tube contains a gas under reduced pressure. Two electrodes are at a voltage just less than that which will allow an electric current to pass. Each time radiation ionises the gas for a fraction of a second, a pulse of electric current flows.

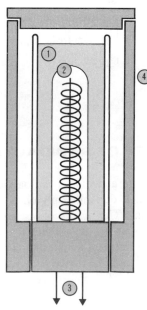

To scale counter

4 Lead 'castle' shields Geiger-Müller tube from background radiation.

5 For α particle emitters, a different design of counter is used. A very thin window is used to allow penetration.

3 The pulses are fed into a loudspeaker to produce a series of clicks, or are counted on an electronic counter.

PHOTOGRAPHIC FILM

...or by a photographic film...

Photographic film is exposed by radioactivity. It was this which led to the discovery of radioactivity by Becquerel. It is used as a protective device in the pieces of photographic film worn by people who work with radioactivity. The 'fogging' of the photographic film is a measure of the amount of radioactivity to which they have been exposed.

SCINTILLATION COUNTER

...or in a scintillation counter

There are substances, such as zinc sulphide, which phosphoresce when affected by radioactivity. As each particle from the radioactive source hits the **phosphor**, a flash of light is emitted. In a scintillation counter, each flash of light gives rise to a pulse of current. A digital counter records the pulses of current.

1.10.7 USES OF RADIOACTIVE ISOTOPES

Radioactivity is used to destroy cancer cells...

1. Cancerous tissue is destroyed by radioactivity in preference to healthy tissue. A cobalt-60 source (a γ emitter, $t_{1/2} = 5$ years) is used to irradiate cancer patients. The dose which the patient receives must be carefully calculated to destroy only the cancer cells without harming the patient's healthy tissues.

...is used in surgery...

2. Surgical instruments can be sterilised more effectively by radioactivity than by boiling.

...and in factories

3. A production line use for radioactivity is to check whether cans have been correctly filled [see Figure 1.18].

FIGURE 1.18
Monitoring a Production Line

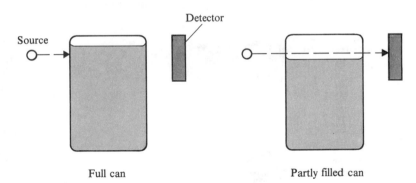

Full can   Partly filled can

Radioactivity reaches the detector because the can is not full. The detector can be made to operate an arm on the production line to reject the can.

4. Figure 1.19 shows a method of using a radioactive source and detector to regulate the thickness of aluminium foil.

Radioactivity is also used for the detection of leaks...

5. Underground leaks in water or fuel pipes can be detected by introducing a short-lived radioisotope into the pipe. The level of radioactivity on the surface can be monitored. A sudden increase of surface radioactivity shows where water or fuel is escaping.

...for measuring engine wear...

6. Engine wear can be measured by using radioactive piston rings. As the piston rings wear away, the lubricating oil becomes radioactive. In this way, the efficiency of various lubricating oils can be tested.

FIGURE 1.19
Monitoring the
Thickness of Metal Foil

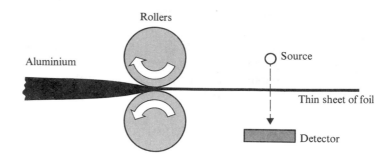

Rollers

Aluminium

Source

Thin sheet of foil

Detector

If the amount of radiation reaching
the detector increases, the detector
operates a mechanism for moving
the rollers further apart, and vice
versa.

...in carbon-14 dating...

7. Carbon-14 dating can be used to calculate the age of plant and animal remains.
Living plants and animals take in carbon, which includes a small proportion of the
radioactive isotope carbon-14. When a plant or animal dies, it takes in no more
carbon-14, and that which is already present decays. The rate of decay decreases
over the years, and the activity that remains can be used to calculate the age of the
plant or animal material [see example, p. 301].

...in medicine...

8. Tracer studies use radioactive isotopes to track the path of an element through
the body. Radioactive iodine (iodine-131) is administered to patients with defective
thyroids to enable doctors to follow the path of iodine through the body. As the
half-life is only 8 days, the radioactivity soon falls to a low level. [See Figure 1.20.]

FIGURE 1.20
Medical Uses of
Radioisotopes. Injection
of a Short Half-life
Radioactive Isotope and
a Miniature Nuclear
Battery for a Heart
Pacemaker

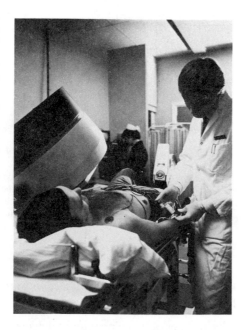

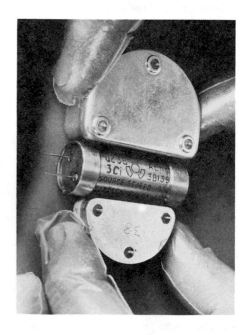

...in analysis...

9. Dilution analysis is the name given to the use of labelled compounds in analysis.

Example   The problem is to find the mass of a substance **X** present in a mixture.
An isotopically labelled form of **X** has an activity of 3×10^6 cpm g^{-1} (counts per
minute per gram). 1 μg of this tracer is added to the mixture. After thorough mixing,

a pure specimen of **X** is isolated from the mixture. It is assayed and found to have an activity of $30\,\text{cpm}\,\text{g}^{-1}$.

The dilution of radioactivity is $1/10^5$; therefore $1\,\mu\text{g}$ is present in $100\,\text{mg}$ of the specimen. The specimen contains $100\,\text{mg}$ of the substance **X**.

...in elucidating structure...

10. Structural studies sometimes call on the use of radioisotopes. A question asked about the structure of the thiosulphate ion, $S_2O_3^{2-}$, was whether the two sulphur atoms occupy equivalent positions in the ion. A radioactive isotope of sulphur can be used in the preparation of thiosulphate:

$$^{35}\text{S(s)} + \text{SO}_3^{2-}(\text{aq}) \rightarrow {}^{35}\text{SSO}_3^{2-}(\text{aq})$$

When dilute acid is added to the thiosulphate formed, sulphur is precipitated, and it is found that all the radioactivity is present in the precipitate of sulphur and none in the sulphur dioxide:

$$^{35}\text{SSO}_3^{2-}(\text{aq}) + 2\text{H}^+(\text{aq}) \rightarrow {}^{35}\text{S(s)} + \text{SO}_2(\text{g}) + \text{H}_2\text{O(l)}$$

The two sulphur atoms in the thiosulphate ion cannot occupy the same type of position in the ion.

Further work has shown that the structure of the thiosulphate ion is

11. Mechanistic studies sometimes employ radioisotopes.

...and in studies of reaction mechanisms

The path of a labelled atom in a molecule can be followed through a sequence of reactions [see esterification, p. 697].

12. People who work with radioactive materials take precautions to ensure that they do not receive a high dose of radiation. A radioactive source is surrounded by a wall of lead bricks, except for an outlet through which a beam of radiation can emerge. To handle a powerful source of radiation, people use long-handled tongs [see Figure 1.21]. Since radioactivity fogs photographic film, workers exposed to radioactivity wear badges containing film, which is examined periodically so that the dose of radiation they are receiving can be monitored.

CHECKPOINT 1C: NUCLEAR REACTIONS I

1. Write the symbols for the isotopes of chlorine (proton number 17, nucleon numbers 35 and 37).

2. If 8 g of a radioactive isotope decay in a year to 4 g, will 6 g of the same isotope decay to 2 g in the same time? Explain your answer.

3. Supply the missing proton numbers and nucleon numbers:

(a) $^{14}_{6}\text{C} \rightarrow \text{N} + {}_{-1}^{0}\text{e}$

(b) $_{10}\text{Ne} \rightarrow {}^{19}\text{F} + {}_{+1}^{0}\text{e}$

(c) $_{88}\text{Ra} \rightarrow {}^{4}_{2}\text{He} + {}^{222}\text{Rn}$

(d) $^{73}\text{As} + {}_{-1}^{0}\text{e} \rightarrow {}_{32}\text{Ge}$

(e) $^{24}_{12}\text{Mg} + {}^{4}_{2}\text{He} \rightarrow \text{Si} + {}^{1}_{0}\text{n}$

(f) $^{19}_{9}\text{F} + \rightarrow {}^{16}_{7}\text{N} + {}^{4}_{2}\text{He}$

1.10.8   MASS CHANGES IN NUCLEAR REACTIONS; BINDING ENERGIES

The mass of the nucleus is less than the sum of the nucleon masses...

The mass of a nucleus is slightly less than the sum of the masses of the protons and neutrons of which it is composed. The difference in mass, which is called the **mass defect**, is transformed into the **binding energy** of the nucleus. The binding energy can be defined as the energy required to separate a nucleus into individual nucleons. The connection between mass and energy is given in Einstein's equation

$$E = mc^2$$

...The mass defect is the source of the binding energy of the nucleus

where E = energy released, m = loss in mass, and c = velocity of light. Since the constant c^2 has a large numerical value, even a very small loss in mass is equivalent to the loss (or release) of a large amount of energy. This is the origin of the substantial binding energies of atomic nuclei, and it is also the reason why nuclear reactions are such an important source of energy.

Example   Calculate the binding energy of the beryllium nucleus in J atom^{-1} and J nucleon^{-1}. ($^9_4$Be = $9.012\,200\,m_u$, $^1_1$H = $1.007\,825\,m_u$, $^1_0$n = $1.008\,665\,m_u$, $c = 2.998 \times 10^8\,\mathrm{m\,s^{-1}}$, $1\,m_u = 1.661 \times 10^{-27}\,$kg. The masses of $^9_4$Be and $^1_1$H are for the whole atoms, including electrons.)

Method

Calculating binding energy

A $^9_4$Be atom contains 4 protons, 4 electrons and 5 neutrons

Mass of 4 protons + 4 electrons = $4\,^1_1$H = $4.031\,300\,m_u$

Mass of 5 neutrons = $5.043\,325\,m_u$

Total mass of particles = $9.074\,625\,m_u$

Mass of $^9_4$Be atom = $9.012\,200\,m_u$

Mass defect = $0.062\,425\,m_u$

Since $E = mc^2$,

$E = 0.062\,425 \times 1.661 \times 10^{-27} \times (2.998 \times 10^8)^2$

$\quad = 9.32 \times 10^{-12}\,$J

The binding energy is $9.32 \times 10^{-12} \mathrm{J\,atom^{-1}}$ or $5.61\,\mathrm{TJ\,mol^{-1}}$.

Binding energy per nucleon $= 9.32 \times 10^{-12}/9 = 1.04 \times 10^{-12}\,\mathrm{J\,nucleon^{-1}}$.

The calculation in the example above can be done for other nuclides. Figure 1.22 shows a plot of binding energy per nucleon against mass number.

FIGURE 1.22 Graph of Binding Energy per Nucleon against Mass Number

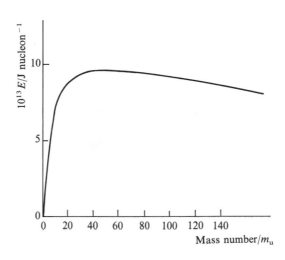

Nuclei of mass greater than $60\,\mathrm{m_u}$ should be able to split...

...lighter nuclei should be able to combine

You can see from the graph that elements with mass numbers of around 60 are the most stable. Elements with nuclei heavier than this should be able to split up to form lighter, more stable nuclei with the release of energy. Elements with nuclei lighter than $60\,m_\mathrm{u}$ should be able to combine, if the repulsion between nuclear charges can be overcome, to form heavier nuclei with the release of energy. These processes are known as **fission** and **fusion** respectively.

1.10.9 NUCLEAR FISSION

The first person to obtain energy from '**splitting the atom**' was O Hahn, in 1937. He bombarded uranium-235 with neutrons. Atoms of ^{235}U split into two smaller atoms and two neutrons, with the release of energy:

Splitting the atom of $^{235}_{92}U$

$$^{235}_{92}\mathrm{U} + {}^{1}_{0}\mathrm{n} \rightarrow {}^{144}_{56}\mathrm{Ba} + {}^{90}_{36}\mathrm{Kr} + 2{}^{1}_{0}\mathrm{n}$$

Mass is converted into energy

The sum of the masses of the fission products is less than the mass of the $^{235}_{92}U$ atom

If the sample of uranium-235 is smaller than a certain size, called the **critical mass**, neutrons will escape from the surface. In a large block of uranium-235, neutrons are more likely to meet uranium-235 atoms and produce fission than to escape. Since each nuclear fission produces two neutrons, as shown in Figure 1.23, a chain reaction is set up. Each time an atom of uranium-235 is split, the mass of the atoms produced is $0.2\,m_\mathrm{u}$ less than the mass of an atom of $^{235}_{92}$U. The lost mass is converted into energy. This is where the energy of the atomic bomb comes from. The atomic bomb consists of two blocks of uranium-235, each smaller than the critical mass. On detonating the bomb, one mass is fired into the other to make a single block larger than the critical mass. The detonation is followed by an **atomic explosion**.

FIGURE 1.23

The Chain Reaction in
Fission of Uranium-235

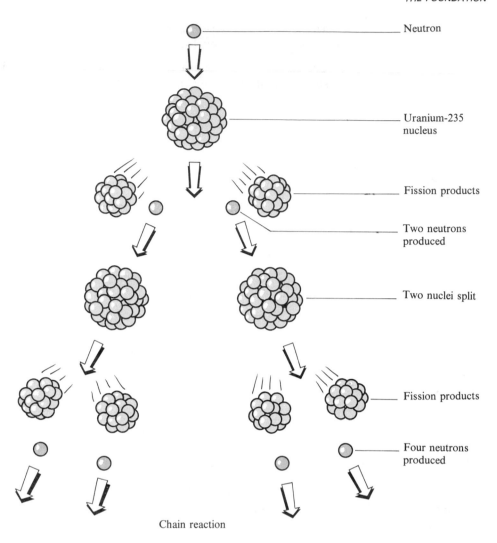

Neutron

Uranium-235
nucleus

Fission products

Two neutrons
produced

Two nuclei split

Fission products

Four neutrons
produced

Chain reaction

Atomic bombs   The only time atomic bombs have been used in warfare was when two cities in Japan, Hiroshima and Nagasaki, were destroyed in 1945. The death and destruction which followed were on such a terrible scale that nations fighting subsequent wars have avoided using atomic weapons.

SEPARATION OF URANIUM ISOTOPES

Separation of ^{235}U from ^{238}U by gaseous effusion of $UF_6(g)$ Natural uranium is a mixture of $^{235}_{92}U$ and $^{238}_{92}U$. Since $^{238}_{92}U$ absorbs neutrons without undergoing fission, the $^{235}_{92}U$ used in atomic bombs must be separated from the other isotopes. This is achieved by means of gaseous **effusion** [p. 135]. Uranium(VI) fluoride, $UF_6(g)$, is prepared and passed along a porous pipe which is surrounded by a larger concentric pipe [see Figure 1.24]. As $^{235}UF_6(g)$ effuses out of the pipe faster than $^{238}UF_6(g)$, the ratio of $^{235}UF_6$ to $^{238}UF_6$ in the outer pipe gradually increases. The rates of effusion are related by the equation [p. 134]

$$\frac{\text{Rate of effusion of } ^{235}UF_6(g)}{\text{Rate of effusion of } ^{238}UF_6(g)} = \sqrt{\frac{M_r(^{238}UF_6)}{M_r(^{235}UF_6)}} = 1.004$$

Miles of piping must be used to give a substantial separation.

FIGURE 1.24
Separation of Uranium
Isotopes

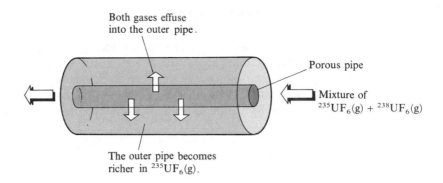

Both gases effuse
into the outer pipe.

Porous pipe

Mixture of
$^{235}UF_6(g) + {}^{238}UF_6(g)$

The outer pipe becomes
richer in $^{235}UF_6(g)$.

FIGURE 1.24
Separation of Uranium
Isotopes

1.10.10 NUCLEAR REACTORS

In a nuclear reactor...

Energy from **nuclear reactors** is obtained by fission of ^{235}U, carried out in a controlled way. The reactors use a mixture of uranium-235 and uranium-238, avoiding the need to separate uranium-235 from the natural mixture of uranium isotopes. Uranium-235 is the only isotope which undergoes fission. Uranium-238 absorbs fast neutrons but not slow neutrons. The neutrons produced from uranium-235 fission are slowed down by passing them through blocks of graphite to prevent them from being absorbed by uranium-238. The method of controlling the rate of fission to avoid overheating is to insert rods of boron, an element which is a very good neutron-absorber, into the reactor. If the fission process speeds up, the rods are pushed further into the reactor; if the chain reaction slows down, the rods are pulled out to allow the number of neutrons to increase and speed up the reaction.

*...the release of nuclear
energy is controlled...*

*...Boron is used to
absorb neutrons*

*The heat evolved is used
to generate electricity*

As heat is generated in the nuclear reactor, it is transferred to a stream of gas which circulates around the reactor. The hot gas is used to boil water; the steam produced is used to drive a turbine and generate electricity. [See Figure 1.25.] Other coolants, e.g. water and liquid sodium, are used in different designs.

FIGURE 1.25
A Nuclear Reactor

4 Rods of boron, a good
neutron-absorber, regulate the
supply of neutrons.

3 Graphite rods slow neutrons
and prevent capture by ^{238}U.

2 Uranium rods.

1 Cool gas is passed into reactor
to absorb heat produced.

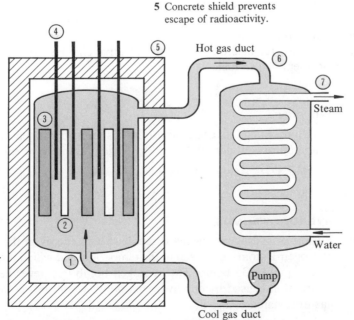

5 Concrete shield prevents
escape of radioactivity.

Hot gas duct

6 Hot gas is used to boil water.

7 Steam is used to drive a
turbine which generates
electricity.

Steam

Water

Pump

Cool gas duct

1.10.11 RADIOISOTOPES

New isotopes are made in reactors

Many radioisotopes have been made by putting stable isotopes into a nuclear reactor. They are bombarded with neutrons and absorb one or two neutrons into the nucleus. The new isotope may well be radioactive because it has an excessive number of neutrons. The number of elements in the universe has been extended from 92 to 105 by the formation of artificially produced radioisotopes [See Periodic Table, p. 752.]

1.10.12 NUCLEAR FUSION

When two atoms of hydrogen-2 (**deuterium**) collide at high speed, they may interact by either of the reactions

In theory, fusion is a source of nuclear energy

$$2\,_1^2\text{H} \rightarrow \,_2^3\text{He} + \,_0^1\text{n}$$

$$2\,_1^2\text{H} \rightarrow \,_1^3\text{H} + \,_1^1\text{H}$$

For the second of these reactions

$$\text{Loss in mass} = 3.016\,05 + 1.007\,825 - 2(2.014\,10) = 0.004\,325\,m_\text{u}$$

$$\text{Release of energy} = mc^2 = 0.004\,325 \times 1.661 \times 10^{-27}(2.998 \times 10^8)^2$$

$$= 6.46 \times 10^{-13}\,\text{J}$$

$$= 3.89 \times 10^{11}\,\text{J}\,\text{mol}^{-1} \text{ of the equation [see p. 196]}$$

This is a release of 389 TJ (3.89×10^8 MJ) for the fusion of 4.00 kg of deuterium. The energy produced by the combustion of 1 tonne of coal is 33 MJ.

Fusion requires hydrogen-2

The enormous release of energy from fusion gives more $_1^2\text{H}$ nuclei the energy they need to fuse, and a chain reaction is set up. Hydrogen-2 (deuterium) is present together with the normal isotope, $_1^1\text{H}$, in all natural compounds of hydrogen, e.g. water. It is therefore an inexhaustible and a more readily accessible starting material than uranium-235. This is why the fusion process is expected eventually to replace fission as the source of nuclear energy. The problem of accelerating the hydrogen-2 atoms

The ^2H atoms must be accelerated before they can fuse

sufficiently to initiate the reaction has still to be solved. Unless the atoms are moving very fast, repulsion between their positive charges prevents the nuclei from getting close enough to fuse.

THE SUN

Fusion takes place in the Sun...

The Sun obtains its energy from the fusion of hydrogen atoms. In the Sun, the temperature is about 10^7 K, and hydrogen atoms have enough energy to fuse:

$$4\,_1^1\text{H} \rightarrow \,_2^4\text{He} + 2\,_{+1}^0\text{e}$$

HYDROGEN BOMBS

...and in a hydrogen bomb

Fusion of hydrogen-2 nuclei is the source of energy in the hydrogen bomb. The hydrogen-2 atoms are raised to the temperature at which they can fuse by the explosion of a uranium-235 bomb. The hydrogen bomb has never been used in war. The destruction caused by one hydrogen bomb would be so catastrophic that no nation has ever dared to use it.

CHECKPOINT 1D: NUCLEAR REACTIONS II

1. How does the size of the nucleus compare with that of the whole atom? What is the binding energy of the nucleus? How can this energy be calculated?

2. How does nuclear fission differ from nuclear decay? How does nuclear fission differ from nuclear fusion? What are the difficulties that have to be overcome before energy from nuclear fusion becomes a commercial proposition?

3. What is a nuclear chain reaction? What is done to stop a nuclear reactor from becoming dangerously hot?

4. What is the source of the energy that can be obtained from uranium-235? Why must uranium-235 be separated from uranium-238 when it is used for making atomic bombs but not when it is used in nuclear power stations?

QUESTIONS ON CHAPTER 1

†**1.** Draw a labelled diagram of a mass spectrometer. Explain how this instrument is used to measure molecular mass. Are there any limits to its use for molecular mass determination?
[For help with second part, see p. 11.]

2. Identify the emitted particles (1) and (2), and state in which groups of the Periodic Table the elements Pb, **X**, **Y** and **Z** occur.

$$^{212}_{82}Pb \xrightarrow{(1)} {}^{212}_{83}X \xrightarrow{(2)} {}^{208}_{81}Y \xrightarrow{\beta \text{ particle}} Z$$

3. In the case of lighter elements, a nuclide in which the ratio neutrons/protons is greater than 1 is likely to emit radiation to bring the ratio nearer to 1. Calculate the ratio of neutrons/protons for the radioactive isotope $^{32}_{15}P$ and the stable isotope $^{31}_{15}P$. Write the equation for the decay of $^{32}_{15}P$ by α particle emission. Calculate the ratio of neutrons/protons in the nuclide formed.

4. Plot the number of neutrons against the number of protons in the nuclides ^1H, ^4He, ^7Li, ^9Be, ^{11}B, ^{12}C, ^{14}C, ^{14}N, ^{16}O, ^{19}F, ^{23}Na, ^{24}Na, ^{24}Mg, ^{27}Al.

In which two nuclides is the neutron/proton ratio greater than the stable ratio? [See Question 3.] Write equations for the decay by β emission of these radioactive isotopes.

5. Write equations for

(a) β particle decay of $^{27}_{13}Al$

(b) α particle decay of $^{27}_{13}Al$

(c) neutron capture by $^{27}_{13}Al$ followed by β emission

(d) a possible path for

$$^{211}_{83}Bi \rightarrow {}^{207}_{83}Bi$$

6. The mass spectrum of C_2H_5Cl shows peaks corresponding to ^1H, ^2H, ^{12}C, ^{13}C, ^{35}Cl and ^{37}Cl. Calculate the mass numbers of the most abundant molecular ion and the heaviest molecular ion. Write the formulae of all the possible ions that contribute to the peak at a mass number of 66.

7. Imagine you have a mixture of hydrogen-1, hydrogen-2 and hydrogen-3, (hydrogen, deuterium and tritium) present as diatomic molecules and that the numbers of atoms of the three species are the same. Sketch the mass spectrum.

8. Give values for a, b, c and d, and the symbols for **X** and **Y** in the equations

(a) $^{35}_{17}Cl + {}^1_0n \rightarrow {}^a_b X + {}^1_1H$

(b) $^7_3Li + {}^2_1H \rightarrow 2{}^c_d Y + {}^1_0n$

†**9.** (a) Define (i) mass number, (ii) mass deficit.

(b) The mass spectra of methylbenzene and butanone are shown in Figures 1.26 and 1.27.
   (i) Draw structural formulae for methylbenzene and butanone.
   (ii) Identify the peak in each spectrum corresponding to the parent molecule.

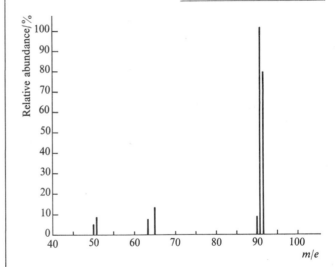

FIGURE 1.26 Mass Spectrum of Methylbenzene

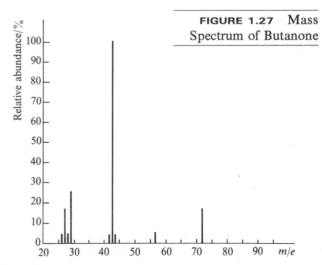

FIGURE 1.27 Mass Spectrum of Butanone

(iii) Deduce from the spectra the *main* fragmentation process that occurs in the mass spectrometer for molecules of methylbenzene, and for molecules of butanone.

[For help with *b*(i), see Chapters 28 and 31.]

(AEB 82)

10. In 1909, Geiger and Marsden reported the amazing results of their experiments on α particles and thin metal foils.

(*a*) What is an α particle?

(*b*) Why did most α particles pass through the foils?

(*c*) Why were some α particles scattered backwards?

(*d*) What did they infer from their results about the structure of metal atoms?

11. A series of radioactive decays can be represented

$$\underset{90}{\overset{232}{}}\text{Th} \xrightarrow{\alpha \text{ emission}} \text{X} \xrightarrow{\beta \text{ emission}} \text{Y} \xrightarrow{\beta \text{ emission}} \text{Z}$$

State the mass number and atomic number of the element **Z**.

†**12.** Define: *isotope*; *relative atomic mass*; *the mole*. Explain whether you think the value of the Avogadro constant is dependent on the latter definition(s).

With the aid of a suitably labelled diagram, describe the operation of a simple mass spectrometer and outline how the instrument may be used to determine (i) the *m/e* value (relative mass) of one isotopic form of chlorobenzene and (ii) the relative atomic mass of an element of your choice.

Naturally occurring carbon, hydrogen and chlorine each contain two isotopes: ^{12}C and ^{13}C (relative abundance 1.11%), ^1H and ^2H (relative abundance 0.015%) and ^{35}Cl and ^{37}Cl (relative abundance 24.23%). Deduce which molecular isotopic species contribute to the peak in the mass spectrum of chlorobenzene at *m/e* = 115, stating which of the species is the most abundant and estimating its natural percentage abundance.

[For help with first part, see Chapter 3.]   (WJEC 82)

†*Note*   A dagger denotes a question which should be tackled on re-reading, as it involves material which is covered in later chapters.

2
THE ATOM: THE ARRANGEMENT OF ELECTRONS

2.1 ATOMIC SPECTRA

Atomic spectra pose a problem which the Rutherford picture of the atom does not solve.

If sunlight or light from an electric light bulb is formed into a beam by a slit and passed through a prism on to a screen, a rainbow of separated colours is seen. The spectrum of colours is composed of visible light of all wavelengths and is called a **continuous spectrum** [see Figure 2.1].

FIGURE 2.1
A Continuous Spectrum

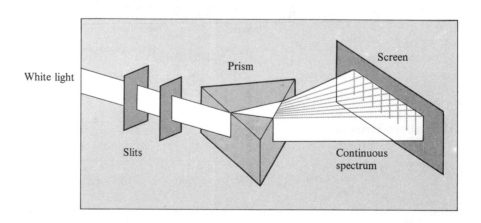

White light

Prism

Screen

Slits

Continuous spectrum

If the source of light is a discharge tube containing a gaseous element, the observed spectrum consists of a number of coloured lines on a black background. The spectrum is called an **atomic emission spectrum** or **line spectrum** [see Figure 2.2].

Elements have emission spectra in the visible and ultraviolet region

All substances give emission spectra when they are excited in some way, by the passage of an electric discharge or by a flame. The atomic emission spectra of elements are in the visible and ultraviolet region of the spectrum. When sodium or a sodium compound is put into a flame, it emits light with a wavelength of 590 nm, and colours the flame yellow. A tube of hydrogen gas which has been excited by an electric discharge glows a reddish-pink colour.

FIGURE 2.2
An Emission Spectrum
(or Line Spectrum)

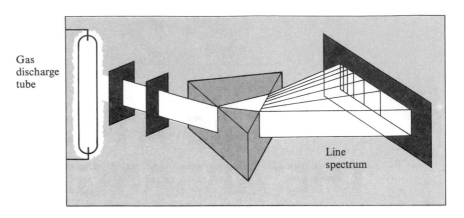

Absorption spectra are black lines on a bright background

An absorption spectrum is seen when white light (light of all visible wavelengths) is passed through a substance. Black lines can be seen where light of some wavelengths has been absorbed by the substance. Spectrometers are instruments for viewing emission and absorption spectra. [See also molecular spectra, p. 727.]

Through a spectrometer, the hydrogen emission spectrum is seen to consist of series of lines

Viewed through a spectrometer, the emission spectrum of hydrogen is seen to be a number of separate sets of lines or **series** of lines. These series of lines are named after their discoverers, as shown in Figure 2.3. The Balmer series, in the visible part of the spectrum, is shown in Figure 2.4.

FIGURE 2.3
The Hydrogen Spectrum

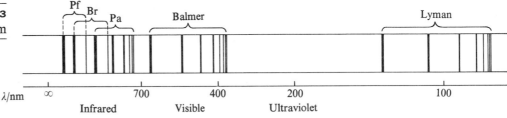

(Pf = Pfund   Br = Brackett   Pa = Paschen   Br overlaps Pf and Pa.)

FIGURE 2.4
The Balmer Series of
Hydrogen

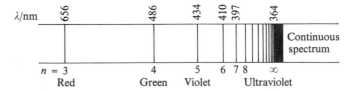

In each series, the lines become closer together as the frequency increases

At very high frequency the lines coalesce

In each series, the intervals between the frequencies of the lines become smaller and smaller towards the high frequency end of the spectrum until the lines run together or **converge** to form a **continuum** of light. The frequencies of the spectral lines fit **Rydberg's equation**:

$$v = cR_H \left(\frac{1}{n_1{}^2} - \frac{1}{n_2{}^2} \right)$$

where v = frequency, n_1 and n_2 = integers, c = velocity of light, and R_H = Rydberg's constant. Wavelength λ and frequency v are related by the equation

$$v\lambda = c$$

For the Lyman series, $n_1 = 1$; for the Balmer series, $n_1 = 2$, and so on.

The Rutherford model of the atom does not explain spectral lines

Why do atomic spectra consist of **discrete** (separate) lines? Why do atoms absorb or emit light of certain frequencies? Why do the spectral lines converge to form a continuum? The Rutherford picture of the atom offers no explanation.

2.2 THE BOHR−SOMMERFELD ATOM

2.2.1 THE BOHR MODEL

Planck theorised that energy is quantised. Bohr suggested that electrons can have only certain amounts of energy...

N Bohr put forward his picture of the atom in 1913 to explain line spectra. He referred to M Planck's recently developed **quantum theory**, according to which energy can be absorbed or emitted in certain amounts, like separate packets of energy, called **quanta**. Bohr suggested that an electron moving in an orbit can have only certain amounts of energy, not an infinite number of values: its energy is **quantised**. The energy that an electron needs in order to move in a particular orbit depends on the radius of the orbit. An electron in an orbit distant from the nucleus requires higher energy than an electron in an orbit near the nucleus. If the energy of the electron is quantised, the radius of the orbit also must be quantised. There is a restricted number of orbits with certain radii, not an infinite number of orbits.

...and their orbits can have only certain radii

Electrons which absorb photons move to higher orbits

An electron moving in one of these orbits does not emit energy. In order to move to an orbit farther away from the nucleus, the electron must absorb energy to do work against the attraction of the nucleus. If an atom absorbs a **photon** (a quantum of light energy), it can promote an electron from an inner orbit to an outer orbit. If sufficient photons are absorbed, a black line appears in the absorption spectrum.

Electrons which fall to lower orbits emit photons of light...

According to the quantum theory, the energy contained in a photon of light of frequency v is hv, h being Planck's constant (6.626×10^{-34} Js). For an electron to move from an orbit of energy E_1 to one of energy E_2, the light absorbed must have a frequency given by **Planck's equation**:

$$hv = E_2 - E_1$$

The emission spectrum arises when electrons which have been excited (raised to orbits of high energy) drop back to orbits of lower energy. They emit energy as light with a frequency given by Planck's equation. [See Figure 2.5.]

FIGURE 2.5 The Origin of Spectral Lines

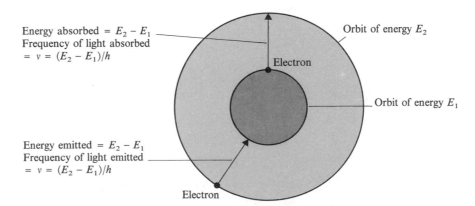

Energy absorbed = $E_2 - E_1$
Frequency of light absorbed
= $v = (E_2 - E_1)/h$

Orbit of energy E_2

Electron

Orbit of energy E_1

Energy emitted = $E_2 - E_1$
Frequency of light emitted
= $v = (E_2 - E_1)/h$

Electron

Planck's equation gives the frequencies of light emitted

Bohr gave orbits of different energy different quantum numbers

Bohr assigned **quantum numbers** to the orbits. He gave the orbit of lowest energy (nearest to the nucleus) the quantum number 1. An electron in this orbit is in its **ground state**. The next energy level has quantum number 2 and so on [see Figure 2.6]. If the electron receives enough energy to remove it from the attraction of the nucleus completely, the atom is **ionised**.

FIGURE 2.6
The Energy Levels at
Various Values of the
Quantum Number, n

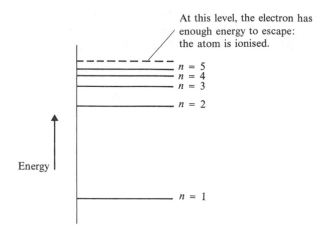

*The hydrogen emission
spectrum arises as
electrons move from orbits
of high quantum number
to orbits of lower quantum
number*

Figure 2.7 shows how the lines in the hydrogen emission spectrum arise from transitions between orbits. The Lyman series in the emission spectrum arise when the electron moves to the $n = 1$ orbit (the ground state) from any of the other orbits. The Balmer series arise from transitions to the $n = 2$ orbit from the $n = 3$, $n = 4$ etc. orbits. The Paschen, Brackett and Pfund series arise from transitions to the $n = 3$, $n = 4$ and $n = 5$ orbits from higher orbits.

*The frequency of the
convergence of spectral
lines can be used to give
the ionisation energy*

FIGURE 2.7   Energy
Transitions in the
Hydrogen atom

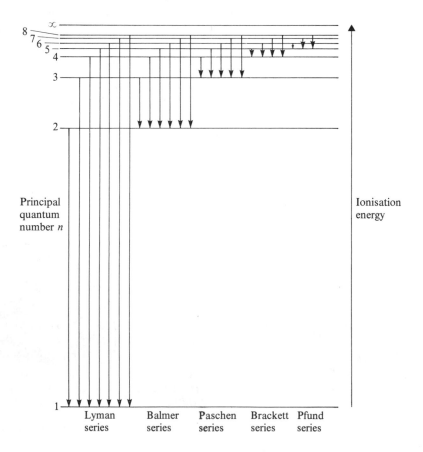

Now consider the absorption spectrum. In each series of absorption lines, as the frequency increases, each line becomes closer to the previous line until the lines converge, and the absorption becomes continuous. The Lyman series in the absorption spectrum arises from transitions from the ground state to higher energy levels. The highest frequency lines relate to the highest energy levels. The limit of the Lyman series (the

When the lines in the absorption spectrum converge it means that the electron has left the atom...

convergence of the lines) corresponds to a transition from the $n = 1$ orbit (the ground state) to $n = \infty$, i.e. to an energy level where the electron has escaped from the atom:

$$A \rightarrow A^+ + e^-$$

...the atom has ionised

The atom has ionised. The same convergence in the emission spectrum arises when an electron collides with an ion and returns to the ground state. The convergence frequency can be used to find the **ionisation energy** of the atom.

2.2.2 DETERMINATION OF IONISATION ENERGY

The definition of the first ionisation energy

The first ionisation energy of an element is the energy required to remove one electron from each of a mole of atoms in the gas phase to form a mole of cations in the gas phase:

$$A(g) \rightarrow A^+(g) + e^-$$

*THE RYDBERG EQUATION

The Rydberg equation is

$$v = cR_{H} \left(\frac{1}{n_1{}^2} - \frac{1}{n_2{}^2} \right)$$

(a) The frequency, v_1, for the first line in the Lyman series must be measured. It fits the equation

$$v_1 = cR_{H} \left(\frac{1}{1^2} - \frac{1}{2^2} \right)$$

so that cR_{H} can be found.

Finding the first ionisation energy...

...from the Rydberg equation...

(b) The frequency at the start of the continuum fits the equation

$$v = cR_{H} \left(\frac{1}{1^2} - \frac{1}{\infty^2} \right) = cR_{H}$$

(c) The energy transition, ΔE, from the ground state ($n = 1$) to the start of the continuum ($n = \infty$) is therefore

$$\Delta E = hv$$

$$= hcR_{H}$$

As the value of h (Planck's constant) is known, and cR_{H} has been found in (a), the equation gives the value of ΔE.

(d) Since ΔE is the energy required to ionise one atom, it must be multiplied by the Avogadro constant ($6.022 \times 10^{23}\,\text{mol}^{-1}$) to give the first ionisation energy of the element.

A GRAPHICAL METHOD

The interval between the frequencies of spectral lines becomes smaller and smaller as they approach the continuum.

(a) The frequencies of the first lines in the Lyman series are measured. If these are v_1, v_2, v_3, v_4, etc., the intervals $\Delta v = (v_2 - v_1)$, $(v_3 - v_2)$, $(v_4 - v_3)$, etc., can be calculated.

...by a graphical method...

(*b*) A graph of v (the lower frequency) against Δv is shown in Figure 2.8. It can be extrapolated back to $\Delta v = 0$. If there is no interval between lines, this is the beginning of the continuum.

FIGURE 2.8 Finding the Convergence Frequency by a Graphical method

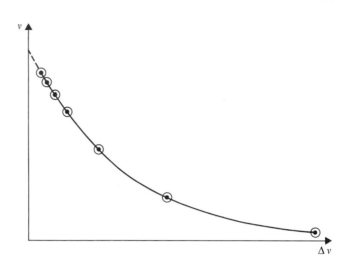

(*c*) The value of v at $\Delta v = 0$ is read off and inserted in Bohr's equation

$$\Delta E = hv$$

(*d*) The value of ΔE is multiplied by Avogadro's constant to give the first ionisation energy for a mole of atoms.

Example The value of the wavelength at the start of the continuum in the sodium emission spectrum is 242 nm. Calculate the first ionisation energy of sodium.

Method

...the calculation Since $\Delta E = hv$

$$= hc/\lambda$$

where $c = 2.998 \times 10^8 \, \mathrm{m \, s^{-1}}$, $h = 6.626 \times 10^{-34} \, \mathrm{J \, s}$ and $L = 6.022 \times 10^{23} \, \mathrm{mol^{-1}}$.

$$\text{Ionisation energy} = L\Delta E$$

$$= Lhc/\lambda$$

$$= 6.022 \times 10^{23} \times 6.626 \times 10^{-34} \times 2.998 \times 10^8/(242 \times 10^{-9})$$

$$= 494\,300 \, \mathrm{J \, mol^{-1}}$$

$$= 494 \, \mathrm{kJ \, mol^{-1}}$$

IONISATION ENERGY FROM AN ELECTRICAL METHOD

Ionisation energy can be determined by measuring the potential difference at which ionisation takes place [see Figure 2.9].

FIGURE 2.9 Method used for the Measurement of First Ionisation Energy

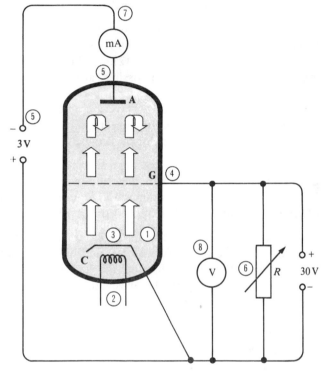

5 Anode, **A**, is made negative by applying a small voltage so that electrons are repelled, (Still called 'anode' because it is *normally* the positive electrode). Anode, **A**, repels electrons, and no current flows through the valve.

4 Grid, **G**. A positive potential is applied between grid and cathode to accelerate electrons towards grid.

3 Electrons stream from heated cathode, **C**, into the gas.

2 Heater heats cathode, **C**.

1 Container of gas at very low pressure.

7 The positive ions are attracted to the anode. Current flowing between cathode and anode is detected by a milliammeter.

6 The accelerating voltage on the grid is increased gradually by means of the variable resistance, *R*, until the electrons are moving fast enough to be able to ionise atoms or molecules with which they collide to form positive ions, e.g.
$$Ar + e^- \rightarrow Ar^+ + 2e^-$$

8 The first ionisation energy is calculated from the potential difference at which ionisation takes place.

9 Calculation: Charge on $1 e^- = 1.602 \times 10^{-19} C$
$1 e^-$ accelerated through a p.d. of 1 V acquires $1.602 \times 10^{-19} J$ of energy.
$1 \, mole^-$ accelerated through 1 V acquire
$1.602 \times 10^{-19} \times 6.022 \times 10^{23} J = 96.5 \, kJ \, mol^{-1}$
First ionisation energy = p.d. at which ionisation takes place $\times 96.5 \, kJ \, mol^{-1}$.

2.2.3 SOMMERFELD'S QUANTUM NUMBERS

Sommerfeld elaborated Bohr's theory in 1916. He proposed that each quantum number governed the energy of a circular orbit and also a set of elliptical orbits of similar energy. He called *n* the **principal quantum number** and introduced a second quantum number which describes the shape of the elliptical orbits (the degree of eccentricity). The **second quantum number**, *l*, can have values from $(n - 1)$ down to 0. If $n = 4$, $l = 3, 2, 1$ and 0.

Sommerfeld's second quantum number

2.3 THE WAVE THEORY OF THE ATOM

According to the wave theory of light, refraction* and diffraction* can be explained by the properties of waves. Other properties of light such as the origin of line spectra and the photoelectric effect*, need a particle or photon theory for their explanation. The success of the dual theory of light led Louis de Broglie to speculate in 1924 on whether particles might have wave properties. He postulated that a particle of mass m moving with velocity v has a wavelength λ associated with its motion. He predicted the value of λ from the **de Broglie equation**

Light is a wave motion with the properties of particles also

$$\lambda = h/mv$$

This equation has the form

Do electrons have the properties of waves as well as being particles?

Wave property = Constant/Particle property

Applying this equation to an electron, and inserting the values $v = 6 \times 10^6 \, \mathrm{m\,s^{-1}}$ for an electron accelerated through 100 volts, $m = 9 \times 10^{-31} \, \mathrm{kg}$, and $h = 6.63 \times 10^{-34} \, \mathrm{J\,s}$ gives $\lambda = 0.12 \, \mathrm{nm}$. This is the sort of distance between the ions in a crystal. It should be possible, according to de Broglie's theory, to send a beam of electrons with this velocity at a crystal and obtain a diffraction pattern. The experiment was tackled by Davisson and Germer. They succeeded in showing that a crystal could act as a diffraction grating for a beam of electrons. This experimental evidence was strong support for de Broglie's bold prediction.

Like waves, electrons can give diffraction patterns

The de Broglie equation can be applied to more massive particles. The wavelengths associated with macroscopic bodies cannot be detected because they are very much less than the spacing in any diffraction grating. This is why de Broglie's equation is of greatest importance when applied to the least massive of particles.

De Broglie theorised...

The Bohr–Sommerfeld picture of the atom specifies the velocity of an electron and the orbit occupied by an electron. To find out the position of an electron, light of short wavelength (comparable with the size of the electron) must be used. Since $E = h\nu = hc/\lambda$, light of short wavelength has photons of high energy. When these interact with an electron, they change its velocity. To avoid changing the velocity of an electron, light of longer wavelength could be used, but the position of the electron would not be found accurately. The idea that it is impossible to measure accurately both the velocity and the position of a particle was expressed by W K Heisenberg and termed the **Uncertainty Principle**. From the viewpoint of Heisenberg's Uncertainty Principle, the Bohr picture of the atom does not appear satisfactory. The electrons are moving in orbits of specified radii at specified velocities, and these quantities cannot both be measured experimentally. A theory which involves quantities which cannot be measured does not follow the tradition of scientific work. Advances in science have never come through theories which cannot be tested by quantitative measurement. Scientists were dissatisfied with the Bohr–Sommerfeld theory of the atom on these grounds.

...a moving particle has a wavelength associated with its motion

Heisenberg's Uncertainty Principle...

...states that it is not possible to measure both the position and the velocity of an electron accurately

ATOMIC ORBITALS

To steer a way around this difficulty, E Schrödinger and W K Heisenberg and P A M Dirac worked out wave theories of the atom. The Schrödinger treatment, which is the best known, sets up a wave equation for the atom. Solutions to the Schrödinger wave equation can be obtained only under certain conditions. If the electron is to be treated as a wave, then an integral number of wavelengths must be fitted into one circuit of the electron. [See Figure 2.10.]

The Schrödinger wave equation...

*See R Muncaster, *A-Level Physics* (Stanley Thornes)

FIGURE 2.10
The Wave Nature of the
Electron

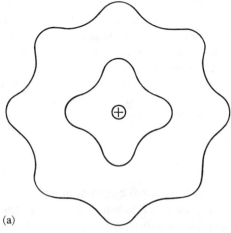

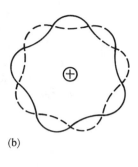

(a)

The waves shown in (a) are standing waves. The
maxima and minima of the waves match exactly
for consecutive circuits of the electron about the
nucleus. (The figure is a two-dimensional
representation of three-dimensional waves.)

The pattern in (b) is not a
solution. A maximum in one
circuit is cancelled by a
minimum in the next.

*...Its solution gives the
probability of finding the
electron at any distance
from the nucleus*

The solution of the wave equation gives the **probability density** of the electron. This
is the probability that the electron is present in a given small region of space. The
probability that the electron is at a distance, r, from the nucleus is plotted against r
for the hydrogen atom in its ground state in Figure 2.11. The maximum probability
of finding the electron is at a distance of 0.053 nm. This is the same as the radius of
the orbit occupied by the electron in its ground state according to the Bohr–Sommerfeld
theory.

FIGURE 2.11
A Probability Density
Diagram for the
Hydrogen Atom in its
Ground State

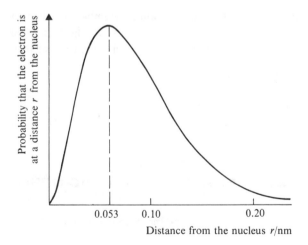

Probability that the electron is
at a distance r from the nucleus

0.053   0.10   0.20

Distance from the nucleus r/nm

*The volume of space in
which there is a 95%
chance of finding the
electron is called the
atomic orbital*

There is a possibility that the electron will be either closer to the nucleus or outside
the radius of 0.053 nm. The probability of finding the electron decreases sharply,
however, as the distance from the nucleus increases beyond $3r$. The volume of
space in which there is a 95% chance of finding the electron is called the **atomic
orbital**. There is a 5% probability that the electron will be outside this volume of
space at a given instant. On this model, the electron is not described as revolving in
an orbit. The electron is said to occupy a three-dimensional space around the nucleus
called an atomic orbital. The nucleus is described as being surrounded by a
three-dimensional 'cloud of charge' or 'electron cloud'.

THE FOUR QUANTUM NUMBERS

Since the Schrödinger wave equation has a limited number of solutions, the total energy of the atom is therefore found to have only certain values. This agrees with Bohr's postulate that the electronic energy levels are quantised. Solutions of the wave equation can be obtained if the orbitals are described by four quantum numbers. The first is Bohr's quantum number, n. The second quantum number, l, corresponds to Sommerfeld's quantum number describing the shape of elliptical orbits. The values of l are assigned letters

Bohr's quantum number ,n.
Sommerfeld's quantum number ,l

$$l = 0 \quad 1 \quad 2 \quad 3 \quad 4$$
$$\quad \text{s} \quad \text{p} \quad \text{d} \quad \text{f} \quad \text{g}$$

If an electron has a principal quantum number $n = 2$ and a second quantum number $l = 0$, it is said to be a 2s electron. For various values of n, the different combinations of the two quantum numbers are

1s
2s   2p
3s   3p   3d
4s   4p   4d   4f
5s   5p   5d   5f   5g

A third quantum number, m_l...

The wave equation leads to a **third quantum number**, m_l. This gives the maximum number of orbitals for the different values of l as

one   s orbital
three   p orbitals
five   p orbitals
seven   p orbitals

...and a spin quantum number, m_s

The **fourth quantum number** is called the **spin quantum number**, m_s. It has values of $+\frac{1}{2}$ and $-\frac{1}{2}$. It represents the spin of an electron on its own axis, which can be clockwise or anticlockwise, relative to the orbital of the electron.

The Pauli Exclusion Principle

In his Exclusion Principle, W Pauli stated that <u>no two electrons in an atom can have the same four quantum numbers</u>. It follows that, if two electrons in an atom have the same values of n, l and m_l, they must have different values of m_s. Their spins must be opposed. Each orbital can hold two electrons with opposed spins.

2.3.1   SHAPES OF ATOMIC ORBITALS

(a) The shape of a 1s orbital   (b) The shape of a 2s orbital

   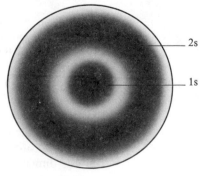

FIGURE 2.12
The Shapes of s Orbitals

(The density of the shading is a measure of the probability of finding an electron at that distance from the nucleus.)

s orbitals are spherical

p orbitals have an hourglass shape

The shape of an s orbital is spherically symmetrical about the nucleus. The orbital has no preferred direction. The probability of finding an electron at a distance r from the nucleus is the same in all directions. [See Figure 2.12.] A p orbital is not symmetrical: it is concentrated in certain directions. The electron density is shaped like an hourglass. [See Figure 2.13.] The shapes of d orbitals are shown in Figure 2.14.

FIGURE 2.13

The Shape and Orientation of p Orbitals

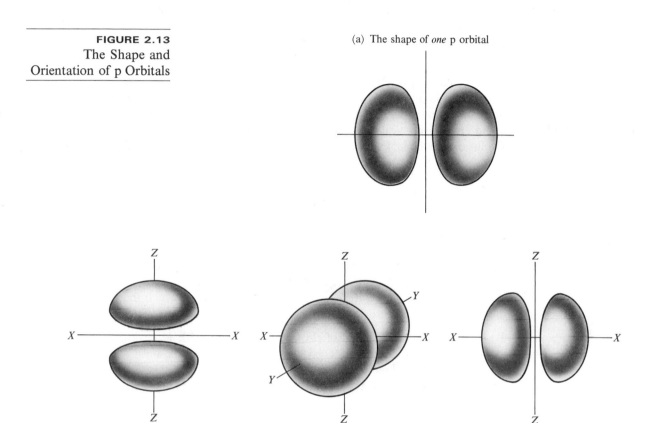

(a) The shape of *one* p orbital

(b) The orientation in space of *three* p orbitals.(Each of the three p orbitals is perpendicular to both of the others. The shape of each orbital is as shown in (a).)

FIGURE 2.14

The Shape and Orientation of d Orbitals

There are four d orbitals of shape (a). The lobes lie between the X–Y axes as shown in (a), between the X–Z axes, between the Y–Z axes and, in the fourth case, along the axes X and Y as shown in (b). The fifth orbital has the shape shown in (c).

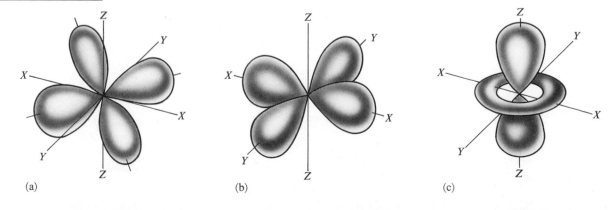

(a)

(b)

(c)

2.4 ELECTRONIC CONFIGURATION OF ATOMS

In atoms with more than one electron, there are shells of orbitals with the same principal quantum number

The energy levels of the orbitals of the hydrogen atom are illustrated in Figure 2.7 [p. 30]. For atoms with more than one electron, there are different values of the quantum number l for each value of the principal quantum number n. The energy levels for each value of n are split between orbitals with different values of l. The relative energy levels of the orbitals are shown in Figure 2.15. The term **shell** is used for a group of orbitals with the same principal quantum number. The $n = 1$ shell is termed the K shell, $n = 2$ the L shell, $n = 3$ the M shell and so on. A **subshell** is a group of orbitals with the same principal and second quantum numbers, e.g. the 3p subshell.

The arrangement of electrons in atomic orbitals is governed by two factors. The first is that in a normal atom the electrons are arranged so that the energy is at a minimum. Any other arrangement would make the atom an excited atom, which could emit energy and pass to its ground state. The second factor is the Pauli Exclusion Principle, i.e. no two electrons can have the same four quantum numbers.

The orbitals of lowest energy are filled first

It is convenient to draw an 'electrons-in-boxes' diagram to show the arrangement of electrons in orbitals. A box represents one orbital and can contain two electrons with opposite spins. The electrons are represented by arrows, pointing upwards for $m_s = +\frac{1}{2}$ and downwards for $m_s = -\frac{1}{2}$:

'Electrons-in-boxes'

$\boxed{\uparrow\downarrow}$

An s subshell consists of one box, a p subshell of *three* boxes, a d subshell of *five* boxes and an f subshell of *seven* boxes. The boxes are arranged in order of energy in Figure 2.16. An aid to help you remember the order in which the levels are filled is shown in Figure 2.17.

Electrons occupy lowest energy 'boxes' first

To work out the arrangement of electrons in an atom with 12 electrons, the electrons must be put into the lowest energy boxes first, two to a box, until all the electrons are accommodated [see Figure 2.16]. It is interesting to do this for the elements in order of atomic number, the order in which they appear in the Periodic Table [p. 518].

FIGURE 2.15
The Relative Energy Levels of Atomic Orbitals (not to scale)

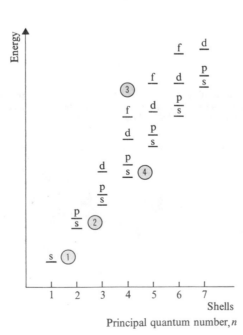

3 Note 4f > 4d > 4p > 4s in energy.

4 Note 4s < 3d in energy. The orbitals of $n = 4$ overlap those with $n = 3$.

2 Note 2p electrons have more energy than 2s.

1 Lowest energy: $n = 1$

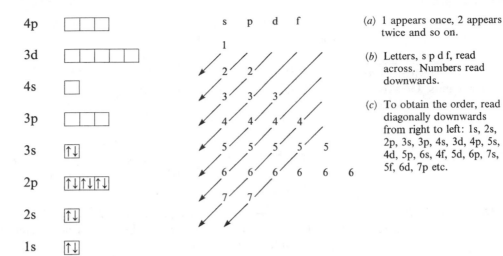

FIGURE 2.16
'Electrons-in-boxes' Diagram for an Atom with 12 Electrons

FIGURE 2.17
How to Remember the Order in which Boxes are Filled

(a) 1 appears once, 2 appears twice and so on.

(b) Letters, s p d f, read across. Numbers read downwards.

(c) To obtain the order, read diagonally downwards from right to left: 1s, 2s, 2p, 3s, 3p, 4s, 3d, 4p, 5s, 4d, 5p, 6s, 4f, 5d, 6p, 7s, 5f, 6d, 7p etc.

Measurements of successive ionisation energies support the idea of shells

Evidence for the arrangement of electrons in shells of different energies is provided by values of successive **ionisation energies** for elements. Figure 2.18 shows a graph of the logarithm of the ionisation energy required for the removal of one electron after another from a potassium atom. A logarithmic plot is used in order to give a condensed graph. You can see that the electrons fall into four groups. The higher the ionisation energy, the more difficult the electrons are to remove and the nearer they must be to the nucleus.

FIGURE 2.18 Graph of lg (Ionisation Energy) against Number of Electron Removed for the Potassium Atom

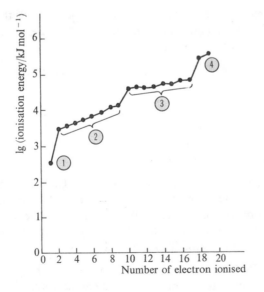

4 The two electrons with the highest ionisation energies are closest to the nucleus, and form the K shell.

3 These eight electrons are in the next shell, the L shell.

2 These eight electrons are in the M shell.

1 This electron has the lowest ionisation energy. It is the easiest to remove. It is in the N ($n = 4$) shell.

2.5 ELECTRONIC CONFIGURATIONS OF ELEMENTS IN THE PERIODIC TABLE

2.5.1 HISTORY OF THE PERIODIC TABLE

Newlands arranged the elements in octaves

There are 92 naturally occurring elements. Chemists have always wanted to find a way of fitting all the diverse elements into a logical pattern. In 1865, J Newlands arranged the elements in order of increasing relative atomic mass. Noticing that the eighth element resembled the first, the ninth resembled the second and so on, he arranged them in columns:

FIGURE 2.19
Newlands' Octaves of
Elements

H	Li	Be	B	C	N	O
F	Na	Mg	Al	Si	P	S
Cl	K	Ca	Cr	Ti	Mn	Fe

...but his ideas were scorned

Newlands compared his chemical 'octaves' with musical octaves, and called the resemblance the **Law of Octaves**. The comparison was unfortunate: people poured scorn on his ideas.

Mendeleev carried on the work on octaves...

Only a few years later, in 1869, a Russian chemist, D I Mendeleev, presented a similar idea. He also arranged the elements in order of increasing relative atomic mass, and put them into vertical groups.

...and drew up his Periodic Table

He called this arrangement the **Periodic Table** [see Figure 2.20]. It included some improvements on Newlands' system.

FIGURE 2.20 Part of Mendeleev's Periodic Table of 1871

	Gp 1	*Gp 2*	*Gp 3*	*Gp 4*	*Gp 5*	*Gp 6*	*Gp 7*	*Gp 8*
Row 1	H							
Row 2	Li	Be	B	C	N	O	F	
Row 3	Na	Mg	Al	Si	P	S	Cl	
Row 4	K	Ca	–	–	–	–	–	Ti V Cr Mn Fe Co Ni
Row 5	Cu	Zn	–	–	As	Se	Br	

Mendeleev's improvements included long periods to accommodate transition metals...

1. Mendeleev introduced long rows or **periods** for the elements we now call **transition metals**. This meant that the metals Ti, Mn, Fe were no longer placed under the non-metals Si, P, S [see Figure 2.19].

...spaces...

2. He left spaces. When he saw that arsenic fitted naturally into group 5 he left two spaces between zinc and arsenic.

...new values of relative atomic mass...

3. When elements did not fit comfortably into the slots in the Periodic Table dictated by their relative atomic masses, Mendeleev instigated new determinations of relative atomic mass. In each case (Cr, In, Pt, Au) the new value justified the arrangement in Mendeleev's Periodic Table.

...predictions about undiscovered elements

4. Where he had left gaps in the Periodic Table, Mendeleev predicted that new elements would be discovered to fill the gaps. He had some outstanding successes in predicting the properties of elements. When elements were discovered and found to have the relative atomic mass and the physical and chemical properties Mendeleev had predicted, faith in the Periodic Table soared.

In his **Periodic Law**, Mendeleev stated that the properties of chemical elements vary with relative atomic mass in a systematic way.

The noble gases were discovered and found to fit into a new group of the periodic table...

The **noble gases** had not been discovered when the Periodic Table was drawn up. As they were discovered one by one, they were found to fit in between the halogens in Group 7 and the alkali metals in Group 1. A separate Group 0 was added to the right-hand side of the table. Argon, however, has a higher relative atomic mass than potassium (A_r(Ar) = 40; A_r(K) = 39) [p. 47]. It made more sense chemically to put potassium with the alkali metals, rather than keep to the order of relative atomic masses. Another example of this kind was the positions of tellurium and iodine. Relative atomic masses placed tellurium under bromine, and iodine under sulphur and selenium; chemical properties placed them in the reverse order.

...Discrepancies were resolved by arranging elements in order of atomic number

Moseley's work on X rays, in 1914, solved this problem. He showed that the atomic numbers (proton numbers) of elements are more significant than their relative atomic masses [see p. 7]. This discovery was the final step in the validation of the Periodic Table. In the modern Periodic Table elements are arranged in order of proton number (atomic number).

2.5.2 FEATURES OF THE PERIODIC TABLE

In the periodic table...

...the alkali metals fall underneath one another in Group 1...

...the vertical groups contain similar elements...

Various types of chemical behaviour are repeated after an interval. The noble gases have proton numbers 2, 10, 18, 36, 54 and 86. The intervals are 2, 8, 8, 18, 18 and 32. Following each noble gas is an extremely reactive metal which forms M^+ ions. These are the **alkali metals**: Li, Na, K, Rb and Cs. Preceding each noble gas is a reactive, non-metallic element which forms a Y^- ion. These are the **halogens**: F, Cl, Br, I and At. The alkali metals in Group 1A are followed by the **alkaline earths**: Be, Mg, Ca, Sr and Ba, in Group 2A. The halogens in Group 7B are preceded by O, S, Se and Te: a set of elements with valency 2 and a gradation in properties from non-metal to metalloid. Groups 3B, 4B and 5B are sets of less similar elements. They all show the group valency and a gradation in properties from non-metal to metal down the group. The Periodic Table is shown on p. 752.

...and the long periods contain transition metals

The method of portraying these chemical similarities is to divide the elements into seven rows or **periods**. The number of elements in each period is: 2 in Period 1, 8 in Period 2, 8 in Period 3, 18 in Period 4, 18 in Period 5, 32 in Period 6 and 17 in Period 7 (which may possibly be still incomplete). In Periods 4 and 5 there are ten metals in between Group 2A and Group 3B. They are sets of rather similar metals called **transition metals**. In Periods 6 and 7, there are sets of transition metals and also **inner transition metals**: sets of 14 extremely similar metals.

Groups which precede the block of transition metals are designated **A** groups, and those which follow the transition metals are termed **B** groups.

2.5.3 THE ELECTRONIC CONFIGURATIONS OF THE ELEMENTS

In the Periodic Table, elements are arranged in order of their proton numbers (atomic numbers). The proton number, Z, is the nuclear charge and also the number of electrons in an atom of the element.

Hydrogen has one electron. It occupies the lowest energy level, 1s.

H 1s $\boxed{\uparrow}$
This can be written as 1s

Helium has two electrons which can occupy the same 1s orbital with opposing spins.

He 1s $\boxed{\uparrow\downarrow}$ written as $1s^2$
The superscript 2 denotes the number of electrons in the subshell.

Lithium has three electrons. Two occupy the 1s box, which is then full. The third must go into a box at the next level.

Li 2s $\boxed{\uparrow}$
   1s $\boxed{\uparrow\downarrow}$ written as $1s^2 2s$

Beryllium has four electrons.

Be 2s $\boxed{\uparrow\downarrow}$
   1s $\boxed{\uparrow\downarrow}$ $1s^2 2s^2$

Boron has five electrons. The fifth electron cannot enter the 1s or 2s boxes: it occupies an orbital in the 2p subshell.

B

2p ↑ ▢ ▢
2s ↑↓
1s ↑↓ $1s^2 2s^2 2p$

Carbon has six electrons. There are three ways in which the fifth and sixth electrons can be accommodated in the 2p subshell. The two 2p electrons may be (*a*) in the same box or in different boxes with the spins (*b*) parallel or (*c*) opposed. In fact arrangement (*b*) is favoured.

(*a*) 2p ↑↓ ▢ ▢
(*b*) 2p ↑ ↑ ▢
(*c*) 2p ↑ ↓ ▢

C

2p ↑ ↑ ▢
2s ↑↓
1s ↑↓ $1s^2 2s^2 2p^2$

According to Hund's Rule, only when all the orbitals in a subshell contain an electron do electrons begin to occupy orbitals in pairs

Hund's Multiplicity Rule comes into play here. F Hund stated that <u>the favoured configuration is the one in which the electrons occupy different boxes and have the same spins</u>. This arrangement puts the electrons further apart than the others. According to Hund's rule, electrons do not pair in an orbital until all the other orbitals in the subshell have been occupied by a single electron.

Nitrogen ($Z = 7$) obeys Hund's rule by accommodating the three 2p electrons in different boxes with the same spins.

N

2p ↑ ↑ ↑
2s ↑↓
1s ↑↓ $1s^2 2s^2 2p^3$

Oxygen ($Z = 8$). The fourth 2p electron pairs with one of the other three 2p electrons.

O

2p ↑↓ ↑ ↑
2s ↑↓
1s ↑↓ $1s^2 2s^2 2p^4$

Fluorine ($Z = 9$) has the arrangement $1s^2 2s^2 2p^5$.

F

2p ↑↓ ↑↓ ↑
2s ↑↓
1s ↑↓ $1s^2 2s^2 2p^5$

Neon ($Z = 10$) $1s^2 2s^2 2p^6$, has a full 2p subshell, thus completing the L shell. The next element, **sodium** ($Z = 11$) has to utilise the M shell, starting with the 3s subshell. The diagrams for sodium and the twelfth element, **magnesium**, are

Na 3s ↑
    2p ↑↓ ↑↓ ↑↓
    2s ↑↓
    1s ↑↓ $1s^2 2s^2 2p^6 3s$

Mg 3s ↑↓
    2p ↑↓ ↑↓ ↑↓
    2s ↑↓
    1s ↑↓ $1s^2 2s^2 2p^6 3s^2$

The following six elements have electrons in the 3p subshell. The configurations of the next six elements are (writing (Ne) for $1s^2 2s^2 2p^6$)

Al ($Z = 13$)   (Ne)$3s^2 3p$

Si ($Z = 14$)   (Ne)$3s^2 3p^2$

P ($Z = 15$)   (Ne)$3s^2 3p^3$

S ($Z = 16$)   (Ne)$3s^2 3p^4$

Cl ($Z = 17$)   (Ne)$3s^2 3p^5$

Ar ($Z = 18$)   (Ne)$3s^2 3p^6$

How do the elements fit into the Periodic Table? So far we have:

Electron configuration and the periodic table

Filling the K shell: H and He: First period

Filling the L shell: Li to Ne: Second period (first short period)

Filling the M shell: Na to Ar: Third period (second short period)

The fourth period   The next elements start the fourth period. They are

$$K \ (Z = 19) \ (Ar)4s$$

Filling the d subshell... $$Ca \ (Z = 20) \ (Ar)4s^2$$

Once the 4s subshell is full, the 3d orbitals are filled [see Figure 2.16, p. 39]. Over the next 10 elements, electrons enter the 3d subshell. These elements are

$$Sc \ (Z = 21) \ (Ar)4s^2 3d$$

$$to \ Zn \ (Z = 30) \ (Ar)4s^2 3d^{10}$$

While the d subshell fills, the chemistry of the elements is not greatly affected. The metals scandium to zinc are a very similar set of metals, called **transition metals** [Chapter 24]. The elements gallium ($Z = 31$) to krypton ($Z = 36$) complete the M shell by filling the 4p orbitals. The 18 elements from potassium to krypton comprise the **first long period**.

TABLE 2.1 Electronic Configurations of the Atoms of the Elements

	Z	1s	2s	2p	3s	3p	3d	4s	4p	4d	4f	5s	5p	5d	5f	6s	6p	6d	6f	7s
H	1	1																		
He	2	2																		
Li	3	2	1																	
Be	4	2	2																	
B	5	2	2	1																
C	6	2	2	2																
N	7	2	2	3																
O	8	2	2	4																
F	9	2	2	5																
Ne	10	2	2	6																
Na	11	2	2	6	1															
Mg	12				2															
Al	13				2	1														
Si	14		10 electrons		2	2														
P	15				2	3														
S	16				2	4														
Cl	17				2	5														
Ar	18	2	2	6	2	6														
K	19	2	2	6	2	6		1												
Ca	20							2												
Sc	21						1	2												
Ti	22						2	2												
V	23						3	2												
Cr	24						5	1												
Mn	25						5	2												
Fe	26						6	2												
Co	27		18 electrons				7	2												
Ni	28						8	2												
Cu	29						10	1												
Zn	30						10	2												
Ga	31						10	2	1											
Ge	32						10	2	2											
As	33						10	2	3											
Se	34						10	2	4											
Br	35						10	2	5											
Kr	36	2	2	6	2	6	10	2	6											

The fifth period   The fifth period is a **second long period**, extending from rubidium ($Z = 37$) to xenon ($Z = 54$). This period includes a **second transition series**.

	Z	1s	2s	2p	3s	3p	3d	4s	4p	4d	4f	5s	5p	5d	5f	6s	6p	6d	6f	7s
Rb	37	2	2	6	2	6	10	2	6			1								
Sr	38											2								
Y	39									1		2								
Zr	40									2		2								
Nb	41									4		1								
Mo	42									5		1								
Tc	43									5		2								
Ru	44									7		1								
Rh	45					36 electrons				8		1								
Pd	46									10		0								
Ag	47									10		1								
Cd	48									10		2								
In	49									10		2	1							
Sn	50									10		2	2							
Sb	51									10		2	3							
Te	52									10		2	4							
I	53									10		2	5							
Xe	54	2	2	6	2	6	10	2	6	10		2	6							

The sixth period

The rare earth elements

The sixth period starts with caesium ($Z = 55$) and ends with radon ($Z = 86$). During this period, the 4f orbitals are filled between cerium ($Z = 58$) and ytterbium ($Z = 70$). The **lanthanons** (from lanthanum ($Z = 57$) to lutetium ($Z = 71$)) are the first set of **rare earth elements**, and are even more similar to one another than are the transition metals. The elements from lutetium to mercury, ($Z = 80$), comprise a second transition series in which the 5d subshell is filled.

	Z	1s	2s	2p	3s	3p	3d	4s	4p	4d	4f	5s	5p	5d	5f	6s	6p	6d	6f	7s
Cs	55	2	2	6	2	6	10	2	6	10		2	6			1				
Ba	56											2	6			2				
La	57											2	6	1		2				
Ce	58										2	2	6			2				
Pr	59										3	2	6			2				
Nd	60										4	2	6			2				
Pm	61										5	2	6			2				
Sm	62										6	2	6			2				
Eu	63										7	2	6			2				
Gd	64										7	2	6	1		2				
Tb	65										8	2	6	1		2				
Dy	66										10	2	6			2				
Ho	67										11	2	6			2				
Er	68										12	2	6			2				
Tm	69										13	2	6			2				
Yb	70					46 electrons					14	2	6			2				
Lu	71										14	2	6	1		2				
Hf	72										14	2	6	2		2				
Ta	73										14	2	6	3		2				
W	74										14	2	6	4		2				
Re	75										14	2	6	5		2				
Os	76										14	2	6	6		2				
Ir	77										14	2	6	7		2				
Pt	78										14	2	6	9		1				
Au	79										14	2	6	10		1				
Hg	80										14	2	6	10		2				
Tl	81										14	2	6	10		2	1			
Pb	82										14	2	6	10		2	2			
Bi	83										14	2	6	10		2	3			
Po	84										14	2	6	10		2	4			
At	85										14	2	6	10		2	5			
Rn	86	2	2	6	2	6	10	2	6	10	14	2	6	10		2	6			

The seventh period The seventh and final period starts at francium ($Z = 87$). The last naturally occurring element, uranium, has $Z = 92$. The electron configurations of all the elements are shown in Table 2.1. The elements which follow uranium are artificially made and are radioactive [p. 12].

	Z	1s	2s	2p	3s	3p	3d	4s	4p	4d	4f	5s	5p	5d	5f	6s	6p	6d	6f	7s
Fr	87	2	2	6	2	6	10	2	6	10	14	2	6	10		2	6			1
Ra	88															2	6			2
Ac	89															2	6	1		2
Th	90															2	6	2		2
Pa	91														2	2	6	1		2
U	92														3	2	6	1		2
Np	93														5	2	6			2
Pu	94														6	2	6			2
Am	95					78 electrons								7	2	6			2
Cm	96														7	2	6	1		2
Bk	97														7	2	6	2		2
Cf	98														9	2	6	1		2
Es	99																			
Fm	100																			
Mv	101																			
No	102																			
Lw	103																			

CHECKPOINT 2A: ELECTRONIC CONFIGURATIONS

1. There are six calcium isotopes, of nucleon number 40, 42, 43, 44, 46 and 48. How many protons and neutrons are there in the nuclei?

2. Draw 'electrons-in-boxes' diagrams of the electronic configurations of the following atoms, given the proton number (Z): boron (5), fluorine (9), aluminium (13) and potassium (19).

3. Draw diagrams to show the electronic configurations of the ions K^+, Cl^-, Ca^{2+}, O^{2-}, Al^{3+}, H^-. (Proton numbers are K = 19, Cl = 17, Ca = 20, O = 8, Al = 13 and H = 1.)

4. Write down the electronic configurations of the atoms with the proton numbers 4, 7, 18, 27, 37. State to which Group of the Periodic Table each element belongs.

5. Write the electronic configurations of the following species (e.g., Li = $1s^2 2s$). Their proton numbers range from Na = 11 to Ar = 18.

Na^+, Mg^{2+}, Al, Si, P, S, S^{2-}, Cl, Cl^-, Ar

QUESTIONS ON CHAPTER 2

1.

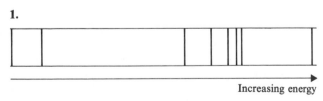

Increasing energy

Figure 2.21

Figure 2.21 represents the atomic emission spectrum of hydrogen. Explain why it is composed of lines, and say what each line indicates. Why do the lines become closer together as you read from left to right?

2. 'In its ground state, the electron in a hydrogen atom is in a 1s orbital.' Explain this statement.

3. (*a*) Describe how the first ionisation energy of an element can be determined experimentally.

(*b*) Results obtained for the ionisation energies of boron are

Electron number	1st	2nd	3rd	4th	5th
Ionisation energy/ kJ mol^{-1}	800	2400	3700	25 000	32 800

On graph paper, plot lg (ionisation energy/kJ mol^{-1}) against the number of electron removed. From the graph, deduce the most likely formula of boron chloride.

4. (*a*) The spectrum of atomic hydrogen includes a series of lines in the ultraviolet region. The separation of these lines decreases with increasing frequency (i.e. decreasing wavelength), and eventually the lines coalesce. There is a similar series which starts in the visible region of the spectrum. Explain these facts.

†(*b*) Describe *one* example of the use of isotopes which *either* throws light on the mechanism of an organic reaction, *or* helps to solve a problem of industrial or technological importance.

(c) The element boron has two isotopes of relative atomic masses 10.01 and 11.01. The relative atomic mass of naturally occurring boron differs slightly according to its origin. The value (as found by a chemical method) for boron obtained from a Californian mineral was 10.84, while that for boron from an Italian mineral was 10.82. Calculate the percentage of the lighter isotope in the two samples of boron. Indicate (without giving any detailed explanation) what experiment might be performed to decide whether or not this difference in the relative atomic mass of boron from the two sources was real.

[For help with (b) see Chapters 14 and 33.] (O 83)

5. (a) The ratio of protons to neutrons in the nuclei of atoms gives a measure of their stability.

Using the following data, plot a graph of *neutron number* (*y* axis) against *proton number* to give the upper and lower bands of stability.

Element	Atomic number	Mass number of isotopes				
Lithium	3	6	7			
Oxygen	8	16	17	18		
Sulphur	16	32	33	34	36	
Chromium	24	50	52	53	54	
Strontium	38	84	86	87	88	
Silver	47	107	109			
Tungsten	74	180	182	183	184	186

Use the graph
 (i) to determine which of the following isotopes are *unstable*

$$^{67}_{30}Zn; \quad ^{101}_{45}Rh; \quad ^{32}_{12}Mg; \quad ^{43}_{20}Ca; \quad ^{141}_{57}La.$$

 (ii) For these unstable isotopes, suggest the numbers and types of radiation they must emit in order to become stable.
 (iii) Give the mass numbers and atomic numbers of the stable isotopes produced in (ii).

(b) The following values represent the successive standard molar ionisation energies/kJ of an element **X**:

$$1400, 2880, 4520, 7450, 9450, 53\,000, 64\,200$$

 (i) What is meant by the term *ionisation energy*?
 (ii) Account for the successive increases in values shown.
 (iii) To what Group in the Periodic Table should element **X** be assigned? Give a reason for your choice.
 (iv) Name *two* elements from this group.
 (v) Name and describe the type of bond formed in the hydride of the element, **X**.

(c) If the radioactive decay ^{63}Ni to ^{63}Cu has a half life of 120 years, how long will it take for three-quarters of the nickel to change into copper?

(AEB 81)

3

EQUATIONS AND EQUILIBRIA

3.1 RELATIVE ATOMIC MASS

The masses of atoms are very small, from 10^{-24} to 10^{-22} grams. Instead of using the actual masses of atoms, **relative atomic masses** (A_r) are used. Originally, they were defined as

Atoms range in mass from $10^{-24}g$ to $10^{-22}g$

$$\text{Original relative atomic mass} = \frac{\text{Mass of one atom of an element}}{\text{Mass of one atom of hydrogen}}$$

Since relative atomic masses are now determined by mass spectrometry, and since volatile carbon compounds are much used in mass spectrometry, the mass of an atom of $^{12}_6C$ is now taken as the standard of reference:

The definition of relative atomic mass...

$$\text{Modern relative atomic mass} = \frac{\text{Mass of one atom of an element}}{1/12 \text{ the mass of one atom of carbon-12}}$$

The difference between the two scales is small. On the carbon-12 scale, the relative atomic mass of $^{12}_6C$ is 12.0000, and the relative atomic mass of ^1_1H is 1.0078. The mass of a $^{12}_6C$ atom is $12.0000\,m_u$, and the mass of a ^1_1H atom is $1.0078\,m_u$ [p. 3].

3.2 THE MOLE

It follows that, if an atom of carbon is 12 times as heavy as an atom of hydrogen, then 12g of carbon contain the same number of atoms as 1g of hydrogen. The same is true of the relative atomic mass expressed in grams for any element, e.g., 4g of helium and 200g of mercury contain the same number of atoms. This number is $6.022 \times 10^{23}\,\text{mol}^{-1}$, the Avogadro constant, L. We call 6.022×10^{23} units of any species a **mole** of that species. Thus, 6.022×10^{23} atoms of aluminium are a mole of aluminium, 6.022×10^{23} molecules of hydrogen are a mole of hydrogen molecules and 6.022×10^{23} electrons are a mole of electrons. The symbol for mole is **mol**.

...and the definition of the mole

The mole is defined in this way. A mole of substance is the amount of that substance which contains as many elementary entities as there are atoms in 12 grams of carbon-12.

3.3 MOLAR MASS

The **relative molecular mass** (M_r) of a compound is defined by the expression

Relative molecular mass and . . .

$$\text{Relative molecular mass} = \frac{\text{Mass of one molecule of the compound}}{1/12 \text{ the mass of one atom of carbon-12}}$$

The relative molecular mass of a compound expressed in grams is a mole of the compound; thus 44 g of carbon dioxide are a mole of carbon dioxide, and contain 6.0×10^{23} molecules. In the case of ionic compounds, which do not consist of molecules, we refer to a **formula unit** of the compound. A formula unit of sodium sulphate is $2Na^+ \ SO_4^{2-}$; thus the **relative formula mass** in grams is 142 g, a mole of sodium sulphate.

. . . relative formula mass

The mass of one mole of an element or compound is referred to as its **molar mass**. The molar mass of sodium is 23 g mol^{-1}. The molar mass of sodium hydroxide is 40 g mol^{-1}. If you have m grams of a substance which has a molar mass of $M \text{ g mol}^{-1}$, then the amount of substance in moles, n, is given by

Molar mass and . . .

$$n/\text{mol} = \frac{m/\text{g}}{M/\text{g mol}^{-1}}$$

The **number of moles** of substance is referred to simply as the **amount of substance**:

. . . the amount of substance

$$\text{Amount of substance (number of moles)} = \frac{\text{Mass}}{\text{Molar mass}}$$

================ CHECKPOINT 3A: THE MOLE ================

†**1.** Calculate the molar masses of the following:
(a) $NH_4Fe(SO_4)_2 \cdot 12H_2O$
(b) $Al_2(SO_4)_3$
(c) $K_4Fe(CN)_6$

2. How many moles of substance are present in the following?
(a) 0.250 g of calcium carbonate
(b) 5.30 g of anhydrous sodium carbonate
(c) 5.72 g of sodium carbonate-10-water crystals

3. Use the value of $6.0 \times 10^{23} \text{ mol}^{-1}$ for the Avogadro constant to find the number of atoms in
(a) 2.0×10^{-3} g of calcium
(b) 5.0×10^{-6} g of argon
(c) 1.00×10^{-10} g of mercury

3.4 EMPIRICAL FORMULAE

The **empirical formula** of a compound is the simplest formula which represents its composition. It shows the elements present and the ratio of the amounts of elements present.

Finding an empirical formula . . .

To find an empirical formula, you need to work out the ratio of the amounts of the elements present.

Example   A 0.4764 g sample of an oxide of iron was reduced by a stream of carbon monoxide. The mass of iron that remained was 0.3450 g. Find the empirical formula of the oxide.

Method

...A worked example

Elements present	Iron	Oxygen
Mass/g	0.3450	0.1314
A_r	56	16
Amount/mol	0.3450/56	0.1314/16
	$= 6.16 \times 10^{-3}$	$= 8.21 \times 10^{-3}$

Ratio of amounts $\quad 1 \qquad : \quad \dfrac{8.21 \times 10^{-3}}{6.16 \times 10^{-3}}$

$\qquad\qquad\qquad\quad 1 \qquad : \quad 1.33$

$\qquad\qquad\qquad\quad 3 \qquad : \quad 4$

Empirical formula is Fe_3O_4.

3.5 MOLECULAR FORMULAE

Finding a molecular formula...

The **molecular formula** is a simple multiple of the empirical formula. If the empirical formula is CH_2O, the molecular formula may be CH_2O, $C_2H_4O_2$, $C_3H_6O_3$ and so on.

The way to find out which molecular formula is correct is to find out which gives the correct molar mass.

Example A polymer of empirical formula CH_2 has a molar mass of $28\,000\,g\,mol^{-1}$. What is its molecular formula?

Method

...A worked example

Empirical formula mass $= 14\,g\,mol^{-1}$

Molar mass $= 28\,000\,g\,mol^{-1}$

The molar mass is 2000 times the empirical formula mass; therefore the molecular formula is $(CH_2)_{2000}$.

3.6 CALCULATION OF PERCENTAGE COMPOSITION

The empirical formula shows percentage by mass composition...

From the formula of a compound and the relative atomic masses of the elements in it, the percentage of each element in the compound can be calculated. This is called the **percentage composition by mass**.

Example Calculate the percentage by mass of water of crystallisation in copper(II) sulphate-5-water.

Method

...A worked example

Formula is $CuSO_4 \cdot 5H_2O$

Relative atomic masses are $Cu = 63.5 \quad S = 32 \quad O = 16 \quad H = 1$

Molar mass $= 63.5 + 32 + (4 \times 16) + (5 \times 18)$

$\qquad\qquad\quad = 249.5\,g\,mol^{-1}$

Percentage of water $= \dfrac{90}{249.5} \times 100$

$\qquad\qquad\qquad\qquad = 36.1\%$

CHECKPOINT 3B: FORMULAE AND PERCENTAGE COMPOSITION

1. Calculate the percentage by mass of the named element in the compound listed:

(a) Mg in Mg_3N_2

(b) Na in NaCl

(c) Br in $CaBr_2$

2. Calculate the empirical formulae of the compounds for which the following analytical results were obtained:

(a) 27.3% C, 72.7% O

(b) 53.0% C, 47.0% O

(c) 29.1% Na, 40.5% S, 30.4% O

(d) 32.4% Na, 22.6% S, 45.0% O

3. Find the empirical formulae of the compounds formed in the reactions described below:

(a) 10.800g magnesium form 18.000g of an oxide.

(b) 3.400g calcium form 9.435g of a chloride.

(c) 3.528g iron form 10.237g of a chloride.

4. Weighed samples of the following crystals were heated to drive off the water of crystallisation. When they reached constant mass, the following masses were recorded. Deduce the empirical formulae of the hydrates:

(a) 0.942g of $MgSO_4 \cdot a\,H_2O$ gave 0.461g of residue

(b) 1.124g of $CaSO_4 \cdot b\,H_2O$ gave 0.889g of residue

(c) 1.203g of $Hg(NO_3)_2 \cdot c\,H_2O$ gave 1.172g of residue

3.7 EQUATIONS FOR THE REACTIONS OF SOLIDS

Equations give us much information

Equations tell us not only what substances react together but also what amounts of substances react together. The equation for the action of heat on sodium hydrogencarbonate

$$2NaHCO_3(s) \rightarrow Na_2CO_3(s) + CO_2(g) + H_2O(g)$$

tells us that 2 moles of sodium hydrogencarbonate give 1 mole of sodium carbonate. Since the molar masses are $NaHCO_3 = 84\,\mathrm{g\,mol^{-1}}$ and $Na_2CO_3 = 106\,\mathrm{g\,mol^{-1}}$, it follows that 168g of sodium hydrogencarbonate give 106g of sodium carbonate.

The stoichiometry of a reaction is the relationship between the amounts of reactants and products

The amounts of substances undergoing reaction, as given by the balanced chemical equation, are called the **stoichiometric** amounts. **Stoichiometry** is the relationship between the amounts of reactants and products in a chemical reaction. If one reactant is present in excess of the stoichiometric amount required to react with another of the reactants, then the excess of one reactant will be left unused at the end of the reaction.

A worked example of a calculation based on a stoichiometric equation

Example What mass of zinc can be obtained from the reduction of 10.00 tonnes of zinc oxide by 10.00 tonnes of charcoal? (1 tonne $= 10^3$ kg)

Method Write the equation

$$ZnO(s) + C(s) \rightarrow Zn(s) + CO(g)$$

Amount of ZnO $= 10.00 \times 10^6/(65.4 + 16.00) = 1.23 \times 10^5\,\mathrm{mol}$

Amount of C $= 10.00 \times 10^6/12.00 = 8.33 \times 10^5\,\mathrm{mol}$

Since zinc oxide is present in the smaller amount, the amount of zinc formed is limited by the amount of zinc oxide. From the equation, you can see that 1 mole of ZnO forms 1 mole of Zn.

Amount of Zn $= 1.23 \times 10^5\,\mathrm{mol}$

Mass of Zn $= 1.23 \times 10^5 \times 65.4 \times 10^{-6}\,\mathrm{tonne}$

Mass of Zn $= 8.04\,\mathrm{tonne}$

CHECKPOINT 3C: MASSES OF REACTING SOLIDS

1. What mass of pure aluminium oxide must be electrolysed to give 50 tonne of aluminium?

2. The sulphur present in 0.1000 g of an organic compound is converted into barium sulphate. A precipitate of 0.1852 g of dry $BaSO_4$ is obtained. Calculate the percentage by mass of sulphur in the compound.

3. What is the maximum mass of 2,4,6-trichlorophenol, $C_6H_2Cl_3OH$, that can be obtained from 10.00 g of phenol, C_6H_5OH? A chemist who carried out this conversion obtained 19.54 g of the product. What percentage yield did he obtain?

4. What is the maximum mass of N-benzoylphenylamine, $C_6H_5NHCOC_6H_5$, that can be obtained from 1.00 g of phenylamine, $C_6H_5NH_2$? A chemist who made this derivative obtained 2.04 g. What percentage yield did she obtain?

3.8 EQUATIONS FOR REACTIONS OF GASES

The volume of 1 mole of any ideal gas is the same

For reactions of gases, it is more usual to consider the volumes of reactants and products, rather than their masses. The volume of 1 mole of any ideal gas is the same; 22.414 dm³ at 0 °C and 1 atm (standard temperature and pressure). The gas molar volume is 22.414 dm³ at stp. (Examples of reactions between gases are given on p. 137.)

3.9 EQUATIONS FOR REACTIONS OF SOLIDS AND GASES

When a reaction involves both solids and gases, the solids are usually measured by mass and the gases by volume.

Example What mass of potassium chlorate(V) must be decomposed to supply 200 cm³ of oxygen (measured at stp)? In the presence of a catalyst, decomposition proceeds according to the equation

A worked example of a reaction of solids and gases

$$2KClO_3(s) \xrightarrow{\text{MnO}_2} 2KCl(s) + 3O_2(g)$$

Method From the equation you can see that

2 mol of $KClO_3$ give 3 mol of O_2

Molar mass of $KClO_3$ = 92.5 g mol^{-1}

Therefore 2×92.5 g $KClO_3 \rightarrow 3 \times 22.4$ dm³ O_2

To supply 200 cm³ O_2 you need $\dfrac{2 \times 92.5}{3 \times 22.4} \times 200 \times 10^{-3}$ g $KClO_3$

Mass of $KClO_3$ decomposed = 0.551 g

CHECKPOINT 3D: REACTING VOLUMES OF GASES

---

1. What volume of hydrogen is formed when 3.00 g of magnesium react with an excess of dilute sulphuric acid?

2. Carbon dioxide is obtained by the fermentation of glucose:

$$C_6H_{12}O_6(aq) \rightarrow 6CO_2(g) + 6H_2O(l)$$

If 20.0 dm³ of carbon dioxide (at stp) are collected, what mass of glucose has reacted?

3. In the preparation of hydrogen chloride by the reaction

$$NaCl(s) + H_2SO_4(l) \rightarrow HCl(g) + NaHSO_4(s)$$

what masses of sodium chloride and sulphuric acid are required for the production of 10.0 dm³ of hydrogen chloride (at stp)?

3.10 EQUATIONS FOR REACTIONS IN SOLUTION

Concentrations of solutions are measured in $mol\,dm^{-3}$

When chemical reactions take place between two solutions, the equation for the reaction may be used to find out the volumes of the solutions that react. These depend on the concentrations of the solutes as well as on the stoichiometry of the reaction. The concentration of a solution is measured in terms of the amount (moles) of a solute present in a cubic decimetre (a litre) of solution.

$$\text{Concentration/mol dm}^{-3} = \frac{\text{Amount of solute/mol}}{\text{Volume of solution/dm}^3}$$

(In strict SI units, concentrations are expressed in $mol\,m^{-3}$.) A solution of known concentration is called a **standard solution**.

Titrimetric analysis...

...and the end-point (equivalence-point)

Standard solutions are used in **titrimetric** (or **volumetric**) analysis. The volume of one solution (e.g., an acid) that will react with a known volume of a standard solution of another reagent (e.g., an alkali) is found. The measured addition of one solution to another until the reaction is complete is called **titration**. The point in a titration when the amount of **titrant** added is the stoichiometric amount needed to react with the amount of the other reagent present is called the **end-point** or **equivalence-point**.

3.10.1 ACID—BASE TITRATION

Titrations often involve using standard solutions...

A common practice in titrimetric analysis is to make a standard solution of sodium carbonate and use it to find the concentration of an acid. Anhydrous sodium carbonate is a **primary standard**. This means that it can be weighed out to make a solution of accurately known concentration. The reasons are that it can be obtained in a high state of purity, and it is stable in air and in solution. Other primary standards are ethanedioic acid, $C_2H_2O_4$, butanedioic acid, $C_4H_6O_4$, and sodium hydrogencarbonate.

A worked example of an acid–base titration

Example   A standard solution was prepared by dissolving 2.6061 g of anhydrous sodium carbonate in distilled water and making up to 250 cm³. A 25.0 cm³ portion of this solution was titrated against hydrochloric acid, using methyl orange as indicator. This indicator changes colour when sodium carbonate has been converted into sodium chloride [p. 259]. 18.7 cm³ of the acid were required for neutralisation. What is the concentration of the acid?

Method

(*a*) Find the concentration of the standard solution:

Molar mass of Na_2CO_3 = 106 g mol⁻¹

Concentration of standard solution = 2.6061/(106 × 0.250)

= 0.0983 mol dm⁻³

(*b*) Write the equation:

$$Na_2CO_3(aq) + 2HCl(aq) \rightarrow 2NaCl(aq) + CO_2(g) + H_2O(l)$$

It shows that $1\,mol\,Na_2CO_3$ neutralises $2\,mol\,HCl$.

(*c*) Find the amount of the standard reagent used in titration:

$$\text{Amount of } Na_2CO_3 = \text{Volume/dm}^3 \times \text{Concentration/mol\,dm}^{-3}$$
$$= 25.0 \times 10^{-3} \times 0.0983\,\text{mol}$$
$$= 2.46 \times 10^{-3}\,\text{mol}$$

Amount of $HCl = 2 \times$ Amount of $Na_2CO_3 = 4.92 \times 10^{-3}\,mol$

But amount of HCl used in titration $= 18.7 \times 10^{-3} \times c$

where $c =$ is the concentration of HCl.

Equating these two values gives $4.92 \times 10^{-3} = 18.7 \times 10^{-3} \times c$

The concentration of the acid $c = 0.263\,mol\,dm^{-3}$

3.10.2 BACK-TITRATION

In back-titration the amount of reactant unused at the end of the reaction is found

In the technique known as **back-titration**, a known excess of one reagent **A** is allowed to react with an unknown amount of **B**. At the end of the reaction, the amount of **A** that remains is found by titration. A simple calculation gives the amount of **A** that has been used and the amount of **B** that has reacted.

Example A sample containing ammonium chloride was warmed with $100\,cm^3$ of $1.00\,mol\,dm^{-3}$ sodium hydroxide solution. After all the ammonia had been driven off, the excess of sodium hydroxide required $50.0\,cm^3$ of $0.250\,mol\,dm^{-3}$ sulphuric acid for neutralisation. What mass of ammonium chloride did the sample contain?

Method The two reactions which have taken place are

(*a*) The reaction between the ammonium salt and alkali:

A worked example of back-titration

$$NH_4^+(aq) + OH^-(aq) \rightarrow NH_3(g) + H_2O(l) \tag{1}$$

(*b*) The neutralisation of the excess alkali:

$$2NaOH(aq) + H_2SO_4(aq) \rightarrow Na_2SO_4(aq) + 2H_2O(l) \tag{2}$$

Amount of NaOH initially present $= 100 \times 10^{-3} \times 1.00 = 0.100\,mol$

The amount of NaOH left unused is titrated against sulphuric acid.

Amount of $H_2SO_4 = 50.0 \times 10^{-3} \times 0.250\,mol = 0.0125\,mol$

From equation [2]

Amount of $NaOH = 2 \times$ Amount of $H_2SO_4 = 0.0250\,mol$

Amount of NaOH used in reaction $=$ Initial amount $-$ Amount left over

$$= 0.100 - 0.0250\,mol = 0.0750\,mol$$

From equation [1]

Amount of $NH_4Cl =$ Amount of $NaOH = 0.0750\,mol$

Mass of ammonium chloride $= 0.0750 \times 53.5 = 4.01\,g$

1. A solution is made by dissolving 5.00 g of impure sodium hydroxide in water and making it up to 1.00 dm³ of solution. 25.0 cm³ of this solution is neutralised by 30.3 cm³ of hydrochloric acid, of concentration 0.102 mol dm⁻³. Calculate the percentage purity of the sodium hydroxide.

2. Sodium carbonate crystals (27.8230 g) were dissolved in water and made up to 1.00 dm³. 25.0 cm³ of the solution were neutralised by 48.8 cm³ of hydrochloric acid of concentration 0.100 mol dm⁻³. Find n in the formula $Na_2CO_3 \cdot n\,H_2O$.

3. A fertiliser contains ammonium sulphate. A sample of 0.500 g of fertiliser was warmed with sodium hydroxide solution. The ammonia evolved was absorbed in 100 cm³ of 0.100 mol dm⁻³ hydrochloric acid. The excess of hydrochloric acid required 55.9 cm³ of 0.100 mol dm⁻³ sodium hydroxide for neutralisation. Calculate the percentage of ammonium sulphate in the sample.

4. When sodium hydroxide is added to copper (II) sulphate solution, a precipitate of $Cu_a(OH)_b(SO_4)_c$ is obtained. A 25.0 cm³ portion of a 0.100 mol dm⁻³ solution of copper(II) sulphate required 3.75 cm³ of a 1.00 mol dm⁻³ solution of sodium hydroxide to precipitate all the copper ions.

(*a*) Find the ratio, moles Cu^{2+} : moles OH^- in the precipitate,

(*b*) By considering the charges on the ions, find the simplest formula of $Cu_a(OH)_b(SO_4)_c$,

(*c*) Write an equation for the reaction between copper(II) sulphate and sodium hydroxide.

*See Footnote.

3.11 EQUATIONS FOR OXIDATION–REDUCTION REACTIONS

3.11.1 REDOX REACTIONS

Oxidising agents accept electrons . . .

Oxidising agents are substances which can accept electrons from other substances. **Reducing agents** are substances which can give electrons to other substances. **Oxidation** and **reduction** occur together. In an **oxidation–reduction** reaction or **redox** reaction, electrons pass from the reducing agent to the oxidising agent.

. . . reducing agents donate electrons

Iron(II) ions are reducing agents, losing electrons to form iron(III) ions:

$$Fe^{2+}(aq) \rightarrow Fe^{3+}(aq) + e^- \tag{1}$$

The half-reaction equations for the oxidising agent and the reducing agent combine to give the equation for the redox reaction

Chlorine is an oxidising agent, accepting electrons to form chloride ions:

$$Cl_2(aq) + 2e^- \rightarrow 2Cl^-(aq) \tag{2}$$

Equations [1] and [2] are described as **half-reaction equations**. Free electrons never occur under ordinary laboratory conditions. To represent the real process, the half-reaction equations must be combined. If equation [1] is multiplied by 2 and added to equation [2], the result is

$$2Fe^{2+}(aq) + Cl_2(aq) + 2e^- \rightarrow 2Fe^{3+}(aq) + 2Cl^-(aq) + 2e^-$$

or, as the electrons cancel out

$$2Fe^{2+}(aq) + Cl_2(aq) \rightarrow 2Fe^{3+}(aq) + 2Cl^-(aq)$$

The technique of combining the half-reaction equations for the **oxidant** and the **reductant** is useful because often the equations for redox are more complicated than this example.

---

*For further practice, see E N Ramsden, *Calculations for A-Level Chemistry* (Stanley Thornes)

REACTION BETWEEN MANGANATE(VII) AND IRON(II)

Obtaining the equation for the redox reaction between acidic MnO_4^- and Fe^{2+}

Potassium manganate(VII), $KMnO_4$, in acid solution is a powerful oxidising agent, widely used in titrimetric analysis. The sudden change from purple MnO_4^- ions to pale pink Mn^{2+} ions at the end-point means that no indicator is needed. In balancing the half-reaction equation, $8H^+(aq)$ are needed to combine with 4O in MnO_4^-:

$$MnO_4^-(aq) + 8H^+(aq) \rightarrow Mn^{2+}(aq) + 4H_2O(l)$$

The charge on the left-hand side (LHS) = $-1 + 8 = +7$ units.
The charge on the RHS = $+2$ units.
To equalise the charge on both sides of the equation, 5 electrons are needed on the LHS:

$$MnO_4^-(aq) + 8H^+(aq) + 5e^- \rightarrow Mn^{2+}(aq) + 4H_2O(l) \qquad [3]$$

Potassium manganate(VII) in acidic solution oxidises iron(II) ions. The equation for the reaction is obtained by combining half-reaction equations [1] and [3]. Since Fe^{2+} gives one electron and MnO_4^- needs five, equation [1] is multiplied by 5 and then added to equation [3]:

$$MnO_4^-(aq) + 8H^+(aq) + 5Fe^{2+}(aq) \rightarrow Mn^{2+}(aq) + 4H_2O(l) + 5Fe^{3+}(aq)$$

REACTION BETWEEN DICHROMATE(VI) AND ETHANEDIOATE

Potassium dichromate(VI), $K_2Cr_2O_7$, is an oxidising agent which is most effective in acidic solution. It is reduced to a chromium(III), Cr^{3+}, salt. Balancing the half-reaction equation with respect to mass gives

The redox reaction between acidic $Cr_2O_7^{2-}$ and $C_2O_4^{2-}$

$$Cr_2O_7^{2-}(aq) + 14H^+(aq) \rightarrow 2Cr^{3+}(aq) + 7H_2O(l)$$

Balancing the equation with respect to charge gives

$$Cr_2O_7^{2-}(aq) + 14H^+(aq) + 6e^- \rightarrow 2Cr^{3+}(aq) + 7H_2O(l) \qquad [4]$$

A check shows that the charge on the LHS = $-2 + 14 - 6 = +6$, and the charge on the RHS = $+6$ units.

Sodium ethanedioate and ethanedioic acid are oxidised to carbon dioxide. The half-reaction equation is

$$\begin{matrix} CO_2^- \\ | \\ CO_2^- \end{matrix}(aq) \rightarrow 2CO_2(g) + 2e^- \qquad [5]$$

To obtain the equation for the redox reaction between acidified potassium dichromate and sodium ethanedioate, equation [5] is multiplied by 3 and added to equation [4]:

$$Cr_2O_7^{2-}(aq) + 14H^+(aq) + 3C_2O_4^{2-}(aq) \rightarrow 2Cr^{3+}(aq) + 7H_2O(l) + 6CO_2(g)$$

REACTION BETWEEN IODINE AND SODIUM THIOSULPHATE

Obtaining the equation for the reaction between I_2 and $S_2O_3^{2-}$

Sodium thiosulphate(VI), $Na_2S_2O_3$, is a reducing agent. It is most often used in titrimetric analysis for reducing iodine to iodide ions, being oxidised in the process to sodium tetrathionate, $Na_2S_4O_6$. When the brown colour of iodine fades as the end-point approaches, a little starch solution is added. This gives an intense blue colour with even a trace of iodine. At the end-point the blue colour vanishes. The two half-reaction equations are

$$2S_2O_3^{2-}(aq) \rightarrow S_4O_6^{2-}(aq) + 2e^- \qquad [6]$$

$$I_2(aq) + 2e^- \rightarrow 2I^-(aq) \qquad [7]$$

Combining the two half-reaction equations gives

$$2S_2O_3^{2-}(aq) + I_2(aq) \rightarrow S_4O_6^{2-}(aq) + 2I^-(aq)$$

1. Write balanced half-reaction equations for the oxidation of each of the following:

(a) $Sn^{2+}(aq) \rightarrow Sn^{4+}(aq)$

(b) $Cl^-(aq) \rightarrow Cl_2(aq)$

(c) $H_2S(aq) \rightarrow S(s) + H^+(aq)$

(d) $SO_3^{2-}(aq) \rightarrow SO_4^{2-}(aq) + H^+(aq)$

(e) $H_2O_2(aq) \rightarrow O_2(g) + H^+(aq)$

(f) $NO_2^-(aq) + H_2O(l) \rightarrow NO_3^-(aq) +$

(g) $MnO_2(s) \rightarrow MnO_4^-(aq) + H^+(aq)$

Check that the equations are balanced with respect to charge as well as mass. Remember that H_2O is present in all solutions; you will need it to balance some of the equations.

2. Write balanced half-reaction equations for the following reductions:

(a) $Br_2(aq) \rightarrow Br^-(aq)$

(b) $MnO_2(s) + H^+(aq) \rightarrow Mn^{2+}(aq)$

(c) $NO_3^-(aq) + 10H^+(aq) \rightarrow NH_4^+(aq)$

(d) $PbO_2(s) + H^+(aq) \rightarrow Pb^{2+}(aq)$

(e) $IO^-(aq) + H^+(aq) \rightarrow I_2(aq)$

(f) $ClO_3^-(aq) + H^+(aq) \rightarrow Cl_2(aq)$

(g) $PbO_2(s) + H^+(aq) \rightarrow Pb^{2+}(aq)$

Balance the equations for mass, using H_2O from the solution if needed, and then balance with respect to charge.

3. By combining half-reaction equations, write balanced equations for the following reactions:

(a) $Fe^{3+}(aq) + I^-(aq) \rightarrow$

(b) $Fe^{3+}(aq) + Sn^{2+}(aq) \rightarrow$

(c) $MnO_4^-(aq) + Cl^-(aq) + H^+(aq) \rightarrow$

(d) $MnO_4^-(aq) + H_2O_2(aq) + H^+(aq) \rightarrow$

(e) $MnO_4^-(aq) + H^+(aq) + I^-(aq) \rightarrow$

(f) $Cr_2O_7^{2-}(aq) + H^+(aq) + I^-(aq) \rightarrow$

(g) $Cr_2O_7^{2-}(aq) + H^+(aq) + Fe^{2+}(aq) \rightarrow$

(h) $Cr_2O_7^{2-}(aq) + H^+(aq) + NO_2^-(aq) \rightarrow$

(i) $MnO_4^-(aq) + H^+(aq) + Sn^{2+}(aq) \rightarrow$

(j) $Cr_2O_7^{2-}(aq) + H^+(aq) + SO_3^{2-}(aq) \rightarrow$

(k) $MnO_4^-(aq) + H^+(aq) + SO_3^{2-}(aq) \rightarrow$

(l) $MnO_4^-(aq) + H^+(aq) + Fe^{2+}(aq) \rightarrow$

(m) $Cr_2O_7^{2-}(aq) + H^+(aq) + Sn^{2+}(aq) \rightarrow$

(n) $PbO_2(s) + H^+(aq) + Cl^-(s) \rightarrow$

(o) $ClO_3^-(aq) + H^+(aq) + I^-(aq) \rightarrow$

(p) $Br_2(aq) + I^-(aq) \rightarrow$

(q) $Cl_2(g) + NO_2^-(aq) + H_2O \rightarrow$

(r) $Cl_2(g) + H^+(aq) + IO^-(aq) \rightarrow$

(s) $MnO_2(s) + H^+(aq) + Cl^-(s) \rightarrow$

(t) $Br_2(g) + H_2S(g) \rightarrow$

3.12 OXIDATION NUMBERS

Redox reactions can be discussed in terms of changes in oxidation numbers

Oxidation–reduction reactions are often discussed in terms of the change in the **oxidation number** or **oxidation state** of each reactant. In the case of ions, the oxidation number of the element is the charge on the ion. The oxidation numbers of iron and chlorine in $FeCl_2$ are $Fe = +2$ and $Cl = -1$. Oxidation results in an increase in the oxidation number: iron is oxidised from an oxidation state of $+2$ in $FeCl_2$ to an oxidation state of $+3$ in $FeCl_3$.

The use of oxidation numbers is extended to covalent compounds. Some elements are assigned positive oxidation numbers and others negative oxidation numbers, in accordance with certain rules.

3.12.1 RULES FOR ASSIGNING OXIDATION NUMBERS

The oxidation numbers of elements...

1. The oxidation number of an uncombined element is zero.

...in ions and compounds...

2. In ionic compounds, the oxidation number is equal to the charge on the ion.

3. The sum of the oxidation numbers of all the atoms or ions in a compound is zero. For example, in $FeCl_3$ the sum of the oxidation numbers is

$$+3 + 3(-1) = 0$$

...the sum of oxidation numbers

4. The sum of the oxidation numbers of all the atoms in an ion is equal to the charge on the ion. In SO_4^{2-}, the sum of the oxidation numbers ($S = +6$, $O = -2$) is

$$+6 + 4(-2) = -2$$

which is the charge on the ion.

5. Some elements nearly always employ the same oxidation number in their compounds. They are used as reference points in assigning oxidation numbers to other elements. The reference elements are

Reference elements...

K	Na	+1	H	+1	except in metal hydrides
Mg	Ca	+2	F	−1	
Al		+3	Cl	−1	except in compounds with O and F
			O	−2	except in peroxides, superoxides, fluorides

Example What is the oxidation number of thallium in $TlCl_3$?

Method Chlorine always has the oxidation number −1.

...Some worked examples

Therefore (Ox. No. of Tl) + 3(−1) = 0

and the oxidation number of thallium is +3.

Example What is the oxidation number of Cl in Cl_2O_7?

Method The exceptions to the rule that the oxidation number of Cl equals −1 are compounds with O and F. Oxygen is the reference point with the oxidation number −2.

Therefore 2 (Ox. No. of Cl) + 7(−2) = 0

and the oxidation number of chlorine is +7.

Example What is the oxidation number of Cr in $Cr(CN)_6^{3-}$?

Method The cyanide ion, CN^-, has a charge of −1.

Therefore (Ox. No. of Cr) + 6(−1) = −3

and the oxidation number of chromium is +3.

CHECKPOINT 3G: OXIDATION NUMBERS

1. What is the oxidation number of the named element in the following species (ions or molecules)?

(a) N in NO, NO_2, N_2O_4, N_2O, NO_2^-, NO_3^-, N_2O_5

(b) Mn in $MnSO_4$, Mn_2O_3, MnO_2, MnO_4^-, MnO_4^{2-}

(c) As in As_2O_3, AsO_2^-, AsO_4^{3-}, AsH_3

(d) Cr in CrO_4^{2-}, $Cr_2O_7^{2-}$, CrO_3

(e) I in I^-, IO^-, IO_3^-, I_2, ICl_3, ICl_2^-

3.12.2 CHANGES IN OXIDATION NUMBER

Oxidation numbers increase on oxidation and decrease on reduction...

When an element is **oxidised,** its oxidation number increases; when an element is **reduced,** its oxidation number decreases. In a redox reaction

$$x\mathbf{A} + y\mathbf{B} \rightarrow$$

if the oxidation number of **A** changes by $+a$ units, and the oxidation number of **B** changes by $-b$ units

then $x(+a) + y(-b) = 0$

Example Consider the reduction of iron(III) ions by a tin(II) salt:

...Some worked examples

$$Sn^{2+}(aq) + 2Fe^{3+}(aq) \rightarrow Sn^{4+}(aq) + 2Fe^{2+}(aq)$$

For tin, change in Ox. No. = $+2$

For iron, change in Ox. No. = -1

And $1(+2) + 2(-1) = 0$

Example

$$3I_2(aq) + 3OH^-(aq) \rightarrow IO_3^-(aq) + 5I^-(aq) + 3H^+(aq)$$

The only element which changes its oxidation number is iodine.

On the LHS, in I_2 Ox. No. of I = 0

On the RHS, in IO_3^- (Ox. No. of I) $+ 3(-2) = -1$

and the Ox. No. of I = $+5$

In I^-, Ox. No. of I = -1

Iodine has changed from oxidation number zero on the LHS to a combination of Ox. No. $+5$ and Ox. No. -1 on the RHS. Part of the iodine has been oxidised and part has been reduced. A reaction of this kind is termed a **disproportionation reaction.**

3.12.3 BALANCING EQUATIONS BY THE OXIDATION NUMBER METHOD

Balancing equations...

The oxidation number method of balancing equations is best explained through an example.

Example Balance the equation

$$a\,KIO_3(aq) + b\,Na_2SO_3(aq) \rightarrow c\,KIO(aq) + d\,Na_2SO_4(aq)$$

Iodine changes from Ox. No. $+5$ in KIO_3 to $+1$ in KIO.

Change in Ox. No. of I = -4

Sulphur changes from Ox. No. $+4$ in Na_2SO_3 to $+6$ in Na_2SO_4.

...A worked example

Change in Ox. No. of S = $+2$

Therefore $a(-4) + b(+2) = 0$

If $a = 1$, $b = 2$ and the equation becomes

$$KIO_3(aq) + 2Na_2SO_3(aq) \rightarrow c\,KIO(aq) + d\,Na_2SO_4(aq)$$

By stoichiometry, it follows that $c = 1$ and $d = 2$, giving

$$KIO_3(aq) + 2Na_2SO_3(aq) \rightarrow KIO(aq) + 2Na_2SO_4(aq)$$

1. Use the oxidation number method to balance the equations:

(a) $MnO_4^-(aq) + H^+(aq) + Fe^{2+}(aq) \rightarrow$
$$Mn^{2+}(aq) + Fe^{3+}(aq) + H_2O(l)$$

(b) $Mn^{2+}(aq) + BiO_3^-(aq) + H^+(aq) \rightarrow$
$$MnO_4^-(aq) + Bi^{3+}(aq) + H_2O(l)$$

(c) $As_2O_3(s) + MnO_4^-(aq) + H^+(aq) \rightarrow$
$$As_2O_5(s) + Mn^{2+}(aq) + H_2O(l)$$

(d) $Sn(s) + HNO_3(aq) \rightarrow SnO_2(s) + NO_2(g) + H_2O(l)$

(e) $Cu^{2+}(aq) + I^-(aq) \rightarrow CuI(s) + I_2(aq)$

(f) $Cl_2(g) + OH^-(aq) \rightarrow Cl^-(aq) + ClO^-(aq) + H_2O(l)$

(g) $Zn(s) + Fe^{3+}(aq) \rightarrow Zn^{2+}(aq) + Fe^{2+}(aq)$

(h) $I_2(aq) + S_2O_3^{2-}(aq) \rightarrow I^-(aq) + S_4O_6^{2-}(aq)$

2. One mole of the compound ICl_x reacts with an excess of potassium iodide solution to give two moles of I_2. Write an equation for the reaction, and state the oxidation number of I in ICl_x.

3. State the oxidation number of the species which is underlined in the following equations. Say whether the species is oxidised or reduced during the reaction. Complete and balance the equations:

(a) $\underline{H_2O_2} + \underline{I}^- + H^+(aq) \rightarrow \underline{H_2O} + \underline{I_2}$

(b) $\underline{Cu} + \underline{NO_3}^- + H^+(aq) \rightarrow \underline{Cu}^{2+} + \underline{N}O + H_2O$

(c) $\underline{Fe}^{2+} + \underline{Cr_2O_7}^{2-} + H^+(aq) \rightarrow \underline{Fe}^{3+} + \underline{Cr}^{3+} + H_2O$

(d) $\underline{S_2O_3}^{2-} + \underline{I_2} \rightarrow \underline{S_4O_6}^{2-} + \underline{I}^-$

(e) $K\underline{I}O_3 + K\underline{I} + H\underline{Cl} \rightarrow KCl + \underline{ICl} + 3H_2O$

(f) $\underline{CrO_4}^{2-} + H^+(aq) \rightarrow \underline{Cr_2O_7}^{2-}$

3.13 OXIDATION NUMBERS AND NOMENCLATURE

3.13.1 SYSTEMATIC NOMENCLATURE

Systems for naming compounds

Oxidation numbers are used in the naming of compounds. Systematic nomenclature is set out by IUPAC (the International Union of Pure and Applied Chemistry) in their *Manual of Symbols and Terminology for Physiochemical Quantities and Units* and also by ASE (the Association for Science Education) in *Chemical Nomenclature, Symbols and Terminology* (2nd edition, 1979). The systems are not quite the same. Although the IUPAC system, coming from an international body, is more widely used, the examination boards follow the ASE system. This book, since it aims to prepare the readers for examinations, also follows the ASE system.

3.13.2 CATIONS

The oxidation state of the element is specified if it is variable

Cations (positive ions) are given the name of the element together with the oxidation number. This system of naming was devised by A Stock:

> e.g. Fe^{2+} iron(II) ion Fe^{3+} iron(III) ion

Names of cations

When there is no doubt about the oxidation state because an element assumes one only, then it is omitted:

> e.g. Na^+ sodium ion Al^{3+} aluminium ion

3.13.3 ANIONS

Elemental **anions** (negative ions) are named after the element, with the ending *-ide*:

> e.g. H^- hydride N^{3-} nitride

Compound anions have names ending in *-ide*, *-ite* or *-ate*:

> e.g. OH^- hydroxide NO_2^- nitrite NO_3^- nitrate

Many elements form more than one **oxoanion**, using more than one oxidation state (e.g., NO_2^-, NO_3^-). The names are derived from the name of the element which is combined with oxygen and the ending *-ate*:

e.g. SO_4^{2-} sulphate ion HCO_3^- hydrogencarbonate ion

Both ClO^- and ClO_3^- are chlorate ions. To distinguish between them, the oxidation number of chlorine is added:

Names of anions

e.g. ClO^- chlorate(I)	ClO_3^- chlorate(V)
also CrO_4^{2-} chromate(VI)	$Cr_2O_7^{2-}$ dichromate(VI)
MnO_4^{2-} manganate(VI)	MnO_4^- manganate(VII)
NO_3^- nitrate(V) or nitrate	NO_2^- nitrate(III) or nitrite
SO_4^{2-} sulphate(VI) or sulphate	SO_3^{2-} sulphate(IV) or sulphite

The Stock names for the last four examples have not been widely adopted, and people prefer to use nitrate, nitrite, sulphate and sulphite. These names date back to their usage before the changes in nomenclature of 1970.

3.13.4 ACIDS

Acids are named after their anions:

Names of acids

e.g. $HClO$ chloric(I) acid

$HClO_2$ chloric(III) acid

$HClO_3$ chloric(V) acid

Again, the names nitrous acid and sulphurous acid are preferred to the Stock names (nitric(III) and sulphuric(IV)) for the acids HNO_2 and H_2SO_3.

3.13.5 SALTS

Salts are named by combining the names of the cation, with its oxidation number if that is variable, and the anion:

Names of salts

e.g. $FeSO_4$ iron(II) sulphate $NaClO$ sodium chlorate(I)

When a salt is hydrated, the number of water molecules per formula unit is stated:

e.g. $CuSO_4 \cdot 5H_2O$ copper(II) sulphate-5-water

3.13.6 STOICHIOMETRIC FORMULAE

Some compounds are named by stoichiometry

The oxides, sulphides and halides of non-metallic elements are usually named, not by the Stock system but according to their stoichiometry:

e.g. NO nitrogen oxide	CS_2 carbon disulphide
N_2O dinitrogen oxide	$SiCl_4$ silicon tetrachloride
NO_2 nitrogen dioxide	$POCl_3$ phosphorus trichloride oxide
N_2O_4 dinitrogen tetraoxide	$SOCl_2$ sulphur dichloride oxide

Phosphorus compounds are sometimes named by stoichiometry, as above, and sometimes by the Stock system:

e.g. PCl_5 phosphorus pentachloride or phosphorus(V) chloride.

3.14 TITRIMETRIC ANALYSIS USING REDOX REACTIONS

Titrimetric analysis using redox reactions...

Redox reactions are used in titrimetric analysis. For example, a solution of unknown concentration of a reductant is titrated against a standard solution of an oxidant. From the volumes of the two solutions and the equation for the reaction, the concentration of the unknown solution can be found.

Example (a) Find the concentration of an iron(II) sulphate solution, given that $25.0\,cm^3$ of the solution, when acidified, required $19.8\,cm^3$ of $0.0200\,mol\,dm^{-3}$ potassium manganate(VII) for oxidation.

Method The equation for the reaction comes first. A combination of the two half-reaction equations, as described on p. 54, gives

...Some worked examples

$$MnO_4^-(aq) + 5Fe^{2+}(aq) + 8H^+(aq) \rightarrow Mn^{2+}(aq) + 5Fe^{3+}(aq) + 4H_2O(l)$$

The equation indicates that $1\,mol$ of MnO_4^- oxidises $5\,mol$ of Fe^{2+}.

Amount of MnO_4^- in $19.8\,cm^3$ = $19.8 \times 10^{-3} \times 0.0200\,mol$
$$= 0.396 \times 10^{-3}\,mol$$

Amount of Fe^{2+} in $25.0\,cm^3$ = $5 \times$ amount of MnO_4^-
$$= 1.98 \times 10^{-3}\,mol$$

Concentration of Fe^{2+} = $(1.98 \times 10^{-3})/(25.0 \times 10^{-3})\,mol\,dm^{-3}$

Concentration of $FeSO_4$ = $7.92 \times 10^{-2}\,mol\,dm^{-3}$

Example (b) A standard solution is prepared by dissolving $1.185\,g$ of 'AnalaR' potassium dichromate(VI) and making up to $250\,cm^3$ of solution. This solution is used to find the concentration of a sodium thiosulphate solution. A $25.0\,cm^3$ portion of the oxidant was acidified and added to an excess of potassium iodide to liberate iodine.

$$Cr_2O_7^{2-}(aq) + 6I^-(aq) + 14H^+(aq) \rightarrow 3I_2(aq) + 2Cr^{3+}(aq) + 7H_2O(l) \quad [1]$$

When the solution was titrated against sodium thiosulphate solution, $17.5\,cm^3$ of 'thio' were required. Find the concentration of the thiosulphate solution.

Method Combining the half-reaction equations

$$I_2(aq) + 2e^- \rightarrow 2I^-(aq)$$

$$2S_2O_3^{2-}(aq) \rightarrow S_4O_6^{2-}(aq) + 2e^-$$

gives

$$2S_2O_3^{2-}(aq) + I_2(aq) \rightarrow S_4O_6^{2-}(aq) + 2I^-(aq) \quad [2]$$

Concentration of $K_2Cr_2O_7$ = $(1.185/294) \times 4 = 0.0161\,mol\,dm^{-3}$

Amount of I_2 in $25.0\,cm^3$ = $0.0161 \times 25.0 \times 10^{-3} \times 3\,mol$
(from equation [1])　　　 = $1.208 \times 10^{-3}\,mol$

Amount of thio in $17.5\,cm^3$ = $1.208 \times 10^{-3} \times 2\,mol$
(from equation [2])　　　 = $2.416 \times 10^{-3}\,mol$

Concentration of thiosulphate = $(2.416 \times 10^{-3})/(17.5 \times 10^{-3})\,mol\,dm^{-3}$
$$= 0.138\,mol\,dm^{-3}$$

CHECKPOINT 3I: REDOX TITRATIONS

1. A 0.1576 g piece of iron wire was converted into Fe^{2+} ions and then titrated against potassium dichromate solution of concentration $1.64 \times 10^{-2} \, mol \, dm^{-3}$. From the fact that 27.3 cm³ of the oxidant were required, calculate the percentage purity of the iron wire.

2. A volume of 27.5 cm³ of a 0.0200 mol dm⁻³ solution of potassium manganate(VII) was required to oxidise 25.0 cm³ of a solution of hydrogen peroxide. Calculate the concentration of hydrogen peroxide and the volume of oxygen (at stp) evolved during the titration.

3. Calculate the percentage purity of an impure sample of sodium thiosulphate from the following data. A 0.2368 g sample of the sodium thiosulphate was added to 25.0 cm³ of 0.0400 mol dm⁻³ iodine solution. The excess of iodine that remained after reaction needed 27.8 cm³ of 0.0400 mol dm⁻³ thiosulphate solution in a titration.

4. What volume of potassium manganate(VII) solution of concentration 0.0100 mol dm⁻³ will oxidise 50.0 cm³ of iron(II) ethanedioate solution of concentration 0.0200 mol dm⁻³ in acid conditions?

5. Mercury is oxidised by potassium manganate(VII) solution, with the formation of manganese(IV) oxide, MnO_2, and potassium hydroxide and an oxide of mercury. 50.0 cm³ of 0.0200 mol dm⁻³ potassium manganate(VII) solution oxidise 0.600 g of mercury. Work out the equation for the reaction.

*See Footnote.

3.15 PHYSICAL EQUILIBRIA

A system is a part of the universe considered in isolation

We have looked at some chemical reactions. Later in this book, we shall look at some physical changes such as evaporation and condensation, melting and dissolution. In a system in which a chemical or physical change is occurring, the properties of the system change. (The word **system** is used to describe a part of the universe which one wants to study in isolation from the rest of the universe.) Systems in which no change in the properties of the system is occurring are said to be **at equilibrium**. There are two kinds of systems: systems **in a state of change** and systems at equilibrium.

Consider what happens when you drop 5 cm³ of the brown liquid, bromine, into a gas jar and replace the lid [see Figure 3.1].

FIGURE 3.1
Vaporisation of Liquid Bromine

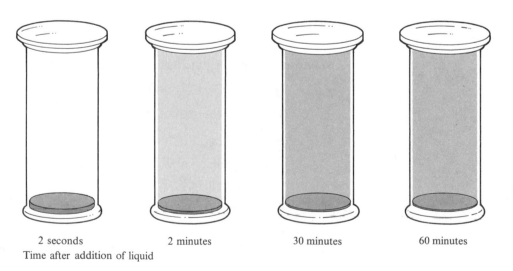

| 2 seconds | 2 minutes | 30 minutes | 60 minutes |

Time after addition of liquid

---

*For further practice, see E N Ramsden, *Calculations for A-Level Chemistry* (Stanley Thornes)

A phase is a physically distinct part of a system

As soon as the liquid enters the gas jar, it begins to **vaporise**: some molecules leave the liquid phase and enter the vapour phase. A **phase** is a part of a system which is physically distinct from other parts of the system. The system we are considering is the contents of the closed gas jar. Two phases are present: liquid (bromine) and gas (bromine vapour and air).

After 2 minutes, the gas in the gas jar is brown because it contains bromine molecules as well as air. Vaporisation (or **evaporation**) continues, and after 30 minutes the brown colour of bromine vapour is even more intense. The colour does not continue to deepen for ever. After 60 minutes, it is no more intense than after 30 minutes. It looks as though vaporisation has ceased, and the system is at equilibrium.

$Br_2(l)$ and $Br_2(g)$ reach equilibrium in a closed system...

If you could see individual molecules of bromine, however, you would see that the population of bromine molecules in the gas phase is constantly changing. Molecules of bromine are still passing from the liquid to the gas phase but, as fast as they do this, molecules of bromine pass from the gas phase to the liquid phase, that is, they **condense**. The system is at equilibrium because

Rate of vaporisation = Rate of condensation

...and the equilibrium is dynamic

This kind of system is described as being in **dynamic equilibrium**. Dynamic means *moving*, and, at a molecular level, the system is in motion. The properties of the system in bulk are unchanging; the volume of liquid bromine and the concentration of bromine in the gas phase are no longer changing:

$$Br_2(l) \rightleftharpoons Br_2(g)$$

If the system were not closed, it would not come to equilibrium. If the gas jar were open, bromine would continue to vaporise until there was no liquid bromine left.

Consider what happens when you stir a scoopful of copper(II) sulphate crystals in a beaker of water. As the salt dissolves, the solution becomes a more and more intense blue colour.[See Figure 3.2.]

FIGURE 3.2   Dissolution

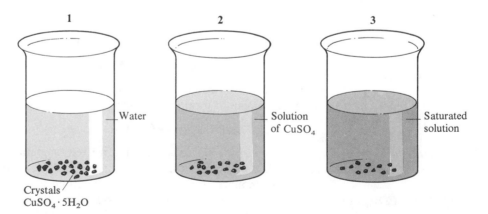

A saturated solution is in a state of dynamic equilibrium

After a while, the intensity of the blue colour remains constant, although (provided you have used an excess of crystals) undissolved copper(II) sulphate remains at the bottom of the beaker. The saturated solution is a system at equilibrium. Although nothing more seems to be happening, in fact copper(II) sulphate is still dissolving but, as fast as it does so, copper(II) sulphate is crystallising from solution:

$$CuSO_4 \cdot 5H_2O(s) + aq \rightleftharpoons Cu^{2+}(aq) + SO_4^{2-}(aq) + 5H_2O(l)$$

A radioactive tracer can be used to demonstrate the dynamic nature of the equilibrium

There is a way of demonstrating that this system is in dynamic equilibrium. It involves the use of a radioactive **tracer**. If some crystals of $Cu^{35}SO_4 \cdot 5H_2O$, which contain radioactive ^{35}S, are added, you might expect that none would dissolve because the solution is already saturated. After a time, however, it is found that the radioactivity is divided between the solution and the undissolved crystals. The reason is that undissolved solid is constantly dissolving, while solute crystallises from the solution at the same rate. [See Figure 3.3.]

FIGURE 3.3
An Experiment using a Radioactive Tracer

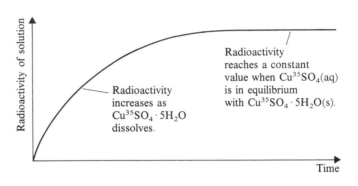

Radioactivity of solution

Radioactivity increases as $Cu^{35}SO_4 \cdot 5H_2O$ dissolves.

Radioactivity reaches a constant value when $Cu^{35}SO_4(aq)$ is in equilibrium with $Cu^{35}SO_4 \cdot 5H_2O(s)$.

Time

3.16 CHEMICAL EQUILIBRIA

The dynamic equilibria described above are physical changes. Chemical reactions can also come to equilibrium.

Chemical reactions, like physical changes, can reach a state of equilibrium

Some chemical reactions take place in one direction almost exclusively. For example, magnesium burns to form magnesium oxide:

$$2Mg(s) + O_2(g) \rightarrow 2MgO(s)$$

The tendency for magnesium oxide to split up to form magnesium and oxygen is negligible at normal temperatures.

Other chemical reactions take place in both directions at comparable rates. For example, when calcium carbonate is heated strongly, it decomposes:

$$CaCO_3(s) \rightarrow CaO(s) + CO_2(g)$$

The products formed are the base, calcium oxide, and the acid gas, carbon dioxide. They recombine to form calcium carbonate:

$$CaO(s) + CO_2(g) \rightarrow CaCO_3(s)$$

In the thermal dissociation of $CaCO_3$...

...$CaO(s) + CO_2(g)$ are in equilibrium with $CaCO_3(s)$ in a closed system...

When calcium carbonate is heated at a fixed temperature in a closed container, at first calcium carbonate decomposes faster than the products recombine. After a while the amounts of calcium oxide and carbon dioxide build up to a level at which the rate of combination of calcium oxide and carbon dioxide is equal to the rate at which calcium carbonate dissociates. The system has reached a state of dynamic equilibrium:

$$CaCO_3(s) \rightleftharpoons CaO(s) + CO_2(s)$$

...if one of the products is removed the equilibrium is disturbed

Equilibrium is reached in a closed system. If the container is open, carbon dioxide can escape. The equilibrium is disturbed, and more calcium carbonate dissociates to try to restore the equilibrium. When limestone is heated in a lime kiln, as the aim is to make plenty of quicklime, the carbon dioxide formed is removed by a powerful through draft of air in order to stop the system coming to equilibrium.

The pressure, the temperature and other external factors affect systems in equilibrium. H L Le Chatelier made a study of the way in which systems at equilibrium adjust when external factors are changed. His work is covered in Chapter 11.

3.17 EQUILIBRIUM CONSTANTS

The equilibrium constant is a measure of the extent of reaction

The extent to which the reactants are converted into the products before equilibrium is reached is measured by the **equilibrium constant** for the reaction. An example is the reaction between ethanoic acid and ethanol:

$$CH_3CO_2H(l) + C_2H_5OH(l) \rightleftharpoons CH_3CO_2C_2H_5(l) + H_2O(l)$$

The equilibrium constant K_c is given by

$$K_c = \frac{[CH_3CO_2C_2H_5][H_2O]}{[CH_3CO_2H][C_2H_5OH]}$$

where $[C_2H_5OH]$ is the concentration of ethanol in $mol\,dm^{-3}$. In general, in an equilibrium

$$mA + nB \rightleftharpoons pC + qD$$

$$K_c = \frac{[C]^p[D]^q}{[A]^m[B]^n}$$

The convention is to put the products in the numerator and the reactants in the denominator.

$$Equilibrium\ constant = \frac{Product\ of\ the\ concentrations\ of\ the\ products\ raised\ to\ the\ appropriate\ powers}{Product\ of\ the\ concentrations\ of\ the\ reactants\ raised\ to\ the\ appropriate\ powers}$$

The appropriate power is the coefficient before that species in the stoichiometric equation for the reaction.

3.17.1 CATALYSTS AND EQUILIBRIUM CONSTANTS

Catalysts increase the speed with which the equilibrium conditions are reached

Catalysts alter the rates of chemical reactions. In the case of a reaction which reaches a state of dynamic equilibrium, a catalyst increases the rates of both the forward reaction and the reverse reaction by the same ratio. The position of equilibrium is therefore unchanged. The equilibrium constant remains the same. What the catalyst does is to decrease the time needed for the system to reach a state of equilibrium. In industrial processes, a catalyst can make a valuable contribution to the economy of the process. In the Haber process, the percentage conversion of nitrogen and hydrogen to ammonia at the temperatures at which plants operate is small. The use of a catalyst to achieve the same percentage conversion in a shorter time increases the productivity of the plant.

3.18 OXIDATION–REDUCTION EQUILIBRIA

Like other reactions, oxidation–reduction reactions do not always go nearly to completion. In many cases, a state of equilibrium is reached. The reactants and

Redox reactions reach a position of equilibrium...

products may be present at equilibrium in comparable amounts. If solutions of the reducing agent, iron(II) sulphate, and the oxidising agent, iodine, are mixed, the resulting solution will contain Fe^{3+}, Fe^{2+}, I_2 and I^-:

...for example

$$I_2(aq) + 2Fe^{2+}(aq) \rightleftharpoons 2I^-(aq) + 2Fe^{3+}(aq) \qquad [1]$$

$Fe^{2+} + I_2$...

When solutions of the oxidising reagent, Fe^{3+}, and the reducing agent, I^-, are mixed, the resulting solution will again contain all four species:

...$Fe^{3+} + I^-$

$$2Fe^{3+}(aq) + 2I^-(aq) \rightleftharpoons 2Fe^{2+}(aq) + I_2(aq) \qquad [2]$$

If the same amounts of iron and iodine have been used to make the two solutions, the compositions of solutions [1] and [2] will be the same. The reason is that equilibrium is established, and the position of equilibrium is the same, whether it is reached by route [1] or route [2]. The concentrations of the four species at equilibrium are related

The equilibrium constant

by the equilibrium constant. For reaction [2], the equilibrium constant is

$$K_c = \frac{[Fe^{2+}]^2[I_2]}{[Fe^{3+}]^2[I^-]^2} = 6 \times 10^7 \, mol^{-1} dm^3$$

The concentration of each reactant and product is raised to the appropriate power, i.e., the coefficient before that reactant in the stoichiometric equation. The value of K_c is high because the position of equilibrium lies far over to the right hand side of equation [2]. In $1 \, dm^3$ of solution made by adding $0.1 \, mol$ of Fe^{3+} and $0.05 \, mol$ of I_2, the concentration of Fe^{3+} remaining at equilibrium is only $5 \times 10^{-5} \, mol \, dm^{-3}$.

With other redox reactions, the **position of equilibrium** (the extent to which the reaction reaches completion) lies closer to the left hand side. The study of equilibria will be continued in Chapter 11.

QUESTIONS ON CHAPTER 3

1. Explain what is meant by
(a) stoichiometric equation
(b) oxidation
(c) reduction
(d) disproportionation

2. State what has been oxidised and what has been reduced in the following reactions:

(a) $Zn(s) + 2HCl(aq) \rightarrow ZnCl_2(aq) + H_2(g)$

(b) $CH_4(g) + 4Cl_2(g) \rightarrow CCl_4(l) + 4HCl(g)$

(c) $NH_4^+NO_3^-(s) \rightarrow N_2O(g) + 2H_2O(l)$

(d) $IO_3^-(aq) + 5I^-(aq) + 6H^+(aq) \rightarrow 3I_2(aq) + 3H_2O(l)$

(e) $2Fe(CN)_6^{4-}(aq) + Cl_2(g) \rightarrow 2Fe(CN)_6^{3-}(aq) + 2Cl^-(aq)$

(f) $2CrO_4^{2-}(aq) + 2H^+(aq) \rightarrow Cr_2O_7^{2-}(aq) + H_2O(l)$

(g) $2CuCl(aq) \rightarrow Cu(s) + CuCl_2(aq)$

3. What volume of $0.250 \, mol \, dm^{-3}$ sodium hydroxide solution is required to neutralise $25.0 \, cm^3$ of $0.150 \, mol \, dm^{-3}$ sulphuric acid?

4. Explain what is meant by the terms
(a) equilibrium
(b) dynamic equilibrium
(c) equilibrium constant

5. A $25.0 \, g$ measure of household ammonia was dissolved in water and made up to $500 \, cm^3$. A $25.0 \, cm^3$ portion of this solution required $29.4 \, cm^3$ of $0.250 \, mol \, dm^{-3}$ sulphuric acid for neutralisation. What is the percentage by mass of ammonia in the cleaning fluid?

6. Arsenic can be oxidised to arsenic(V) acid, H_3AsO_4. This acid oxidises I^- ions to I_2, which can be estimated by titration against a standard thiosulphate solution:

$$As + 5HNO_3 \rightarrow H_3AsO_4 + 5NO_2 + H_2O$$

$$H_3AsO_4 + 2HI \rightarrow H_3AsO_3 + I_2 + H_2O$$

If $0.1058 \, g$ of a sample containing arsenic required $28.7 \, cm^3$ of a $0.0198 \, mol \, dm^{-3}$ solution of sodium thiosulphate in the final titration, what is the percentage of arsenic in the sample?

7. Use the following experimental data to deduce an equation for the reaction between chlorine and sodium thiosulphate in aqueous solution, and explain your reasoning.

$3.10 \, g$ of $Na_2S_2O_3 \cdot 5H_2O$ were dissolved in water to give $250 \, cm^3$ of solution, and chlorine was passed through until reaction was complete. Excess chlorine was then swept from the solution using gaseous nitrogen.

$25.0 \, cm^3$ samples of the resulting solution were found

(*a*) to require 12.5 cm³ of 1.00 M KOH for neutralization

(*b*) to require 10.0 cm³ of 1.00 M AgNO₃ to complete the precipitation of chloride ions, and

(*c*) to give 0.583 g of white precipitate when treated with an excess of aqueous BaCl₂

Iodine reacts with sodium thiosulphate according to the equation

$$I_2 + 2Na_2S_2O_3 \rightarrow 2NaI + Na_2S_4O_6$$

Compare the changes in oxidation number of the chlorine and sulphur in the first reaction with those of iodine and sulphur in the second. What explanation can you offer for this difference between the two halogen elements? (Relative atomic masses: H = 1, O = 16, Na = 23, S = 32, Ba = 137)   (L 83)

8. The following is a method by which the reaction between iron(III) ions and hydroxylammonium chloride, $NH_3OH^+Cl^-$, may be investigated:

25.0 cm³ of a solution containing 3.60 g dm⁻³ of hydroxylammonium chloride is added to a solution containing an excess of Fe^{3+} ions and about 25 cm³ of 1 M sulphuric acid, and the mixture boiled. It is then diluted with water, allowed to cool, and the Fe^{2+} ions titrated with 0.02 M potassium manganate(VII) (potassium permanganate) of which 25.9 cm³ were required.

(*a*) Calculate the molar ratio Fe^{3+}/NH_3OH^+ in the reaction, and hence determine the oxidation number of nitrogen in the product.

(*b*) Using the oxidation number concept, or otherwise, deduce the equation for the reaction.

(H = 1, N = 14, O = 16, Cl = 35.5)   (L80.S)

4

THE CHEMICAL BOND

4.1 THE IONIC BOND

In working out the electronic configurations of atoms, we saw how the configuration ns^2np^6 is associated with neon, argon, krypton and xenon. These gases, together with helium (configuration $1s^2$) are called the noble gases. For many years after their discovery, it appeared that the noble gases were incapable of taking part in any chemical reactions; they are unreactive (see Chapter 16). The chemical stability of the noble gases seemed to be associated with a full outer shell of eight electrons (or a full K shell of two electrons in the case of helium). In 1916, W Kossel and G N Lewis independently put forward theories of chemical bonding. They both explained the formation of chemical bonds in terms of the tendency of atoms to give or receive or share electrons in order to attain a stable, noble gas type of electron configuration.

The noble gases are unreactive...

...Their stability appears to be associated with their full outer shells of electrons

Consider the section of the Periodic Table containing neon:

$$O(2.6) \quad F(2.7) \quad Ne(2.8) \quad Na(2.8.1) \quad Mg(2.8.2) \quad Al(2.8.3)$$

The configurations give the numbers of electrons in the K, L and M shells. Sodium can attain the stable electron configuration of neon by losing one electron:

Na can attain a full outer shell by losing an electron to form Na^+...

$$Na(2.8.1) \rightarrow Na^+(2.8) + e^-$$

With ten electrons and eleven protons (the same number as the sodium atom) the species formed has a positive charge; it is a sodium ion. Fluorine is one electron short of the neon electron configuration. If it obtains one electron (from sodium for example) it can achieve a full outer shell of eight electrons:

...F can attain a full outer shell by gaining an electron to form F^-

$$F(2.7) + e^- \rightarrow F^-(2.8)$$

The species formed has ten electrons and nine protons; it is a negatively charged fluoride ion.

Between the positively charged sodium ion and negatively charged fluoride ion there exists a force of **electrostatic attraction**. The attraction constitutes a chemical bond. This type of bond is called an **ionic bond** or an **electrovalent bond**.

Electrostatic attraction between cations and anions is one type of chemical bond

Magnesium and aluminium have to lose two and three electrons respectively to attain the configuration of neon:

...the ionic bond or electrovalent bond

$$Mg(2.8.2) \rightarrow Mg^{2+}(2.8) + 2e^-$$

$$Al(2.8.3) \rightarrow Al^{3+}(2.8) + 3e^-$$

Oxygen has to gain two electrons to complete its octet:

$$O(2.6) + 2e^- \rightarrow O^{2-}(2.8)$$

Electrovalent or ionic compounds consist of ions

The compounds formed by these **cations** (positive ions) and **anions** (negative ions) are neutral, uncharged substances. They are called **electrovalent** or **ionic** compounds. Magnesium fluoride contains ions in the ratio $Mg^{2+} : 2F^-$: its formula is MgF_2. In aluminium oxide the ratio of ions is $2Al^{3+} : 3O^{2-}$, giving it the formula Al_2O_3.

4.1.1 CRYSTALS

Ionic compounds form crystals

A feature of ionic compounds is that they form **crystals**. The crystals of sodium chloride are perfect cubes. In a dilute solution of sodium chloride, sodium ions and chloride ions are moving about independently of other ions. When the solution is evaporated to the point of crystallisation, the ions are much closer together. A sodium ion attracts chloride ions, as shown in Figure 4.1(a). Each chloride ion attracts other sodium ions, and a three-dimensional arrangement of ions called a **crystal structure** is built up [see Figure 4.1(b)]. There is no pair of Na^+ and Cl^- ions that could be regarded as a molecule of sodium chloride. The formula NaCl represents the ratio in which ions are present in the crystal structure. A pair of ions $Na^+ Cl^-$ is called a **formula unit** of sodium chloride.

FIGURE 4.1
The Arrangement of Ions in a Sodium Chloride Crystal

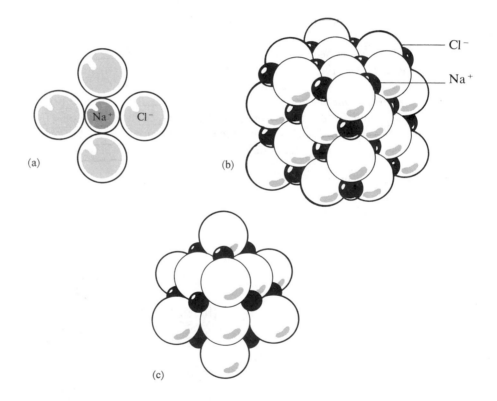

X ray analysis demonstrates the existence of ions

The evidence for the existence of ions is summarised on p. 248. An impressive demonstration of the existence of ions is the use of X ray analysis [p. 111] to obtain electron density maps. The one for sodium chloride is shown in Figure 4.2. It consists of regions of charge which are isolated from other regions of charge. This is the picture one would expect for a structure consisting of separate Na^+ and Cl^- ions. The technique is now sufficiently accurate to show that there are ten electrons, not eleven, associated with each sodium nucleus: the species present is Na^+, not Na. Figure 4.3 shows the electron density map for calcium fluoride.

FIGURE 4.2 Electron
Density Map for Sodium
Chloride

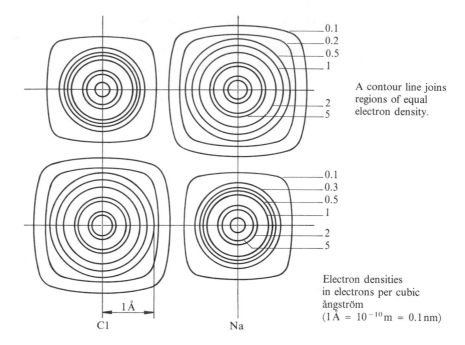

A contour line joins
regions of equal
electron density.

Electron densities
in electrons per cubic
ångström
$(1\,\text{Å} = 10^{-10}\,\text{m} = 0.1\,\text{nm})$

Cl   Na

4.1.2 IONIC RADII

*Ionic radii can be
assigned such that the
sum of the cationic and
anionic radii is equal to
the interionic distance in a
crystal*

The distance between the centres of the ions is the sum of the **cationic radius** and
the **anionic radius**. Linus Pauling was able to apportion the **interionic distance** in
potassium chloride (which equals 0.314 nm) between the K^+ and Cl^- ions. These
ions are **isoelectronic** (have the same number of electrons). He assumed that the
radius of each of the ions is inversely proportional to its **effective nuclear charge**.
This is the nuclear charge modified by the **shielding effect** which the inner shells of
electrons exert on the attraction of the nucleus for the valence electrons [p. 329].

FIGURE 4.3 Electron
Density Map for
Calcium Fluoride

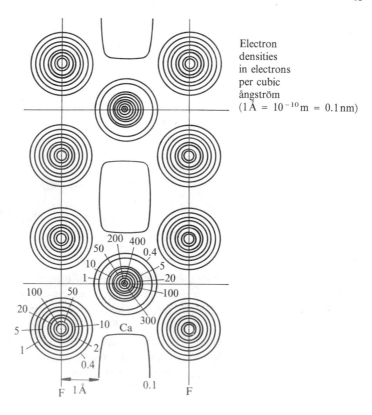

Electron
densities
in electrons
per cubic
ångström
$(1\,\text{Å} = 10^{-10}\,\text{m} = 0.1\,\text{nm})$

Pauling obtained values for the ionic radii: K^+, $0.133\,nm$ and Cl^-, $0.181\,nm$. Ionic radii appear to be *additive*. Pauling could subtract the radius of K^+ from the interionic distance in another potassium salt K^+B^-, to obtain the radius of the anion B^-. He used the radius of Cl^- and interionic distances in M^+Cl^- crystals to obtain values for the ionic radii of a number of cations. Some values of ionic radii are shown in Figure 15.4 [p. 329].

4.1.3 ENERGY CHANGES IN COMPOUND FORMATION

Electrovalent bonds are formed if the reaction between elements to give an ionic solid is exothermic

A detailed picture of the energy changes involved in the formation of an ionic compound was drawn by the theoreticians M Born and F Haber. They considered, for example, the formation of sodium chloride from its elements:

$$Na(s) + \tfrac{1}{2}Cl_2(g) \rightarrow NaCl(s)$$

They analysed the reaction as the sum of five steps. These are

1. Vaporisation of sodium $Na(s) \rightarrow Na(g)$ *Endothermic*
2. Ionisation of sodium $Na(g) \rightarrow Na^+(g) + e^-$ *Endothermic*
3. Dissociation of chlorine $\tfrac{1}{2}Cl_2(g) \rightarrow Cl(g)$ *Endothermic*
4. Ionisation of chlorine $Cl(g) + e^- \rightarrow Cl^-(g)$ *Exothermic*
5. Combination of ions to form a crystalline solid
   $Na^+(g) + Cl^-(g) \rightarrow NaCl(s)$ *Exothermic*

The sum of the five energy changes is exothermic. [See p. 205 for a fuller treatment.]

The driving force behind the reaction is the fact that sodium metal and chlorine molecules can pass to a lower energy level by forming ionic bonds. The formation of sodium chloride is exothermic. This picture of electrovalency proves to be more fruitful than the simple picture of attaining a noble gas electron configuration. Elements will not form an ionic compound if it is at a higher energy level than the elements. They may combine by the formation of covalent bonds.

================ CHECKPOINT 4A: IONS ================

1. Write the formulae of the electrovalent compounds which contain the following pairs of ions:

(*a*) Mg^{2+} and N^{3-}

(*b*) Al^{3+} and F^-

(*c*) Al^{3+} and S^{2-}

(*d*) Fe^{2+} and O^{2-}

(*e*) Fe^{3+} and O^{2-}

(*f*) Co^{3+} and $SO_4{}^{2-}$

(*g*) Ni^{2+} and $NO_3{}^-$

2. Write the electron configuration of each of the following ions, and give the name of the *isoelectronic* noble gas (which has the same electron configuration):

$$Li^+, N^{3-}, Be^{2+}, K^+, S^{2-}$$

(Atomic numbers are: $Li = 3$, $N = 7$, $Be = 4$, $K = 19$, $S = 16$.)

3. Write the electron configurations for the following cations:

$$Mn^{2+}, Cu^+, Cu^{2+}, Zn^{2+}$$

(Atomic numbers are: $Mn = 25$, $Cu = 29$, $Zn = 30$.) With which noble gases are they isoelectronic?

4.2 THE COVALENT BOND

THE VALENCE BOND TREATMENT

In the formation of a covalent bond...

The electrovalent bond is not the only type of chemical bond. In a chlorine molecule, Cl_2, a different type of bond is envisaged. Lewis suggested that the bond involves

...electrons are shared   each of the two chlorine atoms sharing one of its outermost electrons – **valence electrons** as they are termed – with the other chlorine atom. The two atoms have to approach sufficiently closely for their atomic orbitals to overlap. The shared pair of electrons is called a **covalent bond**. They occupy the same orbital with opposing spins [p. 38]. The Cl_2 molecule can be represented as shown in Figure 4.4:

FIGURE 4.4 Ways of Representing the Chlorine Molecule

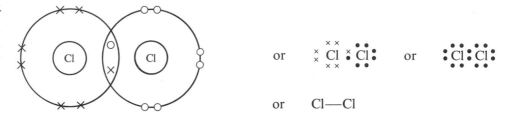

The importance of a noble gas configuration   There are not really two types of electron, but it makes it easier to count the electrons if those from one atom are represented as crosses and those from the other as dots. By sharing a pair of electrons, each of the chlorine atoms has obtained eight electrons in its outer shell: it has 'completed its octet'. Electrons are shared when half-filled atomic orbitals of adjacent atoms overlap in space.

Many carbon compounds are covalent   Covalent bonding is important in carbon compounds. The carbon atom, with four valence electrons, can attain a full octet by sharing one electron with each of four hydrogen atoms. The bonding in methane, CH_4, can be shown by a 'dot-and-cross' diagram [see Figure 4.5]. The hydrogen electrons are shown as dots and the carbon electrons as crosses. Carbon has completed its octet, and hydrogen has attained the noble gas configuration of helium by completing its 1s shell.

In carbon dioxide, the carbon atom shares two electrons with each of two oxygen atoms, in order to give all three atoms a full octet of valence electrons. [See Figure 4.6.]

FIGURE 4.5 Ways of Representing the Bonding in Methane

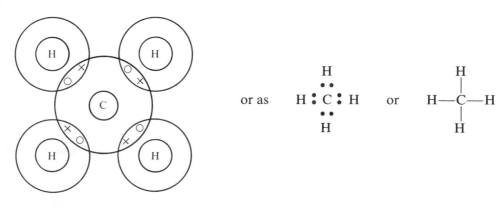

FIGURE 4.6 Ways of Representing the Carbon Dioxide Molecule

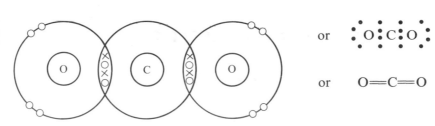

As each shared pair of electrons is a covalent bond, the two pairs of shared electrons between carbon and oxygen constitute a double bond. The pairs of electrons on the oxygen atoms which are not shared are described as **'lone pairs' of electrons**.

In a molecule of nitrogen, each nitrogen atom, with five electrons, needs to share three of its electrons with the other atom of nitrogen in order to complete its octet. The bonding can be written as shown in Figure 4.7.

FIGURE 4.7 Ways of Representing the N_2 Molecule

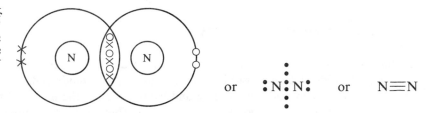

The valence bond method considers the overlapping of atomic orbitals in isolation from the rest of the molecule

This method of describing molecules is called the **valence bond method**. It considers the atoms in a molecule in isolation from the rest of the molecule, except that one or more electrons in the outer shell of one atom are accommodated in the outer shell of another atom by an overlapping of atomic orbitals.

THE MOLECULAR ORBITAL TREATMENT

The molecular orbital method considers the molecule as a whole...

...and calculates the electron density over the whole molecule

Another approach is to consider the *entire* molecule as a unit. Each electron is under the influence of all the nuclei and the electrons in the molecule. The atomic orbitals are replaced by molecular orbitals. The molecular orbital method of viewing the covalent bond uses **quantum mechanics** to calculate the distribution of electron density over the molecule. Figure 4.8 shows the results of the calculation for the hydrogen molecule. The contour lines join regions of the same electron density.

FIGURE 4.8 Electron Density Map for the Hydrogen Molecule

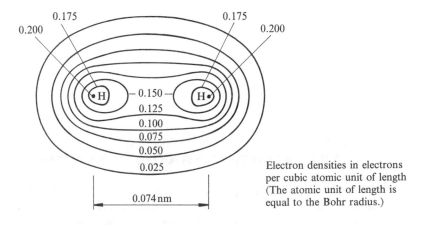

Electron densities in electrons per cubic atomic unit of length (The atomic unit of length is equal to the Bohr radius.)

In the molecular orbital model the 2H in H_2 are held together by the attraction of the nuclei for the electron cloud between the two nuclei

The highest electron density is near each nucleus. There is also a region of high electron density between the two nuclei called the **electron cloud**. The electron density between the nuclei screens the nuclei from one another and prevents repulsion between the two positive nuclei from driving the atoms apart. Although there is a force of repulsion between the positively charged nuclei, the force of attraction between each nucleus and the electron cloud between the two nuclei is greater. It is this attraction that holds the atoms together in the molecule. [See Figure 4.9.]

FIGURE 4.9 Attraction
and Repulsion in the H_2
Molecule

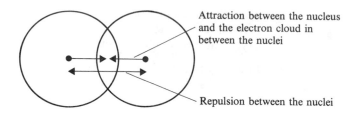

4.2.1 COVALENT RADII

Covalent radii (atomic radii)

The distance between the nuclei of covalently bonded atoms is the sum of their **covalent radii**. Covalent radii, which are also called **atomic radii**, are *additive*. The sum of the covalent radii of chlorine and hydrogen gives the length of the covalent bond in hydrogen chloride [see Figure 4.10]. This figure also shows the **van der Waals radii**, which will be covered on p. 81.

FIGURE 4.10 Covalent
Radii and Van der Waals
Radii for Hydrogen and
Chlorine

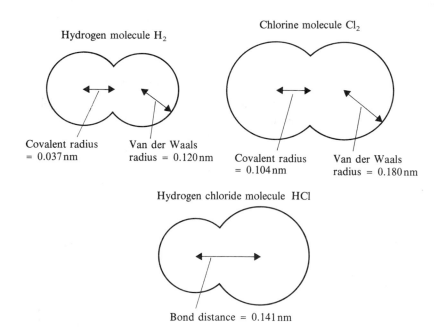

4.2.2 ELECTRONEGATIVITY AND THE COVALENT BOND

Electronegativity is the ability of an atom in a covalent bond to attract electrons

Fluorine is the most electronegative of elements

In a bond between identical atoms, e.g. H—H, the electron density of the bonding orbital is distributed symmetrically between the bonded atoms [see Figure 4.8]. In a bond between different atoms, the bonding electrons may be more attracted to one of the bonded atoms than to the other. In the molecule HF, for example, the electron density of the bonding electrons lies more towards the fluorine atom than towards the hydrogen atom. The ability of an atom in a covalent bond to attract the bonding electrons is called **electronegativity**. Thus, fluorine is more **electronegative** than hydrogen. Electronegativity is not a quantity which can be measured or to which a unit can be assigned. Pauling derived a scale of relative electronegativity values. He assigned a value of 4.00 to fluorine, the most electronegative of elements. Some of his values are:

Various electronegativity *values*	H	Li	Be	B	C	N	O	F
	2.1	1.0	1.5	2.0	2.5	3.0	3.5	4.0

If the bonded atoms differ in electronegativity a covalent bond is polar

The HF molecule is described as **polar**: the centre of the negative charge (due to the electrons) does not coincide with the centre of positive charge (due to the nuclei). There are many other covalent molecules which, like HF, are polar. There is no sharp distinction between an ionic bond and a covalent bond [see Figure 4.11].

FIGURE 4.11
Bond Types

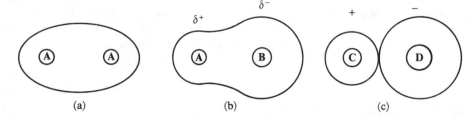

(a)

(b)

(c)

(a) Covalent bond. The electron density is symmetrically distributed.

(b) Polar covalent bond. The bonding electron density is greater in the region of **B**. **A** has a small positive (δ^+) charge, while **B** is slightly negative (δ^-).

(c) Ionic bond. The electron cloud of **C**$^+$ does not come under the influence of **D**$^-$. The electron cloud of **D**$^-$ is not distorted by **C**$^+$

The term **ionic bond** is used for bonds which are predominantly ionic. The term **covalent bond** is used for non-polar bonds, such as C—I, and also bonds in which

Many covalent bonds have some degree of ionic character

there is a considerable degree of polarity. Bonds such as $\overset{\delta_+}{C}$—$\overset{\delta_-}{Cl}$ and $\overset{\delta_+}{C}$—$\overset{\delta_-}{O}$, are termed **polar covalent bonds**. The curve which Pauling drew to relate the percentage of ionic character in a bond to the difference in electronegativity between the bonded atoms is shown in Figure 4.12. Approximate values for some covalent bonds are given:

Bond	C—I	C—H	C—Cl	C—F
Ionic character/%	0	4	6	40

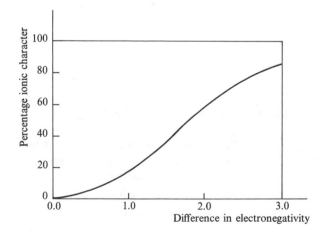

Pauling's electronegativity values can be applied to the consideration of a bond between two atoms, either in a diatomic molecule or between two atoms considered in isolation from other atoms and bonds. They should *not* be applied to a formula unit in a crystal structure as they will give an erroneous impression. If applied to $CaCl_2$, NaCl or LiF, electronegativity values will predict the nature of the bonding

Some ionic compounds show a degree of sharing of electron clouds

in an ion pair considered in isolation, e.g. $Na^+(g)\ Cl^-(g)$. In reality, we are interested in crystalline sodium chloride, in which each Cl^- ion is part of a crystal structure and its electron cloud is influenced symmetrically by six Na^+ ions around it. The net result is that the electron cloud of the Na^+ ion becomes rather cubical in shape, as shown in Figure 4.2, p. 70. The electron cloud is drawn into the spaces between the ions in the structure, but sodium chloride is still overwhelmingly ionic in nature. The electron clouds associated with the cation and the anion influence each other to a more pronounced extent in lithium fluoride [see Figure 4.13]. In contrast, the electron density map of calcium fluoride [see Figure 4.3, p. 70] shows an electron distribution which is, in cross-section, perfectly circular around each ion. Despite the deformation of the electron clouds in sodium chloride and lithium fluoride, these compounds are still ionic in character.

A cation may attract and deform the electron cloud of an anion

FIGURE 4.13 Electron Density Map for Lithium Fluoride

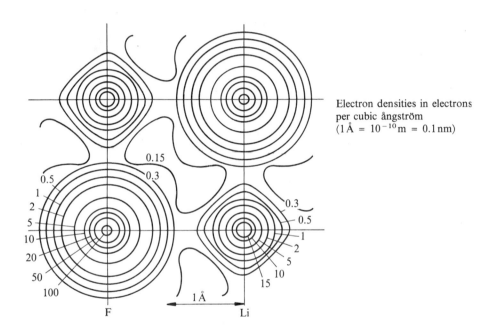

Electron densities in electrons per cubic ångström
$(1\text{Å} = 10^{-10}\text{m} = 0.1\,\text{nm})$

4.2.3 FAJANS' RULES OF BOND TYPE

The distortion of an electron cloud by a neighbouring charged ion or **dipole** is called **polarisation**. Comparing ions with similar nuclear charges, large ions are more easily polarised than small ions. The electrons in small ions are more closely controlled by the positively charged nucleus.

K Fajans formulated two rules to predict the proportions of ionic and covalent character in the bond formed between two atoms.

Fajans' rules predict the ionic and covalent character of bonds

1. Bonds will tend to be ionic if the ions formed are small in charge. For example, sodium chloride is likely to be ionic because Na^+ and Cl^- bear unit charges, whereas aluminium chloride is likely to be covalently bonded because Al^{3+} ions are highly charged.

2. Bonds will tend to be ionic if the radius of the possible cation is large (e.g. the alkali metals) and the radius of the possible anion is small (e.g. the smaller halogens).

The sizes of some ions are shown in Figure 15.4, p. 329. Compare the radii of the ions Na^+, Mg^{2+} and Al^{3+}. The high charge on the Al^{3+} ion results in a small radius because the remaining electrons are drawn in close to the nucleus. The combination of a high charge and a small radius gives Al^{3+} a high **charge density** (i.e. charge/volume ratio), and this enables the cation to attract the electron cloud of a neighbouring anion (or of a molecule). The electron cloud of the anion will be distorted in such a way as to increase the electron density near the cation: the anion has become **polarised**.

A cation with a small radius and a high charge will polarise anions

An anion is larger than the atom from which it was formed, and has one or two more electrons. The nucleus is less able to attract the electrons closely than it is in the parent atom. In a large anion, the electrons are further from the nucleus and less under its control than in a small anion, making the larger anion easier to polarise. If the cation is small and the anion is large, the cation will be able to polarise the anion, and there will be some sharing of the electron cloud of the anion: i.e. the bond will have some covalent character.

A large anion is easily polarised

4.2.4 DIPOLE MOMENTS

Polar molecules have a dipole moment...

The molecule HF is neutral, but, arising from the separation of positive and negative charges in the molecule, it has a **dipole moment**. This means that when the molecule is placed in an electric field between a positive and a negative plate, it will tend to turn so that the more positive end (the hydrogen end) is turned towards the negative plate, and the more negative end of the molecule (the fluorine end) is turned towards the positive plate. Being neutral, the molecule will not move towards either plate [see Figure 4.14]. The tendency of molecules to orient themselves in this way can be measured; it is an indication of the strength of the dipole moment.

...they orient themselves in an electric field so that the positive end of the molecule is towards the negative pole

FIGURE 4.14
Orientation of Polar Molecules in an Electric Field

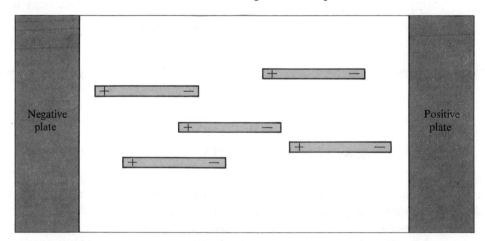

From the measured dipole moment, the **percentage of ionic character** in a bond can be calculated. For HF the value is 45%; in HCl the bonds are 17% ionic in character.

4.3 THE PROPERTIES OF IONIC AND COVALENT COMPOUNDS

Ionic solids dissolve as ions...

Ionic compounds often dissolve in *water*. Their solutions contain the free ions, and can be electrolysed [p. 237]. Covalent compounds are often insoluble in water. If they do dissolve, they dissolve as molecules. Covalent compounds are soluble in *organic* solvents, such as propanone and tetrachloromethane; these too are covalent compounds.

...covalent compounds dissolve in organic solvents

Molecular covalent compounds have lower melting and boiling temperatures than ionic compounds...

Molecular covalent compounds often melt at lower temperatures than ionic compounds. An electrovalent compound in the solid state consists of an array of ions held together by strong electrostatic forces. Many covalent compounds in the solid state consist of individual molecules. The **intramolecular** bonds, the bonds between atoms *within* the molecule, are strong. The **intermolecular** forces, the forces of attraction *between* the molecules (dipole–dipole interactions and others), are easily broken to allow molecules to move about independently, that is to enter the liquid phase. Such covalent compounds have low melting temperatures; many are liquids or gases at room temperature. The boiling temperatures of many covalent compounds are low for the same reason: the forces of attraction between individual molecules are very weak.

...because intermolecular forces are weak

Elements and compounds with macromolecular covalent structures have high melting and boiling temperatures

Other covalent substances do not consist of individual molecules. They are **macromolecular structures** held together by covalent bonds. Examples are: diamond [see Figure 6.14, p. 121], boron nitride (BN), silicon(IV) oxide [see Figure 6.17, p. 122] and silicon(IV) carbide (SiC). The strong covalent bonds holding these macromolecules together give the compounds high melting temperatures (higher than most ionic solids) and make them **involatile** (i.e. they have high boiling temperatures).

CHECKPOINT 4B: THE COVALENT BOND

1. Draw dot-and-cross diagrams to show the bonding in

(a) H—C≡N

(b) CCl_4

2. Distinguish between (a) the intramolecular bonds, and (b) the intermolecular forces in chloroform, $CHCl_3$. What makes you think that the bonds in (a) are strong and the forces in (b) are weak?

3. Sodium chloride melts at 800 °C. Tetrachloromethane, CCl_4, is a liquid at room temperature. Explain how this difference arises.

4.4   THE COORDINATE BOND

In a coordinate bond, a donor atom shares a lone pair of electrons with an acceptor atom

A **coordinate bond** is a covalent bond in which the shared pair of electrons is provided by only *one* of the bonded atoms. One atom is the **donor**, the other is the **acceptor**, and the bond is sometimes called the **dative covalent bond**. Once formed, a coordinate bond has the same characteristics as a covalent bond. For an atom to act as a donor, it must have at least one pair of unshared electrons (a lone pair) in its outermost shell (the valence shell). The acceptor has at least one vacant orbital in its outer shell. It may be a metal cation or a transition metal atom or an atom in a molecule.

Once formed, a coordinate bond is like a covalent bond

In water, H—$\overline{O}$—H, the oxygen atom has two lone pairs of electrons. They can be shared with an atom which needs them to complete its valence shell. A proton H^+ has an empty 1s orbital. By accepting a pair of electrons from oxygen, it achieves a full s shell:

The oxonium ion, H_3O^+, is formed by coordination

$$H\!:\!\overset{\displaystyle\cdot\cdot}{\underset{\displaystyle\cdot\cdot}{O}}\!:\;+\;H^+ \longrightarrow \left[\, H\!:\!\overset{\displaystyle\cdot\cdot}{O}\!:\!H \,\right]^+$$

The species formed is an **oxonium** ion, H_3O^+. The positive charge contributed by the proton is spread over the whole ion. The proton has one unit of positive charge spread over a surface area which is minute compared with other ions. The high charge density makes it extremely reactive: the proton cannot exist by itself. It is stabilised by the coordination of water molecules.

Water coordinates to metal ions...

...an example is $[Ca(H_2O)_6]^{2+}$

Water also coordinates to metal ions. The fact that bonds are formed between metal ions and water is responsible for the solubility of many salts. Energy is required to break the bonds holding ions together in a crystal structure. If energy is given out when coordinate bonds form between metal ions and water, this may swing the balance in favour of solution. Metal ions are hydrated, e.g. $[Ca(H_2O)_6]^{2+}$:

$$Ca^{2+} + 6H\!:\!\overset{\bullet\bullet}{\underset{\bullet\bullet}{O}}\!:\!H \longrightarrow \left[Ca\left(:\!\overset{\bullet\bullet}{\underset{\bullet\bullet}{O}}\!:\!\overset{H}{H}\right)_6 \right]^{2+}$$

The lone pair on N in NH_3 is responsible for the formation of coordination compounds by NH_3

The nitrogen atom in ammonia, $H{-}\!\!\overset{H\backslash}{\underset{H/}{N}}\!|$, has an unshared pair of electrons which it is able to share with an atom that needs two electrons to complete its octet. When ammonia meets the vapour of aluminium fluoride, which consists of covalent AlF_3 molecules, a white solid of formula NH_3AlF_3 is formed. It is formed by the coordination of the lone pair of the nitrogen atom into the valence shell of the aluminium atom as shown below [see also Question 3, p. 80].

$$H\!:\!\overset{H}{\underset{H}{N}}\!: + Al\!:\!\overset{F}{\underset{F}{F}} \longrightarrow H\!:\!\overset{H}{\underset{H}{N}}\!:\!Al\!:\!\overset{F}{\underset{F}{F}}$$

or

$$H{-}\!\overset{H}{\underset{H}{N}}\!: + Al{-}\!\overset{F}{\underset{F}{F}} \longrightarrow H{-}\!\overset{H}{\underset{H}{N}}{\rightarrow}Al{-}\!\overset{F}{\underset{F}{F}}$$

The symbol $\longrightarrow$ is used for a coordinate bond, an arrow pointing from the donor towards the acceptor.

Copper(II) chloride is blue in solution. In the presence of a high concentration of chloride ions, the solution turns a very deep green. The colour is due to the formation of $CuCl_4^{2-}$ ions. Coordinate bonds form between Cl^- ions and Cu^{2+} ions:

Ions such as $CuCl_4^{2-}$ are formed by coordination...

$$4\!:\!\overset{\bullet\bullet}{\underset{\bullet\bullet}{Cl}}\!:^- + Cu^{2+}(aq) \longrightarrow \left[\begin{array}{c} :\!\overset{\bullet\bullet}{\underset{\bullet\bullet}{Cl}}\!: \\ :\!\overset{\bullet\bullet}{\underset{\bullet\bullet}{Cl}}\!:\!\overset{\bullet\bullet}{\underset{\bullet\bullet}{Cu}}\!:\!\overset{\bullet\bullet}{\underset{\bullet\bullet}{Cl}}\!: \\ :\!\overset{\bullet\bullet}{\underset{\bullet\bullet}{Cl}}\!: \end{array} \right]^{2-}$$

or

...they are called complex ions

$$4Cl^- + Cu^{2+} \rightarrow \left[\begin{array}{c} Cl \\ \downarrow \\ Cl \longrightarrow Cu \longleftarrow Cl \\ \uparrow \\ Cl \end{array} \right]^{2-}$$

The copper atom now has eight electrons in its valence shell. Ions such as $CuCl_4^{2-}$ and $Cu(NH_3)_4^{2+}$, which are formed by the combination of an ion with an oppositely charged ion or a molecule are called **complex ions**.

CHECKPOINT 4C: THE COORDINATE BOND

1. Draw a dot-and-cross diagram to show the arrangement of valence electrons in the complex ion $Cu(H_2O)_4^{2+}$. It is formed by coordination of H_2O molecules on to a Cu^{2+} ion.

2. Draw a dot-and-cross diagram to show the bonding in the complex ion $Fe(H_2O)_6^{3+}$. How many electrons has the iron atom in its valence shell? How can it accommodate this number?

3. Explain the bonding in the compound $NH_3 \cdot BF_3$.

4. The formula for carbon monoxide can be written $|C{=}O|$. Draw a dot-and-cross diagram for this molecule.

The lone pair on the oxygen atom can be used to form a coordinate bond to a nickel atom. By means of a diagram, show the bonding in nickel carbonyl, $Ni(CO)_4$.

5. Explain the bonding in the tetraammine copper ion, $Cu(NH_3)_4^{2+}$.

6. The slightly soluble compound lead(II) chloride dissolves in concentrated hydrochloric acid to form a soluble complex ion. What do you think the formula of this ion might be? Explain the bonding in the ion.

4.5 INTERMOLECULAR FORCES

Dipole–dipole interactions, van der Waals forces and hydrogen bonds are all intermolecular forces

On p. 78 a reference was made to the forces of attraction that can exist between covalent molecules. These intermolecular forces and bonds are of a number of types: dipole–dipole interactions, van der Waals forces and the hydrogen bond.

4.5.1 DIPOLE–DIPOLE INTERACTIONS

Attractions between dipoles result in an ordered arrangement of molecules

In the solid state, polar molecules interact to form an ordered arrangement. Dipole–dipole interactions between the molecules lead the molecules to pack in such a way that partial positive charges will be adjacent to partial negative charges [see Figure 4.15].

Solvation helps solids to dissolve

Polar solids dissolve in polar solvents. The energy required to break up the crystal is recouped by the energy released when polar solute molecules interact with polar solvent molecules. [See Figure 4.15.] This interaction is called **solvation**; if the solvent is water, it is called **hydration** [p. 347].

FIGURE 4.15
The Process of
Dissolution

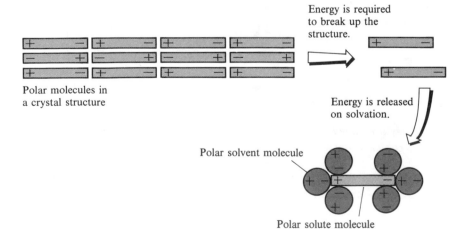

Energy is required to break up the structure.

Polar molecules in a crystal structure

Energy is released on solvation.

Polar solvent molecule

Polar solute molecule

4.5.2 VAN DER WAALS FORCES

There are different types of van der Waals forces...

When molecules pack together in the liquid or solid state, there must be forces of attraction between them. J D van der Waals' work on gases [p. 147] led him to postulate the existence of forces of attraction and repulsion that are neither ionic nor covalent. Such forces arise in a number of ways and are collectively called **van der Waals forces**. Dipole–dipole interactions [p. 80] between polar molecules are one type of van der Waals force. Attractive forces exist also between non-polar molecules. Even atoms of the noble gases are attracted to one another to a slight degree; this is why the noble gases can be liquefied.

...Attractive forces exist between non-polar molecules...

A type of van der Waals forces called **dispersion forces** or **London forces** are thought to operate between non-polar molecules. The theoretician, F London, explained them in the following manner. Consider two non-polar molecules which are very close together. Since they are non-polar, the arrangement of electrons is *on the average* symmetrical. Yet, *at any given instant*, the electron distribution in one molecule may be unsymmetrical. There may be a dipole in the molecule *for an instant*. Figure 4.16 shows how a temporary dipole in one molecule **A**, can attract the electron cloud of a neighbouring molecule **B**. This means that both molecules will have dipoles, and the direction of the dipoles will be such that they attract one another. Since the electrons are moving about at high speed, the attraction has a fleeting existence. In the next instant, the dipole in **A** may be in the opposite direction. Again, the dipole which it **induces** in **B** will result in an attraction. The dipoles are temporary, but the net attraction which they produce is permanent.

...London explained these forces as being due to the momentary polarisation of molecules...

...leading to a permanent attraction

FIGURE 4.16 Attraction between Momentary Dipoles

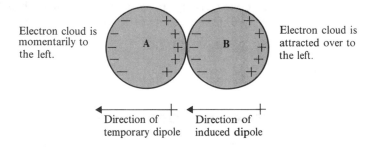

Electron cloud is momentarily to the left.

Electron cloud is attracted over to the left.

Direction of temporary dipole

Direction of induced dipole

The ease with which an electron cloud is distorted is called polarisability

The ease with which an electron cloud is distorted, and therefore the ease with which a dipole is induced, is called **polarisability**. Polarisability increases with the number of electrons in a molecule, and the strength of dispersion forces therefore increases with molar mass. This is why xenon has a higher boiling temperature than argon. The shape of molecules is another factor. Electrons in elongated molecules are more easily displaced than those in small, compact, symmetrical molecules.

Van der Waals forces determine distances between atoms in condensed phases

If a pair of molecules are far apart, there will be no induction of dipoles and no attractive forces between them. Should the molecules move too close together, repulsion between electron shells will predominate over the induction effect and drive the molecules apart. Figure 4.17 shows how closely argon atoms can approach in the liquid state. Half the distance between argon atoms at their closest distance of approach is the **van der Waals radius** of the atom.

FIGURE 4.17
Van der Waals Forces
determine the Distance
between Atoms in the
Liquid State

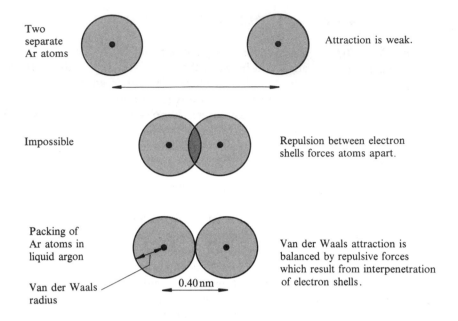

In addition to the dipole–dipole interactions and the London forces described,
van der Waals forces can involve interactions between ions and induced dipoles,
and between permanent dipoles and induced dipoles. All types of van der Waals
forces between small molecules are weak. Between molecules with long chains of
atoms, giving many points of contact, van der Waals forces are stronger. This is

*Van der Waals forces are
stronger between large
linear molecules*

why in the series of alkanes [p. 543] ethane, C_2H_6, is a gas at stp, hexane, C_6H_{14}, is
a liquid, and octadecane, $C_{18}H_{38}$, is a solid. Branched-chain hydrocarbons are
more **volatile** (have lower boiling temperatures) than unbranched-chain hydrocarbons.
Hydrocarbon polymers have molecules which are continuous chains, containing
thousands of repeating units. The long, strand-like molecules of poly(ethene),
$-(CH_2-CH_2-)_n$ [p. 564] can align themselves to give thousands of contacts
between atoms and set up very strong van der Waals forces. Poly(ethene) is an
extremely tough material, which is used for the manufacture of laboratory and kitchen
ware.

4.5.3   THE HYDROGEN BOND

THE NATURE OF THE BOND

*When hydrogen is
combined with an
electronegative atom
intermolecular hydrogen
bonds can be formed*

The bond in hydrogen fluoride is a polar covalent bond [p. 75] which can be written
as $\overset{\delta_+}{H}—\overset{\delta_-}{F}$. If two polar HF molecules are close enough, there is an attraction
between the positive end of one molecule and the negative end of the other molecule:

$$\overset{\delta_+}{H}—\overset{\delta_-}{F}\cdots\cdots\overset{\delta_+}{H}—\overset{\delta_-}{F}$$

Since the attraction between the hydrogen atom in one molecule and the fluorine
atom in the next molecule is stronger than the repulsion between the two hydrogen
atoms and between the two fluorine atoms, the two molecules are bonded together.
The attraction between a hydrogen atom in one molecule and an electronegative
atom, such as fluorine, in another molecule is called a **hydrogen bond**. Hydrogen
bonding holds a number of molecules of hydrogen fluoride together in liquid hydrogen
fluoride [see Figure 4.18]. A molecule of HF forms an even stronger hydrogen bond
to a fluoride ion, F^-. The ion $[F\cdots\cdots H\cdots\cdots F]^-$ is formed. Hydrogen fluoride
forms acid salts, e.g. KHF_2, containing this anion.

FIGURE 4.18 Hydrogen
Bonding in HF(l)

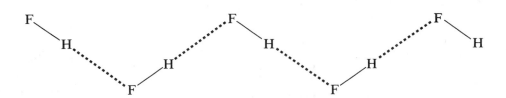

Hydrogen bonds form when a hydrogen atom is covalently bonded to one of the electronegative atoms, fluorine, chlorine, oxygen or nitrogen. Hydrogen bonding is different from the dipole–dipole interactions in other polar molecules. A hydrogen atom has no inner, non-bonding electrons to set up forces of repulsion with the non-bonding electrons of the other atom. The strength of a hydrogen bond is $\frac{1}{10}$ to $\frac{1}{20}$ of that of a covalent bond.

Hydrogen is unique

The strength of H bonds

Evidence for H bonds is given by the melting and boiling temperatures of
HF(l)
H_2O(l)
NH_3(l)

Some evidence for the existence of the hydrogen bond is shown in Figures 4.19 and 4.20. A comparison is made of the melting temperatures and boiling temperatures of hydrogen fluoride, water and ammonia with those of other hydrides in the same groups of the Periodic Table. The melting temperatures and boiling temperatures of these compounds are much higher than those of other hydrides in the same groups. The molecules of HF, H_2O and NH_3 must be held together by intermolecular bonds stronger than those between molecules of the other hydrides. Since fluorine, oxygen and nitrogen are the most electronegative of elements, the intermolecular forces are thought to be hydrogen bonds. [See Figure 4.21.]

FIGURE 4.19 Graph of Melting
Temperature against Period Number

FIGURE 4.20 Graph of Boiling
Temperature against Period Number

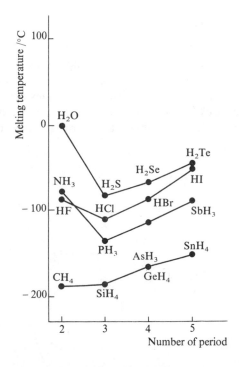

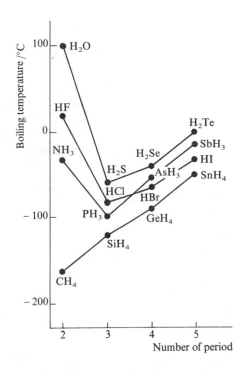

FIGURE 4.21 Hydrogen
Bonding in $H_2O(l)$ and
$NH_3(l)$

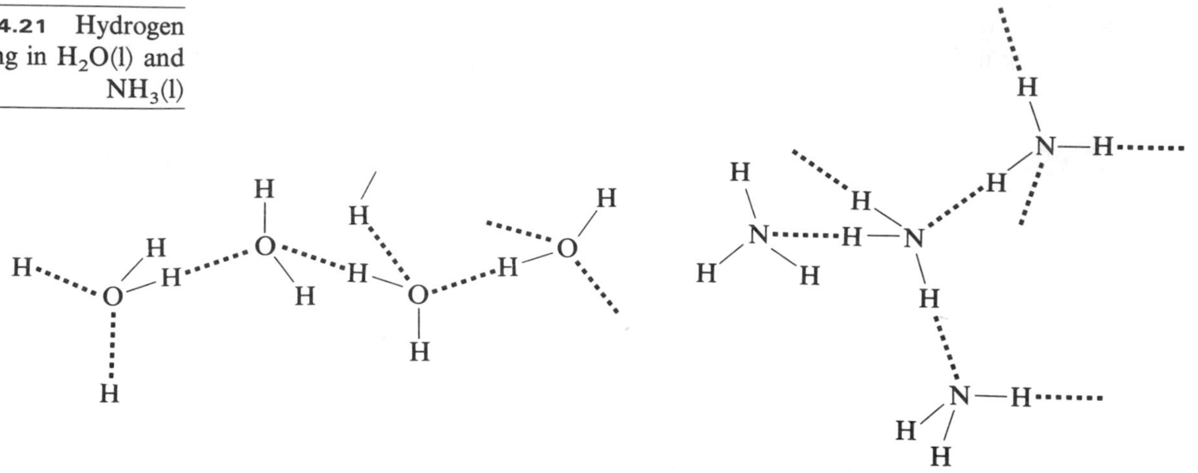

When the molar masses of carboxylic acids are found from measurements in the vapour phase and from solutions in organic solvents, the values are often up to twice the values calculated from the formulae. The molecules are thought to dimerise through the formation of hydrogen bonds [see Figure 4.22].

FIGURE 4.22
H Bonding in RCO_2H

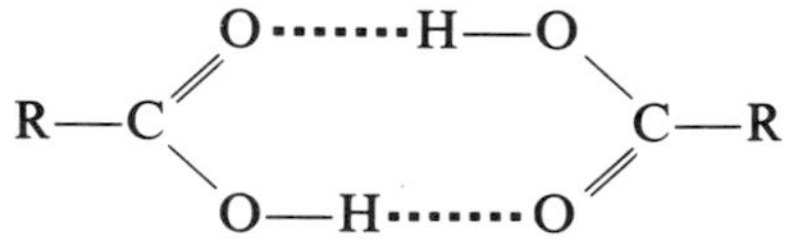

Some compounds can form **intramolecular hydrogen bonds** between two groups in the same molecule. [See Figures 4.25, 4.26, 4.27, 4.28 and Question 5, p. 64.]

SOLUBILITY CONSIDERATIONS

Substances dissolve in water if they can form H bonds with it

Water is a hydrogen-bonded association of water molecules. A substance such as ethanol, C_2H_5OH, will dissolve in water as molecules of ethanol can displace water molecules in the association. New hydrogen bonds form between molecules of ethanol and water [see Figure 4.23]. Halogenoalkanes such as chloroethane, C_2H_5Cl, do not form hydrogen bonds with water and are only slightly soluble. There are more references to solubility and hydrogen bonding in Part 4: Organic Chemistry.

FIGURE 4.23 Hydrogen
Bonds between Alcohols
and Water

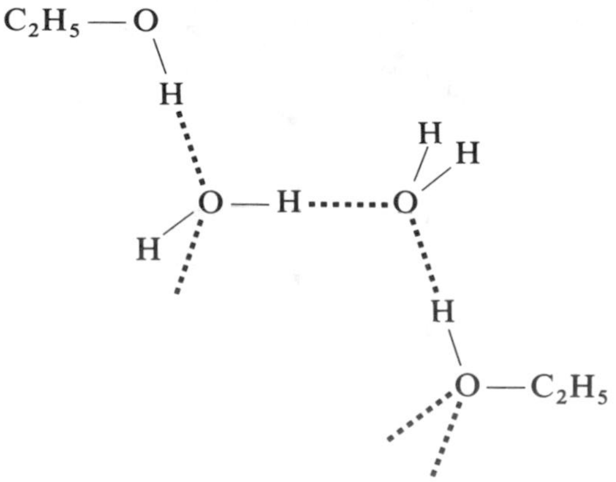

VOLATILITY

Hydrogen bonding occurs in alcohols and amines

In the liquid state, the molecules of alcohols are associated by hydrogen bonding. Energy must be supplied to break these bonds when the liquid is vaporised, making the boiling temperatures of alcohols higher than those of non-associated liquids, e.g. alkanes, of comparable molar mass. Amines also are hydrogen-bonded in the liquid state [p. 671].

THE STRUCTURE OF ICE

The bonds in H_2O are inclined at approximately the tetrahedral angle of 109.5°. The lone pairs occupy the other apices of the tetrahedron [see Figure 5.5, p. 92].

Liquid water is associated by H bonding

Hydrogen bonding extends throughout the whole structure in ice...

Liquid water contains associations of water molecules [see Figure 4.21, p. 84]. In ice, the arrangement of molecules is similar, but the regularity extends thoughout the whole structure [see Figure 4.24]. The structure spaces the molecules further apart than they are in liquid water. This is why, when water freezes, it expands by 9%, and why ice is less dense than water at 0 °C. The underlying structure of ice resembles that of diamond [see Figure 6.14, p. 121].

FIGURE 4.24 Hydrogen Bonding in Ice
Notes Each H_2O molecule uses both its H atoms to form hydrogen bonds and is also bonded to two other H_2O molecules by means of their H atoms. The arrangement of bonds about the O atoms is tetrahedral

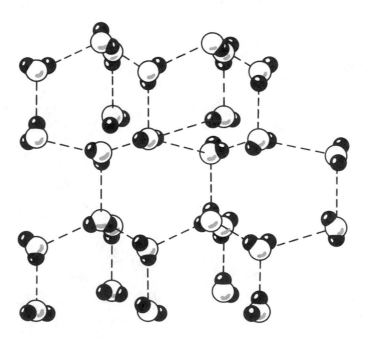

...This explains why ponds freeze from the surface downwards

The fact that ice is less dense than water at 0 °C explains why ponds and lakes freeze from the surface downwards. Water reaches its maximum density at 4 °C. As it cools further, the water at the surface becomes less dense and therefore stays on top of the slightly warmer water until it freezes. The layer of ice on the surface helps to insulate the water underneath from further heat loss. Fish and plants survive under the ice in Canadian lakes and rivers for months.

THE STRUCTURE OF PROTEINS

Protein molecules have H bonds...

Hydrogen bonding is important in protein molecules. Proteins consist of long chains of formula

$$\left(\begin{array}{ccc} & R & \\ & | & \\ -C & -C-N- \\ | & \parallel & | \\ H & O & H \end{array} \right)_n$$

...between C=O *and*

N—H groups

R can be a number of groups. Since both the $\overset{\delta+}{C}=\overset{\delta-}{O}$ group and the $\overset{\delta-}{N}-\overset{\delta+}{H}$ groups are polar, hydrogen bonding can occur between them:

$$\overset{\delta+}{C}=\overset{\delta-}{O}\cdots\cdots\overset{\delta+}{H}-\overset{\delta-}{N}$$

The helical structure of proteins is sustained by H bonds

A single protein molecule contains many hydrogen bonds. They are one of the forms of intramolecular attraction which hold the protein in a three-dimensional arrangement described as the *secondary structure* of the protein. Figure 4.25 shows the α helical structure proposed by Pauling and co-workers as a result of their X ray diffraction studies on protein molecules. An **α helix** is a spiral, which, looking away from you, is spiralling in a clockwise direction.

FIGURE 4.25 A Protein Chain with the α Helical Structure

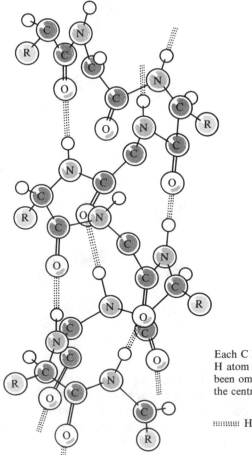

Each C has an R group and a H atom attached; these have been omitted from C atoms in the centre of the helix.

░░░░░ Hydrogen bond

THE DOUBLE HELIX

Nucleic acids contain phosphate groups, sugar groups and bases

Hydrogen bonding is important in the famous **double helix** of DNA. **Chromosomes** are the bodies in the nuclei of the cells of living organisms which carry genetic information. They contain macromolecular substances called **nucleic acids**. These are of two types: ribonucleic acid, RNA, and deoxyribonucleic acid, DNA. The macromolecular chains in DNA are of the type

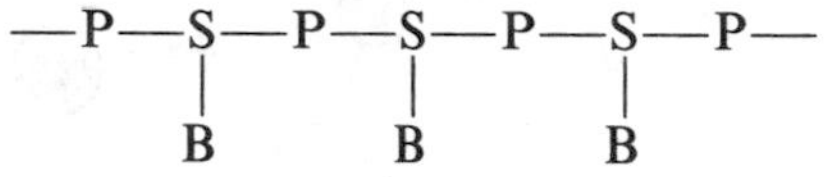

$$-P-S-P-S-P-S-P-$$
$$\quad\ \ |\qquad\ \ |\qquad\ \ |$$
$$\quad\ \ B\qquad\ B\qquad\ B$$

where P is a phosphate group and S is the sugar deoxyribose:

P represents

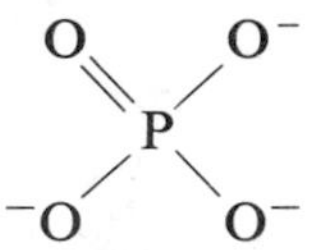

S represents

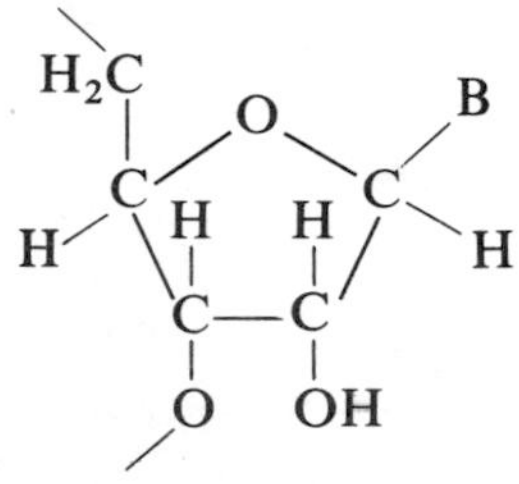

The double helix of DNA is held in this configuration by H bonds

B is one of the four bases: adenine, thymine, cytosine and guanine. DNA consists of two macromolecular strands spiralling round each other in a double helix, as shown in Figure 4.26. The strands are held together by hydrogen bonding between the bases, as shown in Figure 4.27.

FIGURE 4.26
The Double Helix

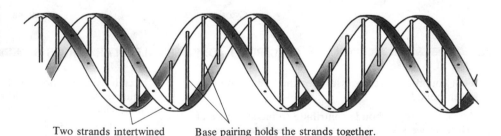

Two strands intertwined   Base pairing holds the strands together.

FIGURE 4.27 Hydrogen Bonding between Strands in DNA

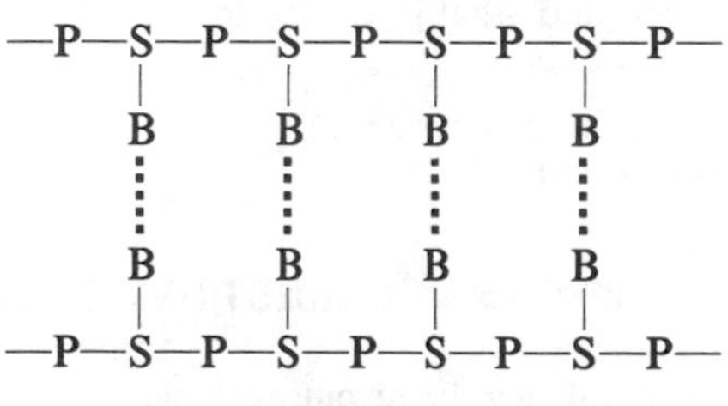

$$-P-S-P-S-P-S-P-S-P-$$

Of the four bases, *thymine* can pair up with *adenine* by hydrogen bonding and *cytosine* can form hydrogen bonds with *guanine*. The double helix brings these base pairs into contact so that they can form the bonds that keep the structure intact. Figure 4.28 shows the details of the hydrogen bonding.

FIGURE 4.28 Hydrogen Bonding of Base Pairs in DNA

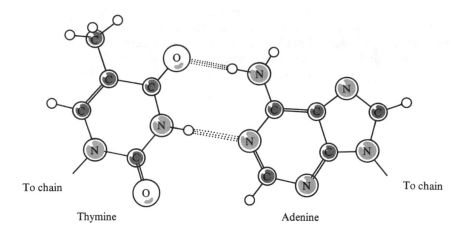

To chain   To chain

Thymine   Adenine

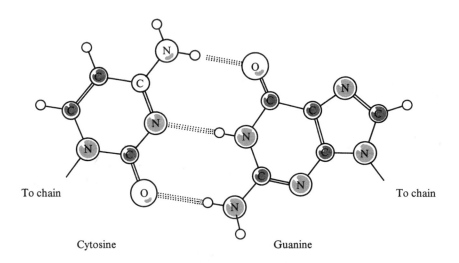

To chain   To chain

Cytosine   Guanine

CHECKPOINT 4D: INTERMOLECULAR FORCES

1. What conditions are necessary for the formation of a hydrogen bond?

2. How do hydrogen bonds contribute to the structure of ice? How do we know that some hydrogen bonding is present in water?

3. What reason do we have for postulating the existence of van der Waals forces? What is responsible for the van der Waals attractive forces, and what gives rise to repulsion?

4. What type of intermolecular forces operate in (a) HBr(g), (b) Br_2(g), (c) ICl(g) and (d) HF(l)?

5. Sketch hydrogen bonding in (a) liquid ethanol, (b) aqueous ethanol, (c) liquid ethanoic acid, (d) aqueous ethanoic acid and (e) liquid ammonia. How many hydrogen bonds can be formed by one molecule of (i) H_2O, and (ii) NH_3?

6. If water were not hydrogen-bonded, what would you expect for its boiling temperature and melting temperature and the relative densities of the liquid and solid states?

QUESTIONS ON CHAPTER 4

1. Explain the 'octet theory' of valency. Point out examples of the success of the theory in explaining the formation of chemical bonds. Discuss cases in which the octet theory is inadequate.

2. Discuss the bonding in (a) NaH, (b) NH_4^+, (c) $BeCl_2$, (d) HF (l), (e) CCl_4 and (f) Cu. [For (f), see p. 113.]

†3. Put the following substances in order of increasing boiling temperatures, giving reasons for your choice:

$$C_4H_9OH, \ CH_3CH_2CH_2CH_2CH_3,$$

$$(CH_3)_3CCH_3, \ N_2$$

†**4.** Which members of the following pairs would you expect to have the higher boiling temperature?

(*a*) C_3H_8 and CH_3OCH_3

(*b*) $CH_3CH_2NH_2$ and CH_3CH_2OH

(*c*) CH_3CH_2OH and C_2H_6

(*d*) C_3H_8 and $(CH_3)_2C{=}O$

Give reasons for your choice.

5. Briefly explain the importance of the hydrogen bond to living creatures.

*†**6.** What do you understand by the 'octet rule' of valency? Summarise the limitations of this rule and indicate the principles on which modern valency theory is based. Illustrate your answer by reference to at least two compounds from each of the following three groups.

(*a*) $ZnCl_2$ \quad $PbCl_2$ \quad $FeCl_2$

(*b*) $BeCl_2$ \quad BF_3 \quad PF_5 \quad SF_6

(*c*) Na_2SO_3 \quad Na_2SO_4 \quad H_3PO_4

(L81, S)

5

THE SHAPES OF MOLECULES

5.1 THE ARRANGEMENT IN SPACE OF COVALENT BONDS

Ionic bonds are not directed in space whereas...

Electrovalent bonds are the electrostatic attractions that exist between oppositely charged ions. Since ions radiate a spherically symmetrical positive or negative field, ionic bonds are *non-directional*.

...covalent bonds have a preferred direction in space

When atoms approach one another closely, their atomic orbitals overlap and molecular orbitals are formed. Covalent bonds are formed when a bonding pair of electrons enters a molecular orbital of low energy. The bonding electrons must have opposing spins, in accordance with the Pauli Exclusion Principle [p. 36]. The more the atomic orbitals overlap, the more stable will be the molecular orbital formed. The strongest bonds will be formed if the atoms approach in such a way that there is maximum overlap between atomic orbitals. It follows that a covalent bond will have a *preferred direction*. A covalent molecule will have a shape which is determined by the angles between the bonds joining the atoms together.

$BeCl_2$ and $SnCl_2$ differ in shape...

...BCl_3 and NH_3 differ in shape

There must be a reason why beryllium chloride, $BeCl_2$, is a linear molecule without a dipole moment, while tin(II) chloride, $SnCl_2$, is a bent molecule with a dipole moment. There must also be a reason why the four atoms in boron(III) chloride, BCl_3, are coplanar, whereas in ammonia the nitrogen atom lies above three coplanar hydrogen atoms. [See Figure 5.1.]

FIGURE 5.1 $BeCl_2$ Compared with $SnCl_2$ and BCl_3 Compared with NH_3

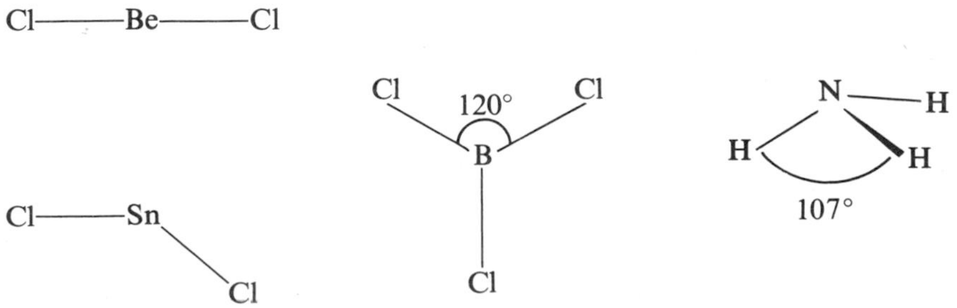

The differences can be explained by the atomic orbital approach or...

The explanation lies in the difference in the atomic orbitals used by beryllium and tin, and by boron and nitrogen. The atomic orbital approach to the question of molecular geometry is covered on p. 96.

A simpler but less precise theory to account for the shapes of molecules was put forward by Sidgwick and Powell in 1940. It is known as the **Valence Shell Electron Pair Repulsion Theory**. Sidgwick and Powell considered the shapes of small molecules and molecular ions, such as $BeCl_2$, BCl_3, NH_3, NH_4^+ and CH_4. They pointed out that the arrangement of electron pairs around the central atom in a molecule depends on the number of electron pairs. Between each electron pair and any other electron pair there is a force of electrostatic repulsion, which forces the orbitals as far apart as possible. Any lone pairs of electrons on the central atom occupy atomic orbitals, and they too repel the bonding pairs of electrons and influence the geometry of the molecule.

...the Sidgwick–Powell theory, which is the Valence Shell Electron Pair Repulsion Theory

5.1.1 LINEAR MOLECULES

The molecules of gaseous beryllium chloride, $BeCl_2$, are linear. Beryllium, in Group 2 of the Periodic Table, has two electrons in its valence shell, and forms two covalent bonds. A **linear** arrangement of the atoms (a bond angle of 180°) puts the two electron clouds as far apart as possible:

$$Cl—Be—Cl$$

Other linear molecules are

Examples of linear molecules are $BeCl_2$, $HC≡CH$, $H—C≡N$ and $O=C=O$

$$H—C≡C—H \qquad H—C≡N \qquad O=C=O$$

The electron pairs in a multiple bond are assumed on the Sidgwick–Powell theory to occupy the position of one electron pair in a single bond.

5.1.2 TRIGONAL PLANAR MOLECULES

The arrangement of 3 pairs of valence electrons is trigonal planar

When there are three pairs of electrons around the central atom, the bonds lie in the same plane at an angle of 120° to one another. Three atoms form a triangle about the central atom, and the arrangement is described as **trigonal planar**. An example is boron trichloride, BCl_3. Boron, in Group 3 of the Periodic Table, has three valence electrons and forms three covalent bonds. Gaseous tin(II) chloride, $SnCl_2$, has a dipole moment, proving that the molecule is not linear. The reason is that tin, in Group 4, is using only two of its four electrons for bond formation. The lone pair of electrons repel the bonding pairs and a trigonal planar arrangement of orbitals results. [See Figure 5.2.] This arrangement maximises the angle between the electron pairs and minimises the repulsion between them.

Lone pairs of electrons can determine the shape of the molecule

FIGURE 5.2
The Trigonal Planar Arrangement of Electron Pairs in BCl_3 and $SnCl_2$

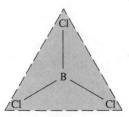
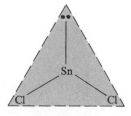

Other structures based on a trigonal planar arrangement are ethene, the nitrate ion and sulphur dioxide [see Figure 5.3].

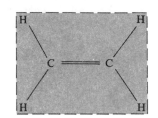
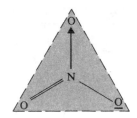
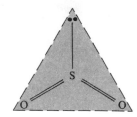

5.1.3   TETRAHEDRAL MOLECULES

*4 electron pairs adopt a
tetrahedral configuration*

The molecules CH_4, NH_3, NH_4^+ and H_2O all have four pairs of electrons around the central atom. Whether they are bonding pairs or lone pairs of electrons, they experience mutual repulsion. To minimise this repulsion, the four electron orbitals adopt the spatial arrangement that maximises the angle between the orbitals. This is the **tetrahedral** arrangement. [See Figures 5.4 and 5.5.]

FIGURE 5.4
The Bonding in CH_4,
NH_3, NH_4^+, H_2O

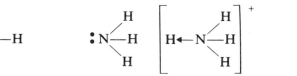

FIGURE 5.5
The Tetrahedral
Arrangement of Valence
Electron Pairs in CH_4,
NH_3, NH_4^+ and H_2O

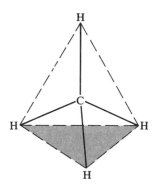
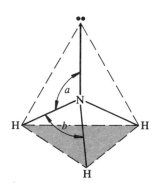

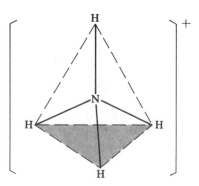
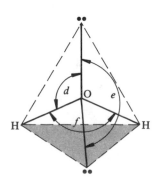

The arrangements of atoms are of course *not* all the same. In CH_4 and NH_4^+, the atoms form a tetrahedron, in NH_3 they form a trigonal pyramid, and in H_2O they form a bent line.

In CH_4 and NH_4^+ all the bonds are the same. Once formed, a coordinate bond is the same as a covalent bond. The structures are perfect tetrahedra with the tetrahedral angle of 109.5° between each pair of bonds. In NH_3 the bond angle is 106°, and in H_2O it is 104.5°.

To account for such departures from the expected bond angle, Gillespie and Nyholm suggested a refinement of the theory of valence shell electron pair repulsion. They suggested that, since lone pairs are closer to the nucleus than bonding pairs, they will exercise a greater force of repulsion. Repulsion between electron pairs decreases in the order

Lone pairs are closer to the nucleus than bonding pairs and exert a greater repulsive force

| Lone pair : lone pair repulsion | ＞ | Lone pair : bonding pair repulsion | ＞ | Bonding pair : bonding pair repulsion |

Repulsion between the lone pair and the bonding pairs in NH_3 makes the angle *a* in Figure 5.5 greater than the tetrahedral angle (109.5°) and consequently the angle *b* less than 109.5°. Similarly in H_2O, angles *d* and *e* are greater than 109.5°, and the angle *f* between the H—O—H bonds is 104.5°. Other structures based on the tetrahedron are the sulphate and sulphite ions [see Figure 5.6].

FIGURE 5.6
The Shapes of SO_4^{2-} and SO_3^{2-}

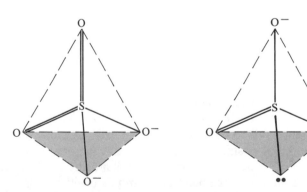

5.1.4 STRUCTURES WITH 5, 6 OR 7 PAIRS OF VALENCE ELECTRONS

Structures with more than four pairs of electrons about the central atom may occur if the element is in the second short period or a later period. This is known as **expansion of the octet**. A molecule of phosphorus(V) chloride, PCl_5, with *five* bonding pairs of electrons, has the shape of a **trigonal bipyramid**. The angles between the bonds are 90° or 120°. Two Cl atoms occupy **axial** positions in the bipyramid, and three occupy **equatorial** positions [see Figure 5.7].

Some atoms have more than 8 electrons in the valence shell

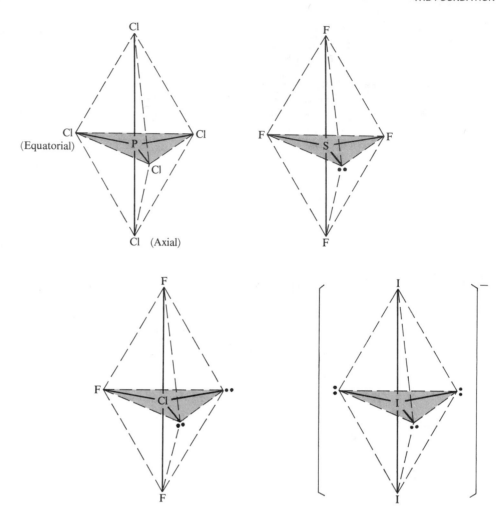

The same type of bond distribution is adopted by SF$_4$. Sulphur, in Group 6, uses four of its six electrons to form the bonds and has a lone pair of electrons. The lone pair could be in an axial position or, as shown in Figure 5.7, in an equatorial position. The choice of an equatorial position fits in with the Gillespie–Nyholm idea of the lone pair orbital being concentrated in a volume closer to the nucleus than the bond pairs. In this position it can have a bond angle of 120° with two orbitals and a bond angle of 90° with two others. The molecule of ClF$_3$ has the shape illustrated in Figure 5.7, with a trigonal bipyramidal arrangement of bonds. In the I$_3^-$ ion, iodine (in Group 7) uses only two of its seven electrons in bond formation. The remaining five plus the electron that gives the ion its negative charge make up three lone pairs. The total five pairs of electrons are distributed in space as shown in Figure 5.7. The arrangement of bonds is trigonal bipyramidal; the arrangement of atoms is linear.

The electron pairs take up positions so as to maximise the angle between electron pairs and minimise the repulsion between them

Structures with *six* pairs of electrons around the central atom are sulphur(VI) fluoride, SF$_6$, iodine(V) fluoride, IF$_5$, and the ICl$_4^-$ ion. The **octahedral arrangement** of electron pairs is shown in Figure 5.8.

The arrangement of *atoms* in IF$_5$ is **square pyramidal**, a lone pair occupying the sixth position in the octahedron. In ICl$_4^-$ the four chlorine atoms are in a **square planar** configuration, with lone pairs occupying the axial positions of the octahedron.

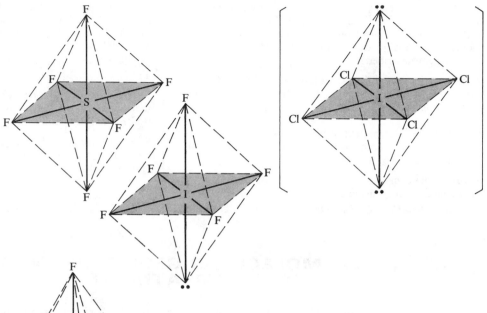

FIGURE 5.9
The Pentagonal
Bipyramidal Shape of
IF_7

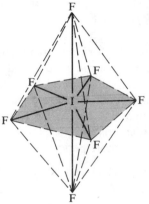

A molecule like IF_7, with *seven*
orbitals around the central atom,
has the **pentagonal bipyramidal**
arrangement of bonds shown in
Figure 5.9.

5.1.5 SUMMARY

TABLE 5.1 A
Summary
of the Shapes
of Molecules

No. of valence electrons	No. of bond pairs	No. of lone pairs	Total electron pairs	Arrangement of orbitals	Arrangement of atoms	Example
4	2	0	2	Linear	Linear	$BeCl_2$
6	3	0	3	Trigonal planar	Trigonal planar	BF_3
8	4	0	4 ⎫		Tetrahedral	CH_4
8	3	1	4 ⎬ Tetrahedral		Trigonal pyramidal	NH_3
8	2	2	4 ⎭		Bent line	H_2O
10	5	0	5 ⎫		Trigonal bipyramidal	PF_5
10	4	1	5 ⎪ Trigonal		⟨	SF_4
10	3	2	5 ⎬ bipyramidal		⊤	ClF_3
10	2	0	5 ⎭		Linear	I_3^-
12	6	0	6 ⎫		Octahedral	SF_6
12	5	1	6 ⎬ Octahedral		Square pyramidal	IF_5
12	4	2	6 ⎭		Square planar	ICl_4^-

═══════ **CHECKPOINT 5A: SHAPES OF MOLECULES** ═══════

1. (*a*) Take three long balloons, blow them up and tie the
ends. Hold the three tied ends between your finger and
thumb. What positions do the three balloons adopt?
(*b*) Add a fourth balloon and notice the positions which the
balloons take up.

2. In BF_3, how many electron pairs are there around the
B atom? Sketch the spatial distribution of bonds. How
would you describe the shape of this molecule?

3. In BrF_3, how many electrons does Br use for bond formation? How many lone pairs does Br possess? What is the total number of electron pairs around the Br atom? Sketch the spatial distribution of bonds. How is this arrangement described?

4. Sketch the spatial distribution of bonds in HOBr. How would you describe the shape of (*a*) the electron orbitals, (*b*) the molecule?

5. Sketch the arrangement of bonds in CCl_4. If the $\overset{\delta+}{C}{\,-\,}\overset{\delta-}{Cl}$ bond has a dipole moment of *x* debyes, what is the dipole moment of the CCl_4 molecule?

6. (*a*) In the compound XeF_4, how many electrons is the noble gas xenon using for bond formation? How

many lone pairs does xenon have? What is the total number of electron pairs around the central Xe atom? Sketch the arrangement of bonds. What shape is the molecule?

(*b*) Sketch the arrangement of bonds in XeF_2, XeF_6 and XeO_3.

7. Explain how the Sidgwick–Powell theory predicts the shape of the following molecules: (*a*) $SnCl_4$, (*b*) PH_3, (*c*) PF_5, (*d*) BH_3 and (*e*) BeH_2.

8. Sketch the spatial arrangement of bonds in the following: (*a*) F_2O, (*b*) $SeCl_4$, (*c*) SO_3, (*d*) ICl_3, (*e*) PF_6^- and (*f*) $COCl_2$.

5.2   MOLECULAR GEOMETRY: A MOLECULAR ORBITAL TREATMENT

An alternative to the Sidgwick–Powell treatment is the molecular orbital approach

The Sidgwick–Powell theory provides a simple treatment of the shapes of covalent molecules. A more precise treatment of the spatial distribution of covalent bonds about a central atom involves a consideration of the atomic orbitals used in bond formation. The shapes of atomic orbitals are described on p. 36 and shown in Figures 2.12, 2.13 and 2.14 [p. 36 and p. 37]. When **atomic orbitals** overlap, **molecular orbitals** are formed. The **molecular orbital approach** is illustrated by the following compounds of elements in the first short period of the Periodic Table:

$$HF \quad BeCl_2 \quad BF_3 \quad CH_4 \quad H_2O \text{ and } NH_3$$

5.2.1   HYDROGEN FLUORIDE

The electronic configurations of the fluorine atom and the hydrogen atom are

2p $\boxed{\uparrow\downarrow\,\uparrow\downarrow\,\uparrow}$
2s $\boxed{\uparrow\downarrow}$
1s $\boxed{\uparrow\downarrow}$ $\boxed{\uparrow}$
F $1s^2 2s^2 2p^5$ H $1s$

In HF, an s orbital and a p orbital overlap

When a molecule of hydrogen fluoride is formed, the s orbital of the hydrogen atom overlaps with the fluorine p orbital containing the unpaired electron. Figure 5.10 shows two of the three p orbitals at right angles to one another and shows how the bonding orbital of the fluorine atom becomes concentrated between the F and H nuclei when the bond is formed.

FIGURE 5.10
Hydrogen Fluoride

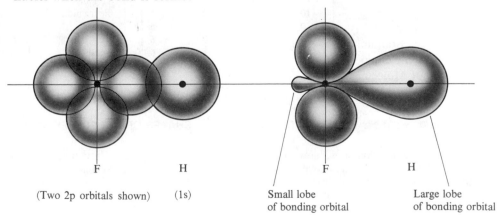

(Two 2p orbitals shown)   (1s)   Small lobe   Large lobe
                                       of bonding orbital   of bonding orbital

5.2.2 BERYLLIUM CHLORIDE

Beryllium has the electron configuration $1s^2 2s^2$, and chlorine has $1s^2 2s^2 2p^6 3s^2 3p^5$.

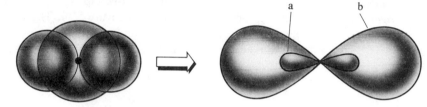

2s ⬆⬇
1s ⬆⬇
Be $1s^2 2s^2$

2p ⬆ □ □
2s ⬆
1s ⬆⬇
Be* $1s^2 2s 2p$

3p ⬆⬇⬆⬇⬆
3s ⬆⬇
2p ⬆⬇⬆⬇⬆⬇
2s ⬆⬇
1s ⬆⬇
Cl $1s^2 2s^2 2p^6 3s^2 3p^5$

Beryllium uses two orbitals, an s and a p orbital . . .

. . . The bonds formed are identical . . .

. . . Two sp hybrid orbitals are used

In order to combine, beryllium needs unpaired electrons: otherwise it would be as unreactive as helium. If one 2s electron is promoted into the 2p shell, then the atom will have two unpaired electrons. The atom must absorb energy in order to promote the electron and is described as an 'excited' atom. The bonding orbitals formed by beryllium are not of two different kinds, derived simply from the s and p atomic orbitals, they are two identical **hybrid** orbitals. The electron densities in the s and p atomic orbitals combine in compound formation to give sp hybrid orbitals of the shape shown in Figure 5.11.

FIGURE 5.11
sp Hybrid Bonds

![sp hybrid bonds diagram]

One s orbital + one p orbital   Two sp hybrid orbitals

In future diagrams, the small lobe *a* will be omitted and only the main lobe *b* in each sp hybrid bond will be shown.

When the two beryllium sp hybrid orbitals overlap with p orbitals of chlorine atoms, the bonding is as shown in Figure 5.12.

FIGURE 5.12 $BeCl_2$, a Linear Molecule

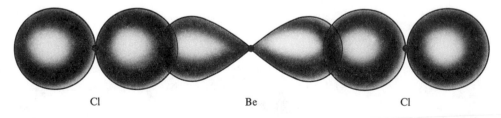

Cl   Be   Cl

The non-bonding p orbitals of Cl are not shown.

This linear arrangement of atoms is as predicted by the Sidgwick–Powell theory.

5.2.3 BORON TRIFLUORIDE

In an atom of boron ($1s^2 2s^2 2p$), an s electron can be promoted to a p orbital.

2p ⬆ □ □
2s ⬆⬇
1s ⬆⬇
B $1s^2 2s^2 2p$

2p ⬆ ⬆ □
2s ⬆
1s ⬆⬇
B* $1s^2 2s 2p^2$

2p ⬆⬇⬆⬇⬆
2s ⬆⬇
1s ⬆⬇
F $1s^2 2s^2 2p^5$

In boron, two p orbitals hybridise with one s orbital to form three sp² hybrid bonds...

The three orbitals which an excited atom of boron, B*, uses in bond formation are identical orbitals formed by a combination of the electron densities in the s and two p orbitals and called sp² hybrid orbitals. Quantum mechanical calculations give them the shape shown in Figure 5.13.

FIGURE 5.13
sp² Hybrid Orbitals

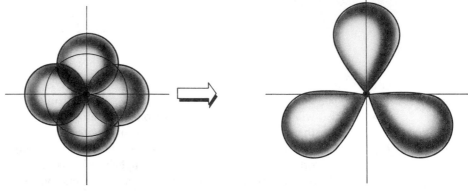

One s and two p orbitals   Three sp² hybrid orbitals

...The arrangement of orbitals is trigonal planar

When three sp² hybrid orbitals overlap with p orbitals from fluorine atoms, the bonding formed is as shown in Figure 5.14. The molecule is described as trigonal planar. Although the B—F bonds are polar, the centre of negative charge coincides with the centre of positive charge, the boron atom, and the molecule has no dipole moment.

FIGURE 5.14 BF₃, a Trigonal Planar Molecule

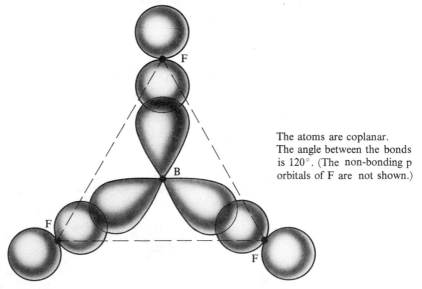

The atoms are coplanar. The angle between the bonds is 120°. (The non-bonding p orbitals of F are not shown.)

5.2.4 METHANE

The electronic configurations of carbon in its normal, C, and excited, C*, states and hydrogen, H, are shown below:

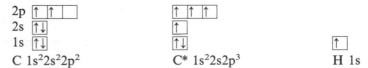

C $1s^2 2s^2 2p^2$ C* $1s^2 2s 2p^3$ H 1s

Each carbon atom, C, has two unpaired electrons, and one might expect carbon to form two bonds. It would not then attain a neon-like structure: it needs to share four electrons to do this. A sharing of four electrons can be achieved by promoting

In carbon, an s electron is promoted to a p orbital...

one of the 2s electrons into the 2p level. The excited carbon atom, C*, might be expected to form two different kinds of bond, using one s orbital and three p orbitals. Actually, the electron density distributes itself evenly through four bonding orbitals, which are called sp^3 hybrid orbitals. The sp^3 atomic orbital is more concentrated in direction than a p orbital [see Figure 5.15]. An sp^3 orbital is therefore able to overlap more extensively and form stronger bonds than a p orbital.

FIGURE 5.15
Comparison of Atomic Orbitals

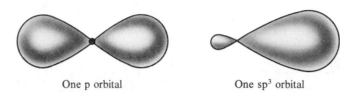

One p orbital   One sp^3 orbital

...four sp hybrid orbitals are formed

The tetrahedral arrangement of the four sp^3 orbitals of carbon is shown in Figure 5.16. The Sidgwick–Powell theory predicts this arrangement of bonds. Quantum mechanical calculations arrive at the same picture. Experimental evidence for the tetrahedral arrangement is gained from X ray diffraction studies of diamond. The angle between the bonds is shown to be 109.5° [see Figure 6.14, p.121].

FIGURE 5.16
Overlapping of Atomic Orbitals in Methane

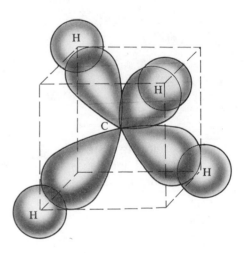

...they are tetrahedrally arranged in space

Six planes of symmetry run through the tetrahedron of hydrogen atoms in CH_4, and the centres of positive charge and negative charge coincide in the carbon atom at the centre of the tetrahedron. [See Figure 5.17.]

FIGURE 5.17
The Symmetry of the Tetrahedron

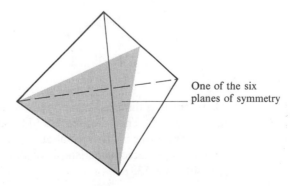

One of the six planes of symmetry

*Four different groups
bonded to carbon destroy
the symmetry of the
tetrahedron...*

If four different atoms or groups are attached to a carbon atom, there is no longer a plane of symmetry in the molecule. Nor is there a centre of symmetry or an axis of symmetry. The carbon atom in CHClBrF is an **asymmetric** carbon atom. There are two ways of drawing a tetrahedral arrangement for this formula. You will see from Figure 5.18 that one molecule is the mirror image of the other; they are called **enantiomers**. The molecules cannot be superimposed, being related in the same way as a left hand and a right hand [see Figure 5.19]. They show the geometric property of **chirality**.

FIGURE 5.18
Enantiomers of
Bromochlorofluoromethane

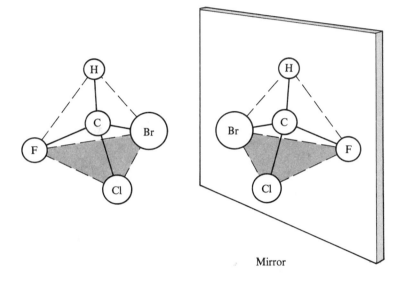

FIGURE 5.19
Left Hand and Its
Mirror Image

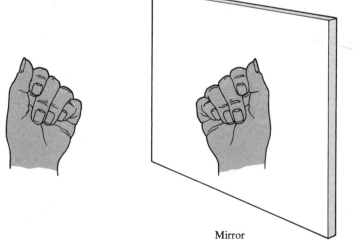

*...such compounds exist
in stereoisomeric forms*

There are two different compounds with the two different kinds of molecules shown in Figure 5.18. They are **isomers** (different compounds with the same formula). Since they differ in the spatial arrangement of atoms, they are called **stereoisomers**. This type of stereoisomerism, involving enantiomers, is called **optical isomerism** [pp. 492 and 538].

5.2.5 WATER

H₂O is not a linear molecule

The H_2O molecule has a dipole moment. If the molecule were linear, the dipoles in the two H—O bonds would cancel out: $\overset{\delta+}{H}—\overset{\delta-}{O}—\overset{\delta+}{H}$, and the molecule would have no dipole moment. The reason for its dipole moment lies in the atomic orbitals used for bonding:

2p	↑↓ ↑ ↑		↑↓ ↑ ↑↓
2s	↑↓		↑↓
1s	↑↓	↓	↑↓
	O $1s^2 2s^2 2p^4$	H 1s	O in H_2O

If O uses two p orbitals for bonding, the bond angle should be 90°...

If oxygen uses two p orbitals for bonding, the bond angle will be 90° [Figure 2.13(b), p. 37]. In fact, X ray studies show the bond angle to be 105°. This is close to the tetrahedral angle of 109.5°. It is believed that hybridisation does occur between the s orbital and the three p orbitals of the oxygen atom. Of the four sp^3 hybrid orbitals, two are occupied by bonding pairs of electrons and the other two by lone pairs [see Figure 5.20]. The difference between the bond angle of 105° and the tetrahedral angle is explained by the greater repulsion between lone pair orbitals than between bonding orbitals [p. 93].

...In fact it uses sp³ hybrid orbitals with bond angles of 105°

FIGURE 5.20
sp³ Bonds in H₂O

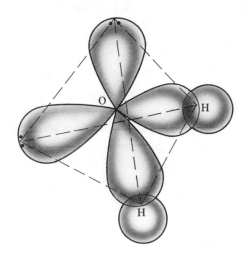

5.2.6 AMMONIA

The electrons used by nitrogen to form bonds are the three unpaired 2p electrons. They are in orbitals which are mutually perpendicular [see Figure 2.13(b), p. 37].

2p	↑ ↑ ↑		↑↓ ↑↓ ↑↓
2s	↑↓		↑↓
1s	↑↓	↓	↑↓
	N $1s^2 2s^2 2p^3$	H 1s	N in NH_3

In NH₃, N uses four sp³ hybrid orbitals...

...one is occupied by a lone pair of electrons

Measurements give a value of 107° for the angle between the bonds in the NH_3 molecule. It is believed that four sp^3 hybrid orbitals are formed, one occupied by a lone pair and the other three by bonding pairs of electrons [see Figure 5.21]. The difference between the bond angle, 107°, and the tetrahedral angle, 109.5°, arises from the repulsion between a lone pair and a bonding pair of electrons being greater than the repulsion between two bonding pairs [p. 93].

FIGURE 5.21
sp³ Hybrid Bonds in
NH₃

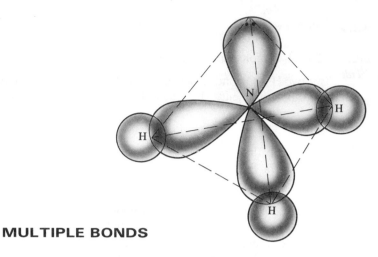

5.2.7 MULTIPLE BONDS

*A double bond is less than
twice as strong as a single
bond...*

Carbon forms double bonds in compounds such as carbon dioxide, $O=C=O$, and ethene, $H_2C=CH_2$. The double bond is not simply two single bonds. The amount of energy required to break a certain bond in a mole of molecules is called the **standard bond enthalpy** [p. 203]. Standard bond enthalpies of carbon–carbon bonds are

*...and a triple bond has
less than three times the
strength of a single bond*

$$C-C \quad 346 \, kJ \, mol^{-1}$$

$$C=C \quad 610 \, kJ \, mol^{-1}$$

$$C\equiv C \quad 837 \, kJ \, mol^{-1}$$

The $C=C$ bond is less than twice as strong as a $C-C$ bond, and the $C\equiv C$ bond is less than three times as strong as a $C-C$ bond.

In a molecule of ethene, each carbon atom uses a 2s orbital and two of the three 2p orbitals to form three sp² hybrid bonds [see Figure 5.13, p. 98]. The electronic configurations of carbon are shown below:

2p ⎡↑⎢↑⎢ ⎤   ⎡↑⎢↑⎢↑⎤   ⎡↑↓⎢↑↓⎢↑⎤
2s ⎡↑↓⎤   ⎡↑⎤   ⎡↑↓⎤
1s ⎡↑↓⎤   ⎡↑↓⎤   ⎡↑↓⎤
C 1s²2s²2p²   C* 1s²2s2p³   C in ⟩C—

*Each C has an
unhybridised p orbital...*

The carbon–carbon bond formed when the sp² orbitals of neighbouring carbon atoms overlap is called a σ, **sigma**, bond. In σ bonds, e.g. any single bond, overlap of atomic orbitals occurs along the line joining the two bonded atoms. There is an unhybridised p orbital at right angles to the plane of the two sp² orbitals, and the p orbitals on adjacent carbon atoms are close enough to overlap. The overlapping occurs at the sides of the orbitals [see Figure 5.22].

FIGURE 5.22
The Ethene Molecule

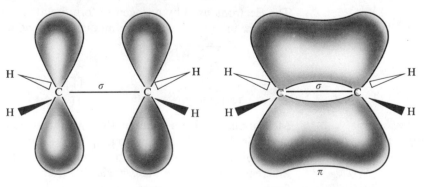

(a) The C atoms have
unhybridised p orbitals.

(b) Sideways overlap between
the two parallel p orbitals
produces one π orbital.

...Sideways overlap between p orbitals is called a π bond...

This type of bond, produced by sideways overlapping of p orbitals above and below the plane of the sp² bonds, is called a π, **pi**, bond. It is not as strong as a σ bond since there is less overlapping of orbitals [see Figure 5.23]. This is why the C=C bond is less than twice as strong as a C—C bond.

...π bonds are less strong than σ bonds

Since overlapping of p orbitals on adjacent carbon atoms can occur only when the p orbitals are parallel, the two

H
\
C—
/
H

structures must be coplanar, i.e., lie in the

For π bonds to be formed the atoms in H₂C=CH₂ must be coplanar

same plane. If one CH₂ group twists with respect to the other, the amount of overlapping of p orbitals will decrease, and the π bond will be partially broken. Since it requires energy to break a bond, the most stable arrangement of the molecule is the one in which all six atoms lie in the same plane [see Figure 27.2(a), p. 554].

FIGURE 5.23
The Difference between σ and π bonds

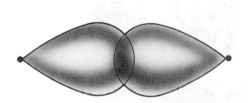

(a) σ Bonding.
The orbitals point towards each other.

(b) π Bonding.
The orbitals are parallel and overlap sideways.

There is an interesting consequence of the coplanar arrangement. The formula of but-2-ene is $CH_3CH=CHCH_3$. There are two structures with this formula:

(a) CH₃—C—H   (b) CH₃—C—H
       ‖   ‖
   CH₃—C—H   H—C—CH₃

Cis- and trans- *forms of but-2-ene*

Structure (a), in which the hydrogen atoms are on the same side of the double bond, is called *cis*-but-2-ene, and structure (b), with the hydrogen atoms on opposite sides of the double bond, is called *trans*-but-2-ene. The existence of *cis*- and *trans*-forms of compounds is covered on p. 534. [See Figure 5.24.]

FIGURE 5.24 Cis– and trans–But-2-ene

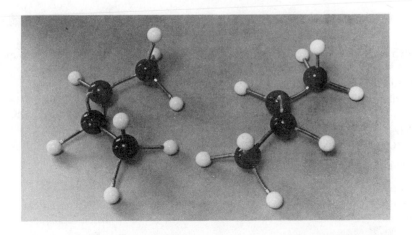

In the —C≡C— bond
each C uses two sp hybrid
orbitals...

...the two unhybridised
p orbitals overlap with
those of the other C atom
to form two π bonds

When a carbon atom is bonded to only two other atoms, as in HC≡CH or
O=C=O the two σ bonds formed use sp hybrid orbitals [p. 97]. In
H—C≡C—H, σ bonds are formed by the overlapping of sp hybrid orbitals of
the two carbon atoms and by the overlapping of sp orbitals of each carbon atom
with the 1s orbital of a hydrogen atom. For maximum overlapping, the four atoms
must lie in a straight line. Two π bonds are formed by overlapping of the remaining
unhybridised p orbitals. There are two of these on each carbon atom at right angles
to the sp hybrid orbitals [see Figure 5.25].

FIGURE 5.25

The Ethyne Molecule

(a) Two unhybridised
p orbitals on one C atom
overlap with two on the other
C atom.

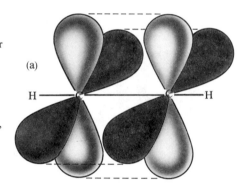 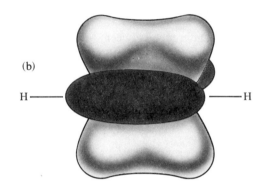

(b) Two π bonds are formed,
one above and below the
line of the σ bonds, the
other in front and at the
back.

There is a π bond in CO_2

In carbon dioxide, there are sp hybrid bonds between the carbon atom and each of
the oxygen atoms. Since two sp hybrid bonds are colinear, the CO_2 molecule is linear.
Sideways overlapping of p orbitals produces single π bonds between each oxygen
atom and the carbon atom.

π bond formation is
restricted to small atoms

The formation of strong π bonds is restricted to members of the first short period:
carbon, nitrogen and oxygen. In larger atoms, strong π bonds are not formed
because, being removed from the line between the centres of the atoms, the π bond
becomes rapidly weaker as the size of the atom increases.

CHECKPOINT 5B: BONDING

1. Match each of these species with one of the hybridisation
schemes sp, sp², sp³: (a) PH_3, (b) PH_4^+, (c) $BeCl_2$,
(d) $SiCl_4$, (e) BrF_3, (f) Al_2Cl_6.

2. Describe the bonding in the molecules CO_2 and CO.

3. State what type of bonding orbitals are employed by
the central atom in the following: (a) PH_3, (b) SCl_2,
(c) HCHO, (d) HCN, (e) F_2O.

4. Draw an electrons-in-boxes diagram to show the
electronic configuration in each of the underlined atoms:
$\underline{Be}H_2$, $\underline{B}F_3$, $\underline{B}F_4^-$, $H\underline{C}N$. What is the nature of the hybrid
bonds formed in each species, and what is the shape of
each species?

5. Deduce the shapes of the following species: AsH_3,
PH_4^+, H_3O^+, CS_2, $CH_2=C=CH_2$, HC≡N.

6. What is the arrangement of bonds around the central
atom in each of the following species: CH_4, BF_3, NF_3,
ICl_4^-, BrO_3^-, ClO_4^-, $CHCl_3$? State the nature of the
hybridisation of the atomic orbitals on the central atom.

7. Write structures which show the arrangement of electrons
in the bonding orbitals of O_2, CO_2, CO, NO_3^- and CN^-

(e.g., $\ddot{:}\overset{..}{O}\ddot{:}\ \ddot{:}\overset{..}{O}\ddot{:}$)

8. What can you deduce from the fact that, whereas water
has a dipole moment, carbon dioxide has none?

9. Write electron structures for PH_3, NH_3, NH_4Cl, H_2O,

H_2O_2, SiH_4, HOCl and NO_2^- (e.g., $H \!:\! \overset{\displaystyle ..}{\underset{\displaystyle H}{P}} \!:\! H$)

10. The ammonium ion and methane are said to be *isoelectronic*. What does this mean? Why do the compounds have different chemical properties?

11. (*a*) Sketch the arrangement of bonds in $CF_2{=}CF_2$.

(*b*) Explain why there are two isomers with the formula $CFCl{=}CFCl$.

5.2.8 *HYBRIDISATION INVOLVING d ORBITALS

Complex ions are formed by means of coordinate bonds

Hybrid bonds formed by s, p and d orbitals are important in the formation of **complexes** or **coordination compounds**. A **complex ion** is formed by the combination of a simple ion with a **ligand** or **complexing agent**. This ligand may be an oppositely charged ion or a neutral molecule. It must be a species containing an unshared pair (or pairs) of electrons that can be donated to the simple ion with the formation of a

A ligand shares a lone pair with an acceptor ion or atom

coordinate bond. Ammonia, $: N \overset{\displaystyle H}{\underset{\displaystyle H}{—}} H$, and the cyanide ion, $[:C{\equiv}N:]^-$, are ligands.

Ammonia coordinates to copper(II) ions to form complex tetraammine copper(II) ions, $Cu(NH_3)_4{}^{2+}$. Cyanide ions coordinate to nickel(II) ions to form complex tetracyanonickel(II) ions, $Ni(CN)_4{}^{2-}$. Nickel atoms complex with carbon monoxide to form the coordination compound nickel carbonyl, $Ni(CO)_4$.

Below are shown the electronic configurations of Ni, Ni^{2+} and Ni in the complex ion $Ni(CN)_4{}^{2-}$:

Ni^{2+} uses a d orbital for complex formation...

...four dsp^2 hybrid bonds are formed in a square planar arrangement

When coordination occurs in Ni^{2+}, the two unpaired 3d electrons pair up to vacate one 3d orbital. This combines with the empty 4s orbital and two 4p orbitals to form four hybrid dsp^2 orbitals. These four hybrid dsp^2 orbitals are directed to the corners of a square. The four pairs of electrons from four cyanide ligands enter these four orbitals to form four bonds in a square planar configuration. Since four lone pairs of electrons are bonded to the central ion, the **coordination number** of the complex is said to be four [see Figure 5.26].

In the hexacyanoferrate ions, $Fe(CN)_6{}^{4-}$ and $Fe(CN)_6{}^{3-}$, cyanide ions coordinate to Fe^{2+} or Fe^{3+} ions. The electron configurations of Fe, Fe^{2+}, Fe^{3+} and the complex ions are as follows:

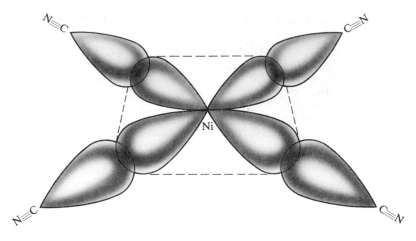

Fe^{2+} and Fe^{3+} use six d^{2}sp^3 hybrid orbitals...

...The bonds formed are octahedrally directed

The unpaired d electrons pair up to vacate orbitals in Fe^{2+} and Fe^{3+} which can be occupied by lone pairs of electrons. Six hybrid d^{2}sp^3 orbitals are formed. They are directed to the apices of an octahedron, as shown in Figure 5.27. Lone pairs of electrons from the cyanide ions coordinate into the empty hybrid orbitals.

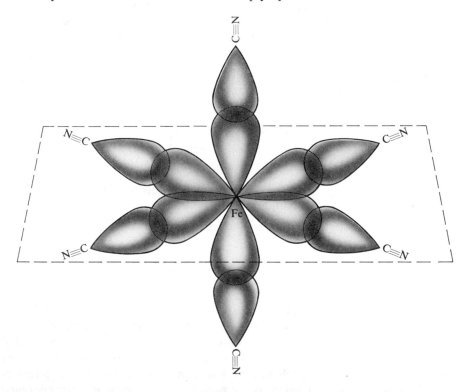

CHECKPOINT 5C: HYBRID BONDS I

*1. The electron configuration of cobalt is (Ar) 3d^{7}4s^2. Draw an electrons-in-boxes diagram of the valence electrons in Co and Co^{3+}. Explain which orbitals are used in the formation of the complex ion Co(CN)$_6$$^{3-}$. What is the type of hybrid bond formed? What is the spatial arrangement of bonds?

*2. Draw an electrons-in-boxes diagram for the outer electrons in Fe, (Ar) 3d^{6}4s^2, and Fe^{3+}. Which orbitals are used for complexing with cyanide ions in Fe(CN)$_6$$^{3-}$? What type of hybrid bond is formed? Sketch the spatial arrangement of bonds.

3. Identify the orbitals used for bonding in CH$_4$ and in BF$_4$$^-$. Sketch the spatial arrangement of bonds.

4. State the types of bonds which operate in (a) HCN and (b) Ag(CN)$_2$$^-$. In each case, specify the nature of any hybrid bonds formed. Sketch the arrangement in space of the atoms.

5. SiF$_4$ is a covalent molecule which can combine with fluoride ions to form SiF$_6$$^{2-}$. Once formed, all six bonds are indistinguishable. Why is this? Describe the arrangement in space of the bonds.

5.3 DELOCALISED ORBITALS

Localised electrons are to be found between the nuclei of two bonded atoms...

In the compounds discussed so far, the electrons in the σ and π bonds have been located in the region *between* the nuclei of the bonded atoms. They are **localised** electrons. In some molecules, some of the electrons are **delocalised**: they do not remain between a pair of atoms.

5.3.1 BENZENE

Benzene, C_6H_6, is an aromatic hydrocarbon [p. 530]. Kekulé [p. 574] proposed the structural formula for benzene which is shown here.

An alternating system of single and double bonds is called a **conjugated double bond system**. Between each pair of adjacent carbon atoms is a σ bond, formed by overlapping of sp^2 hybrid orbitals. Since sp^2 bonds are coplanar, all the carbon atoms lie in the same plane and form a regular hexagon. The unhybridised p orbitals of the carbon atoms are perpendicular to the plane of the hexagonal benzene 'ring'.

FIGURE 5.28
The Benzene Ring

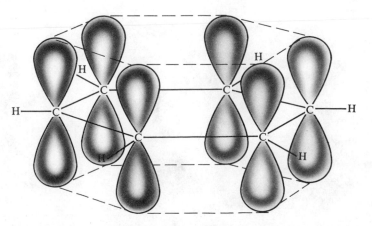

(a) Each of the C atoms has an unhybridised p orbital. These overlap sideways.

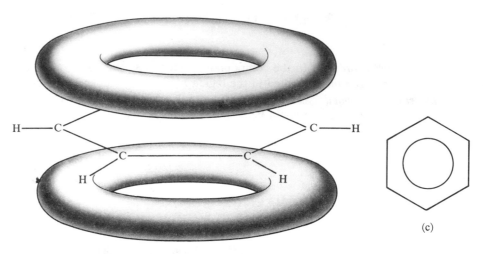

(b) The resulting distribution of π electron charge can be represented by two doughnut-shaped regions, one above and one below the hexagon of carbon atoms.

...Delocalised electrons do not remain between a pair of bonded atoms

As in ethene [see Figure 5.22], overlapping of p orbitals on adjacent carbon atoms gives rise to π bonds. In benzene, the p orbitals are able to overlap all round the ring [see Figure 5.28]. The electrons in the p orbitals cannot be regarded as located between any two carbon atoms: they are free to move between all the carbon atoms in the ring. They are described as **delocalised** and are represented as an annular cloud of electron density above and below the plane of the molecule.

Electrons in the π bonds in benzene are delocalised

The formula of benzene is often written as shown in Figure 5.28(c) to represent the delocalisation of π electrons.

FIGURE 5.29 Models of Ethene and Benzene

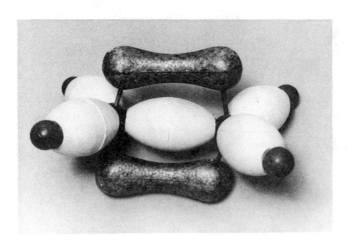

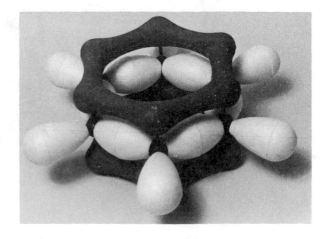

The delocalisation of π electrons confers stability on benzene. One can calculate the standard enthalpy content of benzene on the basis of the structure

by adding the average standard bond enthalpies for the bonds [pp. 102 and 203]. The sum of 3(C—C) bonds plus 3(C=C) bonds plus 6(C—H) bonds is $-5350 \, kJ \, mol^{-1}$. The standard enthalpy content can also be found from measurement

Delocalisation of electrons confers stability on benzene

of the standard enthalpy of combustion [p. 195]. The experimental value of $-5550 \, kJ \, mol^{-1}$ is more negative than the theoretical value. This means that benzene is *more* stable than one would expect it to be on the basis of the formula

The *difference,* $-200 \, kJ \, mol^{-1}$ is called the **delocalisation energy** of benzene.

The structure of benzene can be represented as intermediate between valence bond structures

Another way of representing delocalisation is to visualise the structure of the benzene molecule as intermediate between the two valence bond formulae. The two structures are called **canonical structures** or **canonical forms**. The actual structure of benzene is said to be a **resonance hybrid** of the two, and can be represented as shown below:

The double-headed arrow does *not* mean that benzene resonates or oscillates between the two canonical structures. It cannot possibly do this because neither of the two canonical structures exists. The actual structure of benzene is the delocalised structure.

CHECKPOINT 5D: HYBRID BONDS II

1. How is it known that the four C—H bonds in methane are equivalent?

2. What is meant by the terms: (*a*) localised molecular orbital and (*b*) delocalised molecular orbital?

3. Describe the formation of the second bond between the two carbon atoms in ethene. Explain why the ethene molecule, $H_2C=CH_2$, is planar.

†4. Why was the Kekulé structure for benzene adopted? Why was it superseded? [Refer to Chapter 28.]

*5. The molecular orbital structure for the sulphate ion is given below. It can be represented as a resonance hybrid of six canonical structures, one of which is shown. Draw the other five.

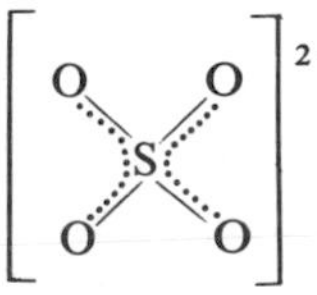

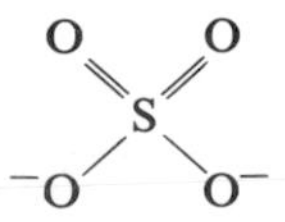

QUESTIONS ON CHAPTER 5

1. State the Sidgwick–Powell theory of electron pair repulsion. Sketch the arrangement of atoms in (*a*) $BeCl_2$, (*b*) BCl_3, (*c*) CCl_4, (*d*) NH_3, (*e*) H_2O and (*f*) IF_5. Explain how the arrangement of bonds is predicted on the Sidgwick–Powell theory.

2. Sketch the shapes of the following species:

$CO_3{}^{2-}$, $NO_2{}^-$, $NO_2{}^+$, $NO_3{}^-$, $PCl_6{}^-$,

$PCl_4{}^+$, $ICl_4{}^-$

3. Give the formula of a molecule whose atoms occupy each of the following shapes: (*a*) linear, (*b*) planar trigonal, (*c*) tetrahedral and (*d*) octahedral. State the angle between the bonds in each structure.

Some molecules which are based on a tetrahedral structure have bond angles different from the regular tetrahedral angle. Give an example, and explain the difference.

4. 'The shapes of simple molecules can be deduced from a consideration of the bonding electrons employed.' Discuss this statement. Apply the principle to the shapes of (*a*) NH_3, (*b*) H_2O, (*c*) CO_3^{2-}, (*d*) $H_2C=CH_2$, (*e*) ICl_3, (*f*) I_3^- and (*g*) SF_6.

5. Explain the following statements:

(*a*) Two different compounds have the molecular formula

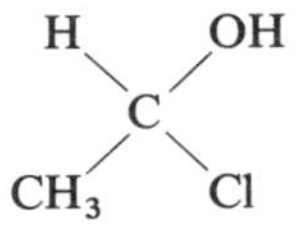

$$H \diagdown \quad \diagup OH$$
$$C$$
$$CH_3 \diagup \quad \diagdown Cl$$

(*b*) The spatial arrangement of atoms in BF_3 is different from that in NH_3.

(*c*) XeF_4 is planar, whereas CCl_4 is tetrahedral.

(*d*) The bond angle in NH_3 is greater than that in H_2O.

6. State the nature and spatial arrangement of the bonds in each of the following species: (*a*) CO_2, (*b*) HCN and (*c*) NO_3^-.

7. (*a*) Write a brief account, with illustrative examples, of the 'electronic theory of valency', in which you explain

(i) the underlying principles;

(ii) the formation of the several kinds of chemical bond, and the differences in properties conferred by the presence of these bonds;

(iii) the concept of 'limiting types' of chemical bond;

(iv) the formation of the hydrogen bond, and its consequences.

(*b*) On the basis of the electron-pair repulsion theory, explain how the shapes of the species in the sequences below may be explained, and sketch both the bond diagrams and the shapes described:

(i) CH_4, NH_3, H_2O;

(ii) NH_4^+, NH_3, NH_2^-.

Which principle is exemplified by the atoms (other than hydrogen) in each sequence? State this principle.

(SUJB 82)

6

CHEMICAL BONDING AND THE STRUCTURE OF SOLIDS

6.1 X RAY DIFFRACTION

A crystal is a regular three-dimensional arrangement of particles

A solid in which the arrangement of atoms or ions or molecules follows a regular three-dimensional design has a crystalline form. The surfaces of the crystal are planes, called **faces**. They intersect at angles that are characteristic for the substance. The ordered arrangement of particles in the crystal is called the **crystal structure**. It is based on a **lattice**, i.e., a geometrical arrangement of points in space. There are fourteen lattices on which crystal structures are based. The crystal structure gives rise to many series of equally spaced planes of particles [see Figure 6.1]. Since the wavelengths of X rays are comparable with the distances between the planes of particles, a crystalline solid acts as a three-dimensional diffraction grating for X rays.*

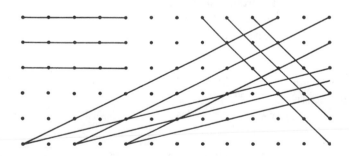

FIGURE 6.1 Some of the Planes in a Crystal (in two dimensions)

A crystal acts as a three-dimensional diffraction grating for X rays

When a beam of X rays meets the solid, X rays interact with electrons, and the beam is scattered. The scattered X rays must be made to produce a visible pattern, e.g., on a photographic film. [See Figure 6.2.] From the pattern of scattering [see Figure 6.3] one can infer the pattern of distribution of electronic charge in the crystal and hence the nature of the crystal structure.

*See R Muncaster, *A-Level Physics* (Stanley Thornes)

111

FIGURE 6.2 Diffraction
of X rays by a Crystal

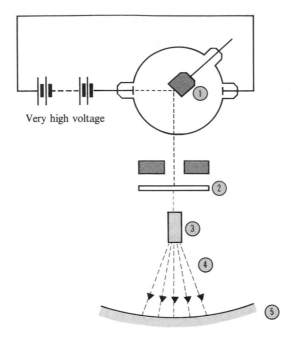

Very high voltage

1 Cathode rays (electrons)
impinge on metal and
generate X rays.

2 Metal filter removes X rays
of all frequencies except one.

3 Narrow beam of X rays of
known frequency strikes
crystal.

4 X rays are diffracted through
various angles.

5 Photographic film detects X
rays.

6.1.1   THE BRAGG EQUATION

The **Bragg equation**, derived by the pioneer X ray crystallographers, W H and
W L Bragg, relates the angle at which a diffraction pattern is obtained to the distance
between planes in the crystal. The equation is

$$n\lambda = 2d \sin \theta$$

where λ = wavelength, n = integer, d = distance between planes in the crystal and
θ = angle of diffraction.

Figure 6.3 shows a developed X ray film. The central spot has been produced by
undeflected X rays, and the circles of spots are produced by X rays which have
been diffracted through various angles by the planes of atoms or ions in the crystal.

FIGURE 6.3 An X ray
Diffraction Pattern

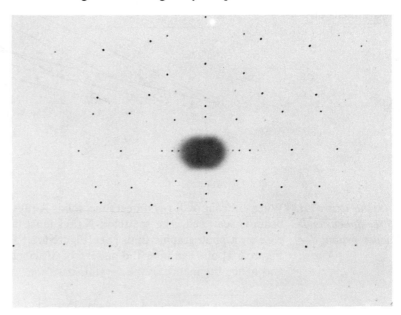

Crystallographers compare results of measurements with results calculated for models

Arriving at a crystal structure from an X ray photograph is not easy. A method which crystallographers employ is to construct a model of the crystal structure and calculate the angles and intensities of diffracted X rays which the model would produce. If the calculations do not agree with the measured results, a different model must be constructed and fresh calculations performed. Eventually, with the aid of a computer, agreement is reached between the model and the measurements. The model is then held to represent the structure of the crystal.

Hydrogen atoms do not show on X ray patterns

Since it is the *electrons* of atoms that scatter X rays, the small atoms in a compound, which have few electrons, especially hydrogen atoms, are difficult to detect. The structures of metals, ionic compounds and macromolecular substances (such as diamond and graphite) have been worked out from X ray diffraction measurements.

6.2 METALLIC SOLIDS

6.2.1 THE METALLIC BOND

The nature of the metallic bond must be responsible for the properties of metals

Modern technology is based on the use of metals. Most of our machines and most of our forms of transport are made of metal. There must be some feature of the bond between metal atoms that gives metals their special properties. Many metals are strong and can be deformed without breaking; many are **malleable** (can be hammered) and **ductile** (can be drawn out under tension). They are shiny when freshly cut and good conductors of heat and electricity. Any theory of the metallic bond must account for all these physical properties.

In a metal...

...Atomic orbitals overlap to form molecular orbitals...

...The valence electrons become delocalised...

...Metal cations are formed...

...These are attracted to the electron cloud

The outer shell electrons of a metal (the valence electrons) are relatively easily removed, with the formation of metal cations. When two metal atoms approach closely, as in a metal structure, their outer shell orbitals overlap to form molecular orbitals. If a third atom approaches, its atomic orbitals can overlap with those of the first two atoms to form another molecular orbital. For a large number of atoms, a large number of molecular orbitals are formed, extending over three dimensions. As a consequence of the multiple overlapping of atomic orbitals, the outer electrons from each atom come under the influence of a very large number of atoms. They are free to move through the structure and are no longer located in the outer shell of any one atom: they are **delocalised**. The removal of the electrons leaves behind metal cations. The reason why the cations are not pushed apart by the repulsion between them is that, in a pair of cations, each cation is attracted to the delocalised electron cloud between them [see Figure 6.4].

FIGURE 6.4
Deformation of Metal Structure

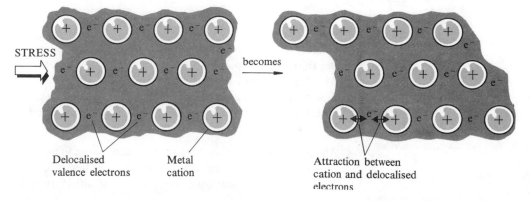

Delocalised valence electrons   Metal cation

Attraction between cation and delocalised electrons

The metallic bond explains the strength of metals...

This theory of the metallic bond explains the physical properties of metals. If a stress is applied to the metal, the structure can change in shape without fracturing [see Figure 6.4]. This contrasts with the effect of stress on an ionic structure [see Figure 6.11, p. 119].

...also their thermal conductivity...

The high thermal conductivity of metals is accounted for. When heat is supplied to one end of a piece of metal, the kinetic energy of the electrons is increased. The increase is transmitted through the system of delocalised electrons to other parts of the metal.

...and electrical conductivity...

Electrical conductivity can also be explained. If a potential difference is applied between the ends of a metal, the delocalised electron cloud will flow towards the positive potential.

...and the shiny appearance of metals

The shiny appearance of metals fits in which the theory of the nature of the metallic bond. The metal contains a large number of molecular orbitals at a large number of different energy levels. When light falls on to the metal, electrons are excited. A large number of transitions between energy levels is possible, with a whole range of frequencies being absorbed. As electrons return to lower energy levels, light is emitted and makes the metal shine.

6.2.2   THE STRUCTURE OF METALS

Types of metal structures

Metal atoms pack closely together in a regular structure. There is no way of packing spheres to fill a space completely without leaving gaps between them. Arrangements in which the gaps are kept to a minimum are called **close-packed** arrangements.
X ray studies have revealed three main types of metallic structures. In the **hexagonal close-packed structure**, and the **face-centred-cubic close-packed structure**, the metal atoms pack to occupy 74% of the space. In the **body-centred-cubic close-packed structure**, the atoms occupy 68% of the total volume.

When identical spheres pack together so as to minimise the space between them, a hexagonal close-packed structure or a face-centred-cubic structure may be formed...

The way in which the structures are built up can be envisaged by considering (or, better, by experimenting with) the stacking of spheres (e.g., marbles, table tennis balls, styrofoam spheres) together. Imagine a layer A of identical spheres packed together on a flat surface. Each sphere is in contact with six others. [See Figure 6.5(a).] There are spaces or holes between the spheres. Then imagine laying a second layer B on top of the first. Each sphere in layer B will rest in a depression between three spheres in layer A. The holes in layer B are of two kinds. [See Figure 6.5(b).] Tetrahedral holes in layer B lie over spheres in layer A, and octahedral holes in layer B lie over holes in layer A:

Tetrahedral   Octahedral
hole   hole

There are two ways in which a third layer can be laid on layer B. The spheres of the third layer can cover the tetrahedral holes; in this case the third layer is identical with layer A. This type of structure is made up of alternating layers ABABA and is described as hexagonal close-packed [see Figure 6.5(c)]. Since every atom is in contact with twelve others (six in the same layer, three in the layer above and three in the layer below), it is said to have a **coordination number** of 12.

...In both these structures the coordination number is twelve

Alternatively, the spheres in layer C can cover the octahedral holes in layer B, so that layer C is not identical with layer A. When a fourth layer of spheres is laid on layer C, it is identical with A. This type of structure, with the sequence ABCABC,

FIGURE 6.5

The Packing of Spheres
in the Hexagonal
Close-packed and the
Face-centred-cubic
Close-packed Structures

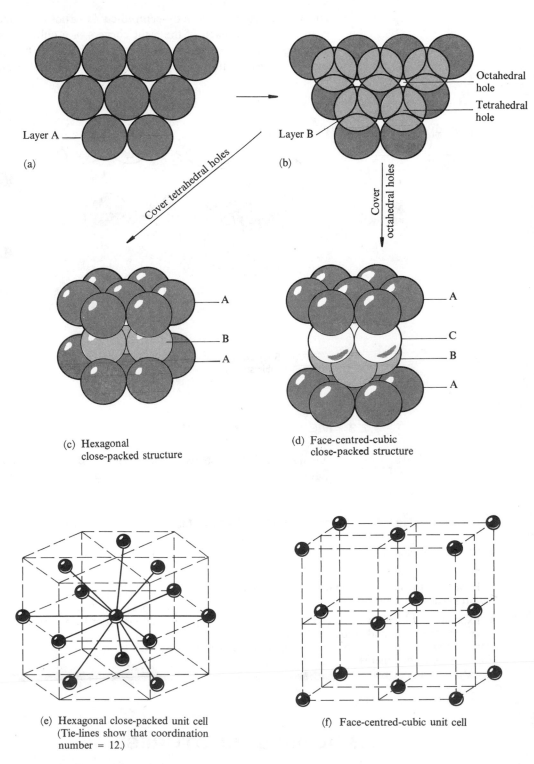

Layer A

(a)

Cover tetrahedral holes

Layer B

Octahedral
hole

Tetrahedral
hole

(b)

Cover octahedral holes

A
B
A

(c) Hexagonal
close-packed structure

A
C
B
A

(d) Face-centred-cubic
close-packed structure

(e) Hexagonal close-packed unit cell
(Tie-lines show that coordination
number = 12.)

(f) Face-centred-cubic unit cell

is a face-centred-cubic close-packed structure. [See Figure 6.5(d).] The coordination number is 12. The high coordination numbers in these structures arise from the non-directed nature of the metallic bond.

A unit cell   Also shown in Figure 6.5 are the **unit cells** of the two types of structure. A unit cell is the smallest part of the crystal that contains all the characteristics of the structure. The whole structure can be generated by repeating the unit cell in three directions.

The less closely packed body-centred-cubic structure is shown in Figure 6.6(a). With one atom at each of the eight corners of a cube and one in the centre touching these eight, the coordination number is 8. Figure 6.6(b) shows an expanded view, and (c) shows the unit cell with tie-lines to show that the coordination number is 8.

FIGURE 6.6
Body-centred-cubic
Structure

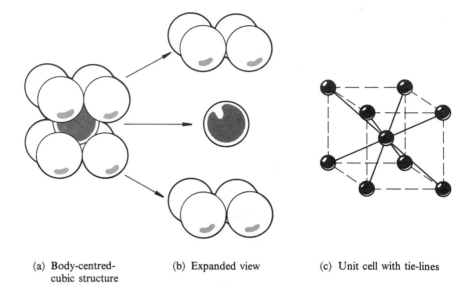

(a) Body-centred-   (b) Expanded view   (c) Unit cell with tie-lines
cubic structure

6.2.3 METALLIC RADIUS

One half the distance between the nuclei of adjacent metal atoms is called the **metallic radius**.

CHECKPOINT 6A: METALLIC STRUCTURES

1. Why might someone describe an entire piece of metal as a large molecule?

2. How does the nature of the metallic bond account for the properties of metals?

3. Explain what is meant by (*a*) a close-packed structure, (*b*) tetrahedral holes and (*c*) octahedral holes. Why are there two arrangements, rather than one, for achieving the closest packing of spheres?

†**4.** Why does a metal give rise to X rays of a limited number of frequencies when excited by electron bombardment as in Figure 6.2, p. 112? The metal filter in Figure 6.2 is a metal with proton number (atomic number) one less than the target metal. Why does this cut out X rays of all frequencies but one? [For help, see p. 7.]

6.3 IONIC STRUCTURES

Ionic structures are regular three-dimensional arrangements of ions

The alkali metal halides are ionic compounds. The ions are arranged in a regular three-dimensional structure. The melting and boiling temperatures of ionic compounds are high, owing to the strong forces of electrostatic attraction between the ions in the crystal. When the salts are melted or dissolved, the ions become free to move, and the salts conduct electricity. The crystals of alkali metal halides are *cubic* in shape, and X ray analysis shows two kinds of structures. The sodium chloride structure is illustrated in Figure 6.7(a).

FIGURE 6.7
The Sodium Chloride
Structure

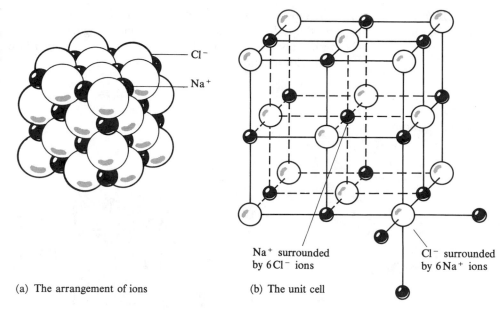

Cl⁻

Na⁺

Na⁺ surrounded
by 6 Cl⁻ ions

Cl⁻ surrounded
by 6 Na⁺ ions

(a) The arrangement of ions

(b) The unit cell

The best arrangement of ions in a structure, being the one with the lowest energy, is that which allows the greatest number of contacts between oppositely charged ions without pushing together ions with the same charge. Many structures are close-packed arrangements of anions, with the smaller cations occupying the octahedral holes. Sodium chloride has a face-centred-cubic close-packed lattice of chloride ions (radius, 0.181 nm), which is expanded to accommodate sodium ions (radius 0.098 nm) in the octahedral holes [Figure 6.7(a)]. In Figure 6.7(b), only the centres of the ions are drawn in order to show the features of the structure clearly. There are six sodium ions surrounding each chloride ion: the coordination number of chlorine is 6. Similarly, there are six chloride ions surrounding each sodium ion: the coordination number of sodium is 6. The structure shows 6 : 6 coordination.

In NaCl the anions form a face-centred-cubic lattice with cations in the octahedral holes...

...and 6 : 6 coordination

In CsCl the Cs⁺ ions and Cl⁻ ions both adopt simple cubic structures with 8 : 8 coordination

The caesium chloride structure [see Figure 6.8(a)] is different. Since the caesium ion (radius 0.168 nm) is larger than the sodium ion, a larger number of chloride ions can surround it. The structure shows 8 : 8 coordination. Both the caesium ions and the chloride ions adopt simple cubic lattices, which interpenetrate [see Figure 6.8(b)] so that each cube of chloride ions has a caesium ion at its centre and vice versa.

FIGURE 6.8
The Caesium Chloride
Structure

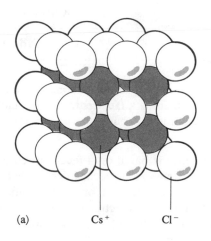

(a) Cs⁺ Cl⁻

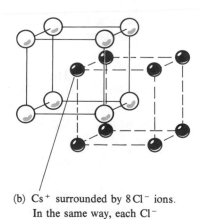

(b) Cs⁺ surrounded by 8 Cl⁻ ions.
In the same way, each Cl⁻
is surrounded by 8 Cs⁺ ions.

The radius ratio r_C/r_A determines the structure

The type of structure adopted depends on the radius ratio, r_C/r_A, where r_C and r_A are the radii of the cation and the anion. Figure 6.9 shows the structure of a crystal with a coordination number of 6. The anions are just in contact with other anions and with a cation.

FIGURE 6.9 Radius Ratio for Coordination Number 6

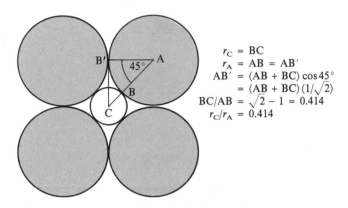

$$r_C = BC$$
$$r_A = AB = AB'$$
$$AB' = (AB + BC)\cos 45°$$
$$= (AB + BC)(1/\sqrt{2})$$
$$BC/AB = \sqrt{2} - 1 = 0.414$$
$$r_C/r_A = 0.414$$

The critical radius ratio for 6-coordination is 0.414. If the radius ratio is less than 0.414, a binary compound C^+A^- must adopt a structure with a coordination number less than 6. A similar calculation shows that if the radius ratio is greater than 0.732, the structure adopted will have a coordination number of 8. Table 6.1 gives examples:

TABLE 6.1 Radius Ratios and Coordination Numbers

Radius ratio r_C/r_A	Coordination number	Example
0.225–0.414	4	ZnS
0.414–0.732	6	Alkali metal halides except caesium halides below
0.732	8	CsCl, CsBr, CsI

The stoichiometric ratio of cations to anions is important...

The stoichiometric ratio of cations to anions also affects the type of structure formed. In calcium chloride, there is a face-centred-cubic close-packed lattice of chloride ions, in which calcium ions occupy alternate octahedral holes, thus maintaining the formula $CaCl_2$.

...cadmium chloride has a layered structure

Cadmium chloride also is based on a face-centred-cubic close-packed lattice of chloride ions. The cadmium ions occupy every other octahedral hole, but the structure formed is different from calcium chloride. The cations arrange themselves to leave alternate layers of anions in direct anion to anion contact. This compound has a **layered structure**. [See Figure 6.10.] Many transition metals form layered oxides and sulphides.

FIGURE 6.10
The CdCl$_2$ Structure

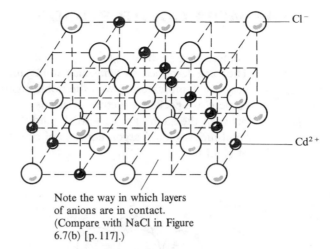

Note the way in which layers
of anions are in contact.
(Compare with NaCl in Figure
6.7(b) [p. 117].)

Unlike metals, ionic crystals are brittle

Ionic crystals are brittle. Figure 6.11 shows what happens when an ionic crystal is subjected to stress. A slight dislocation in the crystal structure brings similarly charged ions together. Repulsion between the like charges fractures the crystal. This is a different picture from the effect of stress on a metallic crystal, where deformation of the structure does not result in fracture [see Figure 6.4, p. 113].

FIGURE 6.11 An Ionic
Structure is Easily
Fractured

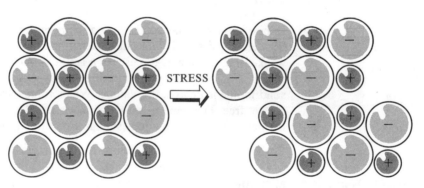

Structure is deformed.

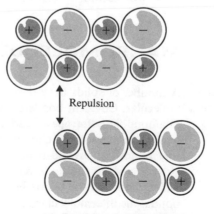

Repulsion shatters structure

6.4 MOLECULAR SOLIDS

Van der Waals forces hold the molecules together in solid argon and in solid iodine...

Some solids are held together by weak attractions between individual molecules. They are described as **molecular solids** and said to have a **molecular structure**. At very low temperatures, even the noble gases can be solidified. Figure 6.12 shows the cubic close-packed structure of atoms in solid argon. The van der Waals forces between the atoms are very weak, and if the temperature rises above − 170 °C the solid melts. Liquid argon consists of separate atoms. Figure 6.13 shows the structure of solid iodine.

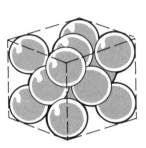

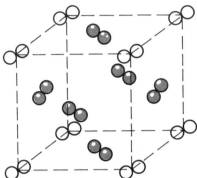

The I_2 molecules in the centre of each face are shaded.

Figure 6.12 Solid Argon (face-centred cube)

Figure 6.13 Solid Iodine (face-centred cube)

Iodine is a molecular solid up to a temperature of 30 °C. The atoms are covalently bonded in pairs as I_2 molecules. Operating between the molecules are the much weaker van der Waals forces. As a result of the regular arrangement of molecules, iodine is a crystalline solid with regular faces, which give a shiny appearance. When solid iodine is heated, the van der Waals forces are broken and individual molecules are set free. The vapour phase, which is purple, consists of individual I_2 molecules. Bromine and chlorine adopt similar structures at lower temperatures.

...solid CO_2 also is a molecular structure

Carbon dioxide is well known in its solid form as 'dry ice' or 'drikold'. Above − 78 °C it sublimes, absorbing heat from its surroundings to do so. From this property arises the widespread use of solid carbon dioxide as a refrigerant, both in laboratory work and in the food industry. Pop singers sometimes like to enhance their performance by having lumps of dry ice on stage. As it sublimes, it cools the moist air, and swirling clouds of water droplets form. Solid carbon dioxide has a face-centred-cubic structure, resembling that of iodine, which is shown in Figure 6.13.

6.5 MACROMOLECULAR STRUCTURES

A macromolecular structure is held together by covalent bonds, e.g. diamond...

A number of solids have the kind of structure described as **macromolecular** or **giant molecular**. Covalent bonds between atoms bind all the atoms into a giant molecule. Diamond, an **allotrope** of carbon [p. 412], is one of the hardest substances known and has a macromolecular structure [see Figure 6.14]. Each carbon atom forms four bonds (sp³ hybrid bonds) to four other carbon atoms. The giant molecular structure which results is very strong. It is different from a molecular structure, in which, although the bonds between the atoms in a molecule are strong covalent bonds, the intermolecular forces of attraction are weak. Diamond remains a solid up to a temperature of 3500 °C, at which it sublimes.

...in which the strong covalent bonds result in a hard, abrasive character

The hard, abrasive character of diamond finds it many uses. Diamond-tipped tools are used for cutting and engraving, and diamond-tipped tools are used by oil prospectors for boring through rock [see Figure 6.15]. The high **refractive index*** of diamond gives it the sparkle that makes it the most prized of jewels [see Figure 6.16].

FIGURE 6.14
The Structure of
Diamond

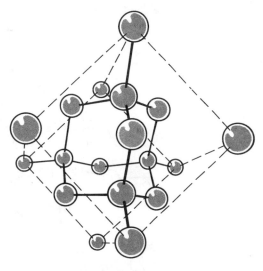

Each C atom is surrounded by 4 others:
the coordination number is 4.

FIGURE 6.15
Diamond-studded
Drill Bits

FIGURE 6.16
A Selection of
Diamonds

*See R Muncaster, *A-Level Physics* (Stanley Thornes)

Other macromolecular structures are SiC, BN and SiO₂

Other solids with a diamond-like structure are silicon carbide $(SiC)_n$, and boron nitride $(BN)_n$. The formula unit, BN, is isoelectronic with the unit CC. Silicon(IV) oxide, SiO_2 (**silica**), also forms a three-dimensional structure. The Si—O bonds about each silicon atom are tetrahedrally distributed and each oxygen atom is bonded to two other silicon atoms [see Figure 6.17]. This structure occurs in **quartz** and other crystalline forms of silica. Quartz remains solid up to a temperature of 1700 °C.

FIGURE 6.17
Silicon(IV) Oxide
Structure

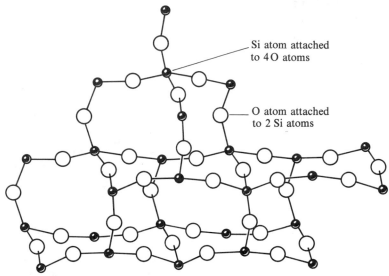

Si atom attached to 4O atoms

O atom attached to 2 Si atoms

6.6 LAYER STRUCTURES

Layer structures have covalent bonds within each layer and weak van der Waals forces between layers e.g. graphite

Graphite, the other allotrope of carbon, has a **layered** structure. Within each layer, every carbon atom uses three coplanar sp^2 hybrid orbitals to bond to two other carbon atoms. A network of coplanar hexagons is formed, with a C—C bond distance of 0.142 nm. Between layers, the distance is 0.335 nm. The weak van der Waals forces of attraction between the layers allow one layer of bonded atoms to slide over another layer. The structure, which is shown in Figure 6.18, accounts for the properties of graphite. It is a lubricant, whereas diamond is abrasive. The unhybridised p electrons form a delocalised cloud of electrons similar to the metallic bond. They enable graphite to conduct electricity and are responsible for its shiny appearance.

FIGURE 6.18
The Structure of
Graphite

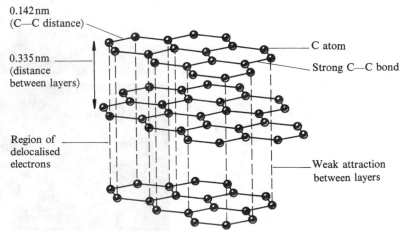

0.142 nm
(C—C distance)

0.335 nm
(distance between layers)

Region of delocalised electrons

C atom

Strong C—C bond

Weak attraction between layers

6.7 CHAIN STRUCTURES

Some substances exist in chain-like structures. One example is sulphur(VI) oxide, which crystallises in long shiny needles with the structure shown in Figure 6.19. Another is beryllium chloride which, in the solid state, has the structure shown in Figure 6.20.

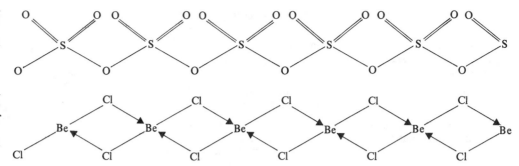

6.8 GLASSES

Glasses are supercooled liquids...

Glasses are **supercooled liquids** i.e., liquids that have been cooled below their freezing temperatures without crystallising. Although they are solids, over a period of years they flow slightly. X ray diffraction patterns of glasses resemble those of liquids. There are ordered arrangements of molecules over short distances, but there is no long-range order. Glasses are non-crystalline.

...they are non-crystalline...

...are formed from some molten oxides...

Glasses are formed when the molten oxides, SiO_2, B_2O_3, GeO_2 and P_4O_{10}, are cooled.

...and have many uses in industry and in the home...

Glasses are transparent to visible light. They have a multitude of uses in the construction industry, in the home and in the manufacture of optical instruments containing lenses, prisms and mirrors.

The manufacture of optical fibres is a new use for glass

Glass has an expanding use in the production of **optical fibres**. Glass can be melted and drawn out to form hair-thin fibres, which are as strong as steel. If a fibre is clad in a material of lower refractive index, light is guided along the fibre by total internal reflection at the surface of the fibre. With a source of light at one end and a light sensor at the other, optical fibres are used for communication. Lasers and light-emitting diodes are used as light sources. The big advantage of fibre optics is that, because the frequency of light is so high, a single fibre can carry a much larger number of channels of information than a coaxial cable.

The chemical inertness of glass has led to its use as a guardian of radioactive waste

The long term storage of liquid radioactive waste from nuclear power stations [p. 23] is a serious problem. Some of the waste needs to be stored for thousands of years. France has tackled the problem by **vitrification** (glass making). Concentrated radioactive waste is combined with glass-forming oxides. The resulting glass blocks are enclosed in steel canisters and stored behind thick concrete walls, which absorb radiation. The chemical inertness of glass should ensure that there is no natural mechanism by which the contents could contaminate the environment. Stores of this kind will have to be supervised for thousands of years. Some people feel that we have no right to impose this task on future generations. Others are concerned that the stores will present dangers from accidents, terrorists and war.

CHECKPOINT 6B: COVALENT STRUCTURES

1. From Figures 6.14, p. 121 and 6.18, p. 122, which would you expect to have the higher density, diamond or graphite? Why is diamond used in cutting tools but not graphite?

2. What is meant by the statement that the electrons in diamond are *localised*, whereas graphite has *delocalised* electrons? Which electrons in graphite are delocalised? How do they affect the properties of graphite?

3. Sketch the structure of silicon carbide, SiC, which resembles diamond. Why do you think that **carborundum** (SiC) is used as an abrasive? Why does silicon carbide not exist in a graphite-type of structure?

*4. Boron nitride, BN, has a structure like that of graphite. Explain the bonding in BN.

QUESTIONS ON CHAPTER 6

1. What type of intramolecular and intermolecular bonds exist in (a) solid argon, (b) solid bromine, (c) diamond, (d) graphite and (e) silica?

2. Crystals of salts fracture easily, but metals are deformed under stress without fracturing. Explain the difference.

3. What is the coordination number of an ion? What is the coordination number of the cation in (a) the NaCl structure and (b) the CsCl structure? What is the reason for the difference?

4. What type of structure would you expect for (a) LiF, and (b) RbBr? Ionic radii/nm

$$Li^+ = 0.074, Rb^+ = 0.149$$

$$F^- = 0.131, Br^- = 0.196$$

5. Say what properties you would expect of substances which are (a) metals, (b) ionic compounds, (c) composed of individual covalent molecules, and (d) macromolecular covalent compounds.

6. Explain the bonding present in the solids sodium chloride, sodium, phosphorus(V) chloride (PCl_5) graphite and ice. Point out how the type of bonding determines the physical properties of the solids.

7. Give two examples of (a) ionic solids, (b) molecular solids and (c) covalent macromolecular solids. What are the factors that determine whether each of these types of solid will dissolve in water?

8. Trace Figure 6.21(a) and (b). The circles represent atoms in a certain layer (layer 1) of a close-packed metallic structure.

(a)

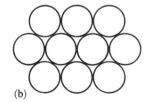
(b)

Figure 6.21

(a) On your tracing, mark with a cross the positions of the centres of the atoms in layer 2, in (a) and (b).

(b) The atoms in layer 3 can be arranged in two different ways. In (a) mark with a circle the positions of the centres of the atoms in layer 3 for a hexagonal close-packed structure.

(c) In (b) mark with a circle the positions of the centres of the atoms in layer 3 of a face-centred-cubic structure.

9. Explain how the *radius ratio* is related to the type of ionic structure formed by an ionic solid.

10. (a) Figure 6.22 illustrates one way in which atoms or ions can be arranged in a crystal. The entire crystal is simply a repeating pattern of this structure which is often known as the unit cell.

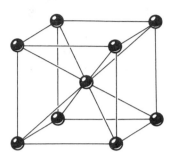

Figure 6.22

(i) What is meant by the term coordination number as applied to crystal structure?

(ii) What is the name given to this arrangement if the particles in the diagram above are atoms of a metal?

(iii) What is the coordination number of the central atom in the diagram?

(iv) How many cubes, of the entire structure, are shared by each atom at the corners of the cube in the diagram?

(v) Hence, what is the coordination number of the atoms at the corners of the cube?

(b) Suppose the central particle of the cube is a *cation* **M**. Suppose the corner particles of the cube are *anions* **X**.

(i) Is the central ion **M** shared by any other cube of the entire crystal other than the cube shown?

(ii) How many cubes, of the entire structure, are shared by each anion **X**?

(iii) Hence, deduce, with some explanation, the empirical formula of the substance containing the cation **M** and the anion **X**.

(iv) Name a metal ionic halide which has the structure shown in the diagram.

(c) A metal halide has a face-centred cubic arrangement of cations and anions, the cations and anions both having a coordination number of 6.

(i) Draw a diagram, like that above, to show this different type of arrangement.

(ii) Name a metal halide from the same group of the periodic table as the halide in (b) (iv) which nevertheless has this different arrangement (c) (i).

(iii) Give some explanation as to why the halides (b) (iv) and (c) (ii) have these two different structural arrangements.

†(iv) The theoretical lattice energies of ionic chlorides are sometimes significantly different from experimental values. Explain why this is so.
[For (c) (iv) see Chapter 10.]

(SUJB 81)

6.9 POSTSCRIPT: LIQUID CRYSTALS

You will read in Chapter 34 that pure solids melt sharply, the temperature remaining constant at the melting temperature until all the solid has melted. There are, however, many crystalline solids which pass at a sharp **transition temperature** to a turbid liquid phase before finally melting to form a clear liquid. Some compounds of this type are listed in Figure 6.23. The turbid liquid phases are liquids in that they can flow as liquids do and in possessing surface tension. Their molecules, however, possess some degree of order, with the result that these turbid liquids resemble crystals in certain optical properties. They are known as **liquid crystals**.

In many crystals, the velocity of light and therefore the refractive index is the same in every direction. Such crystals are described as **isotropic**. (Greek: *isos*, equal; *tropos*, a turn.) Other crystals (e.g., Iceland spar, $CaCO_3$) are **anisotropic**; the velocity of light is *not* the same in all directions. Such crystals are **birefringent**, i.e., an object viewed through the crystal is seen as a double refracted image. Besides this characteristic of birefringence*, anisotropic crystals give rise to interference patterns* in plane polarised light. Liquid crystals are anisotropic.

The liquid crystalline state exists between two temperatures, the melting temperature and the **clearing temperature**.

$$\text{Crystal} \underset{\text{melting temperature}}{\rightleftarrows} \text{liquid crystal} \underset{\text{clearing temperature}}{\rightleftarrows} \text{Isotropic liquid}$$

All liquid crystals are organic and have elongated molecules which end in a polar group,
e.g., $-CN$, $-OR$, $-NO_2$ and $-NH_2$. Many have flat portions, such as a benzene ring. The molecules are linear in conformation with doubly-bonded linking groups, e.g. $C=C$ and $-N=N$, to inhibit rotation and hold the molecule

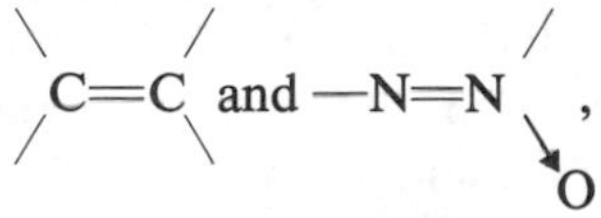

rigid about its long axis. The molecules possess strong dipoles and also easily polarisable groups. You can easily imagine how dipole–dipole interactions might give rise to intermolecular forces of attraction in compounds of this type. Such forces align the molecules side by side with their long axes parallel.

FIGURE 6.23 Some Liquid Crystalline Materials

(a)

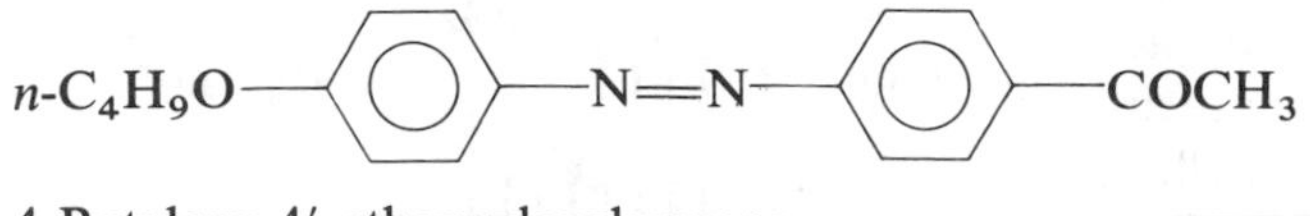

4-Butyloxy-4'-ethanoylazobenzene

(b)

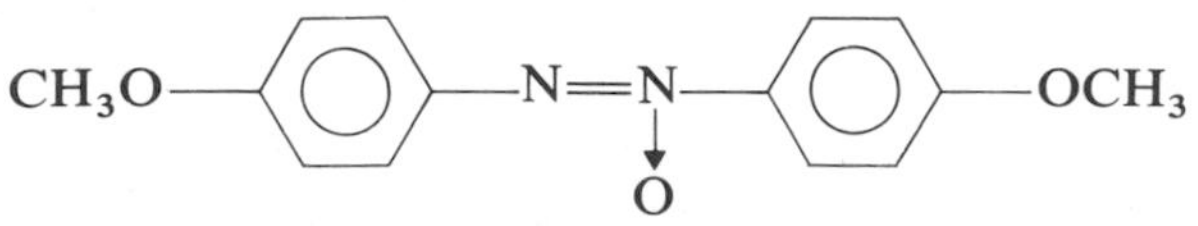

4,4'-Dimethoxyazoxybenzene

(c)

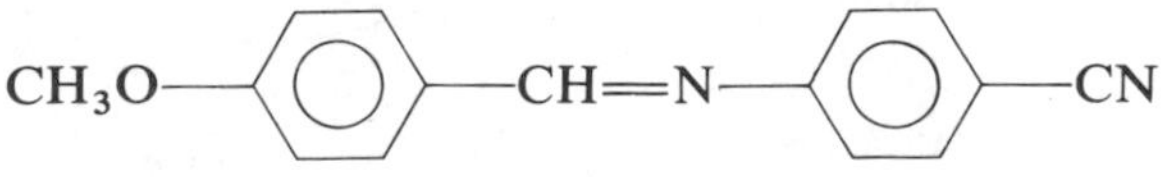

4-Methoxybenzylidene-4'-cyanophenylamine

*See R. Muncaster, *A-Level Physics* (Stanley Thornes)

(d) $n\text{-}C_8H_{17}O$—⟨benzene ring⟩—C⟨O—H······O, O······H—O⟩C—⟨benzene ring⟩—$OC_8H_{17}\text{-}n$

n-Octyloxybenzoic acid dimer

(e)

$CH(CH_3)(CH_2)_3CH(CH_3)_2$

CH_3

CH_3

$C_6H_5CO_2$

Cholesteryl benzoate (the first to be discovered)

(f) $CH_3(CH_2)_4$—⟨biphenyl⟩—CN

4-Cyano-4′-pentylbiphenyl (a liquid crystal at room temperature)

There are three types of liquid crystals: **smectic, nematic** and **cholesteric**. Figure 6.24 shows the relationship between the smectic and nematic phases. Smectic liquids (Greek: *smektikos*: soaplike) do not flow freely: they glide along in one plane. X ray diffraction patterns suggest a structure consisting of a series of planes, which are further apart than the molecules in a crystal. A smectic phase may melt to form an isotropic liquid or may pass at a transition temperature into a nematic phase. Nematic phases flow readily, and their X ray diffraction patterns resemble those of true liquids. Viewed in plane-polarised light under a microscope, liquid crystals show characteristic coloured patterns. Nematic phases show threadlike patterns, which give the phase its name. (Greek: *nema*: thread.)

FIGURE 6.24 Order in Liquid Crystals. (The elongated molecules are shown as ellipses. Except in the isotropic liquid they lie with their long axes parallel. T = transition temperature.)

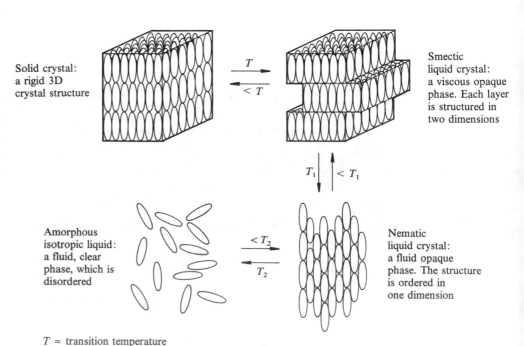

Solid crystal: a rigid 3D crystal structure

T →, ← $< T$

Smectic liquid crystal: a viscous opaque phase. Each layer is structured in two dimensions

T_1 ↓ | ↑ $< T_1$

Amorphous isotropic liquid: a fluid, clear phase, which is disordered

$< T_2$ →, ← T_2

Nematic liquid crystal: a fluid opaque phase. The structure is ordered in one dimension

T = transition temperature

A third type of liquid crystal phase (a modified nematic phase) is described as **cholesteric**. All the compounds which show cholesteric behaviour are **optically active** [p. 538]. Examples are the cholesteryl esters [see Table 6.2]. Cholesteric liquid crystals may reflect different colours when viewed in ordinary light. The phenomenon is associated with a structure of layers 500 to 5000 molecules thick. Each layer is a stack of sheets of molecules. The long axes of the molecules in the same sheet point in the same direction. Figure 6.25 shows how, passing from one sheet to the next, the long axes of the molecules are displaced through a small angle. The displacement progresses in a clockwise direction (or, in some compounds, an anticlockwise direction), giving rise to a helical structure. The distance between one sheet in which the molecular long axes point in a certain direction and the next sheet in which the axes point in the same direction is the **pitch** of the helix. This distance is also the thickness of the layers mentioned above.

FIGURE 6.25

The Structure of a Cholesteric Liquid Crystal

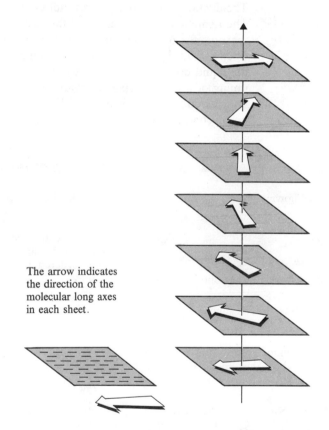

The arrow indicates the direction of the molecular long axes in each sheet.

The structure reflects light of different wavelengths to different extents. If the reflected light is in the visible region, the cholesteric liquid crystal appears coloured. The wavelength of light reflected is proportional to the pitch of the helix. The light reflected is complementary to the light transmitted, and gives rise to the beautiful iridescence shown by such liquid crystals. As the pitch of the helix alters with temperature, the colour of the reflected light alters; this explains why cholesteric liquid crystals can be used as temperature sensors. Some materials give a full range of colour change over as little as 0.1 °C, while others change from violet to red over 40 °C, and some form the true liquid state before the colour range is completed.

A mixture of cholesteric esters which gives a full colour response over about 3 °C in the range of body temperature is used for **skin thermography**. Skin overlying veins and arteries is slightly warmer than in other areas, and the difference in temperature can be detected by cholesteric liquid crystals. Specialists can use the technique of skin thermography to detect blockages in veins and arteries. The technique has been successful in the early diagnosis of breast cancer. When a layer of cholesteric material is painted or sprayed on to the surface of the breast, a tumour shows up as a 'hot area', which is coloured blue.

The electronics industry uses cholesteric liquid crystals to find points of potential failure in circuits by detecting them as 'hot spots'. In the manufacture of aeroplanes, lightness is achieved by the use of laminates with a honeycomb interior. The quality of the bonding can be investigated by applying a film of cholesteric liquid, which will show up any defect in the bonding as a difference in thermal conductivity. Faults in welding and cracks due to metal fatigue can be detected in the same way.

Room thermometers contain cholesteric liquid crystals with a suitable temperature range. Figures show up in different colours as the temperature changes.

The digital displays you see in watches and calculators contain nematic liquid crystals. The nematic phase is fluid, and the liquid crystal molecules are polar. For these two reasons, the orientation of the molecules can be changed by the use of a very small electric field. If the change in orientation results in a change in optical properties, the liquid crystal can be used to display information, e.g., the time, or date. The timing of a watch display is controlled by a quartz crystal. A small electric current induces quartz to resonate* at 32 768 oscillations per second. A quartz watch has no mechanical moving parts, giving it a big advantage over traditional watches.

CHECKPOINT 6C: LIQUID CRYSTALS

1. Explain why the following groups are flat:

2. Explain why the following groups have a dipole:

$$-OCH_3, -NO_2, -NH_2 \text{ and } -N=N^{\nearrow}_{\searrow O}.$$

3. Explain why these linking groups inhibit rotation and hold the molecules in which they are present in linear conformations:

$$-N=N-, -N=N^{\nearrow}_{\searrow O}, -CH=CH-,$$

$$-C\equiv C- \text{ and } -CH=N-.$$

4. Sketch a pair of molecules of the liquid crystal

$$(CH_3)_2NC_6H_4N=NC_6H_4NO$$

(a) end to end and (b) side by side, in order to show how dipole–dipole interactions could hold the molecules in an ordered, parallel arrangement.

†5. Suggest methods of preparation for compounds (a), (b) and (d) on pp. 125–6.
[For help, see Part 4.]

*See R Muncaster, *A-Level Physics* (Stanley Thornes)

Part 2

PHYSICAL CHEMISTRY

7

GASES

7.1 STATES OF MATTER

The three main states of matter are the solid, liquid and gaseous states

There are three important **states of matter**: the gaseous, liquid and solid states. In addition, there are the **liquid crystalline state** [p. 125] and the **plasma state** [p. 150]. In solids and liquids, the molecules are close together, and considerable forces of attraction exist between them. When a solid melts, it expands slightly (with some exceptions) whereas, when a liquid evaporates, it forms many times its own volume of vapour. The gaseous state of a substance at temperatures below its critical temperature [p. 148] is called a **vapour**. Gases have much lower densities and are much more compressible than solids and liquids. Many scientists became interested in the behaviour of gases, and the results of the experimental studies which were made in the seventeenth, eighteenth and nineteenth centuries are embodied in the **gas laws**. A **law** is a concise statement of experimental results; it need not include an explanation of them. A **hypothesis** is an explanation of experimental findings by means of a concept or model. A well-established hypothesis becomes known as a **theory**.

The gas laws state the results of experimental studies on gases

7.2 THE GAS LAWS

7.2.1 BOYLE'S LAW

Boyle's Law deals with the effect of pressure on the volume of a gas

In 1662, Robert Boyle published his work on the **compressibility** of gases. He found that, at a constant temperature, the volume of a fixed mass of a given gas is inversely proportional to its pressure. Boyle's Law can be expressed as

$$PV = \text{Constant}$$

where P = pressure, and V = volume.

Figure 7.1 shows three ways of representing Boyle's Law graphically.

FIGURE 7.1 Graphs illustrating Boyle's Law (Plots of (*a*) P against V, (*b*) P against $1/V$ and (*c*) PV against P)

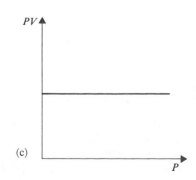

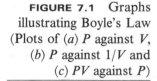

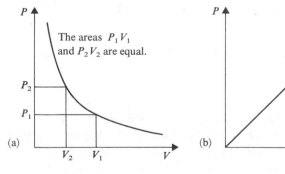

The areas P_1V_1 and P_2V_2 are equal.

PV = Constant

Further work has shown that gases do not obey Boyle's Law accurately under all conditions. All gases come closer to obeying the law at low density.

7.2.2 CHARLES' LAW

Charles' Law deals with the effect of temperature on the volume of a gas

Two French scientists, J L Gay-Lussac and J A C Charles, independently measured the expansion of gases that occurs when the temperature is raised. They found that the variation of volume with temperature was linear at constant pressure [see Figure 7.2].

FIGURE 7.2 Graph illustrating Charles' Law (Plot of volume against temperature (°C) for a gas at constant pressure)

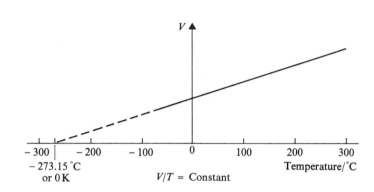

On extrapolation, the line cuts the temperature axis at − 273.15 °C. When similar plots are made for different gases, the extrapolated linear plots always cut the temperature axis at this point. It looks as though all gas volumes would become zero at − 273.15 °C, but in fact gases liquefy or solidify long before this temperature is reached. The temperature − 273.15 °C was adopted as the zero on a new temperature scale called the **absolute temperature scale** or the **Kelvin scale**. Temperatures on this scale are measured in kelvins and are obtained by adding 273.15 to temperatures on the Celsius scale:

Absolute zero and the Kelvin scale of temperature . . .

. . . 273 K = 0 °C

$$\text{Temperature/K} = \text{Temperature/°C} + 273.15$$

$$273.15 \, \text{K} = 0 \, °\text{C}$$

If the graph in Figure 7.2 is referred to zero on the Kelvin scale, the equation for the graph is

V/T = constant . . .

$$V = kT$$

. . . The relationship leads to a statement of Charles' Law

where T is the temperature in kelvins and k is a constant. Thus, Charles' Law (or Charles' and Gay-Lussac's Law) can be stated: <u>the volume of a fixed mass of a given gas at constant pressure is directly proportional to its temperature in kelvins</u>. In fact, gases do not obey Charles' Law closely at all temperatures. A gas which obeys both Boyle's Law and Charles' Law is said to behave **ideally**.

7.2.3   THE EQUATION OF STATE FOR AN IDEAL GAS

By combining Boyle's Law (PV = constant) and Charles' Law (V/T = constant), one finds that PV/T = constant. This relationship is often written as

Combining Boyle's Law and Charles' Law gives the equation of state for an ideal gas...

$$\frac{P_1 V_1}{T_1} = \frac{P_2 V_2}{T_2}$$

...This can be used to calculate the volume of gas at stp

This equation is called **the equation of state for an ideal gas**. Real gases do not show ideal behaviour at all temperatures and pressures. The equation of state enables one to calculate the effect of a change in temperature and pressure on the volume of a gas. If a given mass of gas has a volume V_1 at a temperature T_1 and pressure P_1, then the volume, V_2, which the gas would occupy at a temperature T_2 and pressure P_2, can be found. Gas volumes are usually compared at 0 °C and 1 atmosphere, conditions which are referred to as **standard temperature and pressure** (stp). Sometimes the comparison is made at rtp, a room temperature of 20 °C and a pressure of 1 atm. The SI unit of pressure is the newton per square metre ($N\,m^{-2}$), called the pascal (**Pa**):

Pressure units

$$1 \text{ atmosphere} = 1.0132 \times 10^5\,N\,m^{-2} = 1.0132 \times 10^5\,Pa$$

$$= 760\,mm \text{ mercury}$$

The SI unit of volume is the cubic metre, m^3, but cubic decimetres, dm^3, cubic centimetres, cm^3, and litres, l, are also used.

Volume units

$$1\,m^3 = 10^3\,dm^3 = 10^6\,cm^3$$

$$1\,dm^3 = 1\,litre$$

Temperatures must be in kelvins in the ideal gas equation of state.

Example   If the volume of a gas collected at 60 °C and $1.05 \times 10^5\,N\,m^{-2}$ is $60\,cm^3$, what would be the volume of gas at stp?

Method   The experimental conditions are

A sample calculation of the volume of a gas at stp...

$$P_1 = 1.05 \times 10^5\,N\,m^{-2}$$

$$T_1 = 273 + 60 = 333\,K$$

$$V_1 = 60\,cm^3$$

Standard conditions are

$$P_2 = 1.01 \times 10^5\,N\,m^{-2}$$

$$T_2 = 273\,K$$

Since

$$\frac{P_1 V_1}{T_1} = \frac{P_2 V_2}{T_2}$$

The volume of gas at stp

$$V_2 = \frac{1.05 \times 10^5 \times 60 \times 273}{1.01 \times 10^5 \times 333}\,cm^3$$

$$= 51\,cm^3$$

Alternative Method   When the pressure *decreases* from 1.05×10^5 to 1.01×10^5 the volume *increases* in the same ratio:

$$V_2 = 60 \times \frac{1.05 \times 10^5}{1.01 \times 10^5} \, \text{cm}^3$$

When the temperature *decreases* from $60\,^{\circ}\text{C}$ to $0\,^{\circ}\text{C}$, the volume *decreases* by the ratio of the temperatures in kelvins $273/(273 + 60)$.

$$\therefore \qquad V_2 = 60 \times \frac{1.05 \times 10^5}{1.01 \times 10^5} \times \frac{273}{333} = 51\,\text{cm}^3 \text{ (as before)}$$

CHECKPOINT 7A: CORRECTING GAS VOLUMES

1. Correct the following gas volumes to stp:

(*a*) $400\,\text{cm}^3$ of an ideal gas measured at $200\,^{\circ}\text{C}$ and $9.80 \times 10^4\,\text{N}\,\text{m}^{-2}$

(*b*) $64.0\,\text{cm}^3$ of an ideal gas measured at $35\,^{\circ}\text{C}$ and $1.25 \times 10^5\,\text{N}\,\text{m}^{-2}$

(*c*) $20.0\,\text{dm}^3$ of an ideal gas measured at $200\,\text{K}$ and $201\,\text{kN}\,\text{m}^{-2}$

(*d*) $3.15\,\text{dm}^3$ of an ideal gas measured at $250\,^{\circ}\text{C}$ and $1.95\,\text{atm}$

2. (*a*) A volume $24.0\,\text{dm}^3$ of an ideal gas is collected at $1.00\,\text{atm}$ and $25\,^{\circ}\text{C}$. It is subjected to a pressure of $2.05\,\text{atm}$ at $75\,^{\circ}\text{C}$. What is its new volume?

(*b*) After $625\,\text{cm}^3$ of an ideal gas were collected at $90\,^{\circ}\text{C}$ and $9.67 \times 10^4\,\text{N}\,\text{m}^{-2}$, conditions were changed to $25\,^{\circ}\text{C}$ and $1.13 \times 10^5\,\text{N}\,\text{m}^{-2}$. What did the volume become?

7.2.4   GRAHAM'S LAW OF GASEOUS DIFFUSION AND EFFUSION

Gases diffuse . . .   Gases mix when brought into contact with one another. All gases spontaneously **diffuse** into one another to form a homogeneous mixture. The relative rates at

. . . the rate of diffusion depend on the density of the gas . . .   which two gases diffuse into a third gas (e.g., air) depend on their densities. In 1829, T Graham stated that the relative rates of diffusion of gases, under the same conditions, are inversely proportional to the square roots of their densities. Comparing the rates of diffusion of gases **A** and **B**

. . . the rate is proportional to $\sqrt{1/\rho}$ . . .

$$\frac{r_A}{r_B} = \sqrt{\frac{\rho_B}{\rho_A}}$$

(where r = rate of diffusion and ρ = density). Graham's Law is now written as

. . . The rate is therefore proportional to $\sqrt{1/M}$ . . .

$$\frac{r_A}{r_B} = \sqrt{\frac{M_B}{M_A}}$$

(where M = molar mass). (It will be shown on p. 139 that $\rho = M/V_m$ where V_m = molar volume.)

FIGURE 7.3
Gaseous Effusion

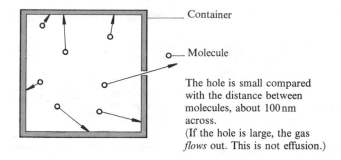

Container

Molecule

The hole is small compared
with the distance between
molecules, about 100 nm
across.
(If the hole is large, the gas
flows out. This is not effusion.)

*... The same relationships
are true of effusion*

Graham's Law applies also to gaseous **effusion**. Effusion is the passage of a gas through a very small hole into a vacuum [see Figure 7.3].

From a practical point of view, it is easier to compare rates of effusion than rates of diffusion. An apparatus is shown in Figure 7.4.

FIGURE 7.4
An Apparatus for
Comparing Rates of
Effusion of Gases

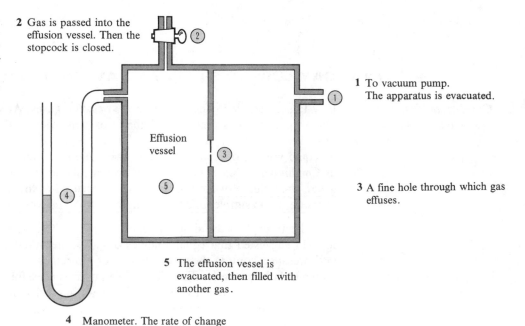

2 Gas is passed into the
effusion vessel. Then the
stopcock is closed.

1 To vacuum pump.
The apparatus is evacuated.

Effusion
vessel

3 A fine hole through which gas
effuses.

5 The effusion vessel is
evacuated, then filled with
another gas.

4 Manometer. The rate of change
of pressure is observed. It is
equal to the rate of effusion.

*Applications of gaseous
effusion*

The application of gaseous effusion to the separation of uranium isotopes is illustrated in Figure 1.24, p. 23. Another application is molar mass determination. The rate of effusion of the gas of unknown molar mass must be compared with that of a gas of known molar mass.

Example $50.0 \, cm^3$ of gas **A** effuse through a tiny aperture in 146 s. The same volume of carbon dioxide effuses under the same conditions in 115 s. Calculate the molar mass of **A**.

Method

*A worked example on
molar mass determination
from rates of effusion*

$$\frac{\text{Rate (CO}_2)}{\text{Rate (A)}} = \frac{50.0/115}{50.0/146} = \sqrt{\frac{M_A}{M_{CO_2}}}$$

$$(1.27)^2 = M_A/44.0$$

The molar mass of **A**, $M_A = 71 \, \text{g mol}^{-1}$.

CHECKPOINT 7B: DIFFUSION AND EFFUSION

1. In 5.00 minutes, 15.00 cm³ of argon effuse through a pinhole. What volume of xenon will effuse through the same pinhole under the same conditions?

2. If it takes 2 minutes 50 seconds for 25.0 cm³ of hydrogen to effuse through a pinhole, what time will be required for the same volume of bromine vapour to effuse through the same pinhole under the same conditions?

†**3.** A gaseous alkane diffuses through a porous partition at a rate of 2.56 cm³ s⁻¹. Helium diffuses through the same partition under the same conditions at a rate of 8.49 cm³ s⁻¹. What is the molar mass of the alkane? What is its molecular formula? [For alkanes, see Chapter 26.]

7.2.5   GAY-LUSSAC'S LAW AND AVOGADRO'S HYPOTHESIS

Gay-Lussac observed that gas volumes combine in a simple ratio

Gay-Lussac studied chemical reactions between gases. Among the observations he made were the facts that one volume of oxygen reacts with exactly twice its volume of hydrogen to form two volumes of steam, and one volume of hydrogen reacts with an equal volume of chlorine to form two volumes of hydrogen chloride. In his **Law of Combining Volumes** in 1809, Gay-Lussac stated that, in a reaction between gases, the volume of the reacting gases, measured at the same temperature and pressure, are in simple ratio to one another and to the volumes of any gaseous products.

Avogadro's explanation of Gay-Lussac's results

In order to explain the simple relationship which Gay-Lussac had found, Avogadro in 1811 suggested that equal volumes of gases, measured at the same temperature and pressure, contain the same number of molecules. This suggestion is called **Avogadro's Hypothesis**. Since it has been used successfully for nearly two centuries, it is often called **Avogadro's Law**. (This is an imprecise use of the term *law*, which should be applied to a statement of experimental observations and not to a theoretical explanation.) On the basis of Avogadro's Hypothesis, one can interpret the observation

Avogadro's Hypothesis can be used to find the formula of a gaseous compound

1 volume hydrogen + 1 volume chlorine → 2 volumes hydrogen chloride

as 1 molecule hydrogen + 1 molecule chlorine → 2 molecules hydrogen chloride

As it must be possible to make 1 molecule of hydrogen chloride

$\frac{1}{2}$ molecule hydrogen + $\frac{1}{2}$ molecule chlorine → 1 molecule hydrogen chloride

The molecules of hydrogen and chlorine must both contain an even number of atoms. The simplest hypothesis is that each molecule contains two atoms. In this case

$$H_2 + Cl_2 \rightarrow 2HCl$$

It has never been found necessary to suggest a number greater than two.

Gas molar volume

It follows from Avogadro's Hypothesis that if equal volumes of gases contain equal numbers of molecules then the volume occupied by one mole of molecules must be the same for all gases. It is called the **gas molar volume** and measures 22.414 dm³ at stp (0 °C and 1 atm) or 24.056 dm³ at rtp (20 °C and 1 atm). This information is used in calculations on the volumes of gases formed in chemical reactions. [See p. 51.]

Eudiometry Avogadro's Hypothesis and the gas molar volume are the basis of the **eudiometric method** of finding the formulae of gases, in which the gas is exploded with an excess of oxygen and the products of combustion are assayed.

Example (a) $10\,cm^3$ of a gaseous hydrocarbon, C_3H_x, were allowed to react with an excess of oxygen at $110\,°C$ and $1\,atm$. There was an expansion of $5\,cm^3$. Deduce the value of x.

Method The equation for the reaction is

Some worked examples of finding the formulae of gases

$$C_3H_x(g) + (3 + x/4)O_2(g) \rightarrow 3CO_2(g) + x/2\,H_2O(g)$$

If the volume of hydrocarbon is $10\,cm^3$, then

   Volume of oxygen $= 10(3 + x/4) = (30 + 5x/2)\,cm^3$

   Volume of carbon dioxide $= 30\,cm^3$

   Volume of steam $= x/2 \times 10\,cm^3 = 5x\,cm^3$

Let $a =$ the unused oxygen. Since there is an expansion of $5\,cm^3$

   Final volume $= 5 +$ initial volume

   $30 + 5x + a = 5 + 10 + 30 + 5x/2 + a$

   $5x/2 = 15$

   $x = 6$

The formula is C_3H_6.

Example (b) When $10\,cm^3$ of a gaseous hydrocarbon, C_aH_b, were exploded with an excess of oxygen, a contraction of $25\,cm^3$ occurred, all volumes being measured at room temperature and pressure. On treatment of the products with sodium hydroxide solution, a contraction of $20\,cm^3$ occurred. Deduce the formula of the hydrocarbon.

Method The equation is

   $$C_aH_b(g) + (a + b/4)O_2(g) \rightarrow aCO_2(g) + b/2\,H_2O(l)$$

   Volume of hydrocarbon $= 10\,cm^3$

   Volume of $CO_2 = a +$ volume of C_aH_b

   From reaction with NaOH, volume of $CO_2 = 20\,cm^2$

$\therefore$ $a = 2$

Let the volume of unused oxygen be c.

   Final volume $=$ Initial volume $- 25\,cm^3$

Note $H_2O(l)$ is a liquid at room temperature and pressure, and does not contribute to the final volume of gas.

   $20 + c = 10 + 10(a + b/4) + c - 25$

   $20 + c = 10 + 20 + 5b/2 + c - 25$

$\therefore$ $b = 6$

The formula is C_2H_6.

Example (c) Figure 7.5 illustrates the measurements that must be made in the determination of the formula of ammonia.

FIGURE 7.5 How to
Determine the Formula
of Ammonia

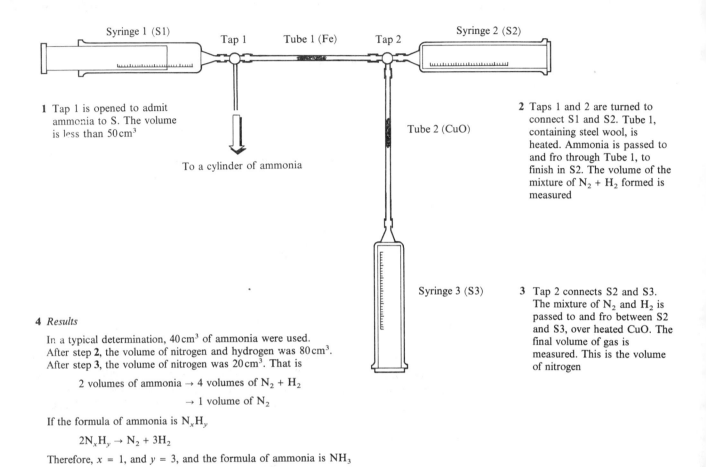

1 Tap 1 is opened to admit ammonia to S. The volume is less than $50\,cm^3$

To a cylinder of ammonia

Tube 2 (CuO)

Syringe 3 (S3)

2 Taps 1 and 2 are turned to connect S1 and S2. Tube 1, containing steel wool, is heated. Ammonia is passed to and fro through Tube 1, to finish in S2. The volume of the mixture of $N_2 + H_2$ formed is measured

3 Tap 2 connects S2 and S3. The mixture of N_2 and H_2 is passed to and fro between S2 and S3, over heated CuO. The final volume of gas is measured. This is the volume of nitrogen

4 *Results*

In a typical determination, $40\,cm^3$ of ammonia were used. After step **2**, the volume of nitrogen and hydrogen was $80\,cm^3$. After step **3**, the volume of nitrogen was $20\,cm^3$. That is

$$2 \text{ volumes of ammonia} \rightarrow 4 \text{ volumes of } N_2 + H_2$$

$$\rightarrow 1 \text{ volume of } N_2$$

If the formula of ammonia is N_xH_y

$$2N_xH_y \rightarrow N_2 + 3H_2$$

Therefore, $x = 1$, and $y = 3$, and the formula of ammonia is NH_3

CHECKPOINT 7C: REACTING VOLUMES OF GASES

1. When $10\,cm^3$ of a gaseous hydrocarbon were sparked with $50\,cm^3$ of oxygen (an excess) and the residual gases cooled to room temperature, a contraction of $20\,cm^3$ occurred. A further contraction of $20\,cm^3$ took place when the residual gases were subjected to aqueous sodium hydroxide. All the volumes were measured at $20\,°C$. Deduce the formula of the hydrocarbon.

2. $20\,cm^3$ of ammonia are burned in an excess of oxygen at $110\,°C$. $10\,cm^3$ of nitrogen and $30\,cm^3$ of steam are formed. Deduce the formula of ammonia, given that the formula of nitrogen is N_2, and the formula of steam is H_2O.

3. Cyanogen is a compound of carbon and nitrogen. On combustion in excess oxygen, $250\,cm^3$ of cyanogen form $500\,cm^3$ of carbon dioxide and $250\,cm^3$ of nitrogen (measured at the same temperature and pressure). What is the formula of cyanogen?

4. $25\,cm^3$ of a mixture of methane and ethane were completely oxidised by $72.5\,cm^3$ of oxygen, measured at the same temperature and pressure. What was the composition of the mixture?

7.2.6 THE IDEAL GAS EQUATION

For ideal gases which obey Boyle's Law and Charles' Law the dependence of volume upon external conditions is given by the equation

The ideal gas equation is
$$PV = nRT$$

$$\frac{P \times V}{T} = \text{Constant (for a given mass of gas)}$$

R is the Universal Gas Constant... It follows from Avogadro's Hypothesis that, if one mole of gas is considered, the constant will be the same for all gases. It is called the Universal Gas Constant, and given the symbol R, so that the equation becomes, when $V = V_m$, the gas molar volume

$$PV_m = RT$$

This equation is called the **ideal gas equation**. For n moles of gas, the equation becomes

$$PV = nRT$$

The value of the constant R can be calculated. Consider one mole of an ideal gas at stp. Its volume V_m is $22.414\,dm^3$. Inserting values of P, V_m and T in SI units into the ideal gas equation gives

...Finding the value of R...

$$P = 1.0132 \times 10^5\,N\,m^{-2}$$

$$T = 273.16\,K$$

$$V_m = 22.414 \times 10^{-3}\,m^3$$

$$n = 1\,mol$$

$$1.0132 \times 10^5 \times 22.414 \times 10^{-3} = 1 \times 273.16 \times R$$

$$R = 8.314$$

...and the unit of R The unit of R is PV/nT i.e., $\dfrac{N\,m^{-2}\,m^3}{mol\,K} = N\,m\,mol^{-1}\,K^{-1} = J\,K^{-1}\,mol^{-1}$.

Thus $R = 8.314\,J\,K^{-1}\,mol^{-1}$ (joules per kelvin per mole).

7.2.7 EXPERIMENTAL DETERMINATION OF THE MOLAR MASS OF A GAS

The density of a gas can be found and used to give its molar mass The gas density is found by weighing a known volume of gas. Since

$$PV = nRT = \frac{m}{M}RT$$

(where m = mass and M = molar mass)
and density

$$\rho = m/V$$

$$M = \rho RT/P$$

An experimental method is illustrated in Figure 7.6.

FIGURE 7.6
Determination of Gas
Density by Direct
Weighing

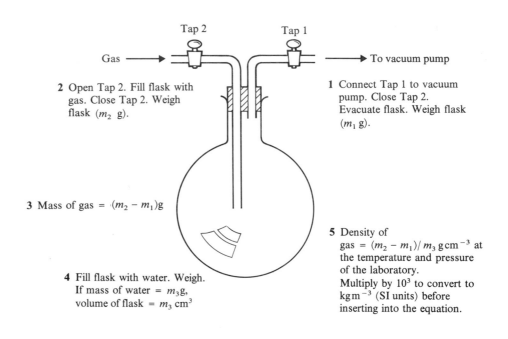

2 Open Tap 2. Fill flask with gas. Close Tap 2. Weigh flask (m_2 g).

1 Connect Tap 1 to vacuum pump. Close Tap 2. Evacuate flask. Weigh flask (m_1 g).

3 Mass of gas = $(m_2 - m_1)$g

5 Density of gas = $(m_2 - m_1)/m_3$ g cm^{-3} at the temperature and pressure of the laboratory. Multiply by 10^3 to convert to kg m^{-3} (SI units) before inserting into the equation.

4 Fill flask with water. Weigh. If mass of water = m_3g, volume of flask = m_3 cm^3

CHECKPOINT 7D: GAS MOLAR VOLUME

$R = 8.31 \, \text{J K}^{-1} \text{mol}^{-1}$,
Gas molar volume = 22.4 dm^3 at stp

1. Calculate the molar mass of a gas which has a density of 2.615 g dm^{-3} at 298 K and 101 kN m^{-2}.

2. At 273 K and 1.01×10^5 N m^{-2}, 6.319 g of a gas occupy 2.00 dm^3. Calculate the molar mass of the gas.

3. An ideal gas occupies 225 cm^3 at 280 K and 4.80×10^5 N m^{-2}. What amount of gas (in moles) is present?

4. Calculate the volume occupied by 1.15 mol of an ideal gas at 1.01×10^5 N m^{-2} and 20 °C.

7.2.8 DALTON'S LAW OF PARTIAL PRESSURES

Definition of partial pressure

Dalton's Law

Dalton considered the pressures exerted by the constituents of a mixture of gases. Since air is approximately $\frac{4}{5}$ nitrogen and $\frac{1}{5}$ oxygen by volume, $\frac{4}{5}$ of the air pressure is due to nitrogen and $\frac{1}{5}$ is due to oxygen. The contribution that each gas makes to the total pressure is called the **partial pressure**. The partial pressure of a gas is the pressure that each gas would exert if it alone occupied the container. Dalton's Law of partial pressures states that, <u>in a mixture of gases which do not react chemically,</u> <u>the total pressure of the mixture is the sum of the partial pressures of the constituent</u> <u>gases.</u> It is an ideal gas law.

Definition of mole fraction

The partial pressure of each gas is equal to the product of the total pressure of the mixture and the **mole fraction** of the gas. Mole fractions are one method of expressing the composition of a mixture. The mole fraction of **A** in a mixture of **A** and **B** is

$$\text{Mole fraction of } \mathbf{A} = \frac{\text{Moles of } \mathbf{A}}{\text{Moles of } \mathbf{A} + \text{Moles of } \mathbf{B}} = \frac{n_A}{n_A + n_B}$$

From Avogadro's Hypothesis, it follows that

$$\frac{n_A}{n_A + n_B} = \frac{\text{Volume of } \mathbf{A}}{\text{Total volume}}$$

Calculation of partial pressure

Example $4.00\,dm^3$ of oxygen at a pressure of $400\,kPa$ and $1.00\,dm^3$ of nitrogen at a pressure of $200\,kPa$ are introduced into a $2.00\,dm^3$ vessel. What is the total pressure in the vessel?

Method When oxygen contracts from $4.00\,dm^3$ to $2.00\,dm^3$, the pressure increases from $400\,kPa$ to

$$400 \times \frac{4.00}{2.00} \qquad \text{i.e. } 800\,kPa.$$

The partial pressure of oxygen in the vessel is $800\,kPa$.
When nitrogen expands from $1.00\,dm^3$ to $2.00\,dm^3$, the pressure decreases from 200 to

$$200 \times \frac{1.00}{2.00} = 100\,kPa.$$

The partial pressure of nitrogen is $100\,kPa$.

$$\text{The total pressure} = p(O_2) + p(N_2)$$
$$= 800 + 100 = 900\,kPa$$

The total pressure in the vessel is $900\,kPa$.

CHECKPOINT 7E: PARTIAL PRESSURES

1. $50.0\,cm^3$ of carbon dioxide at $10^5\,Nm^{-2}$ are mixed with $150\,cm^3$ of hydrogen at the same pressure. If the pressure of the mixture is $1.00 \times 10^5\,Nm^{-2}$, what is the partial pressure of carbon dioxide?

2. A mixture of gases at a pressure $1.01 \times 10^5\,Nm^{-2}$ has the volume composition of 30% CO, 50% O_2, 20% CO_2.
(a) What is the partial pressure of each gas?
(b) If the carbon dioxide is removed by the addition of some pellets of sodium hydroxide, what will be the partial pressures of O_2 and CO?

3. Into a $10.0\,dm^3$ vessel are introduced $4.00\,dm^3$ of methane at a pressure of $2.02 \times 10^5\,Nm^{-2}$, $12.5\,dm^3$ of ethane at a pressure of $3.50 \times 10^5\,Nm^{-2}$ and $1.50\,dm^3$ of propane at a pressure of $1.01 \times 10^5\,Nm^{-2}$. What is the pressure of the resulting mixture?

4. A mixture of 20% NH_3, 55% H_2 and 25% N_2 by volume has a pressure of $9.80 \times 10^4\,Nm^{-2}$.
(a) What is the partial pressure of each gas?
(b) What changes will take place in the partial pressures of hydrogen and nitrogen if the ammonia is removed by the addition of solid phosphorus(V) oxide?

7.2.9 HENRY'S LAW

The solubility of a gas is proportional to its partial pressure

Henry's Law is concerned with the solubility of gases in liquids. It states that the mass of gas dissolved at constant temperature per unit volume of solvent is directly proportional to the partial pressure of the gas. This law may be expressed as

$$m_s = kp$$

where m_s is the mass of gas dissolved p is its partial pressure, and k is a constant.

Definition of solubility...

The **solubility** of a gas is the volume which will dissolve in unit volume of the solvent under stated conditions of temperature and pressure. For example, the solubility of carbon dioxide at 30 °C and 1 atm is 30 cm^3 in 1.00 dm^3 of water.

...and of absorption coefficient

The **absorption coefficient** is the volume of gas (quoted at stp) that will dissolve at a partial pressure of 1 atm and a stated temperature in unit volume of the liquid. For example, the absorption coefficient of nitrogen in water at 20 °C is 0.016.

Henry's Law does not apply to gases which combine chemically with the solvent, for example, hydrogen chloride in solution in water. It applies only to dissolution as a physical process.

A sample calculation of the volume composition of dissolved air

Example   Taking the composition by volume of air as 80% nitrogen; 20% oxygen, calculate the volume composition of dissolved air in water at 20 °C. The absorption coefficients at 20 °C are nitrogen, 0.016; oxygen, 0.031.

Method   Air is composed of 80% by volume of nitrogen and 20% by volume of oxygen. The mole fractions are therefore: nitrogen = 0.80; oxygen = 0.20. The partial pressure of each gas is proportional to its mole fraction. The absorption coefficient must be multiplied by the mole fraction to give the absorption of the gas at its partial pressure.

$$\text{Volume of nitrogen at stp dissolving in } 100 \text{ cm}^3 \text{ water} = 100 \times 0.016 \times 0.80$$
$$= 1.28 \text{ cm}^3$$

$$\text{Volume of oxygen at stp dissolving in } 100 \text{ cm}^3 \text{ water} = 100 \times 0.031 \times 0.20$$
$$= 0.62 \text{ cm}^3$$

The dissolved air has the volume composition
1.28 cm^3 nitrogen : 0.62 cm^3 oxygen = 67% nitrogen : 33% oxygen

---

CHECKPOINT 7F: SOLUBILITY OF GASES

1. At room temperature, the ratio of the solubility of oxygen in water to nitrogen in water is 1.90. The percentage composition by volume of air is taken as 20% oxygen : 80% nitrogen. What is the percentage composition by volume of the mixture of oxygen and nitrogen that is obtained by degassing water saturated with air at room temperature?

2. The absorption coefficients at 280 K for hydrogen, oxygen and nitrogen are 0.021, 0.049 and 0.024 respectively. A 500 cm^3 vessel contains 250 cm^3 of hydrogen, 150 cm^3 of oxygen and 100 cm^3 of nitrogen at 1 atm. What will be the composition by volume of the dissolved gas if the mixture is allowed to come into contact with a water surface at 280 K?

7.3   THE KINETIC THEORY OF GASES

7.3.1   THE BASIC ASSUMPTIONS

The gas laws need an explanation...

All these laws describe the behaviour of gases, and they call for an explanation. Many scientists made a contribution to the development of a theoretical model to represent the behaviour of gases. Even as far back as the seventeenth century, many realised that in a gas the molecules themselves occupy only a tiny fraction of the volume in which the gas is contained. This is evident from the large volume of vapour formed from one volume of liquid, and it explains the low densities and high compressibilities of gases. A number of scientists envisaged the molecules of a gas as

... The kinetic theory of gases offers one

being in motion, but it was R J Clausius (1857) and J C Maxwell (1859) who presented the **kinetic theory of gases** in its present form. (Greek: *Kinein*, to move.) The theory rests on a number of *basic assumptions*.

Molecules of gas are in constant motion...

According to the kinetic theory of gases, the molecules of a gas are in constant motion, moving in straight lines unless they collide with the walls of the container or with another molecule. As the molecules collide with the walls of the container, they exert a pressure on the container. If the volume of the container is decreased, the molecules collide more frequently with the container, and the pressure increases. If the temperature of the gas is raised, the molecules gain more kinetic energy and move faster and collide more often with the container, thus increasing the pressure.

... This explains gas pressure

Molecular collisions are elastic

The collisions are held to be perfectly *elastic*; the molecules bouncing back off the walls with, on average, no loss of kinetic energy. If this were not so, if the molecules rebounded with less energy, they would be giving up energy to the container, and the temperature of the container would rise while the temperature of the gas fell. The volume of the molecules is held to be negligible compared with the space occupied by the gas. The distances between molecules are so great compared with the sizes of the molecules (10^3 or 10^4 times the diameter of a molecule) that the molecules exert no force upon one another except during collisions. The kinetic energy of the molecules is assumed to be proportional to the temperature of the gas on the Kelvin scale.

The volume of the molecules is negligible compared with that of the container

The distances between molecules are great

The kinetic theory equation

From these assumptions, R J Clausius deduced the **kinetic theory equation***:

$$PV = \tfrac{1}{3}mN\overline{c^2}$$

where P = pressure, V = volume, m = mass of one molecule, N = number of molecules, $\overline{c^2}$ = mean square velocity. If there are n_1 molecules with velocity c_1, n_2 with velocity c_2, etc., $\overline{c^2}$ is given by

$$\overline{c^2} = \frac{n_1 c_1{}^2 + n_2 c_2{}^2 + n_3 c_3{}^2 + \ldots}{n_1 + n_2 + n_3 + \ldots}$$

The root mean square velocity

The square root of $\overline{c^2}$, $\sqrt{\overline{c^2}}$, is called the **root mean square velocity** (rms velocity).

The kinetic theory makes a number of assumptions about the behaviour of gaseous molecules. The ideal gas laws describe the behaviour of gases. If the assumptions of the kinetic theory are correct, it should be possible to derive the ideal gas laws from the kinetic theory.

7.3.2 DERIVING THE IDEAL GAS LAWS FROM THE KINETIC THEORY

BOYLE'S LAW

$$\text{Since } PV = \tfrac{1}{3}mN\overline{c^2}$$

Boyle's Law can be derived from the kinetic theory...

for a given temperature, the average kinetic energy of a particle, $\tfrac{1}{2}m\overline{c^2}$, has a constant value. For a certain mass of gas, the number of particles, N, is constant, and

$$PV = \text{Constant}$$

**For a derivation, see R Muncaster, *A-level Physics* (Stanley Thornes)

CHARLES' LAW

...*as can Charles'
Law*...

If the average kinetic energy of the gas molecules is proportional to the temperature/K

$$\tfrac{1}{2}m\overline{c^2} \propto T$$

For a fixed mass of gas, N is constant, and, at constant pressure

$$V \propto T$$

GRAHAM'S LAW OF GASEOUS DIFFUSION/EFFUSION

Density = Mass/Volume

$$\rho = mN/V$$

$$= 3P/\overline{c^2}$$

At constant pressure

$$\overline{c^2} \propto 1/\rho$$

If the rate of diffusion, r, is proportional to $\sqrt{\overline{c^2}}$

...*and Graham's Law*...

$$r \propto \sqrt{1/\rho}$$

AVOGADRO'S HYPOTHESIS

Consider two gases:

...*and Avogadro's
Hypothesis*

$$P_1V_1 = \tfrac{1}{3}m_1N_1\overline{c_1^2}$$

$$P_2V_2 = \tfrac{1}{3}m_2N_2\overline{c_2^2}$$

For equal volumes of the two gases at the same pressure

$$V_1 = V_2 \text{ and } P_1 = P_2$$

Therefore $\tfrac{1}{3}m_1N_1\overline{c_1^2} = \tfrac{1}{3}m_2N_2\overline{c_2^2}$

At the same temperature

$$\tfrac{1}{2}m_1\overline{c_1^2} = \tfrac{1}{2}m_2\overline{c_2^2}$$

$$\therefore \quad N_1 = N_2$$

Thus equal volumes of gases at the same pressure and temperature contain equal numbers of molecules.

THE IDEAL GAS EQUATION

The kinetic energy of a particle is given by

$$\text{KE} = \tfrac{1}{2}m\overline{c^2}$$

*The ideal gas equation
can be derived from the
kinetic theory*

The kinetic energy of all the particles in one mole of a gas is therefore

$$\tfrac{1}{2}mL\overline{c^2}$$

where m = mass of one molecule; L = Avogadro constant. Since kinetic energy is assumed to be proportional to temperature in kelvins

$$\tfrac{1}{2}mL\overline{c^2} \propto T$$

Applying this proportionality to the kinetic theory equation, since

$$PV = \tfrac{1}{3}mL\overline{c^2},$$

$$PV \propto T$$

$$\therefore \quad PV = \text{constant} \times T$$

This is the ideal gas equation.

There is agreement between the kinetic theory equation and the ideal gas laws, which are based on experimental observations

The ideal gas equation was derived from the gas laws. The proponents of the kinetic theory found it very satisfying to see that the same equation can be derived from the kinetic theory, because it shows that not only can the kinetic theory give a *qualitative* explanation of phenomena, such as gaseous pressure and gaseous diffusion, it can also give a *quantitative* explanation of the behaviour of gases. The derivation of experimental laws from the kinetic theory is a validation of the basic assumptions of the kinetic theory.

7.3.3 CALCULATION OF ROOT MEAN SQUARE VELOCITY

Calculation of rms velocity

The kinetic theory can be used to calculate the root mean square velocity of gas molecules, for example, that of oxygen molecules at stp.

Method (a) from the molar mass of oxygen, $32.0\,\text{g mol}^{-1}$

In the equation $PV = \frac{1}{3}mN\overline{c^2}$, substitute $PV = RT$ for 1 mole of gas, which gives

$$\frac{1}{3}mN\overline{c^2} = RT$$

Then, $N = L$, the Avogadro constant, and $mL = M$, the molar mass of gas in kg mol^{-1}, which gives

$$\overline{c^2} = \frac{3RT}{M}$$

$$\overline{c^2} = 3 \times 8.314 \times 273/(32.0 \times 10^{-3})$$

$$\sqrt{\overline{c^2}} = 461\,\text{m s}^{-1}$$

The rms velocity of oxygen molecules at stp is $461\,\text{m s}^{-1}$.

Method (b) from the density of oxygen, $1.428 \times 10^{-3}\,\text{g cm}^{-3}$ at stp.

$$\overline{c^2} = 3PV/mN$$

Since $mN/V = \rho$, the density of the gas,

$$\overline{c^2} = 3P/\rho$$

$\rho = 1.428 \times 10^{-3}\,\text{g cm}^{-3} = 1.428\,\text{kg m}^{-3}$ at stp

$$\overline{c^2} = 3 \times 1.013 \times 10^5/1.428$$

$$\sqrt{\overline{c^2}} = 461\,\text{m s}^{-1}, \text{ as before}$$

7.3.4 CALCULATION OF MOLECULAR ENERGY

The **translational kinetic energy** of a particle is given by $\frac{1}{2}mc^2$. The translational kinetic energy of a number of molecules of gas, E_t, is given by

The value of molecular energy...

$$E_t = \frac{1}{2}mc_1^2 + \frac{1}{2}mc_2^2 + \frac{1}{2}mc_3^2 + \ldots + \frac{1}{2}mc_n^2$$

If N = number of molecules and $\overline{c^2}$ = mean square velocity

$$E_t = \frac{1}{2}mN\overline{c^2}$$

Since $\frac{1}{3}mN\overline{c^2} = RT$

$$E_t = \frac{3}{2}RT$$

...and a calculation For example, the translational kinetic energy of one mole of a gas at 273 K is given by

$$E_t = \tfrac{3}{2} \times 8.314 \times 273 = 3.405\,\text{kJ mol}^{-1}$$

CHECKPOINT 7G: THE KINETIC THEORY AND THE IDEAL GAS EQUATION

(Take Avogadro constant $= 6.022 \times 10^{23}\,\text{mol}^{-1}$, $R = 8.314\,\text{J K}^{-1}\text{mol}^{-1}$, $1\,\text{atm} = 1.013 \times 10^5\,\text{N m}^{-2}$)

1. State the main postulates of the Kinetic Theory of Gases.

2. How does the motion of molecules in a gas differ from that in a liquid? When a liquid is evaporated, heat energy must be supplied. What happens to this energy?

3. Find the ratio of the root mean square velocities of oxygen and hydrogen at the same temperature.

4. Find the ratio of the root mean square velocities of $^{235}\text{UF}_6$ and $^{238}\text{UF}_6$ at the same temperature.

5. Calculate the number of molecules of oxygen in a $0.500\,\text{dm}^3$ flask at a pressure of $101\,\text{N m}^{-2}$ at 273 K.

7.3.5 THE DISTRIBUTION OF MOLECULAR ENERGIES

Molecular energy has a spread of values at any temperature

J C Maxwell and L Boltzmann plotted distribution curves to show how the spread in the values of molecular energy (translational kinetic energy) alters with temperature. They calculated the fraction of the total number of molecules with energy x, and plotted this against the value of molecular energy x, as in Figure 7.7.

FIGURE 7.7
Distribution of
Molecular Energies

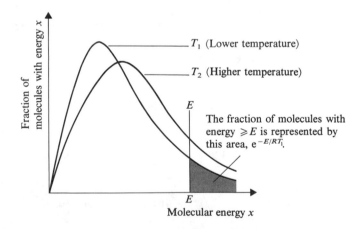

Only a very small fraction of the molecules have very low energies and only a very small fraction have very high energies. The figure shows how the molecules are distributed between different energy levels at two temperatures T_1 and T_2, where T_2 is higher than T_1. In both cases, the area under the curve is the same as all the fractions add up to unity. As the temperature is raised, the average energy of the molecules increases proportionately [p. 144].

At high temperature...

...the average energy increases...

In addition, the shape of the distribution curve alters as the fraction of molecules with low energies decreases and the fraction of molecules with high energies increases. Maxwell and Boltzmann derived an expression for the fraction of molecules possessing high energy. The most important part of this expression is the exponential function $e^{-E/RT}$. For a value of molecular energy E, the fraction of molecules with energy greater than or equal to E is given approximately by $e^{-E/RT}$.

...and the fraction of molecules with high energy increases

7.4 REAL GASES

7.4.1 VAN DER WAALS' EQUATION

Real gases do not show ideal behaviour at high P and low T . . .

Real gases depart from the ideal behaviour illustrated in Figure 7.1, p. 131 at *high pressures* and *low temperatures*. Figure 7.8 shows how the product *PV* departs from a constant value as pressure increases. You will see that the departure from ideal behaviour occurs at very high pressures; many gases approximate to ideal behaviour at pressures in the neighbourhood of 1 atm. The deviation of *PV* from constancy is greatest for easily liquefied gases such as ammonia, and least for gases which are far above their boiling temperatures. At low temperatures, all gases approach their boiling temperatures and deviate from ideal behaviour.

FIGURE 7.8 Plot of *PV* against *P* at a Constant Temperature for one Mole of Gas at 273 K

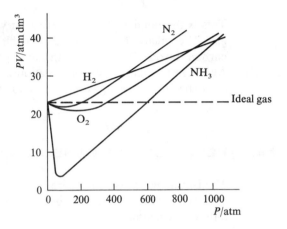

Two of the assumptions of the kinetic theory have to be modified in gases at high pressures or low temperatures. When gases are compressed, the molecules come close enough together for van der Waals forces of attraction [p. 81] to operate between them. These attractions make the gas more compressible than it is at low pressure, and for one mole of gas $PV < RT$. This is shown in the medium pressure range of Figure 7.8. A molecule striking the wall of the container experiences a force of attraction towards other molecules in the gas [see Figure 7.9]. This lessens the force of its impact on the wall.

. . . Van der Waals offered an explanation of this deviation from ideal behaviour

FIGURE 7.9

A Molecule in a Gas at High Pressure Striking the Wall of the Container

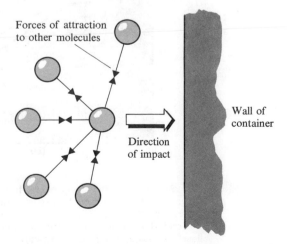

At high pressure, there are forces between the gas molecules...

To the pressure P must be added a term a/V^2, to compensate for the intermolecular attractions.

...and the volume of the molecules cannot be neglected

At still higher pressures, the molecules are pushed so close together that repulsive forces operate between them [see Figure 4.17, p. 82]. These repulsions make the gas less compressible than it is at low pressure, and $PV > RT$. This is shown in the high pressure region of Figure 7.8. Van der Waals postulated that the volume of the actual molecules can no longer be neglected in a highly compressed gas. He subtracted a term b (equal to four times the volume of the molecules) from the volume occupied by the gas, in order to give the volume in which the molecules are free to move.

The van der Waals modified equation of state...

With these modifications to P and V, the equation of state becomes, *for 1 mole of gas*

$$\left(P + \frac{a}{V_m^2}\right)(V_m - b) = RT$$

...represents the behaviour of real gases

This is the **van der Waals equation**, in which a and b are constant for a particular gas. The values of a are found to be highest for the easily liquefiable gases, such as carbon dioxide. The gases with the largest molecules have the highest value of b. Although the van der Waals equation is not completely satisfactory in representing the behaviour of real gases, it does so more faithfully than the ideal gas equation.

7.4.2 BEHAVIOUR OF CARBON DIOXIDE

The **isotherms** shown in Figure 7.10 are the experimental curves which T Andrews obtained for the relationship between pressure and volume for carbon dioxide at different temperatures.

FIGURE 7.10 The Andrews Isotherms for Carbon Dioxide

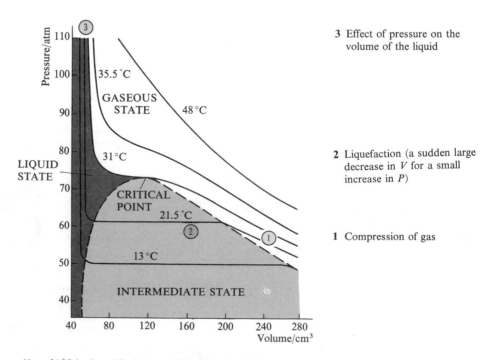

3 Effect of pressure on the volume of the liquid

2 Liquefaction (a sudden large decrease in V for a small increase in P)

1 Compression of gas

Note: 31 °C is the **critical temperature.** It is the highest temperature at which there is evidence of a horizontal portion and therefore the highest temperature at which the gas can be liquefied.
The **critical pressure** is 74 atm. It is the minimum pressure required to liquefy the gas at its critical temperature.

7.4.3 THE WORK DONE IN EXPANSION

The work done when a gas expands can be calculated

Work is done when a gas expands. Consider a gas contained in a cylinder by means of a sliding, frictionless piston [see Figure 7.11]. The gas exerts a pressure, P, on the piston. Since pressure = force/area, the force which the gas exerts on the piston is given by

$$f = PA$$

where A = the cross-sectional area of the piston. The piston is prevented from moving by a weight resting on it. If the temperature of the gas is raised, the gas will expand, and the piston will rise through a distance d.

FIGURE 7.11 The Expansion of a Gas at Constant Pressure

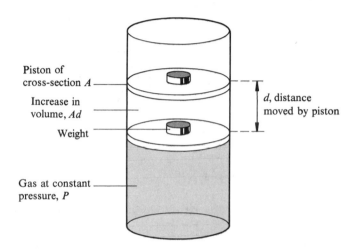

Piston of cross-section A

Increase in volume, Ad

Weight

d, distance moved by piston

Gas at constant pressure, P

Work done by the gas, w = Force × Distance

$$w = PAd$$

Work done is the pressure multiplied by the increase in volume

Since Ad is the increase in volume, ΔV

$$w = P\Delta V$$

You will see in Chapter 10 that work done *on* a system is treated as a positive quantity. Work done *by* a system is therefore a negative quantity. In this case we have

Work done by the gas in expansion = $-P\Delta V$

Work done on the gas by the surroundings = $P\Delta V$

7.4.4 LIQUEFACTION OF GASES

Gases cannot be liquefied above their critical temperatures

Gases cannot be liquefied above their critical temperatures. This is why many early attempts to liquefy gases by compression alone failed. For gases like hydrogen, oxygen and nitrogen, with low critical temperatures (around $-200\,°C$), the problem of cooling them sufficiently to be liquefied was a difficult one. It was solved by using the Joule–Thomson effect.

Sudden expansion results in cooling...

J P Joule and Sir William Thomson noticed that when a gas was allowed to expand through a porous plug its temperature fell. The expanding gas has to do work: it has to push back the gas in front of it, and also the gas molecules become more widely separated against intermolecular forces of attraction. If the expansion takes place too rapidly for energy to be absorbed from the surroundings, the energy must come from the gas itself, and the gas cools.

...This cooling, the Joule–Thomson effect, is used to liquefy air

The industrial liquefaction of air is illustrated in Figure 7.12. After several cycles of compression followed by sudden expansion, the temperature is low enough to allow air to be liquefied at − 200 °C.

FIGURE 7.12
Liquefaction of Air

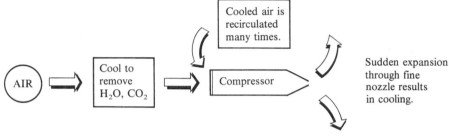

7.4.5 PLASMAS

A **plasma** is an ionised gas. At very high temperatures, 10^4 to 10^5 K, electrons are removed from the atoms to form ions. Plasmas are studied in the atmosphere of the Sun and stars and in research on nuclear fusion.

CHECKPOINT 7H: REAL GASES

1. State two ways in which a real gas differs from an ideal gas. How did van der Waals modify the ideal gas equation in order to take into account these two factors?

2. Which of the gases: hydrogen, propane and hydrogen chloride, would you expect to depart the furthest from ideal gas behaviour? Explain your choice.

3. Figure 7.13 shows the effect of pressure on the volume of a gas at three temperatures, T_1, T_2, and T_3 where $T_1 > T_2 > T_3$.
(*a*) On the basis of these curves, would you say that the gas obeys Boyle's Law?
(*b*) Why do the curves at different temperatures differ in shape?

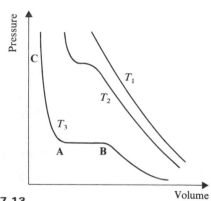

FIGURE 7.13

(*c*) What does the region **AB** signify?
(*d*) What is the significance of the pressure **C**?

QUESTIONS ON CHAPTER 7

1. (*a*) How does the pressure of a given mass of gas in a fixed volume change as the temperature rises? How does the kinetic theory explain this change?

(*b*) At stp, a certain mass of gas has a volume of $1.00\,dm^3$. At 30 atm pressure, the volume is $31.2\,cm^3$. At 60 atm, the volume is $14.9\,cm^3$. Does the gas show ideal behaviour? Explain your answer.

2. State the assumptions that are made in the kinetic theory of gases. Explain how (a) Avogadro's Hypothesis and (b) Dalton's Law of Partial Pressures can be derived from the kinetic theory.

3. If 13 cm³ of nitrogen (at stp) dissolve in 1.00 dm³ of water at room temperature and 1 atm pressure, what is the solubility of nitrogen at room temperature and 60 atm pressure? State any law which you use to arrive at a conclusion, and explain how this law fits in with the kinetic theory. Give examples of gases which do not obey the law.

4. (a) Explain the modifications which van der Waals made to the ideal gas equation.

(b) Explain the principle underlying the methods used for liquefying air.

5. How does the kinetic theory of gases explain (a) the interdiffusion of gases, (b) gas pressure and (c) the compressibility of gases?

6. (a) From the equation

$$pV = \tfrac{1}{3}mN\overline{c^2}$$

derive Graham's Law of Diffusion.

(b) Calculate the kinetic energy of 1 mole of an ideal gas at stp ($R = 8.314\,\mathrm{J\,K^{-1}\,mol^{-1}}$).

†**7.** Refer to Figure 7.4, p. 135. 80 cm³ of oxygen effused from such an apparatus in 70 s, while 80 cm³ of nitrogen dioxide took 100 s. Calculate a value for the molar mass of nitrogen dioxide. After studying pp. 227–30 explain why the value is different from that corresponding to the formula NO_2.

8. Given $R = 8.314\,\mathrm{J\,K\,mol^{-1}}$, $L = 6.022 \times 10^{23}\,\mathrm{mol^{-1}}$, 1 atm = $1.01 \times 10^5\,\mathrm{N\,m^{-2}}$, calculate

(a) the number of molecules of hydrogen in a bulb of capacity 0.500 dm³ which has been evacuated down to a pressure of $1.00\,\mathrm{N\,m^{-2}}$ at 298 K,

(b) the kinetic energy of the molecules in 1.00 mol hydrogen at 298 K,

(c) the ratio of the root mean square velocities of hydrogen and hydrogen bromide at 298 K.

9. Hydrogen diffused 7.94 times faster than a gaseous fluoride of phosphorus (under the same conditions). Calculate the molar mass of the fluoride, and suggest its formula.

10. State Avogadro's Hypothesis.

(a) When 10.0 cm³ of a gaseous hydrocarbon C_4H_x were exploded with an excess of oxygen at 110 °C, under a pressure of 1 atm, there was an expansion of 10.0 cm³. Deduce the value of x.

(b) A gas **X** is known to be an oxide of nitrogen. When 20.0 cm³ of **X** were passed over red hot iron, 10.0 cm³ of nitrogen were obtained. When 20.0 cm³ of **X** were heated with charcoal, 10.0 cm³ of carbon dioxide were obtained. All volumes were measured under the same conditions. Deduce the formula of **X**.

11. (a) Define the terms *relative atomic mass* and *relative molecular mass*. Does *relative molecular mass* have the same significance when applied to potassium chloride as it has when applied to benzene?

(b) What two variables determine the average velocity of the molecules of an ideal gas? State concisely how this velocity depends on these variables. [The derivation of formulae is *not* expected.]

(c) On reacting potassium chlorate(VII) (potassium perchlorate), $KClO_4$, with fluorosulphonic acid, FSO_3H, a gas **X** is evolved.

(i) 0.245 g of **X** are found to occupy 112 cm³ at 293 K and at a pressure of $5.20 \times 10^4\,\mathrm{Pa}$. Calculate to three significant figures the relative molecular mass of **X**.

(ii) By mass spectrometry, **X** is found to contain only the elements Cl, O and F. Suggest a molecular formula for **X**.

(O 82)

12. (a) Discuss what is meant by *an ideal gas* and use this concept to account for

(i) the existence of gas pressure,

(ii) the increase of pressure with increasing temperature,

(iii) the fact that the pressure of a gas is the same in all directions.

(b) State the essential differences between a real gas and an ideal gas and explain how these differences are accounted for in the van der Waals equation for 1 mole of a real gas

$$(p + a/V^2)(V - b) = RT.$$

(c) A 1 litre flask containing gas **A** at a pressure of 1 atmosphere is connected by a tap to a 3 litre flask containing gas **B** at a pressure of 3 atmospheres. The tap is then opened. Assuming that the gases behave ideally and that the temperature remains constant throughout, calculate

(i) the final pressure inside the flasks,

(ii) the partial pressures of **A** and **B**,

(iii) the mole fractions of **A** and **B**.

(JMB 81)

13. (a) Calculate the pressure, p_i, that would be exerted by an ideal gas at a concentration of $1.5\,\mathrm{mol\,dm^{-3}}$ and 400 K.

Suggest reasons why, at this concentration and temperature, hydrogen exerts a pressure of $1.01 \times p_i$ and ethene a pressure of $0.89 \times p_i$.

(b) Discuss the behaviour of gases **A** and **B** using the data below.

(i) Variation of pressure with temperature at constant volume

T/K	300	340	380	420
Pressure of **A**/atm	1.00	1.13	1.27	1.40
Pressure of **B**/atm	1.00	1.46	1.95	2.29

(ii) Variation of density with pressure at constant temperature (340 K)

Pressure/atm	1.00	3.00	5.00
Density of **A**/g dm⁻³	1.43	4.30	7.17
Density of **B**/g dm⁻²	2.04	7.01	12.40

(C 82,S)

8

LIQUIDS

8.1 THE LIQUID STATE

In a liquid, the molecules are closer together than in a gas and less ordered than in a solid

In a liquid, the molecules are closer together than in a gas. This is evident from the fact that one volume of water is formed by the condensation of 1300 volumes of steam. The formation of a liquid from a gas is due to the attractive forces between molecules. These must be strong enough to overcome the kinetic energy of the molecules. When a gas is cooled, the molecules lose kinetic energy until they reach a temperature at which their kinetic energy will no longer overcome the forces of intermolecular attraction, and the gas liquefies.

There are forces of attraction between the molecules in a liquid

The liquid state bears more resemblance to the solid state than to the gaseous state. The particles in a liquid are further apart than in a solid and less ordered. When solids melt, there is usually an expansion of the order of 10%. The amount of heat required to convert 1 mole of a liquid to vapour at 1 atm is called the **standard enthalpy of vaporisation**, $\Delta H^{\ominus}_{\text{Vaporisation}}$. The same amount of heat is released when 1 mole of vapour condenses at 1 atm. The amount of heat required to convert 1 mole of a solid into a liquid at 1 atm is called the **standard enthalpy of melting**, $\Delta H^{\ominus}_{\text{Melting}}$. The same amount of heat is released when 1 mole of a liquid solidifies at 1 atm. Standard enthalpies of melting are much smaller than standard enthalpies of vaporisation because the transition from solid to liquid involves less disruption of intermolecular attractions than the transition from liquid to gas.

FIGURE 8.1
The Distance between Molecules of a Liquid

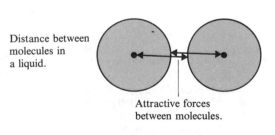

Distance between molecules in a liquid.

Attractive forces between molecules.

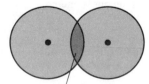

Imaginary compression of liquid forces molecules closer together.

Interpenetration of electron clouds causes repulsion, and molecules move further apart.

Liquids are slightly compressible...

Liquids have a definite volume and are only slightly compressible. The distance between molecules in a liquid is governed by a balance between forces of attraction between molecules and forces of repulsion between neighbouring electron clouds [see Figure 8.1]. The forces of attraction between the molecules, being less strong than those in solids, are not strong enough to hold the liquid in a definite shape. A liquid flows to fit the shape of its container.

...and liquids flow

Diffusion takes place more slowly in liquids than in gases

Gases are able to diffuse because the molecules are in rapid motion. The rates of diffusion of liquids are much less than those of gases, and those of solids are extremely low. Although the molecules of a liquid are in rapid motion, they are not able to move freely from one part of a liquid to another: they are restricted by the attractive forces which bind them into a loose network with neighbouring molecules. Occasionally, a molecule breaks free from the network of molecules that surround it, and moves to another region of the liquid: it **diffuses**.

Molecules with high energy escape from the liquid to the vapour phase

Since molecules of liquid are in constant motion, some of them will have enough energy to escape from the liquid into the vapour state. (The term **vapour** is applied to a gas below its critical temperature.) Those that escape will be molecules with energy considerably above average, sufficient to remove them against the attraction of other molecules of liquid. The molecules that remain in the liquid will be of lower energy than those that escape, and the temperature of the liquid will fall. Since the fraction of molecules with high energy increases as the temperature rises, the rate of evaporation increases with temperature.

At equilibrium, the rate of evaporation equals the rate of condensation

Evaporation will continue until no liquid remains. If the liquid is in a closed container, however, the molecules in the vapour state will collide with the walls of the container, and some will be directed back towards the liquid. Some of these will re-enter the liquid, i.e. **condense**. [See Figure 8.2.] **Equilibrium** will be reached when the rate at which molecules of liquid evaporate is equal to the rate at which molecules of vapour condense.

FIGURE 8.2 A Liquid and its Vapour reach Equilibrium in a Closed Container. The rate of evaporation is equal to the rate of condensation.

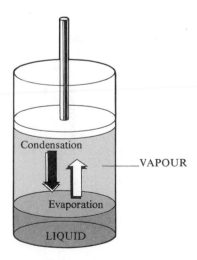

8.2 SATURATED VAPOUR PRESSURE

The maximum vapour pressure developed by a liquid is its saturated vapour pressure at that temperature

As the molecules of vapour collide with the walls of the container they exert a pressure. The maximum vapour pressure that can be developed by a liquid is called the **saturated vapour pressure** of that liquid. Since the fraction of molecules with high energy increases with temperature, evaporation increases as the temperature rises and the saturated vapour pressure increases with temperature. The magnitude of the saturated vapour pressure depends on the *identity* of the liquid and the *temperature*; it does not depend on the amount of liquid present. Solids have vapour pressures too, but, at room temperature, the vapour pressures of most solids are low.

The saturated vapour pressure of a liquid can be measured as shown in Figure 8.3.

FIGURE 8.3 Measuring the Saturated Vapour Pressure of a Liquid

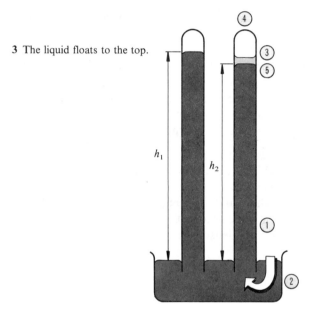

3 The liquid floats to the top.

4 Vapour is formed. The vapour exerts a pressure on the column of mercury.

5 The level of mercury drops. The difference, $(h_1 - h_2)$ mm, is the saturated vapour pressure of the liquid at this temperature.

1 Mercury barometer

2 Some liquid is introduced into the barometer.

A plot of saturated vapour pressure against temperature is shown in Figure 8.4.

FIGURE 8.4 Variation of Saturated Vapour Pressure with Temperature

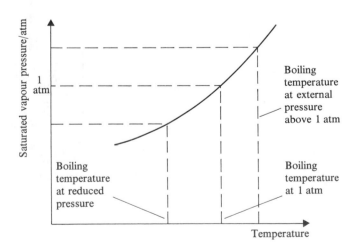

Saturated vapour pressure, svp, increases with temperature

If a liquid is heated in an open container, its saturated vapour pressure increases until it becomes equal to atmospheric pressure. When this happens, bubbles of vapour form in the interior of the liquid and escape into the atmosphere because the vapour pressure is high enough to push the air aside: the liquid boils. The **boiling temperature** of a liquid is defined as the temperature at which its saturated vapour pressure, svp, is equal to the external pressure, normally 1 atm. While a liquid is boiling, the heat taken in is used to produce molecules with enough energy to escape into the vapour phase. The average kinetic energy of the molecules remaining in the liquid does not increase, and its temperature remains constant. Figure 8.4 shows how the boiling temperature increases at pressures greater than 1 atm and decreases at pressures less than 1 atm. Distillation under reduced pressure (sometimes called **vacuum distillation**) is used in the purification of substances which decompose at their boiling temperatures. The apparatus shown in Figure 8.5 could be used. By means of vacuum pumps, pressures down to 10^3 or $10^2 \, \mathrm{N \, m^{-2}}$ can be obtained.

When the svp equals the external pressure, the liquid boils

Distillation under reduced pressure is used for the purification of some substances

The measurement of boiling temperature

A pure liquid can be identified by its boiling temperature at a certain pressure (normally 1 atm). When a boiling temperature is measured, the thermometer must measure the temperature of the vapour in equilibrium with the boiling liquid [see Figure 8.5]. It should not be immersed in the liquid because **superheating** (heating above the boiling temperature) can occur if a liquid is heated rapidly.

FIGURE 8.5 Distillation Apparatus

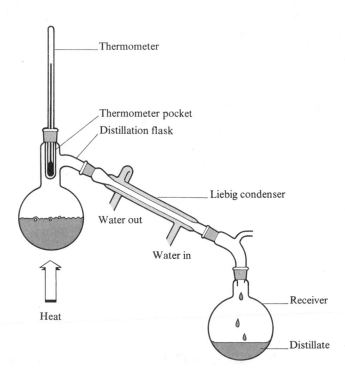

Thermometer

Thermometer pocket
Distillation flask

Liebig condenser

Water out

Water in

Heat

Receiver

Distillate

The saturated vapour pressure of a liquid can be used to calculate the mass of the liquid that is present in the vapour state in a given volume at a stated temperature.

The mass of a liquid that vaporises can be calculated

Example A $1.000 \, \mathrm{dm^3}$ vessel contains air at $1.01 \times 10^5 \, \mathrm{N \, m^{-2}}$ and 0 °C. After $1.000 \, \mathrm{g}$ of water is introduced into the vessel, the temperature is raised to 90 °C. Calculate the mass of water that will vaporise. The saturated vapour pressure of water at 90 °C is $6.99 \times 10^4 \, \mathrm{N \, m^{-2}}$.

Method   Use the Ideal Gas Equation

$$PV = nRT$$

Substitute $P = 6.99 \times 10^4 \, \text{N m}^{-2}$

$R = 8.314 \, \text{J K}^{-1} \text{mol}^{-1}$

$T = 363 \, \text{K}$

$V = 1.000 \, \text{dm}^3 = 1.000 \times 10^{-3} \, \text{m}^3$

$6.99 \times 10^4 \times 1.000 \times 10^{-3} = n \times 8.314 \times 363$

Amount of water $n = 2.32 \times 10^{-2} \, \text{mol}$

Mass of water $= 18.0 \times 2.32 \times 10^{-2} \, \text{g}$

Mass of water vapour $= 0.417 \, \text{g}$

CHECKPOINT 8A: VAPOUR PRESSURE

1. Why are liquids more compressible than solids and less compressible than gases?

2. Explain the term *saturated vapour pressure*. Arrange in order of increasing vapour pressure: $1 \, \text{dm}^3$ water, $1 \, \text{dm}^3$ ethanol, $50 \, \text{cm}^3$ water, $50 \, \text{cm}^3$ ethanol and $50 \, \text{cm}^3$ of ethoxyethane.

3. While a volatile liquid standing in a beaker evaporates, the temperature of the liquid remains the same as that of its surroundings. If the same liquid is contained in an insulated vacuum flask while it vaporises into the atmosphere, its temperature falls below that of its surroundings. Explain the difference in behaviour.

4. Heat must be supplied to boil a liquid. What happens to this heat?

5. Why is distillation under reduced pressure often employed in the purification of chemicals?

6. Explain how each of the following factors influences the vapour pressure of a liquid:
(*a*) intermolecular forces in the liquid
(*b*) the volume of the liquid in the system
(*c*) the volume of vapour in the system
(*d*) the temperature of the liquid

7. A careless student breaks a thermometer. The saturated vapour pressure of mercury at 20 °C is 0.160 Pa. If the spill is not dealt with immediately, what mass of mercury vapour could build up in each cubic metre of laboratory air at this temperature?

8.3   MOLAR MASS DETERMINATION

The volume of vapour formed by a known mass of liquid is measured...

The gas syringe method of determining the molar mass of a volatile liquid is shown in Figure 8.6. A weighed quantity of liquid is injected into a gas syringe, where it vaporises, and the volume occupied by the vapour is measured.

In the method that uses the apparatus designed by Victor Meyer, a known mass of liquid is vaporised, and the volume of air which it displaces is measured. Neither method gives a very accurate result. These methods are used in conjunction with accurately known empirical formulae to establish the molecular formulae of compounds.

FIGURE 8.6 Gas
Syringe Method of
Molar Mass
Determination

*...and its molar mass is
calculated...*

...A sample calculation

2 Furnace keeps the gas syringe
at a temperature 10 °C above
the boiling temperature of
the liquid

1 Gas syringe

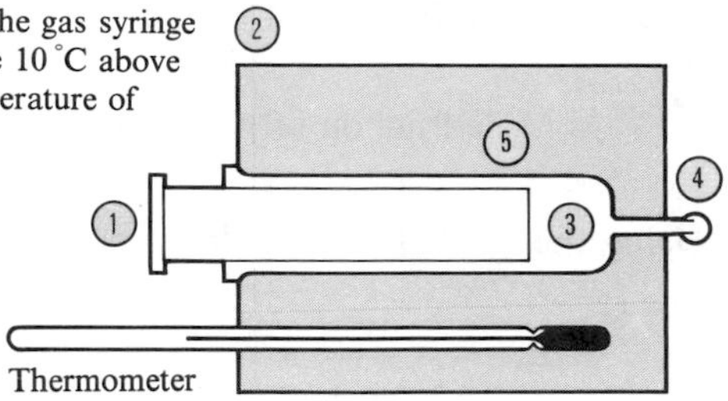

Thermometer

3 The gas syringe contains air.
The volume v_1 cm³ is noted.

4 Self-sealing rubber cap. Liquid
is injected through a
hypodermic needle from a
weighed syringe (m_1 g). The
syringe is reweighed (m_2 g). The
mass of liquid
injected = $(m_1 - m_2)$ g.

5 When the liquid is injected, the
plunger moves backwards. The
volume of gas in the barrel is
read (v_2 cm³). Volume of
vapour = $(v_2 - v_1)$ cm³.

Method of calculation

$$PV = nRT = \frac{m}{M} RT$$

where $V = v_2 - v_1$, and $m = m_1 - m_2$
T = temperature recorded by thermometer,
P = the measured atmospheric pressure
R = the gas constant = $8.314 \, \mathrm{J\,K^{-1}\,mol^{-1}}$
$\therefore M$, the molar mass of the gas, can be found.

Example $m_1 = 20.255 \, \mathrm{g}$, $m_2 = 20.120 \, \mathrm{g}$, $v_1 = 9.80 \, \mathrm{cm^3}$,
$v_2 = 65.8 \, \mathrm{cm^3}$, $T = 363 \, \mathrm{K}$, $P = 1.01 \times 10^5 \, \mathrm{N\,m^{-2}}$

When these experimental results are inserted into the equation

$$PV = \frac{m}{M} RT$$

$$M = \frac{0.135 \times 8.314 \times 363}{1.01 \times 10^5 \times 56.0 \times 10^{-6}}$$

The molar mass, $M = 72 \, \mathrm{g\,mol^{-1}}$

CHECKPOINT 8B: MOLAR MASS OF VOLATILE LIQUIDS

$(R = 8.314 \, \mathrm{J\,K^{-1}\,mol^{-1}}, 1 \, \mathrm{atm} = 1.01 \times 10^5 \, \mathrm{N\,m^{-2}})$

1. When 0.184 g of a liquid was injected into a gas syringe
at 45 °C, the volume of gas formed was 55.8 cm³ at 1 atm.
What value does this indicate for the molar mass of the
liquid? By referring to the empirical formula CH_2 obtain
an accurate value for the molar mass of the liquid.

2. When 0.125 g of a liquid was injected into a gas syringe
at 50 °C, it formed 36.8 cm³ of vapour at atmospheric
pressure. The empirical formula of the compound is
CH_2Cl. What value of molar mass do the experimental data
give? What is the accurate value?

3. A liquid of empirical formula CH_2 was vaporised in a
Victor Meyer apparatus. The volume of air that was displaced
by 0.108 g of vapour was 50.8 cm³, measured at 22 °C and
$9.8 \times 10^4 \, \mathrm{N\,m^{-2}}$. Calculate the molar mass of the liquid.

4. By vaporising 0.100 g of a liquid at 100 °C and
$1.01 \times 10^5 \, \mathrm{N\,m^{-2}}$, 20.0 cm³ of vapour are obtained. Find
the molar mass of the liquid.

8.4 SOLUTIONS OF LIQUIDS IN LIQUIDS

8.4.1 RAOULT'S LAW

A **solution** is a homogeneous (i.e., the same all through) mixture of two substances.
When one liquid dissolves in another, the saturated vapour pressure of the solution
depends on the saturated vapour pressures of the components and on the composition

Definition of mole fraction of the mixture. One way of expressing the composition of a mixture of liquids is to state the **mole fraction** of each constituent. By definition

$$\text{Mole fraction of } \mathbf{A} \text{ in a mixture of } \mathbf{A} \text{ and } \mathbf{B} = \frac{\text{Number of moles of } \mathbf{A}}{\text{Total number of moles}}$$

Using the symbol x for mole fraction

$$x_A = \frac{n_A}{n_A + n_B}$$

Raoult's Law... The vapour above a mixture of liquids **A** and **B** will contain both **A** and **B**. **Raoult's Law** (after F Raoult) states that the saturated vapour pressure of each component in
...gives the vapour the mixture is equal to the product of the mole fraction of that component and the
pressure of a solution saturated vapour pressure of that component when pure. This can be expressed

$$p_A = x_{A(l)} p_A^0$$

where p_A = saturated vapour pressure of **A**, p_A^0 = saturated vapour pressure of pure **A**, and $x_{A(l)}$ = mole fraction of **A** in the solution. Similarly for **B**

$$p_B = x_{B(l)} p_B^0$$

Raoult's Law is obeyed by mixtures of similar compounds. They are said to form
There are conditions for **ideal solutions**. The substances **A** and **B** form an ideal solution if the intermolecular
obeying Raoult's Law forces **A**------**A**, **A**-----**B** and **B**-----**B** are all equal. There is neither a volume change nor an enthalpy (heat) change on mixing.

The vapour above a mixture of liquids does not have the same composition as the liquid. If $x_{A(v)}$ and $x_{B(v)}$ are the mole fractions of **A** and **B** in the vapour phase, then

The vapour is richer than
the liquid in the more
volatile component

$$\frac{x_{A(v)}}{x_{B(v)}} = \frac{p_A}{p_B} = \frac{x_{A(l)} p_A^0}{x_{B(l)} p_B^0}$$

If **A** is more volatile than **B**

$$p_A^0 > p_B^0$$

and

$$\frac{x_{A(v)}}{x_{B(v)}} > \frac{x_{A(l)}}{x_{B(l)}}$$

The vapour is richer than the liquid in **A**, the more volatile component.

Example: how to **Example** At 80 °C, the vapour pressure of benzene is $10.0 \times 10^4 \, \text{N m}^{-2}$; and that
calculate the composition of methylbenzene is $4.00 \times 10^4 \, \text{N m}^{-2}$. Estimate the mole fraction of methylbenzene
of the vapour above a in the vapour in equilibrium with a liquid mixture of benzene and methylbenzene at
solution of two liquids 80 °C in which the mole fraction of methylbenzene is 0.60.

Method

$$\frac{\text{Amount of methylbenzene in vapour}}{\text{Amount of benzene in vapour}} = \frac{0.60}{0.40} \times \frac{4.00 \times 10^4}{10.0 \times 10^4} = 0.60$$

$$\text{Mole fraction of methylbenzene} = \frac{\text{Moles of methylbenzene}}{\text{Total number of moles}}$$

$$= \frac{0.60}{1.60} = 0.375$$

Figure 8.7 shows the results of a number of calculations of this kind on an ideal mixture of **A** and **B**. The composition of vapour in equilibrium with liquid mixtures of different compositions is shown.

FIGURE 8.7 Vapour Pressure–Composition Curve for an Ideal Mixture of Liquids

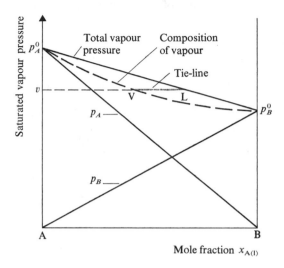

The total saturated vapour pressure is the sum of the partial vapour pressures, $p_A + p_B$. When the total vapour pressure is v, the composition of the liquid is represented by the point **L**, and the composition of the vapour by the point **V**. The line **V·····L** is called a **tie-line**. The vapour is richer than the liquid in the more volatile component, **A**. The difference in composition between the liquid and the vapour is exaggerated in the figure.

Figure 8.8 shows the vapour pressure–composition curves at two temperatures, T_1 and T_2.

FIGURE 8.8 Saturated Vapour Pressure–Composition Curves at Two Temperatures

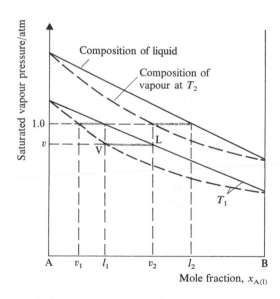

Boiling temperature depends on composition...

...The relationship can be plotted

When the vapour pressure of a liquid reaches atmospheric pressure, the liquid boils. T_1 is the boiling temperature of a liquid with composition l_1 in equilibrium with vapour of composition v_1. T_2 is the boiling temperature of a liquid which has composition l_2, in equilibrium with a vapour of composition v_2. These values can be plotted in the form of a boiling temperature–composition curve. There is a gradual variation in boiling temperature with composition. Neither the liquid curve nor the vapour curve is linear, even for an ideal mixture. The same shape is obtained when boiling temperature is plotted against the mass fraction or percentage composition by mass, instead of the mole fraction (since percentage of **A** by mass = 100 × mass fraction of **A**). A boiling temperature–composition curve is shown in Figure 8.9.

FIGURE 8.9 Boiling Temperature–Composition Curve for an Ideal Mixture of Liquids

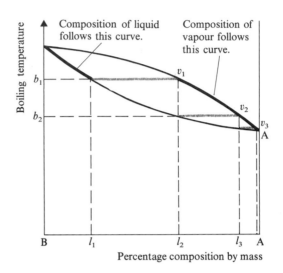

8.4.2 FRACTIONAL DISTILLATION

The process of **fractional distillation** is used to separate the components of a mixture of liquids by means of the difference in their boiling temperatures. An apparatus for fractional distillation is shown in Figure 8.10.

Fractional distillation separates the components of a mixture according to their boiling temperatures

The fractionating column has a large surface area on which ascending vapour and descending liquid come into contact. A mixture rich in the most volatile component distils over at the top of the column, where the thermometer registers its boiling temperature. As distillation continues, the temperature rises towards the boiling temperature of the next most volatile component. The receiver is changed to collect the second component. In this way, the components are distilled over at their boiling temperatures.

The principles underlying fractional distillation are illustrated in Figure 8.9. Imagine that a liquid of composition l_1 is heated until it begins to boil at a temperature b_1.

Each time that equilibrium is established between liquid and vapour, the vapour becomes richer in the more volatile component

Then the vapour in equilibrium with the liquid has composition v_1. If the vapour is condensed by meeting the cold surface of a distillation column, it condenses to form a liquid of composition l_2. This liquid starts to trickle down the column towards the distillation flask. If it is heated, it begins to boil at b_2, to form a vapour of composition v_2. If this vapour is condensed, it forms a liquid of composition l_3. By repeated vaporisation and condensation, the composition of the vapour is made to follow the curve $v_1 v_2 v_3$, becoming richer and richer in **A**, the more volatile component. The liquid is becoming richer in the less volatile component, **B**, and its composition follows the curve from l_1 towards **B**. The longer the column, the more vaporisation followed by condensation steps will be achieved, and the closer to pure **A** and pure **B** will the **distillate** and **residue** become.

FIGURE 8.10 Fractional
                 Distillation

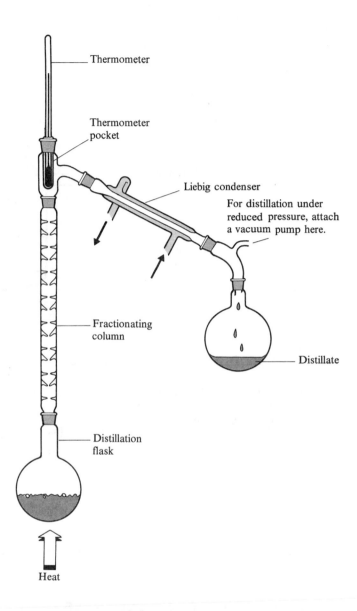

Thermometer

Thermometer
pocket

Liebig condenser

For distillation under
reduced pressure, attach
a vacuum pump here.

Fractionating
column

Distillate

Distillation
flask

Heat

8.4.3 CONTINUOUS FRACTIONAL DISTILLATION

Fuels are obtained from
crude oil by fractional
distillation

The oil industry uses fractional distillation. Crude petroleum oil is vaporised and
fed into a massive fractionating column, which may be 30 to 60 m high and 3 to 6 m
in diameter. Different fractions, such as gasoline and kerosene, are drawn off
continuously from the column at different levels. Figure 8.11 shows how liquid and
vapour attain equilibrium at each level, so that low boiling temperature fractions rise
to the top of the column and high boiling temperature fractions are drawn off from
the bottom of the column.

FIGURE 8.11

Continuous Fractional
Distillation of Petroleum
Oil

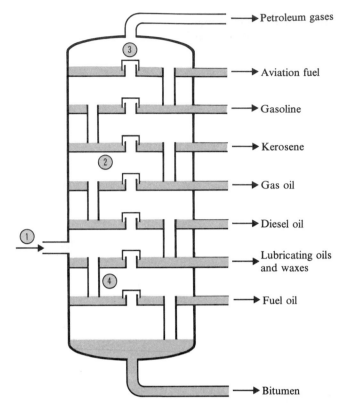

3 Ascending vapour must pass
through *bubble caps*, which
bring it into contact with the
liquid in the plates.

2 Baffle plates,
trays in which
condensate collects.

1 Vaporised petroleum
oil is fed in.

4 Overflow pipes
carry liquid from
baffle plates down
to a lower level.

Petroleum gases

Aviation fuel

Gasoline

Kerosene

Gas oil

Diesel oil

Lubricating oils
and waxes

Fuel oil

Bitumen

CHECKPOINT 8C: VAPOUR PRESSURES OF SOLUTIONS OF TWO LIQUIDS

1. **X** and **Y** are two miscible liquids which form an ideal solution. At 20 °C, the vapour pressures are: $p_X = 25\,kPa$; $p_Y = 45\,kPa$. Calculate the total vapour pressure of a mixture of:

(*a*) 1 mol of **X** and 5 mol of **Y** at 20 °C

(*b*) 5 mol of **X** and 2 mol of **Y** at 20 °C

2. Two liquids **A** and **B** have vapour pressures of $15\,kN\,m^{-2}$ and $40\,kN\,m^{-2}$ at 25 °C. Calculate the vapour pressure of an ideal solution of 1 mol of **A** in 5 mol of **B** at 25 °C.

3. Hexane and heptane form an ideal solution. At 30 °C, the vapour pressures are hexane = $30\,kN\,m^{-2}$; heptane = $12\,kN\,m^{-2}$. Calculate

(*a*) the total vapour pressure

(*b*) the mole fraction of heptane in the vapour above an equimolar mixture of hexane and heptane

4. Explain how fractional distillation separates crude oil into a number of products.
What are these products used for?
How are their uses related to their boiling temperatures?
(See also Chapter 26)

8.4.4 NON-IDEAL SOLUTIONS

*Raoult's Law is obeyed by
ideal solutions...*

*...Non-ideal solutions
have vapour pressures
greater or less than those
predicted*

Solutions which have a vapour pressure greater than that predicted from Raoult's Law are said to show a **positive deviation** from the law. Those with a vapour pressure lower than the calculated value are said to show a **negative deviation**. [See Figure 8.12.] Typical of a pair of liquids showing a slight positive deviation are hexane and ethanol. The molecules are very different, and molecules of ethanol can more easily escape from a mixture of ethanol and hexane than from pure ethanol, where hydrogen bonding holds molecules together. A slight negative deviation from Raoult's Law is shown by a mixture of trichloromethane, $CHCl_3$, and ethoxyethane, $C_2H_5OC_2H_5$. Forces of attraction between the two kinds of molecules tend to prevent the molecules escaping into the vapour phase. When a pair of liquids shows a very large positive deviation from Raoult's Law, the vapour pressure–composition curve has a maximum. A system with a very large negative deviation shows a minimum [see Figure 8.12].

FIGURE 8.12
Saturated Vapour
Pressure–Composition
Curves for Non-ideal
Solutions

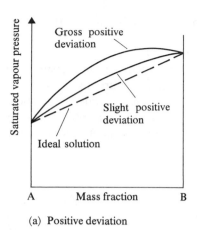

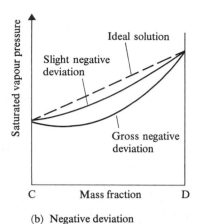

(a) Positive deviation

(b) Negative deviation

Boiling temperature–composition curves for systems which show gross deviations from Raoult's Law are shown in Figure 8.13.

FIGURE 8.13 Boiling
Temperature–Composition
Curves for Non-ideal
Solutions

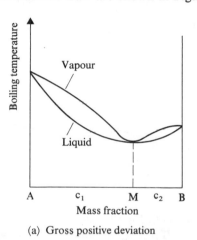

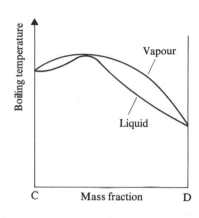

(a) Gross positive deviation

(b) Gross negative deviation

If there is a maximum or minimum in the svp–composition curve, an azeotrope distils

A maximum in the saturated vapour pressure curve results in a minimum in the boiling temperature–composition curve and vice versa. An application of the principles of fractional distillation [p. 160] shows that, in Figure 8.13(a), a liquid of composition c_1 can be separated by distillation into almost pure **A** and a mixture of composition M. A liquid of composition c_2 can be separated by distillation into almost pure **B** and the mixture M. If a liquid of composition M is boiled, the vapour has the same composition as the liquid. This means that the composition, and therefore the boiling temperature, T_b, do not change as distillation is continued. The mixture is therefore called an **azeotropic** or **constant-boiling mixture**. Examples are nitric acid/water (T_b is a maximum) and benzene/ethanol (T_b is a minimum). Azeotropes are *not* classified as compounds because their compositions vary with pressure.

8.5 IMMISCIBLE LIQUIDS

8.5.1 STEAM DISTILLATION

A mixture boils when the combined svp equals the external pressure...

In a system of immiscible liquids, each liquid exerts its own vapour pressure, independently of the other. The saturated vapour pressure of the mixture is equal to the sum of the saturated vapour pressures of the pure components. For a mixture of **A** and **B**,

$$p_{\text{total}} = p_{\text{A}}^0 + p_{\text{B}}^0$$

When the total vapour pressure is equal to the external pressure, the mixture distils.

At T_b, $p_A^0 + p_B^0 = p_{\text{external}}$

...This is the basis of steam distillation

Figure 8.14 shows the saturated vapour pressure–temperature curve for a mixture of phenylamine and water. This mixture distils at 98 °C at 1 atm. When phenylamine is heated at atmospheric pressure, some decomposition occurs before it reaches its boiling temperature of 190 °C. The advantage of steam distillation is that phenylamine will distil at a lower temperature without decomposing. Figure 8.15 shows an apparatus for steam distillation.

FIGURE 8.14 Saturated Vapour Pressure Curve for a Pair of Immiscible Liquids

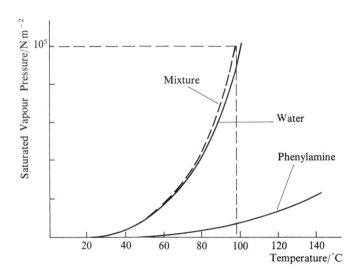

The composition of the steam distillate can be calculated...

The ratio of the amounts of the two liquids in the distillate is equal to the ratio of their vapour pressures. If p_A, p_W are the saturated vapour pressures of phenylamine and water (where $p_A = 3.99 \times 10^3 \, \text{N m}^{-2}$, $p_W = 9.70 \times 10^4 \, \text{N m}^{-2}$ at 98 °C) and n_A, n_W are the amounts of phenylamine and water in a volume V of gaseous distillate, then

$$p_A V = n_A RT$$

$$p_W V = n_W RT$$

and $\dfrac{n_A}{n_W} = \dfrac{p_A}{p_W}$

$$= \frac{3.99 \times 10^3}{9.70 \times 10^4} = 4.11 \times 10^{-2}$$

...Some sample calculations

If m_A and m_W are the masses and M_A and M_W are the molar masses of phenylamine and water respectively (where $M_A = 93 \, \text{g mol}^{-1}$ and $M_W = 18 \, \text{g mol}^{-1}$)

$$\frac{m_A/M_A}{m_W/M_W} = 4.11 \times 10^{-2}$$

$$\frac{m_A}{m_W} = 4.11 \times 10^{-2} \times 93/18 = 0.212$$

The ratio of phenylamine to water by mass in the distillate is 0.212; the percentage by mass of phenylamine is 17.5%.

Example   An organic liquid which does not mix with water distils in steam at 95 °C under a pressure of $1.01 \times 10^5 \, \mathrm{N\,m^{-2}}$. The vapour pressure of water at 95 °C is $8.50 \times 10^4 \, \mathrm{N\,m^{-2}}$. The distillate contains 62.0% by mass of the organic liquid. Calculate its molar mass.

Method   Let M be the molar mass of the organic compound, and n_A and n_W be the amounts of organic liquid and water in the distillate, and p_A and p_W be the vapour pressures of the organic compound and water. Then

$$\frac{n_A}{n_W} = \frac{62.0/M}{38.0/18}$$

$$\frac{n_A}{n_W} = \frac{(1.01 \times 10^5) - (8.50 \times 10^4)}{8.50 \times 10^4} = 0.188$$

hence $M = 156 \, \mathrm{g\,mol^{-1}}$

FIGURE 8.15   Steam Distillation

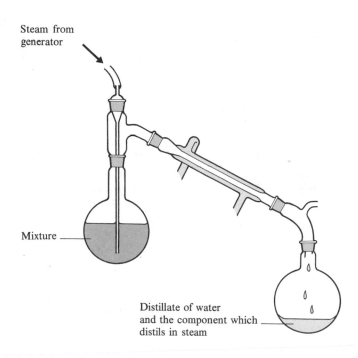

Steam from generator

Mixture

Distillate of water and the component which distils in steam

CHECKPOINT 8D: STEAM DISTILLATION

1. An organic liquid distils in steam. The partial pressures of the two liquids at the boiling point are: organic liquid 5.3 kPa; water 96.0 kPa. The distillate contains the liquids in a ratio of 0.48 g of organic liquid : 1.00 g of water. What is the molar mass of the organic liquid?

2. Chlorobenzene distils in steam at 91 °C at $1.01 \times 10^5 \, \mathrm{N\,m^{-2}}$. At this temperature, the vapour pressure of chlorobenzene is $2.90 \times 10^4 \, \mathrm{N\,m^{-2}}$. Calculate the percentage by mass of chlorobenzene in the distillate.

8.6 DISTRIBUTION OF A SOLUTE BETWEEN TWO SOLVENTS

8.6.1 THE DISTRIBUTION LAW

A solute will distribute itself between a pair of immiscible solvents...

When a solute is added to a pair of immiscible liquids, it may dissolve in both liquids. In this case the solute will distribute itself between the two solvents. It may well be more soluble in one solvent than the other. The way in which a solute is partitioned between two solvents can be studied by shaking the solute with a pair of immiscible liquids in a separating funnel. The two layers can be run off separately and analysed. Provided that there is insufficient solute to saturate either of the two solvents, and provided that the solute is in the same molecular state in both solvents, then it is found that the ratio of the solute concentrations in the two solvents is always the same. If c_U and c_L are the concentrations in the upper and lower layers, then

...in accordance with its distribution coefficient...

$$c_U/c_L = k$$

The constant, k, is called the **partition coefficient** or **distribution coefficient**, and is constant for a particular temperature. Since a dynamic equilibrium exists between

...which is temperature-dependent

Solute in upper layer $\rightleftharpoons$ Solute in lower layer

k is a type of equilibrium constant. Like other equilibrium constants it is temperature-dependent.

8.6.2 EXPERIMENTAL DETERMINATION OF PARTITION COEFFICIENT

The partition coefficient can be found by analysis...

Iodine distributes itself between water and tetrachloromethane. A known mass of iodine (about 1 g) is added to known volumes of water and tetrachloromethane (say 50 cm^3 of each) in a separating funnel. The mixture is shaken well and allowed to stand. The tetrachloromethane layer is run out of the bottom of the separating funnel and a portion of the aqueous layer is titrated against standard thiosulphate solution [p. 55].

Example The mass of iodine used is 0.9656 g and 25.0 cm^3 of the aqueous layer require 4.40 cm^3 of 0.0100 mol dm^{-3} thiosulphate. $A_r(I) = 127$.

25.0 cm^3 aqueous layer require 4.40 cm^3 of 0.0100 mol dm^{-3} thiosulphate
$$= 4.40 \times 10^{-5} \text{ mol thiosulphate}$$

..A sample calculation of partition coefficient

$$[\text{I}_2 \text{ in aqueous layer}] = \tfrac{1}{2} \times 4.40 \times 10^{-5}/(25.0 \times 10^{-3})$$
$$= 8.8 \times 10^{-4} \text{ mol dm}^{-3}$$

Mass of I$_2$ in aqueous layer $= 8.8 \times 10^{-4} \times 254 \times 50.0 \times 10^{-3} = 0.0112$ g

Mass of I$_2$ in CCl$_4$ layer $= 0.9656 - 0.0112 = 0.9544$ g

$[\text{I}_2 \text{ in water}]/[\text{I}_2 \text{ in CCl}_4] = 0.0112/0.9544 = 1.17 \times 10^{-2}$

The partition coefficient between water and tetrachloromethane is 1.17×10^{-2}.

Ethoxyethane extraction is used in the purification of organic products...

Ethoxyethane (ether) extraction is often used in organic preparations. Ethoxyethane is a good solvent for many organic compounds and is immiscible with water. When an aqueous solution of a product is shaken with ethoxyethane in a separating funnel, the product will distribute itself between the two layers. If the partition coefficient is high, a large fraction of product will pass into the ether layer. The ether layer is separated from the aqueous layer and dried. With its low boiling temperature, ethoxyethane is easily distilled off to leave the product. It is more efficient to use the ether in portions for repeated extractions than to use it all in one operation. The following calculation illustrates this point.

Example The product of an organic synthesis, 5.00 g of **X**, is obtained in solution in 1.00 dm^3 of water. Calculate the mass of **X** that can be extracted from the aqueous solution by (a) 50.0 cm^3 of ethoxyethane and (b) two successive portions of 25.0 cm^3 of ethoxyethane. The partition coefficient of **X** between ethoxyethane and water is 40.0 at room temperature.

Method

(a) Let the mass of **X** extracted by 50.0 cm^3 of ether = a_1 g

[**X** in ether]/[**X** in water] = 40.0

$$\frac{a_1/50.0}{(5.00 - a_1)/1000} = 40.0$$

$a_1 = 3.33$ g

The mass of **X** that can be extracted by using all the ether at once is 3.33 g.

It is more efficient to use the solvent in portions (b) Let a_2 g = mass of **X** extracted by the first 25.0 cm^3 of ether, and a_3 g = mass of **X** extracted by the second 25.0 cm^3 of ether

$$\frac{a_2/25.0}{(5.00 - a_2)/1000} = 40.0$$

$a_2 = 2.50$ g

If 2.50 g of **X** are extracted by ether, 2.50 g remain in the aqueous solution. Therefore

$$\frac{a_3/25.0}{(2.50 - a_3)/1000} = 40.0$$

$a_3 = 1.25$ g

Total mass of **X** extracted by ether in two portions is

2.50 + 1.25 = 3.75 g

This is greater than the value of 3.33 g calculated for the mass of **X** extracted by using all the ether at once.

8.6.3 FINDING EQUILIBRIUM CONSTANTS

Partition can be used to investigate an equilibrium in aqueous solution between a covalent species and an ionic species. An example is the equilibrium

$$I_2(aq) + I^-(aq) \rightleftharpoons I_3^-(aq)$$

Partition can be used to study an equilibrium between molecules and ions... for which the equilibrium constant [p. 220] is given by

$$K_{eq} = \frac{[I_3^-(aq)]}{[I_2(aq)][I^-(aq)]}$$

The value of $[I_2(aq)]$ cannot be found by titration because thiosulphate reacts with iodine combined as I_3^- ions as well as with free I_2 molecules. However, only the covalent I_2 molecules will dissolve in an organic solvent, and, as shown in the example below, a partition method can be used to find the concentration of free I_2 molecules in the aqueous layer.

...A sample calculation of an equilibrium constant

Example   A solution of iodine in $0.238\,mol\,dm^{-3}$ aqueous potassium iodide is shaken with tetrachloromethane. The concentrations of iodine in the two layers are: aqueous layer, $0.118\,mol\,dm^{-3}$; organic layer, $0.102\,mol\,dm^{-3}$. The partition coefficient for iodine between water and tetrachloromethane is $1/85$. Calculate the equilibrium constant at room temperature for the reaction

$$I_2(aq) + I^-(aq) \rightleftharpoons I_3^-(aq)$$

Method

$$[I_2 \text{ in } CCl_4] = 0.102\,mol\,dm^{-3}$$

$$[I_2 \text{ free in water}] = 0.102/85 = 1.20 \times 10^{-3}\,mol\,dm^{-3}$$

$$[I_2 \text{ total in water}] = 0.118\,mol\,dm^{-3}$$

$$[I_2 \text{ combined as } I_3^-] = 0.118 - 1.20 \times 10^{-3} = 0.117\,mol\,dm^{-3}$$

$$[I^- \text{ total}] = 0.238\,mol\,dm^{-3}$$

$$[I^- \text{ free}] = 0.238 - 0.117 = 0.121\,mol\,dm^{-3}$$

Inserting these values in the expression

$$K_{eq} = \frac{[I_3^-]}{[I_2][I^-]}$$

gives $K_{eq} = \dfrac{0.117}{1.20 \times 10^{-3} \times 0.121} = 820\,dm^3\,mol^{-1}$

The equilibrium constant is $820\,dm^3\,mol^{-1}$ at room temperature.

8.6.4   FINDING THE FORMULAE OF COMPLEX IONS

To find the formula of the ammine complex formed by copper(II) ions

$$Cu^{2+}(aq) + nNH_3(aq) \rightleftharpoons Cu(NH_3)_n^{2+}(aq)$$

Partition can be used to find the formula of a complex ion...

it is necessary to find out how many moles of ammonia combine with one mole of copper(II) ions. This cannot be done by titration because acid will react with ammonia in the complex as well as with free ammonia. However, the complex ions will not dissolve in an organic solvent: only free ammonia molecules will be partitioned between water and an organic solvent. Partition can therefore be used to find the concentration of free ammonia in the solution and hence the concentration of complexed ammonia and the formula of the complex.

Example $100\,cm^3$ of aqueous copper(II) sulphate of concentration $0.100\,mol\,dm^{-3}$ were mixed with $100\,cm^3$ of ammonia solution of concentration $1.00\,mol\,dm^{-3}$. The mixture was shaken with $100\,cm^3$ of trichloromethane. The concentration of ammonia in the organic layer was found to be $1.17 \times 10^{-2}\,mol\,dm^{-3}$. The partition coefficient of ammonia between water and trichloromethane is 25.0. Find the formula of the $Cu(NH_3)_n^{2+}$ ion.

Method

Since $[\text{NH}_3 \text{ in } \text{CHCl}_3] = 1.17 \times 10^{-2} \text{mol dm}^{-3}$

$[\text{NH}_3 \text{ free in water}] = 1.17 \times 10^{-2} \times 25.0 = 0.293 \text{ mol dm}^{-3}$

Volume of aqueous layer $= 0.200 \text{ dm}^3$

Amount of ammonia free in water $= 0.293 \times 0.200 = 0.0586 \text{ mol}$

Amount of NH_3 in $\text{CHCl}_3 = 1.17 \times 10^{-2} \times 0.100 = 1.17 \times 10^{-3} \text{mol}$

Total amount of $\text{NH}_3 = 100 \times 10^{-3} \times 1.00 = 0.100 \text{ mol}$

Amount of complexed NH_3 in water $= 0.100 - 0.0586 - (1.17 \times 10^{-3}) \text{mol}$

$$= 0.0402 \text{ mol}$$

Amount of Cu^{2+} in water $= 100 \times 10^{-3} \times 0.100 = 0.0100 \text{ mol}$

Ratio, Amount of $\text{Cu}^{2+}/\text{mol}$: Amount of NH_3/mol
$= 0.0100 : 0.0402 = 1 : 4$

The formula of the complex is $\text{Cu}(\text{NH}_3)_4{}^{2+}$.

8.6.5 DISTRIBUTION LAW FOR ASSOCIATION AND DISSOCIATION

When a solute is associated or dissociated in one solvent, the distribution law must be modified

In some solvents, a solute may be **associated**: n molecules of solute **A** may combine to form one molecule of A_n. In other solvents, a solute may be **dissociated**, i.e., split up into smaller molecules or into ions. In these cases, the distribution law has to be modified. For example, ethanoic acid is dimerised in benzene. When ethanoic acid is partitioned between benzene and water, the equilibrium is

$$2\text{CH}_3\text{CO}_2\text{H}(\text{water}) \rightleftharpoons (\text{CH}_3\text{CO}_2\text{H})_2(\text{benzene})$$

Thus $\dfrac{[\text{CH}_3\text{CO}_2\text{H}(\text{benzene})]}{[\text{CH}_3\text{CO}_2\text{H}(\text{water})]^2} = C$

and $\lg[\text{CH}_3\text{CO}_2\text{H}(\text{benzene})] = \lg C + 2\lg[\text{CH}_3\text{CO}_2\text{H}(\text{water})]$

A plot of $\lg[\text{CH}_3\text{CO}_2\text{H}(\text{benzene})]$ against $\lg[\text{CH}_3\text{CO}_2\text{H}(\text{water})]$ gives a straight line with a gradient of 2 and an intercept on the y axis of $\lg C$ [see Figure 8.16].

In general, if a solute is monomeric in solvent 1 and polymeric in solvent 2

$$n\text{A}(\text{solvent 1}) \rightleftharpoons \text{A}_n(\text{solvent 2})$$

A plot of $\lg[\text{A}(\text{solvent 2})]$ against $\lg[\text{A}(\text{solvent 1})]$ gives a straight line of gradient n and intercept $\lg C$ [see Figure 8.16].

FIGURE 8.16 The Distribution Law for an Associated Solute

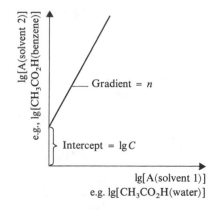

1. The partition coefficient of **S** between ethoxyethane and water is 5.0. A solution containing 10.0 g of **S** in 500 cm³ of water is extracted with 100 cm³ of ether. Which of the following gives the mass of **S** extracted from the water?

A 2.5 g   B 5.0 g   C 7.5 g   D 10.0 g   E 12.5 g

2. A substance **T** of $M_r = 60.0 \text{ g mol}^{-1}$ is present at a concentration of 0.0800 mol dm⁻³ in 500 cm³ of aqueous solution. The distribution coefficient of **T** between water and tetrachloromethane is 2.78×10^{-2}. What mass of **T** will be extracted from the aqueous solution by shaking it with 50.0 cm³ of tetrachloromethane?

3. An aqueous solution contains 5.0 g of **X** in 100 cm³ of solution. The partition coefficient of **X** between water and tetrachloromethane is 0.200. Calculate the mass of **X** extracted by shaking 100 cm³ of the aqueous solution with

(*a*) 50 cm³ of tetrachloromethane

(*b*) two successive 25 cm³ portions of tetrachloromethane.

4. 100 cm³ of aqueous copper(II) sulphate of concentration 0.100 mol dm⁻³ were mixed with 100 cm³ of aqueous ammonia of concentration 1.00 mol dm⁻³. The product, which contained an excess of ammonia, was then shaken with 300 cm³ of trichloromethane. A 50.0 cm³ portion of the trichloromethane layer neutralised 22.5 cm³ of hydrochloric acid of concentration 0.0250 mol dm⁻³. The partition coefficient

$$\frac{[\text{Free ammonia in water}]}{[\text{Free ammonia in trichloromethane}]} = 25.0$$

at the temperature of the experiment. Deduce the formula of the $Cu(NH_3)_x^{2+}$ ion.

8.7 PARTITION CHROMATOGRAPHY

8.7.1 THE THEORY

Chromatographic separations use repeated partition of solutes between two solvents

It has been shown in an example, p. 167 and in Question 3 above, that repeated extractions with organic solvents are effective in removing a considerable quantity of solute from an aqueous solution. The technique of repeated extractions can be applied to the separation of a number of solutes in a solution, provided that the solutes differ in their solubility in a second solvent. Imagine that solutes **S** and **T** are dissolved in water and that **S** is more soluble than **T** in a second solvent **X**. Let the partition coefficient of **S** between **X** and water be 10/1 and that of **T** between **X** and water be 1/10. When an aqueous solution of 10 g of **S** and 10 g of **T** is equilibrated with an equal volume of **X**, the first extraction can be represented by

Water: 10 g **S** + 10 g **T** X: 0 g + 0 g	→	Water: 0.91 g **S** + 9.1 g **T** X: 9.1 g **S** + 0.91 g **T**

If the solvent **X** is now replaced by a fresh portion of **X**, the second extraction can be represented by

Water: 0.91 g **S** + 9.1 g **T** X: 0 g + 0 g	→	Water: 0.083 g **S** + 8.3 g **T** X: 0.83 g **S** + 0.83 g **T**

After a few successive extractions, there will be very little **S** left in the aqueous solution, which will become a solution of almost pure **T**.

In column
chromatography...

In partition chromatography, many extractions are performed in succession in one operation. The solutes are partitioned between a **stationary phase** and a **mobile phase**. The stationary phase is a solvent (often water) **adsorbed** (bonded to the surface) on a solid. This may be paper or a solid such as alumina or silica gel, which has been packed into a column or spread on a glass plate. The mobile phase is a second solvent which trickles through the stationary phase.

...the stationary phase is
adsorbed on a solid which
packs the column...

8.7.2 COLUMN CHROMATOGRAPHY

The *column* is a glass tube packed with an **adsorbent** (e.g., alumina, silica, starch, magnesium silicate). The solid adsorbent is made into a slurry with water, poured in and allowed to settle. [See Figure 8.17.] A solution of the mixture to be analysed is poured on to the top of the column so that the components can be adsorbed on the column. The second solvent, called the **eluant**, is allowed to trickle slowly through the column.

...The moving phase is a
second solvent...

FIGURE 8.17 Column
Chromatography

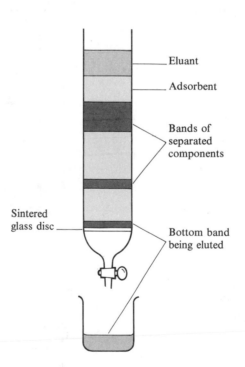

Eluant

Adsorbent

Bands of
separated
components

Sintered
glass disc

Bottom band
being eluted

...Solutes partition
between the solvents...

Partition of the solutes takes place between the moving solvent and the stationary adsorbed water, and the eluant carries the solutes with it, to be readsorbed further down the column. The rates at which the solutes travel down the column depend on their partition coefficients. If these are sufficiently different then, provided the eluant is kept flowing continuously, a complete separation is achieved, and the components become spread out along the column in order of their partition coefficients. The separated bands are called a **chromatogram**. The components can be recovered by dissolving them separately out of the column – **eluting** them – with more solvent, and then evaporating off the solvent.

...The rate at which a
solute travels down the
column depends on its
partition coefficient

8.7.3 PAPER CHROMATOGRAPHY

In paper chromatography...

...the stationary phase is water adsorbed on paper...

...The solvent ascends...

In paper chromatography, the principle is the same as for column chromatography, but the adsorbent material is *paper*. A solution of the mixture to be separated is applied to a strip of chromatography paper. This is hung in a glass tank so that the end dips into solvent in the bottom of the tank [see Figure 8.18]. The solvent may be water, ethanol, glacial ethanoic acid, butanol, etc., or a mixture of any of these. As the solvent rises through the paper, it meets the sample, and the component bands spread out. The separation is stopped when the solvent has travelled nearly to the top of the paper. The distance travelled by the **solvent front** is measured. Then for each component, the **R$_F$ value** is calculated, by

$$R_F = \frac{\text{Distance travelled by the component}}{\text{Distance travelled by the solvent front}}$$

The R$_F$ value can be used to assist in identifying the components of a mixture. The separated components can be obtained by cutting the paper into strips and dissolving out each compound.

FIGURE 8.18 Paper Chromatography

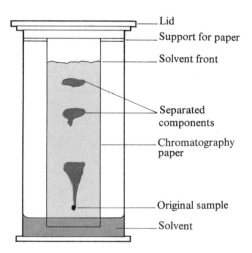

FIGURE 8.19 Thin Layer Chromatography

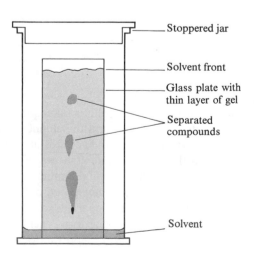

...or descends through the paper

There is a different design of chromatography tank, with a trough at the top of the tank to hold the solvent. The paper hangs down from this trough, and solvent descends through the paper. The two techniques are naturally described as **ascending** and **descending paper chromatography**.

8.7.4 THIN LAYER CHROMATOGRAPHY

A thin layer of adsorbent solid can be used for chromatography

The solid adsorbent may be in the form of a *thin layer* on the surface of a glass plate. The adsorbent (e.g., silica gel or calcium sulphate) is made into a thick paste and spread evenly over a glass plate. The thin layer of paste is allowed to dry and baked in an oven. Spots of the mixture are applied and a chromatogram is developed in the same manner as for paper chromatography. [See Figure 8.19.] Thin layer chromatography has the advantage that a variety of different adsorbents can be used. Since the thin layer is more compact than paper, more equilibrations take place while the solvent front travels over the same distance. Often separations can be achieved in a few centimetres, and coated microscope slides are frequently used for thin layer chromatography.

8.7.5 GAS–LIQUID CHROMATOGRAPHY

In gas–liquid chromatography, the stationary phase is a liquid adsorbed on a solid...

...The mobile phase is a stream of carrier gas

In gas–liquid chromatography, a gas (or volatile liquid or solid) is separated into its components by equilibration with a liquid. The liquid is the stationary phase. It is spread on the surface of inert solid particles which pack the column. Long (5 to 10 m) and narrow (2 to 10 mm bore), the column is often wound into a coil. The mobile phase is a stream of 'carrier gas', nitrogen or a noble gas. When a sample of volatile mixture is injected into the carrier gas, each component sets up a partition equilibrium between the vapour phase and the liquid phase. Some components are less soluble in the liquid phase than others and emerge from the column ahead of the others. A detector records each component as it leaves the column. [See Figure 8.20.] The emerging gases can also be analysed by a mass spectrometer [p. 11].

FIGURE 8.20
Gas–liquid Chromatography

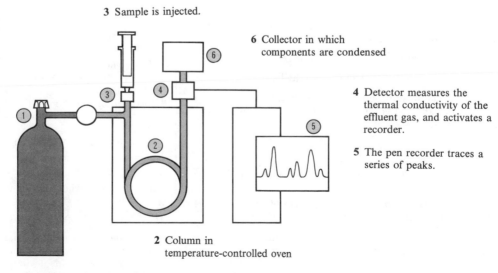

3 Sample is injected.

6 Collector in which components are condensed

4 Detector measures the thermal conductivity of the effluent gas, and activates a recorder.

5 The pen recorder traces a series of peaks.

2 Column in temperature-controlled oven

1 Cylinder of carrier gas with valve to control the flow rate

8.7.6 ION EXCHANGE

Ion exchange resins
replace cations by
$H^+(aq)$...

...and anions by
$OH^-(aq)$

Ion exchange is a type of partition of ionic compounds. The **ion exchange resin** is a polymer which contains at intervals polar groups which can remove undesirable cations (or anions) and replace them with other cations (or anions). The water softener Permutit®, is sodium aluminium silicate, which replaces calcium and magnesium ions in hard water by sodium ions. To purify water, a resin must replace all the cations and anions present by hydrogen ions and hydroxide ions. A combination of a **cation exchanger** and an **anion exchanger** is needed. Cation exchange resins often contain sulphonic acid groups, $-SO_3{}^-H^+$, and anion exchangers often contain quaternary ammonium groups, e.g., $-\overset{+}{N}(CH_3)_3OH^-$. If water passes slowly through a **deioniser**, equilibrium is set up at each level between hydrogen ions attached to the resin and hydrogen ions in solution:

$$-SO_3{}^-H^+(\text{resin}) + Na^+(aq) \rightleftharpoons -SO_3{}^-Na^+(\text{resin}) + H^+(aq)$$

As the water moves on, two of the components in the equilibrium ($H^+(aq)$ and $Na^+(aq)$) are removed and a fresh equilibrium is established. Each successive equilibration increases the replacement of metal cations by hydrogen ions. At the same time, hydroxide ions are replacing other anions. If tap water is run slowly through an ion exchange resin, 'deionised' water of high purity can be obtained.

QUESTIONS ON CHAPTER 8

1. Nitric acid ($T_B = 87\,°C$) and water form a constant boiling mixture, of $T_B = 122\,°C$ and composition 65% by mass nitric acid.

(a) Draw the boiling temperature–composition curve for nitric acid and water.

(b) State what is meant by a constant boiling mixture.

(c) State Raoult's Law. Does this mixture show a positive or negative deviation from this law?

(d) What interaction between nitric acid and water could give rise to this type of deviation?

(e) What happens when a mixture containing 30% by mass of nitric acid is distilled?

2. Explain the principles underlying steam distillation. For what types of compound is it useful?

3.   Mixture A = water + nitrobenzene

    Mixture B = benzene + methylbenzene

State the relationship between the total vapour pressure of the mixture and the vapour pressures of the components in both mixtures.

4. Explain

(a) how liquids mix

(b) why some of a liquid evaporates below its boiling temperature

(c) how the vapour pressure of a system of two immiscible liquids varies with (i) composition and (ii) temperature.

5. Describe practical methods which you have used in the laboratory for column chromatography, paper chromatography and thin-layer chromatography. Give examples of mixtures which can be separated by these methods.

6. Discuss the physical principles involved in

(a) the ethoxyethane extraction of a product from aqueous solution

(b) column chromatography

(c) the separation of two immiscible liquids.

7. Calculate the percentage by volume composition of the vapour over a mixture of 64 g of methanol and 46 g of ethanol at 330 K. The vapour pressures/$N\,m^{-2}$ at this temperature are $CH_3OH = 8.1 \times 10^4$; $C_2H_5OH = 4.5 \times 10^4$.

8. The saturated vapour pressures of phenylamine and water at a number of temperatures are shown below.

TABLE 8.1

Temperature/°C	70	80	90	100
Phenylamine: vapour pressure/$10^5\,N\,m^{-2}$	0.0141	0.0242	0.0384	0.0606
Water: vapour pressure/$10^5\,N\,m^{-2}$	0.311	0.473	0.679	1.010

(a) By means of a graph, find the temperature at which a mixture of phenylamine and water boils at $1.01 \times 10^5\,N\,m^{-2}$ (1 atm).

(b) Calculate the percentage of phenylamine in the distillate.

9. Compare the fractional distillation of an ideal mixture, such as hexane ($T_b = 69\,°C$) and heptane ($T_b = 98.5\,°C$) with that of a non-ideal mixture, such as ethanol ($T_b = 78\,°C$) and water. Draw boiling temperature–composition curves for the mixtures.

10. Define the term *partition coefficient*. A compound **Z** has a partition coefficient of 4.00 between ethoxyethane and water. Calculate the mass of **Z** extracted from 100 cm³ of an aqueous solution of 4.00 g of **Z** by two successive extractions with 50 cm³ of ethoxyethane.

11. Partition coefficients of solutes between immiscible liquids are very seldom equal to one. Explain why you think this is so.

A certain mass of iodine was shaken with carbon disulphide and an aqueous solution containing $0.300 \, mol \, dm^{-3}$ of potassium iodide. On titrating with thiosulphate solution, it was found that the carbon disulphide phase contained $32.3 \, g \, dm^{-3}$ of iodine and the aqueous phase gave a titration equivalent to $1.14 \, g \, dm^{-3}$ of iodine. The partition coefficient for iodine between carbon disulphide and water is 585. Calculate the equilibrium constant for the reaction

$$I_2 + I^- \rightleftharpoons I_3^-$$

$[A_r(I) = 127]$.

Use the principles of the repulsion of electron pairs to suggest a structure for the I_3^- ion. Suggest the highest value of x in the interhalogen compound IF_x.

(WJEC 82)

12. In Figure 8.21, the curve **XY** represents the observed total pressure of the vapour in equilibrium with mixtures of liquids **A** and **B** at constant temperature.

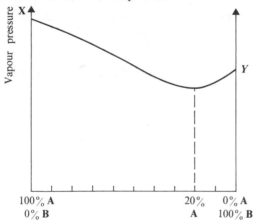

FIGURE 8.21 Composition by mass

(*a*) Sketch the diagram and add to it curves representing

 (i) the change in partial pressure of **A**, in the mixture labelled (i);

 (ii) the change in partial pressure of **B**, in the mixture labelled (ii);

 (iii) the sum of the partial pressures of **A** and **B**, assuming that ideal solutions are formed, labelled (iii).

(*b*) Sketch a second graph to illustrate the variation in composition of vapour and liquid mixtures of **A** and **B** in equilibrium at their boiling points. Use this graph to explain what happens if mixtures of the following compositions are fractionally distilled.

 (i) 80% **A**.

 (ii) 20% **A**.

 (iii) 10% **A**.

†(*c*) Which of the following pairs of chemicals would exhibit behaviour corresponding to the liquids **A** and **B** above? Give reasons for your choice.

 (i) Propan-1-ol and propan-2-ol.

 (ii) Hydrogen chloride and water.

 (iii) Propanone and ethyl ethanoate.

(*d*) How could a knowledge of the enthalpy change accompanying the mixing of **A** and **B** influence your choice in (*c*)?

[For (*c*) see Part 4, for (*d*) see Chapter 10.]

[AEB 81]

13. Copy the graph shown in Figure 8.22, which represents the partial vapour pressures of the miscible liquids dioxane and water above various mixtures of the two liquids at 35 °C.

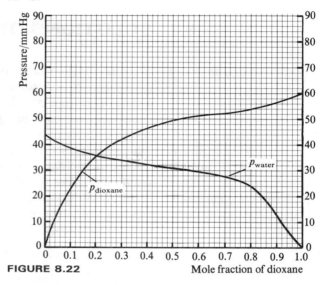

FIGURE 8.22 Mole fraction of dioxane

The structure of dioxane is:

(*a*) (i) Using the values on the graph, plot accurately, on the same axes, a line which would correspond to the variation of total vapour pressure with composition for any mixture of the two liquids. Label this line 'Line A'.

(ii) Also on the same axes, draw accurately a line to represent the variation of total vapour pressure with composition for any mixture of the two liquids, had the liquids behaved ideally and obeyed Raoult's Law. Label this line 'Line B'.

(*b*) What would be the actual vapour pressure of water over a mixture containing a mole fraction of 0.8 of water and 0.2 of dioxane?

(*c*) What would be the actual total vapour pressure over a mixture containing 9 g of water and 132 g of dioxane? (Relative atomic masses: H = 1, C = 12, O = 16)

(*d*) What type of interaction predominates between molecules in:

 (i) pure water?

 (ii) pure dioxane?

(*e*) Explain, in terms of intermolecular forces, the deviation from Raoult's Law that occurs in a mixture of water and dioxane.

(L81, N)

9

SOLUTIONS

9.1 SOLUTIONS OF SOLIDS IN LIQUIDS

Definition of a saturated solution

A solution that contains as much solute as can be dissolved at that temperature in the presence of undissolved solute is called a **saturated** solution. In a saturated solution, a state of dynamic equilibrium exists, with particles of solid constantly dissolving, and solute constantly being deposited. The quantity of solid that dissolves in a certain quantity of liquid to form a saturated solution at a stated temperature is the **solubility** of the solid. There are various ways of expressing solubility, e.g., g solute dm^{-3} solution or mol dm^{-3}; g solute kg^{-1} solvent or mol kg^{-1} and mole fraction : mol solute/(mol solute + mol solvent). The solubility of a solute can be found by (a) titrimetric analysis, (b) evaporating a known mass of solution to dryness and weighing the solute that remains, (c) conductimetric analysis [p. 247] and (d) other physical techniques.

Solubility can be expressed in various units . . .

. . . There are a number of methods of measurement

Figure 9.1 shows a **solubility curve**, a graph of solubility against temperature.

FIGURE 9.1
A Solubility Curve

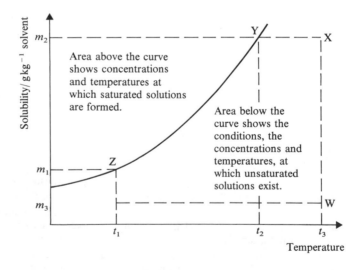

A solubility curve shows the effect of temperature on solubility . . .

. . . and enables one to predict the quantity of solute that will crystallise when a solution is cooled

A solution at point **X** contains m_2 g of solute in 1 kg of solvent at a temperature t_3. It is an unsaturated solution, and when it is cooled it does not deposit solute until it reaches a temperature t_2. At t_2, solute is deposited, and the concentration of the solution decreases. On further cooling, more solute is deposited, and the concentration of the solution again decreases. As the temperature falls, the concentration follows the curve **YZ**. At the temperature t_1, the mass of solute that can remain in solution is m_1 g. The mass of solute that has crystallised between t_2 and t_1 is $(m_2 - m_1)$ g. A solution **W**, containing m_3 g of solute in 1 kg of solvent at t_3, can be cooled from t_3 to t_1 without depositing any solute as the solubility is never exceeded.

9.2 RECRYSTALLISATION

Recrystallisation of a solute is used as a method of purification

Solids are purified by **recrystallisation** from a suitable solvent, e.g., water, ethanol, propanone. The impure material is dissolved in the minimum quantity of hot solvent. The hot solution is filtered to remove insoluble impurities. A heated Buchner funnel and flask [see Figure 9.2] are used to prevent cooling of the solution and crystallisation of the solute. The filtered solution is allowed to cool so that the solute crystallises out. The impurities are present in smaller quantities and remain in solution at the lower temperature. The crystals are filtered, washed in the funnel with a little cold solvent, and dried. The purity of a solid is assessed by finding its melting temperature. [See p. 724 and Figure 34.3, p. 725.]

FIGURE 9.2 Filtration under Reduced Pressure

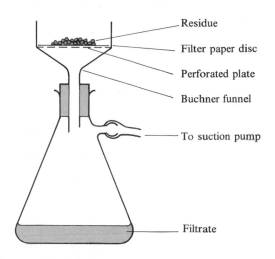

Residue

Filter paper disc

Perforated plate

Buchner funnel

To suction pump

Filtrate

CHECKPOINT 9A: SOLUBILITY

1. A saturated solution of $CaSO_4(aq)$ has some undissolved $CaSO_4(s)$ lying at the bottom of the container. A little $Ca^{35}SO_4$ is mixed with the undissolved $Ca^{32}SO_4$. The isotope ^{35}S is radioactive, and, after a while, both the solution of $CaSO_4(aq)$ and the undissolved $CaSO_4(s)$ are found to be radioactive. Explain how the solution has been able to take up $CaSO_4$ in spite of being saturated.

9.3 LOWERING OF THE VAPOUR PRESSURE OF A SOLVENT BY A SOLUTE

For a solution of two liquids **A** and **B**, Raoult's Law [p. 157] states

$$p = x_A p_A^0 + x_B p_B^0$$

where x_A = mole fraction and p_A^0 = saturated vapour pressure of pure **A** and so on. In a solution of a solid **B** in a liquid **A**, if the solid is involatile, i.e., p_B^0 is negligible, then

A solute lowers the saturated vapour pressure of a solvent...

$$p = x_A p_A^0$$

Since x_A is a fraction, $p < p_A^0$.

The presence of a solute has lowered the saturated vapour pressure of the solvent. It follows that the lowering of the vapour pressure, $p_A^0 - p$, is equal to the mole fraction of the solute. This conclusion holds for ideal solutions, and can be expressed by

$$\frac{p^0 - p}{p^0} = \frac{n_1}{n_1 + n_2}$$

p^0 = vapour pressure of pure solvent

p = vapour pressure of solution

n_1 = number of moles of solute

n_2 = number of moles of solvent

...The molar mass of a solute can be found from the lowering of the vapour pressure that it produces

In dilute solutions $n_2 \gg n_1$ and the expression becomes

$$\frac{p^0 - p}{p^0} = \frac{n_1}{n_2}$$

Notice that the relative lowering of the saturated vapour pressure depends on the molar concentration of the solute, not on its identity. Substituting $n = m/M$, where m = mass, M = molar mass, gives

$$\frac{p^0 - p}{p^0} = \frac{m_1/M_1}{m_2/M_2} = \frac{m_1 M_2}{m_2 M_1}$$

If the masses of solute and solvent are known, and the molar mass of the solvent is known, this expression can be used to give M_1, the molar mass of the solute.

Example   A solution of 100 g of solute in 1.00 dm³ of water has a vapour pressure of 2.27×10^3 Nm^{-2} at 20 °C. The saturated vapour pressure of water at 20 °C is 2.34×10^3 Nm^{-2}. Calculate the molar mass of the solute.

$$\frac{p^0 - p}{p^0} = \frac{n_1}{n_2} = \frac{m_1}{m_2} \times \frac{M_2}{M_1}$$

$$\frac{2.34 - 2.27}{2.34} = \frac{100}{1000} \times \frac{18}{M_1} \quad \therefore \quad M_1 = \frac{18 \times 2.34}{0.07 \times 10} = 60\,\text{g mol}^{-1}$$

Properties such as vapour pressure lowering, which depend on the concentration of dissolved particles and not on their nature, are called *colligative properties*.

CHECKPOINT 9B: VAPOUR PRESSURE LOWERING

1. The vapour pressure of benzene is 9.97×10^3 Nm^{-2} at 20 °C. What is the vapour pressure of a solution of 12.8 g of naphthalene, $C_{10}H_8$, in 100 g of benzene?

2. A solution of 20.0 g of glycerol in 100 g of water has a vapour pressure at 25 °C of 3.04×10^3 Nm^{-2}. Given that the vapour pressure of water at this temperature is 3.17×10^3 Nm^{-2}, calculate the molar mass of glycerol.

9.4   ELEVATION OF BOILING TEMPERATURE

A solute raises the temperature at which a solvent boils

Since the presence of a solute decreases the vapour pressure of a solvent, it also increases the temperature at which the solvent boils. Figure 9.3 shows the relationship between the lowering of the vapour pressure and the **elevation of the boiling temperature** (or boiling point).

FIGURE 9.3 Vapour
Pressure Curves for a
Pure Solvent and for a
Solution of an Involatile
Solute in the Solvent

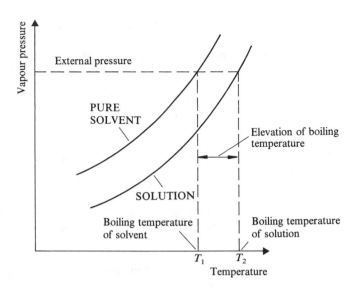

FIGURE 9.3 Vapour Pressure Curves for a Pure Solvent and for a Solution of an Involatile Solute in the Solvent

9.4.1 MEASUREMENT OF BOILING TEMPERATURE ELEVATION

An experimental method for finding the molar mass of a solute from T_b elevation

The boiling temperatures (T_b) of solutions are easier to measure than their vapour pressures. The elevation of the boiling temperature of a solvent produced by a solute can be used to give the molar mass of the solute. Figure 9.4 shows an apparatus which can be used. It employs a special type of thermometer invented by E Beckmann. The capillary tubing is so thin that 20 mm of mercury indicate 1 K, and the range of the thermometer is 5 K. A reservoir of mercury at the top of the thermometer is adjusted to make the thermometer read over the required 5 K of temperature.

FIGURE 9.4 The
Method of Measuring
Boiling Temperature
Elevation devised by
Landsberger

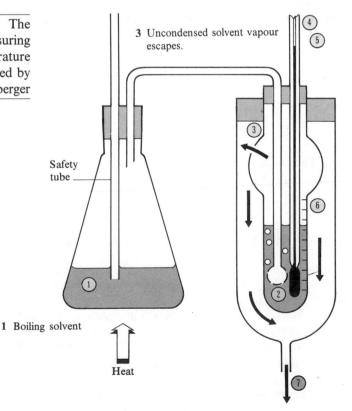

3 Uncondensed solvent vapour escapes.

Safety tube

1 Boiling solvent

Heat

4 Beckmann thermometer registers T_b of solvent, t_1.

5 After T_b of the solvent has been found, a weighed amount of the solute, m_1, is dissolved in it. The procedure is repeated, and T_b of the solution, t_2, is noted.

6 The graduations enable the volume of the solution, v, to be measured.

2 Vapour of boiling solvent condenses, releasing enthalpy of vaporisation and heating the solvent in the apparatus until it boils.

7 Condensed solvent vapour

9.4.2 BOILING TEMPERATURE OR EBULLIOSCOPIC CONSTANT

The boiling temperature elevation is proportional to the concentration of solute and to a constant which has a certain value for each solvent. The elevation produced by 1 mole of solute in 1 kg of water is 0.52 K. This value is called the **ebullioscopic constant** (or boiling temperature constant) for water:

Definition of ebullioscopic
(boiling temperature)
constant

> 1 mole of solute in 1 kg of water raises the boiling temperature by 0.52 K

For other solutions, the boiling temperature elevation depends on the concentration of solute and the magnitude of the ebullioscopic constant for the solvent:

$$\begin{matrix} \text{Boiling} \\ \text{temperature} \\ \text{elevation} \end{matrix} = \begin{matrix} \text{Boiling} \\ \text{temperature} \\ \text{constant} \end{matrix} \times \frac{\text{Amount of solute/mol}}{\text{Mass of solvent/kg}} = k \times \frac{m}{M} \times \frac{1}{W}$$

Method of finding molar
mass from boiling
temperature elevation . . .

where k = the ebullioscopic constant for the solvent (or boiling temperature constant), m = mass and M = molar mass of the solute and W = mass of the solvent.

If W is expressed in kg, then

$$\text{Boiling temperature elevation/K} = k \times \frac{m/\text{g}}{M/\text{g mol}^{-1}} \times \frac{1}{W/\text{kg}}$$

and k has the units K kg mol^{-1}.

Example A solution of 5.00 g of **X** in 100.0 g of water boiled at 100.42 °C. Calculate the molar mass of **X**.

Method Boiling temperature elevation $= k \times \dfrac{m}{M} \times \dfrac{1}{W}$

$$0.42 = 0.52 \times \frac{5.00}{M} \times \frac{1}{0.100}$$

The molar mass $M = 62 \text{ g mol}^{-1}$

CHECKPOINT 9C: BOILING TEMPERATURE ELEVATION

(Take the boiling temperature constant for water = $0.52 \text{ K kg mol}^{-1}$ and for benzene = $2.53 \text{ K kg mol}^{-1}$)

1. Find the molar masses of the solutes in the following solutions:

(a) 25.0 g of **W** in 100 g of water boils at 102.167 °C
(b) 50.0 g of **X** in 1.00 dm³ of water boils at 100.173 °C
(c) 12.5 g of **Y** in 250 cm³ of water boils at 100.144 °C
(d) 20.5 g of **Z** in 300 g of water boils at 100.573 °C

2. If 8.71 g of a compound dissolved in 175 g of benzene raises the boiling temperature by 1.05 °C, what is the molar mass of the compound?

3. Carbon disulphide boils at 46.13 °C and has an ebullioscopic constant of $2.34 \text{ °C kg mol}^{-1}$. If a solution of 3.82 g of sulphur in 100 g of carbon disulphide boils at 46.48 °C, what is the molecular formula of sulphur in this solvent?

9.5 DEPRESSION OF THE FREEZING TEMPERATURE OF A SOLVENT BY A SOLUTE

A solution has a lower freezing temperature (or freezing point) than the pure solvent. Figure 9.5 shows how the **depression of the freezing temperature** arises as a consequence of the lowering of the vapour pressure of the solvent by the solute.

T_1 is the freezing temperature of the solvent, i.e., the temperature at which its vapour pressure–temperature curve intersects that of the solid solvent. T_2 is the freezing temperature of the solution. Since the vapour pressure of the solution is lower than that of the pure solvent, T_2 is lower than T_1: the solute has depressed the freezing temperature of the solvent.

FIGURE 9.5 Vapour Pressure Curves for a Solvent, a Solid Solvent and a Solution

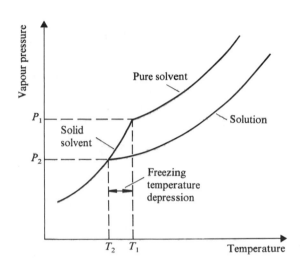

9.5.1 CRYOSCOPIC OR FREEZING TEMPERATURE CONSTANT

Freezing temperature depression is another **colligative property**, i.e., it depends on the *concentration* of particles in solution and not on their nature. The depression of the freezing temperature of 1 kg of solvent produced by 1 mole of solute particles is constant for a particular solvent.

Definition of cryoscopic (freezing temperature) constant

> 1 mole of solute in 1 kg of water lowers the freezing temperature by 1.86 K

The molar depression of the freezing temperature is called the **cryoscopic constant** (or freezing temperature constant) of the solvent.

> $$\begin{array}{l} \text{Freezing} \\ \text{temperature} \\ \text{depression} \end{array} = \begin{array}{l} \text{Freezing} \\ \text{temperature} \\ \text{constant} \end{array} \times \frac{\text{Amount of solute/mol}}{\text{Mass of solvent/kg}} = k \times \frac{m}{M} \times \frac{1}{W}$$

where m = the mass and M = the molar mass of the solute, W = the mass of solvent in kg and k is the cryoscopic constant (freezing temperature constant) which has the units $K\,kg\,mol^{-1}$.

9.5.2 MEASUREMENT OF FREEZING TEMPERATURE DEPRESSION

Determining molar mass from the freezing temperature depression

Measurements of freezing temperature depression can be used to calculate molar masses of solutes in the same way as boiling temperature elevations. The Beckmann method for measuring freezing temperatures, (T_f), is shown in Figure 9.6.

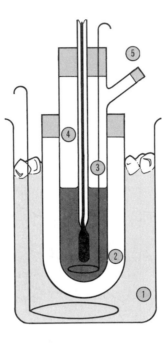

FIGURE 9.6 The Beckmann Apparatus for Freezing Temperature Determination

5 The inner tube is warmed to melt the frozen solvent. A weighed amount of solid is introduced through the side-arm, and allowed to dissolve. The T_f of the solution is found.

4 Beckmann thermometer records T_f of solvent.

3 Stirrer. When the solvent begins to freeze, it is stirred vigorously. The temperature rises a little, and becomes steady at the T_f.

2 Air jacket. This avoids super-cooling of the solvent – cooling below the T_f without freezing.

1 Cooling bath of ice and salt not $> 5\,\mathrm{K}$ below the freezing temperature (T_f) of the solvent

Example A solution of 5.12 g of **Y** in 100 g of water froze at $-0.280\,^\circ\mathrm{C}$. If the cryoscopic constant for water is $1.86\,\mathrm{K\,kg\,mol^{-1}}$, what is the molar mass of **Y**?

Method

$$\text{Freezing temperature depression} = k \times \frac{m}{M} \times \frac{1}{W}$$

$$0.280 = 1.86 \times \frac{5.12}{M} \times \frac{1}{0.100}$$

The molar mass $M = 340\,\mathrm{g\,mol^{-1}}$.

9.5.3 THE MOLTEN CAMPHOR METHOD

Large freezing temperature depressions are observed in molten camphor

Rast introduced the use of molten camphor as a solvent in measurements of freezing temperature depression. As camphor melts at 177 °C, no freezing mixture is needed. The value of the cryoscopic constant is large: it is $40\,\mathrm{K\,kg\,mol^{-1}}$, and the freezing temperature depressions produced by solutes are large enough to be measured with an ordinary thermometer. The method is economical. A mixture of, say, 0.05 g of sample and 0.5 g of camphor could be weighed accurately and then homogenised by melting. After cooling, a portion of the solid mixture could be put into a melting temperature tube, such as that shown in Figure 34.3, p. 725, and its melting temperature found. [See Question 3, p. 183.]

1. (Take the cryoscopic constant for water =
$1.86\,K\,kg\,mol^{-1}$.)
Find the molar masses of the solutes in the following
solutions:

(*a*) 11.0 g of **A** in 100 g of water freezes at $-3.300\,°C$

(*b*) 25.0 g of **B** in 250 g of water freezes at $-3.207\,°C$

(*c*) 22.5 g of **C** in 500 g of water freezes at $-0.244\,°C$

(*d*) 100.0 g of **D** in 1 kg of water freezes at $-2.222\,°C$

2. A silver dollar melts at 875 °C. If it were pure silver, it
would melt at 960 °C. Analysis shows that the only significant
impurity is copper. If the cryoscopic constant for silver is
$48.6\,K\,kg\,mol^{-1}$, what is the percentage by mass of copper
in the coin?

3. A chemist has a small sample of a new compound and
wants to find its molar mass as economically as possible.
She weighs $8.35 \times 10^{-3}\,g$ of the compound, mixes it with
0.1530 g of camphor, and puts the mixture into a melting
temperature tube. In the apparatus shown in Figure 34.3,
p. 725 she finds the melting temperature to be 165.9 °C. If
the melting temperature and cryoscopic constant of camphor
are 179.5 °C and $40.0\,°C\,kg\,mol^{-1}$ respectively, what is the
molar mass of the compound?

9.6 OSMOTIC PRESSURE

*Osmosis is the passage of
solvent from a
concentrated solution to a
more dilute solution*

A **semipermeable membrane** is a film of material which can be penetrated by a solvent
but not by a solute. When two solutions are separated by a semipermeable membrane,
solvent passes from the more dilute to the more concentrated. This phenomenon is
called **osmosis**. Examples of semipermeable membranes are animal and plant cell
walls. Artificial semipermeable membranes are formed by allowing two solutions to
meet in the pores of a porous material.

Copper(II) salts and potassium hexacyanoferrate(II) react in the pores of a porous
pot to produce a semipermeable membrane of copper(II) hexacyanoferrate(II),
$Cu_2Fe(CN)_6$.

*Definition of osmotic
pressure*

When a solution and its solvent are separated by a semipermeable membrane, the
pressure which must be applied to the solution to prevent the solvent from entering
is called the **osmotic pressure** of the solution. There is an analogy with gas pressures.
One mole of a solid **A**, when vaporised, occupies a volume of $22.4\,dm^3$ at 0 °C and
$1.01 \times 10^5\,N\,m^{-2}$. One mole of **A** dissolved in $22.4\,dm^3$ of solvent at 0 °C exerts an
osmotic pressure of $1.01 \times 10^5\,N\,m^{-2}$.

*The osmotic pressure
equation resembles the
Ideal Gas Equation*

The expression which relates osmotic pressure to concentration and temperature is
similar to the Ideal Gas Equation:

$$\pi V = nRT$$

where π = osmotic pressure, V = volume, T = temperature/K, n = amount of
solute/mol, and R = a constant which has the same value as the gas constant,
$8.314\,J\,K^{-1}\,mol^{-1}$. This equation, the van 't Hoff equation, is obeyed by ideal solutions.

9.6.1 MEASUREMENT OF OSMOTIC PRESSURE

The osmotic pressure of a solution depends on the molar concentration of solute
present: it is a colligative property. Measurements of osmotic pressure can be used
to give the molar masses of solutes.

Figure 9.7 shows the Berkeley and Hartley method for the measurement of osmotic
pressure.

Example   Calculate the molar mass of **Z**, given that a solution of 60.0 g of **Z** in
$1.00\,dm^3$ of water exerts an osmotic pressure of $4.31 \times 10^5\,N\,m^{-2}$ at 25 °C.

FIGURE 9.7 Berkeley and Hartley Method for the Measurement of Osmotic Pressure

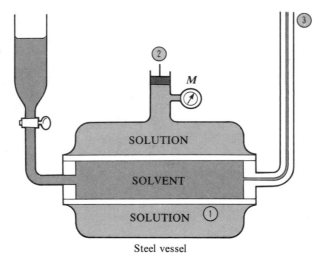

Steel vessel

3 Capillary tube indicates movement of solvent into the solution.

2 Pressure is applied to the solution. It is adjusted until there is no flow of solvent into the solution.

1 Cylindrical tube is porous. In the pores, a semipermeable membrane has been laid down, e.g. $Cu_2Fe(CN)_6$.

Method

A calculation of molar mass from osmotic pressure

$$\pi V = nRT$$

$$\pi = 4.31 \times 10^5 \, \text{N m}^{-2}$$

$$V = 1.00 \times 10^{-3} \, \text{m}^3$$

$$R = 8.314 \, \text{J K}^{-1} \text{mol}^{-1}$$

$$T = 298 \, \text{K}$$

$$\therefore \quad 4.31 \times 10^5 \times 1.00 \times 10^{-3} = \frac{60.0}{M} \times 8.314 \times 298$$

The molar mass $M = 345 \, \text{g mol}^{-1}$

9.6.2 OSMOSIS AND PLANT CELLS

The plasma membrane in a plant cell is a semipermeable membrane. The fluid inside the cell exerts an osmotic pressure. Figure 9.8 shows what happens when a plant cell is placed in solutions of different osmotic pressures. Osmosis is of vital importance in the life of plants, being the means by which the plant roots take in water.

FIGURE 9.8 A Plant Cell

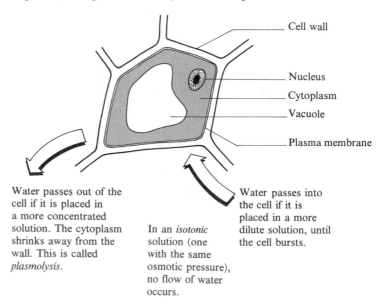

Cell wall

Nucleus

Cytoplasm

Vacuole

Plasma membrane

Water passes out of the cell if it is placed in a more concentrated solution. The cytoplasm shrinks away from the wall. This is called *plasmolysis*.

In an *isotonic* solution (one with the same osmotic pressure), no flow of water occurs.

Water passes into the cell if it is placed in a more dilute solution, until the cell bursts.

9.6.3 OSMOSIS AND BLOOD

Red blood corpuscles behave in the same way. Any solution which is injected into the bloodstream must be **isotonic** with blood, i.e., have the same osmotic pressure.

9.6.4 COMPARISON OF COLLIGATIVE PROPERTIES

Osmotic pressure measurements are used for molar mass determinations on high molar mass solutes...

Measurements of freezing temperature depression and boiling temperature elevation are time-consuming. For substances of high molar mass and limited solubility, these methods are unreliable as the temperature differences are too small to be measured accurately. For substances of high molar mass and limited solubility, these methods are unreliable as the temperature differences are too small to be measured accurately. For instance, a polymer with a molar mass of $5000 \, g \, mol^{-1}$ and a solubility of $10 \, g \, dm^{-3}$ gives a freezing temperature depression of $0.004 \, K$. Osmotic pressure measurement is better here. This solution has an osmotic pressure of $5 \times 10^3 \, N \, m^{-2}$ ($\frac{1}{20} atm$), which can be measured accurately.

...and on temperature-sensitive substances

Osmotic pressure measurements are useful for substances of biological interest. In addition to the greater accuracy, there is the advantage that measurements can be made at room temperature. Many naturally occurring substances, such as proteins, are temperature-sensitive and undergo changes at $0 \, °C$ and $100 \, °C$. Mass spectrometry [p. 11] is the most convenient method for measuring molar masses. Some large molecules, however, fragment and do not give molecular ions. For such compounds, osmotic pressure measurement offers a valuable alternative.

CHECKPOINT 9E: OSMOTIC PRESSURE

1. Calculate (a) the osmotic pressure at $20 \, °C$ of a solution of sucrose, $C_{12}H_{22}O_{11}$, containing $20.0 \, g \, dm^{-3}$ and (b) its boiling temperature.

2. A solution of $2.00 \, g$ of a polymer in $1.00 \, dm^3$ of water has an osmotic pressure of $300 \, N \, m^{-2}$ at $20 \, °C$. Calculate (a) the molar mass of the polymer and (b) the freezing temperature depression in this solution.

3. A solution containing $3.47 \, g$ of **X** in $250 \, cm^3$ of solution at $20 \, °C$ has osmotic pressure $2.06 \times 10^3 \, N \, m^{-2}$. Calculate the molar mass of **X**.

4. What advantages does a measurement of osmotic pressure have over a measurement of the boiling temperature or the freezing temperature of a solution as a method of finding the molar mass of a solute?

9.7 ANOMALOUS VALUES OF MOLAR MASS FROM COLLIGATIVE PROPERTIES

For solutes which ionise, the measured colligative properties are higher than the expected values...

Colligative properties depend on the concentration of particles in solution. If the number of particles increases through dissociation of the solute, the observed colligative property increases. If association occurs, the number of particles is less than expected, and the colligative property is less than expected.

9.7.1 DISSOCIATION

Electrolytes give abnormally high values of colligative properties. J H van 't Hoff discovered this, and the ratio

$$\frac{\text{Observed colligative property}}{\text{Calculated colligative property}}$$

is called the **van 't Hoff factor**, usually represented as *i*.

The reason for the high value of i for electrolytes is dissociation into ions. If a solute, such as sodium chloride, NaCl, is completely dissociated, there are 2 moles of ions for every mole of salt, and the colligative properties have twice the values calculated in the absence of ionisation. If copper(II) nitrate, $Cu(NO_3)_2$, is completely dissociated, there are 3 moles of ions per mole of solute and the colligative properties are three times the values calculated in the absence of ionisation.

...The apparent degree of dissociation can be found

If the solute is incompletely dissociated, the degree of dissociation can be calculated. If a solute **AB** dissociates into two ions, and α is the degree of dissociation, then

| *Species* | **AB** $\rightleftharpoons$ **A**$^+$ + **B**$^-$ | *Total* |

| *Amount* | $(1 - \alpha)$ | α | α | $1 + \alpha$ |

1 mole of **AB** will produce α moles of **A**$^+$ and α moles of **B**$^-$, leaving $(1 - \alpha)$ moles of **AB**. The total number of particles present is $(1 + \alpha)$. Thus

$$\frac{\text{Actual number of particles}}{\text{Expected number of particles}} = \frac{\text{Observed colligative property}}{\text{Calculated colligative property}} = \frac{1 + \alpha}{1}$$

In general, if 1 mole of solute dissociates into n moles of ions

$$\frac{\text{Observed colligative property}}{\text{Expected colligative property}} = \frac{1 + (n - 1)\alpha}{1}$$

It is now believed that strong electrolytes are always completely dissociated in solution. Values of the van 't Hoff factor less than n, where n is the number of ions per formula unit of solute, have been explained in terms of ionic interactions [p. 249]. The value of α calculated from the van 't Hoff factor is referred to as the **apparent degree of dissociation**.

9.7.2 ASSOCIATION

Association of solute molecules gives rise to lower values of colligative properties than expected...

A value for a colligative property lower than the calculated value may indicate **association** of molecules of solute. Consider that n molecules of solute associate to form X_n, and the degree of association is α:

| *Species* | $nX \rightleftharpoons X_n$ | *Total* |

...The degree of association can be found

| *Amount*/mol | $(1 - \alpha)$ | α/n | $1 - \alpha + \dfrac{\alpha}{n} = 1 - \dfrac{(n - 1)\alpha}{n}$ |

and

$$\frac{\text{Observed colligative property}}{\text{Calculated colligative property}} = 1 - \frac{(n - 1)\alpha}{n}$$

If **X** dimerises, $n = 2$ and

$$\frac{\text{Observed colligative property}}{\text{Calculated colligative property}} = 1 - \frac{\alpha}{2}$$

Example (a) A solution of 2.34 g of barium hydroxide in $1.00\,dm^3$ of water freezes at $-0.0726\,°C$. Find the apparent degree of dissociation of the base.

Method

$$\text{Freezing temperature depression} = k \times \frac{m}{M} \times \frac{1}{W}$$

$$= 1.86 \times \frac{2.34}{171} \times \frac{1}{1.00}$$

$$= 0.0254 \,°\text{C}$$

A sample calculation on dissociation The number of ions per formula unit of $Ba(OH)_2$ = 3. Since

$$\frac{\text{Observed } T_f \text{ depression}}{\text{Expected } T_f \text{ depression}} = 1 + (n - 1)\alpha$$

$$\frac{0.0726}{0.0254} = 1 + 2\alpha$$

The apparent degree of dissociation $\alpha = 0.93$.

A sample calculation on association **Example (b)** The dissolution of 5.04 g of ethanoic acid in 100 g of benzene raised the boiling temperature of this solvent by 1.17 K. The boiling temperature constant of benzene is $2.53 \,\text{K kg mol}^{-1}$. What information can you deduce about the molecular condition of ethanoic acid in this solution?

Method

$$\text{Boiling temperature elevation} = k \times \frac{m}{M} \times \frac{1}{W}$$

$$= 2.53 \times \frac{5.04}{60.0} \times \frac{1}{0.100} = 2.13 \,°\text{C}$$

The observed elevation is only just over half the calculated value. The molecules must exist as dimers. The degree of dimerisation can be calculated, using the expression

$$\frac{\text{Observed } T_b \text{ elevation}}{\text{Expected } T_b \text{ elevation}} = 1 - \frac{(n - 1)\alpha}{n}$$

Since the molecules have dimerised, $n = 2$ and

$$\frac{1.17}{2.13} = 1 - \frac{\alpha}{2}$$

The degree of dimerisation $\alpha = 0.90$.

CHECKPOINT 9F: COLLIGATIVE PROPERTIES

Use the following data:

Freezing temperature constant for water	$= 1.86 \,\text{K kg mol}^{-1}$
Boiling temperature constant for water	$= 0.52 \,\text{K kg mol}^{-1}$
Freezing temperature of benzene	$= 5.48 \,°\text{C}$
Freezing temperature constant for benzene	$= 5.12 \,\text{K kg mol}^{-1}$
Boiling temperature constant for benzene	$= 2.53 \,\text{K kg mol}^{-1}$
Boiling temperature of benzene	$= 80.1 \,°\text{C}$

1. (*a*) Explain why 0.1 mol of sodium chloride depresses the freezing temperature of $1 \,\text{dm}^3$ of water twice as much as does 0.1 mol of glucose.

(*b*) Explain why 0.1 mol of ethanoic acid depresses the freezing temperature of 1 kg of benzene half as much as does 0.1 mol of naphthalene.

(*c*) Explain why 0.1 mol of aluminium chloride depresses the freezing temperature of 1 kg of benzene half as much as does 0.1 mol of naphthalene.

2. A solution of 1.00 g of an acid in 100 g of water froze at −0.25 °C. A solution of 1.00 g of the acid in 100 g benzene froze at 5.13 °C. What is the molar mass of the acid?

3. A solution of ethane-1,2-diol, $C_2H_6O_2$, freezes at −10.0 °C. What is the concentration of the solute? This substance is called 'antifreeze', and is used in car radiators. How many kilograms of antifreeze would you have to put into a car radiator containing 5 kg of water to protect the radiator against freezing down to −10 °C?

4. The osmotic pressure of a solution of 1.00 g of chymotrypsin in 100 cm³ of water at 300 K is 994 N m⁻². Calculate the molar mass of chrymotrypsin.

5. The freezing temperature of a liquid decreases when a non-volatile solute is added, yet its boiling temperature

increases. Why is this? If the freezing temperature depression is greater than that calculated from the concentration of solute, what can you deduce about the solute?

6. Which will have the lower freezing temperature, a solution of 100 g of NaCl in 1 dm³ of water or a solution of 100 g of $CaCl_2$ in 1 dm³ of water? Which of these salts is the better de-icing agent, on a weight for weight basis?

7. The cryoscopic constant for ethanoic acid is 3.6 K kg mol⁻¹. A solution of 1.00 g of a hydrocarbon in 100 g of ethanoic acid freezes at 16.14 °C, instead of the usual 16.60 °C. The hydrocarbon contains 92.3% carbon and 7.7% hydrogen. What is its molecular formula?

9.8 COLLOIDS

A colloidal dispersion is a type of solution containing large particles of solute...

Silica, SiO_2, is insoluble in water: the sand on the seashore is evidence of this fact. Yet when an aqueous solution of a silicate is acidified, silica is not precipitated. According to the pH, silica is obtained as a **colloidal dispersion** or a gelatinous precipitate or a solid-like **gel**, in which liquid is trapped. A colloidal dispersion resembles a solution in being clear, apart from a slight opalescence. A colloidal dispersion may contain up to 30% by mass of silica and yet freeze only slightly below 0 °C, indicating that the silica is present as a small concentration of large particles.

9.8.1 OPTICAL PROPERTIES

...The solute particles are large enough to scatter light

The particles of a colloid are not big enough to be visible, but they are big enough to scatter light. This effect, the **Tyndall effect**, was named after its discoverer. It resembles the scattering of light in dust-laden air, as shown in Figure 9.9.

FIGURE 9.9 Scattering of Light by a Colloid

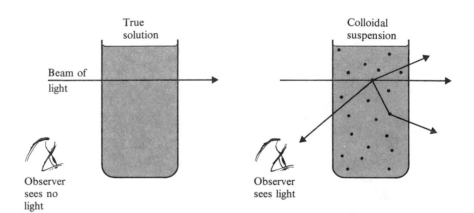

9.8.2 SIZE

Colloidal particles are 100 nm across

Some colloidal particles are illustrated in Figure 9.10, with ions and molecules for comparison.

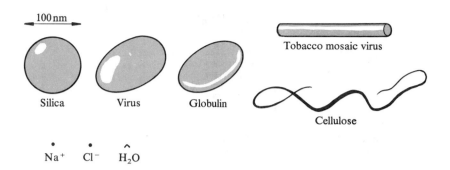

FIGURE 9.10 Size of Colloidal Particles

9.8.3 CHARGE

The reason why colloidal particles do not coalesce is that their surfaces are charged. Colloidal silica particles adsorb hydroxide ions on their surfaces and are negatively charged. Repulsion between like charges keeps the particles in suspension.

Charged colloidal particles migrate in an electric field, as ions do.

9.8.4 PRECIPITATION OF COLLOIDS

Colloids can be precipitated by neutralising their charges

Colloids are precipitated by neutralising the charge on the particles, often by the addition of ions of opposite charge. This is done at one stage in the purification of drinking water, when clay particles and other colloidal suspensions must be removed. Impure water is treated with, e.g., aluminium sulphate. When the negative charges on the clay particles are neutralised by Al^{3+} ions, clay coagulates and settles out of solution.

Colloidal proteins in blood are negatively charged, and styptic pencils, which help blood to clot, contain Al^{3+} ions or Fe^{3+} ions to neutralise the charges on the surfaces of the colloidal particles.

9.8.5 CLASSIFICATION

Colloids are classified as **lyophilic** (solvent-loving; **hydrophilic** if the solvent is water) and **lyophobic** (solvent-fearing; **hydrophobic** if the solvent is water). Table 9.1 lists some types of colloids.

Dispersed phase	Dispersion medium	Type	Example
Liquid	Gas	Aerosol	Fog, mist from aerosol spray can
Solid	Gas	Aerosol	Smoke, dust-laden air
Gas	Liquid	Foam	Soap suds, whipped cream
Liquid	Liquid	Emulsion	Oil in water, milk, mayonnaise
Solid	Liquid	Sol	Clay, colloidal gold
Gas	Solid	Solid foam	Lava, pumice
Liquid	Solid	Solid emulsion	Pearl, opal
Solid	Solid	Solid sol	Some gems, e.g., black diamond, ruby glass

TABLE 9.1 Some Types of Colloids

QUESTIONS ON CHAPTER 9

1. What is a saturated solution? Explain how you could measure the solubility of the crystalline solid, ethanedioic acid-2-water, $(CO_2H)_2 \cdot 2H_2O$, in water at room temperature.

2. Why does the vapour pressure of a liquid increase with temperature? Why does the vapour pressure of a liquid decrease when an involatile solute is dissolved in it?

Assuming that the decrease in vapour pressure is proportional to the concentration of solute, show (with the aid of a diagram) that the increase in the boiling temperature is proportional to the concentration of solute.

3.

	Benzene	*Water*
Freezing temperatures/°C	5.48	0.00
Cryoscopic constants/K kg mol^{-1}	5.12	1.86

TABLE 9.2

(*a*) A solution of 0.1360 g of benzoic acid in 100 g of water freezes at $-0.0207\,°C$. A solution of 2.420 g of the same acid in 100 g of benzene freezes at $4.97\,°C$.

Calculate the molar mass of benzoic acid from the two measurements, and comment on the results.

(*b*) A solution of 0.1255 g of 1-chlorobenzoic acid in 100 g of water freezes at $-0.0150\,°C$. A solution of 2.720 g of the same acid in 100 g of benzene freezes at $4.59\,°C$.

Calculate the molar mass of 1-chlorobenzoic acid from the results in the two solvents. Comment on the values you obtain, and compare the behaviour of the two acids, explaining any difference you find.

4. Explain why osmotic pressure is the only colligative property which offers a practical method for finding the molar masses of polymers with molar masses of the order of $10^5\,g\,mol^{-1}$.

There is a method for finding molar masses which is very much more economical in material and time than the methods which utilise colligative properties. What is it? How much material does this method use? Is it suitable for polymers?

5. (*a*) (i) What is to be found in each of the spaces marked X, Y and Z on Figure 9.11?

(ii) Explain the differences in mercury levels in the three tubes in terms of the properties of dilute solutions.

(iii) What changes, if any, would occur to the mercury levels if the apparatus was placed in a thermostatic bath maintained at a higher temperature? Explain your answer.

(iv) If the pressure in X was reduced by means of a vacuum pump, explain what would happen.

(*b*) Calculate the boiling point of an aqueous solution of urea, $CO(NH_2)_2$, of concentration $12.0\,g\,dm^{-3}$ at a pressure of 101.3 kPa. Assume that the volume of the solute is negligible compared to that of the solution, and the boiling point elevation constant for water is $0.52\,K\,mol^{-1}\,kg$.

(*c*) (i) Write an expression for the mole fraction of a solute in solution.

(ii) Calculate the mole fraction of sodium chloride in an aqueous solution containing 10 g of sodium chloride per 100 g of water.

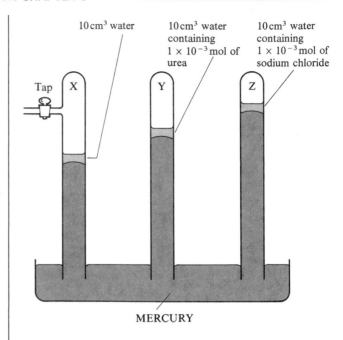

FIGURE 9.11
(Not to scale)   (AEB 81)

6. (*a*) What is meant by the term *colligative properties*? Explain how the measurement of a colligative property can be used to obtain a value of the relative molecular mass of a compound. (Details of apparatus are not required.)

(*b*) Measurement of relative molecular masses by means of colligative properties often gives rise to anomalous results. Discuss the factors which give rise to such results, other than deficiencies in practical techniques.

(*c*) Explain why salt is spread on the roads in winter to prevent formation of ice. Deduce which salt is the more cost effective for this purpose, sodium chloride or calcium chloride.
Cost/1000 kg: sodium chloride £40; calcium chloride £160

(*d*) The relative molecular mass of a protein was determined by measuring the osmotic pressure of solutions of different concentrations. Why was this method selected in preference to, for example, depression of the freezing point?

Calculate the relative molecular mass of the protein from the following data.

TABLE 9.3

Concentration/g l^{-1}	10	15	25	35	45
Pressure/atm	0.0040	0.0063	0.0112	0.0167	0.0228

$R = 0.08205\,litre\,atm\,mol^{-1}\,K^{-1}$ Temperature = 298 K
$\;\;\;\; = 8.314\,J\,mol^{-1}\,K^{-1}$ (JMB, 81, S)

7. Figure 9.12 represents a simple form of osmotic pressure apparatus. **A** is an inert organic solvent and **B** is a solution in **A** of an organic compound **Q** of empirical formula C_7H_6O.

(*a*) The apparatus is set up with the liquid levels **X** and **Y** at equal heights and then left until equilibrium is reached at 27 °C. Explain what happens to the liquid levels.

(*b*) At equilibrium, the osmotic pressure of **B** is 14.0 kPa and the concentration of **Q** in **B** is 1.19 g dm^{-3}. Calculate the relative molecular mass of **Q**.

(C 82)

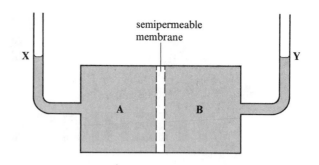

FIGURE 9.12

10

THERMOCHEMISTRY

10.1 ENERGY

Energy is the ability to do work

Work = force × distance

This chapter is about energy. An object which is capable of doing work is said to possess **energy**. The trouble with equating energy with the capacity for doing work is that one then has to explain what work is. When a force is applied to a stationary object, the object moves, and, when this happens, the force is said to be doing **work**. The amount of work done is the product of the force and the distance which the object moves in the direction of the force:

$$\text{Work} = \text{Force} \times \text{Distance}$$

The unit of force is the **newton**. (When 1 newton acts on a mass of 1 kg, it causes the mass to move with an acceleration of $1\,\text{m}\,\text{s}^{-2}$.) The unit of work is the **joule**. (1 joule = 1 newton metre, i.e., $1\,\text{J} = 1\,\text{Nm}$.) Energy and work are measured in the same unit.

10.1.1 ENERGY CHANGES

The energy changes in chemical reactions are important

The energy changes that accompany chemical reactions are of vital importance to us. Our own life processes depend on the energy content of the food we eat. The energy of the chemical bonds in the compounds present in food has come from the Sun. In the process of photosynthesis, plants which contain the catalyst chlorophyll convert carbon dioxide and water and the energy of sunlight into sugars:

Energy comes from the Sun

$$6CO_2(g) + 6H_2O(l) \xrightarrow[\text{chlorophyll}]{\text{sunlight}} C_6H_{12}O_6(aq) + 6O_2(g)$$

A reaction during which energy is absorbed is termed an **endothermic** reaction. Photosynthesis takes in 15 MJ of energy for every kilogram of glucose synthesised.

The energy in fossil fuels can be harnessed

The kind of life we lead depends on harnessing energy from different sources. We harness the fossil fuels, coal, oil and natural gas. These are formed by the slow conversion of plant material over millions of years, and thus derive their energy ultimately from the Sun. In the combustion of octane in a plentiful supply of air, 40 MJ of energy are obtained for every litre of octane burnt.

There are many forms of energy...

...All are either kinetic or potential energy...

Heat is kinetic energy

There are many forms of energy: heat, light, chemical energy, nuclear energy, etc., but basically there are only two kinds of energy, **kinetic energy** and **potential energy**. The energy which an object possesses because it is moving is called kinetic energy. The amount of kinetic energy that an object possesses is the amount of work that it can do before it comes to rest, having used up all its kinetic energy. The energy which an object possesses because of its position or because of the arrangement of its component parts is called potential energy. **Heat** is a form of kinetic energy: it is the kinetic energy associated with the motion of atoms and molecules. The energy

The energy of chemical bonds is potential energy

of chemical bonds is a form of potential energy, arising from the positions of atoms and molecules with respect to one another.

The study of the energy changes that accompany chemical reactions is called **thermochemistry** or **chemical thermodynamics**. It has its origins in the study of heat engines, in which the ideas of heat and movement combined to give the name **thermodynamics**.

The First Law of Thermodynamics states that energy is conserved

Energy can be converted from one form into another. Electrical energy can be converted into heat energy. Our bodies can convert the energy of the chemical bonds in food into other kinds of energy. Calculations on energy conversions show that energy is never created and never destroyed. Observations on physical changes and chemical reactions are summarised in the **First Law of Thermodynamics**. This law states that *energy can be changed from one form into another, but it can neither be created nor destroyed.*

10.1.2 HEAT AND TEMPERATURE

Heat capacity and specific heat capacity

When an object receives heat, its temperature rises. The rise in temperature depends on the **heat capacity** of the object:

$$\Delta T = \Delta Q/C \qquad [1]$$

The relationship between heat absorbed and rise in temperature

ΔT = rise in temperature, ΔQ = heat absorbed and C = heat capacity. You can see that the heat capacity of a mass of substance is the quantity of heat required to raise its temperature by 1 K. The heat capacity of unit mass of a substance is the specific heat capacity of that substance.

$$c = C/m \qquad [2]$$

where c = specific heat capacity and m = mass
Combining equations [1] and [2] gives

$$\Delta Q = mc\Delta T \qquad [3]$$

From equation [3], the rise in temperature that results from the absorption of a certain quantity of heat by a mass of substance can be calculated. For an example, see p. 197.

10.2 INTERNAL ENERGY AND ENTHALPY

Matter possesses energy in the form of...

Matter contains energy. This is in the form of kinetic energy and potential energy. The kinetic energy of matter is the energy of motion at a molecular level. The atoms or ions or molecules in a solid are vibrating and rotating and translating (moving

...kinetic energy and

from one place to another). The potential energy of matter arises from the positions of the atoms relative to one another. Bond-breaking and bond-making involve changes in potential energy. When two solids react, as in the reaction

...potential energy...

$$Ca(s) + I_2(s) \rightarrow CaI_2(s)$$

...the sum of which = internal energy

there is little change in kinetic energy, but there is a big change in potential energy as the bonding in the product is different from the bonding in the reactants. The kinetic energy and potential energy together make up the **internal energy** of matter.

Heat may be given out or taken in during a chemical reaction

Frequently, during the course of a chemical reaction, heat is either given out or taken in from the surroundings. The heat absorbed during a reaction is equal to the internal energy of the products minus the internal energy of the reactants, provided that no work is done by the system on the surroundings [see Figure 10.1].

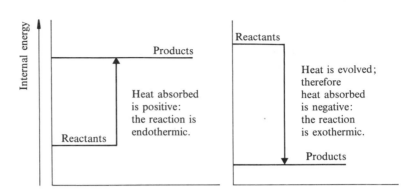

If the reaction takes place at constant pressure, any gases formed are allowed to escape, and they do work in forcing back the atmosphere. The heat absorbed during the reaction is equal to the increase in internal energy plus the work done in expansion.

$$\begin{pmatrix} \text{Heat absorbed at} \\ \text{constant pressure} \end{pmatrix} = \begin{pmatrix} \text{Change in} \\ \text{internal energy} \end{pmatrix} + \begin{pmatrix} \text{Work done on} \\ \text{surroundings} \end{pmatrix} \qquad [a]$$

Since most laboratory work is carried out at constant pressure, it is convenient to define a quantity enthalpy, such that

$$H = U + PV$$

Heat absorbed at constant pressure = change in enthalpy

where H = enthalpy, U = internal energy, P = pressure, and V = volume. Then, at constant pressure, $\Delta P = 0$, and

$$\Delta H = \Delta U + P\Delta V$$

where ΔV is the change in volume that occurs during reaction. Remembering from p. 149 that $P\Delta V$ is the work done by a gas when it expands by a volume ΔV at a pressure P, you can see that

$$\begin{pmatrix} \text{Change in} \\ \text{enthalpy} \end{pmatrix} = \begin{pmatrix} \text{Change in} \\ \text{internal energy} \end{pmatrix} + \begin{pmatrix} \text{Work done on} \\ \text{surroundings} \end{pmatrix} \qquad [b]$$

Comparing equations [a] and [b] shows that <u>the change in enthalpy is equal to the heat absorbed at constant pressure</u>.

If ΔV is positive, as in an expansion, $\Delta H > \Delta U$

If ΔV is negative, as in a contraction, $\Delta H < \Delta U$

If ΔV is zero, i.e. in reaction at constant volume, $\Delta H = \Delta U$

For reactions of gases, ΔH differs from ΔU

Reactions of solids and liquids do not involve large changes in volume, and ΔH is close to ΔU. Reactions in which there is a considerable change in volume are those which involve gases, and ΔH can be calculated from the ideal gas equation. Since

$$PV = nRT$$

then $\qquad P\Delta V = \Delta nRT$

Δn, the increase in the number of moles of gas, is indicated by the equation for the reaction. For example, in the reaction

$$NH_4NO_3(s) \rightarrow N_2O(g) + 2H_2O(g)$$

$\Delta n = 3$.

The standard state of a substance

When you are talking about the enthalpy of a substance, you must state the temperature, pressure and physical state of the substance. It is usual to compare the enthalpies of substances in their **standard states**. The standard state of a substance is the pure substance in a specified state (solid, liquid or gas) at 1 atmosphere pressure.

Standard enthalpy changes under standard conditions

The value of an enthalpy change depends on the temperature, the physical states (s, l, g) of the reactants and products, the pressures of gaseous reactants and products and the concentrations of solutions. Enthalpy changes are therefore stated under standard conditions, and are denoted by the symbol, $\Delta H_T^{\ominus}$, meaning the standard enthalpy change at temperature, *T*. **Standard conditions** are: gases at a pressure of 1 atm, solutions at unit concentration, substances in their normal physical states at the specified temperature. The temperature must be specified. If the temperature is not quoted, as in $\Delta H^{\ominus}$, you can assume that it is 298 K (25 °C). Thus ΔH refers to heat absorbed at constant pressure and $\Delta H^{\ominus}$ refers to heat absorbed under standard conditions.

Definitions of some standard enthalpy changes

$\Delta H_F^{\ominus}$ for an element is zero

Associated with any physical or chemical change is a standard enthalpy change. Some standard enthalpy changes are defined below. **Standard enthalpy of formation**, $\Delta H_F^{\ominus}$, is the heat absorbed when 1 mole of a substance is formed from its elements in their standard states. For example, the standard enthalpy of formation of sodium chloride is calculated for the reaction between solid sodium and gaseous chlorine molecules, Na(s) and $Cl_2(g)$. It follows from this definition that all elements in their standard states are assigned a value of zero for their standard enthalpies of formation. Figure 10.2 illustrates how the value of $\Delta H_F^{\ominus}$ may be negative (as for sodium chloride) or positive (as for ethyne). If the enthalpy absorbed is negative, enthalpy is released, and the reaction is **exothermic**.

FIGURE 10.2 Standard Enthalpy of Formation

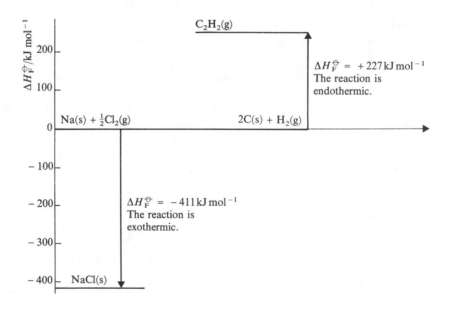

More definitions...

Standard enthalpy of combustion, $\Delta H_C^{\ominus}$, is the heat absorbed when 1 mole of a substance is completely burned in oxygen at 1 atm. Since heat is usually evolved in such a reaction, $\Delta H_C^{\ominus}$ will be negative.

Standard enthalpy of hydrogenation is the heat absorbed when 1 mole of an unsaturated compound is converted to a saturated compound by reaction with gaseous hydrogen at 1 atm.

Standard enthalpy of neutralisation is the heat absorbed when an acid and a base react to form 1 mole of water under standard conditions.

Standard enthalpy of reaction, $\Delta H_R^{\ominus}$, is the heat absorbed in a reaction at 1 atm between the number of moles of reactants shown in the equation for the reaction. In the reaction

$$4H_2O(l) + 3Fe(s) \rightarrow Fe_3O_4(s) + 4H_2(g)$$

there is no reason why the standard enthalpy of reaction should be related to 1 mole of iron or 1 mole of steam or 1 mole of iron oxide. Instead, it is related to the whole reaction, as written, between 4 moles of steam and 3 moles of iron.

Standard enthalpy of dissolution is the heat absorbed when 1 mole of a substance is dissolved at 1 atm in a stated amount of solvent. This may be 100 g or 1000 g of solvent or it may be an 'infinite' amount of solvent, i.e., a volume so large that on further dilution there is no further heat exchange.

Standard enthalpy of atomisation is the enthalpy absorbed when a substance decomposes to form 1 mole of gaseous atoms.

10.3 EXPERIMENTAL METHODS FOR FINDING THE STANDARD ENTHALPY OF REACTION

10.3.1 STANDARD ENTHALPY OF NEUTRALISATION

The reaction to be studied is the formation of water from oxonium (hydrogen) ions and hydroxide ions:

$$H_3O^+(aq) + OH^-(aq) \rightarrow 2H_2O(l)$$

Method of finding standard enthalpy of neutralisation...

The heat released when a known amount of water is formed is found by measuring the temperature rise produced in a calorimeter and its contents [see Figure 10.3]. Any vessel used for determinations of heat changes is called a **calorimeter**, after the old unit of heat, the **calorie** (1 calorie = 4.18 J).

...The two parts of the determination...

Measurements   (*a*) The heat capacity of the calorimeter is found. A known mass of water at a known temperature is poured into the calorimeter, and the temperature rise is noted. (*b*) A neutralisation reaction is carried out in the calorimeter. Known volumes of standard acid and alkali are added to the calorimeter, and the rise in temperature is noted.

FIGURE 10.3
A Calorimeter

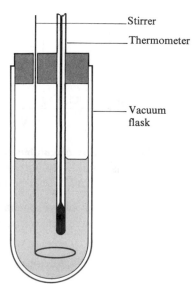

Stirrer

Thermometer

Vacuum flask

...the results... **A Typical Result** (*a*) When 100 g of water at 94.0 °C were added to a calorimeter at 17.5 °C, the temperature rose to 80.5 °C. (*b*) 250 cm^3 of sodium hydroxide (0.400 mol dm^{-3}) were added to 250 cm^3 of hydrochloric acid (0.400 mol dm^{-3}) in the calorimeter. The temperature of the two solutions was 17.5 °C initially and rose to 20.1 °C.

Calculation of $\Delta H^{\ominus}$ (Neutralisation) The assumption is made that the specific heat capacities of the solutions are the same as that of water, 4180 J kg^{-1} K^{-1}.
In (*a*)

...and method of calculation

Heat given out by water = Heat received by calorimeter
of heat capacity C

$$0.100 \times 4180(94.0 - 80.5) = C(80.5 - 17.5)$$

$$C = 90 \, \text{J K}^{-1}$$

In (*b*)

Heat from neutralisation = Heat received by calorimeter + solutions

$$= 90(20.1 - 17.5) + (0.500 \times 4180(20.1 - 17.5))$$

$$= 5670 \, \text{J}$$

Amount of water formed = $250 \times 10^{-3} \times 0.400 = 0.100 \, \text{mol}$

Heat evolved per mole = $5670/0.100 = 56\,700 \, \text{J mol}^{-1}$

The standard enthalpy of neutralisation is $-56.7 \, \text{kJ mol}^{-1}$.

10.3.2 EXPERIMENTAL METHOD FOR FINDING ENTHALPY OF COMBUSTION

Figure 10.4 shows a simple method for obtaining an approximate value for the enthalpy of combustion of a fuel.

FIGURE 10.4
Apparatus for Finding
Enthalpy of Combustion

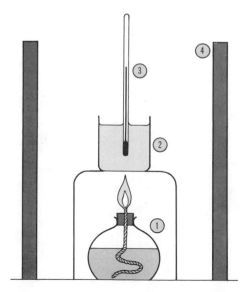

4 Shield reduces heat loss to surroundings.

3 Thermometer records rise in temperature, t °C.

2 Metal calorimeter contains a known mass of water, m_2 g.

1 Spirit burner contains fuel. Weighing before and after burning gives mass of fuel burnt, m_1 g.

A worked example of enthalpy of combustion

Example   When ethanol was burnt in the apparatus shown in Figure 10.4, the results were: $m_1 = 1.50$ g, $m_2 = 500$ g, $t = 13.0$ °C. Find the enthalpy of combustion of ethanol. Compare the experimental value with the listed value of -1368 kJ mol^{-1}.

Method

$$\text{Heat evolved} = 1.50 \times 500 \times 4.18 \times 13.0 = 40.76 \text{ kJ}$$

$$\text{Amount of ethanol burnt} = 1.50/46 = 0.0326 \text{ mol}$$

$$\text{Molar enthalpy of combustion} = 40.76/0.0326 = 1250 \text{ kJ mol}^{-1}$$

The enthalpy of combustion is -1250 kJ mol^{-1}. This is lower than the listed value because some of the heat evolved is lost to the surroundings.

10.3.3   EXPERIMENTAL METHOD FOR HEATS OF COMBUSTION

The bomb calorimeter

Heats of combustion are determined at constant volume in a **bomb calorimeter** [see Figure 10.5]. A substance is burnt completely in oxygen under pressure. The rise in the temperature of a calorimeter is measured. A calibration experiment gives the heat capacity of the calorimeter.

Example   The combustion of 0.725 g of benzoic acid (molar heat of combustion, -3230 kJ mol^{-1}) produced a temperature rise of 1.824 K. The combustion of 0.712 g of ethanamide, CH_3CONH_2, produced a temperature rise of 1.354 K. Find the molar heat of combustion ΔU_C of ethanamide.

Method   (*a*) Find the heat capacity of the calorimeter, C.

$$\text{Heat from combustion of benzoic acid} = (0.725/122) \times 3.23 \times 10^6 \text{ J}$$

$$\text{Heat received by calorimeter} = C \times 1.824 \text{ J}$$

$$\text{Equating these two quantities, } C = 10.52 \text{ kJ K}^{-1}$$

FIGURE 10.5   Bomb Calorimetry

(**a**) The Calorimeter

4 Platinum resistance thermometer measures rise in temperature.

3 Calorimeter contains known mass of water.

2 Platinum crucible contains known mass of substance.

1 The 'bomb', a strong steel cylinder

5 Pipe from oxygen cylinder.   **6** Ignition leads

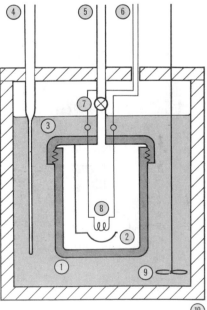

7 Valve admits oxygen to a pressure 25 atm.

8 Platinum ignition coil. A current is passed through the coil for 5 seconds to ignite the sample.

9 Stirrer

10 Insulation

(b) Result of a
Combustion in a Bomb
Calorimeter

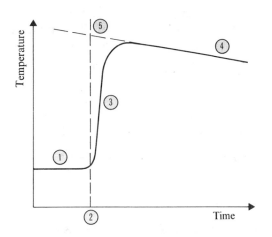

5 Cooling curve is extrapolated
to give the temperature
maximum.

4 Cooling curve records loss of
heat to the surroundings.

3 Temperature rises after
bomb is fired.

1 Temperature is recorded for
10 minutes.

2 Bomb is fired.

(b) Find the molar heat of combustion of ethanamide, ΔU_C.

Heat from combustion of ethanamide = $(0.712/59.1) \times \Delta U_C$ kJ

Heat received by calorimeter = 10.52×1.354 kJ

Equating these two quantities gives $\Delta U_C = -1182$ kJ mol^{-1}

The molar heat of combustion of ethanamide is -1180 kJ mol^{-1}.

CHECKPOINT 10A: COMBUSTION

1. Calculate the amount of energy a woman takes in during a year on a 2200 kcal a day diet (1 calorie = 4.18 J). Calculate the amount of energy she spends in driving 5000 miles a year at 40 miles to the gallon of octane, given 1 gallon of octane weighs 3.12 kg, and $\Delta H_C^{\ominus}$ (octane) = -5512 kJ mol^{-1}.

2. A cocktail contains 33 g of ethanol. If $\Delta H_C^{\ominus}$ (ethanol) is -1370 kJ mol^{-1}, calculate the standard enthalpy released when this amount of ethanol is combusted inside the tissues of the drinker. Your result will be slightly in error because the value of $\Delta H_C^{\ominus}$ which you have been given is not exactly appropriate. What value of $\Delta H_C^{\ominus}$ should be used in the calculation?

3. Compare the liquid fuels propane, butane and iso-octane, and decide which gives the best calorific value for money.

Fuel	Propane (*l*)	Butane (*l*)	Iso-octane (*l*)
Price/pence kg^{-1}	52.5	61.0	57.6
$\Delta H_C^{\ominus}$/kJ mol^{-1}	2220	2880	5510

4. You want to boil a kettle. The kettle contains 2.00 kg of water at 20 °C. What mass of natural gas (methane) must be burned to raise this quantity of water to 100 °C? (Assume no heat is lost.) (Specific heat capacity of water = 4.18 J g^{-1} K^{-1}; $\Delta H_C^{\ominus}$(CH$_4$) = -890 kJ mol^{-1})

10.4 STANDARD ENTHALPY CHANGE FOR A CHEMICAL REACTION

The standard enthalpy change for a chemical reaction can be calculated from the standard enthalpies of formation of all the products and reactants involved. For example, in the reaction of hydrogen chloride with ammonia

Finding $\Delta H^{\ominus}$ of reaction

$$NH_3(g) + HCl(g) \rightarrow NH_4Cl(s)$$

$$(-46) \quad (-92.3) \quad (-315)$$

Written underneath each species is its standard enthalpy of formation in $kJ\,mol^{-1}$. The standard enthalpy of reaction $\Delta H^{\ominus}$ is given by

$$\Delta H^{\ominus} = (-315) - (-46 + (-92.3))$$

$$= -177\,kJ\,mol^{-1}$$

The negative sign means that the reaction is exothermic.

The standard enthalpy of reaction depends only on the difference between the standard enthalpy of the reactants and the standard enthalpy of the products and not on the route by which the reaction occurs. This idea is embodied in **Hess's Law**, which states that, if a reaction can take place by more than one route, the overall change in enthalpy is the same, whichever route is followed.

Hess's Law

Hess's Law is a special case of the First Law of Thermodynamics. In the set of reactions shown in the **enthalpy diagram** in Figure 10.6, the standard enthalpy change in going from **A** to **B** by the direct route 1 is $\Delta H_1^{\ominus}$. The standard enthalpy change in going from **A** to **B** by the indirect route 2 is $\Delta H_2^{\ominus} + \Delta H_3^{\ominus}$.

FIGURE 10.6 An Enthalpy Diagram

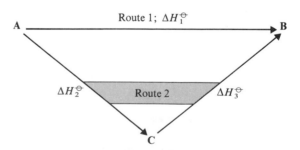

If the sum $(\Delta H_2^{\ominus} + \Delta H_3^{\ominus})$ were less than $\Delta H_1^{\ominus}$, it would be possible to create energy by making **A** from **B** by route 1 and then converting **A** into **B** by route 2. This would be contrary to the First Law of Thermodynamics. It follows that

$$\Delta H_1^{\ominus} = \Delta H_2^{\ominus} + \Delta H_3^{\ominus}$$

and the standard enthalpy change is the same for the different reaction routes. It follows also that the standard enthalpy change for the reaction **B** → **A** is $-\Delta H_1^{\ominus}$.

10.4.1 FINDING THE STANDARD ENTHALPY OF FORMATION OF A COMPOUND INDIRECTLY

$\Delta H_F^\ominus$ of a compound may be found directly by experiment or indirectly, by applying Hess's Law...

Sometimes, the standard enthalpy of formation of a compound can be measured directly by allowing known amounts of elements to combine and measuring the heat evolved. Other reactions are difficult to study, and the standard enthalpy of reaction must be found indirectly.

To find the standard enthalpy of formation of ethyne from practical measurements is impossible as attempts to make ethyne from carbon and hydrogen

$$2C(s) + H_2(g) \rightarrow C_2H_2(g)$$

...that is, by using an enthalpy diagram

will result in the formation of a mixture of hydrocarbons. This is where Hess's Law comes to the rescue. In Figure 10.7, it follows from Hess's Law that the standard enthalpy change via route A is the same as the standard enthalpy change via route B.

FIGURE 10.7 Enthalpy Diagram for Ethyne

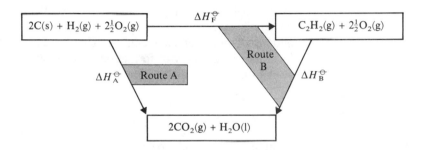

The change in standard enthalpy when carbon and hydrogen burn to form carbon dioxide and water is the same as the sum of the standard enthalpy changes when carbon and hydrogen combine to form ethyne and then ethyne burns to form carbon dioxide and water. Thus

$$\Delta H_A^\ominus = \Delta H_F^\ominus + \Delta H_B^\ominus$$

$$\Delta H_A^\ominus = 2(\Delta H^\ominus \text{ of combustion of C}) + (\Delta H^\ominus \text{ of combustion of H}_2)$$

$$\Delta H_B^\ominus = \Delta H^\ominus \text{ of combustion of C}_2H_2$$

Standard enthalpies of combustion can be measured accurately. They are

$$C(s) + O_2(g) \rightarrow CO_2(g); \qquad \qquad \Delta H_1^\ominus = -394 \, \text{kJ mol}^{-1} \qquad [a]$$

$$H_2(g) + \tfrac{1}{2}O_2(g) \rightarrow H_2O(l); \qquad \qquad \Delta H_2^\ominus = -286 \, \text{kJ mol}^{-1} \qquad [b]$$

$$C_2H_2(g) + 2\tfrac{1}{2}O_2(g) \rightarrow 2CO_2(g) + H_2O(l); \; \Delta H_3^\ominus = -1300 \, \text{kJ mol}^{-1} \qquad [c]$$

Putting these values into the equation

$$\Delta H_F^\ominus = \Delta H_A^\ominus - \Delta H_B^\ominus$$

gives $\quad \Delta H_F^\ominus = 2(-394) + (-286) - (-1300) = +226 \, \text{kJ mol}^{-1}$

A value of $226 \, \text{kJ mol}^{-1}$ is obtained for the standard enthalpy of formation of ethyne. Ethyne is described as an **endothermic compound** since $\Delta H_F^\ominus$ is positive.

A simple method of calculating $\Delta H_F^{\ominus}$

Alternative Method of Calculation　There is a simpler method of calculation. It involves three steps.

1. Write the equation for the combustion of ethyne, since this is the reaction for which $\Delta H^{\ominus}$ can be measured.

2. Under each species, write $\Delta H_F^{\ominus}$. You can see from equations [a] and [b] that the standard enthalpies of formation of carbon dioxide and water are -394 and $-286\,\text{kJ}\,\text{mol}^{-1}$ respectively.

$$C_2H_2(g) + 2\tfrac{1}{2}O_2(g) \rightarrow 2CO_2(g) + H_2O(l); \Delta H_3^{\ominus} = -1300\,\text{kJ}\,\text{mol}^{-1}$$

$$\Delta H_F^{\ominus}(C_2H_2)\ 0 \qquad\quad 2(-394)\quad (-286)$$

3. Since $\Delta H^{\ominus} = $ Total $H^{\ominus}$ of products $-$ Total $H^{\ominus}$ of reactants

$$\Delta H_3^{\ominus} = -1300 = 2(-394) + (-286) - \Delta H_F^{\ominus}(C_2H_2)$$

$$\Delta H_F^{\ominus}(C_2H_2) = +226\,\text{kJ}\,\text{mol}^{-1}$$

The standard enthalpy of formation of ethyne is $226\,\text{kJ}\,\text{mol}^{-1}$, as before.

10.4.2　FINDING THE STANDARD ENTHALPY OF REACTION

When a physical change or a chemical reaction takes place, the standard enthalpy change is equal to the standard enthalpy content of the products minus the standard enthalpy content of the reactants. The standard enthalpy content of a substance is its standard enthalpy of formation. The equation for the hydration of ethene to form ethanol is

Calculating the standard enthalpy of reaction . . .

$$CH_2{=}CH_2(g) + H_2O(l) \rightarrow C_2H_5OH(l)$$

$$(+52) \qquad\qquad (-286)\quad (-278)$$

Written beneath each species is the value of $\Delta H_F^{\ominus}$ in $\text{kJ}\,\text{mol}^{-1}$. The **standard enthalpy of reaction** is given by

$$\Delta H_R^{\ominus} = -278 - (52 - 286)$$

$$= -44\,\text{kJ}\,\text{mol}^{-1}.$$

The reaction is exothermic by $44\,\text{kJ}\,\text{mol}^{-1}$. In short

. . . The method

$$\boxed{\Delta H_{\text{Reaction}}^{\ominus} = \sum \Delta H_{\text{Formation}}^{\ominus} \text{ of products} - \sum \Delta H_{\text{Formation}}^{\ominus} \text{ of reactants}}$$

Alternative Method of Calculation　A different method of finding the standard enthalpy of reaction, $\Delta H_R^{\ominus}$, is illustrated by the physical change from one allotrope of tin to the other:

$$\text{Sn(s, white)} \rightarrow \text{Sn(s, grey)}; \Delta H_R^{\ominus} = ?$$

FIGURE 10.8　Enthalpy Diagram for Tin

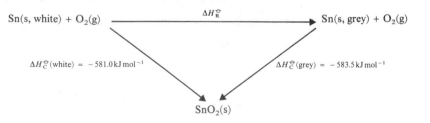

ΔH⊖Reaction can also be found from an enthalpy diagram...

The best way of finding $\Delta H_R^{\ominus}$ for the change is to measure the standard enthalpy of combustion of each allotrope. These are shown in the enthalpy diagram, Figure 10.8.

From Hess's Law

$$\Delta H_C^{\ominus}(\text{white}) = \Delta H_R^{\ominus} + \Delta H_C^{\ominus}(\text{grey})$$

$$\Delta H_R^{\ominus} = -581.0 - (-583.5)$$

$$= +2.5\,\text{kJ mol}^{-1}$$

The method of calculation in this example is simply

...Alternative method

$$\Delta H_{\text{Reaction}}^{\ominus} = \sum \Delta H_{\text{Combustion}}^{\ominus} \text{ of reactants} - \sum \Delta H_{\text{Combustion}}^{\ominus} \text{ of products}$$

CHECKPOINT 10B: ENTHALPY CHANGES

1. State the sign of the standard enthalpy change in the following:

(a) the combustion of octane
(b) the condensation of steam
(c) the freezing of water
(d) the electrolysis of water
(e) the combustion of sodium in chlorine.

2. Why is it necessary to compare the enthalpy content of substances under standard conditions?

3. Arrange the following in order of increasing standard enthalpy:

(a) $1\,\text{mol}\,H_2(g) + \frac{1}{2}\text{mol}\,O_2(g)$ at 25 °C and 1 atm
(b) $1\,\text{mol}\,H_2O(g)$ at 100 °C and 1 atm
(c) $1\,\text{mol}\,H_2O(l)$ at 25 °C and 1 atm
(d) $1\,\text{mol}\,H_2O(s)$ at 0 °C and 1 atm
(e) $1\,\text{mol}\,H_2O(l)$ at 0 °C and 1 atm
(f) $1\,\text{mol}\,H_2O(l)$ at 100 °C and 1 atm

10.5 STANDARD BOND DISSOCIATION ENTHALPY

Standard bond dissociation enthalpies can be measured accurately

The **standard bond dissociation enthalpy** is the energy that must be absorbed to separate the two atoms in a bond. When hydrogen chloride dissociates

$$HCl(g) \rightarrow H(g) + Cl(g); \quad \Delta H^{\ominus} = 429.7\,\text{kJ mol}^{-1}$$

The standard bond dissociation enthalpy of the H—Cl bond in HCl is $429.7\,\text{kJ mol}^{-1}$. The value is accurately known as it is found by precise spectroscopic measurements.

10.6 AVERAGE STANDARD BOND ENTHALPY

Average standard bond enthalpies are approximate values...

When you want to assign a value to the standard enthalpy of dissociation of the C—H bond in methane, the problem is different. The energy required to break the first C—H bond in methane is not the same as that required to remove a hydrogen atom from a $CH_3\cdot$ radical or from CH_2: or CH :. In the complete dissociation

$$CH_4(g) \rightarrow C(g) + 4H(g); \quad \Delta H^{\ominus} = +1662\,\text{kJ mol}^{-1}$$

Dividing the standard enthalpy change equally between the four bonds gives an average value for the C—H bond of $416\,\text{kJ mol}^{-1}$. This value is called the **average standard bond enthalpy** of the C—H bond.

...which give approximate values for $\Delta H^{\ominus}_{Reaction}$ *...*

Tables of average standard bond enthalpies make the assumption that the standard enthalpy of a bond is the same in different molecules. This is only roughly true. Since standard bond enthalpies vary from one compound to another, the use of average standard bond enthalpies gives only approximate values for standard enthalpies of reaction calculated from them. Experimental methods are used to obtain standard enthalpies of reaction whenever possible. Calculations based on average standard bond enthalpies are used only for reactions which have not been studied experimentally, for example, the reactions of a substance which has not been isolated in a pure state.

...and are often called bond energy terms

Average standard bond enthalpy is often called the **bond energy term**. One can say that the bond energy term for the C—H bond is $416\,kJ\,mol^{-1}$. The sum of all the bond energy terms for a compound is the standard enthalpy absorbed in atomising that compound in the *gaseous state*. The standard enthalpy of formation of a hydrocarbon includes the sum of the bond energy terms and also the standard enthalpy of atomisation of the carbon atoms and the standard enthalpy of atomisation of the hydrogen atoms.

Method of calculation of $\Delta H^{\ominus}_F$ *from bond energy terms*

Example Calculate the standard enthalpy of formation of ethane, using the following information. The average standard bond enthalpies are: C—C = 348, C—H = $416\,kJ\,mol^{-1}$. The standard enthalpies of atomisation are: C(s) = $718\,kJ\,mol^{-1}$, and $\frac{1}{2}H_2(g)$ = $218\,kJ$ per mole of H atoms formed.

Method In the C_2H_6 molecule are

one C—C bond of standard enthalpy = $348\,kJ\,mol^{-1}$

six C—H bonds of standard enthalpy = $2496\,kJ\,mol^{-1}$

The total standard enthalpy of the bonds = $2844\,kJ\,mol^{-1}$

This is the energy given out when the atoms combine:

$$2C(g) + 6H(g) \rightarrow C_2H_6(g);\ \Delta H^{\ominus} = -2844\,kJ\,mol^{-1}$$

The standard enthalpy of formation of C(g) from C(s) is $718\,kJ\,mol^{-1}$.
The standard enthalpy of formation of H(g) from $\frac{1}{2}H_2(g)$ is $218\,kJ\,mol^{-1}$.
These values of the standard enthalpy content of each species can be put into the equation:

$$2C(g) + 6H(g) \rightarrow C_2H_6(g);\ \ \Delta H^{\ominus} = -2844\,kJ\,mol^{-1}$$

$$2(718)\quad 6(218)\quad \Delta H^{\ominus}_F$$

$$\Delta H^{\ominus} = \Delta H^{\ominus}_F(\text{product}) - \Delta H^{\ominus}_F(\text{reactants})$$

$$-2844 = \Delta H^{\ominus}_F - 2(718) - 6(218)$$

$$\Delta H^{\ominus}_F = -100\,kJ\,mol^{-1}$$

The standard enthalpy of formation of ethane is $-100\,kJ\,mol^{-1}$.

10.6.1 STANDARD ENTHALPY OF REACTION FROM AVERAGE STANDARD BOND ENTHALPIES

When the standard enthalpy change for a reaction cannot be measured, an approximate value can be obtained by using average standard bond enthalpies. During a reaction, energy must be supplied to break bonds in the reactants, and energy is given out when the bonds in the products form. The standard enthalpy of reaction is the difference between the sum of the average standard bond enthalpies of the products and the sum of the average standard bond enthalpies of the reactants.

Method of calculation of $\Delta H_R^{\ominus}$ from bond energy terms

Example   Calculate the standard enthalpy change of the reaction

$$(CH_3)_2C{=}O(g) + HCN(g) \rightarrow (CH_3)_2C\overset{\displaystyle OH}{\underset{\displaystyle CN}{\diagup\!\!\!\diagdown}}\;(g)$$

Mean standard bond enthalpies/kJ mol^{-1} are C=O, 743; C—H, 412; C—O 360; C—C, 348; H—O, 463.

Method

Bonds broken are   one C=O of $\Delta H^{\ominus} = 743\,\text{kJ mol}^{-1}$
one C—H of $\Delta H^{\ominus} = 412\,\text{kJ mol}^{-1}$

Total standard enthalpy absorbed $= 1155\,\text{kJ mol}^{-1}$

Bonds created are one C—O of $\Delta H^{\ominus} = 360\,\text{kJ mol}^{-1}$
one O—H of $\Delta H^{\ominus} = 463\,\text{kJ mol}^{-1}$
one C—C of $\Delta H^{\ominus} = 348\,\text{kJ mol}^{-1}$

Total standard enthalpy released $= -1171\,\text{kJ mol}^{-1}$

Standard enthalpy change of reaction $= -1171 + 1155$
$= -16\,\text{kJ mol}^{-1}$

10.7   THE BORN–HABER CYCLE

The Born–Haber cycle is a technique for applying Hess's Law to the standard enthalpy changes which occur when an ionic compound is formed. Think of the reaction between sodium and chlorine to form sodium chloride. The reaction can be considered to occur by means of the following steps, even though the reaction itself may not follow this route.

The formation of an ionic compound can be treated as a sequence of separate steps

Vaporisation of sodium
Na(s) → Na(g); $\Delta H_S^{\ominus}$ = standard enthalpy of sublimation or vaporisation of sodium   [1]

Ionisation of sodium
Na(g) → Na$^+$(g) + e$^-$; $\Delta H_I^{\ominus}$ = standard ionisation enthalpy of sodium   [2]

Dissociation of chlorine molecules
$\frac{1}{2}Cl_2$(g) → Cl(g); $\Delta H_D^{\ominus} = \frac{1}{2}$ standard bond dissociation enthalpy of chlorine molecules   [3]

Ionisation of chlorine atoms
Cl(g) + e$^-$ → Cl$^-$(g); $\Delta H_E^{\ominus}$ = electron affinity of chlorine   [4]

Reaction between ions
Na$^+$(g) + Cl$^-$(g) → NaCl(s); $\Delta H_L^{\ominus}$ = standard lattice enthalpy   [5]

Definitions of $\Delta H^{\ominus}$ terms involved in the steps of the Born–Haber cycle

The standard enthalpy changes can be defined as follows:

[1] The standard enthalpy of sublimation or vaporisation is the enthalpy absorbed when 1 mole of sodium atoms is vaporised.

[2] The standard enthalpy of ionisation of sodium is the enthalpy required to remove 1 mole of electrons from 1 mole of gaseous sodium atoms.

[3] The standard bond dissociation enthalpy of chlorine is the enthalpy required to dissociate 1 mole of chlorine molecules into atoms.

[4] The electron affinity of chlorine is the enthalpy absorbed when 1 mole of chlorine atoms accept 1 mole of electrons to become chloride ions. It has a negative value, showing that this reaction is exothermic.

[5] The standard lattice enthalpy is the enthalpy absorbed when 1 mole of sodium chloride is formed from its gaseous ions; its value is negative. (The standard lattice dissociation enthalpy is the enthalpy absorbed when 1 mole of sodium chloride is separated into its gaseous ions; it has a positive value.)

The standard enthalpy changes in steps [1] to [5] are represented in Figure 10.9. The steps in a Born–Haber cycle are represented as going upwards if they absorb energy and downwards if they give out energy.

FIGURE 10.9
Born–Haber Cycle for
Sodium Chloride

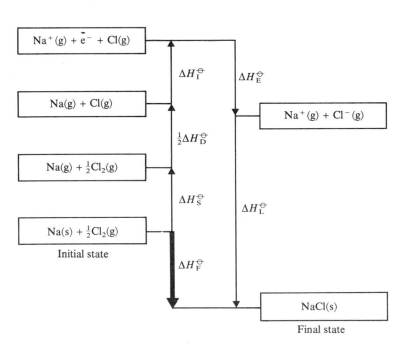

*Application of Hess's Law
to the Born–Haber cycle*

Applying Hess's Law to this cycle, it follows that the sum of the standard enthalpy terms [1] to [5] is equal to the difference in standard enthalpy between the product, sodium chloride, and the reactants, solid sodium and gaseous chlorine molecules, that is the standard enthalpy of formation of sodium chloride. Inserting numerical values gives

$$\Delta H_F^\ominus = \Delta H_S^\ominus + \tfrac{1}{2}\Delta H_D^\ominus + \Delta H_I^\ominus + \Delta H_E^\ominus + \Delta H_L^\ominus$$

$$= +109 + 121 + 494 - 380 - 755 = -411\,\text{kJ}\,\text{mol}^{-1}$$

In practice, it is easier to measure standard enthalpies of formation than to measure some of the other steps. The electron affinity is the hardest term to measure experimentally: Born–Haber cycles are often used to calculate electron affinities.

The Born–Haber cycle for magnesium oxide is shown in Figure 10.10.

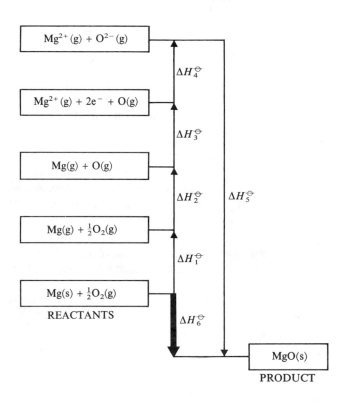

You will be able to identify $\Delta H_1^\ominus$, $\Delta H_2^\ominus$, $\Delta H_3^\ominus$ and $\Delta H_5^\ominus$ by comparison with the sodium chloride cycle. In the case of oxygen, $\Delta H_4^\ominus$ is an endothermic term. The reason is that, although energy is given out in the process

$$O(g) + e^- \rightarrow O^-(g)$$

energy is required for the introduction of a second electron against the repulsion of O^-, in the process

$$O^-(g) + e^- \rightarrow O^{2-}(g)$$

If the lattice enthalpy is strongly exothermic, this may be the term that decides whether an ionic compound is formed

making $\Delta H_4^\ominus$ endothermic. The only exothermic term is the lattice enthalpy. The reason why the formation of the ionic lattice is strongly exothermic is that both ions Mg^{2+} and O^{2-}, are small, and both carry two units of charge. When these opposite charges are brought close together, there is a big release of energy. This example shows the importance of lattice enthalpy in determining whether ionic compounds are formed [see Question 3, p. 208].

10.8 ENTHALPY CHANGES INVOLVED WHEN IONIC COMPOUNDS DISSOLVE

When an ionic solid dissolves in a solvent, two enthalpy terms are involved.

When salts dissolve, energy is required to separate the ions...

1. The ions must be separated from the ionic lattice. The energy required is the lattice dissociation enthalpy.

2. The separate ions interact with the molecules of solvent. If the solvent is polar, a charged ion can be attracted to one end of a polar solvent molecule. The energy

...Solvation of the ions releases energy... released as these attractive forces come into play is compensation for the energy required to dissociate the lattice:

$$\Delta H^{\ominus}_{\text{Dissolution}} = \Delta H^{\ominus}_{\text{Lattice dissociation}} + \Delta H^{\ominus}_{\text{Solvation}}$$

$$\text{(a positive quantity)} \quad \text{(a negative quantity)}$$

...On balance $\Delta H^{\ominus}$ *of dissolution can be positive or negative* If $\Delta H^{\ominus}_{\text{Dissolution}}$ is negative, dissolution is favoured by enthalpy considerations. If $\Delta H^{\ominus}_{\text{Dissolution}}$ is positive, dissolution may still occur endothermically if it is favoured by entropy considerations [p. 209].

Highly negative enthalpies of solvation make water a good solvent As water molecules have large dipoles, powerful interactions are possible between the polar molecules and solute ions. These powerful interactions result in a high negative enthalpy of solvation, which makes water a good solvent. Non-polar solvents, such as hydrocarbons, do not dissolve ionic substances because there is no negative enthalpy of solvation to compensate for the positive lattice dissociation enthalpy.

CHECKPOINT 10C: STANDARD ENTHALPY OF REACTION AND AVERAGE STANDARD BOND ENTHALPIES

The following are standard enthalpies of combustion/kJ mol^{-1} at 298 K:

C(graphite)	− 394	$CH_3CO_2C_2H_5(l)$	− 2246
$H_2(g)$	− 286	C_2H_4	− 1393
$CH_3CO_2H(l)$	− 876	$C_2H_5OH(l)$	− 1400
$CH_4(g)$	− 891	$C_6H_{12}(l)$	− 3924
$C_2H_6(g)$	− 1561	$C_2H_5OH(g)$	− 1444

1. Calculate the standard enthalpy of formation of the following:

(*a*) ethane, $C_2H_6(g)$

(*b*) ethene, $C_2H_4(g)$

(*c*) ethanoic acid, $CH_3CO_2H(l)$

(*d*) ethanol, $C_2H_5OH(l)$

(*e*) ethanol, $C_2H_5OH(g)$

Explain the difference between the values of (*d*) and (*e*).

2. Find the standard enthalpy of formation of ethylethanoate(l), ethanol(l), ethanoic acid(l), water(l). Calculate the standard enthalpy change in the reaction

$$CH_3CO_2H(l) + C_2H_5OH(l) \rightarrow CH_3CO_2C_2H_5(l) + H_2O(l)$$

3. Refer to Figure 10.10. To what do the symbols $\Delta H^{\ominus}_1$ to $\Delta H^{\ominus}_5$ refer? What is the significance of an arrow pointing upwards? Given that the values, in kJ mol^{-1} are: $\Delta H^{\ominus}_1 = +153$, $\Delta H^{\ominus}_2 = +248$, $\Delta H^{\ominus}_3 = +2180$, $\Delta H^{\ominus}_4 = +745$, and $\Delta H^{\ominus}_5 = -3930$, calculate $\Delta H^{\ominus}_6$, the standard enthalpy of formation of magnesium oxide.

4. The following values are $\Delta H^{\ominus}_F$/kJ mol^{-1} at 298 K:

$CH_4(g)$ − 76; $CO_2(g)$ − 394; $H_2O(l)$ − 286; $H_2O(g)$ − 242; $NH_3(g)$ − 46.2; $C_2H_5OH(l)$ − 278; $C_8H_{18}(l)$ − 210; $C_2H_6(g)$ − 85

Calculate the standard enthalpy changes at 298 K for the reactions:

(*a*) $C_2H_6(g) + 3\frac{1}{2}O_2(g) \rightarrow 2CO_2(g) + 3H_2O(l)$

(*b*) $C_2H_5OH(l) + 3O_2(g) \rightarrow 2CO_2(g) + 3H_2O(l)$

(*c*) $H_2(g) + \frac{1}{2}O_2(g) \rightarrow H_2O(l)$

(*d*) $C_8H_{18}(l) + 12\frac{1}{2}O_2(g) \rightarrow 8CO_2(g) + 9H_2O(l)$

5. The Born–Haber cycle for rubidium chloride is shown in Figure 10.11. The letters **A** to **F** represent standard enthalpy changes. Give the names of these quantities. Their values in kJ mol^{-1} are: **A** = − 431, **B** = + 86, **C** = + 122, **D** = + 408, **F** = − 675. Calculate the value of **E**.

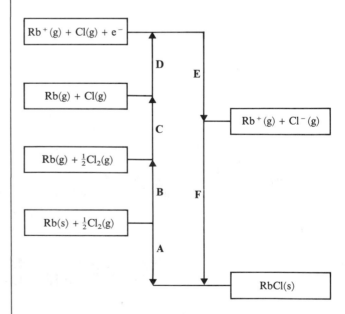

FIGURE 10.11
Born–Haber Cycle for Rubidium Chloride (not to scale)

10.9 FREE ENERGY AND ENTROPY

Some reactions occur spontaneously...

A reaction which happens of its own accord, without any external help, is said to occur **spontaneously**. Some spontaneous reactions are endothermic. Examples are [1] the dissolution of potassium chloride in water, [2] the melting of ice, [3] the evaporation of water, [4] the dissociation of ammonium carbonate:

$$KCl(s) + aq \rightarrow KCl(aq); \Delta H^\ominus = 19\,kJ\,mol^{-1} \qquad [1]$$

$$H_2O(s) \rightarrow H_2O(l); \Delta H^\ominus = 6.0\,kJ\,mol^{-1} \qquad [2]$$

$$H_2O(l) \rightarrow H_2O(g); \Delta H^\ominus = 44\,kJ\,mol^{-1} \qquad [3]$$

$$(NH_4)_2CO_3(s) \rightarrow 2NH_3(g) + CO_2(g) + H_2O(g); \Delta H^\ominus = 68\,kJ\,mol^{-1} \qquad [4]$$

...Such reactions increase the 'disorder' of the system

The difference in enthalpy between the products and the reactants cannot be the only factor which decides whether a chemical reaction takes place. There must be an additional factor involved.

Examples of spontaneous changes

All these four physical or chemical changes have a factor in common. When potassium chloride dissolves, the regular arrangement of the crystal structure is replaced by a random distribution of ions in solution. When ice melts, the regular hydrogen-bonded structure of ice [see Figure 4.24, p. 85] is replaced by the fluid association of water molecules in liquid water. When water evaporates, the association of water molecules is replaced by individual molecules moving independently in the gas phase. (The big difference between $\Delta H^\ominus_{Melting}$ and $\Delta H^\ominus_{Evaporation}$ shows that extensive hydrogen bonding persists in the liquid phase.) In the dissociation of ammonium carbonate, 4 moles of gas are formed from 1 mole of solid. When gases come into contact, they diffuse to form a homogeneous mixture: instead of a tidy arrangement of different molecules in separate containers, the molecules are now mixed up in a random arrangement.

The degree of 'disorder' of a system is measured by its entropy...

All these spontaneous changes involve a transition from an ordered arrangement of particles to a less ordered arrangement. The degree of disorder or randomness in a system is measured by a physical quantity termed the **entropy** of the system. A crystalline solid, with a regular arrangement of ions, has a low entropy. When it melts, the ions are free to move, and it is impossible to describe their positions relative to one another: the degree of order in the system has *decreased*, and the entropy has *increased*. In a gas the molecules move completely independently of one another, and the entropy of a gas is high. Physical and chemical changes in which gases are produced result in an increase in entropy.

...Estimating whether a change involves an increase or a decrease in entropy, S...

Entropy is given the symbol S, standard entropy, $S^\ominus$, and change in entropy, ΔS. An increase in the degree of disorder of a system shows in a positive value of ΔS. It is often possible to tell whether a reaction has a positive or negative value of ΔS by an inspection of the equation. The equation

$$NH_4NO_3(s) \rightarrow N_2O(g) + 2H_2O(g)$$

...and predicting the sign of $\Delta S^\ominus$

shows that 1 mole of the crystalline solid, ammonium nitrate, forms 1 mole of a gas, dinitrogen oxide, and 2 moles of steam, a total of 3 moles of gas. The value of ΔS is positive. Under standard conditions, the water formed is a liquid, and the increase in entropy is less: $\Delta S^\ominus$ has a smaller positive value. Other reactions for which it is easy to predict the sign of ΔS are

$$CaO(s) + H_2O(l) \rightarrow Ca(OH)_2(s); \Delta S \text{ negative}$$

$$CaCO_3(s) \rightarrow CaO(s) + CO_2(g); \Delta S \text{ positive}$$

Both factors, the change in enthalpy and the change in entropy, are important in deciding whether a physical or chemical change will occur. They are combined in the equation

The change in free energy,
ΔG

$$\text{Free energy, } G = \text{Enthalpy, } H - (\text{Temperature/K} \times \text{Entropy, } S)$$

$$G = H - TS$$

from which it follows that, at a constant temperature

$$\Delta G = \Delta H - T\Delta S$$

...must be negative if a change is to occur...

All spontaneous physical and chemical changes take place in the direction of a *decrease* in free energy. They may involve an increase or a decrease in enthalpy; they may involve an increase or a decrease in entropy; they all involve a decrease in free energy. The sign of ΔG for a spontaneous change must be negative. A change is therefore assisted by a decrease in enthalpy (ΔH negative) and by an increase in entropy (ΔS positive).

...If ΔG is negative, the reaction is feasible...

A reaction with a negative value of ΔG is said to be **feasible**. This means that, if the reaction takes place, it will go in the direction of the reactants forming the products and not in the reverse direction. To say that a reaction is feasible does not imply anything about the rate of the reaction. The reacting species may have to surmount an energy barrier, **the activation energy**, before reaction can occur [p. 307].

...The value of $\Delta G^{\ominus}$ can be used to predict the feasibility of a change

If the change takes place under standard conditions, i.e., with each reactant and product at unit concentration (or pressure), then the free energy change is equal to the standard free energy change, $\Delta G^{\ominus}$. When reaction takes place under non-standard conditions, the free energy change, ΔG, differs from $\Delta G^{\ominus}$ as ΔG depends on the concentrations (or pressures) of the reactants and products. It is easy to obtain $\Delta G^{\ominus}$ from tables of standard enthalpies of formation and standard entropies, but one really wants to know the value of ΔG for the real conditions, and this is not easy to compute. However, if $\Delta G^{\ominus}$ has a sufficiently large positive or negative value, $\Delta G^{\ominus}$ may determine the feasibility of reaction over a large range of concentrations (or pressures).

10.9.1 CALCULATION OF CHANGE IN STANDARD ENTROPY

METHOD 1

One method of calculating the standard entropy change of a process is to use the expression

$\Delta S^{\ominus}$

$$\boxed{\Delta S^{\ominus} = \sum S^{\ominus} \text{ of products} - \sum S^{\ominus} \text{ of reactants}}$$

Example Calculate the standard entropy change for the formation of ammonia, given the values of entropy in $JK^{-1}mol^{-1}$: $S^{\ominus}(N_2(g)) = 192$, $S^{\ominus}(H_2(g)) = 131$, $S^{\ominus}(NH_3(g)) = 193$.

Method 1 The equation for the reaction is

A worked example

$$N_2(g) + 3H_2(g) \rightarrow 2NH_3(g)$$

$$S^{\ominus}(\text{products}) = 2 \times 193 = 389\,J\,K^{-1}mol^{-1}$$

$$S^{\ominus}(\text{reactants}) = 192 + (3 \times 131) = 585\,J\,K^{-1}mol^{-1}$$

$$\Delta S^{\ominus} = 386 - 585 = -199\,J\,K^{-1}mol^{-1}$$

The standard entropy change for the reaction is $-199\,J\,K^{-1}\,mol^{-1}$. The negative sign means a decrease in disorder. Since 4 moles of gas have formed 2 moles of gas, this is what one would expect.

Calculation of $\Delta S^\ominus$ for a system at equilibrium

*Method 2 The other method of calculating the change in standard entropy for a process is to derive it from the changes in standard enthalpy and standard free energy. The equation relating these quantities is

$$\Delta G^\ominus = \Delta H^\ominus - T\Delta S^\ominus$$

If a system is at equilibrium, it moves neither in the forward direction, nor the reverse direction, $\Delta G^\ominus = 0$, and the standard entropy change is simply the standard enthalpy change divided by the temperature:

$$\Delta S^\ominus = \Delta H^\ominus/T$$

One process during which equilibrium obtains is the vaporisation of a liquid at its boiling temperature, since this process takes place under reversible conditions at a constant temperature. Another process for which $\Delta G^\ominus = 0$ is the melting of a solid at its melting temperature. For chemical reactions $\Delta G^\ominus \neq 0$, and the change in standard entropy must be found from the equation

$$\Delta S^\ominus = \frac{\Delta H^\ominus - \Delta G^\ominus}{T}$$

*Example Calculate the standard entropy of melting (or fusion) of ice. The standard enthalpy of melting of ice is $6.00\,kJ\,mol^{-1}$.

Method

$$\Delta S^\ominus = \Delta H_M^\ominus/T$$

$$= 6.00/273$$

$$= 2.2 \times 10^{-2}\,kJ\,mol^{-1}\,K^{-1}$$

Note Standard enthalpy of melting of ice was formerly called molar latent heat of fusion. Standard enthalpy of vaporisation (or evaporation) was formerly called molar latent heat of vaporisation.

===== CHECKPOINT 10D: ENTROPY =====

1. Give examples of spontaneous processes which are
(a) exothermic
(b) endothermic
(c) accompanied by an increase in entropy
(d) accompanied by a decrease in entropy.

2. State the sign of the entropy change in the following reactions:
(a) $NH_3(g) + HCl(g) \rightarrow NH_4Cl(s)$
(b) $COCl_2(g) \rightarrow CO(g) + Cl_2(g)$
(c) $PCl_3(g) + Cl_2(g) \rightarrow PCl_5(g)$
(d) $N_2(g) + 3H_2(g) \rightarrow 2NH_3(g)$
(e) $C_6H_{12}(l) + 9O_2(g) \rightarrow 6CO_2(g) + 6H_2O(g)$

3. Arrange the following in order of increasing entropy:
(a) $1\,mol\,H_2O(l)$ 100 °C, 1 atm
(b) $1\,mol\,H_2O(s)$ 0 °C, 1 atm
(c) $1\,mol\,H_2O(l)$ 0 °C, 1 atm
(d) $1\,mol\,H_2O(g)$ 100 °C, 1 atm
(e) $1\,mol\,H_2O(l)$ 25 °C, 1 atm
(f) $1\,mol\,H_2O(g)$ 100 °C, $\frac{1}{2}$ atm

4. When sodium hydroxide is dissolved in water, the temperature of the solution rises. When ammonium nitrate dissolves in water, the temperature of the solution falls. Explain the difference in behaviour.

5. Why is the entropy of a solid less than that of the gaseous form of the same substance?

*6. Do you agree with the statement: 'In a spontaneous chemical process, the system goes to a state of lower energy'? If not, explain why and formulate a correct statement.

*7. Under what conditions does $\Delta H = T\Delta S$? Under what conditions is this statement not true?

8. Is the change in standard entropy for each of the following processes positive or negative?

(a) 1 mol solid ethanoic acid → 1 mol liquid ethanoic acid

(b) 1 mol liquid ethanol → 1 mol gaseous ethanol

(c) $1\,mol\,N_2O_4(g) \rightarrow 2\,mol\,NO_2(g)$

(d) $1\,mol\,O_2(g) + 1\,mol\,N_2(g) \rightarrow 2\,mol\,NO(g)$

(e) $\frac{1}{2}\,mol\,O_2(g) + 1\,mol\,Cu(s) \rightarrow 1\,mol\,CuO(s)$

*9. Calculate the standard entropy of vaporisation of water, given that $\Delta H_{Vaporisation}^{\ominus} = 41.0\,kJ\,mol^{-1}$.

*10. At the boiling temperature of benzene, 80 °C, liquid and vapour are in equilibrium.
If $\Delta H_{Vaporisation}^{\ominus} = 34.3\,kJ\,mol^{-1}$, find $\Delta S_{Vaporisation}^{\ominus}$.

10.9.2   *CHANGE IN STANDARD FREE ENERGY

Change in $G^{\ominus}$ In order to find the change in standard free energy of a reaction, the change in standard enthalpy, the change in standard entropy and the temperature must be known and inserted into the equation

$$\Delta G^{\ominus} = \Delta H^{\ominus} - T\Delta S^{\ominus}$$

Example   Calculate the change in standard free energy at 400 K and 1000 K for the reaction

$$MgCO_3(s) \rightarrow MgO(s) + CO_2(g)$$

... Calculation of $\Delta G^{\ominus}$

	$MgCO_3(s)$	$MgO(s)$	$CO_2(g)$
Standard enthalpy/$kJ\,mol^{-1}$	−1113	−602	−394
Standard entropy/$J\,K^{-1}\,mol^{-1}$	66	27	214

Method

$$\Delta H^{\ominus} = (-602) + (-394) - (-1113) = +117\,kJ\,mol^{-1}$$

$$\Delta S^{\ominus} = 27 + 214 - 66 = 175\,J\,K^{-1}\,mol^{-1} = 0.175\,kJ\,K^{-1}\,mol^{-1}$$

$$\Delta G^{\ominus} = \Delta H^{\ominus} - T\Delta S^{\ominus}$$

$$\Delta G_{400}^{\ominus} = +117 - (400 \times 0.175) = +47\,kJ\,mol^{-1}$$

$$\Delta G_{1000}^{\ominus} = +117 - (1000 \times 0.175) = -58\,kJ\,mol^{-1}$$

$\Delta G^{\ominus}$ varies with
temperature   At 400 K, the value of $\Delta G^{\ominus}$ is $+47\,kJ\,mol^{-1}$, and the reaction will not occur; at 1000 K, the value of $\Delta G^{\ominus}$ is $-58\,kJ\,mol^{-1}$, and the reaction is feasible. (The assumption has been made that $\Delta H^{\ominus}$ and $\Delta S^{\ominus}$ do not vary with temperature.) This example shows how, in the case of a reaction which involves an increase in entropy, the feasibility of reaction increases as the temperature increases.

10.9.3   *ELLINGHAM DIAGRAMS

Ellingham plotted $\Delta G^{\ominus}$
against T   In 1944, T Ellingham published plots of $\Delta G^{\ominus}$ against T for a number of reactions. Figure 10.12 shows an Ellingham diagram for some oxidation reactions. Values of

$\Delta G^{\ominus}$ are compared for reactions written with 1 mole of oxygen on the left-hand side.

For the oxidation of the element, **E**

$$2E(s) + O_2(g) \rightarrow 2EO(s)$$

Reactions of this type involve a decrease in entropy as 1 mole of gas is converted into a solid. It follows from p. 212 that, if $\Delta S^{\ominus}$ is negative, $\Delta G^{\ominus}$ increases (becomes less negative) with temperature.

FIGURE 10.12 An Ellingham Diagram for the Formation of Some Oxides (Values of $\Delta G^{\ominus}$ relate to the reaction of 1 mole of O_2)

The diagram shows why Ag_2O decomposes when heated, but MgO is stable...

...A glance at the diagram also shows that $\Delta G^{\ominus}$ for the reduction of Cr_2O_3 by Al is always negative

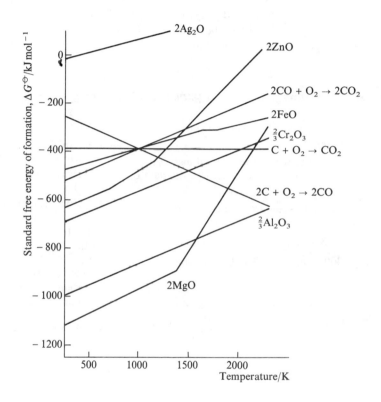

For the formation of silver oxide, $\Delta G^{\ominus}$ is negative up to 440 K. Above this temperature, $\Delta G^{\ominus}$ is positive, and the reverse reaction occurs: silver oxide decomposes into silver and oxygen. For the oxidation of magnesium, $\Delta G^{\ominus}$ is negative over the whole range of temperature shown: magnesium oxide does not decompose when heated to 2500 K. The break in the MgO line occurs at the boiling temperature of magnesium. The value of $\Delta S^{\ominus}$ for the conversion of Mg(g) into MgO(s) is more negative than that for the conversion of Mg(s) into MgO(s).

Compare the aluminium oxide line and the chromium oxide line. At all temperatures, the line for the oxidation of aluminium lies below that for chromium. This shows that the value of $\Delta G^{\ominus}$ for the reaction

$$2Al(s) + Cr_2O_3(s) \rightarrow 2Cr(s) + Al_2O_3(s)$$

is negative over the complete temperature range shown. Aluminium will reduce chromium oxide at any temperature in the range shown [p. 378]. In general, the height of a line on the diagram is a measure of the instability of the oxide.

Ellingham diagrams are useful in explaining the reduction of some metal oxides by carbon and carbon monoxide. In the reaction

The conditions employed in the blast furnace allow CO to be formed and to reduce iron oxides

$$2C(s) + O_2(g) \rightarrow 2CO(g)$$

$\Delta S^\ominus$ is positive, and $\Delta G^\ominus$ therefore decreases (becomes more negative) as T increases. The CO $\rightarrow$ CO_2 line lies below the Fe $\rightarrow$ FeO line at temperatures below 1000 K. The temperature must be *below* 1000 K to allow carbon monoxide to reduce iron oxides. To enable carbon monoxide to be formed, the C $\rightarrow$ CO line must lie below the C $\rightarrow$ CO_2 line. This happens *above* 1000 K. In the blast furnace [see Figure 24.19, p. 499] the combustion of coke at the bottom of the furnace produces a temperature of 2000 K, and carbon monoxide is formed. As it rises up the furnace, carbon monoxide cools until, in the centre of the furnace, a temperature of 1000 K allows it to reduce iron oxides. At the bottom of the furnace, carbon is able to act as a reductant because above 1000 K the lines C $\rightarrow$ CO and C $\rightarrow$ CO_2 both lie below the Fe $\rightarrow$ FeO line.

CHECKPOINT 10E: *FREE ENERGY

1. Explain the difference between a spontaneous reaction and a feasible reaction.

2. Find the temperature at which the thermal dissociation of magnesium carbonate becomes feasible. Use the data on p. 212.

3. Calculate the change in standard free energy and determine whether the reaction

$$Fe_2O_3(s) + 3H_2(g) \rightarrow 2Fe(s) + 3H_2O(g)$$

will take place at 300 K and at 800 K. Use the data:

	$Fe_2O_3(s)$	$H_2(g)$	$Fe(s)$	$H_2O(g)$
Standard enthalpy/kJ mol^{-1}	-822	0	0	-242
Standard entropy/J K^{-1} mol^{-1}	90.0	131	27.0	189

4. Refer to Figure 10.12, p. 213.

(*a*) Explain why specks of carbon appear when magnesium is burnt in a jar of carbon dioxide.

(*b*) Explain the change in gradient in the MgO graph at 1400 K.

(*c*) Give the temperature range over which magnesium oxide can be reduced by coke.

(*d*) When zinc nitrate is heated, the solid product is zinc oxide. When silver nitrate is heated, the solid product is silver.
Explain the reason for this difference in behaviour.

(*e*) Name (i) an oxide which you predict will not be reduced by aluminium,
            (ii) an oxide (other than chromium(III) oxide) which you predict will be reduced by aluminium.

10.10 LINKS WITH OTHER CHAPTERS

10.10.1 WHICH DECIDES WHETHER REACTIONS HAPPEN, THERMODYNAMICS OR KINETICS?

Thermodynamic studies tell us $\Delta G^\ominus$ and therefore whether a change is feasible...

If $\Delta G^\ominus$ for a change is negative, we can state that the reactants are thermodynamically unstable relative to that change. Methane is thermodynamically stable with respect to dissociation into its elements:

$$CH_4(g) \rightarrow C(g) + 2H_2(g); \Delta G^\ominus = +51 \text{ kJ mol}^{-1}$$

Methane is thermodynamically unstable relative to oxidation:

$$CH_4(g) + 2O_2(g) \rightarrow CO_2(g) + 2H_2O(l); \Delta G^\ominus = -580 \text{ kJ mol}^{-1}$$

Thermodynamic instability ($\Delta G^\ominus$ is negative) relative to a certain change is a necessary condition for that change to occur, but it is not the only necessary condition. Methane does not react with air at a measurable rate at room temperature (in the absence of

...*Kinetic studies tell us*
the energy of activation
and therefore how fast a
change will happen

uv light). The reason is that a large amount of energy, called the **activation energy**, must be supplied to enable the molecules to react. The concept of activation energy is covered in Chapter 14. Briefly, the idea is that, although the product molecules have a lower free energy content than the reactant molecules, there is an energy barrier in between the reactant and the product molecules. The reactant molecules must be given enough energy to surmount this barrier in order to react. Very, very few molecules of reactant acquire this activation energy at room temperature. In consequence, the overall rate of reaction is so slow as to be unobservable. The reaction is described as being **under kinetic control**. If the temperature is raised sufficiently, the proportion of energetic molecules increases, and the reaction takes place at a measurable rate. Substances which react slowly because the activation energy of the reaction is high and the temperature is too low are described as **unreactive** or **non-labile** or **inert**.

10.10.2 HOW FAR TOWARDS COMPLETION DO EQUILIBRIUM REACTIONS PROCEED?

The topic of reversible reactions which reach a state of equilibrium was introduced in Chapter 3 and will be extended in Chapter 11. As mentioned in Chapter 3, the extent to which a reaction proceeds from left to right is measured by the equilibrium constant for the reaction, K:

$$K_p = \frac{[\text{Products}]}{[\text{Reactants}]}$$

If $K_p > 1$, the products are favoured over the reactants, and the reaction goes predominantly in the direction

Reactants → Products

The position of equilibrium lies to the right-hand side. If K_p is large, the reactants will be almost completely converted into the products.

The equilibrium constant
tells us how far a reaction
will go in the direction of
forming the products

The relationship between the equilibrium constant for a reaction and the standard free energy change is

$$-\Delta G^{\ominus} = RT\ln K_p$$

For a reaction to be feasible $\Delta G^{\ominus} \leqslant 0$

If

$$\Delta G^{\ominus} = 0$$

$$RT\ln K_p = 0$$

and

$$K_p = 1$$

If $K_p = 1$, the position of equilibrium in a reaction of the simple type

A ⇌ B

lies exactly midway between the reactant and the product. If $\Delta G^{\ominus} < 0$, $K_p > 1$, and the products are favoured over the reactants. The tendency of a reaction of this type to proceed from left to right (from reactants to products) can be measured by the value of $\Delta G^{\ominus}$, which must be negative, or by the value of K_p, which must be greater than unity.

QUESTIONS ON CHAPTER 10

1. (*a*) Describe how you would find the standard enthalpy of neutralisation of a strong acid by a strong base.

[†](*b*) When you have studied Chapter 12.7, explain why the same value is obtained for the neutralisation by sodium hydroxide of nitric, hydrochloric and sulphuric acids, but a different value is obtained with ethanoic acid [p. 253].

2. (*a*) State the first law of thermodynamics.

(*b*) The heat change in a reaction can be expressed as ΔH or ΔU. What is the distinction between these quantities? Write an equation which relates the two quantities.

(*c*) Which of the two, ΔH or ΔU, is the more useful in the study of chemical reactions? Explain your answer.

3. Construct a Born–Haber cycle for the formation of solid potassium chloride from its elements in their standard states. Use the data below, and calculate the standard enthalpy of formation of KCl(s).

$$K(s) \rightarrow K(g); \qquad\qquad \Delta H^{\ominus} = 90\,kJ\,mol^{-1}$$

$$K(g) \rightarrow K^{+}(g) + e^{-}; \qquad \Delta H^{\ominus} = 418\,kJ\,mol^{-1}$$

$$\tfrac{1}{2}Cl_2(g) \rightarrow Cl(g); \qquad \Delta H^{\ominus} = 122\,kJ\,mol^{-1}$$

$$Cl(g) + e^{-} \rightarrow Cl^{-}(g); \qquad \Delta H^{\ominus} = -348\,kJ\,mol^{-1}$$

$$Cl^{-}(g) + K^{+}(g) \rightarrow KCl(s); \; \Delta H^{\ominus} = -718\,kJ\,mol^{-1}$$

4. When 5.000 g of benzene are completely burned in a bomb calorimeter, 209.4 kJ of heat are evolved. Calculate $\Delta U^{\ominus}$ and $\Delta H^{\ominus}$ for the combustion of benzene at 298 K.

5. Use the data below to find (*a*) the C—C bond enthalpy in ethane and (*b*) the value of $\Delta H^{\ominus}$ for
$3C(g) + 8H(g) \rightarrow C_3H_8(g)$

$$C(g) + 4H(g) \rightarrow CH_4(g); \qquad \Delta H^{\ominus} = -1664\,kJ\,mol^{-1}$$

$$2C(g) + 6H(g) \rightarrow C_2H_6(g); \; \Delta H^{\ominus} = -2827\,kJ\,mol^{-1}$$

6. Construct a Born–Haber cycle, and use it to find the standard lattice enthalpy of cadmium(II) iodide.

$$Cd(s) \rightarrow Cd(g); \qquad\qquad \Delta H^{\ominus} = +113\,kJ\,mol^{-1}$$

$$Cd(g) \rightarrow Cd^{2+}(g) + 2e^{-}; \; \Delta H^{\ominus} = +2490\,kJ\,mol^{-1}$$

$$I_2(s) \rightarrow I_2(g); \qquad\qquad \Delta H^{\ominus} = +19.4\,kJ\,mol^{-1}$$

$$I_2(g) \rightarrow 2I(g); \qquad\qquad \Delta H^{\ominus} = +151\,kJ\,mol^{-1}$$

$$I(g) + e^{-} \rightarrow I^{-}(g); \qquad \Delta H^{\ominus} = -314\,kJ\,mol^{-1}$$

$$Cd(s) + I_2(s) \rightarrow CdI_2(s); \; \Delta H^{\ominus} = -201\,kJ\,mol^{-1}$$

7. (*a*) State Hess's Law.

(*b*) Find the standard enthalpy change for the reaction

$$CO(g) + 2H_2(g) \rightarrow CH_3OH(l)$$

Use the data

$$CO(g) + \tfrac{1}{2}O_2(g) \rightarrow CO_2(g); \; \Delta H^{\ominus} = -283\,kJ\,mol^{-1}$$

$$H_2(g) + \tfrac{1}{2}O_2(g) \rightarrow H_2O(l); \; \Delta H^{\ominus} = -286\,kJ\,mol^{-1}$$

$$CH_3OH(l) + 1\tfrac{1}{2}O_2(g) \rightarrow CO_2(g) + 2H_2O(l);$$
$$\Delta H^{\ominus} = -715\,kJ\,mol^{-1}$$

(*c*) Explain how Hess's Law underwrites the method of calculation.

8. For benzene and cyclohexane, the standard enthalpies of combustion/kJ mol^{-1} are $C_6H_6(l)$ -3280; $C_6H_{12}(l)$ -3920. For C(s) and $H_2(g)$ the values of $\Delta H_C^{\ominus}$ are -393 and $-286\,kJ\,mol^{-1}$ respectively. (*a*) Calculate the standard enthalpies of formation of benzene and cyclohexane.

(*b*) Comment on the relative stability of these compounds.

9. Find the standard enthalpy of formation of carbon disulphide, CS_2. It burns in air to form CO_2 and SO_2. Standard enthalpies of combustion/kJ mol^{-1} are $CS_2(l)$ -1075; S(s) -297; C(s), -394.

10. The standard enthalpies of hydrogenation of cyclohexene and benzene are -120 and $-208\,kJ\,mol^{-1}$ respectively.

(*a*) Explain why the value for benzene is not three times that for cyclohexene.

(*b*) Estimate the standard enthalpies of hydrogenation of cyclohexa-1,3-diene and cyclohexa-1,4-diene.

Cyclohexene   Cyclohexa-1,3-diene   Cyclohexa-1,4-diene

11. A pellet of potassium hydroxide weighing 0.166 g is added to 50.0 g of water in a styrofoam cup. The temperature of the water rises from 19.4 to 20.2 °C. Find the standard enthalpy of solution of KOH.

12. The industrial process for the gasification of coal can be represented approximately by the equation

$$2C(s) + 2H_2O(g) \rightarrow CH_4(g) + CO_2(g)$$

From the values of $\Delta H^{\ominus}_{Combustion}$ listed on p. 208 find $\Delta H^{\ominus}$ for this reaction.

13. (*a*) Describe the dissolution of an ionic solid in water, discussing the energetics of the process.

(*b*) Suggest why, in general, $A^{2+}B^{2-}$ compounds are less soluble than $A^{+}B^{-}$ compounds.

(*c*) Suggest why (i) LiF is less soluble than NaF,
(ii) LiCl is more soluble than NaCl,
(iii) NaF is less soluble than NaCl.

	LiCl	LiF	NaCl	NaF
Lattice enthalpies/kJ mol^{-1}	-843	-1029	-775	-968
Hydration enthalpies/kJ mol^{-1}	-883	-1023	-778	-965

14. (i) What do you understand by the term *entropy*?
(ii) For each of the following reactions, say, with reasons, whether the entropy is likely to increase or decrease or stay approximately the same:
(*a*) $N_2(g) + 3H_2(g) \rightarrow 2NH_3(g)$
(*b*) $SO_2(g) + Cl_2(g) \rightarrow SO_2Cl_2(g)$
(*c*) $H_2NCO_2NH_4(s) \rightarrow CO_2(g) + 2NH_3(g)$
(*d*) $CO(g) + H_2O(g) \rightarrow CO_2(g) + H_2(g)$

15. (*a*) What is meant by the statement that a reaction is *feasible*?

(b) Given $\Delta H^{\ominus} = -185\,\text{kJ}\,\text{mol}^{-1}$ and $\Delta S^{\ominus} = 20\,\text{J}\,\text{K}^{-1}\text{mol}^{-1}$ for the reaction

$$H_2(g) + Cl_2(g) \rightarrow 2HCl(g)$$

find out whether the reaction is feasible at room temperature (298 K).

(c) What other factor operates in determining whether the reaction will take place at room temperature?

16. Discuss what is meant by (a) the enthalpy change and (b) the entropy change of a reaction. Discuss how these factors enable the directions of chemical changes to be predicted. What other factor determines whether or not a chemical change will take place?

17. In an experiment to determine the enthalpy of neutralisation of sodium hydroxide with sulphuric acid, 50 cm³ of 0.40 M sodium hydroxide were titrated, thermometrically, with 0.5 M sulphuric acid. The results were plotted as shown in Figure 10.13.

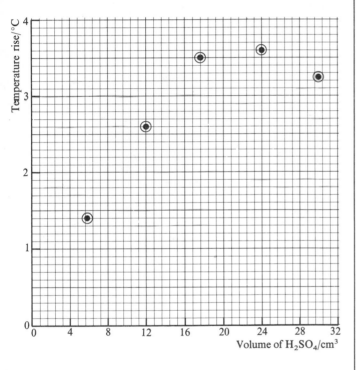

FIGURE 10.13

(a) Define 'enthalpy of neutralisation'.

(b) What is a thermometric titration?

(c) Describe, or draw a labelled diagram of, the apparatus you would use for such a titration.

(d) How do you account for the shape of the graph?

(e) Calculate a value for the enthalpy of neutralisation of sodium hydroxide with sulphuric acid. (The specific heat capacity of water is $4.2\,\text{J}\,\text{K}^{-1}\text{g}^{-1}$.) What assumptions have you made in your calculations?

(f) A similar experiment carried out with hydrocyanic acid, HCN(aq), and ammonia gave a value for the enthalpy of neutralisation of $-5.4\,\text{kJ}\,\text{mol}^{-1}$; the corresponding value for sodium hydroxide with hydrochloric acid was $-57.9\,\text{kJ}\,\text{mol}^{-1}$. Suggest a possible explanation for the difference in these two values.

(L 80)

†**18.** (a) On adding an aqueous solution of ammonia (in sufficient excess) to an aqueous solution of copper(II) sulphate, virtually all the copper ions are converted into the complex ion $Cu(NH_3)_4^{2+}$ according to the equation

$$Cu^{2+}(aq) + 4NH_3(aq) = Cu(NH_3)_4^{2+}(aq)$$

Describe how you would attempt to measure experimentally the enthalpy change for this reaction. What precaution should you take to ensure that your result did in fact apply to the effectively complete conversion of the Cu^{2+} ions into $Cu(NH_3)_4^{2+}$ ions?

(b) The following are the measured enthalpy changes of neutralisation (ΔH) at 20 °C for a mole of the acid of the given formula by a mole of potassium hydroxide, the acid and alkali both being initially in 0.1 M aqueous solutions:

HCl, $\Delta H = -57.6\,\text{kJ}\,\text{mol}^{-1}$;
HNO$_3$, $\Delta H = -57.6\,\text{kJ}\,\text{mol}^{-1}$;
CH$_2$Cl.COOH, $\Delta H = -61.9\,\text{kJ}\,\text{mol}^{-1}$.
[CH$_2$Cl.COOH, chloroethanoic or chloroacetic acid, is a monobasic acid with a dissociation constant of $1.4 \times 10^{-3}\,\text{mol}\,\text{dm}^{-3}$ at 20 °C.]

(i) Explain why the ΔH values for HCl and HNO$_3$ are the same.

(ii) What conclusions may be drawn from the different ΔH value obtained with chloroethanoic acid?

(c) A dibasic acid H$_2$A has two dissociation constants K_1 and K_2, where

$$K_1 = [H^+][HA^-]/[H_2A] \text{ and}$$
$$K_2 = [H^+][A^{2-}]/[HA^-].$$

In the series of dicarboxylic acids of the general formula HOOC(CH$_2$)$_n$COOH, the ratio K_1/K_2 at 25 °C is 730 when $n = 1$, whereas when $n = 3$ this ratio is only 12. Suggest an explanation of this difference between the two acids.

(O 82)

[See also Chapter 12 on acids.]

19. (a) By referring to ammonia, NH$_3$, define the term *average N—H bond enthalpy*.

(b) Use the following bond enthalpies, at 298 K, to estimate the enthalpy change for the reaction:

$$2NH_3(g) \rightarrow N_2(g) + 3H_2(g)$$

Bond	Bond enthalpy/kJ mol^{-1}
H—H	436
N—H	388
N≡N	944

(c) Using the answer you obtained in (b), predict how the proportion of ammonia in an equilibrium mixture of ammonia, nitrogen and hydrogen would change when the temperature is increased.

(d) For the reaction

$$2NH_3(g) \rightarrow N_2(g) + 3H_2(g),$$

the value ΔS in the equation

$$\Delta G = \Delta H - T\Delta S, \text{ is } +199\,\text{J}\,\text{K}^{-1}\text{mol}^{-1}.$$

Assuming that both ΔH and ΔS are independent of temperature, calculate the minimum temperature at which the thermal decomposition of ammonia becomes spontaneous.

(AEB 82)

20. (*a*) (i) Define the term 'internal energy change of a reaction (ΔU)',

(ii) Define the term 'enthalpy change of a reaction (ΔH)'.

(iii) Why is there a difference between ΔU and ΔH?

(*b*) Given the following data:

$$C_2H_4(g) + H_2(g) \rightarrow C_2H_6(g) \qquad \Delta H^{\ominus} = -136 \, kJ \, mol^{-1}$$

$$C_6H_6(g) + 3H_2(g) \rightarrow C_6H_{12}(g) \quad \Delta H^{\ominus} = -208 \, kJ \, mol^{-1}$$

What can you conclude about the bonding in C_6H_6?

(*c*) Calculate the *enthalpy of formation* of methane using the information given in the table opposite.

Substance	Enthalpy of combustion/$kJ \, mol^{-1}$
Methane (gas)	− 890
Hydrogen (gas)	− 286
Carbon (graphite)	− 394

(JMB 82)

11

CHEMICAL EQUILIBRIUM

An introduction to this topic has been made in Chapter 3.

11.1 REVERSIBLE REACTIONS

Some reactions take place in both directions...

Chemical reactions which take place in both directions are called reversible reactions. An example of a reversible reaction between gases is the reaction between hydrogen and iodine to form hydrogen iodide:

...e.g., the reaction between hydrogen and iodine

$$H_2(g) + I_2(g) \rightleftharpoons 2HI(g)$$

M Bodenstein made a detailed study of this reaction from 1890 to 1900. His method was to seal in glass bulbs either a mixture of hydrogen and iodine or pure hydrogen iodide, and place the bulbs in a thermostat bath. After a time interval, he cooled the bulbs rapidly so that chemical reaction stopped. Then he analysed the contents of each bulb to find the amounts of hydrogen, iodine and hydrogen iodide present.

FIGURE 11.1
Bodenstein's Results for a Set of Experiments at 448 °C

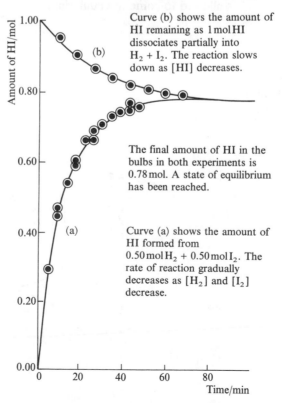

Curve (b) shows the amount of HI remaining as 1 mol HI dissociates partially into $H_2 + I_2$. The reaction slows down as [HI] decreases.

The final amount of HI in the bulbs in both experiments is 0.78 mol. A state of equilibrium has been reached.

Curve (a) shows the amount of HI formed from $0.50 \, mol \, H_2 + 0.50 \, mol \, I_2$. The rate of reaction gradually decreases as [H_2] and [I_2] decrease.

Figure 11.1 shows a set of results at a certain temperature and pressure. If the reaction between 0.50 mol H_2 and 0.50 mol I_2 went to completion, 1.00 mol HI would be formed. It appears that, with 0.78 mol HI formed, no further reaction takes place. The system has reached equilibrium. The same amount of HI is present in the bulbs when equilibrium is reached from the other direction, by the dissociation of hydrogen iodide.

A state of dynamic equilibrium is reached...

...when the forward and reverse reactions occur at the same rate

In an equilibrium mixture of hydrogen, iodine and hydrogen iodide, it is not obvious that chemical reactions are still occurring. In fact, both the forward and the reverse reactions are still taking place. Since the rates of the forward and reverse reactions are equal, the concentration of each species remains constant. The system is said to be in **dynamic equilibrium**. It is possible to prove this by injecting iodine containing a small quantity of radioactive iodine-131 into an equilibrium mixture. Radioactive iodine appears in the hydrogen iodide, showing that the synthesis of hydrogen iodide and the decomposition of hydrogen iodide are still occurring.

The equilibrium concentrations of HI, I_2 and H_2 fit the simple law

$$\frac{[HI]^2}{[H_2][I_2]} = K_c$$

Square brackets represent concentrations in $mol\,dm^{-3}$. The ratio, K_c, has a constant value at a particular temperature, no matter what amounts of HI, I_2 and H_2 are taken initially. K_c is the **equilibrium constant** for the reaction in terms of concentration.

The esterification–hydrolysis reaction comes to equilibrium

An example of a reversible reaction in solution is the esterification of ethanoic acid by ethanol to make ethyl ethanoate. This ester is hydrolysed by water to ethanoic acid and ethanol:

$$CH_3CO_2H(l) + C_2H_5OH(l) \rightleftharpoons CH_3CO_2C_2H_5(l) + H_2O(l)$$

Figure 11.2 shows what happens when ethanoic acid and ethanol are mixed and allowed to come to equilibrium.

FIGURE 11.2
An Esterification
reaches Equilibrium

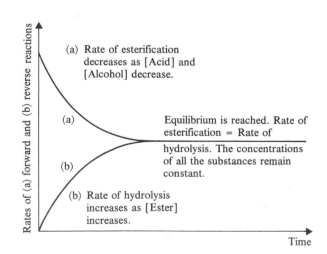

Rates of (a) forward and (b) reverse reactions

(a) Rate of esterification decreases as [Acid] and [Alcohol] decrease.

(a)

Equilibrium is reached. Rate of esterification = Rate of hydrolysis. The concentrations of all the substances remain constant.

(b)

(b) Rate of hydrolysis increases as [Ester] increases.

Time

The concentrations of the reactants and products at equilibrium fit the equilibrium law

The equilibrium concentrations of acid, alcohol, ester and water obey the law

$$\frac{[CH_3CO_2C_2H_5][H_2O]}{[CH_3CO_2H][C_2H_5OH]} = K_c$$

K_c is the equilibrium constant for the esterification.

Esterification is catalysed by inorganic acids. The presence of a catalyst does not alter the equilibrium constant. Its effect is to decrease the time needed for the system to reach a state of equilibrium.

These two reactions are examples of **homogeneous** equilibria, the first in the gas phase and the second in the liquid phase. If the reactants are in different phases, the equilibrium is described as **heterogeneous**. An example of a heterogeneous equilibrium is the reaction between steam and heated iron to form iron(II) iron (III) oxide Fe_3O_4, and hydrogen:

The reaction between iron and steam reaches equilibrium provided that it takes place in a closed container

$$3Fe(s) + 4H_2O(g) \rightleftharpoons Fe_3O_4(s) + 4H_2(g)$$

If the reaction takes place in a closed container, the system reaches equilibrium. In an open container, hydrogen escapes. In an effort to restore equilibrium, more iron reacts with steam to make more hydrogen. When this reaction was used industrially to make hydrogen, the yield was increased by passing a stream of steam over heated iron, and not allowing the system to come to equilibrium [see also p. 226].

11.2 THE EQUILIBRIUM LAW

The equilibrium law summarises the results of a vast amount of research on reactions such as the three mentioned above. For a reaction

$$a\mathbf{P} + b\mathbf{Q} \rightleftharpoons c\mathbf{R} + d\mathbf{S}$$

the equilibrium constant, expressed in terms of concentrations, K_c, is given by the **Equilibrium Law**:

$$K_c = \frac{[\mathbf{R}]^c[\mathbf{S}]^d}{[\mathbf{P}]^a[\mathbf{Q}]^b}$$

A statement of the equilibrium law

The law can be stated as follows. If a reversible reaction is allowed to reach equilibrium, then the product of the concentrations of the products (raised to the appropriate powers) divided by the product of the concentrations of the reactants (raised to the appropriate powers) has a constant value at a particular temperature. The 'appropriate power' is the coefficient of that substance in the stoichiometric equation for the reaction. The dimensions of K_c are concentration$^{(c+d-a-b)}$, and the units vary from one equilibrium to another.

11.3 POSITION OF EQUILIBRIUM

The position of equilibrium is not the same as the equilibrium constant

The proportion of products to reactants in the equilibrium mixture is described as the **position of equilibrium**:

$$a\mathbf{P} + b\mathbf{Q} \rightleftharpoons c\mathbf{R} + d\mathbf{S}$$

If the conversion of **P** and **Q** into **R** and **S** is small, the position of equilibrium lies to the left.
K_c is small.

If the equilibrium mixture is largely composed of **R** and **S**, the position of equilibrium lies to the right.
K_c is large.

The equilibrium constant K_c is not the same as the position of equilibrium. While K_c is constant at a particular temperature, a change in external conditions can alter the position of equilibrium. Chemists are often interested in finding the best conditions for operating a manufacturing process. They want to shift equilibrium reactions in the direction of forming the products. The study of the factors which alter the position of equilibrium is commercially important.

A change in conditions, e.g., pressure, will affect the position of equilibrium

11.4 THE EFFECT OF CONDITIONS ON THE POSITION OF EQUILIBRIUM

11.4.1 LE CHATELIER'S PRINCIPLE

Le Chatelier's Principle describes the effects of external factors on equilibria

Le Chatelier studied the influence of pressure, temperature and concentration on equilibria. His views are known as **Le Chatelier's Principle**. This states that in any equilibrium, when a change is made to some external factor (such as temperature or pressure), the change in the position of equilibrium is such as to tend to change the external factor in the opposite direction. The system cannot completely cancel the change in the external factor, but it moves in the direction that will minimise the change.

11.4.2 CHANGES IN CONCENTRATION

When there is a change in concentration...

An aqueous solution of bismuth(III) chloride is cloudy because of the hydrolysis

$$BiCl_3(aq) + H_2O(l) \rightleftharpoons BiOCl(s) + 2HCl(aq)$$

...the position of equilibrium changes; the equilibrium constant is unchanged

If a little concentrated hydrochloric acid is added, the position of equilibrium shifts in the direction that will absorb acid, i.e., from right to left. The solution cannot absorb *all* the acid added. The hydrolysis of bismuth(III) chloride is much reduced, and a clear solution results.

11.4.3 CHANGES IN PRESSURE

When there is a change in pressure...

It is mainly with gaseous reactions that changes in pressure are important (except for geological reactions). Le Chatelier's Principle can be applied to the formation of ammonia:

$$N_2(g) + 3H_2(g) \rightleftharpoons 2NH_3(g); \quad \Delta H^\ominus = -92\,kJ\,mol^{-1}$$

...the position of equilibrium changes; the equilibrium constant is unchanged

If the pressure of an equilibrium mixture of nitrogen, hydrogen and ammonia is increased, the equilibrium shifts in the direction that tends to decrease the pressure. It does this by decreasing the total number of molecules present, i.e., by moving from left to right of the equation. Although the position of equilibrium has moved towards the ammonia side of the equation, the equilibrium constant has not changed.

11.4.4 CHANGES IN TEMPERATURE

The equilibrium constant of an exothermic reaction decreases with temperature...

The formation of ammonia is exothermic. If the temperature is raised, the system can absorb heat by the dissociation of ammonia into nitrogen and hydrogen. The equilibrium constant for the formation of ammonia is decreased. This is a different kind of effect from the influence of pressure, where the position of equilibrium is altered but the equilibrium constant remains the same.

In the case of an endothermic reaction, such as

$$N_2(g) + O_2(g) \rightleftharpoons 2NO(g); \Delta H^\ominus = 180\,kJ\,mol^{-1}$$

...and that of an endothermic reaction increases with temperature

increasing the temperature increases the equilibrium constant.

There is further discussion of the importance of choosing the most favourable conditions for carrying out reversible reactions in the Haber process [p. 435] and the Contact process [p. 422]. Catalysts do not alter the position of equilibrium [p. 313]. They change only the rate at which equilibrium is attained.

11.4.5 PHYSICAL CHANGES

Le Chatelier's Principle applies to physical changes as well as chemical changes. In the reversible change of state

$$Ice \rightleftharpoons Water; \quad \Delta H^\ominus = 6.00\,kJ\,mol^{-1}$$

Physical changes also obey Le Chatelier's Principle

When pressure is exerted on a mixture of ice and water in equilibrium, the system adjusts in such a manner as to tend to decrease its volume, i.e., ice melts. (The density of water increases from 0 °C to 4 °C.) When the temperature rises, the system adjusts itself in such a manner as to tend to resist an increase in temperature. This can be accomplished by the melting of ice, which is an endothermic process. As long as there is any ice left, the temperature will remain constant at 0 °C (at 1 atm).

CHECKPOINT 11A: LE CHATELIER'S PRINCIPLE

1. Consider the equilibrium

$$N_2(g) + 3H_2(g) \rightleftharpoons 2NH_3(g); \Delta H^\ominus = -90\,kJ\,mol^{-1}$$

At 25 °C and 10 atm, the percentage conversion to ammonia in the equilibrium mixture is 5%. If the pressure is increased to 40 atm, while the temperature remains the same, will the percentage conversion to ammonia be greater or less? If the temperature is increased to 250 °C, while the pressure remains at 10 atm, will the equilibrium amount of ammonia be greater or less? What will happen if some iron and molybdenum are added to the equilibrium mixture?

2. Consider the equilibrium

$$2NO_2(g) \rightleftharpoons N_2O_4(g); \Delta H^\ominus = -54\,kJ\,mol^{-1}$$

How are (i) the extent of conversion of NO_2 to N_2O_4 and (ii) the equilibrium constant affected by (a) an increase in temperature, (b) an increase in pressure and (c) an increase in volume at constant temperature? Explain your answers.

3. In the equilibrium

$$PCl_5(g) \rightleftharpoons PCl_3(g) + Cl_2(g); \Delta H^\ominus = 90\,kJ\,mol^{-1}$$

what is the effect on (i) the position of equilibrium and (ii) the equilibrium constant of (a) increasing the temperature, (b) decreasing the volume of the container, (c) adding a catalyst, (d) adding $Cl_2(g)$ and (e) adding a noble gas? Explain your answers.

4. The freezing of water at 0 °C is represented

$$H_2O(l, \rho = 1.00\,kg\,dm^{-3}) \rightleftharpoons H_2O(s, \rho = 0.92\,kg\,dm^{-3})$$

Explain why the application of pressure to ice at 0 °C makes it melt. Why do most other solids not show the same behaviour?

11.5 EXAMPLES OF REVERSIBLE REACTIONS

11.5.1 REACTION 1: ESTERIFICATION

Example Calculate the amount of ethyl ethanoate formed when 1.0 mol ethanoic acid and 1.0 mol ethanol reach equilibrium. At room temperature, $K_c = 4.0$.

Method Let the amount of ethyl ethanoate at equilibrium $= x$ mol and the volume of the solution $= V$ dm^3. The equilibrium amounts and concentrations are as follows:

$$CH_3CO_2H(l) + C_2H_5OH(l) \rightleftharpoons CH_3CO_2C_2H_5(l) + H_2O(l)$$

Amount/mol	$1 - x$	$1 - x$	x	x
Concentration/mol dm^{-3}	$(1 - x)/V$	$(1 - x)/V$	x/V	x/V

Since

$$K_2 = \frac{[CH_3CO_2C_2H_5][H_2O]}{[CH_3CO_2H][C_2H_5OH]} = 4.0$$

$$x^2/(1 - x)^2 = 4.0$$

$$3x^2 - 8x + 4 = 0$$

Solving this quadratic equation gives $x = 2$ or $\frac{2}{3}$. The value 2 can be excluded because it is higher than the amount of ethanoic acid present initially. The amount of ester formed $= \frac{2}{3}$ mol.

AN EXPERIMENTAL DETERMINATION OF K_c

*A method of finding K_c
for the esterification
reaction between ethanol
and ethanoic acid*

The reaction to be studied is the esterification discussed above. It is catalysed by acid. The stages of the experiment are as follows.

1. A number of mixtures are made up, containing ethanol, ethanoic acid, ethyl ethanoate and water. Each mixture is different. Every mixture contains a small amount of hydrochloric acid to catalyse the reaction.

2. The mixtures are put into stoppered bottles and left for a week in a thermostat bath.

3. At the end of the week, the contents of each flask are titrated against standard sodium hydroxide solution, using phenolphthalein as indicator. The titration gives the amount of CH_3CO_2H + the amount of HCl. The amount of HCl is still the same as that present initially. The amount of CH_3CO_2H is found by subtraction.

4. The amounts of the other substances present are calculated.

A sample calculation A 10.0 cm^3 mixture contains the initial amounts/mol ethanol 0.0515; ethanoic acid 0.0525; water 0.0167; ester 0.0314; H^+(aq) 1.00×10^{-3}.

The equilibrium amount of ethanoic acid $= 0.0255$ mol.
Since the amount of ethanoic acid has decreased by 0.0270 mol, ethanol has decreased by the same amount, and ester and water have both increased by this amount.

Species	CH_3CO_2H +	C_2H_5OH $\rightleftharpoons$	$CH_3CO_2C_2H_5$ +	H_2O
Initial amount/mol	0.0525	0.0515	0.0314	0.0167
Equilibrium amount/mol	0.0255	0.0245	0.0584	0.0437

Since

$$K_c = \frac{[CH_3CO_2C_2H_5][H_2O]}{[CH_3CO_2H][C_2H_5OH]}$$

All the substances are present in the same volume of solution, therefore

$$K_c = \frac{0.0584 \times 0.0437}{0.0255 \times 0.0245} = 4.1$$

From calculations on all the mixtures, an average value of K_c at the chosen temperature is obtained. K_c is dimensionless.

11.5.2 REACTION 2: THE REACTION BETWEEN HYDROGEN AND IODINE

In the reaction between hydrogen and iodine

$$H_2(g) + I_2(g) \rightleftharpoons 2HI(g)$$

For gaseous reactions, it is usual to employ K_p, the equilibrium constant in terms of partial pressures

$$K_c = \frac{[HI]^2}{[H_2][I_2]}$$

Since this is a reaction between gases, the concentration of each gas can be expressed as a partial pressure [p. 140]. Then

$$K_p = \frac{p_{HI}^2}{p_{H_2} \times p_{I_2}}$$

K_p is the equilibrium constant in terms of partial pressures.

You can see that, since the pressure units cancel out in the expression, K_p is dimensionless. K_p predicts the same equilibrium composition, no matter what the pressure may be. This is not the case for all gaseous equilibria [contrast Reaction 3, below. (If the gases are not ideal and do not obey Dalton's Law of Partial Pressures accurately, there will be some discrepancies at high pressures.)

A detailed treatment of the H_2 and I_2 reaction and...

...a worked example

Example   When 1.00 mol hydrogen and 1.00 mol iodine are allowed to reach equilibrium in a 1.00 dm³ flask at 450 °C and 1.01×10^5 N m⁻², the amount of hydrogen iodide at equilibrium is 1.56 mol. Calculate K_p at 450 °C.

Method

	$H_2(g)$	$+ I_2(g) \rightleftharpoons 2HI(g)$		*Total*
P = total pressure				
Initial amount/mol	1.00	1.00	0	2.00
Equilibrium amount/mol	$1-a$	$1-a$	$2a$	2
Since $2a = 1.56$ $a = 0.78$ mol				
Equilibrium amount/mol	0.22	0.22	1.56	2.00
Equilibrium partial pressure	$\left(\frac{0.22P}{2.00}\right)$	$\left(\frac{0.22P}{2.00}\right)$	$\left(\frac{1.56P}{2.00}\right)$	P

$$K_p = \frac{p_{HI}^2}{p_{H_2}p_{I_2}} = \left(\frac{1.56P}{2.00}\right)^2 \Big/ \left(\frac{0.22P}{2.00}\right)^2 = 50$$

11.5.3 REACTION 3: THE HABER PROCESS

In the **Haber process** for making ammonia

$$N_2(g) + 3H_2(g) \rightleftharpoons 2NH_3(g)$$

$$K_p = \frac{p_{NH_3}^2}{p_{N_2} \times p_{H_2}^3}$$

A detailed treatment of   Let the mole fractions of the components in the equilibrium mixture be x_{NH_3}, x_{N_2}
the Haber Process   and x_{H_2}. The total pressure = P. The partial pressures are

$$p_{N_2} = x_{N_2}P; \quad p_{H_2} = x_{H_2}P; \quad p_{NH_3} = x_{NH_3}P$$

$$K_p = \frac{(x_{NH_3}P)^2}{x_{N_2}P(x_{H_2}P)^3}$$

$$= \frac{x_{NH_3}^2}{x_{N_2}x_{H_2}^3 P^2}$$

The unit of K_p is P^{-2}, e.g., atm^{-2} or $N^{-2}\,m^4$.

When P increases, K_p stays constant; therefore x_{NH_3} must increase, and x_{N_2} and x_{H_2} must decrease. The higher the pressure, the greater will be the percentage conversion into ammonia.

...and a worked example   **Example**   Nitrogen and hydrogen are mixed in a molar ratio $1:3$. At equilibrium, at $600\,°C$ and $10\,atm$, 15% of the gases have been converted into ammonia. Find K_p at $600\,°C$.

Method   Let a be the fraction of each mole of N_2 that has reacted at equilibrium. Then the equilibrium amounts of the gases are as listed below.
P = total pressure = $10\,atm$.

Species	$N_2(g)$	$+\,3H_2(g)$	$\rightleftharpoons 2NH_3(g)$	Total
Initial amount/mol	1	3	0	4
Equilibrium amount/mol	$1-a$	$3(1-a)$	$2a$	$4-2a$
Mole fraction	$\dfrac{1-a}{4-2a}$	$\dfrac{3(1-a)}{4-2a}$	$\dfrac{2a}{4-2a}$	

The percentage of $NH_3 = \dfrac{2a}{4-2a} \times 100 = 15 \quad \therefore a = 0.26$

Mole fraction	$\dfrac{0.74}{3.48}$	$\dfrac{2.22}{3.48}$	$\dfrac{0.52}{3.48}$	1
Equilibrium partial pressure	$\dfrac{0.74P}{3.48}$	$\dfrac{2.22P}{3.48}$	$\dfrac{0.52P}{3.48}$	P

$$K_p = p_{NH_3}^2/p_{N_2}p_{H_2}^3$$

$$= \left(\frac{0.52P}{3.48}\right)^2 \bigg/ \left(\frac{0.74P}{3.48}\right)\left(\frac{2.22P}{3.48}\right)^3$$

$$= 0.40P^{-2} = 4.0 \times 10^{-3}\,atm^{-2}$$

11.5.4   REACTION 4: THE REACTION BETWEEN IRON AND STEAM

The heterogeneous   In the reaction between steam and heated iron
reaction between iron and
steam...

$$3Fe(s) + 4H_2O(g) \rightleftharpoons Fe_3O_4(s) + 4H_2(g)$$

$$K_p = \frac{p_{H_2}^4}{p_{H_2O}^4}$$

The solids do not appear in the expression. Their vapour pressures remain constant (at a constant temperature) as long as there is some of each solid present. These constant vapour pressures are incorporated into the value of the constant K_p.

Example A mixture of iron and steam was allowed to reach equilibrium at 600 °C. The equilibrium pressures of hydrogen and steam were 3.2 kPa and 2.4 kPa respectively. Calculate the value of the equilibrium constant in terms of partial pressures.

Method

$$K_p = \frac{p_{H_2}^4}{p_{H_2O}^4} = \left(\frac{3.2}{2.4}\right)^4 = 3.16$$

The equilibrium constant, K_p is 3.2. It is dimensionless.

CHECKPOINT 11B: EQUILIBRIUM CONSTANTS

1. Write an expression for the equilibrium constant, K_p, for each of the following reactions:
(a) $CS_2(g) + 4H_2(g) \rightleftharpoons CH_4(g) + 2H_2S(g)$
(b) $4NH_3(g) + 5O_2(g) \rightleftharpoons 4NO(g) + 6H_2O(g)$
(c) $2NO_2(g) + 7H_2(g) \rightleftharpoons 2NH_3(g) + 4H_2O(g)$

2. Write an expression for the equilibrium constant, K_c, for each of the following reactions:
(a) $Sn^{2+}(aq) + 2Fe^{3+}(aq) \rightleftharpoons Sn^{4+}(aq) + 2Fe^{2+}(aq)$
(b) $Ag^+(aq) + Fe^{2+}(aq) \rightleftharpoons Fe^{3+}(aq) + Ag(s)$
(c) $2Cr^{3+}(aq) + Fe(s) \rightleftharpoons 2Cr^{2+}(aq) + Fe^{2+}(aq)$

3. Equilibrium is established in the reaction

$$A(aq) + B(aq) \rightleftharpoons 2C(aq)$$

If equilibrium concentrations are [A] = 0.25, [B] = 0.40, [C] = 0.50 mol dm^{-3}, what is the value of K_c?

4. The following reaction was allowed to reach equilibrium:

$$2D(aq) + E(aq) \rightleftharpoons F(aq)$$

The initial amounts of the reactants present in 1.00 dm^3 of solution were 1.00 mol D and 0.75 mol E. At equilibrium, the amounts were 0.70 mol D and 0.60 mol E. Calculate the equilibrium constant, K_c.

5. The gases SO_2, O_2 and SO_3 are allowed to reach equilibrium. The partial pressures of the gases are $p_{SO_2} = 0.050$ atm, $p_{O_2} = 0.025$ atm, $p_{SO_3} = 1.00$ atm. Find the values of K_p for the equilibria
(a) $SO_2(g) + \frac{1}{2}O_2(g) \rightleftharpoons SO_3(g)$
(b) $2SO_2(g) + O_2(g) \rightleftharpoons 2SO_3(g)$

6. A mixture contained 1.00 mol of ethanoic acid and 0.500 mol of ethanol. After the system had come to equilibrium, a portion of the mixture was titrated against 0.200 mol dm^{-3} sodium hydroxide solution. The titration showed that the whole of the equilibrium mixture would require 289 cm^3 of the standard alkali for neutralisation. Find the value of K_c for the esterification reaction.

11.5.5 REACTION 5: THERMAL DISSOCIATION

Another type of reaction which reaches an equilibrium position is thermal dissociation. For example, when phosphorus(V) chloride is heated, it dissociates partially into phosphorus(III) chloride and chlorine:

$$PCl_5(g) \rightleftharpoons PCl_3(g) + Cl_2(g)$$

As a result, the pressure of the gas (at constant volume) or, alternatively, the volume of the gas (at constant pressure) is greater than expected.

The fraction of molecules that dissociate is called the **degree of dissociation**, and is represented by the letter α. The amounts of substances in the equilibrium mixture formed from 1 mole of PCl_5 are as follows:

When substances dissociate in the gas phase, the actual pressure is greater than expected...

Species	$PCl_5(g)$	$PCl_3(g)$	$Cl_2(g)$	Total
Amount	$1 - \alpha$	α	α	$1 + \alpha$

$\therefore \quad \dfrac{\text{Actual number of particles}}{\text{Expected number of particles}} = \dfrac{1 + \alpha}{1}$

Since the pressure of a gas is proportional to the concentration of particles present, at constant volume

$$\dfrac{\text{Actual pressure of gas}}{\text{Expected pressure of gas}} = 1 + \alpha$$

For a substance in which one molecule dissociates into n molecules

...From the increase in pressure, the degree of dissociation can be found...

$$\dfrac{\text{Actual number of particles}}{\text{Expected number of particles}} = 1 + (n - 1)\alpha$$

The equilibrium constant for the thermal dissociation of PCl_5 is

...and also the equilibrium constant for the dissociation

$$K_c = \dfrac{[Cl_2][PCl_3]}{[PCl_5]}$$

If V = the volume of the container

$$K_c = \dfrac{(\alpha/V) \times (\alpha/V)}{(1 - \alpha)/V} = \dfrac{\alpha^2}{(1 - \alpha)V} = \dfrac{\alpha^2 c}{1 - \alpha}$$

If c = the initial concentration of PCl_5, then $c = \dfrac{1}{V}$ and

$$K_c = \dfrac{\alpha^2 c}{1 - \alpha}$$

The equilibrium constant K_p, in terms of partial pressures, is

$$K_p = \dfrac{p_{Cl_2} p_{PCl_3}}{p_{PCl_5}}$$

If $p_{Cl_2} = p_{PCl_3} = \left(\dfrac{\alpha}{1 + \alpha}\right) P$, and $p_{PCl_5} = \left(\dfrac{1 - \alpha}{1 + \alpha}\right) P$, where P = total pressure,

$$K_p = \dfrac{\left(\dfrac{\alpha}{1 + \alpha}\right)^2 P^2}{\left(\dfrac{1 - \alpha}{1 + \alpha}\right) P} = \dfrac{\alpha^2 P}{1 - \alpha^2}$$

The unit of K_p is a pressure unit.
If α is small, $1 - \alpha^2 \approx 1$, and $\alpha^2 \approx K_p/P$, i.e., the higher the pressure, the lower is the degree of dissociation.

Example   When 0.100 mol of phosphorus(V) chloride is heated at 150 °C in a 1.00 dm^3 vessel, a pressure of $4.38 \times 10^5\,\mathrm{N\,m^{-2}}$ is measured. Calculate the degree of dissociation and the value of K_p at this temperature.

Method   (a) Find α from a comparison of real and expected pressures. The expected pressure is calculated from $PV = nRT$ [p. 139]. Putting $V = 1.00 \times 10^{-3}\,\mathrm{m^3}$, $T = 423\,\mathrm{K}$, $n = 0.100$ and $R = 8.314\,\mathrm{J\,K^{-1}\,mol^{-1}}$ gives

$$P = 3.51 \times 10^5\,\mathrm{N\,m^{-2}}$$

Since

$$\text{Measured pressure/Expected pressure} = 1 + \alpha$$

$$4.38 \times 10^5/(3.51 \times 10^5) = 1 + \alpha$$

The degree of dissociation, $\alpha = 0.25$

(b) Find K_p.

Since $K_p = \alpha^2 P/(1 - \alpha^2)$

and $\alpha = 0.25$, and $P = 4.38 \times 10^5\,\mathrm{N\,m^{-2}}$

$$K_p = 2.92 \times 10^4\,\mathrm{N\,m^{-2}}$$

Thermal dissociation results in a low measured value for the molar mass...

Another sign of thermal dissociation is an unexpectedly low value for the molar mass. [See method, p. 156.] From the equation

$$PV = \frac{m}{M} RT$$

you can see that, if the volume, V, of a mass of gas, m, is greater than expected, then the value obtained for the molar mass, M, will be less than expected. Since

$$\text{Actual volume of gas/Expected volume of gas} = 1 + \alpha$$

$$\text{Expected molar mass/Measured molar mass} = 1 + \alpha$$

If one molecule dissociates into n particles

$$\text{Expected molar mass/Measured molar mass} = 1 + (n - 1)\alpha$$

Example   A molar mass determination gave a value of 82.5 g mol^{-1} for dinitrogen tetraoxide, N_2O_4, at 25 °C. Calculate the degree of dissociation at this temperature.

Method   The dissociation that occurs is

...An example is the dissociation of N_2O_4

$$N_2O_4(g) \rightleftharpoons 2NO_2(g)$$

Thus, $n = 2$ and

$$\text{Calculated molar mass/Measured molar mass} = 1 + \alpha$$

$\therefore$ $92.0/82.5 = 1 + \alpha$

Degree of dissociation $\alpha = 0.115$

11.5.6   REACTION 6: ASSOCIATION

If association occurs, the pressure of a gas is less than expected...

A pressure less than that expected of a gas and a measured value of molar mass higher than that expected are signs that molecules are associated. If a substance **A** dimerises, and the degree of dimerisation is α, the amounts of monomer and dimer present at equilibrium are as follows:

...and the value of molar mass is higher than expected...

Species		$2A(g) \rightleftharpoons A_2(g)$		Total
Amount/mol		$(1 - \alpha)$	$\alpha/2$	$(1 - \alpha/2)$

$$\frac{\text{Actual volume}}{\text{Expected volume}} = \frac{\text{Actual number of moles}}{\text{Expected number of moles}} = \frac{(1 - \alpha/2)}{1}$$

Since $M = mRT/PV$

$$\frac{\text{Expected molar mass}}{\text{Measured molar mass}} = 1 - \alpha/2$$

In general, if n molecules associate

$$\frac{\text{Expected molar mass}}{\text{Measured molar mass}} = 1 - \frac{(n - 1)\alpha}{n}$$

...An example is the dimerisation of $AlCl_3$

Example   Calculate the degree of dimerisation of aluminium chloride that results in an experimental value of $175\,\text{g}\,\text{mol}^{-1}$ for its molar mass.

Method   The calculated molar mass of $AlCl_3 = 133.5\,\text{g}\,\text{mol}^{-1}$. Since the dimerisation is

$$2AlCl_3(g) \rightleftharpoons Al_2Cl_6(g)$$

$$\frac{\text{Calculated molar mass}}{\text{Measured molar mass}} = 1 - \alpha/2$$

$$\frac{133.5}{175} = 1 - \alpha/2$$

Degree of dimerisation $\alpha = 0.47$.

CHECKPOINT 11C: ASSOCIATION AND DISSOCIATION

1. When 0.100 mol of carbon dichloride oxide, $COCl_2$, was vaporised at 400 °C in a 1.00 dm³ vessel, a pressure of $8.00 \times 10^5\,\text{N}\,\text{m}^{-2}$ developed. Calculate the degree of dissociation of $COCl_2(g)$ into $CO(g) + Cl_2(g)$.

2. The vapour formed from 0.200 g of antimony(V) chloride, $SbCl_5$, occupied 40.4 cm³ at 250 °C and $1.01 \times 10^5\,\text{N}\,\text{m}^{-2}$. Calculate the degree of dissociation into $SbCl_3(g) + Cl_2(g)$ and the value of K_p at 250 °C.

3. The dissociation of nitrogen chloride oxide is represented by the equation

$$2NOCl(g) \rightleftharpoons 2NO(g) + Cl_2(g)$$

If the degree of dissociation is α, how many moles of substance are formed from 1 mole of NOCl? At 200 °C, a value of $54.3\,\text{g}\,\text{mol}^{-1}$ was obtained by a molar mass determination. Calculate the value of α at this temperature, and state what fraction of the molecules in the equilibrium mixture are chlorine molecules.

4. A pressure of $9.20 \times 10^3\,\text{N}\,\text{m}^{-2}$ was measured when 1.00 g of aluminium bromide vaporised in a 1.00 dm³ vessel at 200 °C. Calculate the degree of dimerisation of aluminium bromide molecules.

5. Phosphorus(V) chloride is 30.0% dissociated at a certain temperature and a pressure of $1.01 \times 10^5\,\text{N}\,\text{m}^{-2}$. Find the value of K_p at this temperature for the dissociation,

$$PCl_5(g) \rightleftharpoons PCl_3(g) + Cl_2(g)$$

11.6 *THE DEPENDENCE OF EQUILIBRIUM CONSTANT ON TEMPERATURE

T and K_p are related by the van 't Hoff isochore ...

The quantitative interpretation of the effect of temperature on equilibrium constant was derived by van 't Hoff, and is known as the **van 't Hoff isochore.**

$$\frac{d \ln K_p}{dT} = \frac{\Delta H^{\ominus}}{RT^2}$$

where $\Delta H^{\ominus}$ = the standard enthalpy of reaction at temperature T kelvins, R = the gas constant, and K_p = the equilibrium constant.

On the assumption that $\Delta H^{\ominus}$ is independent of pressure, integration of the equation gives

$$\ln K_p = \frac{-\Delta H^{\ominus}}{RT} + \text{Integration constant}$$

A plot of $\ln K_p$ against T can be used to find $\Delta H^{\ominus}$ for the reaction. The plot gives a straight line with a gradient of $-\Delta H^{\ominus}/R$.

Example The reaction

$$Fe(s) + H_2O(g) \rightleftharpoons FeO(s) + H_2(g)$$

has an equilibrium constant of 2.00 at 1073 K and 1.49 at 1273 K. Calculate the standard enthalpy of the reaction.

Method
Since

$$\ln K_p = \frac{-\Delta H^{\ominus}}{RT} + \text{Constant}$$

$$\ln \frac{K_1}{K_2} = -\frac{\Delta H^{\ominus}}{R} \left(\frac{1}{T_1} - \frac{1}{T_2} \right)$$

where K_1 and K_2 are the values of the equilibrium constant at T_1 and T_2.

$$\Delta H^{\ominus} = \frac{RT_1 T_2 \ln(K_1/K_2)}{T_1 - T_2}$$

$$= \frac{8.314 \times 1073 \times 1273}{-200} \ln \frac{2.00}{1.49}$$

$$= -16\,700 \, \text{J mol}^{-1}$$

$$= -16.7 \, \text{kJ mol}^{-1}$$

This example illustrates how the standard enthalpy of reaction can be found from two values of the equilibrium constant at two temperatures.

11.7 PHASE EQUILIBRIUM DIAGRAMS

Definition of a phase

A **phase** is a homogeneous part of a system which is physically distinct from other parts of the system. A mixture of gases is one phase. A solution is one phase. A saturated solution in the presence of an excess of solute is a two-phase system. A mixture of solids consists of as many phases as there are solids present.

In a phase diagram...

When a substance exists in different physical states, the conditions under which each state exists can be represented by a **phase diagram**. An **area** on a phase diagram represents one phase. A **line** represents the conditions under which two phases can exist in equilibrium. A **triple point** describes the conditions under which three phases can coexist.

11.7.1 WATER

Figure 11.3 shows the phase diagram for water.

...an area = one phase...

...a line = two phases

...a point = three phases, and is named a triple point

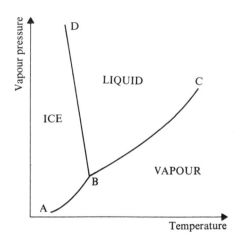

BD = Effect of pressure on freezing temperature of water

BC = Effect of temperature on saturated vapour pressure of water

AB = Effect of temperature on saturated vapour pressure of ice

B = Conditions under which solid, liquid and gaseous forms can coexist — a triple point

FIGURE 11.3 Phase Equilibrium Diagram for Water (not to scale)

11.7.2 SULPHUR

FIGURE 11.4 Phase Equilibrium Diagram for Sulphur (not to scale)

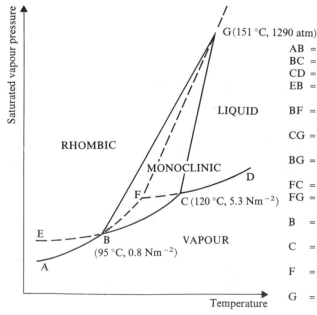

AB = svp curve for rhombic sulphur
BC = svp curve for monoclinic sulphur
CD = svp curve for liquid sulphur
EB = svp curve for metastable monoclinic sulphur
BF = svp curve for metastable rhombic sulphur
CG = Effect of pressure on melting temperature of monoclinic sulphur
BG = Effect of pressure on transition temperature
FC = svp of metastable liquid
FG = Effect of pressure on T_m of metastable rhombic sulphur
B = A triple point, the transition temperature for Rhombic $\rightleftharpoons$ Monoclinic
C = A triple point, T_m of monoclinic sulphur under its own svp
F = A triple point (rhombic, liquid and vapour)
G = A triple point (rhombic, monoclinic and liquid)

Sulphur is allotropic...

...and shows enantiotropy

Sulphur is allotropic...

...and shows enantiotropy

At the transition temperature, both allotropes have the same svp

Figure 11.4 shows the phase equilibrium diagram for sulphur. This element exists in **allotropic** forms. The allotropes, rhombic and monoclinic sulphur, have the same chemical properties but have different crystalline structures. The type of allotropy shown by sulphur is **enantiotropic**. Each allotrope can be changed into the other by changing the temperature. Under certain conditions rhombic sulphur is the stable allotrope; under other conditions the monoclinic form is the stable one. At the transition temperature both allotropes have the same vapour pressure, and can exist in equilibrium. If rhombic sulphur is heated quickly, it may persist as the rhombic form above the transition temperature. In this state it is described as **metastable**, as it is gradually changing into the monoclinic form. Monoclinic sulphur can also exist in a metastable form if it is cooled quickly below the transition temperature.

11.7.3 PHOSPHORUS

Phosphorus shows monotropic allotropy

White phosphorus has a higher vapour pressure than red phosphorus

Figure 11.5 shows the phase equilibrium diagram for phosphorus. It illustrates **monotropic** allotropy. One allotrope (red) is stable; the other (white) is metastable: it is always gradually changing into the stable form at all temperatures. White phosphorus has a higher vapour pressure than red phosphorus. If red phosphorus is heated, its vapour pressure increases, but before it reaches the theoretical transition temperature, E, it reaches its melting temperature, B. To make white phosphorus, the liquid or vapour form must be cooled rapidly.

FIGURE 11.5 Phase Equilibrium Diagram for Phosphorus (not to scale)

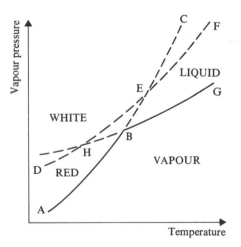

AC = svp curve of red phosphorus
DF = svp curve of white phosphorus
BG = svp curve of liquid phosphorus
HF = svp curve of metastable white phosphorus
BC = svp curve of metastable red phosphorus
E = Hypothetical transition temperature
B = T_m of red phosphorus under its own svp
H = T_m of white phosphorus under its own svp

11.7.4 OXYGEN

O_2 and O_3 show dynamic allotropy

A third type of allotropy is **dynamic allotropy**. Both forms are present at all temperatures, but the proportions in which the forms are present depends on the temperature. Oxygen and ozone are in dynamic equilibrium:

$$3O_2(g) \rightleftharpoons 2O_3(g)$$

CHECKPOINT 11D: PHASE EQUILIBRIA

1. Figure 11.6 shows the pressure–temperature phase diagram for substance **X**.

(a) Trace the figure. On your copy, mark the areas S (solid), L (liquid) and G (gas).

(b) Name the point A, and explain its significance.

(c) Draw in the vapour pressure–temperature curve for a dilute solution in **X** of a non-volatile solute.

(d) Mark the freezing temperature of the solution from (c).

(e) Say, with an explanation, how the freezing temperature of pure **X** varies with an increase in pressure.

FIGURE 11.6

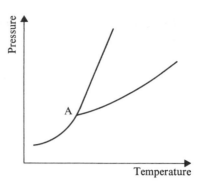

QUESTIONS ON CHAPTER 11

1. Study the gaseous equilibrium

$$N_2O_4(g) \rightleftharpoons 2NO_2(g);$$

$$\Delta H^{\ominus} = 54\,kJ\,mol^{-1} \; Endothermic$$

(a) Write an expression for the equilibrium constant, K_p, in terms of partial pressures.

(b) State, with reasons, the effect on the equilibrium position of (i) an increase in total pressure and (ii) an increase in temperature.

(c) At 60 °C, 1.00 dm³ of the gas weighed 2.585 g under a pressure of $1.01 \times 10^5\,N\,m^{-2}$. Find the degree of dissociation of N_2O_4 and the value of K_p. ($R = 8.314\,J\,K\,mol^{-1}$)

2. The **Contact process** involves the equilibrium

$$2SO_2(g) + O_2(g) \rightleftharpoons 2SO_3(g)$$

(a) Write an expression for K_p.

(b) At 1000 K and 1.00 atm, the equilibrium mixture contains 27.0% SO_2 and 40.0% O_2 by volume. Find the value of K_p.

(c) Use the expression from (a) to show how the ratio of SO_3/SO_2 in equilibrium mixtures is related to the total pressure in the presence of a fixed excess of oxygen.

(d) Explain why industrial plants operate at pressures of around 2 atm.

3. Study the equilibrium

$$H_2O(g) + CO(g) \rightleftharpoons H_2(g) + CO_2(g)$$

(a) Write an expression for K_p.

(b) When 1.00 mol of steam and 1.00 mol of carbon monoxide are allowed to reach equilibrium, 33.3% of the equilibrium mixture is hydrogen. Calculate the value of K_p. State the units of K_p.

(c) Would you expect an increase in total pressure to affect the yield of hydrogen? Explain your answer.

4. An aqueous solution is made by dissolving 1.00 mol of $AgNO_3$ and 1.00 mol of $FeSO_4$ in water and making up to 1.00 dm³. When the equilibrium

$$Ag^+(aq) + Fe^{2+}(aq) \rightleftharpoons Fe^{3+}(aq) + Ag(s)$$

is established, $[Ag^+] = [Fe^{2+}] = 0.44\,mol\,dm^{-2}$, and $[Fe^{3+}] = 0.56\,mol\,dm^{-3}$. Find K_c for the equilibrium.

5. Iron is added to a solution containing the ions, Cr^{3+}, Cr^{2+} and Fe^{2+}. The equilibrium concentrations of the ions present in solution are $[Cr^{3+}] = 0.030\,mol\,dm^{-3}$, $[Cr^{2+}] = 0.27\,mol\,dm^{-3}$, $[Fe^{2+}] = 0.11\,mol\,dm^{-3}$. Find K_c for the equilibrium

$$2Cr^{3+}(aq) + Fe(s) \rightleftharpoons 2Cr^{2+}(aq) + Fe^{2+}(aq)$$

6. For the equilibrium

$$2P(g) + Q(g) \rightleftharpoons 2R(g)$$

K_c is numerically equal to 6.0. Into a 1.00 dm³ flask are introduced 3.0 mol of **P**, 3.0 mol of **Q** and 3.0 mol of **R**.

(a) State the unit in which K_c is expressed.

(b) Is the mixture at equilibrium?

(c) If not, what must the volume of the flask be in order for such a mixture to exist in equilibrium at the temperature for which K_c is given?

7. Into a 1.00 dm³ vessel at 1000 K are introduced simultaneously 0.500 mol of SO_2, 0.100 mol of O_2 and 0.700 mol of SO_3. For the equilibrium

$$2SO_2(g) + O_2(g) \rightleftharpoons 2SO_3(g)$$

the value of K_c at 1000 K is $2.8 \times 10^2\,mol\,dm^{-3}$.

(a) Is the mixture at equilibrium?

(b) If it is not at equilibrium, in which direction will the reaction proceed?

8. Figure 11.7 shows the phase equilibrium diagram for carbon dioxide.

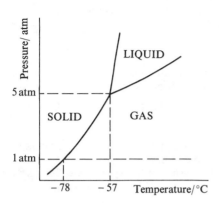

FIGURE 11.7

(a) Using the diagram, predict what will happen when solid CO_2 is warmed above $-78\,°C$ at 1 atm.

(b) Give the conditions under which gaseous CO_2 can be liquefied.

(c) At what temperature is liquid CO_2 stable when the pressure is below 5 atm?

(d) The phase diagram for CO_2 differs in two ways from that for water [see Figure 11.3, p. 232]. What are they?

*9. The following values of K_p were obtained at the stated temperatures in a study of the reaction

$$2SO_2(g) + O_2(g) \rightleftharpoons 2SO_3(g)$$

Use them to obtain a value for the standard enthalpy change of the reaction.

Temperature/°C	856	780	732	684	636
K_p/kPa	21.1	96.5	287	921	3343

*10. Calculate the standard enthalpy of the reaction

$$N_2O_4(g) \rightleftharpoons 2NO_2(g)$$

from the values K_p at 273 K = 159 kPa; K_p at 298 K = 14.4 kPa.

*11. (a) (i) At a temperature of 40 °C and under a pressure of 202.6 kPa the volumes of oxygen and nitrogen dioxide diffusing from an apparatus in a given time were in the ratio of 3 to 2.
Calculate K_p for the reaction:

$$N_2O_4(g) \rightleftharpoons 2NO_2(g)$$

at this temperature and pressure.

(ii) What would be the ratio of the volumes of oxygen to nitrogen dioxide diffusing out of the apparatus if the pressure was reduced to 101.3 kPa, but the temperature remained constant?

(b) For the above reaction the following values for K_p were obtained at constant pressure but at different temperatures:

K_p	Temperature/K
3.89	350
47.90	400
1700.00	507
6030.00	550

(i) Plot a graph of $\lg K_p$ against $\dfrac{1}{T}$.

(ii) It has been found that the following relationship exists between the equilibrium constant K_p and the enthalpy change of reaction:

$$\lg K_p = -\frac{\Delta H}{2.303\,RT} + C$$

where C is a constant,
R is the molar gas constant.
From your graph deduce a value for ΔH.

(c) Write equations for the reactions of nitrogen dioxide with (i) water, (ii) hydrogen sulphide.

(AEB 80,S)

12. This question concerns the gaseous equilibrium reaction:

$$2SO_2(g) + O_2(g) \rightleftharpoons 2SO_3(g)$$

(a) Give an expression for the equilibrium constant K_p specifying clearly the symbols you use.

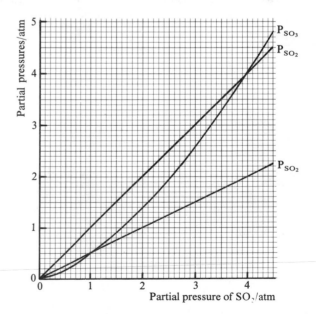

FIGURE 11.8

(b) The graph represents the partial pressures of the gases as a function of the partial pressure of SO_2 when the gases are present in stoichiometric proportions at 700 °C in a constant volume system.

(i) Use the graph to obtain a value for K_p giving your reasoning and any calculation. Give the appropriate units.

(ii) Use the graph to estimate the partial pressure of each gas when the total pressure is 10 atm.

(iii) State the law which has to be assumed to make the estimates in (b) (ii).

(c) The standard enthalpies of formation, $\Delta H_f^{\ominus}$, in kJ mol^{-1} at 298 K of SO$_3$(g) and SO$_2$(g) are -395.8 and -296.8 respectively.

(i) Define $\Delta H_f^{\ominus}$ (298 K).

(ii) Calculate $\Delta H^{\ominus}$ (298 K) for the forward reaction.

(d) Explain how an increase of temperature will affect K_p.

(WJEC 82)

13. (a) State *one* essential characteristic of a chemical equilibrium.

(b) State two ways in which the time taken to establish such an equilibrium may be altered.

(c) The following data indicate the effect of temperature and pressure on the equilibrium concentration of the product, **X**, of the forward reaction in a gaseous equilibrium.

Temperature/°C	Percentage of X present in the equilibrium mixture at		
	1 atm	100 atm	200 atm
550	0.077	6.70	11.9
650	0.032	3.02	5.71
750	0.016	1.54	2.99
850	0.009	0.87	1.68

(i) Use the above data to deduce whether the production of **X** is accompanied by an increase or a decrease in volume and explain your answer.

(ii) Use the above data to determine whether the production of **X** is an exothermic or an endothermic process and explain your answer.

(iii) State qualitatively the theoretical optimum conditions of temperature and pressure suggested by the above data for the commercial production of **X**.

(JMB 80)

14. In the presence of a catalyst, propane may undergo a thermal cracking process represented by the equation

$$CH_3CH_2CH_3(g) \rightleftharpoons C_2H_4(g) + CH_4(g)$$

$$\Delta H = +x \text{ kJ mol}^{-1}$$

(a) At 900 °C, 0.2 mole of propane on attaining equilibrium gave 0.038 mole of ethene.
If the total pressure of the system was 202.6 kPa, calculate

(i) K_p for the reaction at this temperature,

(ii) the mass of the dibromo compound obtainable from the ethene derived from 0.6 mole of propane at the same temperature and pressure.

(b) What is the effect on the composition of the above equilibrium of

(i) reducing the total pressure?

(ii) increasing the temperature?

(iii) adding methane to the equilibrium mixture?

(c) What is the purpose of cracking processes in the Petroleum Industry?

(AEB 81)

12

ELECTROCHEMISTRY

12.1 ELECTROLYTIC CONDUCTION

Electrolytic conduction differs from electronic conduction...

An electric current in a metallic conductor is a stream of electrons. The nature of the metallic bond [p. 113] allows metals to conduct electricity. This process is called **electronic conduction**. It is a physical change: no permanent change occurs in the metal.

...Faraday pioneered this subject

Some compounds conduct electricity when in the molten state or in solution. This process differs from electronic conduction in that chemical changes occur, and new substances are formed. Michael Faraday, who did much of the early work on this phenomenon, described it in 1834 as **electrolytic conduction**. He called the compounds which conduct electricity and are decomposed by it **electrolytes**. The chemical changes which take place are called **electrochemical reactions**. The vessel in which an electrochemical change takes place is called a **cell**, and the conductors which carry electricity in to and out of the cell are called **electrodes**.

12.1.1 FARADAY'S LAWS OF ELECTROLYSIS

Faraday's first law...

Faraday summarised the results of his experiments on electrolysis in two laws. His **first law** states that the mass of a substance liberated at or dissolved from an electrode is proportional to the quantity of electricity passed. The quantity of electricity is measured in **coulombs**. One coulomb of electricity is the electric charge passed by a current of one ampere in one second.

...Faraday's second law

In his **second law**, Faraday stated that, when the same quantity of electricity passes through different electrolytes, **equivalent** masses of the elements are formed at the electrodes. What Faraday meant by *equivalent* is explained below.

12.1.2 COULOMETRY

Coulometry relates electric charge to mass or volume of product

The apparatus shown in Figure 12.1 enables one to repeat Faraday's work on coulometry. A **coulometer** is a cell in which measurements can be made of the quantity of electricity passed and the masses of substances liberated at or dissolved from the electrodes. Also part of the apparatus is a **voltameter**, the name given to a cell which is constructed to allow any gases evolved to be collected.

EXAMPLE OF RESULTS OBTAINED IN COULOMETRY

In an experiment in which 40.0 mA flowed for 2.00 hours, the masses deposited on the cathodes were the same as the masses dissolved from the anodes: 0.322 g silver,

0.0948 g copper, and 0.0518 g chromium. The volumes of gases at room temperature and pressure were 35.8 cm³ hydrogen and 17.9 cm³ oxygen. (Gas molar volume = 24.0 dm³ at rtp.)

FIGURE 12.1
An Apparatus
for Coulometry

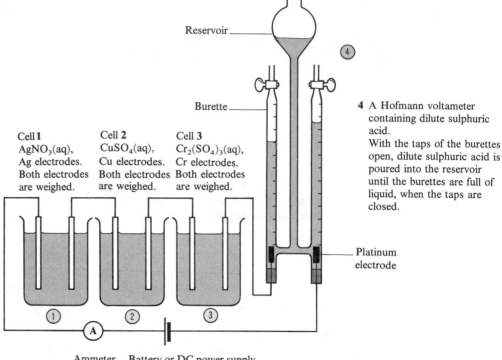

Reservoir

Burette

Cell **1**
AgNO₃(aq),
Ag electrodes.
Both electrodes
are weighed.

Cell **2**
CuSO₄(aq),
Cu electrodes.
Both electrodes
are weighed.

Cell **3**
Cr₂(SO₄)₃(aq),
Cr electrodes.
Both electrodes
are weighed.

4 A Hofmann voltameter
containing dilute sulphuric
acid.
With the taps of the burettes
open, dilute sulphuric acid is
poured into the reservoir
until the burettes are full of
liquid, when the taps are
closed.

Platinum
electrode

Ammeter   Battery or DC power supply

Treatment of results   Quantity of electricity/coulombs

= Current/amperes × Time/seconds

= 0.040 × 2.00 × 60 × 60 = 288 C

The quantity of electricity required to discharge 1 mole of each element can be calculated.

Results obtained by
coulometry

Quantity of electricity required for 1 mol Ag = 288 × 108/0.322   = 96.5 × 10³ C

Quantity of electricity required for 1 mol Cu = 288 × 63.5/0.0948 = 193 × 10³ C

Quantity of electricity required for 1 mol Cr = 288 × 52.0/0.0518 = 289 × 10³ C

Quantity of electricity required for 1 mol H₂ = 288 × 24 000/35.8 = 193 × 10³ C

Quantity of electricity required for 1 mol O₂ = 288 × 24 000/17.9 = 386 × 10³ C

When 96.5×10^3 C of electricity pass through the cells, the amounts of metals deposited are Ag, 1 mol; Cu, $\frac{1}{2}$ mol; Cr, $\frac{1}{3}$ mol; i.e., 1 mol/the valency of the element. These amounts are what Faraday referred to as the **equivalents** of the metals in his second law.

Interpretation of
Faraday's work

Faraday did his work before the structure of the atom and the nature of the chemical bond were elucidated. He postulated that an electric current was conducted through a cell by charged particles, which he called **ions**. Looking at electrolytic conduction from today's viewpoint, we can see that the cathode processes can be represented as

$$Ag^+(aq) + e^- \rightarrow Ag(s)$$

$$Cu^{2+}(aq) + 2e^- \rightarrow Cu(s)$$

$$Cr^{3+}(aq) + 3e^- \rightarrow Cr(s)$$

You can see from these equations that 1 mole of electrons will discharge 1 mole of Ag^+ ions, $\frac{1}{2}$ mole of Cu^{2+} ions or $\frac{1}{3}$ mole of Cr^{3+} ions.

In general

$$\text{Amount of element discharged/mole} = \frac{\text{Number of moles of electrons}}{\text{Number of charges on one ion of the element}}$$

The basis for calculations on coulometry and the Faraday constant

Since coulometry shows that $96\,500\,C$ of charge are required to discharge 1 mol of silver, $96\,500\,C$ must be the charge on 1 mol of electrons. The ratio $96\,500\,C\,mol^{-1}$ (more exactly $96\,485\,C\,mol^{-1}$) is called the **Faraday constant**.

Example   Find how long it will take to deposit $1.00\,g$ of chromium when a current of $0.120\,A$ flows through a solution of chromium(III) sulphate solution.

Method

Since

A worked example

$$Cr^{3+}(aq) + 3e^- \rightarrow Cr(s)$$

3 mol electrons are needed to deposit 1 mol chromium (i.e., 52 g chromium)
3/52 mol electrons deposit 1 g chromium
Number of coulombs required $= 96\,500 \times 3/52\,C$

Since

Number of coulombs = Amperes × Seconds = $0.120 \times t$ (t = time/s)

$0.120 \times t = 96\,500 \times 3/52$

$t = 46\,400\,s$ (approximately)

The current passes for $46\,400\,s$ (12.9 h).

12.2   EXAMPLES OF ELECTROLYSIS

When a molten salt is electrolysed, the products are predictable. When an aqueous solution is electrolysed, hydrogen and oxygen sometimes appear at the cathode and anode. The products formed from a few electrolytes are shown in Table 12.1.

TABLE 12.1   Products of Electrolysis

(a) Using inert electrodes (platinum or graphite)		
Electrolyte	*Cathode*	*Anode*
$PbBr_2(l)$	$Pb(s)$	$Br_2(g)$
$NaCl(l)$	$Na(s)$	$Cl_2(g)$
$CuCl_2(aq)$	$Cu(s)$	$Cl_2(g)$
$NaCl(aq)$	$H_2(g)$	$Cl_2(g)$
$KNO_3(aq)$	$H_2(g)$	$O_2(g)$
$CuSO_4(aq)$	$Cu(s)$	$O_2(g)$
$H_2SO_4(aq)$	$H_2(g)$	$O_2(g)$
$NaOH(aq)$	$H_2(g)$	$O_2(g)$

(b) When the electrodes take part in the reactions		
Electrolyte	Copper cathode	Copper anode
$CuSO_4(aq)$	Cu(s) deposited	Cu(s) dissolves to form Cu^{2+} ions

12.3 EXPLANATION OF ELECTROLYSIS

When a molten salt is electrolysed, the metal ions arrive at the cathode. This negatively charged electrode supplies electrons, which discharge the cations. In the case of lead(II) chloride

Equations for electrode processes

$$Pb^{2+}(l) + 2e^- \rightarrow Pb(l)$$

The anions travel to the anode, which, being positively charged, takes electrons from the anions, thus discharging them.

$$Cl^-(l) \rightarrow Cl(g) + e^-$$

The electrode process is followed by the formation of Cl_2 molecules:

$$2Cl(g) \rightarrow Cl_2(g)$$

The flow of current through the external circuit

The anode takes electrons from the anions. The electrons flow through the external circuit from anode to cathode. At the cathode, the electrons are available to discharge cations. The electric current is conducted through the cell by ions and through the external circuit by electrons.

In aqueous solutions, hydrogen ions may be discharged at the cathode...

When an aqueous solution is electrolysed, the products of electrolysis are not as predictable. Some metal cations are not discharged from aqueous solution. Sodium ions, for example, are present at the cathode during the electrolysis of sodium nitrate solution, but are not discharged. A small concentration of hydrogen ions arises from the dissociation of water:

$$2H_2O(l) \rightleftharpoons H_3O^+(aq) + OH^-(aq)$$

The hydrogen ions accept electrons from the cathode to form hydrogen atoms:

$$H_3O^+(aq) + e^- \rightarrow H(g) + H_2O(l)$$

Subsequently, hydrogen atoms combine rapidly to form hydrogen molecules:

$$2H(g) \rightarrow H_2(g)$$

Although the concentration of hydrogen ions is only $10^{-7}\,mol\,dm^{-3}$ in water, when these hydrogen ions are discharged, more are formed by the dissociation of more water molecules. This gives a vast supply of hydrogen ions, and sodium ions remain in solution while hydrogen is evolved.

...and hydroxide ions may be discharged at the anode

At the anode, both nitrate ions and hydroxide ions are present. Hydroxide ions are easier to discharge than nitrate ions. Nitrate ions remain in solution, while the electrode process is

$$OH^-(aq) \rightarrow OH(aq) + e^-$$

followed by

$$4OH(aq) \rightarrow O_2(g) + 2H_2O(l)$$

12.3.1 THE ELECTROCHEMICAL SERIES

An order can be drawn up for cations to show the relative ease of discharge...

If a solution contains copper(II) ions and zinc ions, during electrolysis zinc ions remain in solution, while copper(II) ions are discharged. If a solution contains bromide ions and iodide ions, iodide ions are discharged, while bromide ions remain in solution. Cations can be arranged in order, according to their relative ease of discharge at a cathode. Anions can be arranged in order of their relative ease of discharge at an anode. The list of ions in order of ease of discharge is called the **Electrochemical Series**. A shortened version of it is shown in Table 12.2 [see also Table 13.1, p. 280].

...The same can be done for anions...

...The result is the Electrochemical Series

Cations				Anions
K^+				NO_3^-
Ca^{2+}				SO_4^{2-}
Na^+			Ease of	Cl^-
Mg^{2+}	Ease of		discharge	Br^-
Al^{3+}	discharge		$A^{n-} \rightarrow A + ne^-$	I^-
Zn^{2+}	$M^{n+} + ne^- \rightarrow M$		increases	OH^-
Fe^{2+}	increases	Chemical		
Sn^{2+}		reactivity		
Pb^{2+}		$M \rightarrow M^{n+} + ne^-$		
H_3O^+		decreases		
Cu^{2+}				
Ag^+				
Au^{3+}				

TABLE 12.2
The Electrochemical Series

If the concentration of one ion is very much greater than the concentration of another, this factor may interfere with the expected order of discharge...

In the electrolysis of a solution containing a mixture of ions, to find out which of the ions will be discharged, you read from the bottom of the Electrochemical Series. When all the ions of that species have been discharged, then the ions next higher up in the series are discharged.

Sometimes, this rule does not predict the observed result. In solutions of halides, the relative concentrations of halide ions and hydroxide ions affect the result. The concentration of hydroxide ions in water is only $10^{-7}\,mol\,dm^{-3}$. In an aqueous solution of a halide of concentration $0.1\,mol\,dm^{-3}$, the concentration of halide ions is 10^6 times greater than the concentration of hydroxide ions. Although, according to the Electrochemical Series, hydroxide ions should be discharged, the concentration of halide ions is so much greater than the concentration of hydroxide ions that halide ions are discharged.

Similarly, inspection of the Electrochemical Series would lead you to expect that hydrogen would be evolved in the electrolysis of a lead(II) salt in aqueous solution. In fact, lead is deposited at the cathode. Since the concentration of lead ions is many powers of ten greater than the concentration of hydrogen ions in an aqueous solution of a lead(II) salt, the effect of concentration overrides the positions of the ions in the Electrochemical Series.

...The nature of the cathode can affect the order of discharge of cations

Another factor which can lead to a departure from the expected order of discharge of ions is the nature of the cathode. At a mercury cathode, the voltage needed to discharge hydrogen ions is much greater than at a platinum or graphite electrode. Sodium ions are discharged in preference to hydrogen ions at a mercury cathode, with the formation of a sodium mercury amalgam. A mercury cathode is the basis of the industrial electrolysis of aqueous sodium chloride [see Figure 18.5, p. 361–2]. The Electrochemical Series reflects the reactivity of metals in reactions in which they form ions:

$$M \rightarrow M^{n+} + ne^-$$

The Electrochemical Series reflects the reactivity of the metals...

This reaction is the opposite to that which occurs when metal ions are discharged:

$$\mathbf{M}^{n+}(\text{aq}) + n\text{e}^- \rightarrow \mathbf{M}(\text{s})$$

...with some exceptions

For this reason, the order of discharge of ions is the reverse of the order of reactivity of the metals. Here again, there are some exceptions. Although sodium is below calcium in the Electrochemical Series, it reacts more vigorously with water than calcium does. The reason is probably that calcium compounds are less soluble than sodium compounds, and the rates at which metals react with water may be governed by the rates at which the compounds formed can dissolve. Aluminium does not appear to be as reactive in practice as its position in the Electrochemical Series would indicate. This is because a film of aluminium oxide forms on the surface of the metal, and this film is much less reactive than the metal itself. If the oxide layer is removed, then aluminium shows its true reactivity [see Chapter 19].

12.4 APPLICATIONS OF ELECTROLYSIS

There are many applications of electrolysis

1. Extraction of sodium by the electrolysis of molten sodium chloride in the Downs process [p. 354].

2. Manufacture of sodium hydroxide by the electrolysis of aqueous sodium chloride in a mercury cathode cell [p. 361] and a diaphragm cell [p. 371].

FIGURE 12.2
Manufacture of Sodium Hydroxide: Diaphragm Cells at ICI, Lostock

3. Manufacture of sodium chlorate(I) and sodium chlorate(V) [p. 401] by the electrolysis of aqueous sodium chloride.

4. Manufacture of chlorine by the electrolysis of sodium chloride as in 1 and 2 above.

5. Manufacture of hydrogen by the electrolysis of brine (aqueous sodium chloride) as in 2 above.

6. Extraction of magnesium and calcium by electrolysis of the fused (molten) chlorides [p. 354].

7. Extraction of aluminium by the electrolysis of fused bauxite [p. 377].

8. Anodisation and dyeing of aluminium [p. 376].

9. Purification of copper by electrolysis, using a lump of impure copper as the anode [p. 508].

10. Electroplating, e.g., chromium plating [p. 503].

FIGURE 12.3
Purification of Copper
by Electrolysis

CHECKPOINT 12A: ELECTROLYSIS

1. The same current was passed through molten sodium chloride and through molten cryolite containing aluminium oxide. If 4.60 g of sodium were liberated in one cell, what mass of aluminium was liberated in the other?

2. What mass of each of the following substances will be liberated by the passage of 0.200 mol of electrons?
(a) Mg(s) from $MgCl_2$(l), (b) Cl_2(g) from NaCl(aq), (c) Cu(s) from $CuSO_4$(aq), (d) Pb(s) from $Pb(NO_3)_2$(aq), (e) H_2(g) from NaCl(aq).

3. During the electrolysis of a 1.00 mol dm^{-3} solution of copper(II) sulphate at 20 °C, a current of 0.100 A was passed for 1.00 h. The mass of the copper cathode increased by 0.118 g. State how the change in mass would be affected if the experiment were repeated

(a) at a current of 0.200 A
(b) at 30 °C
(c) with a 2.00 mol dm^{-3} solution of copper(II) sulphate
(d) for a time of 2.00 h

4. When a metal of relative atomic mass 207 is deposited by electrolysis, a current of 0.0600 A flowing for 66 min increases the mass of the cathode by 0.254 g.

Find the number of moles of metal deposited and the number of moles of electrons that have passed. Deduce the number of units of charge on the cations of this metal.

12.5 ELECTROLYTIC CONDUCTIVITY

Ohm's Law   The conduction of current through an electrolytic cell obeys **Ohm's Law**. This law
states that

$$I \propto V$$

where I = current and V = potential difference between electrodes.
It follows from Ohm's Law that

$$R = V/I$$

where R = resistance.

Resistivity, ρ (rho), is defined by the equation

Definition of resistivity... $$R = \rho l/A \qquad\qquad [1]$$

where l = length and A = cross-sectional area of conductor.
If R is in ohms (Ω), l in metres (m), and A in square metres (m²), then ρ has the
...and its units   unit $\Omega\,\mathrm{m}$. (If l is in cm, and A in cm², then ρ has the unit $\Omega\,\mathrm{cm}$.)

Definition of   The reciprocal of resistance is conductance; the reciprocal of resistivity is **conductivity**,
conductivity... κ (kappa):

$$1/\rho = \kappa \qquad\qquad [2]$$

...and its units κ has the unit $\Omega^{-1}\mathrm{m}^{-1}$ or $\Omega^{-1}\mathrm{cm}^{-1}$.

Combining equations [1] and [2] gives

$$R = l/A\kappa \qquad\qquad [3]$$

12.5.1 MEASUREMENT OF CONDUCTIVITY

The uses of a conductivity   To find the conductivity of a solution, it is necessary to measure its resistance in a
cell   cell such as that in Figure 12.4. Solutions are made up in conductivity water, especially
pure water which has been deionised by passing through an ion-exchange resin. The
method of measuring resistance is to make the conductivity cell one arm of a
Wheatstone bridge circuit.*

FIGURE 12.4
A Conductivity Cell

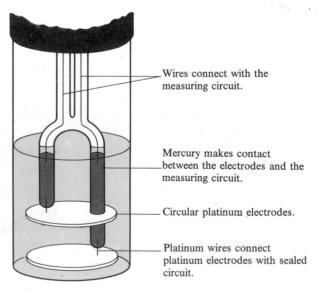

Wires connect with the
measuring circuit.

Mercury makes contact
between the electrodes and the
measuring circuit.

Circular platinum electrodes.

Platinum wires connect
platinum electrodes with sealed
circuit.

---

*See, e.g., R Muncaster, *A-level Physics* (Stanley Thornes)

The cell constant
Once the resistance of the cell has been measured, the conductivity of the solution is found from equation [3]. For any cell, the ratio l/A is a constant, called the **cell constant**. It can be found by measuring l and A, but is usually found by a calibration experiment. A solution of known conductivity is put into the cell, and its resistance is measured. These values of κ and R are inserted into equation [3] to give l/A.

Example   The resistance of a cell containing $0.100 \, mol \, dm^{-3} \, KCl$ solution, which has conductivity $1.29 \, \Omega^{-1} m^{-1}$, is $1.16 \, \Omega$. A second electrolyte in the same cell was found to have a resistance of $15.0 \, \Omega$. Find the conductivity of this electrolyte.

Method   (*a*) Find the cell constant. From equation [3]

$$\text{Cell constant}, \ l/A = R\kappa = 1.16 \times 1.29 = 1.50 \, m^{-1}$$

(*b*) Use the cell constant to find the unknown conductivity.

Since

$$R = l/A\kappa$$

$$15.0 = 1.50/\kappa$$

$$\text{Conductivity } \kappa = 0.100 \, \Omega^{-1} m^{-1}$$

12.5.2   MOLAR CONDUCTIVITY

Molar conductivity is defined by the equation

Definition of molar conductivity

$$\Lambda = \kappa/c$$

where Λ (lambda) = molar conductivity, κ = conductivity, and c = concentration. If κ is in $\Omega^{-1} m^{-1}$ and c is in $mol \, m^{-3}$, then $\Lambda = \Omega^{-1} m^2 mol^{-1}$, and Λ is numerically equal to the conductivity of 1 mole of the electrolyte.

Example   Calculate the molar conductivity of hydrochloric acid of concentration $0.100 \, mol \, dm^{-1}$, given that its conductivity is $3.91 \, \Omega^{-1} m^{-1}$.

Method
Since

$$\Lambda = \kappa/c$$

and $c = 0.100 \, mol \, dm^{-3} = 100 \, mol \, m^{-3}$, and $\kappa = 3.91 \, \Omega^{-1} m^{-1}$

$$\Lambda = 3.91/100 = 3.91 \times 10^{-2} \Omega^{-1} m^2 mol^{-1} = 391 \, \Omega^{-1} cm^2 mol^{-1}$$

12.5.3   MOLAR CONDUCTIVITY AND CONCENTRATION

Molar conductivity increases with dilution. For strong electrolytes Λ soon reaches a maximum...
The molar conductivity of a solute depends on its concentration. When values of molar conductivity, Λ (lambda), are plotted against the volume of solution containing 1 mole of solute, that is the **dilution**, the graphs obtained for different electrolytes fall into two categories. These are shown in Figure 12.5. **Strong electrolytes** are those which are completely ionised in solution. They give graphs of shape **A**, swiftly rising to a maximum. The maximum value of Λ is called the **molar conductivity at infinite dilution**, as further dilution has no effect on its value. It is represented as Λ_{∞} or Λ_0.

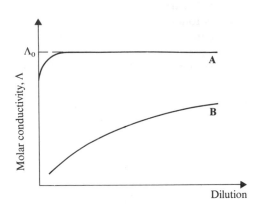

Weak electrolytes, which are incompletely ionised, give graphs similar in shape to **B**. The value of Λ increases as the solution becomes more and more dilute. There is a limit to the range of concentration over which measurements can be made as, in very dilute solutions, the conductivity of the solution is very little greater than the conductivity of water, and no worthwhile values can be obtained.

...For weak electrolytes, Λ cannot be followed to a maximum

S Arrhenius, in 1873, explained the shape of curve **B** as arising from an increase in the ionisation of the solute with dilution. He suggested that the fraction α of the solute which is ionised gradually increases with dilution. He postulated that, if measurements could be extended to even more dilute solutions, α would approach a value of 1, and Λ would level off at a maximum value, Λ_0. The value of the degree of ionisation at any concentration is given by

The degree of ionisation of a weak electrolyte increases with dilution...

$$\alpha = \Lambda/\Lambda_0$$

The ratio Λ/Λ_0 is called the **conductivity ratio**.

The shape of curve **A** suggested to Arrhenius that strong electrolytes are completely ionised in solution, except at very high concentrations. In the light of the twentieth century view of the nature of the chemical bond, this idea is unacceptable. We cannot imagine that an electrolyte such as copper(II) sulphate can be other than completely ionised: there is no possibility of covalent bonding.

Strong electrolytes show low Λ at high concentration

According to the modified ionic theory put forward by P Debye and E Hückel and improved by L Onsager, strong electrolytes are always completely ionised in solution. The decrease in molar conductivity in concentrated solutions must have another explanation. This is discussed on p. 249. The value of α obtained from Λ/Λ_0 is called the **apparent degree of ionisation** or the **conductivity ratio**.

α, the conductivity ratio

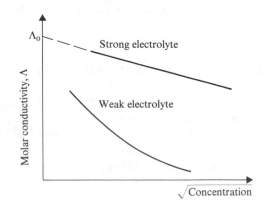

12.5.4   FINDING MOLAR CONDUCTIVITY AT INFINITE DILUTION

Finding Λ for strong electrolytes...

For strong electrolytes, the graph of Λ against $\sqrt{c}$ is a straight line, which can be extrapolated to cut the Λ axis at Λ_0 [see Figure 12.6]. For weak electrolytes, the graph of Λ against $\sqrt{c}$ is of the shape shown in Figure 12.6, and cannot be used to find Λ_0. F Kohlrausch's Law enables us to find Λ_0 for weak electrolytes.

...and for weak electrolytes

Kohlrausch's Law of Independent Mobilities of Ions states that <u>the molar conductivity at infinite dilution of an electrolyte is the sum of two independent contributions, one characteristic of the cation and one characteristic of the anion</u>:

$$\Lambda_0 = \Lambda_0(\text{cation}) + \Lambda_0(\text{anion})$$

Kohlrausch's Law

For an electrolyte $C_n A_m$,

$$\Lambda_0 = n\Lambda_0(C) + m\Lambda_0(A)$$

For every cation and every anion, values of Λ_0 can be looked up in tables and added to give Λ_0 for the salt. Alternatively, measurements on strong electrolytes can be used to give values of Λ_0 which can be computed to give Λ_0 for the required weak electrolyte.

Example   Calculate the molar conductivity at infinite dilution, Λ_0, at 25 °C of thallium(I) chloride, TlCl. Values of Λ_0 at 25 °C in $\Omega^{-1}\text{m}^2\text{mol}^{-1}$ are TlOH = 2.73×10^{-2}, NaOH = 2.48×10^{-2}, NaCl = 1.26×10^{-2}.

Method   According to Kohlrausch's Law

$$\Lambda_0(\text{TlOH}) = \Lambda_0(\text{Tl}^+) + \Lambda_0(\text{OH}^-)$$

$$\Lambda_0(\text{NaOH}) = \Lambda_0(\text{Na}^+) + \Lambda_0(\text{OH}^-)$$

$$\Lambda_0(\text{NaCl}) = \Lambda_0(\text{Na}^+) + \Lambda_0(\text{Cl}^-)$$

$$\Lambda_0(\text{TlOH}) + \Lambda_0(\text{NaCl}) - \Lambda_0(\text{NaOH}) = \Lambda_0(\text{Tl}^+) + \Lambda_0(\text{Cl}^-) = \Lambda_0(\text{TlCl})$$

$$\Lambda_0(\text{TlCl}) = (2.73 \times 10^{-2}) + (1.26 \times 10^{-2}) - (2.48 \times 10^{-2})\,\Omega^{-1}\text{m}^2\text{mol}^{-1}$$

$$= 1.51 \times 10^{-2}\,\Omega^{-1}\text{m}^2\text{mol}^{-1}$$

12.5.5   CALCULATION OF SOLUBILITY FROM MEASUREMENTS OF CONDUCTIVITY

The conductivity of a solution tells the concentration of dissolved salt

The solubility of a sparingly soluble salt can be found by measuring the conductivity of a saturated solution of the salt. The conductivity depends on the concentration of ions in solution. Since the solution is saturated, the concentration is equal to the solubility of the salt. To find the concentration c from the equation

$$\Lambda_0 = \kappa/c$$

where κ = conductivity, and Λ_0 = molar conductivity at infinite dilution, the value of Λ_0 must be known. It is the sum of the molar conductivities at infinite dilution of the cation and anion, which can be looked up in tables.

Example   After subtracting the conductivity of water, the conductivity of a saturated solution of thallium(I) chloride, TlCl, at 25 °C is $2.40 \times 10^{-4}\,\Omega^{-1}\text{m}^{-1}$. Using the value of $\Lambda_0(\text{TlCl}) = 1.51 \times 10^{-2}\,\Omega^{-1}\text{m}^2\text{mol}^{-1}$, find the solubility of TlCl at 25 °C.

Method
Since

$$c = \kappa/\Lambda_0$$

$$c = (2.40 \times 10^{-4})/(1.51 \times 10^{-2})\,\mathrm{mol\,m^{-3}}$$

$$= 1.59 \times 10^{-2}\,\mathrm{mol\,m^{-3}} = 1.59 \times 10^{-5}\,\mathrm{mol\,dm^{-3}}$$

CHECKPOINT 12B: CONDUCTIVITY

1. Explain what is meant by (*a*) the electrolytic conductivity and (*b*) the molar conductivity of an electrolyte. Outline how (*a*) and (*b*) could be found. Sketch curves to show how (*a*) and (*b*) vary with dilution for (i) a strong electrolyte and (ii) a weak electrolyte.

2. An electrolytic cell is 2.50 cm long and has a cross-sectional area of 1.50 cm^2. If the resistance of an electrolyte measured in the cell is 14.2 Ω, what is the conductivity of the electrolyte?

3. The resistance of a cell containing $1.00 \times 10^{-2}\,\mathrm{mol\,dm^{-3}}$ KCl(aq) at 18 °C is 1.23 Ω. The conductivity of KCl at this concentration at 18 °C is 1.22 $\Omega^{-1}\mathrm{cm^{-1}}$. The resistance of the same cell containing potassium nitrate solution of concentration $1.00 \times 10^{-2}\,\mathrm{mol\,dm^{-3}}$ at 18 °C is 1.13 Ω. Find the molar conductivity of potassium nitrate.

4. Calculate the molar conductivity at infinite dilution at 25 °C of ethanoic acid, given the following values of Λ_0 in $\Omega^{-1}\mathrm{m^2\,mol^{-1}}$: HCl 4.26 $\times$ 10^{-2}; NaCl 1.26 $\times$ 10^{-2}; CH_3CO_2Na 9.10 $\times$ 10^{-3}.

5. The conductivity of a saturated solution of silver chloride (corrected for the conductivity of water) is $1.50 \times 10^{-4}\,\Omega^{-1}\mathrm{m^2\,mol^{-1}}$ at 25 °C. If

$$\Lambda_0(Ag^+) = 6.2 \times 10^{-3} \text{ and}$$

$$\Lambda_0(Cl^-) = 7.6 \times 10^{-3}\,\Omega^{-1}\mathrm{m^2\,mol^{-1}} \text{ at 25 °C}$$

what is the solubility of silver chloride in $\mathrm{mol\,dm^{-3}}$?

12.6 THE IONIC THEORY

Arrhenius put forward the theory in 1887 that, when an electrolyte dissolves, a certain fraction of it dissociates into positively and negatively charged particles called ions. His theory explained an enormous volume of work on electrolysis and electrolytic conduction. Evidence for the existence of ions is summarised below.

12.6.1 SUMMARY OF THE EVIDENCE FOR THE EXISTENCE OF IONS

Evidence for the existence of ions...

...from electrolysis...
...and Ohm's Law...

1. Electrolysis is explained. Metals are deposited only at the cathode because their ions are positively charged. Non-metallic elements are liberated only at the anode because their ions are negatively charged. The fact that electrolytes obey Ohm's Law means that ions are already present in the solution or the melt before the current passes. If the current served to dissociate the electrolyte, energy would be consumed, and Ohm's Law would not be obeyed.

...Faraday's Laws are explained...

2. Faraday's Laws of Electrolysis are explained. The amount of a metal deposited on the cathode depends on the number of moles of electrons which flow through the circuit and on the number of charges on one ion of the metal.

...Evidence from colligative properties...

3. The high values of colligative properties [p. 185] observed for electrolytes arise from ionisation. The degree of ionisation, α, can be calculated from the ratio of (measured value of colligative property)/(value expected in the absence of ionisation). For weak electrolytes, there is good agreement between values of α calculated from colligative properties and values derived from the conductivity ratio.

4. The standard enthalpy of neutralisation has a constant value of approximately $-58\,\mathrm{kJ\,mol^{-1}}$, regardless of which strong acid and base are used for its measurement.

...and from $\Delta H^{\ominus}$ of neutralisation...

This is because the neutralisation reaction is in each case the reaction between hydrogen ions and hydroxide ions:

$$H_3O^+(aq) + OH^-(aq) \rightarrow 2H_2O(l)$$

...from the increase of Λ with dilution...

5. The increase in molar conductivity with dilution can be explained. In the case of weak electrolytes, it is attributed to an increase in the degree of ionisation with increasing dilution. The variation of molar conductivity with dilution for strong electrolytes is explained by the Debye–Hückel and Onsager Theory [see below].

...and from Kohlrausch's Law...

6. Kohlrausch's Law of Independent Mobilities of Ions fits into the picture. At infinite dilution, there is no interaction between ions. They move independently, and their separate contributions to Λ_0 can be added to give the total Λ_0, the molar conductivity at infinite dilution of the electrolyte.

...from chemical properties...

7. The chemical properties of an electrolyte are the sum of the properties of its ions. All chlorides give a white precipitate with silver nitrate solution. All iron(II) salts give a green precipitate with sodium hydroxide solution.

...and from X ray crystallography

8. X ray crystallography has shown that salts consist of ions, even in the solid state. X ray diffraction patterns of sodium chloride show a three-dimensional structure of alternate sodium and chloride ions [see Section 6.3, p. 116, Figure 6.7, p. 117 and Figure 4.2, p. 70].

The eighth piece of evidence was not available to Arrhenius. Since his day, the structure of the atom has been elucidated [see Chapters 1 and 2]. The nature of the chemical bond in electrolytes is believed to be an electrostatic attraction between oppositely charged ions. These developments led to a modification of the Arrhenius Theory by Debye and Hückel and by Onsager.

12.6.2 THE DEBYE–HÜCKEL AND ONSAGER THEORY: INTERIONIC ATTRACTIONS

The low values of molar conductivity for strong electrolytes at high concentrations are explained

Debye and Hückel (1923) and Onsager (1927) modified the Ionic Theory as presented by Arrhenius. They said that strong electrolytes are always completely ionised. They explained the low molar conductivities of ions in concentrated solutions in a different way. They postulated that ions move more slowly in concentrated solutions. Since ions are closer together in concentrated solutions than in dilute solutions, the electrical forces between ions are greater. As unlike charges attract, a cation will have more anions than cations in its immediate neighbourhood. The attractive forces between these anions and the cation reduce the velocity of the cation [see Figure 12.7]. The anions are similarly retarded in concentrated solutions.

FIGURE 12.7 Ions are slowed down in Concentrated Solutions

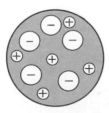

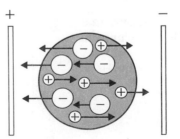

(a) In a concentrated solution, the distribution of ions is not completely random. A cation has more anions than cations surrounding it.

(b) As a cation moves towards the cathode, it is slowed down by the forces of attraction between it and the anions.

12.7 IONIC EQUILIBRIA

12.7.1 ACIDS AND BASES

Early descriptions of acids and alkalis

Before the work of W Ostwald and S Arrhenius on the dissociation of electrolytes, there were attempts to define acids and bases. The sour taste and the effect on vegetable colourings, such as litmus, characterised acids. The soapy feel and detergent power characterised alkalis. It was recognised that acids react with alkalis and also with some other compounds to give salts. The term **base** superseded the term **alkali** as meaning the opposite of an acid. A base was defined as a substance which would react with an acid to form a salt. The study of acids progressed from Boyle

In the dissociation theory...

(1663) through Lavoisier (1780) and Davy (1810) to Liebig (1838). Liebig stated that acids are compounds which contain hydrogen that can be displaced by metals. The explanation of how the possession of hydrogen confers acidic properties was

...acids give H^+ ions and bases give OH^- ions

not forthcoming until the Ostwald–Arrhenius **Theory of Electrolytic Dissociation** in 1880. They identified the hydrogen atoms which give rise to acidic properties as being those which form hydrogen ions in solution. Bases were said to produce hydroxide ions in solution and to neutralise acids by the reaction

$$H^+ + OH^- \rightarrow H_2O$$

Some difficulties arose with these definitions

The Dissociation Theory definitions of acids and bases ran into difficulties. As pure hydrogen chloride does not conduct electricity, should it be classified as an acid, or does it only become an acid in contact with water? A solution of sodium ethoxide in ethanol has strongly basic properties. It contains no OH^- ions, but it does contain $C_2H_5O^-$ ions. Bases such as ammonia neutralise acids by picking up a proton, rather than by providing hydroxide ions:

$$NH_3 + H^+ \rightarrow NH_4{}^+$$

H^+ ions cannot exist in solution: they are so small that the charge density is high, and they must be solvated...

As work proceeded, doubt was cast on the existence of the hydrogen ion, H^+, in solution. The proton, H^+, is very small (10^{-15} m diameter) compared with other cations (around 10^{-10} m diameter). The electric field in its neighbourhood is so intense that it attracts any molecule with unshared electrons, such as H_2O. The reaction

$$H^+ + H_2O \rightarrow H_3O^+$$

...In water, H_3O^+ ions are formed

was shown (by spectroscopic measurements) to liberate $1300\,kJ\,mol^{-1}$. As the reaction is so exothermic, unhydrated protons do not exist in solution. (They are produced in gaseous reactions and in nuclear reactions.) The hydrated proton, H_3O^+, is called the **oxonium ion**, and is also referred to as the hydrogen ion. In this text, H_3O^+ will be called a hydrogen ion. Its structure is shown in Figure 12.8 with that of the ammonium ion for comparison.

FIGURE 12.8
The Oxonium and Ammonium ions (H_3O^+ and $NH_4{}^+$)

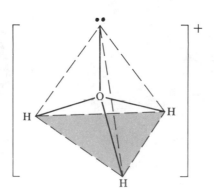 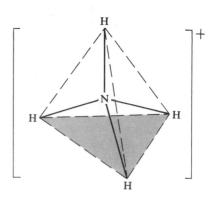

The H_3O^+ ion, like other cations, is solvated by water molecules, and is sometimes written as $H_9O_4^+$. In solution in ethanol, the hydrogen ion is $C_2H_5OH_2^+$, and in liquid ammonia it is NH_4^+.

THE BRÖNSTED–LOWRY DEFINITION OF ACIDS AND BASES

Brönsted and Lowry described acids as proton-donors and bases as proton-acceptors

Acknowledging that the proton does not exist in solution made it necessary to review the definition of acids. The best definition of acids and bases is that proposed by T M Lowry and also, independently, by J N Brönsted in 1923. They said that an acid is a substance which can donate a proton to another substance. A base is a substance which can accept a proton from another substance. The relationship is

$$Acid \rightleftharpoons Base + H^+$$

This equation does not represent an actual reaction in solution since the proton, H^+, cannot exist in solution. The acid and base which are related in this way, by the exchange of a proton, are called a **conjugate acid–base pair**. Since

$$Acid\ 1 \rightleftharpoons Conjugate\ base\ 1 + H^+$$

and $\quad Base\ 2 + H^+ \rightleftharpoons Conjugate\ acid\ 2$

a reaction between an acid and a base is

Conjugate acid–base pairs

$$Acid\ 1 + Base\ 2 \rightleftharpoons Conjugate\ base\ 1 + Conjugate\ acid\ 2$$

Reactions between acids and bases

Acid 1 is transformed into its conjugate base, and Base 2 is transformed into its conjugate acid. No substance can act as an acid in solution unless a base is present to accept a proton: the reactions of acids are reactions between acids and bases. Similarly, all reactions of bases in solution are acid–base reactions.

Examples of **Brönsted–Lowry acids** are

$$HCl + H_2O \rightarrow H_3O^+ + Cl^- \qquad [1]$$

$$HSO_4^- + H_2O \rightleftharpoons H_3O^+ + SO_4^{2-} \qquad [2]$$

Examples of Lowry–Brönsted acids and bases

$$CH_3CO_2H + H_2O \rightleftharpoons H_3O^+ + CH_3CO_2^- \qquad [3]$$

Examples of **Brönsted–Lowry bases** are:

$$NH_3 + H_2O \rightleftharpoons NH_4^+ + OH^- \qquad [4]$$

$$RNH_2 + H_2O \rightleftharpoons RNH_3^+ + OH^- \qquad [5]$$

$$HSO_4^- + H_3O^+ \rightleftharpoons H_2SO_4 + H_2O \qquad [6]$$

$$CH_3CO_2^- + H_2O \rightleftharpoons CH_3CO_2H + OH^- \qquad [7]$$

In [1], [2] and [3], water is acting as a proton-acceptor, a base; in [4], [5], [6] and [7], water is acting as a proton-donor, an acid. Water is described as an **amphoteric** solvent. The HSO_4^- ion is an amphoteric species.

THE LEWIS DEFINITION OF ACIDS AND BASES

There are reactions which appear to us, on common-sense grounds, to be acid–base reactions, and which do not come within the scope of the Brönsted–Lowry definition. Such reactions are

$$CaO + SO_3 \rightarrow CaSO_4$$

$$NH_3 + BF_3 \rightarrow NH_3BF_3$$

A Lewis base gives a lone pair of electrons to a Lewis acid...

To accommodate reactions of this type, G N Lewis (during 1930 to 1940) proposed a fresh definition of acids and bases. He described as acid–base reactions those

...with the formation of a reactions in which an unshared electron pair in the base molecule is accepted by the
covalent bond acid molecule, with the formation of a covalent bond. For example, in the reaction

$$H\underset{\displaystyle H}{\overset{\displaystyle H}{\diagdown}}\!\!N: + B\underset{\displaystyle F}{\overset{\displaystyle F}{\diagup}}\!\!F \rightarrow H\underset{\displaystyle H}{\overset{\displaystyle H}{\diagdown}}\!\!N\!\!\rightarrow\!\!B\underset{\displaystyle F}{\overset{\displaystyle F}{\diagup}}\!\!F$$

A comparison of the two ammonia is the base, and boron trifluoride is the acid. The Lewis definition of a
definitions base includes the Brönsted–Lowry bases because a species with a lone pair of electrons
will accept a proton from a Brönsted–Lowry acid. Lewis acids, such as BF_3 and
SO_3, are not acids in the Brönsted–Lowry sense, and acids such as HCl, H_2SO_4
and CH_3CO_2H are not acids according to the Lewis definition.

The Brönsted–Lowry description of acids and bases lends itself readily to a quantitative
treatment of the strengths of acids and bases [p. 253]. No such quantitative treatment
is possible for Lewis acids and bases.

12.7.2   THE IONIC PRODUCT FOR WATER

The hydrogen ion concentration in a solution can be denoted by means of the **hydrogen
ion exponent** or **pH** of the solution. The pH of a solution is the negative logarithm
to the base ten of the hydrogen ion concentration in $mol\,dm^{-3}$. **pOH** is related in
the same way to the hydroxide ion concentration:

pH and pOH...
$$pH = -lg[H_3O^+/mol\,dm^{-3}]$$

$$pOH = -lg[OH^-/mol\,dm^{-3}]$$

Water is only slightly ionised:

$$2H_2O \rightleftharpoons H_3O^+ + OH^-$$

The product of the concentrations of hydrogen ions and hydroxide ions is equal to
$1.00 \times 10^{-14}\,mol^2\,dm^{-6}$ at 25 °C. This product is called the **ionic product** for water,
K_w.

...The ionic product K_w
and pK_w
$$[H_3O^+][OH^-] = K_w = 1.00 \times 10^{-14}\,mol^2\,dm^{-6}$$

$$pH + pOH = pK_w = 14 \text{ at } 25\,°C$$

This relationship is true of aqueous solutions as well as water.

12.7.3   CALCULATION OF pH AND pOH FOR STRONG ACIDS AND BASES

The pH of a solution of a strong acid or strong base is simply calculated.

Example   What is the pH of a solution of hydrochloric acid of concentration
$0.1\,mol\,dm^{-3}$?

Method
Since

Calculation of pH...
$$[H_3O^+] = 0.1\,mol\,dm^{-3} = 10^{-1}\,mol\,dm^{-3}$$

$$lg[H_3O^+] = -1 \text{ and } pH = 1$$

Example   What is the pH of a solution of sodium hydroxide of concentration
$0.01\,mol\,dm^{-3}$ at 25 °C?

Method

...Calculation of pOH

Since

$$[OH^-] = 0.01 = 10^{-2} \, mol\, dm^{-3}$$

$$pOH = 2$$

$$pH = 14 - pOH = 12 \text{ at } 25 \,°C$$

At 25 °C, solutions with a pH of 7 are neutral, solutions with a pH less than 7 are acidic, and solutions with a pH greater than 7 are alkaline. The relationship between pH and $[H_3O^+]$ and $[OH^-]$ is illustrated in Table 12.3.

$[H_3O^+]/$ $mol\,dm^{-3}$	1	10^{-1}	10^{-2}	10^{-3}	10^{-4}	10^{-5}	10^{-6}	10^{-7}	10^{-8}	10^{-9}	10^{-10}	10^{-11}	10^{-12}	10^{-13}	10^{-14}
$[OH^-]/$ $mol\,dm^{-3}$	10^{-14}	10^{-13}	10^{-12}	10^{-11}	10^{-10}	10^{-9}	10^{-8}	10^{-7}	10^{-6}	10^{-5}	10^{-4}	10^{-3}	10^{-2}	10^{-1}	1
pH	0	1	2	3	4	5	6	7	8	9	10	11	12	13	14
	Strongly acidic			Weakly acidic			Neutral		Weakly alkaline			Strongly alkaline			

TABLE 12.3
pH Values of Acidic and Alkaline Solutions

=========== **CHECKPOINT 12C: ACIDS AND BASES** ===========

1. Define the terms *Brönsted acid* and *Brönsted base*. Give two examples of each, explaining how they fit the definitions. Define the terms *Lewis acid* and *Lewis base*, and give one example of each.

†**2.** Explain the following statements:

(*a*) $[Al(H_2O)_6]^{3+}$ is classified as an acid [p. 379].

(*b*) $ClCH_2CO_2H$ is a stronger acid than CH_3CO_2H [p. 254].

(*c*) The Friedel–Crafts catalyst, $AlCl_3$, is classified as a Lewis acid [p. 584].

3. How are Lewis acids and bases different from Lowry–Brönsted acids and bases and from Arrhenius acids and bases? In the equilibrium

$$NH_3 + H_2O \rightleftharpoons NH_4^+ + OH^-$$

which species is the acid and which the base, according to the definitions of (*a*) Arrhenius and (*b*) Lowry and Brönsted?

4. According to the Lowry–Brönsted theory, water can function as an acid and as a base. Quote one reaction in which water acts as an acid and one in which it acts as a base.

5. At 0 °C, $K_w = 1.14 \times 10^{-15} \, mol^2\, dm^{-6}$. Find:

(*a*) pK_w at 0 °C

(*b*) the pH at 0 °C of a $0.01 \, mol\, dm^{-3}$ solution of sodium hydroxide

6. Name the Lewis acid and the Lewis base in the following reactions:

(*a*) $RCOBr + FeBr_3 \rightleftharpoons RCO^+FeBr_4^-$

(*b*) $Ag^+ + 2NH_3 \rightarrow Ag(NH_3)_2^+$

(*c*) $CH_3CO_2^- + 2HF \rightarrow CH_3CO_2H + HF_2^-$

7. The $H_9O_4^+$ ion consists of an oxonium ion to which $3H_2O$ molecules are attached by hydrogen bonds. Sketch the bonding in this species.

12.7.4 WEAK ELECTROLYTES

The degree of ionisation of a weak electrolyte

Weak electrolytes consist of molecules, some of which dissociate to form ions. The fraction of molecules which dissociate is called the **degree of ionisation** or **degree of dissociation**. As the concentration of a solution of a weak electrolyte decreases, the degree of ionisation increases. **Ostwald's Dilution Law** gives a relationship between the degree of ionisation α of a weak electrolyte and its concentration c. Consider a weak acid HA. An equilibrium is set up between undissociated molecues HA and the ions H_3O^+ and A^-:

$$HA + H_2O \rightleftharpoons H_3O^+ + A^-$$

The equilibrium constant [p. 219] is called the **acid dissociation constant**, K_a, and is given by

The dissociation constant
of a weak acid

$$K_a = \frac{[H_3O^+][A^-]}{[HA]}$$

The square brackets represent concentration in $mol\,dm^{-3}$, e.g., $[H_3O^+]$ is the concentration of hydrogen ions. The expression for K_a does not include $[H_2O]$ because the concentration of water remains constant in all reasonably dilute solutions. If 1 mole of acid is dissolved in $V\,dm^3$ of solution, the amounts and concentrations of each species at equilibrium are as follows:

$$HA \quad + H_2O \rightleftharpoons H_3O^+ + A^-$$

Amount/mol	$1 - \alpha$	α	α
Concentration/$mol\,dm^{-3}$	$(1 - \alpha)/V$	α/V	α/V

Putting these concentrations into the expression for K_a gives

$$K_a = \frac{(\alpha/V)^2}{(1 - \alpha)/V} = \frac{\alpha^2}{(1 - \alpha)V}$$

Since V = volume containing 1 mol of solute, $1/V = c$, and

Ostwald's Dilution Law

$$K_a = \frac{\alpha^2 c}{(1 - \alpha)}$$

This expression embodies Ostwald's Dilution Law.

For many weak electrolytes, α is so small that the error involved in putting $(1 - \alpha) = 1$ is negligible. Then

$$K_a = c\alpha^2$$
$$\alpha = \sqrt{K_a/c}$$

Ostwald's Dilution Law can be applied to weak bases. In the case of a weak base **B**, which is partially ionised in solution

$$B + H_2O \rightleftharpoons BH^+ + OH^-$$

The base dissociation constant, K_b, is given by the equation

$$K_b = \frac{[BH^+][OH^-]}{[B]}$$

$$K_b = c\alpha^2/(1 - \alpha)$$

If $\alpha \ll 1$ $K_b = c\alpha^2$

The value of K_a *for an*
acid is a quantitative
measure of the strength of
the acid in all its typical
reactions . . .

The value of its dissociation constant tells you how strong an acid is and how vigorously it will take part in the reactions which are typical of acids. It enables you to calculate the conductivity and the pH of a solution of the acid [p. 252]. It is a measure of the effectiveness of an acid in acid-catalysed reactions. All these aspects of acid behaviour are covered by one physical constant. From the values of K_a, you can say that chloroethanoic acid is 80 times stronger than ethanoic acid. The value of the dissociation constant of a base is equally important. From the values of K_b, you can say that methylamine is 23 times as strong a base as ammonia. The quantitative measure of the strengths of acids and bases provided by the K_a and K_b values is a splendid feature of the Brönsted–Lowry treatment of acids and bases. No such quantitative treatment can be made of Lewis acids and bases.

. . . K_b *for a base is*
equally important

Tables often list pK values of acids and bases. These are defined as

$$pK_a = -\lg K_a$$

$$pK_b = -\lg K_b$$

The higher the value of K_a (or K_b), the lower the value of pK_a (or pK_b), and the stronger is the acid (or base).

12.7.5 CONJUGATE ACID–BASE PAIRS

The relationship between an acid and its conjugate base

Some tables list the base dissociation constants K_b of bases such as amines. Others list the acid dissociation constants K_a of their conjugate acids. The conjugate acid **BH**$^+$ of a base **B** dissociates according to the equilibrium

$$\mathbf{BH^+ + H_2O \rightleftharpoons B + H_3O^+}$$

The acid dissociation constant of **BH**$^+$ is

$$K_a = \frac{[\mathbf{B}][\mathbf{H_3O^+}]}{[\mathbf{BH^+}]}$$

Multiplying K_a by K_b [p. 254] gives $K_a K_b = [\mathrm{H_3O^+}][\mathrm{OH^-}]$ that is

$$K_a K_b = K_w$$

This is a useful relationship between the dissociation constants of acid–base pairs. It can also be expressed as

$$pK_a + pK_b = pK_w$$

12.7.6 HOW TO CALCULATE THE pH OF A SOLUTION OF A WEAK ACID OR A WEAK BASE

In order to calculate the pH of a solution of a weak acid or base, you must know the concentration of the solution and the dissociation constant of the acid or base. The converse is also true: if the pH of a solution is measured, you can use it to find the dissociation constant of the weak electrolyte.

A sample calculation of the pH of a solution of a weak acid…

Example   Calculate the pH of a $1.00 \times 10^{-2}\,\mathrm{mol\,dm^{-3}}$ solution of butanoic acid, for which $K_a = 1.51 \times 10^{-5}\,\mathrm{mol\,dm^{-3}}$.

Method   Use the equation

$$K_a = \frac{[\mathrm{H_3O^+}][\mathrm{C_3H_7CO_2^-}]}{[\mathrm{C_3H_7CO_2H}]}$$

The concentrations, $[\mathrm{H_3O^+}]$ and $[\mathrm{C_3H_7CO_2^-}]$ are equal.

The concentration $[\mathrm{C_3H_7CO_2H}]$ is very little less than $1.00 \times 10^{-2}\,\mathrm{mol\,dm^{-3}}$. Since the degree of ionisation is small, the approximation $[\mathrm{C_3H_7CO_2H}] = 1.00 \times 10^{-2}\,\mathrm{mol\,dm^{-3}}$ is made to simplify the calculation. Then

$$[\mathrm{H_3O^+}]^2 = 1.51 \times 10^{-5} \times 1.00 \times 10^{-2}$$

$$[\mathrm{H_3O^+}] = 3.89 \times 10^{-4}\,\mathrm{mol\,dm^{-3}}$$

$$pH = 3.42$$

How valid was the approximation $[C_3H_7CO_2H] = 1.00 \times 10^{-2}\,mol\,dm^{-3}$?
Since

$$[H_3O^+] = 3.89 \times 10^{-4}\,mol\,dm^{-3}\ (approximately)$$

$$[C_3H_7CO_2H] = (1.00 \times 10^{-2}) - (3.89 \times 10^{-4}) = 0.96 \times 10^{-2}\,mol\,dm^{-3}$$

If this approximate value is used in a new calculation, a new value of pH = 3.42 is obtained. This is the same as in the first case, and shows that the approximation was justified*. For most weak acids and bases, the approximation can be safely made.

...and a calculation of a dissociation constant from the pH of a solution of a weak base

Example   A solution of dimethylamine of concentration $1.00 \times 10^{-2}\,mol\,dm^{-3}$ has a pH of 7.64 at 25 °C. Calculate (*a*) the dissociation constant of the base and (*b*) the degree of dissociation.

Method   The dissociation of the base can be represented by

$$(CH_3)_2NH + H_2O \rightleftharpoons (CH_3)_2NH_2^+ + OH^-$$

Thus $K_b = \dfrac{[(CH_3)_2NH_2^+][OH^-]}{[(CH_3)_2NH]}$

$\qquad = [OH^-]^2/(1.00 \times 10^{-2})$

Since

$$pH = 7.64,\ pOH = 14.0 - 7.64 = 6.36\ at\ 25\,°C$$

$$[OH^-] = antilg(-6.36) = 4.37 \times 10^{-7}\,mol\,dm^{-3}$$

$$K_b = (4.37 \times 10^{-7})^2/(1.00 \times 10^{-2}) = 1.91 \times 10^{-11}\,mol\,dm^{-3}$$

From the Ostwald Dilution Law, if α = degree of dissociation

$$K_b = \frac{\alpha^2 c}{1 - \alpha}$$

For a weak base

$$K_b = \alpha^2 c$$

$$1.91 \times 10^{-11} = \alpha^2 \times 1.00 \times 10^{-2}$$

Degree of dissociation

$$\alpha = 4.37 \times 10^{-5}$$

12.7.7   CALCULATION OF DEGREE OF DISSOCIATION AND DISSOCIATION CONSTANT OF A WEAK ELECTROLYTE FROM CONDUCTANCE MEASUREMENTS

A dissociation constant can be found from the conductance of a solution of a weak acid or base

Dissociation constants can also be calculated from a measurement of molar conductivity. From the relationship

$$\alpha = \Lambda/\Lambda_0$$

where Λ = molar conductivity and Λ_0 = molar conductivity at infinite dilution, the degree of ionisation α can be found. This value of α can be substituted in the Ostwald equation to give the ionisation constant of the electrolyte.

---

*See E N Ramsden, *Calculations for A-level Chemistry* (Stanley Thornes)

Example The conductivity of a $0.100 \, mol \, dm^{-3}$ solution of ethylamine at $25 \, °C$ is $0.150 \, \Omega^{-1} m^{-1}$. The value of Λ_0 is $2.04 \times 10^{-2} \Omega^{-1} m^2 mol^{-1}$. Find (*a*) the degree of dissociation and (*b*) the dissociation constant of the base.

Method

$$\Lambda = \kappa/c = 0.150/100 = 1.50 \times 10^{-3} \Omega^{-1} m^2 mol^{-1}$$

$$\alpha = \Lambda/\Lambda_0 = 1.50 \times 10^{-3}/(2.04 \times 10^{-2}) = 7.35 \times 10^{-2}$$

From the Ostwald equation

$$K_b = \frac{c\alpha^2}{1 - \alpha} = \frac{0.100(7.35 \times 10^{-2})^2}{0.9265} = 5.83 \times 10^{-4} mol \, dm^{-3}$$

Thus

$$\alpha = 7.35 \times 10^{-2} \text{ and } K_b = 5.83 \times 10^{-4} mol \, dm^{-3}$$

12.7.8 HOW SUBSTITUENTS AFFECT THE STRENGTH OF ACIDS AND BASES

If X is more electronegative [p. 74] than carbon, the acid XCH_2CO_2H is a stronger acid than CH_3CO_2H. An example is chloroethanoic acid, $ClCH_2CO_2H$, which is 100 times stronger than ethanoic acid. The chlorine nucleus in the anion $ClCH_2CO_2^-$ attracts the electrons in the Cl—C bond, enabling the charge to be spread through the anion more than it is in $CH_3CO_2^-$:

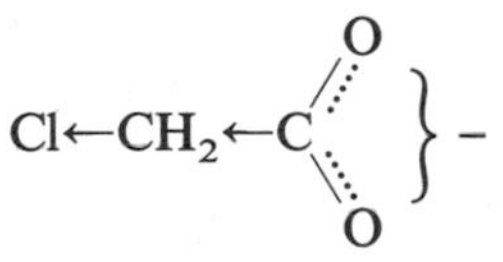

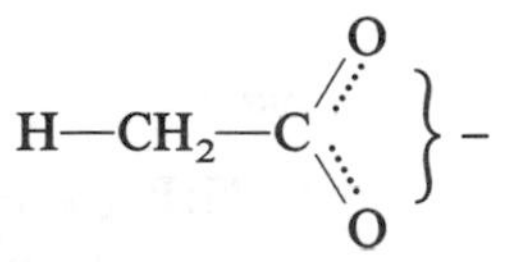

Electron withdrawing substituents make acids stronger, bases weaker...

The reduction of the charge located on the oxygen atoms makes $ClCH_2CO_2^-$ a weaker proton acceptor (a weaker base) than $CH_3CO_2^-$, and therefore makes $ClCH_2CO_2H$ a stronger acid than CH_3CO_2H. Dichloroethanoic acid is stronger still, and trichloroethanoic acid is as strong as some mineral acids. Values of K_a are given below:

CH_3CO_2H $\quad 1.8 \times 10^{-5} mol \, dm^{-3}$ $\quad Cl_3CCO_2H$ $2.2 \times 10^{-1} mol \, dm^{-3}$

$ClCH_2CO_2H$ $1.4 \times 10^{-3} mol \, dm^{-3}$ $\quad HCO_2H$ $\quad 1.7 \times 10^{-4} mol \, dm^{-3}$

...Electron donating substituents make acids weaker, bases stronger

Methanoic acid HCO_2H, is a stronger acid than ethanoic acid. It is inferred from this that the group —CH_3 is less electronegative than hydrogen:

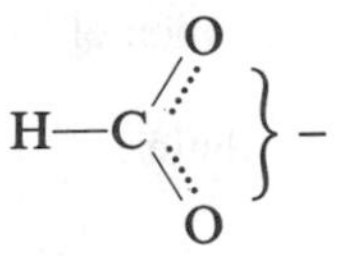

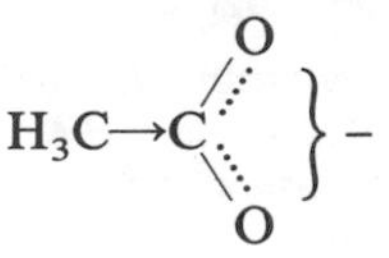

By donating electrons to the —CO_2^- group, —CH_3 makes it a stronger base (a better proton acceptor) and makes CH_3CO_2H a weaker acid than HCO_2H.

Bases are proton acceptors. If a group X is substituted for hydrogen in methylamine, then XCH_2NH_2 will have a different value of K_b from CH_3NH_2. Groups such as —CH_3, which increase the availability of electrons at the nitrogen atom, increase its power to attract electrons, i.e., its basicity. Thus trimethylamine, $(CH_3)_3N$, is a stronger base than methylamine. Values of basic dissociation constants are

CH_3NH_2 $4.4 \times 10^{-4} mol \, dm^{-3}$ $\quad (CH_3)_3N$ $6.3 \times 10^{-3} mol \, dm^{-3}$

A group (e.g. Cl—) which is more electronegative than carbon is described as having a negative inductive effect (a − I effect). A group which is more electropositive than carbon is said to have a positive inductive (+ I) effect. Inductive effects are permanent polarisations of molecules. They affect the physical properties of the compounds, e.g. the dipole moment [p. 77].

12.7.9 *LIQUID AMMONIA

Acids and bases ionise in liquid ammonia...

Ammonia liquefies at − 33 °C at 1 atm. Liquid ammonia is a basic solvent, and many acids which are weakly ionised in water react almost completely with ammonia:

$$HA + NH_3 \rightleftharpoons NH_4^+ + A^-$$

Solutions of the same concentrations of methanoic acid, ethanoic acid, benzoic acid and hydrochloric acid in liquid ammonia have roughly the same catalytic power. This means that all these acids must be almost completely ionised, although they differ greatly in their acidic strengths in aqueous solution.

...and auto-ionisation is very slight...

Liquid ammonia has little acidic character. The auto-ionisation

$$2NH_3 \rightleftharpoons NH_4^+ + NH_2^-$$

takes place to a very small extent. The ionic product, $[NH_4^+][NH_2^-]$, is $10^{-22} \, mol^2 \, dm^{-6}$ at − 33 °C. This is much lower than the ionic product for water, which is $10^{-14} \, mol^2 \, dm^{-6}$ at 25 °C.

Alkali metals react with liquid ammonia to form metal amides:

$$2Na(s) + 2NH_3(l) \rightarrow 2NaNH_2(s) + H_2(g)$$

...thus acid–base reactions can be carried out in liquid ammonia

The acid NH_4^+ will neutralise the base NH_2^-:

$$NH_4^+ + NH_2^- \rightarrow 2NH_3$$

Thus neutralisation reactions can be carried out in liquid ammonia. An example is the reaction between ammonium chloride and sodium amide:

$$NH_4^+Cl^- + Na^+NH_2^- \rightarrow Na^+Cl^- + 2NH_3$$

CHECKPOINT 12D: pH AND DISSOCIATION CONSTANTS

1. The value of K_w, the ionic product for water, increases with temperature. Will the pH of pure water be greater or less than 7 at 100 °C?

2. The pH values of $0.100 \, mol \, dm^{-3}$ solutions of hydrochloric acid and ethanoic acid are 1 and approximately 3 respectively. How does this difference arise?

3. Give an expression for the dissociation constant, K_a, of a weak acid. How is pK_a related to K_a? If two acids, HA and HB, have pK_a values of 3.4 and 4.4, what can you say about the relative strengths of the two acids?

4. Explain what is meant by a conjugate acid–base pair. Give two examples.

5. Find the pH of the following solutions at 25 °C:
(a) $0.00100 \, mol \, dm^{-3}$ HCl(aq)
(b) $2.50 \times 10^{-2} \, mol \, dm^{-3}$ $HClO_4$(aq) (a strong acid)
(c) $3.60 \times 10^{-5} \, mol \, dm^{-3}$ HNO_3(aq)
(d) $2.50 \times 10^{-2} \, mol \, dm^{-3}$ NaOH(aq)
(e) $3.00 \times 10^{-3} \, mol \, dm^{-3}$ $Ca(OHB)_2$(aq)

6. Find the pH of the following solutions:
(a) $1.00 \times 10^{-2} \, mol \, dm^{-3}$ ethanoic acid (pK_a = 4.76)
(b) $0.100 \, mol \, dm^{-3}$ methanoic acid (pK_a = 3.75)

7. Find the dissociation constants of the following acids from the data:
(a) A solution of $2.00 \, mol \, dm^{-3}$ hydrogen cyanide has a pH of 4.55.
(b) A solution of $1.00 \, mol \, dm^{-3}$ iodic(I) acid has a pH of 5.26.

8. Calculate pK_b for the following weak bases at 25 °C from the data:
(a) A solution of $1.00 \times 10^{-2} \, mol \, dm^{-3}$ $C_6H_5CH_2NH_2$ has a pH of 10.7.
(b) A solution of $1.00 \times 10^{-2} \, mol \, dm^{-3}$ $(C_2H_5)_2NH$ has a pH of 11.5.

12.7.10 INDICATORS

Indicators are weak acids or weak bases...
The indicators used in acid–base titrations are weak acids or weak bases. The ions are of a different colour from the undissociated molecules. Litmus is a weak acid, which can be represented by the formula HL. In solution

$$HL + H_2O \rightleftharpoons H_3O^+ + L^-$$

...Litmus
The molecules HL are red, and the anions L^- are blue. The dissociation constant of the indicator is

$$K_a = \frac{[H_3O^+][L^-]}{[HL]}$$

In acid solution, $[H_3O^+]$ is high, and H_3O^+ ions combine with L^- ions to form HL molecules, which are red. In alkaline solution, H_3O^+ ions are removed to form molecules of water, and HL molecules react with OH^- ions to form L^- ions, which are blue:

Their molecules and ions differ in colour...
$$HL + OH^- \rightleftharpoons L^- + H_2O$$

If $[HL] = [L^-]$, the indicator appears purple. Since

$$[H_3O^+] = K_a \frac{[HL]}{[L^-]}$$

this happens when

$$[H_3O^+] = K_a$$

and $pH = pK_a$

...They change colour over 2 units of pH...
If the ratio $[HL]/[L^-] \geqslant 10/1$, the solution appears red. If the ratio $[HL]/[L^-] \leqslant 1/10$, the solution appears blue. Thus litmus changes from red to blue over a range of hydrogen ion concentration from

$$[H_3O^+] = K_a \times 10/1 \text{ to } [H_3O^+] = K_a \times 1/10$$

which gives a range of pH from $pH = pK_a + 1$ to $pH = pK_a - 1$. The colour change takes place over a range of about 2 pH units.

The indicator methyl orange is a weak base, which can be represented as B [p. 686]:

...Methyl orange
$$B + H_3O^+ \rightleftharpoons BH^+ + H_2O$$

The molecules B are yellow, and the cations BH^+ are red. If the ratio $[B]/[BH^+] \geqslant 10/1$, the indicator appears yellow; if the ratio $\leqslant 1/10$, the indicator appears red. The indicator changes from yellow to red over the range of pH given by $pK_b + 1$ to $pK_b - 1$, i.e., 2 pH units. At a pH equal to pK_b, the ratio $[B]/[BH^+] = 1$, and the indicator is orange.

12.7.11 ACID–BASE TITRATIONS

To understand how indicators show the end-point in acid–base titrations, it is necessary to calculate the way in which pH changes during the course of a titration.

TITRATION OF A STRONG BASE INTO A STRONG ACID

Consider the titration of $25.0 \, cm^3$ of $0.100 \, mol \, dm^{-3}$ hydrochloric acid with $0.100 \, mol \, dm^{-3}$ sodium hydroxide solution. At the beginning of the titration

$$[H_3O^+] = 1.00 \times 10^{-1} \, mol \, dm^{-3}; \, pH = 1.00$$

When 24.0 cm^3 of sodium hydroxide have been added

$$[H_3O^+] = (1.0\,cm^3\ of\ 0.100\,mol\,dm^{-3}\ acid)/(49.0\,cm^3\ of\ solution)$$

$$[H_3O^+] = (1.0 \times 10^{-3} \times 0.100)/(49.0 \times 10^{-3}) = 2.04 \times 10^{-3}\,mol\,dm^{-3}$$

$$pH = 2.69$$

Calculation of pH changes during titration of a strong acid by a strong base...

Similar calculations have been employed to give all the pH values plotted in Figure 12.9(a). The figure shows how pH varies during the course of the titration. [See Question 4, p. 266.]

FIGURE 12.9 Changes in pH during Titration (of base (0.100 mol dm^{-3}) into 25.0 cm^3 of acid (0.100 mol dm^{-3}))

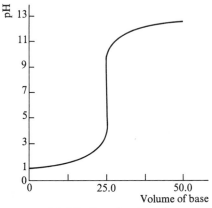

(a) Titration of a strong base into a strong acid

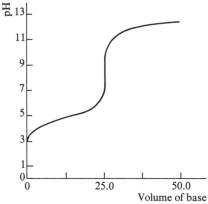

(b) Titration of a strong base into a weak acid

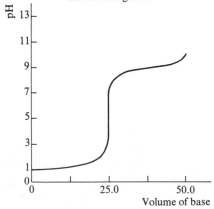

(c) Titration of a weak base into a strong acid

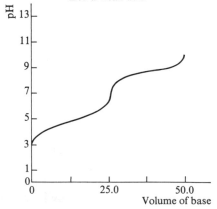

(d) Titration of a weak base into a weak acid

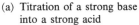

...pH changes rapidly at the end-point

The end-point of the titration occurs when 25.0 cm^3 of alkali have been added. The pH changes rapidly from 3.5 at 24.9 cm^3 of alkali to 10.5 at 25.1 cm^3 of alkali. Any indicator which changes colour over the range of pH 3.5 to 10.5 can be used to show the end-point of the titration.

TITRATION OF A STRONG BASE INTO A WEAK ACID AND A WEAK BASE INTO A STRONG ACID

The weak acid ethanoic acid can be titrated with sodium hydroxide solution. At the beginning of the titration, $25.0 \, cm^3$ of $0.100 \, mol \, dm^{-3}$ ethanoic acid have a pH given by

$$[H_3O^+]^2 = K_a[CH_3CO_2H]$$

$$\therefore \quad pH = 2.88$$

Calculation of pH changes during the titration of a weak acid by a strong base...

When $20.0 \, cm^3$ of alkali have been added

$$[CH_3CO_2H] = 5.0 \, cm^3 \times 0.100 \, mol \, dm^{-3}/45.0 \, cm^3$$

$$[CH_3CO_2^-] = 20.0 \, cm^3 \times 0.100 \, mol \, dm^{-3}/45.0 \, cm^3$$

$$[H_3O^+] = \frac{K_a[CH_3CO_2H]}{[CH_3CO_2^-]} = K_a \times \frac{5.0}{20.0}$$

$$pH = pK_a - \lg(5.0/20.0) = 5.35$$

...of a weak base by a strong acid...

The results of calculations of pH over the range of the titration are shown in Figure 12.9(b). Similar calculations on the course of the titration of the weak base ammonia with a strong acid are shown in Figure 12.9(c). You can see in Figure 12.9(b) that the pH changes from 5 to 10.5 rapidly at the end-point. Indicators which change in this region are methyl orange, litmus and phenolphthalein. In Figure 12.9(c), you can see that the pH changes rapidly from 3 to 7 at the end-point. Indicators which can be used are methyl orange and litmus. Phenolphthalein cannot be used in the titration of a weak base as it changes at pH 9.

TITRATION OF A WEAK ACID AND A WEAK BASE

...of a weak acid with a weak base

In the titration of a weak acid and a weak base, a titration curve like that in Figure 12.9(d) is obtained. The change in pH at the end-point is gradual, and indicators will change colour gradually. No indicator will give a sharp end-point. The way out of this difficulty is to titrate the weak acid against a strong base and the weak base against a strong acid.

TITRATION OF A POLYPROTIC ACID

FIGURE 12.10 Changes in pH during Titration (of $50.0 \, cm^3$ of Phosphoric(V) acid ($0.100 \, mol \, dm^{-3}$) with sodium hydroxide ($0.100 \, mol \, dm^{-3}$)

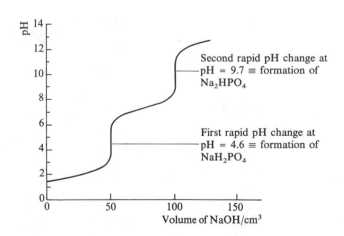

Second rapid pH change at pH = 9.7 ≡ formation of Na₂HPO₄

First rapid pH change at pH = 4.6 ≡ formation of NaH₂PO₄

Volume of NaOH/cm³

...of a polyprotic acid... Figure 12.10 shows the pH changes during the titration of phosphoric(V) acid, H_3PO_4. The titration can be done with an indicator that changes at the first equivalence-point (NaH_2PO_4) or the second equivalence-point (Na_2HPO_4). Figure 12.12, p. 263 shows indicators which could be used. In the third stage of the titration (Na_3PO_4), the curve is very flat, and no indicator will give an end-point. The three stages can be identified because the pK_a values for the three acids, H_3PO_4, $H_2PO_4^-$ and HPO_4^{2-}, are well separated.

TITRATION OF A CARBONATE

FIGURE 12.11 Changes in pH during Titration of sodium carbonate solution $(0.100 \, mol \, dm^{-3})$ with hydrochloric acid $(0.100 \, mol \, dm^{-3})$

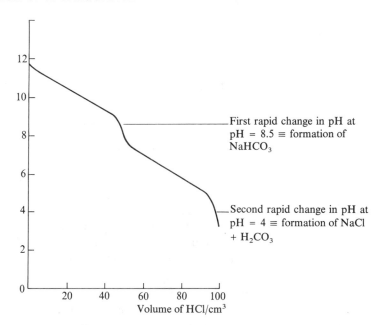

...and of a carbonate... Figure 12.11 shows the pH changes during the titration of sodium carbonate with hydrochloric acid. The two stages in the titration are

$$Na_2CO_3(aq) + HCl(aq) \rightarrow NaHCO_3(aq) + NaCl(aq)$$

$$NaHCO_3(aq) + HCl(aq) \rightarrow NaCl(aq) + H_2O(l) + CO_2(g)$$

Figure 12.12 shows indicators which will change at the two equivalence-points. Use can be made of this two-stage titration to estimate sodium carbonate in a mixture of sodium carbonate and sodium hydroxide or sodium carbonate and sodium hydrogen-carbonate.

Example A $25.0 \, cm^3$ portion of a solution containing sodium carbonate and sodium hydrogencarbonate needed $22.5 \, cm^3$ of a solution of hydrochloric acid of concentration $0.100 \, mol \, dm^{-3}$ to decolourise phenolphthalein. On addition of methyl orange, a further $28.5 \, cm^3$ of the acid were needed to turn this indicator to its neutral colour.

Method In the first stage

Amount of HCl needed to convert CO_3^{2-} to HCO_3^-

$= 22.5 \times 10^{-3} \times 0.100 = 2.25 \times 10^{-3} \, mol$

Amount of $CO_3^{2-} = 2.25 \times 10^{-3} \, mol$

Concentration of $CO_3^{2-} = (2.25 \times 10^{-3})/(25.0 \times 10^{-3})$

$= 9.00 \times 10^{-2} \, mol \, dm^{-3}$

The volume of acid needed to neutralise the total HCO_3^- (aq) is $28.5\,cm^3$. Of this, $22.5\,cm^3$ are needed to neutralise the HCO_3^- (aq) formed from CO_3^{2-} (aq) in the first stage. The remaining $6.00\,cm^3$ neutralise the HCO_3^- (aq) present in the original solution.

$$\text{Amount of HCl needed for } HCO_3^- = 6.00 \times 10^{-3} \times 0.100\,mol$$
$$= 6.00 \times 10^{-4}\,mol$$

$$\text{Amount of } HCO_3^- = 6.00 \times 10^{-4}\,mol$$

$$\text{Concentration of } HCO_3^- = (6.00 \times 10^{-4})/(25.0 \times 10^{-3})\,mol\,dm^{-3}$$
$$= 2.40 \times 10^{-2}\,mol\,dm^{-3}$$

The concentrations are sodium carbonate, $9.00 \times 10^{-2}\,mol\,dm^{-3}$; sodium hydrogen-carbonate, $2.40 \times 10^{-2}\,mol\,dm^{-3}$.

Example $25.0\,cm^3$ of a solution containing sodium hydroxide and sodium carbonate was titrated against $0.100\,mol\,dm^{-3}$ hydrochloric acid, using phenolphthalein as indicator. After $30.0\,cm^3$ of acid had been used, the indicator was decolorised. Methyl orange was added, and a further $12.5\,cm^3$ of hydrochloric acid were needed to turn the indicator orange. Calculate the concentrations of sodium hydroxide and sodium carbonate in the solution.

Method   (a) $30.0\,cm^3$ of acid were needed for the first stage:

$$OH^-(aq) + H^+(aq) \rightarrow H_2O(l)$$
$$CO_3^{2-}(aq) + H^+(aq) = HCO_3^-(aq)$$

(b) $12.5\,cm^3$ of acid were needed for the second stage:

$$HCO_3^-(aq) + H^+(aq) \rightarrow CO_2(g) + H_2O(l)$$

$$\text{Amount of HCl used} = 12.5 \times 10^{-3} \times 0.100 = 1.25 \times 10^{-3}\,mol$$

This is equal to the amount of HCO_3^- formed in the first stage, which is equal to the amount of CO_3^{2-} in $25.0\,cm^3$ of solution.
Therefore

$$[CO_3^{2-}] = 1.25 \times 10^{-3}/(25.0 \times 10^{-3}) = 0.0500\,mol\,dm^{-3}$$

$$\text{Volume of HCl needed by NaOH} = 30.00 - 12.50 = 17.50\,cm^3$$

$$\text{Amount of NaOH in } 25.0\,cm^3 = 17.50 \times 10^{-3} \times 0.100\,mol$$

FIGURE 12.12 Changes in pH for some Common Indicators

pH / Indicator	1	2	3	4	5	6	7	8	9	10	11
Thymol blue	Red	Change			Yellow				Change	Blue	
Methyl orange		Red		Change	Yellow						
Methyl red				Red	Change		Yellow				
Litmus					Red		Change		Blue		
Bromothymol blue					Yellow		Change	Blue			
Phenolphthalein						Colourless			Change	Red	
Universal Indicator		Red		Orange	Yellow		Green	~Blue		Violet	

Therefore

$$[NaOH] = 1.75 \times 10^{-3}/(25.0 \times 10^{-3}) = 0.0700 \, mol \, dm^{-3}$$

The concentrations are $[Na_2CO_3] = 0.050 \, mol \, dm^{-3}$

$$[NaOH] \quad = 0.070 \, mol \, dm^{-3}$$

THE pH RANGE OF INDICATORS

The range over which different indicators can be used is shown in Figure 12.12. Universal Indicator is a mixture of several indicators.

12.7.12 CONDUCTIMETRIC TITRATION

The conductance of an acid changes during the course of a titration

Indicators are not the only means of following a change in pH. During the course of a titration of hydrochloric acid by sodium hydroxide, hydrogen ions are removed from solution to form molecules of water and are replaced by sodium ions. After the end-point, if addition of alkali continues, hydroxide ions accumulate. Since hydrogen ions have a higher value of molar conductivity than any other ions, the graph of conductivity against the volume of alkali added has the form shown in Figure 12.13(b). An apparatus which could be used for **conductimetric titration** is shown in Figure 12.13(a). [For potentiometric titration, see p. 284.]

FIGURE 12.13
Conductimetric Titration

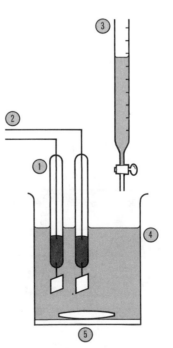

3 Burette measures volume of alkali ($1 \, mol \, dm^{-3}$) added.

2 Leads to conductance meter. This measures the conductance of the solution between the electrodes.

4 Measured volume of $0.1 \, mol \, dm^{-3}$ acid contains H_3O^+ ions of high molar conductivity.

1 Two electrodes. The current flowing between these electrodes depends on the concentration of ions present in the solution.

5 Magnet and magnetic stirrer mix the solution after each addition of alkali.

6 The change in the conductance of the solution as the titration proceeds is shown in (b).

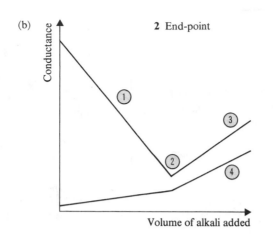

(b)

2 End-point

1 Conductance falls as H_3O^+ ions are replaced by cations from the alkali, e.g. Na^+, of lower conductivity.

3 Conductance rises as OH^- ions are added.

4 Curve for weak acid and strong base. If a weak base is used, the increase in conductance after the end-point is much smaller.

Conductance

Volume of alkali added

12.7.13 MOLAR CONDUCTIVITY OF THE HYDROGEN ION

Λ_0 for H^+ (aq) is high...

The value of molar conductivity at infinite dilution, Λ_0, for the hydrogen ion is much greater than for other ions. The explanation lies in the hydrogen bonding of water molecules [p. 83]. Since long chains of water molecules are linked by hydrogen bonds [see Figure 12.14] if a hydrogen ion enters into the bonding arrangement at one end of a chain, a rearrangement of bonds can take place to release a hydrogen ion at the other end of the chain. These changes can take place very much more quickly than the movement of an ion.

...an explanation in terms of hydrogen bonding

FIGURE 12.14 Water Molecules linked by Hydrogen Bonding

(*a*) H_3O^+ is becoming bonded to one end of a chain of water molecules

(*b*) H_3O^+ is released at the other end of the chain

CHECKPOINT 12E: TITRATION

1. A vegetable extract contains a dye which is a weak acid with $K_a = 10^{-5}\,mol\,dm^{-3}$. At pH 2 the dye is yellow, and at pH 8 it is blue. Explain how this dye could be used for acid–base titrations. Give an example of a titration for which it could be used, and one for which the indicator would be unsuitable.

2. Explain why it is possible to use one indicator for a number of different titrations which reach equivalence-point at different values of pH.

By means of a diagram, explain why phenolphthalein is not used in the titration of a weak base into a strong acid. What kind of indicator should be used for such a purpose?

3. Sketch the change in pH that occurs during the titration of aqueous ammonia into hydrochloric acid of the same concentration.

Suggest an indicator which could be used for the titration, and name one that should be avoided.

4. This question refers to Figure 12.9(a) p. 260. Refer to the worked example on p. 262 if necessary.

Calculate the pH of a solution produced by adding to $25.0 \, cm^3$ of $0.100 \, mol \, dm^{-3}$ hydrochloric acid the following volumes of $0.100 \, mol \, dm^{-3}$ sodium hydroxide solution: (a) $24.5 \, cm^3$, (b) $24.7 \, cm^3$, (c) $24.9 \, cm^3$, (d) $25.1 \, cm^3$, (e) $25.5 \, cm^3$. Check to see whether your results fit the shape of Figure 12.9(a).

5. A $1.000 \, g$ sample of a mixture of Na_2CO_3(anhydrous) and $NaHCO_3$ was dissolved in water. With phenolphthalein as indicator, $17.5 \, cm^3$ of $0.200 \, mol \, dm^{-3}$ hydrochloric acid were required to neutralise the solution. Calculate the percentage by mass composition of the mixture.

6. Explain the high value of Λ_0 of the OH^- ion. Refer to Figure 12.14 for help.

12.7.14 BUFFER SOLUTIONS

A buffer solution absorbs small amounts of $H^+(aq)$ and $OH^-(aq)$ without a change in pH

A **buffer solution** is one which will resist changes in pH due to the addition of small amounts of acid and alkali. An effective buffer can be made by preparing a solution containing both a weak acid and also one of its salts with a strong base, e.g., ethanoic acid and sodium ethanoate. This will absorb small amounts of hydrogen ions because they react with ethanoate ions to form molecules of ethanoic acid:

$$CH_3CO_2^- + H_3O^+ \rightleftharpoons CH_3CO_2H + H_2O$$

Hydroxide ions are absorbed (in small amounts) by combining with ethanoic acid molecules to form ethanoate ions and water:

$$OH^- + CH_3CO_2H \rightleftharpoons CH_3CO_2^- + H_2O$$

A solution of a weak base and one of its salts formed with a strong acid, e.g., ammonia solution and ammonium chloride, will act as a buffer. If hydrogen ions are added, they largely combine with ammonia, and, if hydroxide ions are added, they largely combine with ammonium ions:

The combination of a weak acid and its salt with a strong base acts as a buffer...

$$NH_3 + H_3O^+ \rightleftharpoons NH_4^+ + H_2O$$

$$OH^- + NH_4^+ \rightleftharpoons NH_3 + H_2O$$

The pH of a buffer solution consisting of a weak acid HA and its salt with a strong base is calculated from the equation

$$K_a = \frac{[H_3O^+][A^-]}{[HA]}$$

$$[H_3O^+] = K_a \frac{[HA]}{[A^-]}$$

$$pH = pK_a + lg \frac{[A^-]}{[HA]}$$

Since the salt is completely ionised and the acid only slightly ionised, we can assume that all the anions come from the salt, and put

$$[A^-] = [Salt]$$

$$[HA] = [Acid]$$

$$\therefore \quad pH = pK_a + lg \frac{[Salt]}{[Acid]}$$

...A weak base and its salt with a strong acid act as a buffer...

An effective buffering action is obtained at pH values fairly close to pK_a.

For a buffer made from a weak base $\mathbf{B}$ and its salt with a strong acid $\mathbf{BH^+ X^-}$

$$K_b = \frac{[\mathbf{BH^+}][\mathbf{OH^-}]}{[\mathbf{B}]}$$

$$pOH = pK_b + lg\frac{[\mathbf{BH^+}]}{[\mathbf{B}]}$$

$$pH = pK_w - pK_b + lg\frac{[\mathbf{B}]}{[\mathbf{BH^+}]}$$

Since the weak base is only slightly ionised, we can put

$$[\mathbf{B}] = [\text{Base added}]$$

$$[\mathbf{BH^+}] = [\text{Salt added}]$$

$$\therefore \qquad pH = pK_w - pK_b + lg\frac{[\text{Base}]}{[\text{Salt}]}$$

CALCULATION OF THE pH OF A BUFFER SOLUTION

Example (a)   Three solutions contain ethanoic acid
($K_a = 1.80 \times 10^{-5}\,\text{mol}\,\text{dm}^{-3}$) at a concentration $0.10\,\text{mol}\,\text{dm}^{-3}$ and sodium
ethanoate at a concentration (a) $0.10\,\text{mol}\,\text{dm}^{-3}$, (b) $0.20\,\text{mol}\,\text{dm}^{-3}$,
(c) $0.50\,\text{mol}\,\text{dm}^{-3}$. Calculate the pH values of the three solutions.

Method

$$pH = pK_a + lg\frac{[\text{Salt}]}{[\text{Acid}]}$$

...Calculation of pH values of buffers

In solution (a)   pH $= 4.75 + lg(0.10/0.10)$

$= 4.75 + lg\,1.0$

$= 4.75$

In solution (b)   pH $= 4.75 + lg(0.20/0.10)$

$= 4.75 + lg\,2.0$

$= 5.05$

In solution (c)   pH $= 4.75 + lg(0.50/0.10)$

$= 5.45$

The pH values are (a) 4.75, (b) 5.05, (c) 5.45

*CALCULATION OF THE CHANGE IN pH OF A BUFFER SOLUTION PRODUCED BY ADDITION OF ACID OR ALKALI

Example (b)   Calculate the effect of adding (a) $10\,\text{cm}^3$ of hydrochloric acid of
concentration $1.0\,\text{mol}\,\text{dm}^{-3}$ and (b) $10\,\text{cm}^3$ of sodium hydroxide of concentration
$1.0\,\text{mol}\,\text{dm}^{-3}$ to $1.0\,\text{dm}^3$ of a buffer (a) in Example (1).

Method

In (a)

Amount of hydrochloric acid added = $10 \, cm^3$ of $1.0 \, mol \, dm^{-3}$ solution = $0.010 \, mol$. This amount of hydrogen ion combines with ethanoate ions to form undissociated molecules of ethanoic acid. The amount of salt is therefore decreased by $0.010 \, mol$, and the amount of acid is increased by $0.010 \, mol$. Thus

$$pH = 4.75 + \lg \frac{(0.10 - 0.010)}{(0.10 + 0.010)}$$

$$= 4.66$$

The pH has decreased by 0.09.

In (b)

Amount of OH^- ion added = $10 \times 10^{-3} \times 1.0 = 0.010 \, mol$.

The hydroxide ions react with $0.010 \, mol$ of the weak acid. This results in the formation of an additional $0.010 \, mol$ of anions. Thus, [Acid] is decreased by $0.010 \, mol$, and [Salt] is increased by $0.010 \, mol$.

$$pH = 4.75 + \lg \frac{(0.10 + 0.010)}{(0.10 - 0.010)}$$

$$= 4.84$$

The pH has increased by 0.09.

Addition of $10 \, cm^3$ of $1.0 \, mol \, dm^{-3}$ HCl decreases the pH by 0.09; addition of $10 \, cm^3$ of $1.0 \, mol \, dm^{-3}$ NaOH increases the pH by 0.09.

CHECKPOINT 12F: BUFFERS

1. How does a buffer maintain an almost constant pH, even when small amounts of acid or alkali are added to a solution? What two components are present in a buffer?

2. Solution (a) contains $0.100 \, mol \, dm^{-3}$ potassium chloride solution. Solution (b) contains $0.100 \, mol \, dm^{-3}$ ammonium ethanoate. Both solutions are neutral. Explain why the addition of $1.00 \, cm^3$ of $1.00 \, mol \, dm^{-3}$ hydrochloric acid to $1.00 \, dm^3$ of solution (a) changes its pH to 3, but the same treatment has very little effect on the pH of solution (b).

3. Given a $0.100 \, mol \, dm^{-3}$ solution of dimethylamine, $(CH_3)_2NH$, what would you need in order to prepare a buffer solution? Explain how the buffer would react to the addition of (a) hydrogen ions and (b) hydroxide ions.

*4. A solution of $0.100 \, mol \, dm^{-3}$ ethanoic acid and $0.400 \, mol \, dm^{-3}$ sodium ethanoate has a pH of 5.35. Find the dissociation constant of the acid.

*5. Ethanoic acid, CH_3CO_2H, has $pK_a = 4.76$ at room temperature. Find the pH of the solutions which result from the addition of $50 \, cm^3$ of $0.040 \, mol \, dm^{-3}$ NaOH(aq) to (a) $50 \, cm^3$ of $0.080 \, mol \, dm^{-3} \, CH_3CO_2H$(aq), (b) $50 \, cm^3$ of $0.060 \, mol \, dm^{-3} \, CH_3CO_2H$(aq).

Explain why the pH values of the solutions do not change much when small amounts of acid or alkali are added.

*6. What is the pH of a solution that has been prepared by the addition of $50.0 \, cm^3$ of $0.200 \, mol \, dm^{-3}$ sodium hydroxide to $50.0 \, cm^3$ of $0.400 \, mol \, dm^{-3}$ ethanoic acid $(K_a = 1.75 \times 10^{-5} \, mol \, dm^{-3})$?

What change in pH will take place if $1.00 \, cm^3$ of $1.00 \, mol \, dm^{-3}$ hydrochloric acid is added to the above mixture?

12.7.15 SALT HYDROLYSIS

Many salts dissolve in water to give neutral solutions. Some salts, however, react with water to form acidic or alkaline solutions. These reactions are described as **salt hydrolysis**.

Some salts react with water...

Sodium carbonate is the salt of a strong base and a weak acid. The carbonate ion, CO_3^{2-}, is a base, a proton abstractor. The reaction

$$CO_3^{2-}(aq) + H_2O(l) \rightleftharpoons OH^-(aq) + HCO_3^-(aq)$$

...Their solutions are acidic or alkaline...

makes a solution of sodium carbonate alkaline. Solutions of sodium hydrogen-carbonate are less strongly alkaline because HCO_3^- is a weaker base than CO_3^{2-}:

$$HCO_3^-(aq) + H_2O(l) \rightleftharpoons H_2CO_3(aq) + OH^-(aq)$$

...Such reactions are called salt hydrolysis

Other basic anions are S^{2-}, HS^-, CN^-, $CH_3CO_2^-$ and the anions of many other organic acids.

The hydrolysis of a salt of a weak base and a strong acid, e.g., ammonium chloride, results in an increase in the concentration of hydrogen ions:

$$NH_4^+(aq) + H_2O(l) \rightleftharpoons NH_3(aq) + H_3O^+(aq)$$

The hydrolysis of aluminium salts is covered in Chapter 18 and transition metal salts in Chapter 24.

CHECKPOINT 12G: SALT HYDROLYSIS

1. Predict whether the pH of the following solutions will be 7 or > 7 or < 7. Give your reasons.

(a) $0.10\,mol\,dm^{-3}$ ammonium chloride
(b) $0.010\,mol\,dm^{-3}$ methylammonium chloride
(c) $0.10\,mol\,dm^{-3}$ potassium cyanide
(d) $0.10\,mol\,dm^{-3}$ sodium methanoate

2. The pH readings below refer to the titration of sodium hydroxide solution into $25.0\,cm^3$ of $0.100\,mol\,dm^{-3}$ ethanoic acid.

V/cm^3	0	4.0	6.0	8.0	10.0	12.0	14.0	14.4	14.6	14.8	15.0	15.2	15.4	16.0
pH	2.8	3.8	4.2	4.6	5.1	5.5	6.2	6.5	6.8	7.6	9.0	9.8	10.5	11.4

Plot V, the volume of sodium hydroxide solution, against pH. Use your graph to find the following quantities:

(a) the pH at the end-point
(b) the concentration in $mol\,dm^{-3}$ of sodium hydroxide.
Name an indicator which could be used for this titration.

12.7.16 SOLUBILITY AND SOLUBILITY PRODUCT

Definition of solubility product...

Many salts which we refer to as insoluble do in fact dissolve to a small extent. In a saturated solution, dissolved ions and undissolved salt are in equilibrium. When a saturated solution of nickel(II) sulphide is in contact with solid nickel(II) sulphide, an equilibrium is established:

$$NiS(s) \rightleftharpoons Ni^{2+}(aq) + S^{2-}(aq)$$

The product of the concentrations of nickel ions and sulphide ions is called the solubility product (K_{sp}) of nickel(II) sulphide.

$$K_{sp} = [Ni^{2+}][S^{2-}]$$

The **solubility product** of a salt is the product of the concentrations of the ions in a saturated solution of the salt, raised to the appropriate powers. It applies only to sparingly soluble salts. For lead(II) hydroxide, $Pb(OH)_2$

$$K_{sp} = [Pb^{2+}][OH^-]^2$$

For calcium phosphate(V), $Ca_3(PO_4)_2$

$$K_{sp} = [Ca^{2+}]^3[PO_4^{3-}]^2$$

Like other equilibrium constants, solubility products vary with temperature.

...and solubility   The solubility of a salt is expressed in various ways [p. 176]. They include the mass of solute dissolved in $1 \, dm^3$ of solution and the amount/mol of solute in $1 \, dm^3$ of solution at a stated temperature. The relationship between solubility and solubility product can be seen in the following examples.

Example (a)   Derive an expression for the solubility product of lead(II) fluoride, which has a solubility of $a \, mol \, dm^{-3}$.

Method

$$PbF_2(s) \rightleftharpoons Pb^{2+}(aq) + 2F^-(aq)$$

$$K_{sp} = [Pb^{2+}][F^-]^2$$

If $[PbF_2(aq)] = a$ then $[Pb^{2+}] = a$ and $[F^-] = 2a$

$$K_{sp} = a \times (2a)^2 = 4a^3$$

Example (b)   If the solubility of PbF_2 is $0.64 \, g \, dm^{-3}$, what is the value of the solubility product?

Method

$$\text{Solubility of } PbF_2 = 0.64 \, g \, dm^{-3} = 0.64/245 \, mol \, dm^{-3}$$

$$= 2.61 \times 10^{-3} \, mol \, dm^{-3}$$

As shown in Example (a)

$$K_{sp} = 4a^3$$

$$= 4(2.61 \times 10^{-3})^3 = 7.1 \times 10^{-8} \, mol^3 \, dm^{-9}$$

You will see that the unit in which the solubility product is expressed is the result of multiplying three concentrations together.

Example (c)   Given that the solubility product for calcium sulphate at $25 \, °C = 2.4 \times 10^{-5} \, mol^2 \, dm^{-6}$, calculate the solubility at this temperature.

Method

$$K_{sp} = [Ca^{2+}][SO_4^{2-}] = [Ca^{2+}]^2$$

$$[Ca^{2+}] = \sqrt{K_{sp}} = \sqrt{2.4 \times 10^{-5}}$$

$$= 4.9 \times 10^{-3} \, mol \, dm^{-3}$$

Since each mole of $CaSO_4$ that dissolves puts 1 mole of Ca^{2+} into solution

$$[CaSO_4] = [Ca^{2+}] = 4.9 \times 10^{-3} \, mol \, dm^{-3}$$

The solubility of $CaSO_4$ is $4.9 \times 10^{-3} \, mol \, dm^{-3} = 0.67 \, g \, dm^{-3}$.

12.7.17   THE COMMON ION EFFECT

A saturated solution of a salt **MA** is in contact with solid **MA**.

The solubility of MA is reduced by the addition of $M^{2+}(aq)$ or $A^{2-}(aq)$

$$MA(s) \rightleftharpoons M^{2+}(aq) + A^2(aq)$$

$$K_{sp} = [M^{2+}][A^{2-}]$$

If a solution containing M^{2+} ions is added, $[M^{2+}]$ is increased. Even when the ions are not present in equimolar concentrations, the solubility product, K_{sp}, remains the same. So that the product $[M^{2+}][A^{2-}]$ shall not exceed K_{sp}, M^{2+} ions will be removed from solution as solid **MA**. The addition of a solution containing M^{2+} ions will therefore result in the precipitation of solid **MA**. The addition of a solution containing A^{2-} ions will have the same effect. The precipitation of a solute from solution on addition of an electrolyte solution which has an ion in common with the solute is an example of the **common ion effect**.

Another example of the common ion effect is the change in the concentration of ions produced by the dissociation of a weak acid in the presence of a solution of one of its ions. [See buffer solutions, p. 266.]

Example (d)   Calculate the solubility of manganese(II) sulphide (a) in water and (b) in a $1.0 \times 10^{-2}\,mol\,dm^{-3}$ solution of sulphide ion. $K_{sp}(MnS)$ at $25\,°C = 2.5 \times 10^{-13}\,mol^2\,dm^{-6}$.

Method

(a) $K_{sp} = [Mn^{2+}][S^{2-}]$

$[Mn^{2+}]^2 = 2.5 \times 10^{-13}\,mol^2\,dm^{-6}$

$[Mn^{2+}] = 5.0 \times 10^{-7}\,mol\,dm^{-3}$

$[MnS] = 5.9 \times 10^{-7}\,mol\,dm^{-3}$

(b) The concentration of S^{2-} ions due to dissolved MnS can be neglected in comparison with $1.0 \times 10^{-2}\,mol\,dm^{-3}$.

$[S^{2-}] = 1.0 \times 10^{-2}\,mol\,dm^{-3}$

$K_{sp} = [Mn^{2+}](1.0 \times 10^{-2})$

$[Mn^{2+}] = 2.5 \times 10^{-13}/(1.0 \times 10^{-2})$

$= 2.5 \times 10^{-11}\,mol\,dm^{-3}$

$[MnS] = 2.5 \times 10^{-11}\,mol\,dm^{-3}$

Comparison of the results of (a) and (b) show that the solubility of MnS has been reduced by a factor of 5.0×10^3 by the presence of $1.0 \times 10^{-2}\,mol\,dm^{-3}$ sulphide ions.

12.7.18   APPLICATIONS OF SOLUBILITY PRODUCTS

PRECIPITATION TITRATIONS

Chloride solutions are estimated by titration against silver nitrate solution...

Chlorides are titrated by running silver nitrate solution into a solution of a chloride. Silver chloride is precipitated. The indicator potassium chromate gives a red precipitate of silver chromate at the end-point. It is essential that silver chromate does not precipitate until all the chloride ions have precipitated as silver chloride. At the end-point, solid silver chloride is present, and the value of $[Ag^+]$ can be found.

Solubility products are $K_{sp}(AgCl) = 1.2 \times 10^{-10}\,mol^2\,dm^{-6}$; $K_{sp}(Ag_2CrO_4) = 2.4 \times 10^{-12}\,mol^3\,dm^{-9}$.

$[Ag^+][Cl^-] = [Ag^+]^2 = 1.2 \times 10^{-10}\,mol^2\,dm^{-6}$

$[Ag^+] = 1.1 \times 10^{-5}\,mol\,dm^{-3}$

...Potassium chromate is used as the indicator

The concentration of chromate ions needed to form a precipitate with this concentration of Ag^+ ions is given by

$[Ag^+]^2[CrO_4^{2-}] = 2.4 \times 10^{-12}\,mol^3\,dm^{-9}$

$[CrO_4^{2+}] = 2.4 \times 10^{-12}/(1.1 \times 10^{-5})^2$

$= 2.1 \times 10^{-2}\,mol\,dm^{-3}$

The amount of indicator needed for the titration can be calculated. To give a concentration of $2 \times 10^{-2}\,mol\,dm^{-3}$ in about $50\,cm^3$ of solution it is necessary to add about $0.5\,cm^3$ of $2\,mol\,dm^{-3}$ potassium chromate to the chloride before titration.

QUALITATIVE ANALYSIS

Schemes for the qualitative analysis of mixtures of ions employ methods of separating cations on the basis of the differences between the solubility products of their salts. For example, when hydrochloric acid is added to a solution containing a number of metal cations, only metal chlorides with low solubility products are precipitated:

$$K_{sp}(AgCl) = 1.2 \times 10^{-10}\,mol^2\,dm^{-6}$$

The precipitation of insoluble salts is a useful method for identifying cations

If the concentration of hydrochloric acid is $0.10\,mol\,dm^{-3}$,

$$[Ag^+](0.10) = 1.2 \times 10^{-10}$$

$$[Ag^+] = 1.2 \times 10^{-9}\,mol\,dm^{-3}$$

Only this tiny concentration of silver ions can remain in solution, and if silver ions are present in the mixture, a precipitate of silver chloride is obtained.

CHECKPOINT 12H: SOLUBILITY PRODUCTS

1. Distinguish between the terms *solubility* and *solubility product*. Name three substances to which the term *solubility product* can be applied.

2. Write expressions for the solubility products of the following salts, given their solubilities:

$BaCO_3$, $a\,mol\,dm^{-3}$; BaI_2, $b\,mol\,dm^{-3}$;
Ag_3PO_4, $c\,mol\,dm^{-3}$

3. Explain why, when hydrogen sulphide is passed into an acidic solution containing Cu^{2+} ions and Fe^{2+} ions, CuS is precipitated but FeS is not precipitated.
$(K_{sp}(CuS) = 6 \times 10^{-36}, K_{sp}(FeS) = 6 \times 10^{-18}\,mol^2\,dm^{-6})$.

4. $K_{sp}(CaSO_4) = 2.0 \times 10^{-5}\,mol^2\,dm^{-6}$ at $25\,°C$. Calculate the solubility of $CaSO_4$ in $mol\,dm^{-3}$

(*a*) in water,

(*b*) in $0.100\,mol\,dm^{-3}$ aqueous sodium sulphate,

(*c*) in $0.200\,mol\,dm^{-3}$ aqueous calcium nitrate.

5. $K_{sp}(AgCl) = 1.2 \times 10^{-10}\,mol^2\,dm^{-6}$,
$K_{sp}(AgI) = 8.3 \times 10^{-17}\,mol^2\,dm^{-6}$

A solution of potassium iodide is shaken with solid silver chloride. Describe what you would expect to see. Explain the changes that occur.

12.7.19   COMPLEX IONS

Complex ions are formed by the combination of a cation with a neutral molecule (or molecules) or an oppositely charged ion (or ions). Coordinate bonding is involved [p. 105]. Examples are $[Al(H_2O)_6]^{3+}$, $[Ag(NH_3)_2]^+$, $[CuCl_4]^{2-}$ and $[Al(OH)_4]^-$.

Complex ions have dissociation constants...

The formation of a complex ion is an equilibrium reaction. In the formation of tetracyanozincate ions, $[Zn(CN)_4]^{2-}$, there is set up an equilibrium:

$$Zn^{2+}(aq) + 4CN^-(aq) \rightleftharpoons Zn(CN)_4^{2-}(aq)$$

The dissociation constant of the complex ion is

$$K_d = \frac{[Zn^{2+}][CN^-]^4}{[Zn(CN)_4^{2-}]} = 2.0 \times 10^{-15}\,mol^4\,dm^{-12}$$

The low value of the dissociation constant shows that the complex ion is very stable. The units are $(concentration)^5/(concentration)$, i.e. $mol^4\,dm^{-12}$.

...The reciprocal is the stability constant

The reciprocal of the dissociation constant is called the **stability constant** of the complex.

Edta can be used in complexometric titration...

A number of metal ions form complexes with edta [p. 105]. These complexes are so stable that they can be used for the estimation of the metal ions by **complexometric titration**. The disodium salt of edta is a primary standard. It is used in alkaline solution so that all the carboxyl groups are ionised as

$$(^-O_2CCH_2)_2NCH_2CH_2N(CH_2CO_2{}^-)_2$$

...with an indicator

The end-point in the titration is shown by an indicator which forms a coloured complex with the metal ion which is being titrated. If Eriochrome Black T is used as indicator, the metal-indicator complex colour of red is seen at the beginning of the titration. As the edta solution is added, metal ions are removed from the indicator and complex with edta. At the end-point, the blue colour of the free indicator is seen:

Metal-indicator (red) + Edta → Metal-edta + Indicator (blue)

A worked example of complexometric titration

Example Hardness in water is caused by calcium and magnesium ions. Both these ions complex strongly with edta. To a $200\,cm^3$ sample of tap water were added an alkaline buffer and a few drops of Eriochrome Black T. A volume $3.50\,cm^3$ of $0.100\,mol\,dm^{-3}$ edta was used in titration. Find the concentration of calcium and magnesium ion in the water. The complexes have the formulae Ca(edta) and Mg(edta).

Method

Amount of edta in titration $= 3.50 \times 0.100 \times 10^{-3} = 3.50 \times 10^{-4}\,mol$

Amount of $Ca^{2+} + Mg^{2+} = 3.5 \times 10^{-4}\,mol$

$[Ca^{2+}] + [Mg^{2+}] = 3.50 \times 10^{-4}/(200 \times 10^{-3})$

$= 1.75 \times 10^{-3}\,mol\,dm^{-3}$

CHECKPOINT 12I: COMPLEX IONS

1. Explain these observations, with the help of your knowledge of inorganic chemistry:

(*a*) The solubility of $PbCl_2$ in water decreases on the addition of dilute hydrochloric acid and increases on the addition of concentrated hydrochloric acid.

(*b*) The addition of aqueous ammonia to aqueous magnesium sulphate gives a white precipitate which is soluble in aqueous ammonium chloride.

(*c*) The solubility products of CdS and CuS are 1.6×10^{-28} and $6.3 \times 10^{-36}\,mol^2\,dm^{-6}$ respectively. Explain why, when hydrogen sulphide is bubbled through solutions of (i) Cd^{2+} ions and an excess of CN^- ions, (ii) Cu^{2+} ions and an excess of CN^- ions, CdS is precipitated but CuS is not.

(*d*) A solution of $FeCl_3$ is yellow. Addition of ammonium thiocyanate, NH_4CNS, produces a red colour, which can be discharged by the addition of sodium fluoride.

2. A solution containing silver ions at a concentration of $1.00\,mol\,dm^{-3}$ was added to an equal volume of aqueous ammonia of concentration $1.00\,mol\,dm^{-3}$. If the dissociation constant of the complex ion, $Ag(NH_3)_2{}^+$, is $5.9 \times 10^{-8}\,mol^2\,dm^{-6}$, what is the concentration of free ammonia in the solution?

3. The dissociation constant is $8.0 \times 10^{-16}\,mol^4\,dm^{-12}$ for the equilibrium

$$HgCl_4{}^{2-}(aq) \rightleftharpoons Hg^{2+}(aq) + 4Cl^-(aq)$$

Calculate the concentration of chloride ions in a solution made by dissolving $0.50\,mol$ of K_2HgCl_4 in $1.00\,dm^3$ of solution.

QUESTIONS ON CHAPTER 12

1. Write equations for the reactions that occur when each of the following is separately dissolved in water:
(*a*) HCl(g), (*b*) NH_3(g), (*c*) CH_3CO_2H(l), (*d*) $C_2H_5NH_2$(g), (*e*) $H_2NCH_2CO_2H$(s).

2. Classify the following species as oxidising agent/reducing agent/Brönsted acid/Brönsted base. Quote one reaction

(with its equation) for each species to illustrate its typical behaviour. Some species fall into more than one category:
(*a*) NH_3(aq), (*b*) $NH_4{}^+$(aq), (*c*) $HSO_3{}^-$(aq), (*d*) $(CO_2H)_2$, (*e*) $Fe(H_2O)_6{}^{3+}$(aq).

3. Define the terms *conductivity* and *molar conductivity*. Explain why the molar conductivity of potassium nitrate remains constant over a wide range of dilution. How do the slightly lower values of molar conductivity at high concentrations arise?

4. Find the molar conductivity at infinite dilution, Λ_0, for AgBr. Values of $\Lambda_0/\Omega^{-1}m^2mol^{-1}$ are KBr, 1.52×10^{-2}; KNO_3, 1.45×10^{-2}; $AgNO_3$, 1.33×10^{-2}. Explain the principle on which the calculation is based.

5. What is an ion-exchange resin? Give an example of the use of such a resin, and explain the principles underlying its use [see also p. 174].

6. Explain the significance of the Faraday constant, $96478\,C\,mol^{-1}$. Write an equation for the electrode process that results in the evolution of oxygen at the anode during the electrolysis of an aqueous solution. Calculate the time in minutes for which a current of $0.500\,A$ would need to be passed in order to yield $500\,cm^3$ (at stp) of oxygen.

7. (a) Describe an experimental method for finding the molar conductivity at infinite dilution, Λ_0, of aqueous hydrochloric acid at $20\,°C$.

(b) Explain what data you would need to obtain a value for Λ_0 of ethanoic acid, CH_3CO_2H, at $20\,°C$.

(c) $\Lambda_0(CH_3CO_2H) = 3.90 \times 10^{-2}\Omega^{-1}m^2mol^{-1}$ and $K_a(CH_3CO_2H) = 1.76 \times 10^{-5}mol\,dm^{-3}$. Calculate the electrolytic conductivity of a $0.100\,mol\,dm^{-3}$ solution of ethanoic acid at $20\,°C$.

8. Sketch titration curves showing the change in pH which occurs during the addition of an excess of sodium hydroxide of concentration $1.0\,mol\,dm^{-3}$ to $25\,cm^3$ of (a) $1.0\,mol\,dm^{-3}$ hydrochloric acid, (b) $1.0\,mol\,dm^{-3}$ sulphuric acid, (c) $1.0\,mol\,dm^{-3}$ ethanoic acid ($pK_a = 4.75$).

Explain why the pH at the end-point in (c) is different from that in (a).

9. (a) Explain why an aqueous solution of ammonium chloride is acidic.

(b) Explain why a mixture of ammonium chloride and ammonia in solution has a buffering action.

10. Describe an experiment which you could carry out to find the number of units of charge on a metal cation.

11. A current of $0.200\,A$ is passed for $2.00\,h$ through $200\,cm^3$ of $0.0500\,mol\,dm^{-3}$ aqueous silver nitrate. Calculate the volume (at stp) of hydrogen evolved.

12. Three voltameters were connected in series. They contain silver nitrate solution, copper(II) sulphate solution and acidified chromium(III) sulphate solution. After a current was passed for $30.0\,min$, $0.216\,g$ of silver had been deposited on the cathode of the first cell. Calculate:

(a) the current
(b) the mass of copper deposited in the second cell
(c) the volume (at stp) of oxygen evolved in the second cell
(d) the mass of chromium deposited in the third voltameter

13. A current of $1.75\,A$ passed for $1.00\,h$ through a solution of copper(II) sulphate deposited $1.245\,g$ of copper, and some hydrogen was evolved. Calculate the volume of hydrogen at stp.

14. Silver chloride is sparingly soluble in water. It is even less soluble in dilute hydrochloric acid, more soluble in concentrated hydrochloric acid and much more soluble in aqueous ammonia. Explain the effects of (a) dilute hydrochloric acid, (b) concentrated hydrochloric acid and (c) aqueous ammonia on the solubility.

15. A solution of $0.100\,mol\,dm^{-3}$ sodium hydroxide was added to $25.0\,cm^3$ of $0.100\,mol\,dm^{-3}$ hydrochloric acid:

(a) In one experiment, the pH was measured at intervals during the titration.

(b) In a second experiment, the electrical conductivity was measured throughout the titration.

(c) In the third experiment, the temperature of the acid was recorded as titration proceeded.

Sketch the changes that occur in (a), (b) and (c) as alkali is added to the acid.

16. A $50\,cm^3$ portion of aqueous ammonia ($1.00\,mol\,dm^{-3}$) is titrated with $1.00\,mol\,dm^{-3}$ hydrochloric acid. Draw a graph to show roughly (without doing detailed calculations) the change in pH that occurs as titration proceeds, until $60\,cm^3$ of the acid have been added.

What kind of indicator would be suitable for use in this titration ($K_b(NH_3) = 1.8 \times 10^{-5}mol\,dm^{-3}$)?

17. (a) Describe concisely, with essential experimental detail, how you would determine the electrolytic conductivity (specific conductance) of a $0.100\,M$ aqueous solution of hydrochloric acid.

If the value obtained was $3.5 \times 10^{-2}ohm^{-1}cm^{-1}$, what would the molar conductivity be?

(b) It is easier to obtain an accurate value for the molar conductivity of $0.1\,M$ hydrochloric acid than for a much more dilute solution, say $10^{-5}\,M$. Explain briefly why this is so.

(c) At sufficiently low concentrations, the molar conductivity Λ of a strong electrolyte is proportional to the square root of its concentration c. Use this information and the following data to obtain a value for Λ^∞, the molar conductivity of sodium chloride in water at $25\,°C$ at infinite dilution.

$c/mol\,dm^{-3}$	4.27×10^{-4}	13.4×10^{-4}	21.2×10^{-4}	31.9×10^{-4}
$\Lambda/ohm^{-1}cm^2mol^{-1}$ (for NaCl in water at $25\,°C$)	124.6	123.3	122.5	121.7

(d) What is the purpose of a buffer solution?

(O 83)

18. (a) Explain fully what is meant by the term *solubility product*, indicating any limitations on its use.

If the solubilities at $298\,K$ of calcium sulphate and silver chloride are respectively $1.09\,g\,dm^{-3}$ and $1.79 \times 10^{-3}\,g\,dm^{-3}$, calculate their solubility products, and what their solubilities would be in the presence of $0.100\,M$ calcium chloride solution at this temperature. ($CaSO_4 = 136.1$, $AgCl = 143.5$).

(b) Define the term pH.

If the acid ionization constant (acidity constant, dissociation constant) of ethanoic acid is $1.85 \times 10^{-5}mol\,dm^{-3}$ at $298\,K$, calculate the pH of $0.05\,M$ solution of the acid in water.

(c) Explain why calcium oxalate* is soluble in a strongly ionised acid while calcium sulphate is not.

(*Oxalic acid, ethanedioic acid, $H_2C_2O_4$, is soluble in water.)

(NI 82)

19. (*a*) What is the *Brönsted-Lowry theory* of acids and bases?

(*b*) Explain carefully the meaning of the terms *concentration* and *strength* as applied to acids or bases.
Comment on the relative merits of pH and dissociation constant as measures of the strength of an acid or base.

(*c*) What do you understand by the term *buffer solution*? Calculate the approximate pH of

(i) $0.010 \, mol \, dm^{-3}$ hydrochloric acid,

(ii) $0.010 \, mol \, dm^{-3}$ ethanoic acid,

(iii) a buffer solution which contains $0.010 \, mol \, dm^{-3}$ of ethanoic acid and $0.10 \, mol \, dm^{-3}$ of sodium ethanoate.

[The dissociation constant of ethanoic acid is $1.6 \times 10^{-5} \, mol \, dm^{-3}$.]

(C 83)

20. (*a*) Define, or explain, the term *molar conductance of a solution* (Λ_c). Why is the idea of conductance rather than resistance favoured in Chemistry? What is the special significance of molar conductance at infinite dilution (zero concentration) ($\Lambda_\infty, \Lambda_0$)? Express the effect of dilution on molar conductance graphically, defining the term *dilution*.

(*b*) Sketch graphs to show how molar electrical conductance of weak and strong electrolytes depends on the square root of the concentration, *c* (molarity).

An empirical equation

$$\Lambda_c = \Lambda_0 - k_0 c^{\frac{1}{2}}$$

(where *k* is an experimental constant) has been suggested for use in conductance work. Comment briefly, and relate this equation to the graphs you have sketched.

(*c*) Derive an expression relating the degree of dissociation of a *very weak* acid to its acidity constant (acid ionization constant, dissociation constant) and molar concentration. Without giving experimental details, explain briefly how the validity of the expression might be tested experimentally, and say why this law is of importance in the development of the ionic theory.

Illustrate your discussion with the following data collected for ethanoic acid (acetic acid) at $298 \, K$, the molar conductance at infinite dilution (zero concentration) being $387.9 \, \Omega^{-1} \, cm^2 \, mol^{-1}$.

$c/mol \, dm^{-3}$	$\Lambda_c/\Omega^{-1} \, cm^2 \, mol^{-1}$
6.323×10^{-2}	6.561
3.162×10^{-2}	9.260
1.581×10^{-2}	13.03
3.952×10^{-3}	25.60
1.976×10^{-3}	35.67

Deduce graphically, or by calculations at three or more concentrations, the acidity constant, stating the units in which it is measured.

(*d*) Kohlrausch discovered an interesting relationship between the values of molar conductance at infinite dilution for different electrolytes.

Illustrate what Kohlrausch found, name the law he stated, and show that the molar conductance at infinite dilution of ethanoic acid (acetic acid) is $387.9 \, \Omega^{-1} \, cm^2 \, mol^{-1}$, by using the following data collected at $298 \, K$.

$\Lambda_0/\Omega^{-1} \, cm^2 \, mol^{-1}$

HCl	425.0				
		CH_3COONa	91.0		
		$NaNO_3$	123.0	KNO_3	145.5
		NaCl	128.1	KCl	149.8
		NaOH	246.5	KOH	271.0

(SUJB 81)

21. (*a*) Discuss, with appropriate examples, the Lowry-Brönsted theory of acids and bases.

(*b*) Answer the following in terms of this theory.

(i) What is used as a measure of acid strength? How would this be determined experimentally for a weak acid? Explain the calculation.

(ii) Explain why the relative acid strengths of the following strong acids are in the order

$$HClO_4 > HBr > H_2SO_4 > HCl > HNO_3$$

even though all are strong acids in water.

(iii) Why do all bases stronger than phenylamine (*aniline*) appear to be equally strong in pure ethanoic (*acetic*) acid?

(*c*) (i) Explain why the pH of pure water is 7.0 at 25 °C but at 75 °C it is 6.4. What can be deduced qualitatively from these facts?

(ii) The pH of solution which is 0.02 M with respect to carbonic acid is 4.0. Calculate the concentration of carbonate ion in the solution. Explain the basis of your calculation.

Carbonic acid.

1st dissociation constant $4.2 \times 10^{-7} \, mol \, l^{-1}$

2nd dissociation constant $4.8 \times 10^{-11} \, mol \, l^{-1}$

(JMB 80,S)

22. In order to determine the solubility product, K_{sp}, of lead(II) chloride, $25 \, cm^3$ of a saturated solution of the lead chloride were passed through a cation exchange column, and the column washed through with water. The emergent liquids were combined and titrated with 0.03 M sodium hydroxide of which $26.5 \, cm^3$ were required. The function of the cation exchange column is to replace all the lead(II) ions by H^+ ions so that hydrochloric acid emerges from the column.

$$Pb^{2+}(aq) + 2H^+ \rightarrow Pb^{2+} + 2H^+(aq)$$
from   remains
column   on column

(*a*) Show the equilibria which exist in a saturated solution of lead(II) chloride and give an expression for the solubility product.

(*b*) Calculate a value of the solubility product of lead(II) chloride, stating the units.

(*c*) Using the value of K_{sp} obtained in (*b*), calculate the solubility, in $mol \, dm^{-3}$, of lead(II) chloride in

(i) 2 M hydrochloric acid,

(ii) 1 M lead(II) nitrate solution.

(L 82)

13

OXIDATION–REDUCTION EQUILIBRIA

13.1 ELECTRODE POTENTIALS

If a strip of metal is placed in a solution of its ions, atoms of the metal may dissolve as positive ions, leaving a build-up of electrons on the metal:

$$M(s) \rightarrow M^{2+}(aq) + 2e^-$$

The metal will become negatively charged. Alternatively, metal ions may take electrons from the strip of metal and be discharged as metal atoms:

$$M^{2+}(aq) + 2e^- \rightarrow M(s)$$

The electrode potential of a metal may be positive or negative

In this case, the metal will become positively charged. The potential difference between the strip of metal and the solution depends on the nature of the metal and on the concentration of the ions involved in the equilibrium at the metal surface. Zinc acquires a more negative potential than copper, since it has a greater tendency to dissolve as ions and a smaller tendency to be deposited as metal.

The two metals, zinc and copper in solutions of their ions, may be combined as shown in Figure 13.1 to make an electrochemical cell. The solutions of zinc sulphate and copper(II) sulphate are separated by a porous partition. The metals form the electrodes of the cell, and are connected through a voltmeter. Since the electrode reactions are

$$Zn(s) \rightarrow Zn^{2+}(aq) + 2e^-$$

$$Cu^{2+}(aq) + 2e^- \rightarrow Cu(s)$$

Two metals and solutions of their ions combine to make a cell which can produce a current...

zinc is negatively charged, and copper is positively charged, and electrons flow through the external circuit from zinc to copper. Zinc dissolves from the zinc electrode, and copper is deposited on the copper electrode. Ions flow through the porous partition. The overall cell reaction is

$$Zn(s) + Cu^{2+}(aq) \rightarrow Zn^{2+}(aq) + Cu(s)$$

...and is called a galvanic or voltaic cell

The type of cell in which a chemical reaction results in the production of an electric current is called a **galvanic** or **voltaic** cell. The electromotive force (emf) of the cell is a measure of the tendency of electrons to flow through the external circuit. Under **reversible conditions**, the emf is equal to the difference between the potentials of the two electrodes. To achieve reversible conditions, an electronic voltmeter is used to measure emf. This has a resistance so high that the current which it takes from the cell is negligible. When no current flows, the cell is operating reversibly.

FIGURE 13.1
A Zinc–Copper
Electrochemical Cell

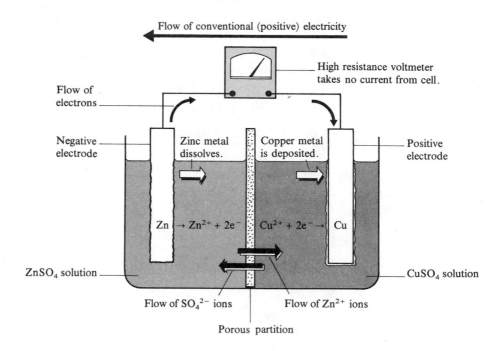

By convention, the emf, E, is taken to act from left to right through the cell. Thus

Convention for the sign of the emf of a cell

$$E_{cell} = E_{RHS\,electrode} - E_{LHS\,electrode}$$

Then a positive emf indicates that electrons are flowing from left to right through the external circuit [in Figure 13.1, from Zn to Cu]. The flow of conventional electricity (positive electricity) is therefore from right to left through the external circuit (from Cu to Zn in Figure 13.1).

The cell in Figure 13.1 can be represented as

$$Zn(s)\,|\,ZnSO_4(aq)\,|\,CuSO_4(aq)\,|\,Cu(s)$$

Its emf is given by

$$E = E_{Cu} - E_{Zn}$$

13.1.1 STANDARD ELECTRODE POTENTIAL

Definition of the standard electrode potential, $E^{\ominus}$, of a metal

If the metal is immersed in a solution of its ions of concentration $1\,mol\,dm^{-3}$ at $25\,°C$, then the potential acquired under these standard conditions is the **standard electrode potential** of that metal, $E^{\ominus}$. In the cell shown in Figure 13.1, if each electrode is immersed in a $1\,mol\,dm^{-3}$ solution of its ions at $25\,°C$, then its potential will be its standard electrode potential. The values are $E^{\ominus}_{Cu} = +0.34\,V$, $E^{\ominus}_{Zn} = -0.76\,V$. The voltmeter will then register an emf of $1.1\,V$. This emf is the difference, $E^{\ominus}_{Cu} - E^{\ominus}_{Zn}$.

13.1.2 FINDING ELECTRODE POTENTIALS

If a cell is constructed from a standard electrode (i.e., one of known potential) and an electrode of unknown potential, the emf of the cell can be measured and used to find the unknown electrode potential. For a cell with the standard electrode on the left-hand side

$$E_{cell} = E_{electrode\,of\,unknown\,potential} - E_{standard\,electrode}$$

The potential of an electrode is found by combining it with a standard electrode to make a cell, and finding the emf of the cell...

The standard hydrogen electrode is the reference electrode with which other electrodes are compared. It consists of a platinised platinum electrode immersed in a solution of $1 \, mol \, dm^{-3}$ hydrogen ions. Hydrogen gas at a pressure of 1 atm is bubbled over the platinum electrode. [Figure 13.2.] On the surface of the platinum, equilibrium is established between hydrogen gas and hydrogen ions.

$$H_2(g) + 2H_2O(l) \rightleftharpoons 2H_3O^+(aq) + 2e^-$$

A potential develops on the surface of the platinum. It is assigned a value of zero volts.

FIGURE 13.2
The Standard Hydrogen
Electrode

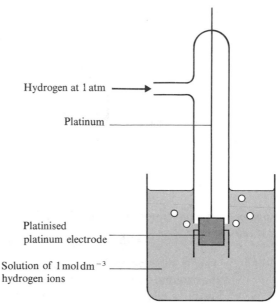

Hydrogen at 1 atm →

Platinum

Platinised
platinum electrode

Solution of $1 \, mol \, dm^{-3}$
hydrogen ions

...$E^{\ominus}$ for another electrode can be found by combination with a standard hydrogen electrode...

The standard electrode potentials of other systems can be found by combining them with a standard hydrogen electrode and measuring the emf of the cell formed. A voltaic cell which combines a standard zinc electrode and a standard hydrogen electrode is shown in Figure 13.3.

FIGURE 13.3 Apparatus
for finding the Standard
Electrode Potential of
the $Zn(s) \rightarrow Zn^{2+} + 2e^-$
system

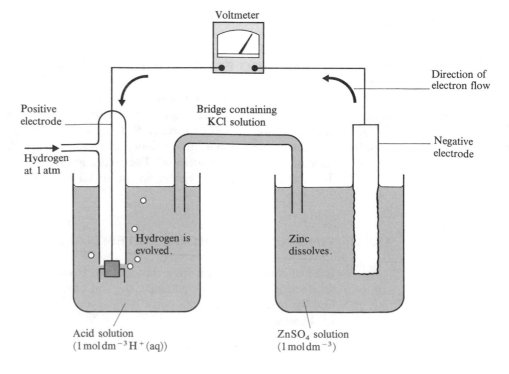

Voltmeter

Direction of
electron flow

Positive
electrode

Bridge containing
KCl solution

Negative
electrode

Hydrogen
at 1 atm

Hydrogen is
evolved.

Zinc
dissolves.

Acid solution
($1 \, mol \, dm^{-3} \, H^+(aq)$)

$ZnSO_4$ solution
($1 \, mol \, dm^{-3}$)

The two compartments in the figure are connected by a **salt bridge**. This contains an electrolyte such as potassium chloride, which conducts electricity but does not allow mixing of the two solutions in the half-cells. The emf of this cell is $-0.76\,V$. The voltmeter shows that electrons flow through the external circuit from zinc to the hydrogen electrode, showing that zinc has a standard electrode potential of $-0.76\,V$.

...but the hydrogen electrode is difficult to operate

The hydrogen electrode is not a convenient reference electrode to use in measurements: maintaining a stream of hydrogen at 1 atm takes careful management. Another standard electrode is the calomel electrode. Calomel is a name for mercury(I) chloride. When mercury and solid mercury(I) chloride are both in contact with a solution of chloride ions, the equilibrium

$$Hg_2Cl_2(s) + 2e^- \rightleftharpoons 2Hg(l) + 2Cl^-(aq)$$

More convenient is the calomel electrode

is set up. The electrode potential of this system depends on the concentration of chloride ions. The easiest way of keeping the concentration constant is to keep the solution saturated by having some undissolved potassium chloride present. The potential of a saturated calomel electrode at $25\,°C$ is $+0.244\,V$. Once the electrode potential has been determined by measuring the emf of a cell composed of a saturated calomel electrode and a standard hydrogen electrode, the calomel electrode can be used as a reference electrode in other measurements. [See Figure 13.4.]

FIGURE 13.4 Two Forms of the Calomel Electrode (in (a) contact is made through the side arm; in (b) contact is made through the sintered glass disc)

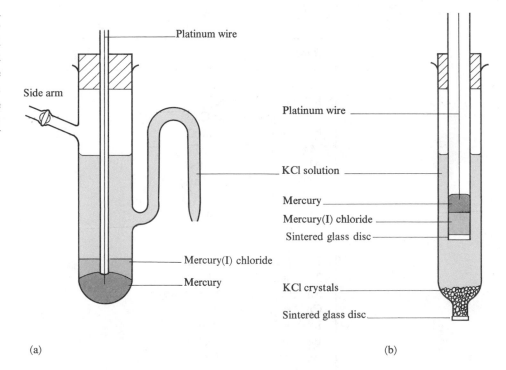

(a)   (b)

In the electrode process

$$Zn(s) \rightarrow Zn^{2+}(aq) + 2e^-$$

Redox systems have electrode potentials

zinc is acting as a reducing agent, supplying electrons. Oxidation–reduction systems other than metals in equilibrium with their ions also have electrode potentials. If a piece of platinum wire is immersed in an acidic solution of manganate(VII) ions, the equilibrium

$$MnO_4^-(aq) + 8H^+(aq) + 5e^- \rightleftharpoons Mn^{2+}(aq) + 4H_2O(l)$$

Definition of standard electrode potential for a redox half-cell

takes electrons from the piece of platinum, which therefore becomes positively charged. If each of the species involved in the equilibrium has a concentration of $1\,mol\,dm^{-3}$, the potential acquired by the platinum is the **standard electrode potential** or **standard reduction potential** of the redox system.

The convention is to write each half-cell reaction as a reduction process:

$$\text{Oxidant} + ne^- \rightleftharpoons \text{Reductant}$$

The standard electrode potentials quoted in tables refer to the reaction written in this way, i.e., as a reduction reaction. They are therefore reduction potentials. Reactions which proceed from left to right more readily than the reduction of hydrogen ions

$$H^+(aq) + e^- \rightleftharpoons \tfrac{1}{2}H_2(g)$$

are given a positive standard electrode potential. Oxidising agents have high positive standard electrode potentials. Reducing agents, such as metals, have highly negative standard electrode potentials.

When all the redox systems are arranged in order of their standard electrode potentials, the Electrochemical Series is obtained. Table 13.1 shows some common redox systems:

Reaction	$E^{\ominus}/V$ (at 298 K)
$Li^+(aq) + e^- \rightleftharpoons Li(s)$	-3.04
$K^+(aq) + e^- \rightleftharpoons K(s)$	-2.92
$Ca^{2+}(aq) + 2e^- \rightleftharpoons Ca(s)$	-2.87
$Na^+(aq) + e^- \rightleftharpoons Na(s)$	-2.71
$Mg^{2+}(aq) + 2e^- \rightleftharpoons Mg(s)$	-2.38
$Al^{3+}(aq) + 3e^- \rightleftharpoons Al(s)$	-1.66
$Zn^{2+}(aq) + 2e^- \rightleftharpoons Zn(s)$	-0.76
$Fe^{2+}(aq) + 2e^- \rightleftharpoons Fe(s)$	-0.44
$Cr^{3+}(aq) + e^- \rightleftharpoons Cr^{2+}(aq)$	-0.41
$Ni^{2+}(aq) + 2e^- \rightleftharpoons Ni(s)$	-0.25
$Sn^{2+}(aq) + 2e^- \rightleftharpoons Sn(s)$	-0.14
$Pb^{2+}(aq) + 2e^- \rightleftharpoons Pb(s)$	-0.13
$2H^+(aq) + 2e^- \rightleftharpoons H_2(g)$	0.00
$Sn^{4+}(aq) + 2e^- \rightleftharpoons Sn^{2+}(aq)$	0.15
$Cu^{2+}(aq) + 2e^- \rightleftharpoons Cu(s)$	0.34
$I_2(s) + 2e^- \rightleftharpoons 2I^-(aq)$	0.54
$Fe^{3+}(aq) + e^- \rightleftharpoons Fe^{2+}(aq)$	0.77
$Ag^+(aq) + e^- \rightleftharpoons Ag(s)$	0.80
$Br_2(l) + 2e^- \rightleftharpoons 2Br^-(aq)$	1.07
$MnO_2(s) + 4H^+(aq) + 2e^- \rightleftharpoons Mn^{2+}(aq) + 2H_2O(l)$	1.23
$Cr_2O_7^{2-}(aq) + 14H^+(aq) + 6e^- \rightleftharpoons 2Cr^{3+}(aq) + 7H_2O(l)$	1.33
$Cl_2(g) + 2e^- \rightleftharpoons 2Cl^-(aq)$	1.36
$Ce^{4+}(aq) + e^- \rightleftharpoons Ce^{3+}(aq)$ (in $H_2SO_4(aq)$)	1.45
$PbO_2(s) + 4H^+(aq) + 2e^- \rightleftharpoons Pb^{2+}(aq) + 2H_2O(l)$	1.47
$MnO_4^-(aq) + 8H^+(aq) + 5e^- \rightleftharpoons Mn^{2+}(aq) + 4H_2O(l)$	1.52
$Ce^{4+}(aq) + e^- \rightleftharpoons Ce^{3+}(aq)$ (in $HNO_3(aq)$)	1.61
$H_2O_2(aq) + 2H^+(aq) + 2e^- \rightleftharpoons 2H_2O(l)$	1.77
$F_2(g) + 2e^- \rightleftharpoons 2F^-(aq)$	2.87

TABLE 13.1 Standard Electrode Potentials

The criterion for a spontaneous cell reaction is that $E^{\ominus}_{cell}$ is positive

When two electrodes combine to form a cell, the value of $E^{\ominus}$ for the cell must be positive if the cell reaction is to happen spontaneously. For example, when copper and silver are in contact with solutions of their ions, two equilibria are set up:

$$Cu^{2+}(aq) + 2e^- \rightleftharpoons Cu(s); \quad E^{\ominus} = +0.34\,V$$

$$Ag^+(aq) + e^- \rightleftharpoons Ag(s); \quad E^{\ominus} = +0.80\,V$$

So that $E^{\ominus}$ shall have a positive value, the reactions that take place are

$$Ag^+(aq) + e^- \rightarrow Ag(s); \quad E^{\ominus} = +0.80\,V$$

$$Cu(s) \rightarrow Cu^{2+}(aq) + 2e^-; \quad E^{\ominus} = -0.34\,V$$

$$Total : Cu(s) + 2Ag^+(aq) \rightarrow Cu^{2+}(aq) + 2Ag(s); \quad E^{\ominus} = +0.46\,V$$

If solutions containing Ce^{4+}, Ce^{3+}, Fe^{3+} and Fe^{2+} are mixed, the redox equilibria in the solution are

$$Fe^{3+}(aq) + e^- \rightleftharpoons Fe^{2+}(aq); \quad E^{\ominus} = +0.77\,V$$

$$Ce^{4+}(aq) + e^- \rightleftharpoons Ce^{3+}(aq); \quad E^{\ominus} = +1.45\,V$$

The redox reaction that takes place is that for which $E^{\ominus}$ is positive, i.e.

$$Ce^{4+}(aq) + Fe^{2+}(aq) \rightarrow Ce^{3+}(aq) + Fe^{3+}(aq); \quad E^{\ominus} = +0.68\,V$$

A redox reaction will go almost to completion between two redox systems which differ by 0.3 V or more in their electrode potentials.

13.1.3 *THE NERNST EQUATION

The Nernst equation relates electrode potentials and standard electrode potentials

If the conditions are not standard, then the electrode potential will not be exactly equal to the standard electrode potential. Its value is given by the **Nernst equation**:

$$E = E^{\ominus} + \frac{RT}{nF} \ln \left(\frac{\text{Product of concentrations in oxidised states raised to the appropriate powers}}{\text{Product of concentrations in reduced states raised to the appropriate powers}} \right)$$

where R = the gas constant, T = temperature in kelvins, F = the Faraday constant and n = the number of electrons transferred from the oxidised state to the reduced state.

If a strip of aluminium dips into a solution of aluminium ions at a concentration of $0.100\,mol\,dm^{-3}$, the potential of the strip is

$$E = E^{\ominus} + \frac{RT}{3F} \ln \frac{[Al^{3+}]}{[Al]}$$

The 'concentration' of a solid remains constant, and may be omitted from the expression since it always has the same value as that under standard conditions. Therefore

since $\quad E^{\ominus} = -1.66\,V$

$$E = -1.66 + 2.303\,\frac{RT}{3F}\,lg\,0.100$$

The constant $2.303\,RT/F$ has the value 0.592 V at 25 °C so that

$$E = -1.66 + \frac{0.0592}{3}\,lg\,0.100 = -1.68\,V$$

The potential of the aluminium electrode is slightly different from its standard value. For the dichromate redox system

$$Cr_2O_7^{2-}(aq) + 14H^+(aq) + 6e^- \rightleftharpoons 2Cr^{3+}(aq) + 7H_2O$$

$$E = E^{\ominus} + \frac{0.0592}{6} \lg \frac{[Cr_2O_7^{2-}][H^+(aq)]^{14}}{[Cr^{3+}]^2}$$

Under standard conditions

$$[Cr_2O_7^{2-}] = [Cr^{3+}] = [H^+(aq)] = 1.0\,mol\,dm^{-3}$$

Dichromate is a powerful oxidising agent under standard conditions... and the potential of the redox system has its standard value of 1.33 V. If dichromate is used in neutral solution instead of acid solution

$$[H^+(aq)] = 10^{-7}\,mol\,dm^{-3}$$

If

$$[Cr_2O_7^{2-}] = [Cr^{3+}] = 1.0\,mol\,dm^{-3}$$

$$E = 1.33 + \frac{0.0592}{6} \lg \left(\frac{1.0 \times (10^{-7})^{14}}{1.0^2} \right)$$

$$= 1.33 + 0.00983\,(\lg 1.0 + 14 \lg 10^{-7})$$

$$= 0.37\,V$$

...but only a weak oxidising agent in neutral solution In neutral solution, the electrode potential is reduced from 1.33 V to 0.37 V, and dichromate is an oxidising agent of the same potency as Cu^{2+} ions at unit concentration.

The same treatment applied to the manganate(VII) system shows that it is a much stronger oxidising agent in acid solution than in neutral solution.

For the hydrogen electrode

$$2H^+(aq) + 2e^- \rightleftharpoons H_2(g)$$

$$E = E^{\ominus} + \frac{RT}{2F} \ln \frac{[H^+(aq)]^2}{[H_2]}$$

The hydrogen electrode can be used to measure pH $[H_2]$ can be omitted from the expression provided that the hydrogen pressure remains the same as that under standard conditions, i.e., 1 atm. Therefore

$$E = 0 + \frac{0.0592}{2} \lg [H^+(aq)]^2$$

$$= 0.0592\,pH$$

The potential of a hydrogen electrode is thus a measure of the pH of the solution in which the electrode is immersed.

CHECKPOINT 13A: ELECTRODE POTENTIALS

1. Does a high positive standard electrode potential for a redox system indicate that the system acts as an oxidising agent or a reducing agent?

2. Explain the terms (*a*) electrode potential, (*b*) standard electrode potential.

Explain how you could find the standard electrode potential for the system

$$Fe^{3+}(aq)\,|\,Fe^{2+}(aq)\,|\,Pt$$

3. This cell is set up:

$$Ag(s)\,|\,Ag^+(aq, 1\,mol\,dm^{-3})\,|\,Cu^{2+}(aq, 1\,mol\,dm^{-3})\,|\,Cu(s)$$

(*a*) State the emf of the cell. Refer to Table 13.1, p. 280.

(*b*) Write the equation for the chemical reaction that takes place in the cell when the copper and silver electrodes are connected by an external circuit.

(*c*) State the direction in which electrons flow through the external circuit.

4. Which electrode, anode or cathode, is associated in an electrochemical cell with oxidation?

5. Refer to Table 13.1, p. 280. Which of the following reactions will occur spontaneously? (Assume all concentrations are $1 \, mol \, dm^{-3}$.)

(a) $Fe(s) + Zn^{2+}(aq) \rightarrow Fe^{2+}(aq) + Zn(s)$

(b) $Fe(s) + Sn^{2+}(aq) \rightarrow Fe^{2+}(aq) + Sn(s)$

(c) $Sn^{4+}(aq) + 2I^{-}(aq) \rightarrow Sn^{2+}(aq) + I_2(s)$

(d) $Zn(s) + Mg^{2+}(aq) \rightarrow Zn^{2+}(aq) + Mg(s)$

(e) $Zn(s) + Sn^{2+}(aq) \rightarrow Zn^{2+}(aq) + Sn(s)$

(f) $Sn^{4+}(aq) + 2Fe^{2+}(aq) \rightarrow Sn^{2+}(aq) + 2Fe^{3+}(aq)$

(g) $Cr_2O_7^{2-}(aq) + 14H^{+}(aq) + 6Cl^{-}(aq) \rightarrow$
$2Cr^{3+}(aq) + 7H_2O(l) + 3Cl_2(g)$

(h) $2Ce^{4+}(aq) + 2Br^{-}(aq) \rightarrow 2Ce^{3+}(aq) + Br_2(l)$

13.1.4   THE GLASS ELECTRODE

The potential of the glass electrode depends on pH

There is another electrode which has a potential that depends on pH. The potential difference between a thin glass membrane and the solution in which it is immersed varies in a regular way with the pH of the solution. This phenomenon is utilised in the glass electrode [see Figure 13.5].

FIGURE 13.5
The Glass Electrode

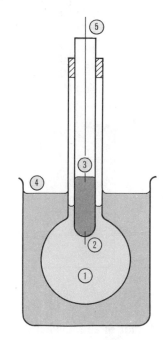

5. Potential of electrode is determined by the difference in pH between the inside and the outside of the bulb.

3. Mercury makes contact with outside circuit.

4. Solution of unknown pH.

2. Platinum wire sealed into an inner tube makes electrical contact with the buffer solution.

1. Bulb of very thin glass is filled with buffer solution.

For pH measurement, a glass electrode is combined with a reference electrode

The glass electrode is coupled with a calomel electrode to complete the cell. As the glass electrode has a high resistance, an electronic voltmeter is used to measure the emf of the cell.

The potential of the glass electrode at 25 °C is given by

$$E = K - 0.0592 \, pH$$

where K = a constant. (K is found by immersing the glass electrode in a buffer solution of known pH, coupling with a calomel electrode, and measuring the emf of the cell.)

13.2 POTENTIOMETRY

13.2.1 ACID–BASE TITRATIONS

The potential of a glass electrode changes during the course of a titration

A potentiometric method can be used to follow the course of a titration. To perform an acid–base titration, a combination of a glass electrode and a calomel electrode is used. The potential of the glass electrode depends on the pH of the solution. As the pH changes during the titration, the potential of the glass electrode changes, and the emf of the cell changes. An apparatus for potentiometric titration is shown in Figure 13.6.

FIGURE 13.6
A Potentiometric Method for Acid–Base Titration

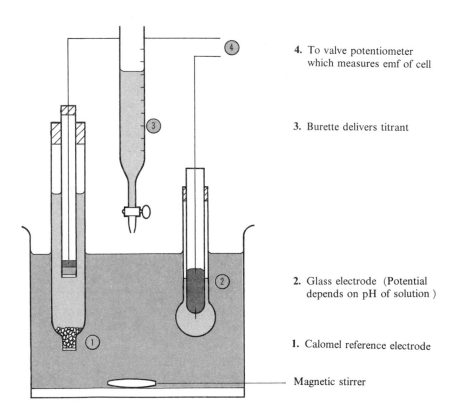

4. To valve potentiometer which measures emf of cell

3. Burette delivers titrant

2. Glass electrode (Potential depends on pH of solution)

1. Calomel reference electrode

— Magnetic stirrer

A plot of emf against the volume of titrant gives the end-point

A plot of emf against the volume of titrant added is plotted. Such plots have the same shapes as the titration curves in Figure 12.9, p. 260: at the end-point, there is a sharp change in emf. Automatic titrators have been developed to add titrant and record the course of the titration automatically.

13.2.2 OXIDATION–REDUCTION TITRATIONS

The potential of a platinum electrode varies during the course of a redox titration

A platinum electrode dipping into the redox solution measures the potential of the system. The usual reference electrode is a calomel electrode. Figure 13.7(a) shows apparatus which could be used. The variation of electrode potential during the course of the titration follows the curve shown in Figure 13.7(b).

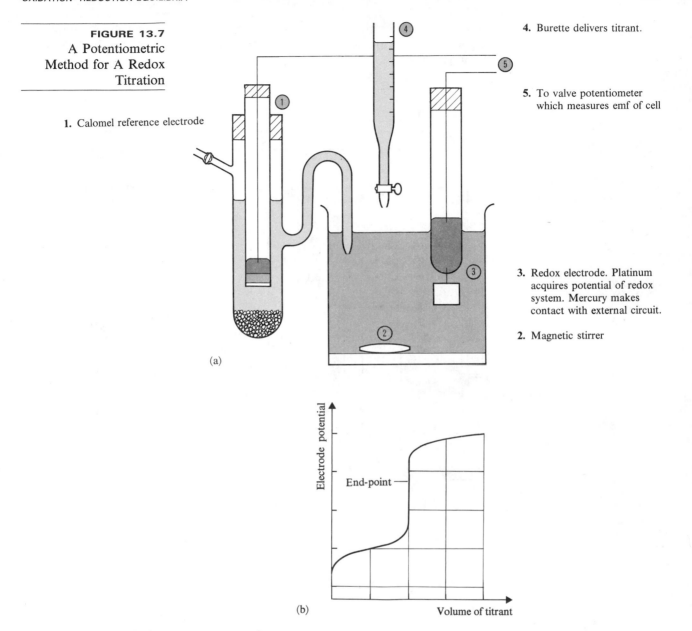

FIGURE 13.7
A Potentiometric
Method for A Redox
Titration

1. Calomel reference electrode

4. Burette delivers titrant.

5. To valve potentiometer
which measures emf of cell

3. Redox electrode. Platinum
acquires potential of redox
system. Mercury makes
contact with external circuit.

2. Magnetic stirrer

(a)

(b)

13.2.3 PRECIPITATION TITRATIONS

*A potentiometric method
can be used to follow
precipitation titrations...*

Precipitation titrations can be performed by a potentiometric method [see Figure 13.8]. If silver nitrate solution is titrated into a neutral solution of sodium chloride, chloride ions are removed from solution:

$$Ag^+(aq) + Cl^-(aq) \rightarrow AgCl(s)$$

The potential of a silver–silver chloride electrode is governed by the equilibrium

$$AgCl(s) + e^- \rightleftharpoons Ag(s) + Cl^-(aq)$$

for which

$$E = E^{\ominus} - 0.059\lg[Cl^-]$$

*...using, for example, the
Ag/AgCl electrode*

If a silver–silver chloride electrode is immersed in the chloride solution, its potential will change during the course of the titration. If it is coupled with a reference electrode, such as a calomel electrode, connected by a bridge containing ammonium nitrate (not potassium chloride for chloride titrations), the emf of the cell can be measured as titration proceeds.

FIGURE 13.8
A Potentiometric
Method for Precipitation
Titration

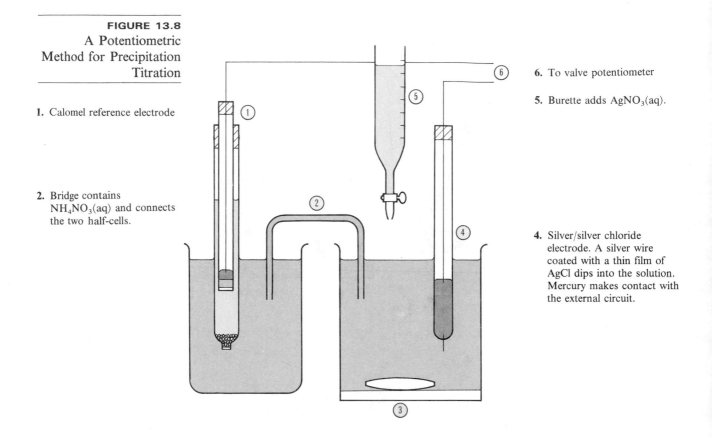

1. Calomel reference electrode

2. Bridge contains $NH_4NO_3(aq)$ and connects the two half-cells.

6. To valve potentiometer

5. Burette adds $AgNO_3(aq)$.

4. Silver/silver chloride electrode. A silver wire coated with a thin film of AgCl dips into the solution. Mercury makes contact with the external circuit.

13.3 VOLTAIC CELLS

In galvanic or voltaic cells, a chemical reaction produces an electric current.

13.3.1 DANIELL CELL

The Daniell cell utilises zinc and copper in solutions of their ions as shown in Figure 13.1, p. 277. The emf of the cell is 1.1 V.

13.3.2 DRY CELLS

Dry cells were invented to overcome the difficulty of electrolyte solution leaking out of cells such as the Daniell cell. In dry cells, the electrolyte is made into a paste. An example is shown in Figure 13.9.

FIGURE 13.9
A Dry Cell

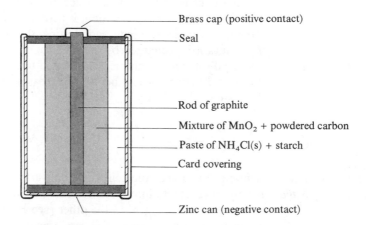

Brass cap (positive contact)
Seal
Rod of graphite
Mixture of MnO_2 + powdered carbon
Paste of $NH_4Cl(s)$ + starch
Card covering
Zinc can (negative contact)

Portable dry cells...

This type of cell is used in radios, flashlights and clocks as it is portable. The emf of the cell shown in Figure 13.9 is 1.5 V. The initial electrode processes are

...batteries...

Anode $Zn(s) \rightarrow Zn^{2+}(aq) + 2e^-$

Cathode $2NH_4^+(aq) + 2e^- \rightarrow 2NH_3(g) + 2H_2(g)$

13.3.3   THE LEAD–ACID ACCUMULATOR

...accumulators

The lead-acid accumulator is charged by passing a direct electric current through it...

This cell stores or accumulates electric charge. It consists of two lead plates dipping into a 30% solution of sulphuric acid. Both plates become covered with an insoluble film of lead(II) sulphate. First, the cell must be charged. A direct electric current is passed through the cell. The processes which take place are

CHARGE

Positive plate
$PbSO_4(s) + 2e^- \rightarrow Pb(s) + SO_4^{2-}(aq)$

Negative plate
$PbSO_4(s) + 2H_2O(l) \rightarrow PbO_2(s) + 4H^+(aq) + SO_4^{2-}(aq) + 2e^-$

The plates are now different and therefore have different potentials, so that, when they are connected, an electric current will flow between them. When the cell supplies electric current, i.e., discharges, the processes which take place are

...During discharge, it supplies an electric current

DISCHARGE

Negative plate
$Pb(s) + SO_4^{2-}(aq) \rightarrow PbSO_4(s) + 2e^-$

Positive plate
$2PbO_2(s) + 4H^+(aq) + SO_4^{2-}(aq) + 2e^- \rightarrow PbSO_4(s) + 2H_2O(l)$

When all the PbO_2 and Pb have been converted to $PbSO_4$, there is no difference between the plates, and the cell can no longer give a current.

The polarity of the plates reverses between charge and discharge

You will notice that the Pb plate is the positive plate during charge, and the negative plate during discharge. The polarity of the PbO_2 plate is also reversed, from negative during charge to positive during discharge. The reactions can be summarised as

$$Pb(s) + PbO_2(s) + 2H_2SO_4(aq) \xrightleftharpoons[\text{charge}]{\text{discharge}} 2PbSO_4(s) + 2H_2O(l)$$

Car batteries

Car batteries consist of six lead accumulator cells joined in series to give an emf of 12 V. When the car is in motion, it drives a generator which charges the battery. If there is too much stopping and starting, the battery loses its charge and becomes 'flat', until it is recharged by the passage of a direct current from a transformer. During discharge, sulphuric acid is used up, and the density of the liquid in the cells drops. A **hydrometer** can be used to measure the density of the liquid and assess the state of the battery.

13.3.4 FUEL CELLS

Fuel cells are a promising source of energy for the future

A fuel cell is a galvanic cell which converts the chemical energy of a continuous supply of reactants into electrical energy. Fuel is supplied to one electrode and an oxidant, usually oxygen, to the other [see Figure 13.10]. A great deal of research is being done on fuel cells as they are a promising source of energy for the future. The American Gemini space probes and Apollo moon probes used hydrogen–oxygen fuel cells. The astronauts used the product of the reaction to supplement their drinking water.

FIGURE 13.10

A Hydrogen–Oxygen Fuel Cell

1 Stream of hydrogen

2 Hydrogen diffuses through the porous cathode (e.g., of Ni). When it comes into contact with the electrolyte, KOH(aq), adsorbed H_2 is oxidised:

$\frac{1}{2}H_2 + OH^- \rightarrow H_2O + e^-$

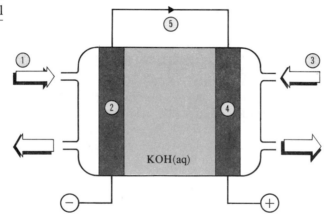

5 Electrons flow through an external circuit from cathode to anode.

3 Stream of oxygen

4 Oxygen diffuses through a porous anode (e.g., of nickel). Adsorbed oxygen is reduced to OH^- ions:

$\frac{1}{2}O_2 + H_2O + 2e^- \rightarrow 2OH^-(aq)$

The overall reaction is

$2H_2(g) + O_2(g) \rightarrow 2H_2O(l)$

QUESTIONS ON CHAPTER 13

1. Refer to Table 13.1, p. 280.
Which of the following species are oxidised by manganese(IV) oxide?
Br^-, Ag, I^-, Cl^-

2. Which of the following species are reduced by Sn^{2+}?
I_2, Ni^{2+}, Cu^{2+}, Fe^{2+}

3. The rusting of iron is prevented by (*a*) a coating of paint or (*b*) a layer of zinc or (*c*) a layer of tin. How do these different methods work? What is meant by 'sacrificial protection'? Give examples of its use [p. 504].

4. Refer to Table 13.1, p. 280. Relate the differences between the chemical properties of the elements mentioned below to the differences between the standard electrode potentials of the following systems:

(*a*) $K^+(aq)|K(s)$ (*d*) $Fe^{2+}(aq)|Fe(s)$

(*b*) $Mg^{2+}(aq)|Mg(s)$ (*e*) $Cu^{2+}(aq)|Cu(s)$

(*c*) $Ca^{2+}(aq)|Ca(s)$ (*f*) $Cl_2(g)|Cl^-(aq)$

5. Two half-cells are

(a) $Co^{2+}(aq, 1\,mol\,dm^{-3})\,|\,Co(s)$

(b) $Cl_2(g, 1\,atm)\,|\,Cl^-(aq, 1\,mol\,dm^{-3})\,|\,Pt$

State which will be the positive and which the negative electrode when the two half-cells are connected. Write the equation for the cell reaction. If the emf of the cell is 1.63 V, what is $E^{\ominus}$ for the cobalt half-cell? [See Table 13.1, p. 280 for $E^{\ominus}(Cl_2/Cl^-)$.]

6. Calculate the standard emf's of the following cells at 298 K:

(a) $Ni(s)\,|\,Ni^{2+}(aq)\,|\,Sn^{2+}(aq), Sn^{4+}(aq)\,|\,Pt(s)$

(b) $Pt(s)\,|\,I_2(s), I^-(aq)\,|\,Ag^+(aq)\,|\,Ag(s)$

(c) $Pt(s)\,|\,Cl_2(g), Cl^-(aq)\,|\,Br_2(l), Br^-(aq)\,|\,Pt(s)$

(d) $Sn(s)\,|\,Sn^{2+}(aq)\,|\,Ag^+(aq)\,|\,Ag(s)$

(e) $Ag(s)\,|\,Ag^+(aq)\,|\,Cu^{2+}(aq)\,|\,Cu(s)$

(f) $Fe(s)\,|\,Fe^{2+}(aq)\,|\,Cu^{2+}(aq)\,|\,Cu(s)$

(g) $Zn(s)\,|\,Zn^{2+}(aq)\,|\,Pb^{2+}(aq)\,|\,Pb(s)$

7. Refer to Table 13.1, p. 280 and to the standard reduction potentials listed here:

$$E^{\ominus}(VO_2^+(aq)\,|\,VO^{2+}(aq)) = +1.00\,V$$

$$E^{\ominus}(Cd^{2+}(aq)\,|\,Cd(s)) = -0.40\,V$$

$$E^{\ominus}(BrO_3^-(aq)\,|\,Br_2(g)) = +1.52\,V$$

$$E^{\ominus}(S_4O_6^{2-}(aq)\,|\,S_2O_3^{2-}(aq)) = 0.090\,V$$

(a) State which of the species MnO_4^-, Ce^{4+}, $Cr_2O_7^{2-}$, VO_2^+, Fe^{3+} are able to liberate chlorine from an acidic solution of sodium chloride.

(b) Write a balanced equation for the reaction between MnO_4^- and VO_2^+ in acid solution.

(c) Find $E^{\ominus}$ for the cell

$Zn(s)\,|\,ZnSO_4(1\,mol\,dm^{-3})\,|\,CdSO_4(1\,mol\,dm^{-3})\,|\,Cd(s)$

(d) Put the following into order of their power as oxidising agents in acid solution: $Cr_2O_7^{2-}$, Cl_2, MnO_4^-, I_2, BrO_3^-, $S_4O_6^{2-}$

8. A student conducted an experiment to determine the solubility product of silver chloride using the apparatus represented in Figure 13.11. The emf measured was $-0.12\,V$.

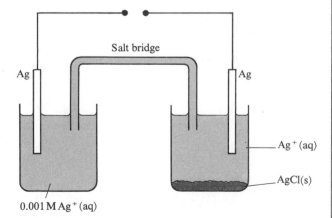

0.001 M Ag^+(aq)

FIGURE 13.11

The potential of the silver electrode in a solution of silver ions is given by the equation:

$$E = E^{\ominus} + 0.06\log_{10}[Ag^+]$$

(a) The standard electrode potential of silver is $+0.80\,V$. What do you understand by this?

(b) Using the above equation, derive an expression for the emf found in this experiment.

(c) Calculate the concentration of silver ions in the right-hand half-cell.

(d) (i) Write an expression for the solubility product of silver chloride. Calculate its value.

(ii) What is the purpose of the solid silver chloride in the right-hand half-cell?

(e) How would you prepare a salt bridge for this experiment?

(f) How would you expect the emf to change if a few drops of aqueous sodium chloride were added to the right-hand half-cell? Explain your answer.

(L 81)

9. Figure 13.12 represents part of an apparatus to be used for the determination of the standard electrode potential of zinc, $E^{\ominus}_{Zn^{2+}_{(aq)}|Zn_{(s)}}$. The apparatus is maintained at 25 °C.

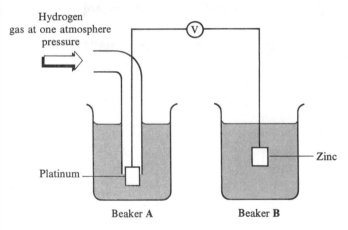

FIGURE 13.12

(a) Name a solution which could be placed in beaker **A** and specify its concentration.

(b) Name a solution which could be placed in beaker **B** and specify its concentration.

(c) One essential part of the cell has been omitted. Give a short description of the missing part.

(d) The reading on the voltmeter was found to be -0.76.

(i) Give the standard electrode potential for the zinc electrode.

(ii) State the direction in which the electrons flow in the external circuit (through the voltmeter, V).

(e) Give the half-cell reactions and the overall cell reaction.

(f) (i) State whether the reading of the voltmeter would become more positive, more negative or remain unchanged if aqueous sodium hydroxide were added to beaker **B**.

(ii) Give equation(s) for the reactions which occur on addition of the aqueous sodium hydroxide. (The electrochemical processes are *not* required.)

(*g*) The standard electrode potential for the $Cu^{2+}(aq)/Cu(s)$ electrode is $+0.34\,V$. If the hydrogen electrode is replaced by the copper electrode what would be the new reading of the voltmeter?

(*h*) In the three half-cells mentioned in this question, state which species is the most powerful oxidising agent. Give reasons.

(WJEC 82)

10. Define *standard electrode potential* and outline the essential features of a method for measuring the standard electrode potential of the zinc ion/zinc electrode.

A cell is constructed as shown in Figure 13.13. Zinc and carbon electrodes are partly immersed in aqueous potassium hydroxide. Some zinc passes into solution as zincate(II) ions, $ZnO_2^{2-}(aq)$.

Use the data below to determine

(*a*) the standard emf of the cell,

(*b*) the cell reaction and its direction when the cell is delivering current.

$$ZnO_2^{2-}(aq) + 2H_2O(l) + 2e^- \rightleftharpoons Zn(s) + 4OH^-(aq);$$

$$E_{298}^{\ominus} = -1.216\,V$$

$$O_2(g) + 2H_2O(l) + 4e^- \rightleftharpoons 4OH^-(aq);$$

$$E_{298}^{\ominus} = +0.401\,V$$

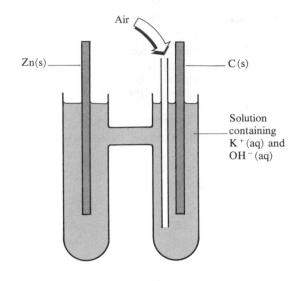

FIGURE 13.13

What advantages and/or disadvantages might influence the industrial use of this cell?

(C 80, S)

14

REACTION KINETICS

14.1 INTRODUCTION

The rates of chemical reactions vary greatly

Chemical reactions vary enormously in speed. Some proceed slowly over a period of months (e.g., the rusting of iron); others take weeks to reach completion (e.g., the fermentation of ethanol). Some reactions are fast (e.g., the precipitation of insoluble salts); others are so fast as to be explosive (e.g., the reaction between hydrogen and oxygen). In Chapter 10, we saw how the sign of $\Delta G^{\ominus}$, the standard free energy change, indicates whether it is possible for a reaction to occur. If $\Delta G^{\ominus}$ is negative, the reaction is feasible, but the study of thermodynamics does not tell us how fast a reaction will occur. This is something that must be found out by experiment. Industrial chemists are interested in knowing how fast a reaction takes place as the speed is a factor in deciding whether a manufacturing process can be carried out profitably. Many factors influence the rate of a chemical reaction, and it is important to discover the conditions under which a reaction will proceed most economically. The rate at which the product is formed is only one factor. The cost of the energy consumed if a high temperature is needed must be computed. If the process requires high pressure, the cost of a plant which is robust enough to withstand the conditions will be high.

The rate of a reaction cannot be predicted from the standard free energy change

Many factors affect the rate of a reaction ...

... Reaction kinetics is the study of these factors

The study of the factors that affect the rates of chemical reactions is called **reaction kinetics**. Such studies throw light on the **mechanisms** of reactions. All reactions take place in one or more simple steps, and the sequence of steps is called the mechanism. The number of reacting species (molecules, atoms, ions or free radicals) that take part in a reaction step is the **molecularity** of that step. Reaction steps are described as **unimolecular**, **bimolecular** or **trimolecular** [p. 580]. Examples of reaction mechanisms are covered on pp. 557, 559, 580, 584, 604, 662 and 706.

14.2 AVERAGE RATE

The rate of a reaction is the rate of change of concentration of a reactant or a product

The average rate of a chemical reaction over a certain interval of time is equal to the change in the concentration of a reactant or product that occurs during that time divided by the time. When an ester is hydrolysed, if the concentration of the ester decreases from $1.00\,mol\,dm^{-3}$ to $0.50\,mol\,dm^{-3}$ in 1.00 hour, then the average rate of reaction over this time interval can be given:

$$\text{Average rate} = (1.00 - 0.50)\,mol\,dm^{-3}/(1.00 \times 60 \times 60)\,s$$

$$= 1.39 \times 10^{-4}\,mol\,dm^{-3}\,s^{-1}.$$

In the example in Figure 14.1, the rate of reaction slows down as the concentration of reactant decreases

The dimension of reaction rate is concentration time^{-1}. Consider a reaction of the type

A → B

where 1 mole of the reactant produces 1 mole of the product. Figure 14.1 shows how the concentration of product increases and the concentration of reactant decreases as the time which has passed since the start of the reaction increases.

FIGURE 14.1 Variation of Concentrations of Reactant and Product with Time

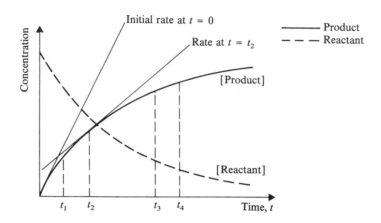

You can see that less product has been formed between t_3 and t_4 than in an equal time interval between t_1 and t_2. The rate of reaction decreases as the reaction proceeds and the reactant is gradually used up. One can only state the rate of reaction at a certain time. At time t_2 the rate of reaction is the gradient of the tangent to the curve at this point [see Figure 14.1]. The rate at the start of the reaction, when an infinitesimally small amount of the reactant has been used up, is called the **initial rate** of the reaction. In Figure 14.1, the gradient of the tangent to the curve at $t = 0$ gives the initial rate.

The rate at the very beginning of the reaction is called the initial rate

The rate of the reaction

A → B

is the rate of decrease in concentration of **A** or the rate of increase in concentration of **B**:

$$\text{Rate} = -\frac{d[A]}{dt} = \frac{d[B]}{dt}$$

where $[A]$ = concentration/mol dm^{-3} of **A**. In the reaction

$$BrO_3^-(aq) + 5Br^-(aq) + 6H^+(aq) \rightarrow 3Br_2(aq) + 3H_2O(l)$$

$$\text{Rate} = -\frac{d[BrO_3^-]}{dt} = -\frac{1}{5}\frac{d[Br^-]}{dt} = \frac{1}{3}\frac{d[Br_2]}{dt} \text{ and so on}$$

The rate of a reaction is found by measuring some property of a reactant or a product at various times after the start of the reaction. Some of the methods of 'following' the reaction in this way will now be described.

14.3 METHODS OF FINDING THE RATES OF CHEMICAL REACTIONS

14.3.1 CHEMICAL METHODS

Titration can be used to follow the change in the concentration of the reactant or the product

The progress of a reaction can often be followed by chemical analysis. The reaction is carried out in a thermostatically controlled water bath. Solutions of the reactants of known concentrations are mixed, and a stop clock is started. A sample of the reacting mixture is withdrawn with a pipette, and the reaction is stopped. This may be done by removing one of the reactants by a chemical reaction. Alternatively, the reaction may be suddenly slowed down by cooling or by dilution. This is done by pipetting a sample into a freezing mixture or into an excess of the solvent. A titration is then performed to find the concentration of one of the reactants or one of the products.

Example (a) The alkaline hydrolysis of an ester

$$CH_3CO_2C_2H_5(aq) + NaOH(aq) \rightarrow CH_3CO_2Na(aq) + C_2H_5OH(aq)$$

The course of an alkaline hydrolysis of an ester is followed by measuring the concentration of alkali at various times after the start of the reaction

Solutions of ester and alkali of known concentrations are allowed to reach the temperature of a thermostat bath. The solutions are mixed, and the time of mixing is noted. A sample of the reaction solution is withdrawn by pipette, and run into about four times its volume of ice-cold water. The dilution and cooling reduce the rate of reaction almost to zero. The alkali that remains is titrated against standard acid, using phenolphthalein as indicator. Ethanoate ions do not affect the colour change of this indicator. The analysis is repeated at various intervals of time after the start of the reaction.

Example (b) The chlorination of aromatic compounds, e.g., 4–chloro-phenoxyethanoic acid

4-Chlorophenoxyethanoic acid   2,4-Dichlorophenoxyethanoic acid

A chlorination reaction is followed by means of thiosulphate titrations

Solutions of known concentrations of ether and chlorine are mixed at the required temperature, and the time is noted. A sample of the reacting mixture is withdrawn and run into an excess of potassium iodide solution. All the chlorine remaining reacts very rapidly with potassium iodide to form iodine, and the chlorination stops. The amount of iodine formed can be found by titration against a standard solution of sodium thiosulphate.

14.3.2 PHYSICAL METHODS

A CHANGE IN GAS VOLUME

The volume of gas evolved can be measured after various time intervals

In a reaction in which a gas is formed, the volume of gas can be recorded at various times. Examples are the reaction of a metal with an acid and the decomposition of hydrogen peroxide:

$$Mg(s) + 2HCl(aq) \rightarrow H_2(g) + MgCl_2(aq)$$

$$2H_2O_2(aq) \rightarrow 2H_2O(l) + O_2(g)$$

FIGURE 14.2
Measuring the Evolution
of Gas in a Reaction

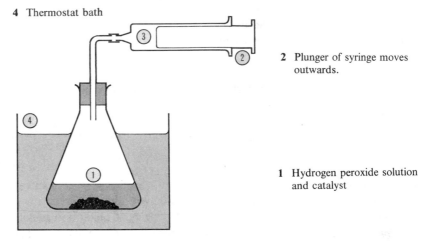

3 Oxygen. Volume is recorded
at certain times after the
start of the reaction.

4 Thermostat bath

2 Plunger of syringe moves
outwards.

1 Hydrogen peroxide solution
and catalyst

A CHANGE IN GAS PRESSURE

An increase or decrease in gaseous pressure can be used to follow many gaseous reactions

Some reactions between gases involve an increase in the number of moles of gas, e.g.

$$2N_2O_5(g) \rightarrow 2N_2O_4(g) + O_2(g)$$

If the reaction takes place at constant volume, the resulting increase in pressure can be followed. The reaction

$$2H_2(g) + O_2(g) \rightarrow 2H_2O(g)$$

is accompanied by a decrease in the number of moles of gas and by a decrease in pressure at constant volume.

Such reactions can be followed by the method shown in Figure 14.3.

FIGURE 14.3 Apparatus
for Following Changes in
Gas Pressure

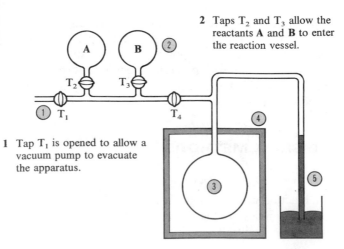

2 Taps T_2 and T_3 allow the
reactants **A** and **B** to enter
the reaction vessel.

1 Tap T_1 is opened to allow a
vacuum pump to evacuate
the apparatus.

3 Reaction vessel. Tap T_4
traps gases **A** and **B** inside.

4 Thermostatically controlled
furnace

5 Manometer records pressure
inside reaction vessel.

THE ROTATION OF THE PLANE OF POLARISATION OF PLANE-POLARISED LIGHT

A change in optical activity is used to follow some reactions

If a change in optical activity [p. 538] occurs during a reaction, the change of optical rotation can be measured. An example is the hydrolysis of sucrose:

$$C_{12}H_{22}O_{11}(aq) + H_2O(l) \xrightarrow{\text{acid}} C_6H_{12}O_6(aq) + C_6H_{12}O_6(aq)$$

$(+)$-Sucrose $\qquad\qquad$ $(+)$-Glucose $\quad$ $(-)$-Fructose

As the reaction proceeds, the direction of rotation changes as the strong $(-)$-rotation of fructose outweighs the weaker $(+)$-rotation of glucose. An **inversion** of rotation occurs.

FIGURE 14.4
A Polarimeter

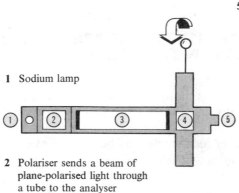

5. The angle of rotation on a graduated scale is read through a microscope. This is the angle through which the solution has rotated the plane of polarisation of plane-polarised light.

1 Sodium lamp

2 Polariser sends a beam of plane-polarised light through a tube to the analyser

3 Optically active solution in cell rotates the plane of polarisation of the light so that light can no longer pass through the analyser.

4 Analyser. *First*, in the absence of solution, it is rotated until the light reaching the eyepiece is of minimum intensity. *Second*, with the solution in the cell, it is rotated to restore the light to minimum intensity.

CHANGES IN THE ABSORPTION SPECTRUM

The use of a spectrophotometer enables light absorption to be measured

Many substances absorb light, either in the visible region or, more frequently, in the ultraviolet region. A **spectrophotometer** is an instrument for measuring the absorption of light at various wavelengths. If a reaction is carried out in a cell inside a spectrophotometer, the change in the absorption spectrum can be used to follow the course of the reaction. Figure 14.5 shows the change in the absorption spectrum during the reaction

$$A \to C$$

FIGURE 14.5 Change in Absorption Spectrum during Reaction

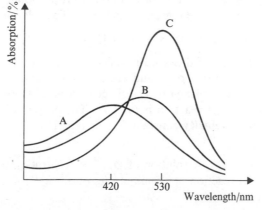

Curve A = absorption spectrum of **A**
Curve C = absorption spectrum of **C**
Curve B = absorption spectrum at an intermediate stage in the reaction

The height of the peak at 530 nm could be used to follow the reaction, because **A** does not absorb at this wavelength.

ELECTRICAL PROPERTIES

A change in conductance indicates a change in the concentration of ions...

A change in the conductance of a solution occurs if ions are used up or created during a reaction. The method of measuring conductance is described on p. 244.

...as does a change in electrode potential

The potential of an electrode in contact with a solution of its ions changes if the concentration of ions changes. Potentiometric methods [see p. 284] can often be used to follow the course of a reaction which involves ions.

THERMAL CONDUCTIVITY

Thermal conductivity

A gaseous reaction can be followed by measuring the thermal conductivity of the mixture of reacting gases.

14.4 THE RESULTS OF MEASUREMENTS OF REACTION RATES

Methods such as those described enable us to measure the rates of chemical reactions. It is interesting to use such methods to find out how the rate of reaction depends on the concentrations of the reactants, the temperature and other factors. The results of such studies can be interpreted to give a detailed picture of what happens during a chemical reaction.

14.4.1 THE EFFECT OF PARTICLE SIZE ON REACTION RATE

Solids react faster in finely divided form

Reactions of solids take place faster when the solids are in a finely divided state. This is because the ratio of surface area to mass is greater in small particles than in large particles, and the area over which the solid can come into contact with liquid or gaseous reactants is greater. Examples are the reaction of powdered zinc and granulated zinc with acids and the reaction of powdered calcium carbonate and marble chips with acid.

14.4.2 THE EFFECT OF CONCENTRATION ON REACTION RATE

Consider a reaction between **A** and **B**:

$$\text{A} + \text{B} \rightarrow \text{Products}$$

The way in which rate is related to concentration is governed by the order of the reaction

The rate of reaction depends on the concentrations of **A** and **B**, but one cannot simply say that the rate of reaction is proportional to the concentration of **A** and proportional to the concentration of **B**. The relationship is

$$\text{Reaction rate} \propto [\text{A}]^m[\text{B}]^n = k[\text{A}]^m[\text{B}]^n$$

An expression of this kind is called a **rate equation**. The indices m and n are usually integers, often 0, 1 or 2, and are characteristic of the reaction. One says that the reaction is of **order** m with respect to **A** and of order n with respect to **B**. The overall order of reaction is $(m + n)$. The proportionality constant k is called the **rate constant** or **velocity constant** for the reaction.

The rate constant for a reaction relates the rate to the concentrations of the reactants

Example (a)   A solution of **Q**, of concentration $0.20\,\text{mol}\,\text{dm}^{-3}$ undergoes a **first-order** reaction at an initial rate of $3.0 \times 10^{-4}\,\text{mol}\,\text{dm}^{-3}\,\text{s}^{-1}$. Calculate the rate constant.

Method   Since, for a first-order reaction

$$\text{Initial rate} = k[\mathbf{Q}]_0$$

$$3.0 \times 10^{-4}\,\text{mol dm}^{-3}\text{s}^{-1} = k \times 0.20\,\text{mol dm}^{-3}$$

$$\text{The rate constant } k = 1.5 \times 10^{-3}\,\text{s}^{-1}$$

The dimension of a first-order rate constant is time^{-1}.

First-order and second-order rate constants have different dimensions

Example (b)   A **second-order** reaction takes place between the reactants **P** and **Q**, which are both initially present at concentration $0.20\,\text{mol dm}^{-3}$. If the initial rate of reaction is $1.6 \times 10^{-4}\,\text{mol dm}^{-3}\text{s}^{-1}$, what is the rate constant?

Method

$$\text{Initial rate} = k[\mathbf{P}]_0[\mathbf{Q}]_0$$

$$\therefore \qquad 1.6 \times 10^{-4}\,\text{mol dm}^{-3}\text{s}^{-1} = k \times (0.20\,\text{mol dm}^{-3})^2$$

$$\text{The rate constant, } k = 4.0 \times 10^{-3}\,\text{dm}^3\,\text{mol}^{-1}\text{s}^{-1}$$

A second-order rate constant has the dimensions concentration^{-1} time^{-1}.

14.5 ORDER OF REACTION

The stoichiometric equation for a reaction does not reveal the order of the reaction

The order of a reaction does not follow from its stoichiometric equation. The reaction between bromate(V) ions, bromide ions and hydrogen ions to give bromine is represented by the equation

$$\text{BrO}_3^-(\text{aq}) + 5\text{Br}^-(\text{aq}) + 6\text{H}^+(\text{aq}) \rightarrow 3\text{Br}_2(\text{aq}) + 3\text{H}_2\text{O}(\text{l})$$

The results of kinetic measurements give the following rate equation:

$$-\frac{d[\text{BrO}_3^-]}{dt} \propto [\text{BrO}_3^-][\text{Br}^-][\text{H}^+(\text{aq})]^2$$

The reaction is first-order with respect to bromate(V), first-order with respect to bromide, second-order with respect to hydrogen ion and fourth-order overall. The negative sign means that $[\text{BrO}_3^-]$ decreases with time.

14.5.1 ZERO-ORDER REACTION

If the rate is independent of concentration, the reaction is zero order...

In a **zero-order** reaction, the rate is independent of the concentration of the reactant. A plot of the concentration of the reactant, $[\text{A}]$, against time has the form shown in Figure 14.6.

FIGURE 14.6 A plot of Concentration against Time for a Zero-Order Reaction

---

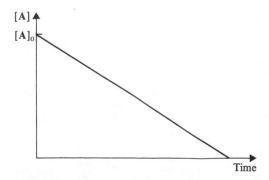

The rate equation for a zero-order reaction is

$$-\frac{d[A]}{dt} = k$$

and the rate constant has the dimension concentration time^{-1}.

...one example is the reaction of iodine and propanone...

In the iodination of propanone [p. 316], the reaction rate does not change if the concentration of iodine is changed:

$$CH_3COCH_3(aq) + I_2(aq) \xrightarrow[\text{buffer}]{\text{acid}} CH_3COCH_2I(aq) + HI(aq)$$

The reaction is said to be zero-order with respect to iodine.

...Another example is the adsorption of reactants in gaseous reactions

Sometimes, reactions between gases are zero-order with respect to one of the reactants. This often indicates that this reactant has been adsorbed on the surface of the vessel. The rate of reaction then depends on the frequency with which molecules of the non-adsorbed gas collide with the inside of the vessel. This frequency is proportional to the concentration of the non-adsorbed reactant.

14.5.2 ORDER OF REACTION FROM INITIAL RATE

In a reaction

$$A \to X$$

The order of a reaction can be found by comparing the initial rates of two reactions at known initial concentrations

the rate of reaction = $k[A]^n$, where n = the order of reaction. Two experiments to find the rate of reaction, two 'runs', are done at different concentrations of A. $[A_0]_1$ = initial concentration of A in Run 1, and $(v_0)_1$ = the initial rate in Run 1.

$$(v_0)_1 = k[A_0]_1{}^n$$

$$(v_0)_2 = k[A_0]_2{}^n$$

The ratio

$$\frac{(v_0)_1}{(v_0)_2} = \left(\frac{[A_0]_1}{[A_0]_2}\right)^n$$

The order of reaction may be found by comparing the initial rates of reactions at different concentrations. This can be done by inspection, as in the example below, or by taking logarithms:

$$\lg\left(\frac{(v_0)_1}{(v_0)_2}\right) = n \lg\left(\frac{[A_0]_1}{[A_0]_2}\right)$$

Example   The following results were obtained for a reaction between A and B:

A worked example: how to find the order by comparing initial rates

Run	Concentrations/mol dm^{-3}		Initial rate/mol dm^{-3} s^{-1}
	[A]	**[B]**	
(a)	0.50	1.0	2.0
(b)	0.50	2.0	8.0
(c)	0.50	3.0	18
(d)	1.0	3.0	36
(e)	2.0	3.0	72

What is the order of reaction with respect to **A** and with respect to **B**? What is the rate equation for the reaction? Calculate the rate constant.

Method   Let the rate equation be

$$\text{Rate} = k[\mathbf{A}]^m[\mathbf{B}]^n$$

Compare runs (d) and (e), in which [**B**] is constant:

$$\frac{\text{Rate (e)}}{\text{Rate (d)}} = \left(\frac{2.0}{1.0}\right)^m = \frac{72}{36} \quad \text{therefore } m = 1$$

Compare runs (b) and (a), in which [**A**] is constant

$$\frac{\text{Rate (b)}}{\text{Rate (a)}} = \left(\frac{2.0}{1.0}\right)^n = \frac{8}{2} \quad \text{therefore } n = 2$$

The reaction is first-order with respect to **A** and second-order with respect to **B**. The rate equation is

Once the order has been found, the rate constant can be found

$$\text{Rate} = k[\mathbf{A}][\mathbf{B}]^2$$

The rate constant can now be calculated from the results of any run, for example (c)

$$18 = k \times 0.5 \times (3.0)^2$$

$$k = 4.0\,\text{dm}^6\,\text{mol}^{-2}\,\text{s}^{-1}$$

This third-order rate constant has the dimension concentration^{-2} time^{-1}.

CHECKPOINT 14A: INITIAL RATES

1. Explain the terms: *rate of reaction*, *order of reaction*, *stoichiometry of reaction* and *rate constant*.

2. Describe three physical methods which can be used to follow the course of a chemical reaction. What are the advantages of physical methods over chemical methods?

3. For the reaction

$$\mathbf{A} + \mathbf{B} \to \mathbf{C}$$

the following results were obtained for kinetic 'runs' at the same temperature:

$[A]_0$/mol dm^{-3}	$[B]_0$/mol dm^{-3}	Initial rate/mol dm^{-3} s^{-1}
0.20	0.10	0.20
0.40	0.10	0.80
0.40	0.20	0.80

Find
(a) the rate equation for the reaction
(b) the rate constant
(c) the initial rate of a reaction, when
$[A]_0 = 0.60\,\text{mol dm}^{-3}$ and $[B]_0 = 0.30\,\text{mol dm}^{-3}$

4. Tabulated are values of initial rates measured for the reaction

$$2\mathbf{A} + \mathbf{B} \to \mathbf{C} + \mathbf{D}$$

Experiment	$[A]$/mol dm^{-3}	$[B]$/mol dm^{-3}	Initial rate/ mol dm^{-3} min^{-1}
1	0.150	0.25	1.4×10^{-5}
2	0.150	0.50	5.6×10^{-5}
3	0.075	0.50	2.8×10^{-5}
4	0.075	0.25	7.0×10^{-6}

(a) Find the order with respect to **A**, the order with respect to **B** and the overall order of the reaction.
(b) Find the value of the rate constant.
(c) Find the initial rate of reaction when
$[A]_0 = 0.120\,\text{mol dm}^{-3}$ and $[B]_0 = 0.220\,\text{mol dm}^{-3}$.

14.6 FINDING THE ORDER OF REACTION FROM THE INTEGRATED RATE EQUATION

Another way of using experimental results is to put them into an integrated form of the rate equation.

14.6.1 *DERIVING THE INTEGRATED RATE EQUATION FOR A FIRST-ORDER REACTION

If the reaction

$$A \rightarrow Products$$

is a first-order reaction, the rate equation will be

$$-d[A]dt = k[A]$$

If $[A]_0$ = initial concentration of A, the integrated form of this equation is

$$\int_{[A]_0}^{[A]} \frac{d[A]}{[A]} = \int_0^t k dt$$

$$-[\ln[A]]_{[A]_0}^{[A]} = k[t]_0^t$$

$$-\ln \frac{[A]}{[A]_0} = kt$$

$$kt = \ln \frac{[A]_0}{[A]} = 2.303 \lg \frac{[A]_0}{[A]}$$

14.6.2 USING THE INTEGRATED FIRST-ORDER RATE EQUATION

A linear plot shows the reaction to be first-order

The integrated form of the rate equation can be used to obtain a linear plot from the results of experimental measurements. If $\ln([A]_0/[A])$ is plotted against t, a linear plot is obtained, and the gradient is k. A simpler method is to plot $\ln[A]$ against t, to give a linear plot with a gradient of $-k$. If $\lg[A]$ is plotted against t, the gradient is $-k/2.303$. If values of $[A]$ obtained from experimental measurements over a considerable extent of reaction (say 0 to 90% completion) give a straight line when plotted in this way, then the reaction is first-order.

14.6.3 HALF-LIFE

Definition of half-life, $t_{1/2}$...

The time taken for the reaction to go to half-completion is called the **half-life** of the reaction, $t_{1/2}$. In the first-order reaction

$$A \rightarrow Products$$

...For first-order reactions, $t_{1/2}$ is independent of concentration

at time $t_{1/2}$

$$[A] = [A]_0/2$$

$$\therefore \qquad kt_{1/2} = \ln 2 = 2.303 \lg 2 = 0.693$$

This shows that the half-life of a first-order reaction is independent of the initial concentration. Radioactive decay is an example of a reaction showing first-order kinetics [p. 15].

Example   The half-life of radium is 1590 years. How long will it take for a sample of radium to decay to 10% of its original radioactivity?

Method   This problem is solved in two steps:

A worked example on radioactive decay

(a) Use $t_{1/2}$ to find k, the rate constant, from the equation

$$kt_{1/2} = 0.693$$

Then

$$k = 0.693/1590 = 4.36 \times 10^{-4}\,year^{-1}$$

(b) Insert this value for k into the equation

$$kt = \ln\frac{[A]_0}{[A]}$$

$$4.36 \times 10^{-4} \times t = \ln(100\%/10\%)$$

$$t = 5280\ years$$

After 5280 years, 10% of the original radioactivity remains.

Example   Carbon-14 dating shows that a piece of ancient wood gives 10 counts per minute per gram of carbon, compared with $15\,cpm\,g^{-1}$ of carbon from a sample of new wood. The half-life of ^{14}C is 5600 years. What is the age of the ancient wood?

Method   Again, there are two steps in the calculation:

A worked example on C-14 dating

(a) Use $t_{1/2}$ to find k

$$k = 0.693/5600 = 1.24 \times 10^{-4}\,year^{-1}$$

(b) Insert this value of k into the equation:

$$kt = \ln\frac{[A]_0}{[A]}$$

Since $$\frac{[A]_0}{[A]} = \frac{^{14}C\ content\ in\ new\ wood}{^{14}C\ content\ in\ ancient\ wood} = \frac{15\,cpm}{10\,cpm}$$

$$t = \ln\frac{15}{10}\Big/(1.24 \times 10^{-4}) = 3270\ years$$

The wood is 3270 years old.

14.6.4   PSEUDO-FIRST-ORDER REACTIONS

The acid-catalysed hydrolysis of an ester, e.g., ethyl ethanoate

$$CH_3CO_2C_2H_5(l) + H_2O(l) \xrightarrow{acid} CH_3CO_2H(l) + C_2H_5OH(l)$$

If the concentration of one reactant is very large, the reaction appears to be zero order with respect to that reactant

is first-order with respect to ester and first-order with respect to water. If water is present in large excess, only a small fraction of the water will be used up in the reaction. The concentration of water is practically constant, and the rate depends on the concentration of ester alone:

$$- d[CH_3CO_2C_2H_5]/dt = k'[CH_3CO_2C_2H_5]$$

k' = a first-order rate constant. The reaction appears to be zero-order with respect to water.

CHECKPOINT 14B: FIRST-ORDER REACTIONS

1. A radioactive element decays with a rate constant of $2.0 \times 10^{-4}\,s^{-1}$. How long will it take for 0.50 g of the substance to decay to 0.10 g?

2. If the half-life of a radioactive element is 150 s, what percentage of the isotope will remain after 600 seconds?

3. The results listed were obtained for a 'run' on the reaction

A → B + C

Plot [A] against t.

(*a*) From the graph, find the order of the reaction with respect to **A**.

(*b*) By drawing a tangent, find the initial rate of reaction.

(*c*) Calculate the rate constant for the reaction.

Time/s	[A]/mol dm^{-3}
0	0.800
400	0.580
800	0.400
1200	0.280
1600	0.200
2000	0.140
2400	0.100

4. If a radioisotope loses 95% of its activity in 110 minutes, what is its half-life?

5. The decomposition of benzene diazonium chloride is first-order:

$$C_6H_5N_2Cl(aq) \rightarrow C_6H_5Cl(aq) + N_2(g)$$

The following results give the volume of nitrogen at 50 °C, measured at various time intervals after the start of the reaction, obtained in the decomposition of 500 cm³ of a $1.10 \times 10^{-3}\,mol\,dm^{-3}$ solution at 50 °C:

Time/min	2.0	4.0	6.0	9.0	12.0	16.0	22.0	28.0
Volume of N$_2$/cm³	1.7	3.4	4.9	6.6	8.1	9.5	11.2	12.2

(*a*) Plot the volume of nitrogen evolved against the time interval.

(*b*) Calculate the volume of nitrogen at 50 °C that will be formed at $t = \infty$ from the amount of benzene diazonium chloride specified. Enter this on the graph.

(*c*) From the graph, estimate the half-life of the reaction, $t_{1/2}$.

(*d*) What evidence have you that the reaction is first-order?

(*e*) Use the value of $t_{1/2}$ to find the rate constant k.

(*f*) Obtain the initial rate of reaction.

(*g*) From it, calculate the rate constant.

(*h*) Method (*e*) relies largely on the accuracy of the $t = t_{1/2}$ point, and method (*g*) relies largely on the $t = 2.0$ point. In order to use all the results to obtain the best possible value for the rate constant, calculate values of $[C_6H_5N_2Cl]$ and plot the ln or lg of these against t. Calculate k.

14.6.5 *DERIVING THE INTEGRATED RATE EQUATION FOR A SECOND-ORDER REACTION

In the second order reaction

A + B → Products

the rate equation is

$$- d[A]/dt = - d[B]/dt = k[A][B]$$

We will consider the simplest case only, when **A** and **B** are present at the same concentration, i.e.

[A] = [B] at all times

$$\frac{-d[A]}{dt} = k[A]^2$$

$$\int_{[A]}^{[A]_0} \frac{-d[A]}{[A]^2} = \int_0^t k dt$$

$$\left[\frac{1}{[A]}\right]_{[A]_0}^{[A]} = k[t]_0^t$$

The integrated rate equation for a second-order reaction between reactants of equal concentration

$$\therefore \quad \frac{1}{[A]} - \frac{1}{[A]_0} = kt$$

14.6.6 *USING THE INTEGRATED SECOND-ORDER RATE EQUATION

A linear plot for 1/[A] against time

The value of k can be obtained by plotting $1/[A]$ against t, to give a linear plot with a gradient of k. If values of $[A]$ obtained over a considerable extent of reaction (say 0–90% completion) give a straight line when plotted in this way, then the reaction is second-order.

Example The results listed were obtained in a kinetic study of the alkaline hydrolysis of an ester:

$$RCO_2R' + NaOH \rightarrow RCO_2Na + R'OH$$

| Ester | Sodium salt of + Alcohol carboxylic acid |

The initial concentrations of ester and alkali were both $0.100 \, mol \, dm^{-3}$. Find the order of reaction and the rate constant.

Time/s	0	100	200	300	400	600
$[RCO_2Na]/10^{-2} \, mol \, dm^{-3}$	0	2.95	4.42	5.55	6.22	7.03

Method If the reaction is first-order, a plot of $\ln[Ester]$ against t will be linear, with a gradient of $-k$.

If the reaction is second-order, a plot of $1/[Ester]$ against t will be linear with a gradient of k.

Tabulated below are values of [Salt formed] and [Ester remaining].

Testing to find out whether a reaction is first- or second-order by plotting ln [Reactant] against t and 1/[Reactant] against t to see which gives a straight line

Time/s	$[Salt]/10^{-2} \, mol \, dm^{-3}$	$[Ester]/10^{-2} \, mol \, dm^{-3}$	$\ln[Ester]$	$1/[Ester]/dm^3 \, mol^{-1}$
0	0	10.0	−2.30	10
100	2.95	7.05	−2.65	14.2
200	4.42	5.58	−2.89	17.9
300	5.55	4.45	−3.11	22.5
400	6.22	3.78	−3.28	26.5
600	7.03	2.97	−3.52	33.7

The plots in Figure 14.7 show that the second-order plot is linear, with a gradient of $6.4 \times 10^2 \, dm^3 \, mol^{-1} \, s^{-1}$. The reaction is therefore second-order with a rate constant of $6.4 \times 10^2 \, dm^3 \, mol^{-1} \, s^{-1}$.

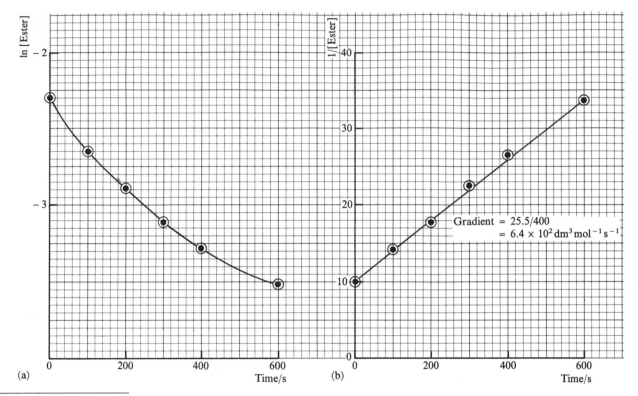

FIGURE 14.7 Plots of
(*a*) ln [Ester] and
(*b*) 1/[Ester] against
Time

14.7 SUMMARY OF THE DEPENDENCE OF THE CONCENTRATIONS OF REACTANT AND PRODUCT ON TIME

The type of linear plot tells the order of reaction.
The gradient of the linear plot gives the rate constant.
(c_0 = initial concentration, c_t = concentration of reactant after time t)

TABLE 14.1 Summary
of the Dependence of the
Concentrations of
Reactant and Product
on Time

Order	Rate Equation	Linear Plot	Gradient	Units of k
0	$kt = [\text{Product}]$	[Product] against t	k	$\text{mol dm}^{-3}\text{s}^{-1}$
1	$kt = \ln\dfrac{c_0}{c_t}$	$\ln\dfrac{c_0}{c_t}$ against t	k	s^{-1}
		$\ln c_t$ against t	$-k$	s^{-1}
		$\lg c_t$ against t	$-k/2.303$	s^{-1}
2	$kt = \dfrac{1}{c_t} - \dfrac{1}{c_0}$	$1/c_t$ against t	k	$\text{dm}^3\text{mol}^{-1}\text{s}^{-1}$

====== **CHECKPOINT 14C: ORDER OF REACTION** ======

1. In a reaction

$$R \rightarrow S$$

the initial rate is $1.7 \times 10^{-4} \, mol \, dm^{-3} s^{-1}$ when $[R]_0 = 0.25 \, mol \, dm^{-3}$. Predict the initial rate when $[R]_0 = 0.75 \, mol \, dm^{-3}$ if the reaction is (*a*) zero-order, (*b*) first-order and (*c*) second-order.

2. The results tabulated refer to the isomerisation

$$trans\text{-}CHCl\!\!=\!\!CHCl \rightarrow cis\text{-}CHCl\!\!=\!\!CHCl$$

Time/s	0	600	900	1200	1500	1800
Trans-*isomer*/mol	1.00	0.90	0.85	0.81	0.77	0.73

From a suitable plot, find the order of reaction and the rate constant.

*3. A second-order reaction between two reactants present at initial concentrations of $0.100 \, mol \, dm^{-3}$ is 20% complete in 20 minutes. If the equation for the reaction is

$$A + B \rightarrow Products$$

calculate
(*a*) the rate constant
(*b*) the time taken for the reaction to be 80% complete
(*c*) the time taken for the reaction to be 20% complete when the reactants are initially present at $0.0200 \, mol \, dm^{-3}$.

4. A first-order reaction is 40% complete at the end of 40 minutes. Calculate
(*a*) the rate constant
(*b*) the time taken for the reaction to reach 80% completion

5. The following results were obtained for the acid-catalysed hydrolysis of methyl ethanoate:

Time/s	0	1150	2050	3600	5050	8000
[Ester]/mol dm^{-3}	0.500	0.375	0.300	0.216	0.150	0.071

Calculate the order of the reaction and the rate constant.

14.8 THE EFFECT OF LIGHT ON REACTION RATES: PHOTOCHEMICAL REACTIONS

Some reactions take place faster in the presence of light

Reactions with very high rates often involve free radicals. When a covalent bond splits **homolytically**, each of the bonded atoms or groups takes one of the bonding pair of electrons, and each product is termed a **free radical**:

$$A : B \rightarrow A \cdot + \cdot B$$

(In **heterolytic** fission

$$A : B \rightarrow A^+ + :B^-$$

one of the bonded atoms takes both bonding electrons, and ions are formed.)

Evidence for the existence of free radicals is obtained from mass spectrometry and other methods.

Light energy may split bonds to form free radicals

Energy must be supplied to break the bonds and produce free radicals. In thermal reactions, this energy comes from collisions with other molecules. Those reactions which are started by the absorption of light energy are called photochemical reactions.

A photochemical reaction takes place between H_2 and Cl_2

Hydrogen and chlorine react slowly in the dark, unless heated above 200 °C, to form hydrogen chloride. In the presence of sunlight, however, reaction takes place rapidly at room temperature:

$$H_2(g) + Cl_2(g) \rightarrow 2HCl(g)$$

It is believed that the first step in the photochemical reaction is the absorption of light, leading to the dissociation of chlorine molecules:

$$Cl_2 + h\nu \rightarrow 2Cl\cdot \qquad [1] \; Initiation \; reaction$$

The chlorine atoms (or radicals) formed react with hydrogen molecules:

$$Cl\cdot + H_2 \rightarrow HCl + H\cdot \qquad [2] \; \textit{Propagation reaction}$$

The hydrogen atoms formed in [2] react with chlorine molecules:

$$H\cdot + Cl_2 \rightarrow HCl + Cl\cdot \qquad [3] \; \textit{Propagation reaction}$$

These two steps are repeated many times, setting up a **chain reaction**. For the absorption of one quantum of light by one chlorine molecule, many thousands of molecules of hydrogen chloride are formed. The chain does not go on for ever; it is brought to an end by the combination of free radicals:

$$2Cl\cdot \rightarrow Cl_2 \qquad\qquad [4]$$
$$2H\cdot \rightarrow H_2 \qquad\qquad\; [5] \qquad \textit{Chain termination reactions}$$
$$H\cdot + Cl\cdot \rightarrow HCl \qquad [6]$$

The chlorination of methane is a photochemical reaction

Another photochemical reaction is the chlorination of methane in sunlight to chloromethane and other derivatives [p. 548].

The absorption of radiation from radioactive elements, for example X rays and γ rays, is sometimes used to initiate chemical reactions.

14.9 THE EFFECT OF TEMPERATURE ON REACTION RATES

Reaction rates increase with temperature...

An increase in temperature increases the rate of a reaction by increasing the rate constant. Figure 14.8 shows plots of rate constant k against temperature T and of $\ln k$ against $1/T$.

FIGURE 14.8
(a) Plot of k against T,
(b) Plot of k against 1/T

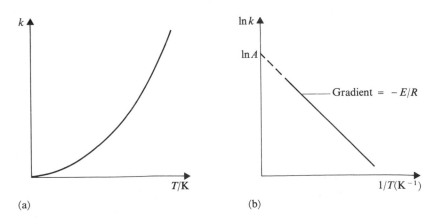

(a)   (b)

The variation of rate constant with temperature was studied by Arrhenius and found to fit the equation

$$k = A e^{-E/RT}$$

...The dependence of rate on temperature fits the Arrhenius equation

This equation is known as the **Arrhenius equation**. R is the gas constant [p. 139] and A and E are constant for a given reaction. The constants can be found by using the equation in logarithmic form:

$$\ln k = \ln A - E/RT$$

ln k against 1/T gives a linear plot

A plot of $\ln k$ against $1/T$ is a straight line of gradient $-E/R$ and an intercept on the y-axis of $\ln A$. If $\lg k$ is plotted against $1/T$, the gradient of the line is $-E/2.303R$, and the intercept is $\lg A$.

The significance ~~ ~~ne temperature dependence is discussed below, and the method of determining E ~~...~~ A is exemplified on p. 308.

14.10 THEORIES OF REACTION RATES

14.10.1 THE COLLISION THEORY

Gas molecules must collide in order to react...

...The rate of reaction is much less than the rate of collision

From the Kinetic Theory of Gases, was developed the **Collision Theory of bimolecular reactions in the gas phase**. In a reaction between two gaseous substances **A** and **B**, a molecule of **A** must collide with a molecule of **B** before reaction can occur. It has been shown that the collision frequency Z is proportional to the product $[A][B]$. If every collision results in reaction, the rate of reaction will equal the collision frequency. Theoreticians have calculated the collision frequency, and have found that the rate constant for a bimolecular gaseous reaction should be of the order of $10^{11} \, \text{dm}^3 \, \text{mol}^{-1} \, \text{s}^{-1}$. Although there are reactions which have rate constants of this magnitude, e.g.

$$H\cdot + Br_2 \rightarrow Br\cdot + HBr$$

such reactions are few. The reason why most reactions are slower than this is that only a small fraction of collisions results in reaction.

You can see from the Arrhenius equation

$$k = A e^{-E/RT}$$

Molecules must not only collide; they must have enough energy to react

that E/RT must be a pure number, without dimension, and E must consequently have the same dimension as RT, i.e., $\text{J K}^{-1} \text{mol}^{-1} \times \text{K} = \text{J mol}^{-1}$, which is an energy unit. E is termed the **energy of activation**. Since the unit of A is the same as that of k, A is a rate constant. It is possible that A is the rate of collision between molecules, and E is the energy which the colliding molecules must possess before a collision will result in reaction.

The Arrhenius equation ties in with the Maxwell–Boltzmann equation

The fraction $e^{-E/RT}$ appears in the Maxwell-Boltzmann calculations [p. 146]. They calculated that for a value of molecular energy $E/\text{J mol}^{-1}$, the fraction of molecules at temperature T that possess energy $\geqslant E$ is given approximately by the value of $e^{-E/RT}$.

The term $e^{-E/RT}$ has values between 0 and 1.
If $E = 0$, $e^{-E/RT} = 1$, then $k = A$, i.e.

Rate constant = Collision frequency

If $E = 50 \, \text{kJ mol}^{-1}$, at 298 K, $e^{-E/RT} = 1.7 \times 10^{-9}$, and

Rate constant $\ll$ Collision frequency

If $E = 50 \, \text{kJ mol}^{-1}$, at 308 K, $e^{-E/RT} = 3.3 \times 10^{-9}$, and

Fraction of molecules with energy $\geqslant E$ is twice that at 298 K

Thus, one would expect a reaction with an activation energy of $50 \, \text{kJ mol}^{-1}$ approximately to double in rate over 10 K. This behaviour is observed for many such reactions.

The pre-exponential term in the Arrhenius equation, A

Returning to the Arrhenius equation, it was suggested that the pre-exponential term, A, might be the collision frequency. Calculated values of collision frequencies, however, are often higher than values obtained from plots such as Figure 14.9, p. 306. The explanation must be that molecules must collide not only with sufficient energy for reaction to occur, but also in a favourable orientation in space. For example, in the reaction

$$CH_3\cdot + CHCl_3 \rightarrow CH_4 + CCl_3\cdot$$

the approach [1] will be more favourable than [2]

Collisions must be favourably oriented

$$CH_3\cdot \rightarrow \leftarrow H-\underset{\underset{Cl}{\big\backslash}}{\overset{\overset{Cl}{\big/}}{C}}-Cl \qquad\qquad\qquad [1]$$

$$CH_3\cdot \rightarrow \leftarrow Cl-\underset{\underset{Cl}{\big\backslash}}{\overset{\overset{H}{\big/}}{C}}-Cl \qquad\qquad\qquad [2]$$

The term A is therefore thought to be the rate at which molecules collide in an orientation which is favourable to reaction.

The Collision Theory applies to reactions in solution as well as to reactions in the gas phase

The Collision Theory was developed for bimolecular reactions in the gas phase, but it has been found to apply to reactions in solution. Although solvent molecules prevent molecules of reactant coming together as frequently as they would in the gas phase, once they have come together, reactant molecules are less able to escape, and will collide repeatedly. These repeated collisions between a pair of molecules may well lead to reaction. Some reactions which have been studied both in the gas phase and in solution are found to have similar rate constants in the two phases.

To summarise: the Collision Theory of bimolecular reactions states that, in order to react, molecules must (*a*) collide, (*b*) collide in a favourable orientation and (*c*) collide with enough energy to react.

*THE METHOD OF FINDING THE ACTIVATION ENERGY

The activation energy and the pre-exponential factor for a reaction can be found from measurements of the rate constant at different temperatures.

Example   Find the activation energy and the pre-exponential factor for the reaction

$$2HI(g) \rightarrow H_2(g) + I_2(g)$$

A worked example...

Values of the rate constant k at various temperatures are given below:

...finding the activation energy from rate constants at different temperatures

T/K	556	629	700	781
$k/dm^3\,mol^{-1}\,s^{-1}$	7.04×10^{-7}	6.04×10^{-5}	2.32×10^{-3}	7.90×10^{-2}
$10^3/T$	1.80	1.59	1.43	1.28
$ln\,k$	-14.2	-9.71	-6.07	-2.54

Method Figure 14.9 shows the linear plot of $\ln k$ against $10^3/T$.

$$\text{Gradient} = -2.35 \times 10^4 \, \text{K} = -E/R$$

$$\text{Activation energy } E = 195 \, \text{kJ mol}^{-1}$$

From the graph when

$$\ln k = -4.00, \, 1/T = 1.34 \times 10^{-3}$$

Since $\ln A = \ln k + E/RT$

$$\ln A = -4.00 + \frac{195 \times 10^3 \times 1.34 \times 10^{-3}}{8.314}$$

Pre-exponential factor $A = 7.9 \times 10^{11} \, \text{dm}^3 \, \text{mol}^{-1} \, \text{s}^{-1}$.

FIGURE 14.9 Graph of $\ln k$ against $10^3/T$

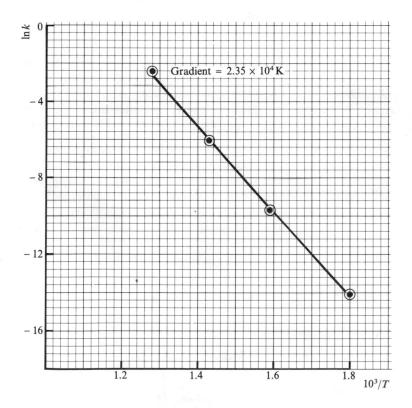

14.10.2 *UNIMOLECULAR REACTIONS

Unimolecular reactions also require collisions before molecules can react...

The Collision Theory is also applied to unimolecular reactions. In a reaction such as

$$C_2H_6 \rightarrow 2CH_3 \cdot$$

it appears as though only one molecule is required for dissociation to occur, and the importance of collisions is not obvious. It is thought that, in a unimolecular reaction

$$A \rightarrow B$$

...Energy is transferred in collisions...

before **A** has enough energy to react it must form an activated species **A*** by obtaining additional energy through collision with another molecule **X**. **X** could be a molecule of **A** or **B** or some other substance:

$$\mathbf{A} + \mathbf{X} \xrightleftharpoons{\text{fast}} \mathbf{A}^* + \mathbf{X} \qquad\qquad [1]$$

$$\mathbf{A}^* \xrightarrow{\text{slow}} \mathbf{B} \qquad\qquad [2]$$

...In this way, some molecules gain enough energy to react

It is assumed that step [1], the transfer of energy in a collision, is a much faster process than [2], the transformation of activated **A*** into **B**. Then

Rate of reaction $= k[\mathbf{A}^*]$

In this way, the Collision Theory can be applied to unimolecular reactions, as well as bimolecular reactions.

There are, however, some shortcomings in the Collision Theory. In particular, no adequate explanation is offered for the enormous variation in the values of the pre-exponential factor.

14.10.3   THE TRANSITION STATE THEORY

The Transition State Theory looks at the course of a collision in detail

The **Transition State Theory** is concerned with what actually happens during a collision. It follows the energy and orientation of the reactant molecules as they collide and seeks an explanation of why such a small fraction of collisions results in reaction.

When two molecules approach each other in a collision, the electron clouds experience a gradual increase in their mutual repulsions, and the molecules begin to slow down. While this is happening, the kinetic energy of the molecules is being converted into potential energy. If the molecules had little kinetic energy to begin with, i.e., if they were not moving very fast, they will come to a stop before their electron clouds have interpenetrated very much, and then fly apart again without reacting [see Figure 14.10(a)].

FIGURE 14.10

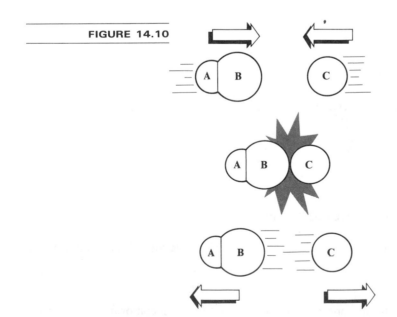

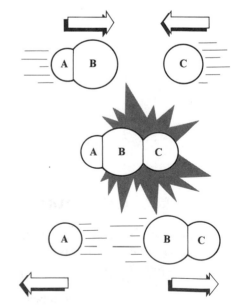

(a) Two slow-moving molecules collide. The electron clouds do not interpenetrate.

(b) Two fast-moving molecules collide. Atoms approach closely, and electron clouds interpenetrate. This leads to reaction.

Kinetic energy is converted into potential energy as reacting molecules collide

When two fast-moving molecules collide, they have a lot of kinetic energy that can be converted into potential energy. They are able to overcome the forces of repulsion between their electron clouds and to approach each other closely. The interpenetration of the electron clouds that occurs permits a rearrangement of valence electrons, with the breaking of old bonds and the formation of new bonds, i.e., a chemical reaction [see Figure 14.10(b)].

A methyl radical and a molecule of hydrogen chloride may react if they collide:

$$CH_3\cdot + HCl \rightarrow CH_4 + Cl\cdot$$

As the reactant molecules move along the reaction coordinate towards becoming the products, the potential energy passes through a peak...

The change in potential energy that takes place during the course of one reactive collision (i.e., one that changes the reactant molecules into the product molecules) is shown in Figure 14.11. The horizontal axis is called the **reaction coordinate**. Positions along the reaction coordinate represent the distance that the reacting species have moved towards forming the products. As $CH_3\cdot$ and HCl approach one another, their potential energy increases to a maximum. The arrangement of atomic nuclei and bonding electrons at the potential energy maximum is called the **activated complex**. It can be represented as

... The arrangement of atomic nuclei and bonding electrons at this peak is called the activated complex

$$H_3C\text{-----}H\text{------}Cl$$

As the new bond, H_3C—H, forms, it assists the breaking of the old bond, H—Cl. The activated complex exists in a **transition state** along the reaction coordinate. Once formed, the transition state is transformed into the products. The difference in potential energy between the activated complex and the reactants is called the activation energy E_A. The number of reacting species that take part in the formation of the transition state is the **molecularity** of the reaction step [p. 291]. This reaction step is **bimolecular**.

FIGURE 14.11 Potential Energy Diagram for a Single Reactive Collision

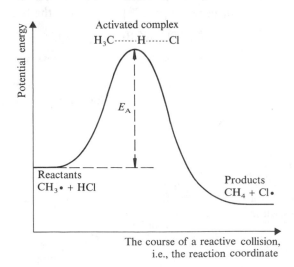

Now consider 1 mole of reactants. Multiplying the potential energy of a molecule by the Avogadro constant gives the internal energy of 1 mole of such molecules [p. 193]. This is approximately equal to the enthalpy of 1 mole of molecules. Figure 14.12 shows the relationship between the standard enthalpy $H^\ominus$ of a mole of the reactants, a mole of the products and a mole of the activated complex. The horizontal axis represents the movement from the reactants to the products along the reaction coordinate. Such a diagram is described as an **enthalpy profile** (or **energy profile**) of the reaction.

FIGURE 14.12 An Enthalpy Profile for a Reaction

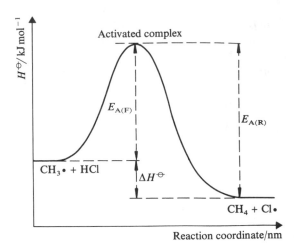

$E_{A(F)}$ and $E_{A(R)}$ in Figure 14.12 represent the activation energies of the forward and reverse reactions. $\Delta H^{\ominus}$ is the standard enthalpy of reaction:

$$\Delta H^{\ominus} = H^{\ominus}_{\text{Products}} - H^{\ominus}_{\text{Reactants}}$$

As can be seen from Figure 14.12

$$\Delta H^{\ominus} = E_{A(F)} - E_{A(R)}$$

The Transition State Theory applies to reactions in solution as well as to gaseous reactions. The alkaline hydrolysis of primary halogenoalkanes (e.g., C_2H_5Br) is discussed on p. 604.

Some reactions take place via a reactive intermediate

Some reactions take place via a **reactive intermediate**. One example is the nitration of benzene [p. 579]. Another is the hydrolysis of tertiary halogenoalkanes [p. 605]:

$$(CH_3)_3CBr \rightarrow (CH_3)_3C^+ + Br^-$$

$$(CH_3)_3C^+ + H_2O \rightarrow (CH_3)_3COH + H^+(aq)$$

The reactive intermediate is the carbocation, $(CH_3)_3C^+$. It is preceded by and followed by a transition state. The enthalpy profile for a reaction of this kind is shown in Figure 14.13.

FIGURE 14.13 Enthalpy Profile of a Reaction which involves a Reactive Intermediate

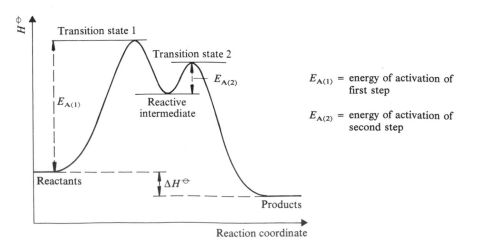

14.11 CATALYSIS

A catalyst lowers the activation energy of a reaction by providing a different route from reactants to products

A **catalyst** is a substance which alters the rate of a chemical reaction without being consumed in the reaction. Thus, a small amount of catalyst is able to catalyse the reaction of a large amount of reactant. A catalysed reaction has a lower activation energy than an uncatalysed reaction. It is believed that a catalyst provides a different mechanism for the reaction, with a lower activation energy [see Figure 14.14].

FIGURE 14.14 Energy Profiles for a Catalysed and an Uncatalysed Reaction

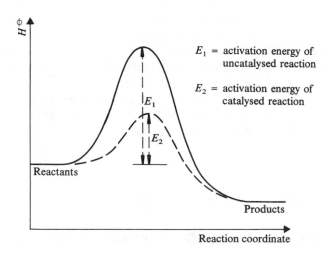

E_1 = activation energy of uncatalysed reaction

E_2 = activation energy of catalysed reaction

Catalysis may be homogeneous or heterogeneous

If the catalyst and the reactants are in the same phase, the process is described as **homogeneous catalysis**. **Heterogeneous catalysis** takes place at the surface of a catalyst which is in a different phase from the reactants (e.g., a solid catalysing a reaction between gases). The reactants are adsorbed on to the surface of the catalyst, where bonds are broken and new bonds are formed. The products are then desorbed from the surface. Catalysis is an extremely widespread phenomenon, and only a few examples are given here.

14.11.1 HOMOGENEOUS CATALYSIS

ACID–BASE CATALYSIS

Ester hydrolysis is an example of homogeneous catalysis

Many reactions are catalysed by acids and bases. An example is the acid-catalysed hydrolysis of esters to give a carboxylic acid and an alcohol or a phenol:

$$RCO_2R' + H_2O \xrightleftharpoons{H^+(aq)} RCO_2H + R'OH$$

ENZYME CATALYSIS

Most biological processes are catalysed by proteins called **enzymes**. For many enzymes there is a specific reaction which that particular enzyme is ideally suited to catalyse. The digestive enzyme, papain, catalyses the hydrolysis of the peptide bond

$$-\overset{\displaystyle \|}{\underset{\displaystyle O}{C}}-\overset{\displaystyle }{\underset{\displaystyle H}{N}}-$$

Enzymes are proteins which act as catalysts for specific reactions

which occurs in proteins:

$$—CONH—\overset{\overset{\displaystyle R'}{|}}{CH}—CONH—\overset{\overset{\displaystyle R''}{|}}{CH}—CONH— + H_2O →$$

$$—CONH—\overset{\overset{\displaystyle R'}{|}}{CH}—CO_2H + H_2N—\overset{\overset{\displaystyle R''}{|}}{CH}—CONH—$$

Enzyme-catalysed reactions do not obey the Arrhenius equation

This is one of the steps which occurs in the hydrolysis of proteins to amino acids. Papain will work only on peptide bonds where R′ and R″ are certain amino acid groups, but not on others. An enzyme and its substrate (the substance which it enables to react) fit together in a three-dimensional arrangement which allows the enzyme to work effectively on the substrate, withdrawing electrons from one bond and supplying electrons to another bond. Any factor which alters the three-dimensional structure of the enzyme is said to **denature** the enzyme, and this destroys its catalytic activity. Heat is one such factor. This is why enzyme-catalysed reactions do not obey the Arrhenius equation.

14.11.2 HETEROGENEOUS CATALYSIS

TRANSITION METALS

Many examples of heterogeneous catalysis involve transition metals. Their empty d orbitals allow them to bond with many substances to form reactive intermediates. In the Haber Process for the manufacture of ammonia, the catalyst is a mixture of iron and vanadium:

Many transition metals act as heterogeneous catalysts...

$$N_2(g) + 3H_2(g) \xrightarrow{Fe/V} 2NH_3(g)$$

...Reaction takes place at the surface

In the hydrogenation of alkenes, a process which is used in the conversion of liquid oils into solid fats, nickel is used. It is finely divided to increase the area of surface over which the reactants can come into contact.

$$R_2C{=}CR_2(g) + H_2(g) \xrightarrow{Ni} R_2CH—CHR_2(g)$$

The oxidation of sulphur dioxide in the Contact Process is catalysed by platinum or by vanadium(V) oxide. If platinum is used, it readily absorbs impurities present in the sulphur dioxide and loses its activity: it is easily 'poisoned'.

'CRACKING'

Catalysis is important in the petroleum industry

The 'cracking' of hydrocarbons in the petroleum industry is catalysed by a mixture of silica and alumina:

$$C_8H_{18}(g) \xrightarrow{Al_2O_3/SiO_2} C_4H_{10}(g) + C_4H_8(g)$$

Octane   Butane   Butene

14.11.3 AUTOCATALYSIS

There are reactions in which one of the products formed in the reaction catalyses the reaction. An example is the oxidation of hydrogen peroxide by acidified manganate(VII):

$$2MnO_4^-(aq) + 6H^+(aq) + 5H_2O_2(aq) \rightarrow 2Mn^{2+}(aq) + 5O_2(g) + 8H_2O(l)$$

The Mn^{2+} ions produced catalyse the reaction. The reaction starts slowly, and then speeds up after a small amount of catalyst has been formed.

CHECKPOINT 14D: REACTION KINETICS

1. Explain the following statements:

(a) The rate of a bimolecular reaction cannot be calculated from the collision frequency alone.

(b) The increase in the rate of a chemical reaction with an increase in temperature is much greater than the corresponding increase in the collision frequency.

(c) The presence of a catalyst can make a big change in the rate of a chemical reaction.

*2. The Collision Theory postulates that collisions between molecules are necessary before reaction can occur. A unimolecular reaction involves the dissociation or isomerisation of a single molecule.
How does the Collision Theory apply to unimolecular reactions?

3. What is an activated complex? Give an example to illustrate your answer. On the Transition State Theory, what is it assumed that an activated complex will do?

4. What is a spontaneous reaction? Do spontaneous reactions always take place rapidly? Give examples. In what way does a catalyst affect the spontaneous reaction?

5. Describe what is meant by a *chain reaction*. By referring to a chosen example, explain what are chain-initiating steps, chain-propagating steps and chain-terminating steps.

6. The root mean square speed of gaseous molecules [p. 145] is given by

$$\tfrac{1}{3}mL\overline{c^2} = RT$$

Calculate the ratio

     rms speed at 308 K/rms speed at 298 K

For a certain reaction, the ratio

     reaction rate at 308 K/reaction rate at 298 K = 2.0

How can you explain the difference between the two ratios?

7. According to thermodynamics, an exothermic reaction is *feasible*. Why do some exothermic reactions proceed very slowly?

8. Why does the probability that a collision will result in reaction depend on the orientation of the colliding molecules? Illustrate your answer with reference to the reactions

$$2HI(g) \rightarrow H_2(g) + I_2(g)$$

$$CH_3Br + I^- \xrightarrow{\text{in propanone}} CH_3I + Br^-$$

$$(CH_3)_3N + C_2H_5I \xrightarrow[\text{solvent}]{\text{in an organic}} (CH_3)_3\overset{+}{N}C_2H_5 + I^-$$

*9. (a) The reaction

     $P \rightarrow Q$

doubles its rate constant between 25 °C and 35 °C. What is its activation energy?

(b) Calculate the activation energy of the reaction

     $R \rightarrow S$

which triples its rate constant over the same temperature range.

*10. By means of a suitable graph, use the following results of kinetic 'runs' on the reaction

$$H_2(g) + I_2(g) \rightarrow 2HI(g)$$

to obtain a value for the activation energy:

T/K	555	606	645	714	769
$k/mol^{-1} dm^3 s^{-1}$	3.72×10^{-5}	7.32×10^{-4}	5.41×10^{-3}	0.111	0.819

14.12   A DETAILED KINETIC STUDY

In a kinetic study of the iodination of propanone...

The iodination of propanone

$$CH_3COCH_3(aq) + I_2(aq) \rightarrow CH_3COCH_2I(aq) + HI(aq)$$

is a reaction with interesting kinetics. The reaction is acid-catalysed. From the equation, one might postulate that the rate equation would be

$$-\frac{d[I_2]}{dt} = k[CH_3COCH_3]^a[I_2]^b$$

A study of the reaction will involve finding a and b.

First: Find the order with respect to iodine

...the concentration of propanone is kept approximately constant

It is arranged that the propanone concentration is much greater than the iodine concentration, e.g., $[CH_3COCH_3] = 1.00\,mol\,dm^{-3}$ and $[I_2] = 0.00500\,mol\,dm^{-3}$ so that, at the end of a run, $[CH_3COCH_3] = 0.995\,mol\,dm^{-3}$, a decrease of 0.5%. One can say that $[CH_3COCH_3]$ is effectively constant, and

$$-\frac{d[I_2]}{dt} = k_1[I_2]^b \text{ so that } b \text{ can be found.}$$

Solutions of known concentration of (a) propanone, (b) iodine in potassium iodide, and (c) an acid buffer of known pH are prepared and brought to the required temperature in a thermostat bath. The reaction is started by pipetting volumes of the three solutions into a flask, and a stop watch is started. After a few minutes, a sample of the reacting mixture is pipetted from the solution into a sodium hydrogencarbonate solution. This stops the reaction instantly by neutralising the acid. The time at which the reaction stops is recorded. The iodine that remains is determined by titration against a standard solution of sodium thiosulphate. The analysis is repeated at intervals of a few minutes. The volume of thiosulphate required is plotted against the time elapsed since the start of the reaction. Figure 14.15 shows the plot obtained. It is a straight line: the gradient of the graph does not change as the concentration of iodine decreases. This shows that the rate of reaction remains constant as the iodine concentration decreases:

The reaction is started by mixing the reagents...

...A sample of the solution is pipetted...

...The reaction is stopped...

...Titration gives the iodine concentration

It is found that the rate of the reaction remains constant as the iodine concentration decreases

$$-\frac{d[I_2]}{dt} = \text{Constant}$$

The reaction is zero-order with respect to iodine, and the rate equation becomes

$$-\frac{d[I_2]}{dt} = k[CH_3COCH_3]^a$$

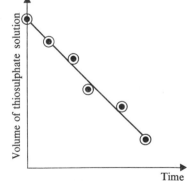

FIGURE 14.15   Graph of Results of Iodine titration

Second: Find the order with respect to propanone

Runs at different propanone concentrations

The procedure is repeated with different concentrations of propanone. It is found that doubling the concentration of propanone doubles the rate of the reaction: the reaction is first-order with respect to propanone:

$$-\frac{d[I_2]}{dt} = k[CH_3COCH_3]$$

Third: Find the effect of acid concentration

Runs at different acid concentrations

A set of runs in buffers with different values of pH and fixed concentrations of propanone and iodine shows that the rate of reaction is proportional to the hydrogen ion concentration. The rate equation is

$$-\frac{d[I_2]}{dt} = k[CH_3COCH_3][H^+(aq)]$$

Interpretation of the rate equation

Iodine must be present for iodination to occur, but its concentration does not appear in the rate equation. To explain this, it is suggested that the reaction takes place in steps. It is the rate of the slowest step that determines the rate of the overall reaction. If iodine is involved in a step which is too fast to be rate-determining, it will not appear in the rate equation. The suggested mechanism for the reaction is

From the results of kinetic studies, a mechanism for the reaction can be worked out

$$CH_3\!-\!\underset{O}{\overset{\|}{C}}\!-\!CH_3 + H^+(aq) \underset{slow}{\rightleftharpoons} CH_3\!-\!\underset{O-H}{\overset{\|}{C}}\!-\!CH_3 \underset{fast}{\rightleftharpoons} CH_3\!-\!\underset{O-H}{\overset{\|}{C}}\!=\!CH_2 + H^+(aq) \quad [1]$$

This is the slow, rate-determining step. An enol is formed. Since halogens are electrophiles [p. 557], iodine reacts rapidly with the $\underset{/}{\overset{\backslash}{C}}\!=\!\underset{\backslash}{\overset{/}{C}}$ bond through its π-electrons:

$$CH_3\!-\!\underset{O-H}{\overset{|}{C}}\!=\!CH_2 + I\!-\!I \xrightarrow{fast} CH_3\!-\!\underset{O-H}{\overset{+}{\underset{|}{C}}}\!-\!CH_2\!-\!I + I^- \quad [2]$$

The intermediate formed has a positive charge on a carbon atom, and fast loses a proton to form the iodoketone:

$$CH_3\!-\!\underset{O-H}{\overset{+}{\underset{|}{C}}}\!-\!CH_2\!-\!I \xrightarrow{fast} CH_3\!-\!\underset{O}{\overset{\|}{C}}\!-\!CH_2\!-\!I + H^+(aq) \quad [3]$$

(For curly arrows [p. 559].)

QUESTIONS ON CHAPTER 14

1. Explain what is meant by the terms *rate constant, order of reaction, molecularity of reaction*. Under what conditions are order and molecularity different? What are the differences between zero-, first- and second-order reactions?

2. In the reaction

$$A \rightarrow B + C$$

what are the units of the rate constant for (a) a zero-order reaction, (b) a first-order reaction and (c) a second-order reaction?

3. The results of experiments on the hydrolysis of ethyl ethanoate in aqueous solution show first-order kinetics. Explain why.

4. What is a chain reaction? Identify the steps described as chain-initiating, chain-propagating and chain-terminating in (a) the chlorination of methane and (b) the fission of ^{235}U.

5. The radioactive isotope $^{24}Na^+(aq)$ is injected into an animal. If the half-life is 15 h, how many hours will elapse before the radioactivity has fallen to 10% of the original dose?

6. A freshly-felled piece of wood gives 15.0 cpm (g of carbon)$^{-1}$. If a piece of wood from an Egyptian mummy case gives 9.5 cpm g^{-1}, how old is the case? The half-life of ^{14}C is 5600 years.

7. Ethanal decomposes thermally to form methane and carbon monoxide:

$$CH_3CHO(g) \rightarrow CH_4(g) + CO(g)$$

Standard enthalpies of formation/kJ mol^{-1} [$\Delta H_F^{\ominus}$, p. 195] are: $CH_3CHO(g) = -166$, $CH_4(g) = -75$, $CO(g) = -110$. The energy of activation, $E_A = 190$ kJ mol^{-1}. When the decomposition is catalysed by iodine, $E_A = 136$ kJ mol^{-1}. Draw the reaction profiles for the catalysed and uncatalysed reactions, indicating the values of E_A and the values of $\Delta H_{Reaction}^{\ominus}$.

8. The following results were obtained in a study of the reaction

$$S_2O_8^{2-}(aq) + 2I^-(aq) \rightarrow 2SO_4^{2-}(aq) + I_2(aq)$$

between peroxodisulphate and iodide ions:

Experiment	$[S_2O_8^{2-}]$ /mol dm^{-3}	$[I^-]$ /mol dm^{-3}	Initial rate, $\left\|\dfrac{d[S_2O_8^{2-}]}{dt}\right\|$ mol dm^{-3} s^{-1}
1	0.040	0.040	9.6×10^{-6}
2	0.080	0.040	1.92×10^{-5}
3	0.080	0.020	9.6×10^{-6}

Find the order of reaction with respect to (a) $S_2O_8^{2-}$, (b) I^- and (c) find the rate constant. What is the initial rate of the reaction when $[S_2O_8^{2-}]_0 = 0.12$ mol dm^{-3}, and $[I^-]_0 = 0.015$ mol dm^{-3}?

9. The nitration of the aromatic compound ArH is zero-order with respect to ArH. Suggest an explanation for this observation.

10. Results are given below for three reactions of a substance **A**:

Reaction 1	
Time/min	$[A]$/mol dm^{-3}
0	1.00
2.0	0.82
4.0	0.67
7.0	0.49
10.0	0.37
14.0	0.24
20.0	0.14
25.0	0.08

Reaction 2	
Time/min	$[A]$/mol dm^{-3}
0	1.00
2.0	0.79
4.0	0.59
7.0	0.30
10.0	0.00

Reaction 3	
Time/min	$[A]$/mol dm^{-3}
0	1.00
2.0	0.84
4.0	0.72
7.0	0.58
10.0	0.50
15.0	0.40
20.0	0.33
25.0	0.29

(a) Construct plots of [**A**] against time. From an inspection of the plots, say which reaction is (i) first-order, (ii) second-order and (iii) zero-order.

(b) Read off the half-life for each reaction.

(c) Give the concentration of **A** remaining after 9.5 min in (i) the first-order, (ii) the second-order and (iii) the zero-order reaction.

(d) From suitable linear plots of concentration against time, find the three rate constants.

11. Give examples of reactions which you could *follow* by means of (a) a conductimetric method, (b) a colorimetric method, (c) pressure measurements and (d) volume measurements.

12. Describe how you would attempt to follow the decomposition of nitrogen(V) oxide (a) in the gas phase and (b) in solution in tetrachloromethane, in which oxygen is insoluble:

(a) $2N_2O_5(g) \rightarrow 4NO_2(g) + O_2(g)$

(b) $2N_2O_5(CCl_4) \rightarrow 4NO_2(CCl_4) + O_2(g)$

13. Ethanal decomposes at 800 K according to the equation

$$CH_3CHO(g) \rightarrow CH_4(g) + CO(g)$$

Explain what is meant by these statements:

(a) The reaction is second-order.

(b) Iodine catalyses the reaction.

(c) The reaction in the presence of iodine is first-order with respect to ethanal and first-order with respect to iodine.

(d) The reaction that takes place in the presence of iodine has a lower activation energy than the reaction of ethanal alone.

14. What is meant by the *mechanism* of a chemical reaction? Explain why, when chemical reactions take place in more than one step, the overall rate of the reaction is determined by the rate of the slowest step. Give an example of a reaction of this kind.

15. Explain the following statements:

(a) At room temperature, a mixture of hydrogen and chlorine will not react until irradiated with ultraviolet light.

(b) The calculated frequency of collisions between molecules in the gas phase is often 10^6 to 10^{10} times greater than the measured rate at which molecules react (expressed in the same unit).

(c) The collision frequency between molecules in the gas phase is proportional to the square root of the temperature in kelvins, but a rise in temperature of 10 K often doubles the rate of reaction.

(d) It is not possible to tell the order of a reaction from its stoichiometric equation.

*16. Chemists use two models to interpret the results of measurements of reaction rates. These are the Collision Theory and the Transition State Theory. Describe the theories briefly. Point out the similarities between the two theories. What are the differences?

17. Describe the effect of temperature on the Maxwell–Boltzmann distribution of molecular speeds. How does a knowledge of this effect help us to understand the effect of change of temperature on the rate constants of reactions in the gas phase?

*18. The Arrhenius equation states

$$k = A\,e^{-E/RT}$$

State what the letters k, R, e and T represent. Explain what quantities are represented by A and E. The following results were obtained for the decomposition of N_2O_5 at different temperatures:

Temperature/K	k/s^{-1}
298	1.74×10^{-5}
308	6.61×10^{-5}
318	2.51×10^{-4}
328	7.59×10^{-4}
338	2.40×10^{-3}

From a suitable plot, find the value of E. Calculate a value for A.

19. The curve represents the variation of energy with reaction coordinate for the reaction

$$A(g) + B(g) \rightleftharpoons C(g) + D(g)$$

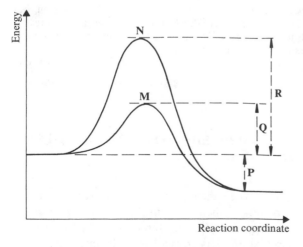

Reaction coordinate

FIGURE 14.16

(a) Of the curves, **M** and **N**, one represents the uncatalysed reaction, and one the reaction in the presence of a catalyst. Which is which?

(b) What are the quantities marked **P**, **Q** and **R**?

(c) How does a catalyst work?

20. (a) Write an equation for the decomposition of hydrogen peroxide.

(b) Explain how you could find the rate of decomposition of hydrogen peroxide in aqueous solution, in the presence of a catalyst, using (i) a gas syringe, (ii) a standard solution of potassium manganate(VII).

(c) Tabulated below are values of the initial rate of decomposition at different concentrations of hydrogen peroxide. Plot a graph of initial rate against concentration.

$[H_2O_2]/mol\,dm^{-3}$	0.100	0.175	0.250	0.300
$Rate/10^{-4}\,mol\,dm^{-3}\,s^{-1}$	0.593	1.04	1.48	1.82

(d) From the graph, find the order of reaction and the rate constant.

21. State *three* of the major assumptions underlying the kinetic-molecular theory of gases and comment on their validity. Describe with a sketch the distribution of velocities among gas molecules and the effect of temperature change on this distribution. Explain why a small rise of temperature usually has a very marked effect on the rate of a homogeneous gas reaction.

Nitrogen(II) oxide (nitric oxide), NO, reacts with hydrogen at 1100 K according to the following equation:

$$2NO + 2H_2 = N_2 + 2H_2O$$

The following are the initial rates of reaction at various partial pressures of reactants at 1100 K.

P_{H_2}/mm^*	400	400	289	147
P_{NO}/mm	300	152	400	400
$rate/mm\,s^{-1}$	1.03	0.25	1.60	0.79

Deduce the rate equation and calculate an *average* value for the rate constant, stating its units.

Note: the non-SI unit, mm Hg, is used for pressure.

(NI 82)

22. What do you understand by the terms (a) order of reaction, (b) rate constant?

Explain, with the aid of suitable diagrams, why a relatively small increase in temperature can cause a large increase in the rate of a chemical reaction.

Figure 14.17 shows the concentration of iodine remaining after a given time when iodine is allowed to react with an organic compound **X** in acidified aqueous solution. Graph **A** refers to a solution which is $1.0\,mol\,dm^{-3}$ with respect to hydrogen ion and graph **B** refers to a solution which is $2.0\,mol\,dm^{-3}$ with respect to hydrogen ion, all other conditions being identical. The reaction is known to be first order with respect to **X**.

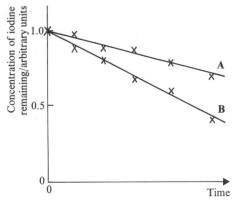

FIGURE 14.17

How does the rate of reaction of iodine with **X** depend on the concentration of (i) iodine, (ii) hydrogen ion?

Explain carefully how you arrive at your answers and write a rate equation for the reaction.

What conclusions can you draw about the mechanism of the reaction from the rate equation?

(C 81)

23. The hypothetical gaseous reaction **A** + **B** → products is exothermic and proceeds 100 times faster at 400 K than at 300 K.

(*a*) With the aid of suitable sketch graphs, discuss the effect of temperature on the reaction.

(*b*) The reaction is catalysed by a metal **M**. Sketch and label the energy profiles for the reaction of **A** with **B** with and without a catalyst. Attempt to explain the role of the catalyst.

(*c*) Using the following logarithmic expression of the Arrhenius equation

$$2.303 \log_{10} k = \text{constant} - E_a/RT$$

(where k = rate constant, R is the gas constant = $8.31 \, \text{J K}^{-1} \text{mol}^{-1}$ and T = temperature/K), determine graphically or otherwise the activation energy E_a of the reaction, assuming that **A** and **B** both have a concentration of $1.0 \, \text{mol dm}^{-3}$.

(L 81, S)

24. Suggest explanations for *three* of the following observations:

(*a*) Bromide ions react with concentrated nitric acid in accordance with the equation:

$$2Br^-(aq) + 4HNO_3(aq) \rightarrow$$
$$Br_2(aq) + 2NO_2(aq) + 2H_2O(l) + 2NO_3^-(aq)$$

During experiments to investigate this reaction, graphs of the form shown in Figure 14.18 were obtained when the amount of bromine formed was plotted against time since the reactants were mixed.

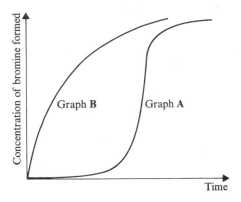

FIGURE 14.18

Graph **A** resulted from the use of colourless concentrated nitric acid. Graph **B** was produced when another sample of concentrated nitric acid was used; this sample was yellow in colour.

(*b*) In 1938 Benford and Ingold showed that if the arenes, benzene, methylbenzene (toluene) and ethylbenzene, were nitrated with an excess of concentrated nitric acid in glacial ethanoic acid as solvent, the same rate of nitration was observed in each case and, in all three cases, the rate of nitration was independent of the concentration of the arene.

(*c*) A solution of potassium peroxodisulphate(VI), $K_2S_2O_8$, is added to an aqueous mixture of potassium iodide, sodium thiosulphate and starch. After a short while a dark blue colour suddenly develops.

The addition of iron(III) ions accelerates the appearance of this blue colour but the addition of chromium(III) ions does not markedly alter the time taken for it to appear.

(*d*) A sample of milk is made alkaline by the addition of a small quantity of sodium carbonate solution. A few drops of phenolphthalein solution are also added and the mixture turns pink. The mixture is divided into two and kept at 40 °C. A dilute solution of pancreatin is added to each, but in one case the pancreatin solution is boiled before the addition. One of the mixtures remains pink but the other loses its pink colour.

(L 82, SN)

14.13 POSTSCRIPT: WORK CAN BE AS MUCH FUN AS GOING TO THE RACES

The great team of Harcourt and Esson devoted their working lives to research on reaction kinetics. They published the results of their work in the *Proceedings of the Royal Society* between 1866 and 1912. Augustus Vernon Harcourt worked in the laboratory in Christ Church College in Oxford. His co-worker, William Esson, worked out the mathematical interpretations of the results of Harcourt's experiments.

One of the reactions they studied was that between hydrogen peroxide and iodide ions, which is still known as the Harcourt–Esson reaction.

This book does not include instructions for practical work. The reason is lack of space, not a lack of respect for the importance of practical work. The author assumes that you, the student, are spending a good deal of time in the laboratory. The Harcourt–Esson reaction is an interesting investigation for you to make during the course of a practical session.

The overall reaction is

$$H_2O_2(aq) + 2H_3O^+(aq) + 2I^-(aq) \rightarrow I_2(aq) + 4H_2O(l)$$

Kinetic measurements showed that the reaction is second-order. The rate is proportional to the iodide ion concentration and to the hydrogen ion concentration. To explain the order of reaction, Harcourt and Esson suggested that the reaction takes place in stages.

1. The first step is a slow reaction between H_2O_2 and I^-:

$$H_2O_2(aq) + I^-(aq) \xrightarrow{\text{slow}} IO^-(aq) + H_2O(l) \qquad [1]$$

2. The second step is the establishment of equilibrium by the weak acid, iodic(I) acid, HIO:

$$IO^-(aq) + H_3O^+(aq) \xrightleftharpoons{\text{fast}} HIO(aq) + H_2O(l) \qquad [2]$$

This happens rapidly.

3. Lastly, iodine is formed in a fast reaction between iodic(I) acid, hydrogen ions and iodide ions:

$$HIO(aq) + H_3O^+(aq) + I^-(aq) \xrightarrow{\text{fast}} I_2(aq) + 2H_2O(l) \qquad [3]$$

This mechanism agrees with the stoichiometry of the overall reaction. You can check up by adding [1] + [2] + [3]. The kinetics and overall order of reaction are those of the slowest step, [1], which is the rate-determining step for the whole reaction. This research work of Harcourt and Esson illustrates one of the important applications of reaction kinetics, which is to throw light on the mechanisms of chemical reactions.

These workers also studied the effect of temperature on the rates of chemical reactions. By extrapolation from measured rates, they predicted the temperature at which, in theory, reaction rates would become zero. They called this temperature the 'zero of chemical action'. It is intriguing to note that their zero was close to the absolute zero of temperature based on the gas laws and on thermodynamics!

After an innings of 46 years, the partnership came at last to a close, and in 1912 Harcourt wrote the last paper of the series. It was published in the *Philosophical Transactions of the Royal Society* in 1912. He ended his last paper in this way.

'Having carried thus far an experimental inquiry which is suggestive of much further work, the author ventures to express a hope that it may attract the attention and pass into the hands of some younger chemist. The mode of working adds to the usual interest of research the particular excitement which attaches to all observations and predictions of time, sporting and scientific, whether it be of the time of a race or of the moment of an occultation.'

Occultation is the eclipsing of one heavenly body by another. People become very excited, waiting to watch an eclipse of the Sun or Moon. As for timing a race, you know how excited people become over watching their horse pass the finishing post or waiting for the first marathon runner to finish the course. This is how Harcourt felt about reaction kinetics. He got the same thrill out of timing reactions, plotting his experimental points and seeing them fall on a straight line. As well as the intellectual satisfaction of thinking out explanations for his results and formulating theories, he got a real thrill out of being in the laboratory. After 46 years of research, his enthusiasm still comes through in this passage. Perhaps you would like to consider his recommendation of the scientific life to 'some younger chemist'?

APPENDIX 1

A SELECTION OF QUESTIONS ON PHYSICAL CHEMISTRY

1. (*a*) State Graham's law of diffusion and explain its theoretical basis.

Under the same conditions equal volumes of the gases **A**, **B** and oxygen each separately diffuse through a small pinhole in 42.4 s, 42.4 s and 119.9 s respectively. A mixture of **A** and oxygen reacts explosively when a spark is passed but there is no reaction when a spark is passed through a mixture of **B** and oxygen. Use these observations to identify gas **A** and gas **B**.

(*b*) Why do all the molecules in a gas not have the same velocity at a given temperature?

Discuss the way the magnitude of the velocities is distributed and explain the following observations.

(i) A liquid can evaporate below its boiling point.

(ii) A small change in temperature can cause a large change in the rate of a reaction   (JMB 81)

2. Suggest explanations for the following observations.

(*a*) Diamond has a high electrical resistivity whereas graphite is a moderately good conductor of electricity.

(*b*) Silver chloride is much less soluble than potassium chloride although Ag^+ and K^+ have closely similar ionic radii.

(*c*) The platinum compound $Pt(py)(NH_3)BrCl$ has been prepared in three isomeric forms, none of which is optically active.

$$py = \;:N \begin{array}{c} CH\!=\!\!=\!\!CH \\ \diagup \qquad\qquad \diagdown \\ \qquad\qquad\qquad CH \\ \diagdown \qquad\qquad \diagup \\ CH\!-\!\!-\!CH \end{array}$$

(*d*) The complexes formed when $Al^{3+}(aq)$ is treated with β-diketones ($RCOCH_2COR$) are insoluble in water but soluble in many organic solvents.   (O & C 82, S)

3. Discuss *each* of the following statements, which *may or may not be correct*.

Note Any numerical data quoted should be taken to be correct.

(*a*) The ionisation energy of hydrogen may be determined from the convergence limit of the series of lines in the visible region of the emission spectrum of atomic hydrogen.

(*b*) The carbon dioxide molecule is linear, but the sulphur dioxide molecule is non-linear.

(*c*) The equilibrium constant K_p for the reaction

$$N_2O_4(g) \rightleftharpoons 2NO_2(g) \qquad \Delta H = +57.2\,kJ\,mol^{-1}$$

increases as either the temperature or the pressure rises.

(*d*) The enthalpy change of solution for an ionic solid such as lithium chloride is often negative ($\Delta H_{solution}(LiCl) = -37.2\,kJ\,mol^{-1}$), although the lattice energy of an ionic solid is large and positive ($LiCl(s) \rightarrow Li^+(g) + Cl^-(g)$; $\Delta H = +846\,kJ\,mol^{-1}$).   (C 83)

4. Account for and explain FIVE of the following.

(*a*) The pH of pure water at 20 °C is 7, but at 45 °C it is 6.7.

(*b*) The boiling point of cis-but-2-ene is higher than that of trans-but-2-ene.

(*c*) Although boron and fluorine have different electro-negativities, boron trifluoride does not have a dipole moment.

(*d*) The crystal lattice of sodium chloride breaks down in water at room temperature, but a temperature of 800 °C is required for melting.

(*e*) Butan-1-ol and propan-1-ol are both primary alcohols, but the former is less soluble in water than the latter.

(*f*) Although iodine has low solubility in water, it is readily soluble in both aqueous potassium iodide and trichloro-methane.

(*g*) The lattice energy of silver iodide determined experimentally is significantly different from the theoretical value.   (L 82)

5. (*a*) (i) State Henry's Law.

(ii) At 293 K, 1 cm^3 of water dissolves 0.12 cm^3 of the rare gas xenon (measured at s.t.p.) when the pressure of this gas is one atmosphere. Assuming that in air one molecule in every 1.7×10^7 molecules is a xenon atom, calculate to two significant figures the number of xenon atoms in one cm^3 of water saturated with air at atmospheric pressure at 293 K.

(*b*) A compound **A** (m.p. 100 °C) and a compound **B** (m.p. 200 °C) are completely miscible in the liquid state. They are completely immiscible in the solid state, and do not combine to form any new substance. In the liquid mixture with the lowest freezing-point (50 °C) the mole fraction of **A** is 0.30.

(i) Sketch the temperature-composition diagram for this system, and label each area on your sketch.

(ii) Describe what will happen on cooling a liquid mixture in which the mole fraction of **A** is 0.50.

(c) Purified water for use in a chemistry laboratory can be obtained by using ion exchange. Explain the principles involved in this method of purifying water.　　　(O 83)

6. Give explanations for the following observations.

(a) The shape of the methane molecule is tetrahedral.

(b) Very tall chimney stacks are used in conventional (coal or oil-fired) power stations.

(c) In the sodium chloride structure each ion is surrounded by six ions of opposite charge, but in caesium chloride each ion has eight nearest neighbours of opposite charge.

(d) Iodine is only slightly soluble in water but readily dissolves in aqueous potassium iodide; it also dissolves in trichloromethane to give a purple solution.

(e) Phenylamine (boiling point 189 °C) can be steam distilled but ethane-1,2-diol (boiling point 197 °C), $HOCH_2CH_2OH$, cannot.　　　(AEB 82)

7. (a) What is meant by the term *standard enthalpy change of solution*?

(b) (i) Indicate what enthalpy changes are involved when solid anhydrous calcium chloride is dissolved in water.

(ii) The enthalpy change of solution of anhydrous calcium chloride is exothermic whilst that of calcium chloride hexahydrate is endothermic. Explain this difference.

(c) How is the enthalpy change of solution of a substance related to its variation of solubility with temperature?

(d) The following data show how the solubility of lead(II) chloride varies with temperature.

Temperature/K	288	293	298	318	328	333	353
Solubility in g/100 g water	8.75×10^{-1}	1.00	1.10	1.55	1.80	1.90	2.35

(i) Plot a graph of *solubility* against *temperature*.

(ii) Calculate the *solubility product* of lead(II) chloride at 308 K.

(iii) Apart from the lowering of temperature, give **two** ways in which the solubility of lead(II) chloride in aqueous solution could be reduced.

(e) (i) How does the change of solubility with temperature enable a lead(II) chloride precipitate to be distinguished from a silver chloride precipitate?

(ii) Give **two** chemical tests, together with the relevant observations, which can be used to distinguish between lead(II) and silver ions.　　　(AEB 81)

8. Answer any two of the options (a), (b) and (c), and one of the options (d) and (e).

(a) Sketch the vapour pressure/temperature diagram for red and white phosphorus, and use it to explain how the relation between these two allotropes differs from that between rhombic and monoclinic sulphur.

(b) Say briefly how steam distillation is carried out in order to separate a substance **X** from a mixture. (A diagram of the apparatus is not required.) Explain what factors determine the relative masses of **X** and of water in the distillate.

(c) For any **one** method of chromatographic separation, describe briefly how the experiment is carried out, and explain the principles underlying this technique of separation.

(d) 'The oxidizing or reducing properties of an element can be assessed by the value of its standard electrode potential.' Illustrate and discuss this statement.

(e) Describe in outline Moseley's experiment, and the reasoning applied to the results of this experiment, which led to the concept of atomic number.　　　(O 82)

Part 3

INORGANIC CHEMISTRY

15

PATTERNS OF CHANGE IN THE PERIODIC TABLE

15.1 PHYSICAL PROPERTIES

The Periodic Table

The Periodic Table was introduced on p. 39. The reason why this arrangement of the elements was proposed is that the elements show **periodicity** in their physical and chemical properties. **Periodic** means repeating after a regular interval. The way in which alkali metals, halogens and noble gases occur at regular intervals of 8 or 18 elements was described on p. 41. The regular intervals are a consequence of the way in which electron shells are filled [p. 42].

The physical properties of elements show periodicity, i.e., they repeat after an interval of 8 or 18 elements

Some of the physical properties which show periodicity are illustrated in Figures 15.1 to 15.6. The melting temperature [see Figure 15.1] of an element depends on the strength of the bonds and also on the structure of the element. When metals melt, some metallic bonding remains in the liquid phase. When macromolecular substances, such as carbon, melt, nearly all the bonds have to be broken to melt the solid. The standard enthalpy of melting varies in a similar way with the proton number (atomic number) of the element.

FIGURE 15.1
Periodicity of Melting Temperatures of the Elements

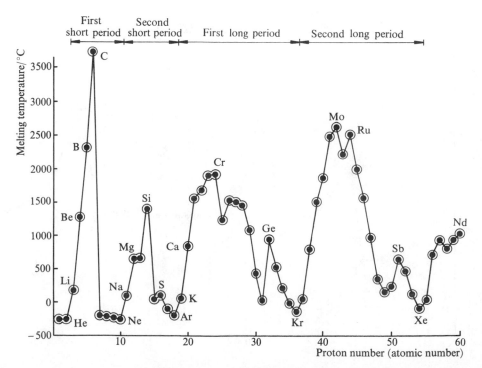

The variation of boiling temperature with proton number is shown in Figure 15.2. The standard enthalpies of vaporisation [p. 152] vary in a similar manner.

FIGURE 15.2

Periodicity of the Boiling Temperatures of the Elements

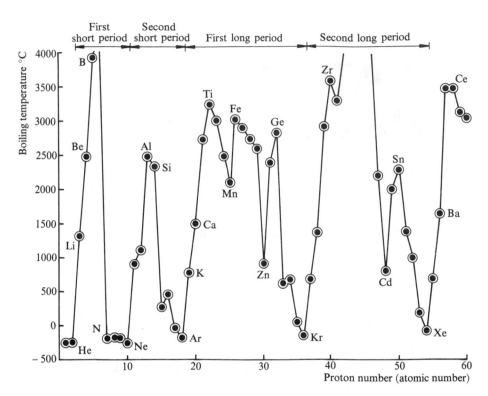

Atomic radius is plotted against proton (atomic) number in Figure 15.3. The sizes of some atoms and ions are compared in Figure 15.4.

FIGURE 15.3   Variation of Atomic Radius with Proton Number. (The values for noble gases are van der Waals radii. The values for other elements are single covalent bond radii.)

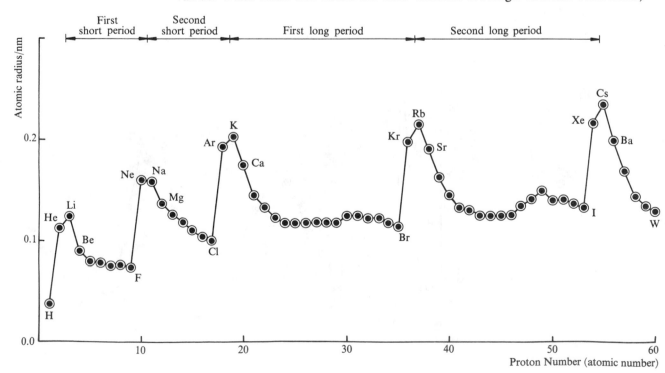

FIGURE 15.4 Atomic and Ionic Radii ($\bigcirc$ = 0.10 nm = 100 pm)

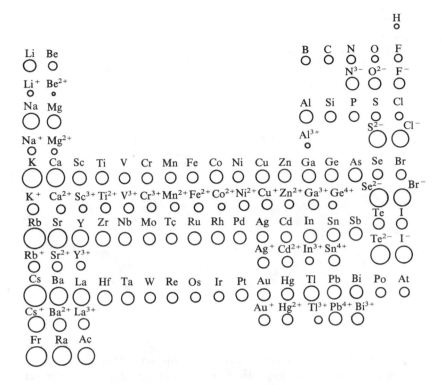

The radii of atoms and ions increase from top to bottom of a group...

From Figure 15.4, you can see that ionic and atomic radii increase from top to bottom of any group. Both the nuclear charge and the number of electron shells increase down a group. Although the increasing nuclear charge decreases the sizes of individual shells, the effect of adding a shell is the dominating effect.

...and decrease from left to right across a period as the effective nuclear charge increases

Covalent and ionic radii both decrease from left to right across any period of the Periodic Table. In the first short period, (Li–F), the nuclear charge increases from 3 to 9. As the nuclear charge increases, it pulls the K electrons closer to the nucleus, and the radius of the K shell decreases. The effect on the L electrons is complicated by the fact that they are *screened* or shielded from the nucleus by the K shell, so that the effective nuclear charge is less than the actual nuclear charge. For example, in lithium the outermost electron is attracted by a nucleus with a charge of +3 screened by two electrons. The net nuclear charge is closer to +1 than +3. In beryllium, the L electrons are attracted by a nucleus which has a charge of +4, and is screened by 2 electrons, which make its effective charge close to +2. Nevertheless, reading from left to right across a period, the effective nuclear charge increases and causes a steady decrease in atomic radius across a period [see Figure 15.4]. A comparison of ions with the same numerical charge, e.g., M^{2+}, shows that ionic radii follow the same pattern.

A series of transition metals differ little in atomic radius

Transition metals show a very small change in atomic radius across a series. In the first transition series, beginning with scandium, the size of the atoms is governed by the N shell, while the additional electrons are entering the M shell.

Ionisation energy varies in a periodic manner

The first ionisation energy [see Figure 15.5] shows periodic variation. It reaches a peak at each noble gas. From helium to lithium and from neon to sodium, the ionisation energy decreases sharply. The additional electrons (a 2s electron in Li and a 3s electron in Na) are much more easily removed than the electrons in the noble gases. Across the periods from lithium to neon and from sodium to argon, the increasing nuclear charge makes it more difficult to remove an electron, and the ionisation energy increases. The increasing ionisation energies for the removal of successive electrons are shown in Figure 2.18, p. 39.

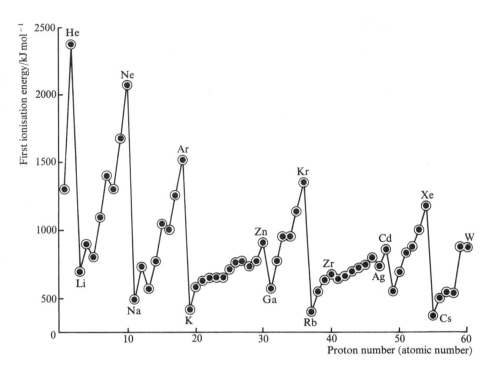

Electron affinity varies in a regular manner

Electron affinity [p. 206]is an important quantity in determining whether a non-metallic element will form ionic compounds. Across a period, as nuclear charge increases, electron affinity increases (becomes more negative, more exothermic). If an electron enters a shell close to the nucleus it becomes more tightly bound, with the release of more energy, than if it enters a shell distant from the nucleus. Electron affinity therefore decreases (becomes less negative, less exothermic) from top to bottom of a group.

Electronegativity increases across a period from left to right and decreases down a group

The property of elements called electronegativity was covered on p. 74. The difference in electronegativity between bonded atoms determines the percentage of ionic character in a covalent bond [see Figure 4.11, p. 75]. Across a period, as nuclear charge increases, electronegativity increases. It decreases down a group as the number of electron shells increases. An element with a very low electronegativity is described as **electropositive**. Figure 15.6 shows how the electronegativity of an element is related to its position in the Periodic Table.

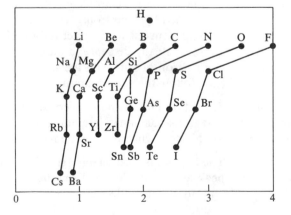

Other physical properties which illustrate periodicity are the standard enthalpy of atomisation, the standard enthalpy of hydration of ions and atomic volume. The last-mentioned was the first property to be studied. Lothar Meyer plotted atomic volume against proton (atomic) number in 1867.

15.2 TRENDS ACROSS THE PERIODIC TABLE

The polarising power of cations and the polarisability of anions vary in a regular manner across the periods and down the groups of the Periodic Table

The trends in ionic radius, first ionisation energy, first electron affinity and electronegativity are summarised in Figure 15.7. These quantities determine the type of bonds formed by an element. Fajans' Rules, which predict the ratio of ionic/covalent character in a bond between two elements are listed on p. 76. They describe the power of small cations and highly charged cations to **polarise** anions, introducing a degree of covalent character into an ionic bond. It follows from Fajans' Rules that the degree of electrovalent character in the compounds formed by cationic elements and anionic elements will vary as shown in Figure 15.7. Electropositive elements will show their most electrovalent behaviour in the top right-hand corner of the table, where oxygen, fluorine and chlorine are to be found.

FIGURE 15.7
Gradations in Bond Type across the s and p Blocks

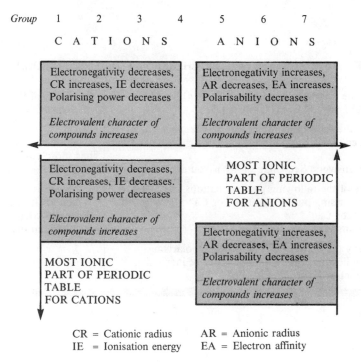

CR = Cationic radius   AR = Anionic radius
IE = Ionisation energy EA = Electron affinity

Trends in chemical properties can be observed across sections of the Periodic Table

In practice, one cannot consider a set of cations across a whole period; for example, the cations from Li^+ to F^{7+} do not all exist. Nor can one follow anionic behaviour across a whole period. The gradations in bond type are applied to sections of periods. One can say that

Period 2:   Li   Be   B   C   N   O   F

            form cations   form covalent bonds   form anions

One can predict that from Li^+ to Be^{2+}, there will be an increase in the covalent character of the bonds, and from O^{2-} to F^-, there will be an increase in the electrovalent character of the bonds.

Diagonal relationships are explained in terms of the polarising power of cations and the polarisability of anions

Figure 15.8 explains the existence of diagonal relationships. The first element in Group 1A, lithium, is diagonally related to the second element in Group 2A, magnesium. Since the polarising power of the cations decreases from lithium down to sodium and increases across from sodium to magnesium, lithium and magnesium form compounds with a similar degree of covalent character. In fact, the two elements are similar in their chemistry [p. 367]. There is a diagonal relationship between beryllium in Group 2A and aluminium in Group 3B [p. 382] and between boron in Group 3B and silicon in Group 4B [p. 382].

FIGURE 15.8 Diagonal Relationships in the Periodic Table

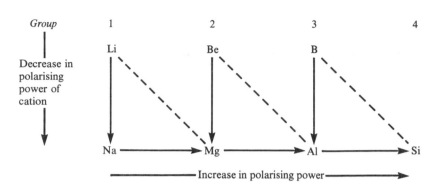

1. Which of the following bonds are polar: (*a*) S—O, (*b*) Cl—Cl, (*c*) I—I, (*d*) C—H, (*e*) C—Cl.

†**2.** Which of the following anions will be the more easily polarised? (*a*) Cl^- or I^-, (*b*) Cl^- or O^{2-}, (*c*) O^{2-} or S^{2-}.
Explain your answers. See Chapter 4.2 if you need help.

3. Which one of the following pairs of cations has the greater polarising power? (*a*) Na^+ or Cs^+, (*b*) Mg^{2+} or Al^{3+}, (*c*) Be^{2+} or Ca^{2+}.
Explain your answers.

4. Explain why the cation Fe^{3+} is strongly polarising. Arrange the following compounds in order of increasing covalent character: $FeBr_3$, $FeCl_3$, FeF_3, FeI_3.

5. Which of the following salts has the highest percentage of electrovalent character?
LiCl, CsCl, CsF, CsI, LiF, LiI.

6. Which one of the following pairs of compounds has more covalent character? (*a*) $AlCl_3$ or Al_2O_3, (*b*) SnF_4 or $SnCl_4$.

7. Which is the more polarising cation, Li^+ or Cs^+? Which of the chlorides, LiCl or CsCl, has the larger degree of covalent character? One of these chlorides is soluble in organic solvents. Which do you suppose it is?

15.3 CHEMICAL PROPERTIES AND BOND TYPE

The chemical behaviour of compounds is determined by the character of the bonds in the compound.

15.3.1 OXIDES

Ionic oxides are basic

Electrovalent oxides are basic. The high negative charge/volume ratio of the O^{2-} ion makes it act as a proton-acceptor, e.g.

$$2Na^+O^{2-}(s) + H_2O(l) \rightarrow 2Na^+(aq) + 2OH^-(aq)$$

Covalent oxides are acidic or neutral

Covalent oxides are either acidic or neutral. An oxide like SO_3 cannot act as a proton-donor, but it can give rise to oxonium ions by acting as an electron-acceptor (a Lewis acid, see p. 251).

$$SO_3(s) + 2H_2O(l) \rightarrow H_3O^+(aq) + HSO_4^-(aq)$$

Oxides of intermediate bond character are amphoteric.

15.3.2 HALIDES

Many ionic chlorides dissolve in water

Electrovalent chlorides simply dissolve in water, with the exception of a few sparingly soluble chlorides, e.g.

$$Na^+Cl^-(s) + aq \rightarrow Na^+(aq) + Cl^-(aq)$$

Many covalent chlorides are hydrolysed

Covalent chlorides may, like CCl_4, neither dissolve in nor react with water or they may, like PCl_3, be hydrolysed to an acid and hydrogen chloride:

$$PCl_3(l) + 3H_2O(l) \rightarrow H_3PO_3(aq) + 3HCl(aq)$$

The phosphorus atom has empty d orbitals, into which electrons from the oxygen atoms in a water molecule can coordinate. Coordination is followed by hydrolysis.

Chlorides of polarising cations are acidic in solution

Chlorides of polarising cations may be partially hydrolysed:

$$BiCl_3(aq) + H_2O(l) \rightleftharpoons BiOCl(s) + 2HCl(aq)$$

The solutions of chlorides and other salts of highly polarising cations are acidic on account of salt hydrolysis [p. 268].

$$[Al(H_2O)_6]^{3+}(aq) + H_2O(l) \rightleftharpoons [Al(OH)(H_2O)_5]^{2+}(aq) + H_3O^+(aq)$$

15.3.3 HYDRIDES

Electrovalent hydrides have a H^- ion, which acts as a proton-acceptor, and they therefore react readily with water to form a base:

$$Na^+H^-(s) + H_2O(l) \rightarrow H_2(g) + Na^+(aq) + OH^-(aq)$$

Covalent compounds of hydrogen can be proton-donors (HCl, H_2S), proton-acceptors (NH_3), amphoteric (H_2O) or neutral (CH_4).

15.3.4 OXIDES, HALIDES AND HYDRIDES IN RELATION TO THE PERIODIC TABLE

Figures 15.9 and 15.10 relate the properties of oxides, chlorides and hydrides to the positions of elements in the Periodic Table. [See p. 344 for hydrides, p. 397–8 for halides and p. 417 for oxides.]

FIGURE 15.9 Bonding in Oxides, Chlorides and Hydrides

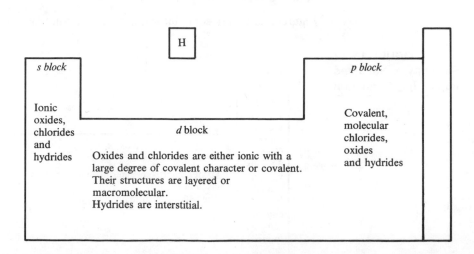

FIGURE 15.10 Acidic
and Basic Compounds

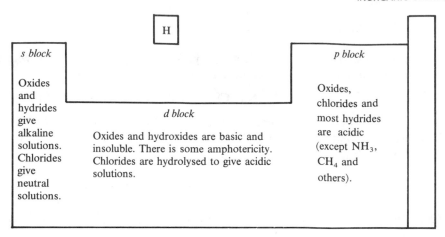

The relationship of electronegativity to the Periodic Table is shown in Figure 15.6, p. 330. The oxidising or reducing properties of the elements are related to their electronegativity, as shown in Figure 15.11.

FIGURE 15.11 Redox
Properties and the
Periodic Table

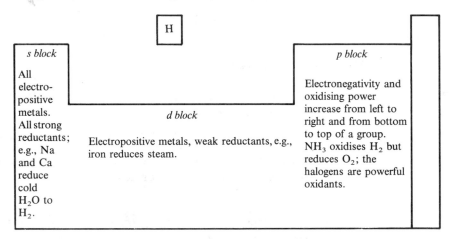

15.3.5 SALTS

*Thermal stability is
related to bond character*

Nitrates, sulphates and carbonates in which the electrovalent character of the bonds is high are stable to heat. Salts with more polarising cations are more easily decomposed by heat: calcium carbonate is decomposed, whereas sodium carbonate is stable. Hydrogencarbonates do not exist after Group 2A, and solid hydrogencarbonates are obtained only in Group 1A.

Figure 15.12 gives some information about solubility of salts and complex formation.

FIGURE 15.12
Solubility of Salts and
Complex Ion Formation

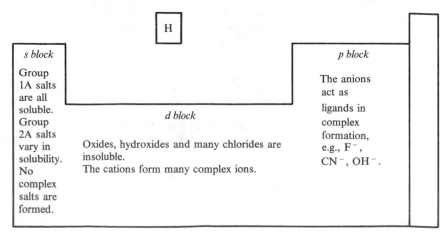

15.3.6 VARIABLE OXIDATION STATE

The compounds of a metal in a high oxidation state show more covalent character than compounds of the metal in a lower oxidation state

In this section an attempt has been made to relate the properties of the compounds of an element to the position of the element in the Periodic Table. The oxidation state of the element also influences the nature of its compounds. Since the polarising power of a cation M^{4+} is greater than that of M^{2+}, $M(IV)$ compounds will show more covalent character than $M(II)$ compounds. For example, lead(II) chloride is a crystalline solid which dissolves sparingly in water to form a solution of Pb^{2+} (aq) and Cl^- (aq) ions. Lead(IV) chloride is a liquid, which is rapidly hydrolysed by water to lead(IV) dichloride oxide and hydrogen chloride:

$$PbCl_4(l) + H_2O(l) \rightarrow PbOCl_2(aq) + 2HCl(aq)$$

Lead(II) oxide, PbO, is basic, while lead(IV) oxide is amphoteric, reacting with alkalis to form plumbates(IV), containing the anion, $Pb(OH)_6{}^{2-}$.

QUESTIONS ON CHAPTER 15

1. A modified statement of Meyer's First Periodic Law (1869) is: 'The properties of the elements are a periodic function of their proton (atomic) numbers.'

What properties of the elements justify this statement? Refer to both physical and chemical properties in your answer.

Meyer originally related properties to relative atomic mass, rather than proton (atomic) number. Why was his statement modified? [p. 7.]

2. (*a*) Which of the cations, Ca^{2+} or Al^{3+}, has the greater polarising power?

(*b*) Which of the chlorides, $CaCl_2$ or $AlCl_3$, is the more covalent in character?

One of these chlorides sublimes when it is heated. Which one do you suppose it is?

3. (*a*) Compare the polarising power of Be^{2+} and Ba^{2+}.

(*b*) What would you predict about the degree of covalent character in $BeCl_2$ and $BaCl_2$?

One of the chlorides is hydrolysed in aqueous solution. Which one do you suppose it is?

4. Of the compounds $SnCl_2$ and $SnCl_4$, which one has the more covalent character? One of the chlorides is a crystalline solid, and one is a fuming liquid. Which is which?

16

GROUP 0: THE NOBLE GASES

16.1 MEMBERS OF THE GROUP

The outermost shell of electrons is full with two electrons in the case of helium and eight for the other elements [see Table 16.1]. There is stability associated with a full valence shell, and the noble gases are very unreactive.

Name	Symbol	Z	Electron Configuration
Helium	He	2	$1s^2$
Neon	Ne	10	$(He)2s^22p^6$
Argon	Ar	18	$(Ne)3s^23p^6$
Krypton	Kr	36	$(Ar)3d^{10}4s^24p^6$
Xenon	Xe	54	$(Kr)4d^{10}5s^25p^6$
Radon	Rn	86	$(Xe)5d^{10}6s^26p^6$

TABLE 16.1
The Noble Gases

16.1.1 HELIUM

Uses of helium Helium forms 5.0 ppm (parts per million) of air. It is obtained from some natural gas wells where the gas may contain up to 5% of helium. It is used to inflate airships, weather balloons and aeroplane tyres, being safer than hydrogen for these purposes. A mixture of oxygen and helium is used by divers instead of air. If they use air, nitrogen dissolves in the blood under pressure and then comes out of the blood when they surface, causing painful spasms which they call 'the bends'.

16.1.2 ARGON

Uses of argon Argon forms 1% of air and is obtained during the fractional distillation of liquid air. It is used to provide an inert atmosphere for gas–liquid chromatography, for risky welding jobs and for some chemical reactions.

16.1.3 NEON, KRYPTON AND XENON

Neon and krypton lights Neon, krypton and xenon are also obtained from air, their concentrations being: Ne, 20 ppm; Kr, 1.0 ppm; Xe, 0.08 ppm. Neon gives out a red glow when an electrical discharge is passed through the gas at low pressure, and finds widespread use in neon lights. Krypton also is used in discharge tubes.

16.1.4 RADON

Radon is a radioactive gas which is formed when radium decays.

16.2 COMPOUNDS OF THE NOBLE GASES

Compounds of xenon...

...XeF₂, XeF₄, XeF₆...

The first noble gas compound ever made was the ionic solid xenon hexafluoroplatinate, $XePtF_6$, which resulted from a reaction between xenon and the powerful oxidising agent platinum(VI) fluoride. This discovery, in 1962, was followed by the synthesis of xenon(II) fluoride, XeF_2; xenon(IV) fluoride, XeF_4; and xenon(VI) fluoride, XeF_6. They were made by warming xenon and fluorine together and, in the case of XeF_6, increasing the pressure. The fluorides are white, crystalline solids. The bonding is covalent, with one, two and three of xenon's 5p electrons being promoted to 5d orbitals. The energy required for promotion is great and can be compensated for only by reaction with the most electronegative of elements, fluorine.

...XeOF₄, XeO₃

Other compounds, xenon(VI) tetrafluoride oxide, $XeOF_4$ and xenon(VI) oxide, XeO_3, have been made by the hydrolysis of xenon(VI) fluoride.

Compounds of krypton...

...KrF₂, KrF₄

The krypton compounds krypton(II) fluoride, KrF_2, and krypton(IV) fluoride, KrF_4, have been synthesised. These require more severe conditions for their synthesis than the xenon fluorides.

This spatial arrangement of the bonds and the lone pairs of electrons about the xenon atom in xenon compounds is shown in Figure 16.1.

FIGURE 16.1 Structures of Some Xenon Compounds

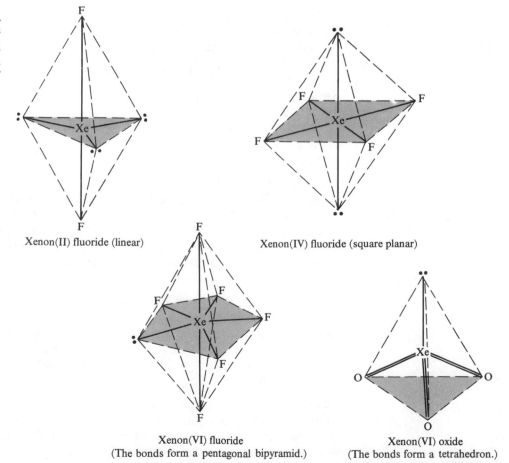

Xenon(II) fluoride (linear)

Xenon(IV) fluoride (square planar)

Xenon(VI) fluoride
(The bonds form a pentagonal bipyramid.)

Xenon(VI) oxide
(The bonds form a tetrahedron.)

1. The electron configurations of the noble gases were, until recently, thought to be perfectly stable. Explain how this belief was the starting-point in the first explanations of the nature of the chemical bond.

2. How did the discovery of the noble gases confirm people's acceptance of the Periodic Table? [See Postscript.]

3. What shapes would you predict for the molecules KrF_2 and KrF_4?

4. Why do you think that it is more difficult for krypton to form compounds than for xenon? Would you expect argon to be more reactive or less reactive than krypton?

5. Write an account of the noble gases (inert gases) with particular reference to their discovery, their importance in the Periodic Classification, and their part in the development of ideas on bonding.

(NI 82)

16.3 POSTSCRIPT: THE 'NEW GAS'

The relative atomic masses of the elements were found by accurate, painstaking work done by scientists in the nineteenth century. Lord Rayleigh, professor at the Cavendish Laboratory in Cambridge from 1880 to 1900, was one of these meticulous workers. One of his methods of preparing nitrogen was the oxidation of ammonia. Another method was to start with air and remove from it oxygen, carbon dioxide and water vapour. Rayleigh found that nitrogen from air was always 0.5% denser than nitrogen from ammonia. He put forward four possible explanations:

1. Nitrogen from air might contain oxygen, which is denser than nitrogen.

2. Nitrogen from ammonia might contain the less dense gas, hydrogen, as impurity.

3. Nitrogen from air might be in the molecular form N_3 as well as the usual form of N_2 molecules. He favoured this theory because he knew that dioxygen, O_2, could be converted into trioxygen or ozone, O_3.

4. Nitrogen from ammonia might contain some N atoms as well as N_2 molecules.

Rayleigh tested these theories. He could not detect oxygen or hydrogen as impurities. Since he knew that ozone, O_3, could be made by passing a silent electric discharge through oxygen, he did the same thing with nitrogen, and then measured the density of the gas to see whether any N_3 had been formed. The density was exactly the same, and this undermined his confidence in the N_3 molecules theory.

His next line of investigation was to take $50 \, cm^3$ of air and pass an electric discharge through it. Nitrogen and oxygen combined to form nitrogen dioxide, which, being an acidic gas, could be absorbed in a solution of an alkali. He added more oxygen, and sparked the mixture again. Since air is 80% nitrogen, he had to keep on adding extra oxygen to convert all the nitrogen into nitrogen dioxide, which he absorbed in alkali. Eventually, of the $50 \, cm^3$ of air, $0.3 \, cm^3$ remained. This gas would not combine with oxygen: it was even less reactive than nitrogen.

Professor William Ramsey of University College, London, became interested in the problem. He passed nitrogen, which he had obtained from the air by the Rayleigh method, over red-hot magnesium. After the gas had been passed to and fro over the metal many times, the volume decreased to $\frac{1}{80}$ of the original volume. This residue of gas would not combine with magnesium as nitrogen did. It was 19.075 times more dense than hydrogen.

Ramsey did scores of experiments with the 'new gas' as he called it. He heated it with metals and non-metallic elements, he mixed it with other gases and passed electrical discharges through the mixtures. All his efforts had the same result: the 'new gas' would not combine with anything.

Sir William Crookes, the inventor of the discharge tube [p. 3], was asked to look at the emission spectrum of the gas [p. 27]. He was amazed to find that he could not identify the spectrum: it was different from the spectra of all the known elements. The 'new gas' was a new element!

Rayleigh and Ramsey discussed the new element. The density they had found for it gave it a relative atomic mass of 40. It was difficult to fit the new element into the Periodic Table. It seemed best to put it in between Cl = 35.5 and K = 39. This was one of the anomalies that led to the arrangement of elements in order of atomic number (proton number) rather than relative atomic mass.

Group	1	2	3	4	5	6	7	
Element A_r	Li 7	Be 9	B 11	C 12	N 14	O 16	F 19	
Element A_r	Na 23	Mg 24	Al 27	Si 28	P 31	S 32	Cl 35.5	? 40
Element A_r	K 39	Ca 40						

TABLE 16.2 Part of the Periodic Table

Rayleigh, who had started the work, and Ramsey, who had finished it, announced their discovery together at the British Association for the Advancement of Science in 1894. They claimed to have discovered a new element, which did not fit into any group of the Periodic Table. When the chairman of the meeting suggested calling the gas *argon* (Greek: *argos*, lazy) this was the name they adopted.

Ramsey went on to discover helium, neon, krypton and xenon. Their relative atomic masses and their lack of chemical reactivity placed them in a group with argon, and they formed a new Group 0 of the Periodic Table. They were called the **inert gases**, but are now generally called the **noble gases**.

CHECKPOINT 16A: ARGON

1. Why was Rayleigh looking for impurities in nitrogen?

2. A sequence of reactions which Ramsey did was

$$\text{Air} \rightarrow \text{Nitrogen} \xrightarrow{\text{Mg}} \text{Magnesium nitride}$$
$$\downarrow \text{H}_2\text{O}$$
$$\text{Nitrogen} \xleftarrow{\text{CuO}} \text{Ammonia}$$

Write equations for the last three reactions. Would the density of the nitrogen obtained at the end of this sequence of reactions be the same as Rayleigh's value for nitrogen from the air or nitrogen from ammonia?

3. What made Ramsey believe that he had discovered a new element?

17

HYDROGEN

17.1 OCCURRENCE

Hydrogen in the Sun   Hydrogen is the most abundant element in the universe. The Sun and the stars derive their energy from the nuclear fusion reaction

$$4{}_1^1H \rightarrow {}_2^4He + 2{}_1^0e \ (\textit{positron}) + \gamma\,(\textit{radiation})$$

There is very little free hydrogen in the Earth's atmosphere, but there is plenty of hydrogen on the Earth, combined as water and organic compounds.

17.2 MANUFACTURE

1. Hydrogen is manufactured from natural gas, which is largely methane. The process is illustrated in Figure 17.1.

FIGURE 17.1   Hydrogen from Natural Gas

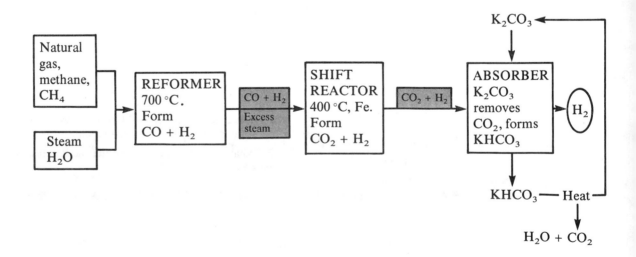

The reactions that take place are

Hydrogen is manufactured from natural gas...

Reforming:	$CH_4(g) + H_2O(g) \rightarrow CO(g) + 3H_2(g)$
Shift Reaction:	$CO(g) + H_2O(g) \rightarrow CO_2(g) + H_2(g)$
Absorption:	$K_2CO_3(aq) + CO_2(g) + H_2O(l) \rightleftarrows 2KHCO_3(aq)$

...from brine... **2.** Hydrogen is produced during the electrolysis of brine for the manufacture of sodium hydroxide [see Figure 18.5, p. 361].

...from cracking of hydrocarbons **3.** It is also a by-product of the cracking of hydrocarbons in a refinery [p. 545].

17.3 INDUSTRIAL USES

Figure 17.2 shows the main industrial uses of hydrogen.

FIGURE 17.2 Some Industrial Uses of Hydrogen

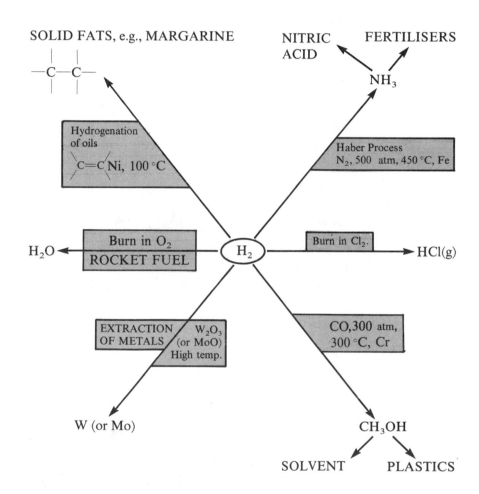

SOLID FATS, e.g., MARGARINE

NITRIC ACID   FERTILISERS

NH_3

Hydrogenation of oils

$C=C$ Ni, 100 °C

Haber Process
N_2, 500 atm, 450 °C, Fe

H_2O Burn in O_2 ROCKET FUEL

H_2

Burn in Cl_2.   HCl(g)

EXTRACTION OF METALS W_2O_3 (or MoO) High temp.

CO,300 atm, 300 °C, Cr

W (or Mo)

CH_3OH

SOLVENT   PLASTICS

17.4 LABORATORY PREPARATION AND REACTIONS

Hydrogen is prepared in the laboratory from a metal + HCl(aq) or H_2SO_4(aq)

The most convenient laboratory preparation is the action of a dilute acid (not nitric acid) on zinc or another metal which is fairly high in the Electrochemical Series. Made in this way, hydrogen contains some hydrogen sulphide, which comes from the zinc sulphide present as an impurity in zinc.

Other laboratory preparations are shown in Figure 17.3. Also shown are some of the reactions of hydrogen. Equations for some of the reactions shown are:

$$2Na(s) + 2H_2O(l) \rightarrow 2NaOH(aq) + H_2(g)$$

$$Mg(s) + H_2O(g) \rightarrow MgO(s) + H_2(g)$$

$$CaH_2(s) + 2H_2O(l) \rightarrow Ca(OH)_2(aq) + 2H_2(g)$$

$$2Al(s) + 2OH^-(aq) + 6H_2O \rightarrow 2Al(OH)_4^-(aq) + 3H_2(g)$$

FIGURE 17.3 Some Reactions of Hydrogen

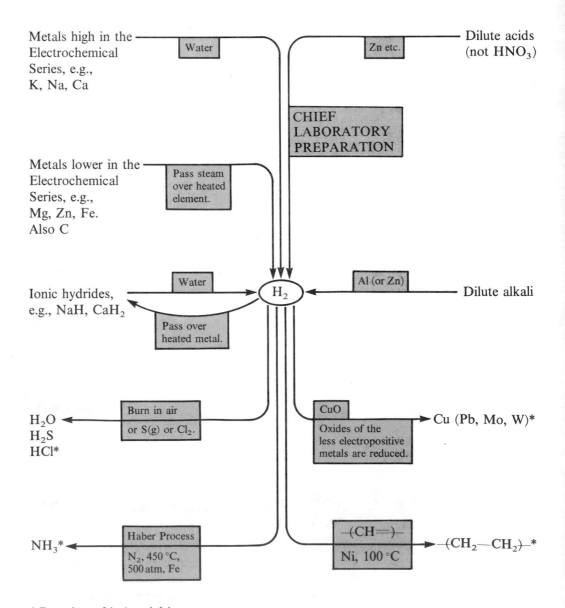

* Reaction of industrial importance

CHECKPOINT 17A: REACTIONS OF HYDROGEN

1. Write equations for the reactions between

(*a*) calcium and water

(*b*) zinc and dilute sulphuric acid

(*c*) magnesium and dilute hydrochloric acid

(*d*) zinc and sodium hydroxide solution to give the zincate ion, $Zn(OH)_4^{2-}$

2. Write equations for the reactions of hydrogen with

(*a*) ethene

(*b*) tungsten(III) oxide

(*c*) oxygen

(*d*) sulphur

(*e*) chlorine

(*f*) nitrogen

Which of these reactions are important in industry? What conditions are employed to give good yields?

3. 'Water gas' was an important fuel twenty years ago. It is a mixture of $CO + H_2$ formed by the reaction of steam and red-hot coke. Write an equation for its formation. Why do you think it is no longer produced? How might you obtain hydrogen from this mixture in the laboratory?

4. State the conditions under which hydrogen will react with the following, and name the products:

(*a*) AgO

(*b*) Pb_3O_4

(*c*) W_2O_3

(*d*) Al_2O_3

(*e*) CaO

17.5 HYDROGEN IONS AND HYDRIDE IONS

The formation of the oxonium ion, H_3O^+, is endothermic

The hydrogen atom, H (1s), can lose one electron to form a proton, H^+, or gain one electron to fill the outer shell in H^- ($1s^2$), or it can fill the outer shell by the formation of a covalent bond.

In aqueous solution, the proton exists as the hydrated oxonium ion, H_3O^+ (aq) [p. 250]. The oxonium ion is often referred to as a hydrogen ion, and its formula is often written as H^+ (aq). The formation of H_3O^+ (aq) is endothermic:

$$\tfrac{1}{2}H_2(g) + H_2O \rightarrow H_3O^+(aq); \quad \Delta H^{\ominus} = +320\,kJ\,mol^{-1}$$

The cation H_3O^+ is formed in combination with an anion, e.g.

$$HCl(aq) + H_2O(l) \rightarrow H_3O^+(aq) + Cl^-(aq)$$

The formation of an anion is usually exothermic. If this is the case, then the formation of $H_3O^+ A^-$ may well be exothermic overall.

The formation of the hydride ion, H^-, is exothermic

The formation of H^- is exothermic:

$$\tfrac{1}{2}H_2(g) \rightarrow H^-(g); \quad \Delta H^{\ominus} = -150\,kJ\,mol^{-1}$$

The anion H^- is formed in combination with cations, e.g.

$$Ca(s) + H_2(g) \rightarrow CaH_2(s)$$

The formation of a hydride M^+H^- or $M^{2+}2H^-$ is exothermic only if the ionisation enthalpy of the cation is small. Only with metals in Groups 1A and 2A is this true.

17.6 REACTIONS OF HYDROGEN IONS

The reactions which characterise acids are those of the hydrogen ion, H_3O^+ (aq) or H^+ (aq). They are summarised in Figure 17.4.

FIGURE 17.4 Some
Reactions of the
Hydrogen Ion

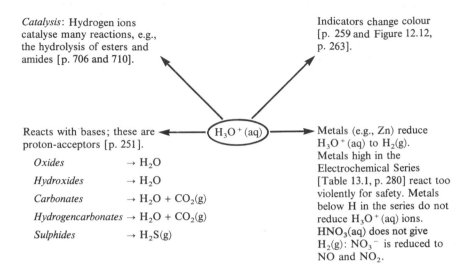

Catalysis: Hydrogen ions catalyse many reactions, e.g., the hydrolysis of esters and amides [p. 706 and 710].

Indicators change colour [p. 259 and Figure 12.12, p. 263].

$H_3O^+(aq)$

Reacts with bases; these are proton-acceptors [p. 251].

Oxides	$\rightarrow H_2O$
Hydroxides	$\rightarrow H_2O$
Carbonates	$\rightarrow H_2O + CO_2(g)$
Hydrogencarbonates	$\rightarrow H_2O + CO_2(g)$
Sulphides	$\rightarrow H_2S(g)$

Metals (e.g., Zn) reduce $H_3O^+(aq)$ to $H_2(g)$. Metals high in the Electrochemical Series [Table 13.1, p. 280] react too violently for safety. Metals below H in the series do not reduce $H_3O^+(aq)$ ions. $HNO_3(aq)$ does not give $H_2(g)$: NO_3^- is reduced to NO and NO_2.

17.7 HYDRIDES

17.7.1 IONIC HYDRIDES

Ionic hydrides are formed by Group 1A and 2A metals

Ionic hydrides are formed by the metals of Groups 1A and 2A. They are made by passing hydrogen over the heated metal. These hydrides are crystalline solids with structures similar to the corresponding halides. They react vigorously with water to give hydrogen, and burn vigorously in air:

The hydrides react with water and burn in air

$$H^-(s) + H_2O(l) \rightarrow H_2(g) + OH^-(aq)$$

$$2H^-(s) + O_2(g) \rightarrow O^{2-}(s) + H_2O(g)$$

Ionic hydrides, including complex hydrides, are reducing agents

In these reactions the hydride is acting as a reducing agent. Complex hydrides, such as lithium tetrahydridoaluminate, $Li^+[AlH_4]^-$, and sodium tetrahydridoborate, $Na^+[BH_4]^-$, are preferred as reducing agents because they are more stable. Both are valued reducing agents in organic chemistry. Sodium tetrahydridoborate is stable in cold water, but lithium tetrahydridoaluminate must be used in solution in dry ethoxyethane because it reacts with water.

Figure 17.5 shows where in the Periodic Table ionic hydrides are formed.

17.7.2 COVALENT HYDRIDES

Covalent hydrides are molecular and often gaseous

Covalent hydrides are molecular. Except for water and hydrogen fluoride, which are associated by hydrogen bonding, covalent hydrides are gaseous.

Hydrogen combines directly with all other non-metallic elements, but the yield of hydride is sometimes low (e.g., NH_3). Hydrolysis is often used to prepare hydrides, for example

Methods of preparing covalent hydrides

(a)   Zinc sulphide + dilute acid $\rightarrow$ hydrogen sulphide, H_2S

(b)   Calcium nitride + water or dilute acid $\rightarrow$ ammonia, NH_3

(c)   Calcium carbide + water or dilute acid $\rightarrow$ ethyne, C_2H_2

(d)   Magnesium silicide + dilute acid $\rightarrow$ silane, SiH_4

FIGURE 17.5   Hydrides

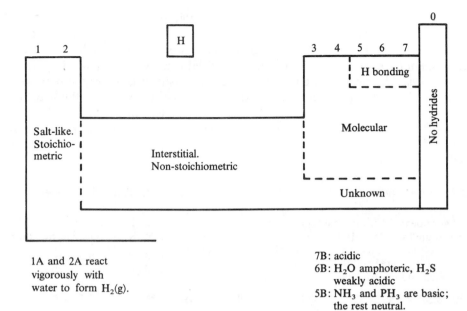

1A and 2A react
vigorously with
water to form $H_2(g)$.

7B: acidic
6B: H_2O amphoteric, H_2S
     weakly acidic
5B: NH_3 and PH_3 are basic;
     the rest neutral.

17.7.3   INTERSTITIAL HYDRIDES

Table 17.1 summarises the hydrides of the elements in Periods 2 and 3 of the Periodic
Table.

	Hydride	Physical state	Reaction with air	Reaction with water
Gp 1	LiH NaH	Solid Solid	Burn vigorously	React violently → $H_2(g)$ + $OH^-(aq)$
Gp 2	BeH_2 MgH_2	Solid Solid	Burn vigorously	React → $H_2(g)$ + $OH^-(aq)$
Gp 3	BH_3 AlH_3	Gas Gas	Burns Burns	Neutral solution Reacts → $H_2(g)$
Gp 4	CH_4 SiH_4	Gas Gas	Burns Ignites	Do not dissolve or react
Gp 5	NH_3 PH_3	Gas Gas	Burns in O_2 Ignites in air	Forms weak alkali Insoluble
Gp 6	H_2O H_2S	Liquid Gas	None Burns	— Forms weak acid
Gp 7	HF HCl	Liquid Gas	None None	Forms acid Forms strong acid

TABLE 17.1   The
Properties of Hydrides

Interstitial hydrides are formed by metals

In a close-packed metallic structure [see Figure 6.5, p. 115] there are holes in between the metal atoms. The tetrahedral holes are small enough to retain hydrogen atoms. Since there are two tetrahedral holes for every metal atom, if all these holes are filled, an **interstitial hydride** of formula MH_2 will be formed. Interstitial hydrides are unusual compounds as their composition depends on the hydrogen pressure. If this is reduced, some hydrogen atoms leave the structure, and a hydride with a formula between MH_2 and MH_0 is left. Interstitial hydrides are used as a means of storing hydrogen and releasing it gradually, for example, for use in fuel cells.

CHECKPOINT 17B: HYDRIDES

1. What appears at (*a*) the cathode and (*b*) the anode, when sodium hydride is electrolysed? Why is sodium hydride not electrolysed in aqueous solution? Electrolysis is often carried out in molten sodium chloride. Why do you suppose that molten sodium hydride is not used alone?

2. Predict from Figure 17.5 the properties of the hydrides of K, Ar, Ge, Sn, Se and Pd.

3. Write equations for hydrolysis reactions which can be used to prepare (*a*) H_2S, (*b*) NH_3, (*c*) ethyne, C_2H_2, and (*d*) silane, SiH_4.

17.8 ISOTOPY

Hydrogen, deuterium and tritium...

There are two stable isotopes of hydrogen, ^1_1H and ^2_1H. The heavier isotope is called **deuterium**. It constitutes 0.01% of naturally occurring hydrogen. The isotope ^3_1H, called **tritium**, is radioactive, and does not occur naturally.

...differ more in mass, relatively speaking, than other isotopes...

...There are some chemical differences between them

The difference in mass between hydrogen and deuterium is, relatively speaking, greater than for any other pair of isotopes (except for hydrogen and tritium). In consequence, the differences between hydrogen and deuterium are greater than the differences between the isotopes of other elements. Deuterium is usually less reactive than hydrogen. Deuterium compounds react at different rates from hydrogen compounds, sometimes differing by a factor of ten in rate. During the electrolysis of water, deuterium ions are discharged more slowly than hydrogen ions. As electrolysis proceeds, the water gradually becomes richer in deuterium. Pure D_2O, **heavy water**, can eventually be obtained. This is how deuterium oxide is obtained commercially. Other deuterium compounds are made from it [see Figure 17.6]. Deuterium oxide has $T_m = 3.8\,°C$, $T_b = 101.8\,°C$ and density $= 1.10\,g\,cm^{-3}$.

Tritium is artificially made, and is radioactive

Tritium is obtained by bombarding deuterium compounds with deuterons:

$$^2_1H + {}^2_1H \rightarrow {}^3_1H + {}^1_1H$$

It decays by β particle emission.

$$^3_1H \rightarrow {}^3_2He + {}_{-1}^0e$$

17.9 ATOMIC HYDROGEN

When an electric discharge is passed through hydrogen gas, atoms of hydrogen are formed:

$$H_2(g) \rightarrow 2H(g); \quad \Delta H^\ominus = +436\,kJ\,mol^{-1}\text{ of }H_2$$

FIGURE 17.6 Routes to some Deuterium Compounds

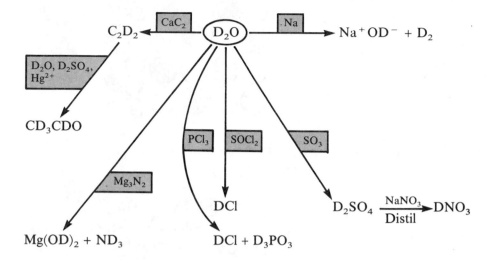

The atomic hydrogen blowlamp

In the **atomic hydrogen blowlamp**, a metal catalyst enables the hydrogen atoms to recombine in a strongly exothermic reaction. The heat evolved enables the blowlamp to cut and weld metals. The reducing atmosphere protects metals from oxidation.

CHECKPOINT 17C: HYDROGEN

1. Explain the following statements:

1a) There is very little hydrogen in the Earth's atmosphere.

(b) Methane, CH_4, burns in air at 500 °C, whereas silane, SiH_4, ignites spontaneously in air.

2. In what ways is the chemistry of hydrogen similar to that of Group 1A and in what ways does it differ?

3. How does hydrogen (a) resemble and (b) differ from the halogens?

4. State how the following could be prepared from D_2O:

$$NaD, \ ND_3, \ DCl, \ C_2D_2, \ LiAlD_4$$

17.10 WATER

The physical properties of water

Some of the physical properties of water have been mentioned. These are: bond angle in H_2O [p. 101 and Figure 5.5, p. 101], hydrogen bonding [p. 83], melting and boiling temperatures [p. 83], ice [p. 85] and the dissolution of organic solutes [p. 84].

The polar nature of water molecules

$$\overset{\delta+}{H}-\overset{\delta-}{O}-\overset{\delta+}{H}$$

explains why water is a good solvent for ionic compounds. Figure 17.7 shows how a salt dissolves in water. More detail is given on pp. 207–9.

FIGURE 17.7
Dissolution of a Salt in
Water

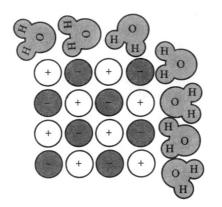
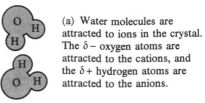

(a) Water molecules are
attracted to ions in the crystal.
The $\delta-$ oxygen atoms are
attracted to the cations, and
the $\delta+$ hydrogen atoms are
attracted to the anions.

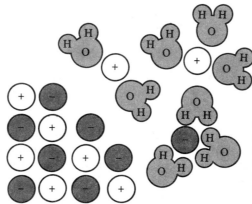

(b) As forces of attraction
come into play, energy is given
out. This compensates for the
energy required to break up the
crystal structure. Ions leave the
structure.

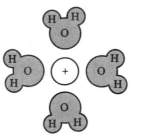
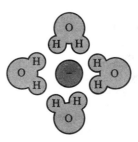

(c) These hydrated ions are
called *aqua* ions. Many are
surrounded by six water
molecules.

Aqua ions   Hydrated ions (ions surrounded by water molecules) are called **aqua ions**. Figure 17.8
shows the aquasodium ion, $[Na(H_2O)_6]^+$ or $Na^+(aq)$ for short.

FIGURE 17.8
The Aquasodium Ion,
$[Na(H_2O)_6]^+$ or
$Na^+(aq)$

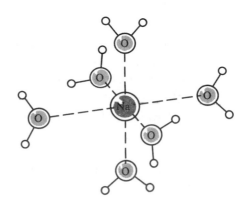

Water hydrolyses some salts, covalent inorganic compounds and organic compounds

The acid–base behaviour of water has been described on p. 250. When salts dissolve in water, their solutions are often neutral. Some salts, however, react with water to form acidic or alkaline solutions [see Salt Hydrolysis, p. 268].

Water hydrolyses many inorganic compounds (e.g., $SiCl_4$, PCl_3) and organic compounds (e.g., acid chlorides, amides, anhydrides and esters). Unlike the hydrolysis of salts, many of these reactions go to completion, sometimes slowly and sometimes rapidly.

═══════════════ **CHECKPOINT 17D: WATER** ═══════════════

1. Explain the following statements:

(*a*) Ponds freeze from the top downwards.

(*b*) Oil and water do not mix.

(*c*) Whisky and water do mix.

(*d*) Water is a good solvent for ionic compounds.

(*e*) Many classes of organic compounds dissolve in water.

(*f*) A solution of sodium sulphide has an unpleasant smell.

(*g*) It is dangerous to dissolve potassium cyanide in water.

(*h*) Aluminium sulphate solution is used in fire-extinguishers to generate carbon dioxide.

17.11 HARD WATER

Calcium and magnesium ions form 'scum' with soap

The presence of calcium and magnesium ions makes water hard. A soap such as sodium octadecanoate, $C_{17}H_{35}CO_2Na$, reacts with calcium and magnesium ions to form an insoluble 'scum' of calcium and magnesium octadecanoate:

$$Ca^{2+}(aq) + 2C_{17}H_{35}CO_2^-(aq) \rightarrow (C_{17}H_{35}CO_2)_2Ca(s)$$

Detergents lather, even in hard water

when all the calcium and magnesium ions have been precipitated, then the soap will lather. Detergents do not form scum in hard water. They are the sodium salts of sulphonic acids, and the calcium and magnesium salts are soluble:

$$C_{12}H_{25}-\langle\bigcirc\rangle-SO_3Na$$

Sodium dodecylbenzenesulphonate (a detergent)

17.11.1 TEMPORARY HARDNESS

Temporary hardness is caused by the hydrogen carbonates of calcium and magnesium...

Limestone reacts with rain water that contains dissolved carbon dioxide to form the soluble salt calcium hydrogencarbonate:

$$CaCO_3(s) + H_2O(l) + CO_2(aq) \xrightarrow[\text{boil hard water}]{\text{limestone and rainwater}} Ca(HCO_3)_2(aq)$$

...and is removed by boiling

The reaction is reversed by boiling. When hard water containing calcium hydrogen-carbonate is boiled, calcium carbonate is deposited, and the water becomes soft. Hardness caused by hydrogencarbonates is therefore called **temporary hardness**.

17.11.2   PERMANENT HARDNESS

Permanent hardness is caused by calcium and magnesium sulphates...

When water trickles over rocks containing calcium or magnesium sulphates, these minerals dissolve. The resulting hard water cannot be softened by boiling, and is described as **permanently hard water**. It can be softened by the following methods, which also remove temporary hardness.

WASHING SODA

...and is removed by washing soda...

Washing soda, $Na_2CO_3 \cdot 10H_2O$, is added to precipitate calcium and magnesium ions as insoluble carbonates.

ION EXCHANGE

Hard water can be softened by passing it slowly through a column containing sodium aluminium silicate (Permutit®). An exchange of ions takes place:

$$Ca^{2+}(aq) + 2NaAlSilicate(s) \underset{\text{Regenerating spent Permutit}^{\circledR}}{\overset{\text{Permutit}^{\circledR}\text{ softening hard water}}{\rightleftarrows}} 2Na^+(aq) + Ca(AlSilicate)_2(s)$$

When all the sodium ions in the Permutit® have been replaced by calcium and magnesium ions, the Permutit® must be regenerated. This is done by passing through it a concentrated solution of sodium chloride. After this solution has been washed out of the column, the Permutit® is ready for use. [See p. 174 for the theory of ion exchange.]

CALGON

...and by complexing agents

Calgon® and similar products are the sodium salts of polyphosphate ions. They are able to form insoluble complexes with calcium ions and magnesium ions, releasing sodium ions as they do so.

17.11.3   ESTIMATION OF TOTAL HARDNESS

See p. 273.

17.11.4   PURE WATER

Pure water

Pure water is made by distillation [see Figures 8.5, p. 155 and 8.10, p. 161]. Many laboratories find it more convenient to purify water by running it through a column containing ion-exchange resins [p. 174].

CHECKPOINT 17E: HARD WATER

1. Explain the following statements:

a) Washing soda softens water.

b) Washing soda makes water alkaline.

c) Stalactites and stalagmites and the 'scale' in kettles are all formed by the same chemical reaction.

(d) Mixed ion-exchange resins do not neutralise each other. (I.e., $R^+OH^- + H^+P^- \rightarrow R^+P^- + H_2O$ does not happen.)

(e) Biochemists prefer to use distilled water in their work, rather than de-ionised water.

1. Deuterium is an isotope of hydrogen. From your knowledge of the chemistry of hydrogen compounds, suggest methods for the preparation from D_2O of (a) D_2SO_4, (b) D_2S, (c) ND_3, (d) CD_4, (e) C_2D_2, (f) D_2, (g) DCl, (h) $Ca(OD)_2$.

2. 'In some ways, water is an abnormal liquid.' Give three properties which support this statement, and explain how these properties arise.

3. Distinguish between *hydration* and *hydrolysis* with examples. If you have started your study of organic chemistry, include some examples from this area.

†4. Discuss the acid–base reactions of water. (Do this question after studying Chapter 12.)

5. Describe with the aid of diagrams the bonding in (a) NaH, (b) CaH_2, (c) CH_4, (d) NH_3, (e) HCl, (f) HF. What reactions take place when these hydrides are added to water?

†6. (a) Briefly describe the chemical properties of hydrogen and explain why the chemistry of hydrogen is unique among the elements in some respects.

(b) Discuss the properties of the hydrides of *three* typical elements which are selected one from each of the s-, p- and d-blocks of the periodic table.

(c) Comment on the observation that the mass spectrum of CD_3OH, made by the reaction between CD_3ONa and HCl, shows only *one* major peak (at mass number 35) whereas the mass spectrum of a sample of HDO, made by the interaction of NaOD and HCl, shows *three* major peaks (at mass numbers 18, 19 and 20).

(O & C 82)

7. Consider the following processes involving hydrogen:

$$2H^+(aq) \underset{2}{\overset{1}{\rightleftharpoons}} H_2(g) \underset{4}{\overset{3}{\rightleftharpoons}} 2H^-(s) \text{ (in NaH)}$$

$$\downarrow 5$$

$$2H(g)$$

Thermochemical values ($kJ\,mol^{-1}$) for hydrogen are as follows: Electron affinity (ΔH_{EA}) = 78; Enthalpy of atomisation = 218; Ionisation energy = 1314; Enthalpy change of hydration of $H^+(g)$ = −1532.

(a) Suggest how each of the processes, 1 to 5, may be effected.

(b) Calculate the enthalpy changes for:

(i) Process 5.

(ii) Process 1, and comment on the value you obtain.

(iii) The process

$$H_2(g) \rightarrow 2H^-(g)$$

(c) Two of the single (monatomic) hydrogen species have radii of 148 and 37 pm. Explain which species have these radii.

(WJEC 81)

18

THE s BLOCK METALS: GROUPS 1A AND 2A

18.1 THE MEMBERS OF THE GROUPS

s block:
Group 1A; (core)ns
Group 2A; (core)ns²

The s block metals are the metals in Group 1A and Group 2A of the Periodic Table. They are called the s block elements because they occupy an area of the Periodic Table following the noble gases, an area in which the s orbitals are being filled. The s block metals have much in common, and aluminium in Group 3 shares many of their properties. The s block metals are listed in Table 18.1.

Group 1A			T_m/°C	T_b/°C	AR/nm	IR/nm	IE/kJ mol⁻¹	$E^\ominus$/V
Lithium	Li	3 1s²2s	180	1330	0.15	0.06	519	− 3.05
Sodium	Na	11 (Ne)3s	98	892	0.19	0.10	494	− 2.71
Potassium	K	19 (Ar)4s	64	760	0.23	0.13	418	− 2.93
Rubidium	Rb	37 (Kr)5s	39	688	0.24	0.15	402	− 2.92
Caesium	Cs	55 (Xe)6s	39	690	0.26	0.17	376	− 2.92
Francium	Fr	87 (Rn)7s	A radioactive element which is artificially made					
Group 2A								
Beryllium	Be	4 1s²2s²	1280	2770	0.11	0.03	2660	− 1.85
Magnesium	Mg	12 (Ne)3s²	650	1110	0.16	0.07	2186	− 2.37
Calcium	Ca	20 (Ar)4s²	840	1440	0.20	0.10	1740	− 2.87
Strontium	Sr	38 (Kr)5s²	768	1380	0.21	0.11	1608	− 2.89
Barium	Ba	56 (Xe)6s²	714	1640	0.22	0.13	1468	− 2.91
Radium	Ra	88 (Rn)7s²	A radioactive element					

TABLE 18.1 Physical Properties of s block metals

(Where T_m/T_b = melting/boiling temperature, AR = atomic radius, IR = ionic radius, IE = ionisation energy; for Group 2A, the sum of the first and second ionisation energies, $E^\ominus$ = standard electrode potential.)

The members of Groups 1A and 2A are all metals. They are silvery coloured and tarnish rapidly in air. They show relatively weak metallic bonding because they have only one valence electron. They differ in a number of ways from metals later in the Periodic Table:

In these metals, the metallic bond is relatively weak

1. They are soft: they can be cut with a knife.

2. Their melting and boiling temperatures are low.

3. They have low standard enthalpies (heats) of melting and vaporisation.

4. They have low densities. (Li, Na, K are less dense than water.) Group 2A, with two valence electrons, show stronger metallic bonding, which is reflected in their physical properties.

The types of metallic structure

Group 1A have a body-centred cubic structure, beryllium and magnesium have a hexagonal close-packed structure, barium has a body-centred cubic structure, and calcium and strontium have a face-centred cubic structure. [p. 114.]

The outer electron or electrons can be excited to a higher energy level. When they fall to a lower energy level, energy is emitted. For these metals, the energy is sufficiently low to have a wavelength in the visible spectrum [p. 27]. These elements therefore colour flames: Li—red, Na—yellow, K—lilac, Rb—red, Cs—blue, (Be—colourless), Mg—brilliant white, Ca—brick red, Sr—crimson, Ba—apple green.

The flame colours

Ionisation energies are low

The elements show constant oxidation numbers of +1 in Group 1A and +2 in Group 2A. The ionisation energy required for the process

$$M(g) \rightarrow M^{n+}(g) + ne^-$$

is low. The s electrons are shielded from the attraction of the nucleus by the noble gas core and are easily removed. As the size of the atoms increases down the groups, the electrons to be removed become more distant from the nuclear charge, and the ionisation energy decreases.

Metals are reducing agents. The highly negative $E^{\ominus}$ values show that s block metals are powerful reducing agents

Metals are reducing agents. The power of s block metals as reducing agents is shown by the vigour with which they reduce water to hydrogen. The standard electrode potential, $E^{\ominus}$ [p. 277], measures the tendency for the reduction process

$$M^{n+}(aq) + ne^- \rightarrow M(s)$$

to occur. A highly negative value for $E^{\ominus}$ indicates that the reverse process will take place: the metal atoms will form ions and electrons.

FIGURE 18.1 Trends in Ionisation Energy and Standard Electrode Potential

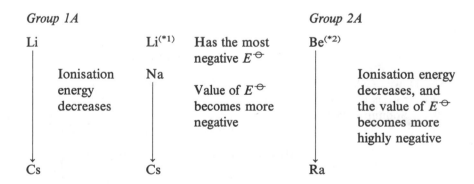

(*1) The small size of Li^+ gives it a high standard enthalpy of hydration. This is why Li^+ has the most negative value of $E^{\ominus}$ in the group. The small size of Li^+ enables it to polarise anions, and its compounds have some covalent character.

(*2) Be^{2+} is small and highly charged. It polarises anions, and beryllium compounds are mainly covalent.

18.2 USES

SODIUM

Sodium is used in some nuclear reactors

1. Molten sodium is used as a coolant in some types of nuclear reactor. Its high thermal conductivity and low melting temperature and the fact that its boiling temperature is much higher than that of water make sodium suitable for this purpose [p. 23].

...in electrical circuits...

2. Sodium wire is used in electrical circuits for special applications. It is very flexible and has a high electrical conductivity. The wire is coated with plastics to exclude moisture.

...in lamps...

3. Sodium vapour lamps are used for street lighting.

...as a reducing agent...

4. Sodium amalgam and sodium tetrahydridoborate, $NaBH_4$ [p. 741], are used as reducing agents.

5. Sodium cyanide is used in the extraction of silver and gold.

BERYLLIUM

Beryllium is used for making containers for uranium-238 as it does not absorb neutrons, and therefore does not become radioactive.

MAGNESIUM

Magnesium is used in alloys and in flares

1. Magnesium is alloyed with aluminium to make Duralumin® [p. 372].

2. Magnesium is used as a sacrificial anode to prevent iron from rusting [p. 502].

3. The intense white light of burning magnesium is used in flares and distress signals.

18.3 OCCURRENCE AND EXTRACTION

Extraction is by electrolysis

The s block metals are too reactive to occur uncombined. Group 1A are found as chlorides. Group 2A are found as chlorides, carbonates and sulphates. The metals are obtained by electrolysis of the molten chlorides [see Figures 18.2 and 18.3].

FIGURE 18.2

The Downs Cell for the Electrolysis of Molten Sodium Chloride

---

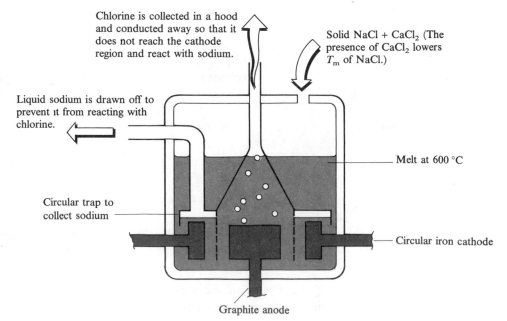

Chlorine is collected in a hood and conducted away so that it does not reach the cathode region and react with sodium.

Solid $NaCl$ + $CaCl_2$ (The presence of $CaCl_2$ lowers T_m of $NaCl$.)

Liquid sodium is drawn off to prevent it from reacting with chlorine.

Melt at 600 °C

Circular trap to collect sodium

Circular iron cathode

Graphite anode

*Chemical reducing agents
cannot reduce the oxides
of s block metals*

Electrolysis is an expensive way of obtaining metals. When possible, a reducing agent, such as carbon monoxide or carbon, is employed to reduce a metal oxide to the metal. The metals of Groups 1A and 2A are themselves such powerful reducing agents that their oxides cannot be reduced by chemical reducing agents. The metallurgist must resort to electrolysis [see Ellingham diagrams, p. 212].

FIGURE 18.3 Extraction
of Calcium by
Electrolysis

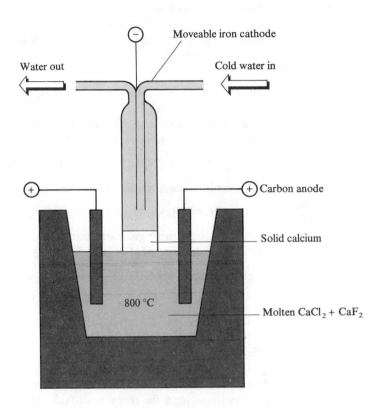

Moveable iron cathode

Water out

Cold water in

+ Carbon anode

Solid calcium

800 °C

Molten $CaCl_2$ + CaF_2

═══════════ **CHECKPOINT 18A: THE METALS** ═══════════

1. Refer to Table 18.1, p. 352.

(*a*) Considering Group 1A, explain why the radius of each ion is smaller than that of the atom.

(*b*) Explain why the ionic radius increases from Li^+ to Cs^+.

(*c*) Explain why, although Na^+ and Mg^{2+} have the same electron configuration, the radius of Mg^{2+} is only 0.07 nm.

(*d*) Which of the ions listed in Table 18.1 is likely to have the highest enthalpy (energy) of hydration? Explain your choice.

2. Explain why Group 2A metals have higher boiling temperatures and melting temperatures than Group 1A metals. What other physical properties differ?

3. For the process

$$M(s) \rightarrow M^{n+}(aq) + ne^-$$

list the individual steps by which this process can be considered to occur [see Born–Haber cycle, p. 205]. Name the enthalpy changes associated with each step.

4. Magnesium is mined as magnesium carbonate. Explain what is wrong with the following suggested method for extracting magnesium:

$$MgCO_3(s) \xrightarrow{\text{heat}} MgO(s) + CO_2(g)$$

$$MgO(s) + CO(g) \xrightarrow{\text{heat}} Mg(s) + CO_2(g)$$

5. Calcium chloride solution is the waste-product of the ammonia–soda process [p. 363]. Explain why calcium cannot be made by the electrolysis of this solution.

18.4 REACTIONS

18.4.1 REACTION WITH HYDROGEN

*Hydrides contain H⁻,
except for those of Be and
Mg*

The highly electropositive metals in Groups 1A and 2A react with hydrogen to form **hydrides**, in which hydrogen is the anion, H^- [p. 344]. Sodium hydride has a structure of the sodium chloride type. Beryllium and magnesium in Group 2A are less electropositive than the rest of the s block and form covalent hydrides.

18.4.2 REACTION WITH AMMONIA

Apart from beryllium and magnesium, all the s block metals react with liquid ammonia to form amides, e.g., $Na^+NH_2^-$, and hydrogen:

$$2Na(s) + 2NH_3(l) \rightarrow 2NaNH_2(s) + H_2(g)$$

*All (except Be and Mg)
form amides, containing
NH_2^-*

Metal amides react readily with water to give ammonia:

$$NaNH_2(s) + H_2O(l) \rightarrow NaOH(aq) + NH_3(g)$$

Group 1A react with gaseous ammonia in the same way, but Group 2A react to form a nitride (e.g., Mg) or a hydride (e.g., Ca):

$$3Mg(s) + 2NH_3(g) \rightarrow Mg_3N_2(s) + 3H_2(g)$$

$$3Ca(s) + 2NH_3(g) \rightarrow 3CaH_2(s) + N_2(g)$$

18.4.3 REACTION WITH WATER

*All (except Be and Mg)
react with cold water...*

Except for beryllium and magnesium, Groups 1A and 2A react with cold water to give hydrogen and the metal hydroxide. Group 1A metals are kept under oil to prevent them from reacting with water vapour in the air:

$$2Na(s) + 2H_2O(l) \rightarrow 2NaOH(aq) + H_2(g)$$

Magnesium reacts only slowly with cold water but with steam rapidly forms the oxide and hydrogen:

$$Mg(s) + H_2O(g) \rightarrow H_2(g) + MgO(s)$$

The vigour of the reaction with water is a good illustration of the increase in reactivity in passing down the groups. In short:

*...Reactivity increases
down each group*

Lithium reacts slowly with water, violently with acids.

Sodium reacts vigorously with water, violently with acids.

Potassium reacts so vigorously with water that the hydrogen formed catches fire.

Rubidium and caesium react even more violently.

Beryllium does not react with water, but reacts rapidly with acids.

Magnesium reacts slowly with water, fast with steam and fast with acids.

Calcium undergoes a steady reaction with water and a vigorous reaction with acids.

Strontium and barium are more reactive than calcium and about equal in reactivity to lithium.

18.4.4 REACTION WITH NON-METALLIC ELEMENTS

All the s block metals react with oxygen. They tarnish in air as they soon become coated with a film of oxide. They burn readily in air to form oxides, peroxides and superoxides [pp. 360 and 418] and caesium and rubidium inflame spontaneously in air.

They all react with halogens on heating to form halides.

FIGURE 18.4 Some Reactions of Sodium and its Compounds

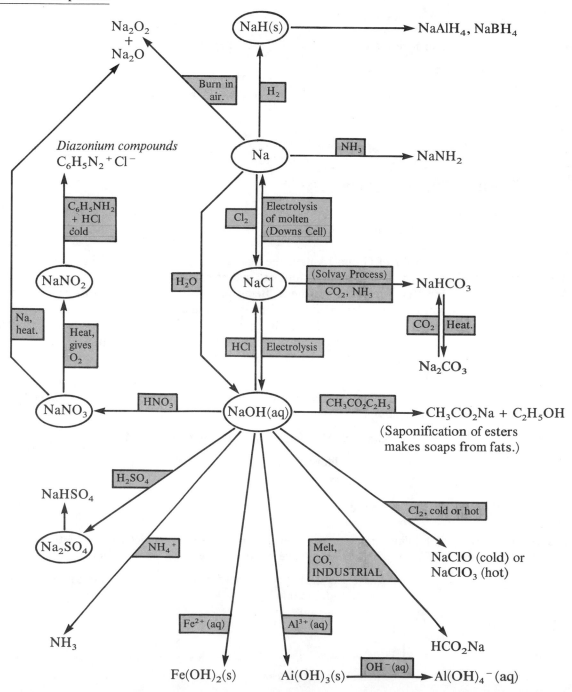

When heated in nitrogen, the metals that react are lithium in Group 1A and all of Group 2A except beryllium. Those nitrides are formed that have sufficiently high lattice enthalpies to compensate for the high ionisation enthalpies of the metal ions:

All react readily with oxygen and with the halogens...

$$6Li(s) + N_2(g) \rightarrow 2Li_3N(s)$$

On addition of water, nitrides give ammonia:

...some react with nitrogen...

$$Mg_3N_2(s) + 6H_2O(l) \rightarrow 3Mg(OH)_2(s) + 2NH_3(g)$$

...Group 2A react with carbon

Group 2A metals react with carbon at high temperatures to form carbides; Group 1A do not react. Calcium dicarbide, CaC_2, is the important carbide. Its structure is

$$Ca^{2+}(C\equiv C)^{2-}$$

and it reacts with water to give ethyne [p. 567]

$$H-C\equiv C-H$$

$$CaC_2(s) + 2H_2O(l) \rightarrow Ca(OH)_2(s) + C_2H_2(g)$$

18.5 COMPOUNDS

18.5.1 IONIC CHARACTER

The ions are hydrated, not hydrolysed, and form few complex ions

The salts of Group 1A have the highest amount of electrovalent character of any salts [p. 331]. The metal ions are hydrated in solution. The ions of Group 1A form very few complex compounds. Being small with high charge densities, the ions Li^+ and Be^{2+} polarise anions, and form compounds with some degree of covalent character [p. 331]. The cations formed by the rest of the elements in the s block, being larger, have smaller charge densities and do not polarise anions appreciably. The Born–Haber treatment [p. 205] explains why these metals prefer to form electrovalent compounds.

Li and Be compounds have some covalent character

Lithium and magnesium [see Fig. 29.6, p. 608] form covalent organometallic compounds:

$$2Li(s) + C_2H_5Cl(ether) \rightarrow C_2H_5Li(ether) + LiCl(s)$$

Chloroethane   Ethyllithium

18.5.2 SOLUBILITY

Solubilities are determined by two factors

Solubilities are not easy to explain because they are determined by two factors. Differences in ionic size have opposing effects on the two factors:

Small ions

> *Factor 1* Crystal lattices are hard to break up. (The lattice dissociation enthalpy is highly endothermic.)

> *Factor 2* Much energy is released when the ions are hydrated. (The enthalpy of hydration is highly exothermic.)

The salts of Group 1A are soluble because, since the cations are singly charged, the lattice dissociation enthalpies are not high. The trends in solubility down the group are shown below.

GROUP 1A   SALTS AND HYDROXIDES

Li

The solubilities of salts and hydroxides in Group 1A . . .

(a) Salts of small anions (e.g., F^-) and hydroxides *increase* in solubility because of the decreasing lattice dissociation enthalpy.

(b) Salts of large anions (e.g., SO_4^{2-}) *decrease* in solubility because of the decreasing enthalpy of hydration.

(c) The only insoluble salts are $NaZn(UO_2)_3(CH_3CO_2)_9$ and $K_2Na[Co(NO_2)_6]$.

Cs

GROUP 2A   SALTS AND HYDROXIDES

Compared with Group 1A

(a) Lattice enthalpies are higher.

(b) Enthalpies of solvation are higher.

These two factors have opposing effects on solubility.

Sparingly soluble compounds			Soluble salts	
MSO_4 and MCO_3		$M(OH)_2$	$M(NO_3)_2$	
Mg	*Decrease* in solubility is due to the decrease in the enthalpy of hydration (*Note 1*).	*Increase* in solubility is due to the decrease in the lattice dissociation enthalpy (*Note 2*).	Mg	*Increase*
Ca			Ca	
Sr			Sr	
Ba			Ba	*Decrease*

. . . and in Group 2A

TABLE 18.2

Note 1 The lattice enthalpies do not vary much. Since in size

Anion ≫ Cation

the differences in cation size do not greatly affect the lattice enthalpy.

Note 2 This outweighs the change in the enthalpy of hydration.

18.5.3   THERMAL STABILITY OF COMPOUNDS

Lattice enthalpies depend on two factors

The thermal stability of an ionic solid is measured by its standard lattice enthalpy [p. 205]. This depends on two factors:

Standard lattice enthalpy

Factor 1 The greater the charges on the ions, the greater is the attraction between them, and the greater is the lattice enthalpy.

Factor 2 The smaller the ions, the more closely they can approach in the lattice, and the greater is the lattice enthalpy.

NITRATES

Nitrates decompose when heated. Those of Group 1A (except Li) decompose to form nitrites (nitrates(III)) and oxygen:

$$2KNO_3(s) \rightarrow 2KNO_2(s) + O_2(g)$$

Nitrates of Group 2A and lithium decompose further to form the metal oxide, nitrogen dioxide and oxygen:

$$2Mg(NO_3)_2(s) \rightarrow 2MgO(s) + 4NO_2(g) + O_2(g)$$

Since the NO_2^- ion is smaller than the NO_3^- ion, the solid lattice of $Na^+NO_2^-$ is more stable than that of $Na^+NO_3^-$. The nitrite lattice is sufficiently stable to avoid further decomposition to the oxide. This is not the case for Group 2A nitrates. The O^{2-} ion is smaller and more highly charged than the NO_3^- ion. Oxides $M^{2+}O^{2-}$ have high lattice enthalpies, and nitrates of Group 2A metals therefore decompose to form oxides.

CARBONATES AND HYDROGENCARBONATES

The carbonates of Group 1A (except Li) are thermally stable. Those of Group 2A (and Li) decompose when heated to form the oxides:

$$CaCO_3(s) \rightarrow CaO(s) + CO_2(g)$$

The hydrogencarbonates of Group 1A are solids which decompose at 100 °C:

$$2NaHCO_3(s) \xrightarrow{100\,°C} Na_2CO_3(s) + CO_2(g) + H_2O(g)$$

This is why sodium hydrogencarbonate is used in baking powder.

The hydrogencarbonates of Group 2A are less stable. They exist only in solution. When their solutions are boiled, they decompose to form the carbonates [p. 348].

HYDROXIDES

Hydroxides follow the same pattern as carbonates. Those of Group 1A (except Li) are stable. Those of Group 2A and lithium are decomposed by heat to form oxides:

$$Mg(OH)_2(s) \xrightarrow{heat} MgO(s) + H_2O(l)$$

18.5.4 OXIDES

TABLE 18.3 Methods of Preparing Oxides

Group 1A		Group 2A	
$Li + O_2 \xrightarrow{heat} Li_2O$	*Normal oxide*	$M + O_2 \xrightarrow{heat} MO_2$ *Peroxide*	
$Na + O_2 \xrightarrow{heat} Na_2O_2$	*Peroxide*	$M + air$	
		MCO_3	
$K/Rb/Cs + O_2 \xrightarrow{heat} M_2O_2$ *Superoxide*			$\xrightarrow{heat} MO$ *Normal oxide*
(A large M^+ ion is more able to form a crystal structure with the large O_2^- ion.)		$M(NO_3)_2$	
		$M(OH)_2$	
To make the normal oxide			
$MNO_3 + M \xrightarrow{heat} M_2O + N_2$			

Normal oxides, peroxides and superoxides...

...all react with water...

...and all give OH⁻ (aq)

The s block metals form three kinds of oxides [see Table 18.3]. These are (a) **normal oxides** containing the anion O^{2-}, (b) **peroxides** containing the anion O_2^{2-} and (c) **superoxides** (or **hyperoxides**) containing the anion O_2^{-}. All the oxides react with water to give solutions of hydroxide ions. Peroxides give hydrogen peroxide in addition. Superoxides give hydrogen peroxide and oxygen in addition to hydroxide ions:

$$O^{2-}(s) + H_2O(l) \rightarrow 2OH^{-}(aq)$$

$$O_2^{2-}(s) + 2H_2O(l) \rightarrow 2OH^{-}(aq) + H_2O_2(aq)$$

$$2O_2^{-}(s) + 2H_2O(l) \rightarrow 2OH^{-}(aq) + H_2O_2(aq) + O_2(g)$$

18.5.5 HYDROXIDES

Group 1A hydroxides are strongly basic and are soluble...

...Group 2A hydroxides are less soluble

The manufacture of NaOH

All Group 1A hydroxides are soluble [p. 358]. All (except LiOH) are **deliquescent**, i.e., they absorb water vapour from the air and dissolve in it. The strongly alkaline solutions give Group 1A their name of the **alkali metals**.

The hydroxides of Ca, Sr and Ba dissolve in water [p. 359], but their lower solubilities make them weaker alkalis. The hydroxides of Be and Mg are insoluble in water. Beryllium hydroxide is amphoteric.

The manufacture of sodium hydroxide by the electrolysis of brine is illustrated in Figure 18.5. With most electrodes, the electrolysis of brine yields hydrogen at the cathode and chlorine at the anode [p. 239]. With a mercury cathode, hydrogen has a high overvoltage which raises the negative potential required for the discharge of hydrogen ions. At the same time, the discharge potential of sodium ions is lowered beçause discharged sodium atoms combine with mercury to form an amalgam. The amalgam passes to a chamber where it can react with water to form sodium hydroxide and hydrogen, with the regeneration of mercury, which is returned to the electrolysis cell:

$$Na^{+}(aq) + e^{-} + Hg(l) \rightarrow Na \cdot Hg(l)$$

$$2Na \cdot Hg(l) + 2H_2O(l) \rightarrow 2NaOH(aq) + H_2(g) + 2Hg(l)$$

FIGURE 18.5(a)
Manufacture of Sodium Hydroxide in the Mercury Cathode Cell (top) and Decomposer (bottom)

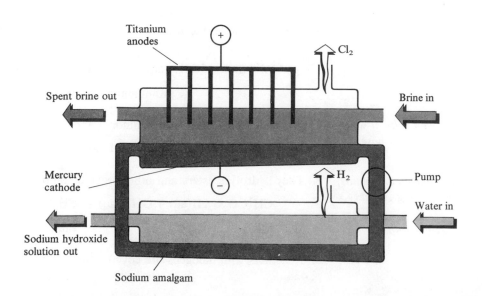

FIGURE 18.5(b)
Mercury Cathode Cells
at ICI, Runcorn

The reactions of alkalis are:

Alkalis react with acids...   **1.** They neutralise acids to form salts:

...metal cations...
$$Na^+(aq)\,OH^-(aq) + H^+(aq)\,Cl^-(aq) \rightarrow Na^+(aq)\,Cl^-(aq) + H_2O(l)$$

2. They precipitate insoluble metal hydroxides from solution:
$$Cu^{2+}(aq) + 2OH^-(aq) \rightarrow Cu(OH)_2(s)$$

3. They precipitate and then dissolve amphoteric hydroxides:

$$Al^{3+}(aq) + 3OH^-(aq) \rightarrow Al(OH)_3(s) \xrightarrow{\;OH^-(aq)\;} Al(OH)_4{}^-(aq)$$
$$\text{Aluminium hydroxide} \qquad \text{Aluminate ion}$$

...with salts of weak   **4.** They displace weak bases (ammonia and amines) from their salts:
bases...
$$NH_4{}^+(s) + OH^-(aq) \rightarrow NH_3(g) + H_2O(l)$$

...organic compounds,   **5.** They hydrolyse esters and other organic compounds:
e.g., esters...
$$CH_3CO_2C_2H_5(aq) + OH^-(aq) \rightarrow CH_3CO_2{}^-(aq) + C_2H_5OH(aq)$$

...and with halogens   Ethyl ethanoate   Ethanoate ion   Ethanol

6. Reactions with halogens are covered on p. 395; reactions with phosphorus on p. 434.

Compound	Uses
Sodium hydroxide (Potassium hydroxide is similar.)	(*1*) Titrimetric analysis of acids [p. 52]. (*2*) Hydrolysis of organic compounds, e.g., esters [p. 604], amides [p. 710] and nitriles [p. 712]. **Saponification**, the hydrolysis of the esters of glycerol to give soaps [p. 707]. (*3*) Manufacture of sodium methanoate, HCO_2Na [p. 694].
Calcium hydroxide (Limestone is heated in a lime kiln to give calcium oxide, *quicklime*, which is 'slaked' with water to give *slaked lime*, $Ca(OH)_2$.)	Called *slaked lime*; the aqueous solution is called *limewater*. (*1*) Treatment of fields which have become too acidic for healthy plant growth. (*2*) Mortar = Slaked lime + Sand + Water The paste sets and then hardens as it reacts with carbon dioxide in the air to form calcium carbonate. (*3*) Manufacture of calcium hydrogensulphite. The paper industry needs this to remove lignin from wood and leave cellulose, ready to be made into paper. (*4*) Reaction with chlorine to form bleaching powder, $Ca(OCl)_2 \cdot CaCl_2$. This is a useful source of chlorine, which it liberates readily on treatment with acid.

TABLE 18.4 Uses of Hydroxides

18.5.6 CARBONATES

Thermal stability has been mentioned on p. 358.

Hydrates of sodium carbonate

When sodium carbonate crystallises from solution, it appears as crystals of sodium carbonate-10-water, $Na_2CO_3 \cdot 10H_2O$. These crystals **effloresce** (give off water vapour) to form the monohydrate, $Na_2CO_3 \cdot H_2O$. When heated at 100 °C, the hydrates are converted into the anhydrous salt.

Sodium carbonate is used as a primary standard

Anhydrous sodium carbonate is used as a primary standard in titrimetric analysis [pp. 52 and 262]. Since it is deliquescent, anhydrous potassium carbonate cannot be used as a primary standard.

Commercial uses of sodium carbonate Manufacture of sodium carbonate by the ammonia–soda process

Sodium carbonate crystals (washing soda) are used in softening hard water [p. 350]. Anhydrous sodium carbonate is required in great quantities by the glass and paper industries and for the manufacture of soaps and detergents. The manufacture of sodium carbonate by the ammonia–soda process or Solvay process is illustrated in Figure 18.7. The process uses the materials coke, limestone, sodium chloride and ammonia. It employs a clever recycling of materials so that the only by-product is calcium chloride. Although some of this is used as a drying agent, it is largely a waste-product.

FIGURE 18.6 Reactions
of Calcium and
Magnesium

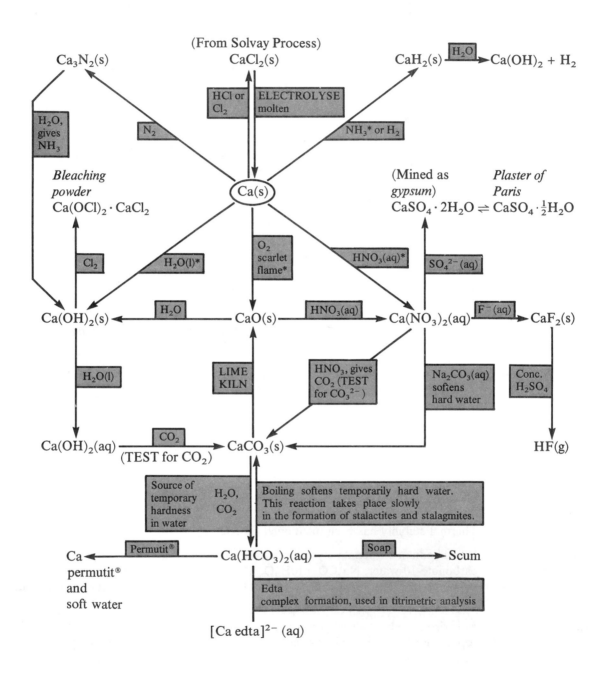

The reactions of magnesium are similar, but

* $Mg + NH_3 \rightarrow Mg_3N_2$
* Mg burns with a bright white flame.
* Mg reacts with steam to form MgO, but only slowly
  with water.
* Mg + cold, dilute $HNO_3(aq)$ gives $H_2(g)$, not $NO_2(g)$
  as in common.
  Mg forms Grignard reagents.

The Solvay process begins with heating limestone in a kiln. The energy required is supplied by burning coke. Limestone and coke are mixed in a kiln, and the coke is ignited:

The raw materials are plentiful...

$$CaCO_3(s) \rightarrow CaO(s) + CO_2(g) \qquad [1]$$

The carbon dioxide produced is passed up the Solvay tower, down which trickles ammoniacal brine (a solution of ammonia in brine). Cooling coils in the tower remove the heat of reaction. This reaction occurs:

$$NaCl(aq) + CO_2(g) + NH_3(g) + H_2O(l) \rightarrow NaHCO_3(s) + NH_4Cl(s) \qquad [2]$$

...Clever recycling of materials reduces costs

The precipitated sodium hydrogencarbonate is filtered off and washed free of ammonium chloride. It is heated to give sodium carbonate and carbon dioxide, which is sent to the Solvay tower:

$$2NaHCO_3(s) \rightarrow Na_2CO_3(s) + CO_2(g) + H_2O(g) \qquad [3]$$

The calcium oxide produced in [1] is slaked:

$$CaO(s) + H_2O(l) \rightarrow Ca(OH)_2(s) \qquad [4]$$

Calcium hydroxide produced in [4] and ammonium chloride from [2] react to give ammonia, which is dissolved in brine, and calcium chloride:

$$2NH_4Cl(aq) + Ca(OH)_2(s) \rightarrow 2NH_3(g) + CaCl_2(aq) + 2H_2O(l) \qquad [5]$$

FIGURE 18.7(a)

The Solvay Process for the Manufacture of Sodium Carbonate

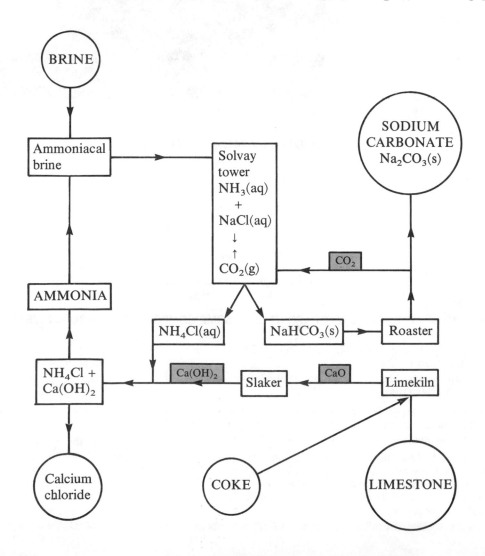

KHCO₃ is more soluble than NaHCO₃

The Postscript [p. 370] tells you more about the ammonia–soda process. Potassium carbonate is not made in the same way because potassium hydrogencarbonate is too soluble to crystallise at the bottom of the Solvay tower.

MgCO₃…

…CaCO₃ and its uses

Calcium and magnesium are mined as the carbonates, $MgCO_3$, *magnesite*; $MgCO_3 \cdot CaCO_3$, *dolomite*; and $CaCO_3$, *calcite* (Iceland spar); *marble*; *limestone*; *chalk* and *aragonite* in coral shells. Calcium carbonate is used in the ammonia–soda process, in the iron and steel industry, in the glass industry and in the manufacture of cement.

FIGURE 18.7(b)
Winnington ICI, including the Solvay Plant

18.5.7   HYDROGENCARBONATES

Hydrogencarbonates of Group 1A: preparation

The hydrogencarbonates of Group 1A (except Li) are obtained as solids by treating a solution of the carbonate with carbon dioxide or with a stoichiometric amount of acid, and then evaporating the solution:

$$CO_3^{2-}(aq) + H_2O(l) + CO_2(g) \rightarrow 2HCO_3^-(aq)$$

$$CO_3^{2-}(aq) + H^+(aq) \rightarrow HCO_3^-(aq)$$

The hydrogencarbonates decompose at 100 °C [p. 360].

CHECKPOINT 18B: SOME COMPOUNDS OF s BLOCK METALS

1. (*a*) Explain why CaO(s) has a higher standard lattice enthalpy than CaCO₃(s).

(*b*) Explain why MgCO₃(s) is more easily decomposed by heat than BaCO₃(s).

*2. What are an *oxide*, a *peroxide* and a *superoxide*? Why do you think that, on burning, lithium forms an oxide, sodium a peroxide and potassium a superoxide?

3. A pellet of sodium hydroxide weighing 0.254 g was left to stand in the air. It changed to a colourless liquid and then to colourless, transparent crystals. After some days it formed a white solid, weighing 0.394 g. Explain the changes that have occurred.

4. Explain why Mg(HCO₃)₂ and Ca(HCO₃)₂ make water hard, and how the water can be softened.

5. What is the net reaction in the Solvay process? Comment on the cost and availability of the raw materials used. What techniques are used to keep running costs to a minimum?

†Explain why the presence of NaCl reduces the solubility of NaHCO₃ in the Solvay tower [p. 270].

18.5.8 HALIDES

FLUORIDES

Halides... *Group 1A*
e.g., KF
KHF_2

Prepared from aqueous HF and aqueous carbonate or hydroxide in a polyethene vessel.
An excess of HF gives a hydrogenfluoride, e.g., KHF_2, containing the ion $[F\cdots\cdots H\cdots\cdots F]^-$.

Group 2A
e.g., CaF_2

Prepared by precipitation because of their low solubilities. *Fluorspar*, CaF_2, is mined, and is used as a source of fluorine [p. 392].

CHLORIDES

Group 1A
NaCl

Common salt, NaCl, is mined, e.g., in Cheshire.
It is used in the manufacture of sodium [p. 354], sodium carbonate [p. 363], sodium hydroxide [p. 361] and sodium chlorate(I) and sodium chlorate(V) [p. 395].

Group 2A
$CaCl_2$

Calcium chloride is the by-product of the ammonia–soda process. It is deliquescent, and is used as a drying agent except for ammonia and ethanol, with which it forms complexes. Crystallises as $CaCl_2 \cdot 6H_2O$.

$MgCl_2$

Crystallises as $MgCl_2 \cdot 6H_2O$. For hydrolysis, see p. 368.

$BeCl_2$

The structure is described on p. 97.

BROMIDES AND IODIDES

KBr

These salts are similar to chlorides.
Potassium bromide is used as a sedative and also in the production of silver bromide for photographic film.

KI

Potassium iodide is used for making solutions of iodine as the soluble complex ion, I_3^-(aq).

18.5.9 NITRATES

...nitrates... Sodium nitrate and potassium nitrate are mined. They are used as fertilisers.

The nitrates of Group 2A are made by the reaction of dilute nitric acid on the metal, metal oxide, hydroxide or carbonate.

The thermal stability of nitrates has been covered on p. 360.

18.5.10 SULPHATES

...sulphates **SODIUM SULPHATE**
Na_2SO_4

Prepared by titrating aqueous sodium hydroxide with sulphuric acid to pH 7. Crystallises as $Na_2SO_4 \cdot 10H_2O$.

$NaHSO_4$

If a second, equal volume of sulphuric acid is added, the hydrogensulphate is formed. It is acidic, dissociating partially to give H^+(aq) and SO_4^{2-} ions.

MAGNESIUM SULPHATE
$MgSO_4$

Mined as $MgSO_4 \cdot 7H_2O$.
Present in tap water, causing permanent hardness.
Used as the laxative Epsom salts.

CALCIUM SULPHATE
$CaSO_4$
$CaSO_4 \cdot 2H_2O$
$(CaSO_4)_2 \cdot H_2O$

Mined as *anhydrite*, $CaSO_4$, and as *gypsum*, $CaSO_4 \cdot 2H_2O$.
The latter give: laster of Paris when heated at 100 °C:

$$2CaSO_4 \cdot 2H_2O(s) \underset{\text{Add water}}{\overset{100\,°C}{\rightleftharpoons}} (CaSO_4)_2 \cdot H_2O(s) + 3H_2O(l)$$

Calcium sulphate-2-water   Plaster of Paris

On the addition of water, plaster of Paris expands slightly and sets to form calcium sulphate-2-water. Calcium sulphate is used to make the fertiliser ammonium sulphate.
Present in tap water, it is a cause of permanent hardness [p. 350].

18.6 HYDROLYSIS OF SALTS

Sodium carbonate solutions are strongly alkaline

The larger cations of Groups 1A and 2A are hydrated in solution. Their salts do not undergo hydrolysis unless they are the salts of weak acids, e.g., H_2S, HCN, H_2CO_3 [p. 269 and p. 349]. The strongly alkaline nature of sodium carbonate solutions gives them their detergent power. When a solution of sodium carbonate is added to a solution of copper(II) sulphate, a precipitate of basic copper(II) carbonate, $CuCO_3 \cdot Cu(OH)_2 \cdot xH_2O$, is obtained. A less strongly alkaline sodium hydrogencarbonate solution must be used to give a precipitate of copper(II) carbonate, $CuCO_3$.

Mg^{2+} salts are hydrolysed

When magnesium chloride solutions are evaporated, they do not give anhydrous magnesium chloride. Hydrolysis occurs, with the formation of the basic chloride:

$$MgCl_2(aq) + H_2O(l) \rightleftharpoons Mg(OH)Cl(s) + HCl(aq)$$

If evaporation is carried out in a stream of hydrogen chloride, the equilibrium is reversed, and magnesium chloride can be obtained.

The hydrolysis of aluminium salts is mentioned on p. 379 and that of transition metal salts on p. 507.

18.7 COMPARISON OF LITHIUM WITH MAGNESIUM

The diagonal relationship between lithium and magnesium

The reasons for this diagonal relationship are outlined on p. 331. Lithium ions, Li^+, being extremely small, are able to polarise anions and give compounds a high degree of covalent character. The features in which lithium resembles magnesium and differs from other alkali metals are listed in Table 18.5. Beryllium and aluminium are compared in Table 19.2, p. 383.

CHECKPOINT 18C: SALTS

1. Explain the following statements:
(a) Sodium carbonate solutions are alkaline.
(b) Calcium dicarbide is a compound of industrial importance.
(c) Calcium sulphate is a compound with (i) medical and (ii) agricultural importance.

(d) Potassium forms two fluorides.
(e) Sodium forms two sulphates.
(f) Beryllium chloride solutions are acidic.
(g) Beryllium chloride is a covalent substance in the vapour state.

Reaction	Group 1A	Lithium	Magnesium
Combination with O_2	Peroxides and hyperoxides	Li_2O	MgO
Combination with N_2	No reaction	Li_3N	Mg_3N_2
Action of heat on carbonate	No reaction	$Li_2O + CO_2$	$MgO + CO_2$
Action of heat on hydroxide	No reaction	$Li_2O + H_2O$	$MgO + H_2O$
Action of heat on nitrate	$MNO_2 + O_2$ Nitrite	$Li_2O + NO_2 + O_2$	$MgO + NO_2 + O_2$
Hydrogen-carbonates	Exist as solids	Only in solution	
Solubility of salts in water	Most salts are more soluble than those of Li, Mg.	Fluoride, hydroxide, carbonate, phosphate, ethanedioate are sparingly soluble.	
Solubility of salts in organic solvents	Halides only slightly soluble in organic solvents	Chloride, bromide, iodide dissolve in organic solvents.	

TABLE 18.5
A Comparison of Lithium with Magnesium and Group 1A

━━━━━━━━ **QUESTIONS ON CHAPTER 18** ━━━━━━━━

1. From your knowledge of the chemistry of Group 2A, predict what you can of the properties of radium. In particular, comment on

(a) the reaction of radium with water

(b) the solubility of its hydroxide and a likely value for the pH of its solution

(c) the solubility of the sulphate, chloride and carbonate

(d) the action of heat on the nitrate and carbonate.

2. Francium, the last member of Group 1A, is a short-lived radioactive element. From what you know of the chemistry of Group 1A, deduce what the properties of francium are likely to be, with respect to

(a) the nature of its hydride and the reaction of the hydride with water

(b) combination with nitrogen

(c) combination with oxygen

(d) the action of heat on the carbonate, hydrogencarbonate and nitrate

(e) the solubility of its salts in water and in organic solvents.

3. Comment on the statement, 'The metals of Group 1A are a similar set of elements, yet some gradation in properties can be observed from top to bottom of the group.'

4. Comment on the statement. 'In some ways, the members of Group 2A are a very similar set of elements; yet one can also look at them as a pair of similar elements (Be, Mg) and a trio of elements (Ca, Sr, Ba).'

5. 'The properties of the first members of Groups 1A and 2A are not typical of the groups as a whole.' Discuss this statement, pointing to some of the differences in the chemistry of the elements and also to the reasons for these differences.

6. Give an account of Groups 1A and 2A. Point out the similarities and differences between the two groups.

7. 'In some ways hydrogen resembles the alkali metals.' Discuss this statement.

8. Explain the meaning of the terms 'first electron affinity', 'first ionisation energy' and 'electronegativity'.
Discuss the trends in first ionisation energy and electronegativity in Groups IA and IIA of the Periodic Table.
In each case, relate your answer to the chemical properties of the elements.
Mention and explain three other trends which occur in one or both of these periodic groups.
Discuss briefly the relationships which exist between the two groups of elements.   (L80)

9. Some physical properties of the alkaline earth metals (Group II of the Periodic Table) are tabulated below.

Element	Atomic number	Electronic structure	Ionic radius M^{2+}/nm	First ionisation energy/ kJ mol^{-1}	Standard electrode potential $E^{\ominus}$/V
Beryllium	4	2,2	0.031	900	−1.85
Magnesium	12	2,8,2	0.065	736	−2.38
Calcium	20	2,8,8,2	0.099	590	−2.87
Strontium	38	2,8,18,8,2	0.113	548	−2.89
Barium	56	2,8,18,18,8,2	0.135	502	−2.90

Discuss the importance of *each* of these physical properties. Explain any observable trends in their magnitude and show how they may be used to interpret the chemical properties of the alkaline earth metals and their compounds.

The element radium, Ra (atomic number 88), is also an alkaline earth metal. Predict the approximate magnitude of the ionic radius, first ionisation energy and standard electrode potential for this element.

The radioactive decay of the isotope radium-226 can be represented by the equation

$$^{226}_{88}Ra \rightarrow X + \alpha.$$

State, giving reasons, in which group of the Periodic Table element **X** occurs.

(C 81)

10. The following is a simple account of the production of magnesium from sea water:
'Sea water is concentrated and calcium removed by the controlled addition of carbonate ions (**A**). The mixture is filtered and the clear solution treated by controlled addition of hydroxide ions (**B**). The precipitate is thermally decomposed (**C**) and the residue converted into anhydrous magnesium chloride (**D**). Magnesium is then obtained by electrolysis (**E**).

(*a*) Write equations, as simply as possible, for reactions **A**, **B**, **C** and **D**.

(*b*) What is meant by 'controlled addition'?

(*c*) Outline the process **E**, emphasising what you consider to be the important points (details of plant are *not* required).

(*d*) Suggest why it is difficult to produce magnesium by heating the residue from the thermal decomposition with carbon.

Mention a metallic ore reducible by carbon.

(*e*) Write *two* commercial uses for magnesium or its compounds.

(*f*) Group II metals can only form one oxidation state. Using a suitable electronic configuration diagram (outer shells only), show how they achieve this.

(*g*) Arrange the atoms of the Group II metals in order of decreasing atomic size (i.e. put the *smallest* last).

Comment on the stability of the metal cation as the group is ascended.

(*h*) Suggest why beryllium chloride hydrolyses completely, magnesium chloride hydrolyses partially, but barium chloride does not hydrolyse at all.

(*i*) State the 'flame' colours for calcium, strontium and barium. Suggest why magnesium compounds fail to produce a visible 'flame' colour.

(SUJB 82)

18.8 POSTSCRIPT: TWO INDUSTRIES BASED ON SALT

18.8.1 THE AMMONIA–SODA PROCESS

The ammonia–soda process [p. 363] was established in the UK in 1872. John Brunner and Ludwig Mond acquired the right to make sodium carbonate by the process which had been patented by Alfred and Ernest Solvay in Belgium in 1861. The Brunner–Mond plant, built at Winnington in Cheshire [see Figure 18.7(b)], is now a part of ICI. At nearby Northwich, Middlewich and Nantwich, there are vast salt deposits underground. Salt was mined to such an extent that Northwich was suffering from subsidence by 1890. Another method of extracting salt is employed now. A hole is drilled in the ground, and water is pumped in. After the water has been left underground long enough to become saturated with salt, brine is pumped out. Pillars of salt are left intact at intervals. The holes are left full of saturated brine so that the ground above will not subside.

The brine that is pumped out of the ground contains sulphates. These need to be removed before the brine goes to the ammonia–soda plant. 'Milk of lime', a suspension of calcium hydroxide, is added to precipitate sulphate ion as calcium sulphate:

$$Na_2SO_4(aq) + Ca(OH)_2(aq) \rightarrow CaSO_4(s) + 2NaOH(aq)$$

The solid calcium sulphate is dumped down a bore hole into one of the underground brine-filled caverns.

The ammonia–soda process discharges no pollutant gases into the air. The only by-product is calcium chloride. Some of this is sold for use as a drying agent. Most of it is waste. It is discharged into a short stretch of river which flows into the sea. The amount of chloride ion dumped in the sea annually is negligible compared with that present in sea water. Calcium ion is constantly added to sea water by the action of rainwater on limestone [p. 349]. Marine creatures remove calcium ion from the sea to build their shells. The contribution which the ammonia–soda process makes to the calcium content of the sea is negligible.

FIGURE 18.8
Winnington, Runcorn and Environment

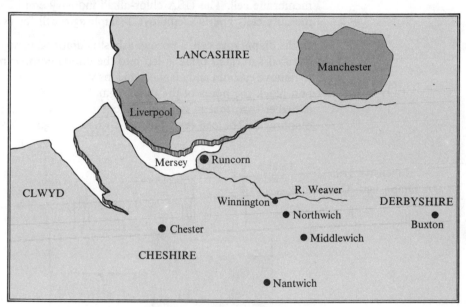

In Japan, the ammonia–soda process is operated to produce sodium carbonate, ammonium chloride and calcium oxide. Reaction [5] on p. 365 is eliminated. Ammonium chloride is a good fertiliser for rice. It is not used as a fertiliser on dry land crops because it leads to a build-up of chloride ion, which crops will not tolerate. Rice is grown in paddy fields which are continuously irrigated by large volumes of water. Ammonium ions are absorbed by the plants, and chloride ions are carried away by the flow of water. The calcium oxide produced is sold for use in the iron and steel industry and in the cement industry.

Japan has no salt deposits. It obtains salt from Australia. With plenty of sunshine and large areas of land available, Australia uses the method of solar evaporation to obtain salt from sea water. Japan also has access to sea water and has a sunny climate, but the rainfall is too high for solar evaporation to yield salt from sea water.

18.8.2 THE CHLOR–ALKALI INDUSTRY

Mercury cathode cells [see Figure 18.5, pp. 361–2] produce a large fraction of the chlorine and caustic soda that industry uses. ICI has a chlor-alkali plant at Runcorn [see Figure 18.8]. There are disadvantages to the process. Mercury is expensive, and it is poisonous. Although mercury is recirculated, some mercury escapes with the spent brine. This converts it into the soluble salt, mercury(II) chloride. It has been the practice to discharge the effluent into lakes and rivers. High levels of mercury build up in fish, which take in the mercury compounds but cannot excrete them. In 1970, 700 lakes in Canada and the USA were closed to fishermen because of the high mercury content of fish caught there. It was feared that the Minamata tragedy might be repeated. In 1958, 100 people died and thousands were maimed in Minamata in Japan. They had eaten fish with a very high mercury content caught in Minamata Bay. The source of mercury there was the effluent from a PVC plant, which used mercury(II) chloride as a catalyst in the polymerisation of ethene [p. 564]. The mercury in the North American lakes was traced to the chlor-alkali industry.

The industry immediately took steps to reclaim mercury from the effluent by precipitation as mercury(II) sulphide. Since this date, a larger share in the production of chlorine and sodium hydroxide and hydrogen has been taken by two types of electrolytic cell which do not use mercury. These are the diaphragm cell and the membrane cell. The USA chlor-alkali industry employs 2 diaphragm cells for every mercury cell. The UK employs 1 diaphragm cell for every 20 mercury cells.

In the **diaphragm cell**, a porous asbestos diaphragm separates the cathode and anode. Purified saturated brine is fed into the anode compartment. Purification is necessary to remove calcium and magnesium ions which would precipitate as insoluble hydroxides and block the pores of the diaphragm. The level of liquid in the anode compartment is higher than that in the cathode compartment so that brine will seep through the diaphragm. The cell reactions are shown in Figure 18.9.

FIGURE 18.9
A Diaphragm Cell

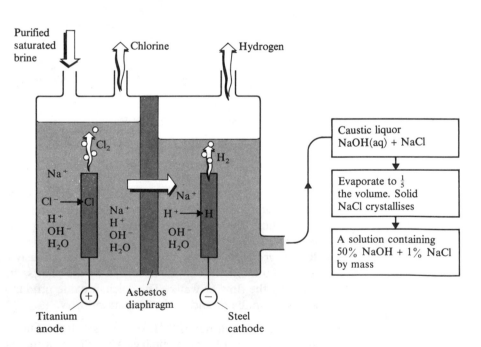

A newer type of cell is the **membrane cell**. The electrode reactions are the same as those in the diaphragm cell. The electrodes are separated by an ion-exchange membrane. Since chloride ions cannot pass through this membrane, the sodium hydroxide formed in the cathode compartment is free from sodium chloride. It is sold as a 35% by mass solution of NaOH or evaporated to give a 50% solution or solid sodium hydroxide.

Installations contain 50 to 100 single cells joined in series. The advantages of each type of cell are summarised in Table 18.6.

Type	Construction	Operation	Product
Mercury	Expensive to construct and fill	Mercury must be reclaimed from any effluent to avoid pollution.	High-purity NaOH(aq) is 50% by mass NaOH.
Diaphragm	Simpler and lower in cost	The diaphragm is replaced frequently. No pollution	Must be evaporated to remove NaCl and to concentrate the product
Membrane	Simple, low cost. Single cells are small and easily transported.	Needs brine of high purity. The membrane lasts for 2–3 years. No pollution	High-purity product. Evaporation increases the concentration.

TABLE 18.6 The Cells used for the Electrolysis of Brine

CHECKPOINT 18D: THE CHLOR-ALKALI INDUSTRY

1. Where do you think Winnington ICI obtains (a) limestone and (b) coal? [See Figure 18.8, p. 371]. Why is their product important?

2. What effluent is produced by the ammonia–soda process? Is it an environmental hazard? How well is the Winnington plant situated for effluent disposal?

3. (a) Why does Australia not mine salt?
(b) Why does Japan not produce calcium chloride as a by-product?

4. How does the electrolysis of brine in the mercury cathode cell compare with the ammonia–soda process with respect to pollution?

5. What are the two strategies that are being adopted by the chlor-alkali industry to reduce pollution by mercury?

6. What are the three important products of the chlor-alkali industry? What are they used for?

7. If nant is an old English word for south, what does wich mean? [See Figure 18.8, p. 371.]

19

GROUP 3B

19.1 THE MEMBERS OF THE GROUP

The elements in Group 3B of the Periodic Table are listed in Table 19.1.

Name	Symbol	Z	Electron Configuration
Boron	B	5	$1s^2 2s^2 2p$
Aluminium	Al	13	$(Ne)3s^2 3p$
Gallium	Ga	31	$(Ar)3d^{10}4s^2 4p$
Indium	In	49	$(Kr)4d^{10}5s^2 5p$
Thallium	Tl	81	$(Xe)5d^{10}6s^2 6p$

Table 19.1
The Elements
of Group 3B

The elements of Group 3 With the exception of boron, they are metals. With the exception of boron, they form ionic compounds by losing the s and p electrons to form an ion M^{3+}. They also form covalent compounds, through the promotion of an s electron to an unoccupied p orbital and the formation of three sp^2 hybrid bonds [p. 98]. Thallium forms some Tl^+ compounds. Aluminium is by far the most important of the metals.

19.1.1 BORON

Boron is non-metallic

Uses of boron

Boron is a non-metallic element, which is a poor electrical conductor. It is extracted from borates, e.g., sodium tetraborate, $Na_2B_4O_7 \cdot 10H_2O$ (borax). Boron is used as an absorber of neutrons in nuclear reactors [p. 23], and in the manufacture of steels. It is a reactive element, combining readily with oxygen, halogens and other elements to form covalent compounds.

19.1.2 GALLIUM, INDIUM AND THALLIUM

Gallium, indium and thallium are metals which resemble aluminium.

19.2 ALUMINIUM

19.2.1 USES

Aluminium has a low density and it is not corroded

Aluminium is the most abundant metal in the surface of the earth; yet it has been extracted in quantity only since the end of the nineteenth century. Every month new uses are being found for this metal which resists corrosion and which has a low density. Being completely non-toxic, it is ideal for packaging food [see Figure 19.2].

It is a good thermal
conductor and also a
reflector of heat

It is used in headlights...

...in electrical cables...

...and for the
construction of boats and
planes
Some parts of cars are
made of aluminium

Aluminium is amazing in being of a good thermal conductor which can also be used as a thermal insulator. As a thermal conductor, it is used for the manufacture of saucepans and cooking foil. The insulating property of aluminium arises from its ability to reflect radiant heat (i.e., infrared rays). Prematurely born babies are sometimes wrapped in aluminium foil, which keeps them warm by reflecting heat lost from the body. Firefighters in the USA wear suits which are coated with aluminium to reflect the heat from the fire and keep them cool [see Figure 19.3]. The polished surface of aluminium finds it a use in the reflectors of car headlights. Aluminium is a good electrical conductor and is replacing copper in overhead cables: to support aluminium cables, which are lighter, the pylons can be spaced at longer intervals.

Since it has a low density and is not corroded, aluminium has obvious advantages over iron as a manufacturing material. Pure aluminium is too soft for construction purposes, but alloys (e.g., Al/Mg and Al/Mg/Cu) have a higher tensile strength and are used for the construction of aeroplanes and small boats [see Figure 19.4]. More and more parts of cars are being made of aluminium: engine blocks can be cast from aluminium; piston heads are made of aluminium and encircled by steel rings, and rocker covers are made of aluminium. Some vehicles have an aluminium body, but the chassis needs the strength of steel, and the engine is of cast iron. When a car is scrapped because the iron in it has rusted, the aluminium parts are as good as new and can be recycled. Another advantage of incorporating aluminium parts is that the vehicle becomes lighter and consumes less petrol.

FIGURE 19.1
The Tanker is made of
Aluminium with an Inner
Tank of Stainless Steel.
The Door of the Inner
Tank can be seen

FIGURE 19.2
The Can is made of
Mild Steel or
Aluminium. The
Ring-pull Cap is of
Aluminium

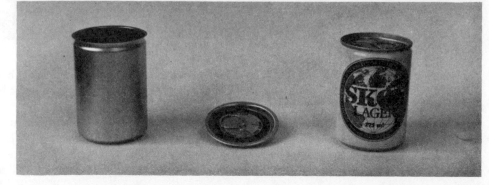

FIGURE 19.3 Firefighter
in an Aluminium Fabric
Suit

FIGURE 19.3 Firefighter
in an Aluminium Fabric
Suit

FIGURE 19.4 The A300
Airbus made from
Aluminium Alloys

*The metal is coated with
a film of aluminium
oxide, which is unreactive*

The use of aluminium alloys in construction is possible because the metal is coated
with a thin film of aluminium oxide, which resists attack by corrosive reagents.

*Anodised aluminium can
be dyed, and is used for
construction purposes*

When aluminium is **anodised**, that is, made the anode in an electrolytic cell of sulphuric
acid or chromic acid, the layer of oxide is thickened. When formed in this way, the
oxide is hydrated and can absorb dyes. Dyed anodised aluminium is used for door
frames and window frames, which are decorative as well as weatherproof.

19.2.2 EXTRACTION OF ALUMINIUM

Purification of Al_2O_3 requires separation from Fe_2O_3 and SiO_2

Aluminium is mined as the ore bauxite, aluminium oxide-2-water, $Al_2O_3 \cdot 2H_2O$, which contains silicon(IV) oxide and iron(III) oxide as impurities. Pure aluminium oxide is obtained from the ore by utilising the fact that it is amphoteric, whereas, of the impurities, silicon(IV) oxide is acidic and iron(III) oxide is basic. After being ground, the ore is treated as shown in Figure 19.6.

FIGURE 19.6
Purification of Aluminium Oxide

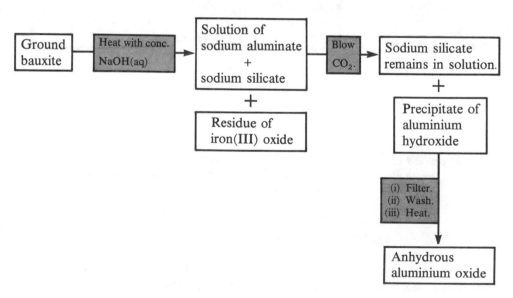

Electrolysis of Al_2O_3 in molten Na_3AlF_6 is used to extract aluminium

Aluminium is obtained from its oxide by electrolysis. Since the melting temperature of aluminium oxide (2050 °C) is so high that electrolysis of the molten oxide cannot be accomplished, a solvent must be used. The cell shown in Figure 19.7 contains a molten mixture of the ore cryolite, sodium hexafluoroaluminate, Na_3AlF_6, with calcium fluoride and aluminium fluoride added to lower its melting temperature.
Aluminium oxide is dissolved in this melt, and electrolysed at 850 °C to give aluminium and oxygen.

The electrode processes It is postulated that the equilibrium

$$Al_2O_3 \rightleftharpoons Al^{3+} + AlO_3^{3-}$$

gives rise to the electrode processes

Cathode $Al^{3+} + 3e^- \rightarrow Al$

Anode $4AlO_3^{3-} \rightarrow 2Al_2O_3 + 3O_2 + 12e^-$

FIGURE 19.7
A Hall–Héroult Cell
(5 m × 3 m × 1 m,
30 000 A, 5 V)

---

3 Carbon anode blocks replaced often because of oxidation to CO_2 by the O_2 evolved.

2 Molten Al is syphoned off.

1 Electrolyte: molten cryolite, Na_3AlF_6 ($+ CaF_2 + AlF_3$ to lower T_m) $+ Al_2O_3$. More Al_2O_3 is added periodically.

7 Steel case

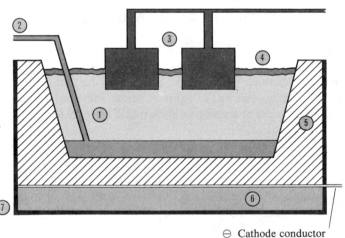

⊕ Anode conductor

4 Crust of solid Al_2O_3 protects molten Al from oxidation.

5 Carbon cathode

6 Insulation

⊖ Cathode conductor

The process consumes much electricity Electrolysis is an expensive method of obtaining metals from their ores. The reasons why it is used for aluminium follow from Section 9.3 on p. 212 and are described in the Postscript on p. 385.

19.2.3 THE METAL

A fresh surface of aluminium reacts rapidly with water vapour in the air Aluminium is high in the Electrochemical Series [see Table 12.2, p. 241], but its true reactivity is masked by the presence of a layer of aluminium oxide on its surface. This can be removed by reaction with mercury or mercury(II) chloride. When a fresh aluminium surface is exposed, it reacts immediately with water vapour in the air to form strands of aluminium hydroxide.

Aluminium reacts with non-metallic elements... Aluminium reacts directly with the non-metallic elements oxygen, sulphur, nitrogen, carbon (at a high temperature) and the halogens to form compounds Al_2O_3, Al_2S_3, AlN, Al_4C_3 and AlX_3. The high electropositivity of aluminium leads to its use in the thermit process for extracting other metals (e.g., chromium) from their oxides:

...with metal oxides... $$2Al + Cr_2O_3 \rightarrow 2Cr + Al_2O_3$$

Nitric acid makes aluminium 'passive' by increasing the thickness of the oxide film. Hydrochloric acid and sulphuric acid in fairly concentrated solutions react with aluminium to form salts:

... with HCl(aq) and

$$2Al(s) + 6HCl(aq) \rightarrow 2AlCl_3(aq) + 3H_2(g)$$

$H_2SO_4(aq)$ ...

$$2Al(s) + 6H_2SO_4(aq) \rightarrow Al_2(SO_4)_3(aq) + 3SO_2(g) + 6H_2O(l)$$

Alkalis react with aluminium to form an aluminate with the evolution of hydrogen:

... and with alkalis

$$2Al(s) + 2OH^-(aq) + 6H_2O(l) \rightarrow 2Al(OH)_4^-(aq) + 3H_2(g)$$

The reaction starts slowly and speeds up as the metal is stripped of its protective film.

19.2.4 THE Al³⁺ ION

The small size of the Al³⁺ ion gives its compounds a high degree of covalent character

The charge/radius ratio of the Al^{3+} ion is 3 units of charge/0.05 nm = 60 units of charge nm^{-1}. This ratio is high compared with the ratios Na^+ = 10 and Mg^{2+} = 30 units of charge nm^{-1}. It is similar to that of Be^{2+} = 66 units of charge nm^{-1}. The Al^{3+} ion is able to polarise the electron cloud of an anion [p. 76 and p. 331] to form a bond with a high degree of covalent character. Aluminium fluoride is ionic, the oxide is largely ionic, with some covalent character, and the anhydrous chloride, bromide and iodide have polar covalent bonds.

In aqueous solution, H_2O molecules coordinate to the Al³⁺ ion

In aqueous solution, the Al^{3+} ion is stabilised by the coordination of water molecules to form the complex hexaaquaaluminium(III) ion, $[Al(H_2O)_6]^{3+}$. The high ionisation energy required to form Al^{3+} ions is largely balanced by the energy released when Al—OH_2 bonds are formed. Six water molecules are distributed octahedrally about an Al^{3+} ion [see Figure 19.8(a)]. The coordination of water molecules occurs through the donation by the oxygen atom of a lone pair of electrons [see Figure 19.8(b)].

FIGURE 19.8
(a) $[Al(H_2O)_6]^{3+}$
(b) Polarity of O—H Bonds

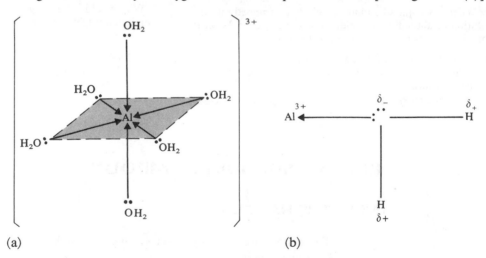

(a)

(b)

Coordinated water molecules tend to donate protons to free water molecules

In consequence, the electrons in the O—H bond move closer to the oxygen atom, and the hydrogen atoms have a greater degree of positive charge than in free water molecules. The partial positive charges on the hydrogen atoms attract bases, which may abstract protons from the coordinated water molecules. Water molecules of the solvent itself may function as bases:

This makes aluminium salts acidic in solution

$$[Al(H_2O)_6]^{3+} + H_2O \rightleftharpoons [Al(OH)(H_2O)_5]^{2+} + H_3O^+$$

$$[Al(OH)(H_2O)_5]^{2+} + H_2O \rightleftharpoons [Al(OH)_2(H_2O)_4]^+ + H_3O^+$$

As they do so, oxonium ions are formed, and the solution becomes acidic.

The process is called salt hydrolysis

These equilibria are examples of salt hydrolysis [p. 268]. If stronger bases are present (e.g., OH^-, CO_3^{2-}, S^{2-}), H_3O^+ ions are removed, and the hydrolysis equilibria move to the right. The stronger base can also remove the third proton:

$$[Al(OH)_2(H_2O)_4]^+ + OH^- \rightleftharpoons [Al(OH)_3(H_2O)_3](s) + H_2O$$

A precipitate of hydrated aluminium hydroxide appears. If an excess of hydroxide ions is added, protons are removed from the precipitate:

$$[Al(OH)_3(H_2O)_3](s) + OH^- \rightleftharpoons [Al(OH)_4(H_2O)_2]^-(aq) + H_2O$$

The solution formed contains diaquatetrahydroxoaluminate(III) ions:

$$[Al(OH)_4(H_2O)_2]^-$$

usually written $Al(OH)_4^-(aq)$ and called tetrahydroxoaluminate(III) ions or simply aluminate ions. If an acid is added to the solution, the above equilibrium is reversed, and hydrated aluminium hydroxide is precipitated.

$Al^{3+}(aq)$ ions are used as coagulating agents

The high charge/radius ratio of aluminium ions makes them useful coagulating agents. Aluminium ions are adsorbed on to the surface of negatively charged colloidal particles. The charges on the surfaces of the colloidal particles are reduced, and they are able to join to form a solid precipitate. Aluminium sulphate is used in water treatment plants to remove colloidal organic material from water [p. 189].

CHECKPOINT 19A: ALUMINIUM

1. (a) What are the advantages and disadvantages of iron and aluminium for use in car engines, car bodies and car bumpers?
(b) Why have aluminium saucepans become more popular than iron cooking pans?

2. When a sample of sodium sulphide is dropped into a solution of aluminium chloride, a gas is detected. What is this gas? Explain how it comes to be formed.

3. Two products are formed when hydroxide ions react with diaquatetrahydroxoaluminate(III) ions. Give their names and formulae.

4. Which of the following solutions do you think will be acidic?

$$FeSO_4, Fe_2(SO_4)_3, ZnSO_4, Cr_2(SO_4)_3$$

Explain your answer.

5. Why would it be dangerous to allow a solution of aluminium nitrate to come into contact with sodium cyanide?

6. Explain why aluminium finds use in (a) mirrors, (b) overhead electrical cables, (c) milk-bottle tops and (d) window frames.

19.3 ALUMINIUM COMPOUNDS

19.3.1 THE HALIDES

The halides are made by direct synthesis or by heating aluminium in a stream of hydrogen halide:

The synthesis of halides

$$2Al + 3X_2 \rightarrow 2AlX_3$$

$$2Al + 6HX \rightarrow 2AlX_3 + 3H_2$$

An apparatus which could be used for the preparation of anhydrous aluminium chloride is shown in Figure 19.9.

The dimerisation of AlX_3

Aluminium chloride sublimes as the dimer Al_2Cl_6. This reacts readily with water to give $Al^{3+}(aq)$ and $Cl^-(aq)$. The bromide and iodide also exist as dimers in the gaseous state. They dissociate into the monomers on further heating. Figure 19.10 shows the structures of the monomers and the dimers.

FIGURE 19.9
Preparation of
Aluminium Chloride

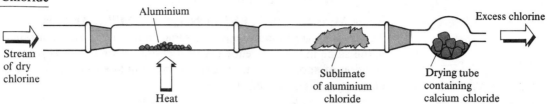

The chloride, bromide and iodide dissolve in covalent solvents like benzene. They are important catalysts in the Friedel–Crafts reactions [p. 582], because of their ability to act as Lewis acids (electron-acceptors) [p. 251 and p. 584].

FIGURE 19.10
The Trigonal Planar
Structure of $AlBr_3$ and
the Tetrahedral
Arrangement of Bonds
in Al_2Br_6

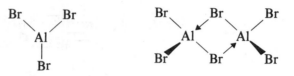

19.3.2 THE OXIDE AND HYDROXIDE

Aluminium oxide is amphoteric...

Aluminium oxide is mined as bauxite. It is amphoteric. With acids, it reacts to form salts of the $[Al(H_2O)_6]^{3+}$ ion (usually written Al^{3+} (aq)), and with alkalis it forms salts of the tetrahydroxoaluminate ion (the aluminate ion), $Al(OH)_4^-$.

...It is used as a catalyst...

Aluminium oxide (often called alumina) catalyses the cracking of alkanes in the petrochemical industry [p. 545] and the dehydration of ethanol to ethene [p. 623]. It is frequently used as the stationary phase in column chromatography [p. 171].

...and in chromatography

The hydroxide also is amphoteric...

Aluminium hydroxide is a white, gelatinous precipitate formed when ammonia solution is added to a solution of an aluminium salt. It is amphoteric. On standing, it loses water to become hydrated aluminium oxide.

...It is used as a mordant

Aluminium hydroxide is used in the dyeing industry. It is called a *mordant* (Latin: *mordere*, to bite) as it helps the dye to 'bite' the cloth. The cloth is soaked in a solution of aluminium sulphate, and alkali is added so that a gelatinous precipitate of the hydroxide is deposited in the fibres of the cloth. When the cloth is dipped into a vat of dye, the dye is adsorbed by the water occluded in the gelatinous precipitate and held by the charge on the Al^{3+} (aq) ion.

19.3.3 ALUMS

Alums are double salts

Aluminium sulphate crystallises as the hydrate, $Al_2(SO_4)_3 \cdot 18H_2O$. If equimolar amounts of aluminium sulphate and potassium sulphate are allowed to crystallise together, a **double salt**, aluminium potassium sulphate-24-water

$$K_2SO_4 \cdot Al_2(SO_4)_3 \cdot 24H_2O$$

crystallises. It is not a complex salt: in solution, it behaves as a mixture of the two salts. The double salt has a high lattice enthalpy (energy), making it less soluble than the simple salts from which it forms.

Crystals of alums are   Crystals of the same octahedral shape are obtained with other double salts. They
isomorphous   are called **alums** and have the general formula

$$M_2^I SO_4 \cdot M_2^{III}(SO_4)_3 \cdot 24H_2O$$

where M^I is Na^+, K^+, Rb^+ or NH_4^+ and M^{III} is Al^{3+}, Fe^{3+}, Cr^{3+} or Mn^{3+}.
The crystals are **isomorphous** (the same shape) because the arrangement of the ions
is the same in the different salts. If a crystal of one alum is placed in a saturated
solution of another alum, the crystal will continue to grow in size as the second
alum crystallises around the first.

FIGURE 19.11 Some
Reactions of Aluminium
and its Compounds

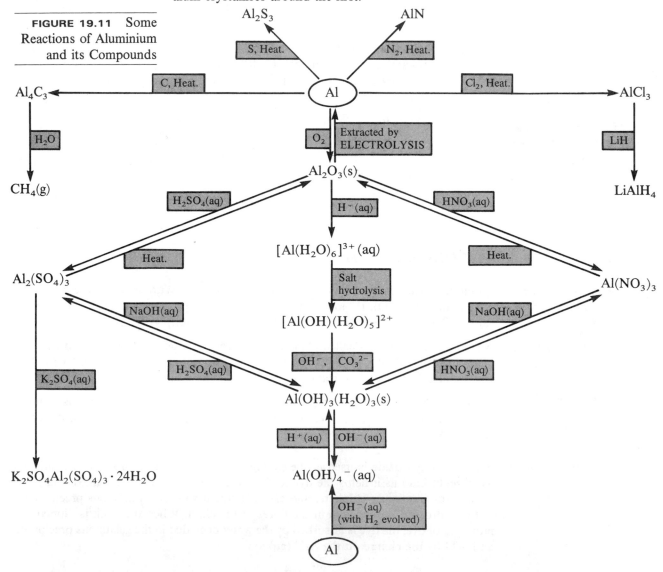

CHECKPOINT 19B: ALUMINIUM COMPOUNDS

1. From the reaction of aluminium with hydrochloric acid
can be obtained crystals of $AlCl_3 \cdot 6H_2O$. Why is it not
possible to obtain anhydrous aluminium chloride by heating
these crystals?

2. Why do aluminium halides act as Lewis acids? The ion
$AlBr_4^-$ is an intermediate in a Friedel–Crafts reaction.
Name the ion and describe the arrangement of bonds
about the central atom.

3. If you need to prepare aluminium hydroxide, why is it
better to add a solution of ammonia to a solution of an

aluminium salt, rather than to add a solution of sodium
hydroxide?

4. What do you think would happen if a crystal of aluminium
ammonium sulphate-24-water were suspended in a saturated
solution containing equimolar amounts of chromium(III)
sulphate and sodium sulphate?
Explain your answer.

5. *Potash alum*, $K_2SO_4 \cdot Al_2(SO_4)_3 \cdot 24H_2O$, can be made
by allowing an excess of aluminium to react with warm
potassium hydroxide solution until the evolution of hydrogen

ceases, decanting, acidifying with dilute sulphuric acid solution and setting the solution aside to crystallise. Explain what reactions take place and write equations for them.

6. Why do you think aluminium salts are used in 'styptic pencils' to assist blood to clot?

19.4 DIAGONAL RELATIONSHIPS

Beryllium, at the beginning of Group 2, resembles aluminium, the second element in Group 3. Boron, at the head of Group 3, resembles silicon, the second element in Group 4.

	Beryllium	Aluminium
Charge/radius ratio	66 charge units nm^{-1}	60 charge units nm^{-1}
Oxide and hydroxide	Covalent and amphoteric	Oxide has some covalent character. Both are amphoteric.
Halides	Covalent when anhydrous; ionise in water. Dimerise, e.g., Be_2Cl_4	AlF_3 is ionic. The rest are covalent when anhydrous and ionise in water. Dimerise, e.g., Al_2Cl_6
Reaction with alkali	Forms the complex $[Be(OH)_4]^{2-}$, the beryllate ion	Forms the complex $[Al(OH)_4]^-$, the aluminate ion
Complex formation with F^- ion	Forms the complex $[BeF_4]^{2-}$	Forms the complex $[AlF_6]^{3-}$

TABLE 19.2 Comparison of Beryllium and Aluminium

	Boron	Silicon
Elements	Both are non-metallic elements. Silicon is well-known as a semi-conductor. Boron is a very poor electrical conductor. The melting temperatures are high.	
Oxides	B_2O_3	SiO_2
	The oxides are acidic. They are both covalent, macromolecular, glassy solids.	
Halides	BX_3	SiX_4
	All the halides are covalent and completely hydrolysed by water.	
Hydrides	B_2H_6, B_4H_{10}, etc.	SiH_4, Si_2H_6, etc.
	The hydrides are reactive, flammable gases.	
Metals	Both elements form some binary compounds with metals, e.g., Ca_3B_2	e.g., Mg_2Si

TABLE 19.3 Comparison of Boron and Silicon

1. Describe the manufacture of aluminium, explaining the reasons for the conditions employed. What advantages does aluminium have over steel?
Give examples of the uses to which aluminium is put.

2. Aluminium fluoride boils at 1270 °C and aluminium bromide at 265 °C. Can you explain the large difference in boiling temperatures?

3. Describe the preparation of $Al_2(SO_4)_3 \cdot 12H_2O$ from aluminium. Why are solutions of this salt acidic?

4. How does aluminium react with (a) chlorine, (b) hydrochloric acid, (c) sodium hydroxide solution?

5. What is 'potash alum'? How can this salt be made, using aluminium as a starting material?

6. (a) Show, by means of a fully labelled diagram, how anhydrous aluminium chloride can be prepared in the laboratory. Use sodium chloride and aluminium as your starting materials, plus any other laboratory chemicals. Give equations for the reactions involved.

(b) Describe and explain what happens when the solution obtained by adding aluminium sulphate to water is treated with the following reagents added dropwise until in excess:
 (i) aqueous sodium carbonate,
 (ii) aqueous sodium hydroxide.

†(c) Describe and explain the mechanism of a reaction in aromatic organic chemistry which uses a halogen carrier. [See Chapter 28.]

(JMB 82)

7. Giving reasons for your choices, decide which of the following first ionisation energies and ionic radii correspond to sodium, magnesium and aluminium.

 First ionisation energies (kJ mol^{-1}) 494, 577, 736
 Ionic radii (nm) 0.050, 0.068, 0.100

Compare (a) the base strengths and solubilities of the hydroxides of these metals and (b) the bonding of the chlorides of sodium and aluminium.

Explain what constitutes an alum and outline a laboratory preparation of a pure crystalline sample of an alum starting from aluminium.

4.74 g of an alum which were found to contain 0.39 g of potassium and 0.27 g of aluminium were dissolved in 250 cm^3 of water. 20 cm^3 of the solution produced 0.373 g of $BaSO_4$ on quantitative precipitation. Calculate the number of molecules of water of crystallisation in the alum.
[$A_r(H) = 1.0$; $A_r(O) = 16$; $A_r(Al) = 27$; $A_r(S) = 32$; $A_r(K) = 39$; $A_r(Ba) = 137$]

(WJEC 82)

8. (a) Describe and explain an electrolytic process for the extraction of either sodium *or* magnesium *or* aluminium from a named raw material.

(b) Aluminium chloride finds many uses in preparative chemistry.
 (i) Describe how a sample of anhydrous aluminium chloride could be prepared in the laboratory.
 (ii) Using balanced equations and stating the appropriate conditions, give *two* organic reactions in which aluminium chloride is used as a catalyst.
 (iii) Explain the part played by this compound as a catalyst in organic reactions. Why is it advisable to use stoichiometric quantities of aluminium chloride under appropriate conditions for this catalytic role?

(c) (i) 0.30 g of gaseous aluminium chloride was found to occupy a volume of 61.0 cm^3 at a temperature of 315 °C and a pressure of 90.0 kPa. Calculate the relative molecular mass of the aluminium chloride at this temperature and pressure. Draw a structure for the aluminium chloride which is consistent with this value.
(ii) Explain with suitable diagrams, how the structure changes when the compound is vaporised beyond 400 °C.

(d) (i) What is the formula of the hydrated aluminium ion?
(ii) Why is the hydrated aluminium ion referred to as a Brönsted acid? Why is the hydrated aluminium ion a stronger acid than the hydrated magnesium ion?

(AEB 81)

[For (b)iii, see Chapter 28.]
[For (c), see Chapter 7.]
[For (d), see Chapter 12.]

9. From a consideration of the vertical trends in Group III of the Periodic Table and of the data given below, discuss the chemistry of thallium. Pay particular attention to the structure of, the bonding in, and the chemical properties of the halides and oxides.

	Aluminium	Thallium	Potassium
$M(g) \rightarrow M^+(g) + e^-$; $\Delta H/kJ\,mol^{-1}$	578	589	419
$M^+(g) \rightarrow M^{2+}(g) + e^-$; $\Delta H/kJ\,mol^{-1}$	1816	1970	3069
$M^{2+}(g) \rightarrow M^{3+}(g) + e^-$; $\Delta H/kJ\,mol^{-1}$	2700	2866	4400
Ionic radius of M^+/nm	–	0.140	0.133
Ionic radius of M^{3+}/nm	0.050	0.090	–
Covalent radius of M/nm	0.126	0.148	0.203
$M^{3+}(aq) + 3e^- = M(s)$; $E^{\ominus}/V$	−1.66	+0.72	–

(JMB 80, S)

19.5 POSTSCRIPT: THE ALUMINIUM PROBLEM

Aluminium is the most abundant metal in the surface of the earth, yet the metal was extracted from its ore only 150 years ago and remained a rarity for another 60 years. The Tsar of Russia gave his baby son an aluminium rattle to play with as it was more expensive than gold and he wanted the little fellow to have only the best.

Two compounds of aluminium have been known for hundreds of years. One is a beautiful crystalline substance which chemists called *alum* and which we know as aluminium potassium sulphate. The other compound known to chemists of old was a basic substance from which they could make alum. They called it *alumina* and deduced that it was the oxide of a metal, the metal aluminium, which no one had ever seen.

Sir Humphrey Davy tried in 1807 to obtain aluminium by electrolysing alumina, but he failed. In 1825, the Danish scientist, H C Oersted, decided to try a chemical method of isolating aluminium. He made aluminium chloride from alumina, charcoal and chlorine. Then he allowed aluminium chloride to react with potassium amalgam. Potassium was the most reactive metal he knew. He hoped that it would displace aluminium from aluminium chloride and leave him with potassium chloride and aluminium amalgam. He distilled what he hoped was aluminium amalgam under reduced pressure. Mercury distilled over. Left in the distillation flask was 'a lump of metal which in colour and lustre somewhat resembles tin'.

Wöhler was a German chemistry professor. He was unable to get Oersted's method to work and in 1827 he tried an extraction from alum. He obtained aluminium hydroxide by adding alkali to a solution of alum, and then converted it into aluminium chloride. Wöhler put aluminium chloride and potassium into a crucible, and, when he just touched it with a flame, a violent reaction occurred. To get rid of the excess of potassium, he plunged the crucible into a trough of water. To his delight, a grey powder floated to the surface, and after several experiments he had enough grey powder to melt it and get a lump of metal. This was the first certain extraction of aluminium.

A French chemist called Henri Sainte-Claire Déville decided to try to improve on Wöhler's extraction of aluminium, and turn it into a commercial process. Déville took aluminium oxide, charcoal and salt and heated the mixture in a stream of chlorine. The compound sodium hexachloroaluminate, Na_3AlCl_6, was formed. When he melted this compound with an excess of sodium, he obtained sodium chloride and, as he had hoped, molten aluminium, which he was able to run off and cast into ingots. Iron had been known for 5000 years, but this was the first time, in 1860, that aluminium had been obtained in quantity.

Now that there was a commercial process for making aluminium on a large scale, people began to discover what a useful metal aluminium is and to invent new uses for it. It was still an expensive metal because of the high cost of the large quantities of sodium used in its extraction. Since the reactive metals (Na, K, Ca, Mg) are extracted from their ores by electrolysis, it was natural that people should return to the possibility of using an electrolytic method for the extraction of aluminium. The big problem was that the oxide could not be melted to give a conducting liquid.

The problem was solved by a young American student called Charles Martin Hall. Hall's professor had worked under Wöhler, and his accounts of the early research work had made Hall impatient to contribute his chapter to the aluminium story. Hall bought some batteries and found an outhouse where he could experiment. He discovered that he could melt the ore *cryolite*, sodium hexafluoroaluminate, Na_3AlF_6, at 1000 °C, and then dissolve aluminium oxide in the molten cryolite to form a conducting solution. Hall electrolysed the melt, using graphite electrodes and, to his joy, he obtained aluminium at the cathode. The year was 1886, and Hall was just 21 years old! The Aluminium Company of America was founded to develop the process he had discovered.

Across the Atlantic, another young man was obsessed with the same problem. In France, Paul Héroult (23 years old) was working away in another makeshift laboratory. He arrived at the same conclusion as Hall. The electrolytic cell used in aluminium plants is called the Hall–Héroult cell [see Figure 19.7, p. 378].

Aluminium is mined in many parts of the world. The ore is called *bauxite*, as it has been mined for a long time in Baux in France. The production of aluminium uses a great deal of electricity. It takes 15 000 kilowatt hours to make 1 tonne (10^3 kg) of aluminium. Hydroelectric power is the most economical form of electricity, and aluminium plants are built close to a waterfall or dam which can be used as a source of power. Hydroelectric power stations are often in regions of great natural beauty; aluminium plants are unsightly. There can be a conflict of interest between conservationists, who want to preserve the landscape, and industrialists, who want to provide us with more and more of this amazingly useful metal.

CHECKPOINT 19C: ALUMINIUM

1. Write equations for (*a*) the sequence of reactions by which Oersted obtained aluminium from alumina and (*b*) the sequence of reactions by which Wöhler obtained aluminium from alum.

2. How did the Hall–Héroult method improve on Déville's method of extracting aluminium?

3. In the Hall–Héroult cell, cryolite, Na_3AlF_6, is present. Why is sodium not formed at the cathode? Why does aluminium not form at the cathode when molten cryolite is electrolysed without added alumina?

4. Calculate how many coulombs of electricity are required to produce 1 tonne of aluminium (1 tonne = 10^3 kg, $A_r(Al) = 27$, Faraday constant = $96\,500\,C\,mol^{-1}$).

20

GROUP 7B: THE HALOGENS

20.1 THE MEMBERS OF THE GROUP

The appearance and physical state of the halogens

The elements of Group 7B are fluorine, chlorine, bromine, iodine and astatine. Fluorine is a poisonous pale yellow gas, chlorine is a poisonous dense green gas (Greek: *chloros*, green), bromine is a caustic and toxic brown volatile liquid (Greek: *bromos*, stench) and iodine is a shiny black solid which sublimes to form a violet vapour on gentle heating. Astatine is radioactive and does not occur naturally. Some properties of the elements are shown in Table 20.1.

TABLE 20.1 Physical Properties of the Halogens

Property	*Fluorine* F	*Chlorine* Cl	*Bromine* Br	*Iodine* I
Proton number (Atomic number)	9	17	35	53
Outer electron configuration	$2s^22p^5$	$3s^23p^5$	$3d^{10}4s^24p^5$	$4d^{10}5s^25p^5$
Atomic radius/nm	0.072	0.099	0.114	0.133
Ionic radius/nm	0.136	0.181	0.195	0.216
Boiling temperature/°C	− 187	− 35	59	183
Standard enthalpy of dissociation/kJ mol^{-1} *of X*	79.1	122	111	106
Electron affinity/kJ mol^{-1}	333	348	340	297
Standard electrode potential/V	+ 2.87	+ 1.36	+ 1.07	+ 0.54
Electronegativity	4.00	2.85	2.75	2.20
Oxidation states	− 1	− 1,1,3,5,7	− 1,1,3,5,7	− 1,1,3,5,7
Standard enthalpy of formation of NaX/kJ mol^{-1}	− 573	− 414	− 361	− 288
Standard lattice enthalpy of NaX/kJ mol^{-1}	− 902	− 771	− 733	− 684

Van der Waals forces between halogen molecules

The halogens are a very similar set of non-metallic elements. Their name 'halogens' is derived from the Greek for 'salt formers'. They exist as diatomic molecules, X_2. The strength of the van der Waals forces between X_2 molecules increases as the number of electrons in the molecule of X_2 increases, i.e.

$$F_2 < Cl_2 < Br_2 < I_2$$

This explains the order of melting and boiling temperatures. In iodine, the van der Waals forces are strong enough to sustain a solid structure of iodine molecules (see Figure 6.13, p. 120].

20.2 IONIC BOND FORMATION

Halogens react to form X^- ions...

...in the formation of metal halides

The halogens form salts by accepting one electron to complete an octet of valence electrons, with the formation of a halide ion X^-. The steps involved in the formation of ionic halides are described in the Born–Haber cycle on p. 205. The relative ease with which the halogens form ionic halides is determined by three factors. They are

Factor (a) The standard bond dissociation enthalpy of $X_2(g)$:

Energy is absorbed in $X_2(g) \rightarrow 2X(g)$

Factor (b) The first electron affinity of X:

Energy is released in $X(g) + e^- \rightarrow X^-(g)$

Factor (c) The standard lattice enthalpy of the halide formed:

Energy is released in $M^+(g) + X^-(g) \rightarrow MX(s)$

Values of (a), (b) and (c) are listed in Table 20.1.

The X—X bond strength decreases down the group...

Factor (a) One would expect the X—X bond to decrease in strength as the size of X increases [see Figure 20.1]. This happens for the bond strength in

$$Cl_2 > Br_2 > I_2$$

[See Figure 20.2.]

FIGURE 20.1
The Bonding in a Halogen Molecule

Attraction of nucleus of X_b for electrons of X_a decreases with increasing bond length.

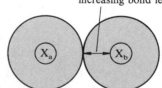

FIGURE 20.2 Standard Bond Dissociation Enthalpy for $X_2(g)$

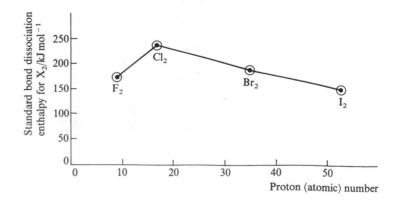

<table>
<tr><td>... with the exception that the F—F bond is weaker than expected</td><td>The F—F bond is unexpectedly weak [see Figure 20.3]. The reason for this is thought to be that the small size of the fluorine atom brings the lone pairs of electrons in F_2 closer together than in other halogen molecules.</td></tr>
</table>

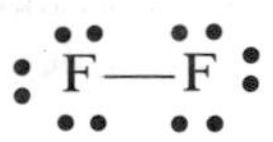

Repulsion between lone pairs of electrons weakens the F—F bond.

Electron affinity

Factor (b) Values of first electron affinities are shown in Table 20.1. You will find it interesting to plot electron affinity against atomic number.

Standard lattice enthalpy of halides

Factor (c) The standard lattice enthalpies of NaX are shown in Table 20.1. The F^- ion is so small that ionic fluorides have highly negative standard lattice enthalpies: much energy is given out when metal cations and fluoride ions form ionic lattices.

Fluorine brings out the highest oxidation states of metals [see Table 20.2]. Factor (*a*) is less endothermic for fluorine than for other halogens. Factors (*b*) and (*c*) are highly exothermic for fluorine and compensate for the ionisation enthalpy of the metal. Metals can therefore use their highest oxidation states in combination with fluorine. People working with fluorine use containers of copper or nickel steel. Although fluorine attacks these metals, the coating of metal fluoride which is formed resists further attack.

TABLE 20.2 The Reactions of some Metals with the Halogens

Metal	Fluorine	Chlorine	Bromine	Iodine
Most metals	Catch fire	Combine when heated		
Gold	Reacts	Does not react		
Platinum	PtF_6	$PtCl_4$	Does not react	
Silver	AgF_2	AgCl	Does not react	
Iron	FeF_3	$FeCl_3$	$FeBr_3$	FeI_2
Tin	SnF_4	$SnCl_4$	$SnBr_2$	SnI_2
Uranium	UF_6	UF_4	UBr_4	UI_4

20.3 COVALENT BOND FORMATION

Halogens form covalent bonds in X_2 and in compounds with other non-metallic elements

Halogens form covalent bonds in the diatomic molecules X_2 and in compounds with other non-metallic elements [see Table 20.3]. Fluorine differs from the rest of the group in being restricted to the L shell. The other elements have empty d orbitals which are close enough in energy to the occupied p orbitals to allow promotion of electrons from p orbitals to d orbitals [see Figure 20.4].

FIGURE 20.4 Electron Configurations in F, Cl and Cl*

2p $\boxed{\uparrow\downarrow\,\uparrow\downarrow\,\uparrow}$
2s $\boxed{\uparrow\downarrow}$
$1s^2$

F ground state can form one covalent bond.

3d $\boxed{}$
3p $\boxed{\uparrow\downarrow\,\uparrow\downarrow\,\uparrow}$
3s $\boxed{\uparrow\downarrow}$
$1s^2 2s^2 2p^6$

Cl ground state can form one covalent bond.

3d $\boxed{\uparrow}$
3p $\boxed{\uparrow\downarrow\,\uparrow\,\uparrow}$
3s $\boxed{\uparrow\downarrow}$
$1s^2 2s^2 2p^6$

Cl* first excited state can form three covalent bonds.

TABLE 20.3
The Reactions of some Non-metallic Elements with the Halogens

Element	Fluorine	Chlorine	Bromine	Iodine
He, Ne, Ar, N_2	Do not react			
Kr, Xe	React when heated	Do not react		
S	Reacts	Reacts when heated		
C	Reacts	Does not react		
O_2	Reacts	Does not react		
H_2	Reacts explosively, even in the dark at $-200\,°C$	Reacts explosively in sunlight; slowly in the dark below 200 °C	Reacts above 200 °C and at lower temperatures with Pt catalyst	Reacts to form an equilibrium mixture of H_2, I_2, HI
Other non-metallic elements	React readily. Fluorine brings out the highest oxidation state. Examples of fluorides which have no corresponding chlorides are SF_6, SiF_6, IF_7, XeF_6.			

Elements employ their highest oxidation states in combination with fluorine

Fluorine brings out the highest oxidation states of elements with which it combines. The reason is the high standard bond enthalpies of covalent bonds between fluorine and other elements. Much energy is given out when these bonds are formed. Atoms can promote electrons from shared orbitals to unoccupied orbitals because the energy required for promotion will be repaid when covalent bonds are formed.

CHECKPOINT 20A: BONDING

1. Why is fluorine more reactive than the other halogens?

2. Why do metals show their highest oxidation states in combination with fluorine? Explain why uranium forms UF_6 but the highest chloride is UCl_4.

3. Explain why sulphur combines with fluorine to form SF_6 but with chlorine forms SCl_4; why iodine forms IF_7 with fluorine but forms ICl_3 with chlorine; why bromine forms BrF_5 but the highest chloride is $BrCl_3$.

4. For iodine, draw electrons-in-boxes diagrams for the ground state and the first, second and third excited states. Say how many covalent bonds can be formed by iodine in each state. Explain why electron promotion occurs more readily in iodine than in chlorine.

5. For the halogens, referring to Table 20.1, plot:
(a) covalent bond radius against the proton (atomic) number of the halogen,
(b) ionic radius against proton number.

20.4 OXIDISING REACTIONS

Fluorine is the most powerful oxidant of the halogens

The oxidising power of the halogens is measured by the value of the standard reduction potential for the process

$$X_2(aq) + 2e^- \rightleftharpoons 2X^-(aq)$$

From the values in Table 20.1, p. 387 and Figure 20.5, you can see that fluorine is the most powerful oxidant of the group. Electron affinity is a less reliable measure of oxidising power as it relates to the process

$$X(g) + e^- \rightarrow X^-(g)$$

In most of their oxidising reactions, the halogens are reacting as X_2 molecules and forming hydrated ions, $X^-(aq)$.

TABLE 20.4 Some Oxidising Reactions of the Halogens

Oxidant	Reaction
All halogens	Sulphite, $SO_3^{2-} \rightarrow$ Sulphate, SO_4^{2-}
All halogens	Hydrogen sulphide, $H_2S \rightarrow$ Sulphur, S
Cl_2, Br_2	Thiosulphate, $S_2O_3^{2-} \rightarrow$ Sulphate, SO_4^{2-}
I_2	Thiosulphate, $S_2O_3^{2-} \rightarrow$ Tetrathionate, $S_4O_6^{2-}$ (This reaction is used for titrimetric analysis of iodine [p. 55].)
Cl_2, Br_2	Organic compounds are oxidised, e.g., methane, CH_4, ethyne, C_2H_2.
F_2	Reacts explosively with organic compounds
I_2	Does not react with organic compounds

FIGURE 20.5 Standard Electrode Potentials for the Halogens

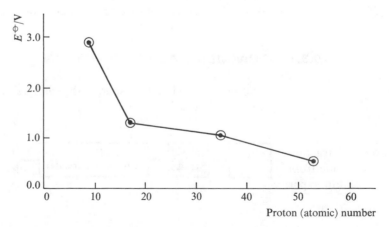

The values of $E^{\ominus}$ for fluorine and chlorine are

$$F_2(aq) + 2e^- \rightleftharpoons 2F^-(aq); \; E^{\ominus} = 2.87\,V$$

$$Cl_2(aq) + 2e^- \rightleftharpoons 2Cl^-(aq); \; E^{\ominus} = 1.36\,V$$

Standard electrode potentials measure oxidising power

Combining the values for $E^{\ominus}$ shows that the oxidation of chloride ions by fluorine will have a positive value of $E^{\ominus}$:

$$F_2(aq) + 2Cl^-(aq) \rightarrow 2F^-(aq) + Cl_2(aq); \; E^{\ominus} = 1.51\,V$$

This reaction therefore occurs: fluorine oxidises chlorides to chlorine. It also oxidises bromides and iodides to the halogens. Similarly, inspection of $E^{\ominus}$ values shows that chlorine will oxidise bromides and iodides, while bromine will oxidise iodides.

Chlorine water is often used as an oxidising agent

Fluorine is rarely used as an oxidising reagent as it is difficult to handle. Chlorine and its aqueous solution, 'chlorine water', are often used as oxidising agents. In chlorine water, there are two oxidising agents: chlorine and chloric(I) acid, HClO [p. 395].

CHECKPOINT 20B: REACTIVITY

1. Explain why bromine oxidises iodides, but bromides are oxidised by chlorine.

2. Some half-reaction equations are listed below:

(*a*) $SO_3^{2-}(aq) + H_2O(l) \rightarrow SO_4^{2-}(aq) + 2H^+(aq) + 2e^-$

(*b*) $H_2S(aq) \rightarrow S(s) + 2H^+(aq) + 2e^-$

(*c*) $X_2(aq) + 2e^- \rightarrow 2X^-(aq)$

(*d*) $S_2O_3^{2-}(aq) + 5H_2O(l) \rightarrow 2SO_4^{2-}(aq) + 10H^+(aq) + 8e^-$

(*e*) $2S_2O_3^{2-}(aq) \rightarrow S_4O_6^{2-}(aq) + 2e^-$

By combining half-reaction equations, obtain equations for the reactions of chlorine with (i) sulphites, (ii) hydrogen sulphide, (iii) a thiosulphate and (iv) obtain an equation for the reaction of iodine with a thiosulphate.

3. Choose two reactions which illustrate the gradation in reactivity down Group 7B.

20.5   OCCURRENCE AND EXTRACTION

The halogens are too reactive to occur free. The methods used for the commercial extraction of the halogens and the laboratory preparation of the halogens are related to their oxidising power. There is no oxidant that will oxidise fluorides to fluorine, and electrolysis is used. Chlorine also is made by electrolysis industrially. Sufficiently powerful oxidants will oxidise bromides and iodides to the halogens.

20.5.1   COMMERCIAL EXTRACTION

Fluorine is mined as fluorspar, CaF_2, and as cryolite, Na_3AlF_6. It is obtained as shown in Figure 20.6

FIGURE 20.6   Extraction of Fluorine from Fluorspar

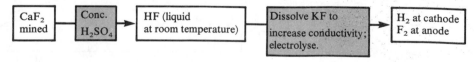

(The carbon anodes are replaced often because F_2 attacks them.)

Chlorine is obtained by the electrolysis of molten sodium chloride [p. 354] and the electrolysis of brine [p. 361].

Bromine is obtained from seawater as shown in Figure 20.7.

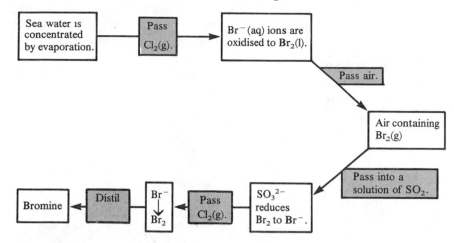

FIGURE 20.7 Extraction of Bromine from Sea Water

Iodine is mined as sodium iodate(V), $NaIO_3$, which is present in Chile saltpetre, $NaNO_3$. Sodium hydrogensulphite is employed to reduce iodate(V) ions to iodide ions. A reaction between iodide ions and iodate(V) ions produces iodine:

$$IO_3^-(aq) + 3HSO_3^-(aq) \rightarrow I^-(aq) + 3HSO_4^-(aq)$$

$$IO_3^-(aq) + 5I^-(aq) + 6H^+(aq) \rightarrow 3I_2(s) + 3H_2O(l)$$

20.5.2 LABORATORY PREPARATION

In the laboratory, chlorine, bromine and iodine are obtained by oxidation of the halides:

$$2X^- \rightleftharpoons X_2 + 2e^-$$

Figure 20.8 shows the preparation of chlorine by the action of concentrated sulphuric acid and manganese(IV) oxide on sodium chloride.

FIGURE 20.8
Laboratory Preparation of dry Chlorine (in fume cupboard)

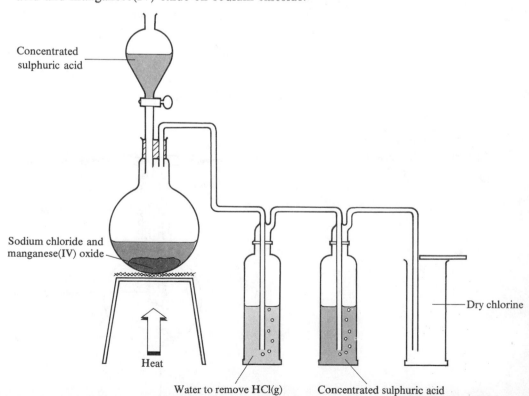

Concentrated sulphuric acid

Sodium chloride and manganese(IV) oxide

Heat

Water to remove HCl(g)   Concentrated sulphuric acid

Dry chlorine

Alternatively, concentrated hydrochloric acid can be used as a source of chlorine. It can be run from a tap funnel on to potassium manganate(VII) or warmed with manganese(IV) oxide or lead(IV) oxide. If the chlorine is not required dry, it can be collected over water, in which it is only slightly soluble.

FIGURE 20.9
Laboratory Preparation
of Bromine

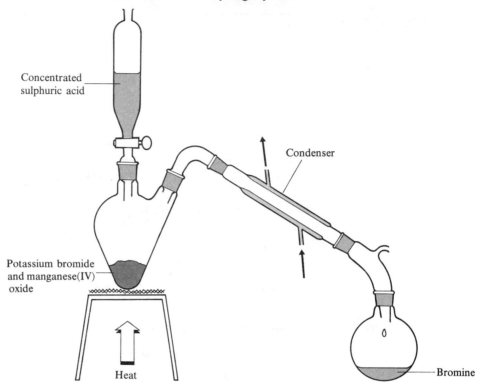

Concentrated
sulphuric acid

Condenser

Potassium bromide
and manganese(IV)
oxide

Heat

Bromine

The laboratory preparation of bromine is shown in Figure 20.9 and that of iodine in Figure 20.10. Concentrated sulphuric acid displaces the hydrogen halide from its salt, and manganese(IV) oxide oxidises it to the halogen [see Question 3, Checkpoint 20C].

A solution of chlorine in water is readily made by dissolving bleaching powder, $CaCl_2 \cdot Ca(ClO)_2$, in water and adding dilute hydrochloric acid:

$$Ca(ClO)_2 + 4H^+(aq) + 2Cl^-(aq) \rightarrow Ca^{2+}(aq) + 2H_2O(l) + 2Cl_2(aq)$$

FIGURE 20.10
Laboratory Preparation
of Iodine

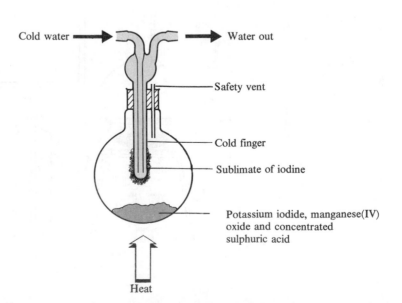

Cold water   Water out

Safety vent

Cold finger

Sublimate of iodine

Potassium iodide, manganese(IV)
oxide and concentrated
sulphuric acid

Heat

1. Draw an apparatus in which concentrated hydrochloric acid is run from a tap funnel on to potassium manganate(VII) in a flask fitted with a delivery tube. Illustrate how chlorine can be collected over water. What are the advantages of collecting chlorine (a) over water and (b) downwards?

2. What do the laboratory preparations of chlorine, bromine and iodine have in common? Why cannot fluorine be prepared in this way?

3. Concentrated sulphuric acid displaces HX from KX, where X = Cl, Br, I. Write equations for the reactions of concentrated H_2SO_4 with KCl, KBr and KI. One product of the reaction is $KHSO_4$.
Write a half-reaction equation for the oxidation of X^- to X_2.
Write a half-reaction equation for the reduction of acidified MnO_2 to Mn^{2+}.

Combine the half-reaction equations to give the equation for the oxidation of X^- by MnO_2 and acid.

4. Fluorine has an oxidation state of -1 only. Iodine has oxidation states of -1, $+1$, $+3$, $+5$ and $+7$ in I^-, ICl, ICl_3, IF_5 and IF_7. What is the reason for this difference between the two halogens?

5. Refer to Table 20.4, p. 391. Write equations for the reactions between (a) Cl_2 and Br^-, (b) SO_3^{2-} and Br_2. (This may be obtained by combining the half-reaction equations for $Br_2 \rightarrow Br^-$ and $SO_3^{2-} \rightarrow SO_4^{2-}$.)

6. What is the quickest way of preparing chlorine water?

7. Little is known of the chemistry of astatine. Describe the physical characteristics which you would expect for this halogen. What do you think would be formed in a reaction between sodium astatide and concentrated sulphuric acid?

20.6 REACTION WITH WATER

The value of $E^{\ominus}$ measures the strength of an oxidising agent

The standard electrode potentials of the halogens are shown in Table 20.1, p. 387 and in Figure 20.5, p. 391. The standard electrode potential (reduction potential) for oxygen is $+1.23\,V$.

$$O_2(g) + 4H^+(aq) + 4e^- \rightleftharpoons 2H_2O(l); \; E^{\ominus} = +1.23\,V$$

Fluorine and chlorine are thus capable of oxidising water, while bromine and iodine are not. Fluorine oxidises water to oxygen (and some hydrogen peroxide and ozone):

$$2F_2(g) + 2H_2O(l) \rightarrow 4HF(aq) + O_2(g)$$

Chlorine reacts slowly with water to form hydrochloric acid and chloric(I) acid, HClO:

$$Cl_2(g) + H_2O(l) \rightarrow HClO(aq) + H^+(aq) + Cl^-(aq)$$

Fluorine oxidises water

Chlorine slowly oxidises water, with the formation of HClO

Chloric(I) acid decomposes to give oxygen slowly, unless sunlight is present to accelerate the decomposition:

$$2HClO(aq) \rightarrow 2H^+(aq) + 2Cl^-(aq) + O_2(g)$$

In the presence of a reducing agent, chloric(I) acid acts as an oxidising agent:

$$HClO(aq) + H^+(aq) + 2e^- \rightleftharpoons Cl^-(aq) + H_2O(l)$$

20.7 REACTION WITH ALKALIS

Chlorine reacts with alkalis, to form ClO^- and ClO_3^- ...

Chlorine reacts faster with dilute alkalis than with water:

$$Cl_2(g) + 2OH^-(aq) \rightarrow Cl^-(aq) + ClO^-(aq) + H_2O(l)$$

The chlorate(I) that is formed may decompose to form a chloride and a chlorate(V):

$$3ClO^-(aq) \rightarrow 2Cl^-(aq) + ClO_3^-(aq)$$

... The reaction has commercial use

The decomposition is slow at room temperature but fast at 70 °C. A reaction like this, in which a species is simultaneously oxidised and reduced, is called a **disproportionation** reaction. Part of the ClO^- is oxidised to ClO_3^-, while the rest is reduced to Cl^-. These reactions are used commercially in the manufacture of sodium chlorate(I), NaClO, a widely used mild antiseptic (Milton®), and sodium chlorate(V), $NaClO_3$, a powerful weedkiller (Tandar®). Both chlorine and sodium hydroxide are products of the electrolysis of brine. If they are allowed to come into contact, sodium chlorate(I) is produced; if the temperature is raised, they can be made to form sodium chlorate(V).

Br_2 and I_2 react with alkalis to form the halate(V) ions

Bromine and iodine react with dilute alkalis, either cold or warm, to give a mixture of halide and halate(V):

$$3Br_2(aq) + 6OH^-(aq) \rightarrow BrO_3^-(aq) + 5Br^-(aq) + 3H_2O(l)$$

Alkalis + F_2 give F_2O or O_2

Fluorine reacts with cold, dilute alkalis to give oxygen difluoride, OF_2, and with warm concentrated alkalis to give oxygen:

$$2F_2(g) + 2OH^-(aq) \xrightarrow{\text{cold, dil.}} OF_2(g) + 2F^-(aq) + H_2O(l)$$

$$2F_2(g) + 4OH^-(aq) \xrightarrow{\text{warm, conc.}} O_2(g) + 4F^-(aq) + 2H_2O(l)$$

Arenes are chlorinated in the ring or in the side chain.
Alkynes, $HC\equiv CH \rightarrow C + HCl$; *Explosive*
Alkenes, $CH_2{=}CH_2 \rightarrow CHCl{=}CHCl \rightarrow CHCl_2CHCl_2$
Alkanes, $CH_4 \rightarrow CH_3Cl \rightarrow CH_2Cl_2 \rightarrow$ etc.

FIGURE 20.11　Some Reactions of Chlorine

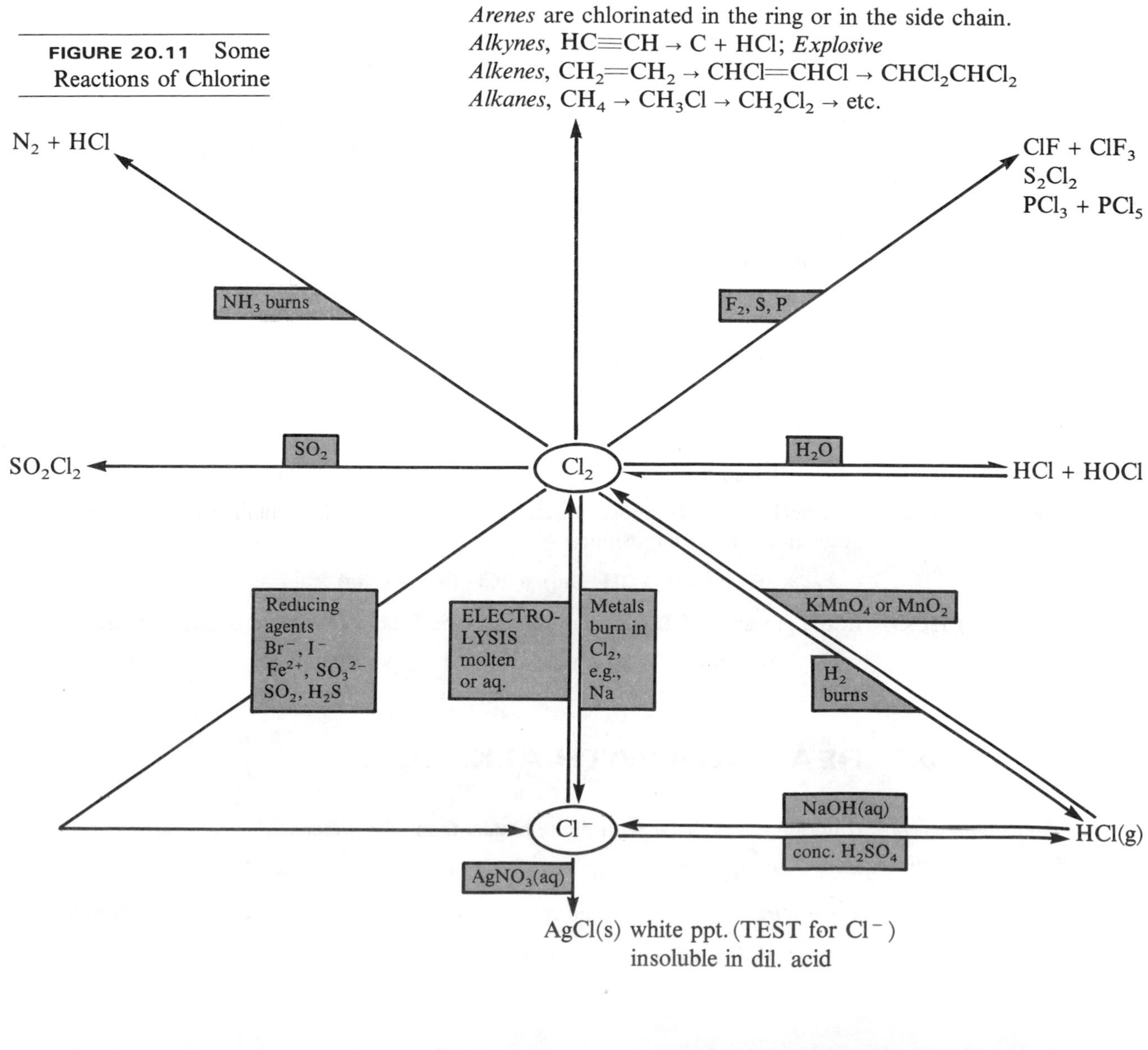

▰▰▰▰▰▰▰▰▰▰▰ CHECKPOINT 20D: REACTIONS ▰▰▰▰▰▰▰▰▰▰▰

1. Write the equation for the reaction of chlorine with water. Explain how the addition of an alkali speeds up the reaction.

2. Write the oxidation number of chlorine in each of the species

$$3\underline{Cl}O^-(aq) \rightarrow 2\underline{Cl}^-(aq) + \underline{Cl}O_3{}^-(aq)$$

Why is this reaction described as *disproportionation*?

3. Write the oxidation numbers of oxygen and fluorine in each of the species

(a) $2F_2(g) + 2\underline{O}H^-(aq) \rightarrow \underline{O}F_2(g) + 2\underline{F}^-(aq) + H_2O(l)$

(b) $2F_2(g) + 4\underline{O}H^-(aq) \rightarrow \underline{O}_2(g) + 4\underline{F}^-(aq) + 2H_2O(l)$

4. How do the standard electrode potentials of the X_2/X^- half-cells determine (a) the reactions of the halogens with water and (b) the methods of preparation of the elements?

20.8 METAL HALIDES

Metal + X_2 or HX(g) gives metal halide

Anhydrous metal halides are prepared by heating the metal in a stream of dry halogen or hydrogen halide [see Figure 19.9, p. 381]. Metals with variable oxidation states usually give the halide of a higher oxidation state with the halogen and the halide of a lower oxidation state with the hydrogen halide. Iron gives iron(III) chloride, $FeCl_3$, with chlorine, and iron(II) chloride, $FeCl_2$, with hydrogen chloride. Exceptions are the reaction of iodine with iron to form iron(II) iodide, FeI_2, and with copper to form copper(I) iodide, CuI.

Metal + HX(aq) gives hydrated metal halide

Hydrated metal halides are prepared by the reaction of a hydrohalic acid with a metal or its oxide, hydroxide or carbonate:

$$Zn(s) + 2HCl(aq) \rightarrow ZnCl_2(aq) + H_2(g)$$
$$CuO(s) + 2HCl(aq) \rightarrow CuCl_2(aq) + H_2O(l)$$

Test reactions of solid halides . . .

Reactions of solid halides				
Reagent	*Fluoride*	*Chloride*	*Bromide*	*Iodide*
Conc. H_2SO_4	HF(g)	HCl(g)	HBr(g) Br$_2$(g)	I$_2$(g)
Conc. H_2SO_4 + MnO_2	HF(g)	Cl$_2$(g)	Br$_2$(g)	I$_2$(g)

. . . and of solutions

Reactions of halide ions in aqueous solution				
Reagent	*Fluoride*	*Chloride*	*Bromide*	*Iodide*
$Pb(NO_3)_2$(aq)	PbF$_2$(s) white	PbCl$_2$(s) white	PbBr$_2$(s) cream	PbI$_2$(s) yellow
$AgNO_3$(aq) + HNO_3(aq)	No reaction	AgCl(s) white soluble in dil. NH$_3$(aq)	AgBr(s) pale yellow soluble in conc. NH$_3$(aq)	AgI(s) yellow insoluble in NH$_3$(aq)
Effect of light on AgX(s)	Not precipitated	Turns black	Turns yellow	No change

TABLE 20.5 Reactions of Metal Halides

Many halides crystallise with water of crystallisation, e.g., $MgCl_2 \cdot 6H_2O$, $AlCl_3 \cdot 6H_2O$. When the hydrates are heated in an attempt to obtain the anhydrous salt, they are often hydrolysed to give a basic chloride or hydroxide:

$$MgCl_2 \cdot 6H_2O(s) \rightleftharpoons MgCl(OH)(s) + HCl(g) + 5H_2O(g)$$

$$AlCl_3 \cdot 6H_2O(s) \rightleftharpoons Al(OH)_3(s) + 3HCl(g) + 3H_2O(g)$$

Some hydrates are hydrolysed on heating

The degree of hydrolysis depends on the degree of covalent character in the bonds [p. 333]. Anhydrous magnesium chloride can be obtained by heating the crystals of the hydrate in a stream of hydrogen chloride gas. This moves the equilibrium over to the left-hand side.

Most metal halides are soluble

Metal halides are soluble, except for the lead halides and the chloride, bromide and iodide of mercury(I) and silver. Some ionic fluorides differ in solubility from the corresponding chlorides. Calcium fluoride is insoluble, and silver fluoride is soluble. Some tests for metal halides are summarised in Table 20.5.

20.9   NON-METAL HALIDES

Combination with non-metallic elements

Halogens combine with many non-metallic elements. Fluorine brings out the highest oxidation states of elements with which it combines. In the case of other halides, the use of an excess of the halogen often results in the formation of the halide of the element in its highest oxidation state. For example, phosphorus reacts with chlorine to form phosphorus(III) chloride PCl_3, or phosphorus(V) chloride, PCl_5, depending on the proportion of chlorine used.

The halides of non-metallic elements are covalent and are hydrolysed by water, with the exception of tetrachloromethane, CCl_4 [p. 463]:

Halides of non-metallic elements are hydrolysed with the formation of HX(g)

$$SiCl_4(l) + 2H_2O(l) \rightarrow SiO_2(aq) + 4HCl(aq)$$

Silicon(IV) chloride   Silicon(IV) oxide

$$PCl_3(l) + 3H_2O(l) \rightarrow H_3PO_3(aq) + 3HCl(g)$$

Phosphorus(III) chloride   Phosphonic acid

20.9.1   HYDROGEN HALIDES

The hydrogen halides, hydrogen fluoride, hydrogen chloride, hydrogen bromide and hydrogen iodide, can be made by direct synthesis. The method works well for hydrogen chloride, which is manufactured industrially by burning a stream of hydrogen in chlorine:

$$H_2(g) + Cl_2(g) \rightarrow 2HCl(g)$$

Industrial manufacture of HCl(g)...

The reaction between hydrogen and fluorine is dangerously fast, and the reactions between hydrogen and bromine or iodine do not give a good yield.

Hydrogen fluoride is toxic and caustic. It is made by the action of concentrated sulphuric acid on calcium fluoride:

...and HF(g)

$$CaF_2(s) + 2H_2SO_4(l) \rightarrow 2HF(g) + Ca(HSO_4)_2 \text{ (in solution)}$$

Laboratory preparation of hydrogen chloride...

A similar displacement reaction between sodium chloride and concentrated sulphuric acid can be used for the laboratory preparation of hydrogen chloride [see Figure 20.12].

FIGURE 20.12
Laboratory Preparation
of Hydrogen Chloride

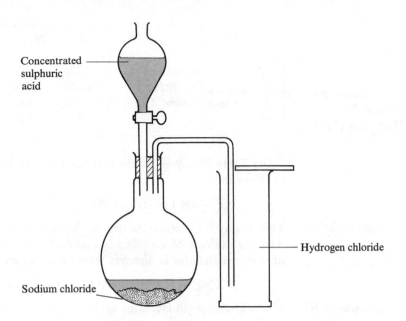

Concentrated
sulphuric
acid

Hydrogen chloride

Sodium chloride

. . . hydrogen iodide . . .   Hydrogen bromide and iodide cannot be made in this way because concentrated
sulphuric acid oxidises them to the halogens. Figure 20.13 shows the laboratory
preparation of hydrogen iodide. Phosphorus triiodide is formed and then hydrolysed.
Since water is added dropwise, the hydrogen iodide formed is in its gaseous state:

$$PI_3(l) + 3H_2O(l) \rightarrow H_3PO_3(aq) + 3HI(g)$$
Phosphonic acid

FIGURE 20.13
Laboratory Preparation
of Hydrogen Iodide

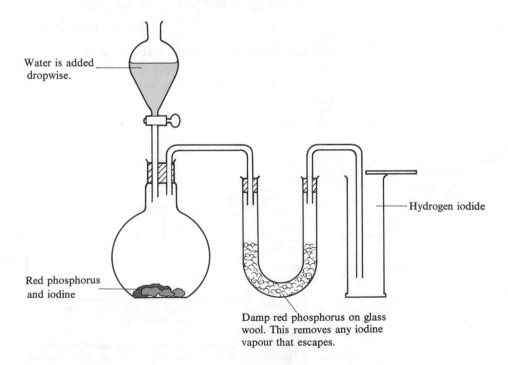

Water is added
dropwise.

Hydrogen iodide

Red phosphorus
and iodine

Damp red phosphorus on glass
wool. This removes any iodine
vapour that escapes.

. . . and hydrogen bromide   The preparation of hydrogen bromide is similar. Bromine is added dropwise to a
paste of red phosphorus and water. The apparatus shown in Figure 20.13 can be
used

	HF	HCl	HBr	HI
$\Delta H^{\ominus}_{Formation}/\text{kJ mol}^{-1}$	-270	-92	-36	$+26$
$\Delta H^{\ominus}_{Bond\ dissociation}/\text{kJ mol}^{-1}$	$+560$	$+430$	$+370$	$+300$
Boiling temperature/°C	20	-85	-67	-35
pK_a	3.25	-7.4	-9.5	-10

TABLE 20.6
The Physical Properties
of the Hydrogen Halides

From Table 20.6, you can see that the standard bond dissociation enthalpy decreases in the order

$$HF > HCl > HBr > HI$$

Thermal stability of HX   This is why the thermal stability of the compounds decreases in the same order. Another example of the effect of standard bond dissociation enthalpies is the ability of hydrogen halides in aqueous solution to act as proton-donors:

$$HX(aq) + H_2O(l) \rightleftharpoons H_3O^+(aq) + X^-(aq)$$

Acid strength of HX   The acidic strength increases in the order

$$HF \ll HCl < HBr < HI$$

Hydrogen fluoride is a much weaker acid than the rest. In concentrated solutions, the complex ion, HF_2^-, is formed [see Question 1, Checkpoint 20E].

HX(aq) show typical acid reactions   The hydrohalic acids react with metals above hydrogen in the Electrochemical Series and with metal oxides, hydroxides and carbonates. Hydrofluoric acid reacts with silicon(IV) oxide to form hexafluorosilicic(IV) acid:

HF reacts with glass

$$SiO_2(aq) + 4HF(aq) \rightarrow SiF_4(g) + 2H_2O(l)$$

$$SiF_4(g) + 2HF(aq) \rightarrow H_2SiF_6(aq)$$

$$\text{Hexafluorosilicic acid}$$

For this reason, hydrofluoric acid attacks glass, and glass apparatus cannot be used for work with hydrogen fluoride or fluorine. Fluorine does not attack perfectly dry glass, but since fluorine reacts with water to form hydrogen fluoride, it will attack damp glassware.

The characteristics of chlorides are summarised in Figure 20.14.

FIGURE 20.14 The
Nature of Chlorides

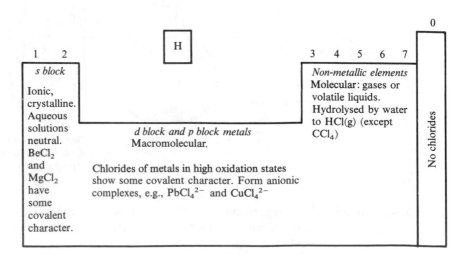

1. Write an equation for (a) the ionisation of HF in water and (b) the formation of the complex ion, HF_2^-. How does equilibrium (b) affect equilibrium (a)? Will the fraction of HF molecules that ionises be greater in dilute or concentrated solutions? Why is it that other hydrohalic acids do not take part in equilibria such as (b)?

2. What is meant by describing the hydrohalic acids as 'fuming' gases? Why do they behave in this way?

†3. The boiling temperatures of the hydrogen halides depend on intermolecular forces.

(a) Explain the order of T_b:

$$HCl < HBr < HI$$

(b) Explain why T_b of HF is so much greater than those of the other hydrogen halides. [p. 82].

4. Would you expect a solution of HAt to be a strong acid or a weak acid? What would you expect to see when a solution of silver nitrate was added to a solution of HAt(aq)?

5. Tetrachloromethane, CCl_4, does not react with water, but silicon(IV) chloride, $SiCl_4$, is rapidly hydrolysed. Explain this difference.

6. (a) In standard bond dissociation enthalpy

$$HF > HCl > HBr > HI$$

(b) In thermal stability

$$HF > HCl > HBr > HI$$

(c) In acid strength

$$HF \ll HCl < HBr < HI$$

Explain how statements (b) and (c) follow from statement (a).

7. Discuss the hydrogen halides with respect to their power as reducing agents.

8. Which of the halogens has (a) the smallest electron affinity, (b) the greatest oxidising power, (c) the greatest ability to form hydrogen bonds and (d) the weakest acid HX?

20.10 OXIDES

OF₂

Oxygen difluoride, OF_2, is made by passing fluorine into cold, dilute sodium hydroxide solution:

Oxides of chlorine and bromine

$$2F_2(g) + 2OH^-(aq) \rightarrow OF_2(g) + 2F^-(aq) + H_2O(l)$$

I₂O₅

Chlorine forms the acidic oxides, Cl_2O, ClO_2, Cl_2O_6 and Cl_2O_7, which are all unstable and explosive. The oxides of bromine are unstable above $-40\,°C$. Iodine forms the oxide, I_2O_5.

20.10.1 OXO-ACIDS AND THEIR SALTS

The oxo-acids of chlorine

The oxo-acids of chlorine are:

HClO	chloric(I) acid	
$HClO_2$	chloric(III) acid	Thermal stability increases
$HClO_3$	chloric(V) acid	Acid strength increases
$HClO_4$	chloric(VII) acid	Oxidising power increases

Some salts...

...NaClO   NaClO₃
KClO₃   KBrO₃
KIO₃

Some of the salts of these acids have been mentioned, e.g., NaClO and $NaClO_3$. Potassium chlorate(V), $KClO_3$, is used in matches and fireworks. Potassium bromate(V) is used as a source of bromine in volumetric analysis. Potassium iodate(V) is used as a primary standard in volumetric analysis [p. 52].

Chloric(VII) acid, HClO₄

Chloric(VII) acid, $HClO_4$, is a strong acid. It is prepared by the action of concentrated sulphuric acid on potassium chlorate(VII), and distilled off at reduced pressure to avoid decomposition.

20.11 INTERHALOGEN COMPOUNDS

A compound of two halogens is called an **interhalogen** compound. Some interhalogen compounds are listed below:

$$ClF(g) \qquad ClF_3(g)$$

$$BrF(g) \qquad BrF_3(l) \qquad BrF_5(l)$$

Interhalogen compounds $BrCl(g)$

$$ICl(l) \qquad ICl_3(s) \qquad IF_5(l) \qquad IF_7(g)$$

$$IBr(s)$$

They are covalent compounds and are made by direct synthesis.

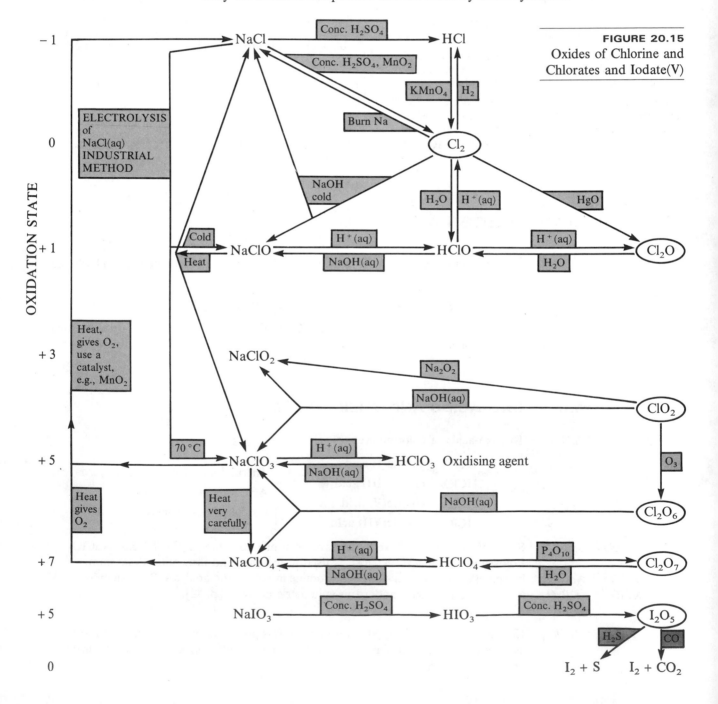

FIGURE 20.15
Oxides of Chlorine and
Chlorates and Iodate(V)

Iodine forms **polyhalide** ions. The triiodide ion, I_3^-, is formed when iodine dissolves in a solution of iodide ions:

Polyhalide ions such as I_3^-

$$I_2(aq) + I^-(aq) \rightleftharpoons I_3^-(aq)$$

If the free iodine present in the solution is used up in a reaction, triiodide ions dissociate to form more free iodine. A solution of triiodide ions therefore behaves like a solution of free iodine. Salts of polyhalide ions are stable if the metal cation is large: KI_3, RbI_3 and CsI_3 are formed, but not LiI_3 and NaI_3.

CHECKPOINT 20F: OXIDES AND INTERHALOGEN COMPOUNDS

1. Write the equation for the reaction of hydroxide ions with

(a) $Cl_2O \rightarrow ClO^-$

(b) $ClO_2 \rightarrow ClO_2^- + ClO_3^-$

(c) $Cl_2O_6 \rightarrow ClO_3^- + ClO_4^-$

(d) $Cl_2O_7 \rightarrow ClO_4^-$

Name each of these species.

2. Write the equation for the reaction of I_2O_5 with

(a) H_2S to give $S + I_2$

(b) CO to give $CO_2 + I_2$

Explain how reaction (b) could be used for the quantitative estimation of carbon monoxide.

3. Draw the spatial arrangement of bonds in ClF, ClF_3, BrF_5, IF_7 and ICl_4^-, ICl_2^-, HOI.

4. Write the formulae of three ions containing astatine.

5. Classify the following compounds as ionic or molecular or macromolecular: KBr, $AlBr_3$, BCl_3, CCl_4, $PbCl_2$, $PbCl_4$, $SiCl_4$, PCl_3, PCl_5. What tests would you make on the substances to substantiate your classification?

20.12 SUMMARY OF GROUP 7B

Summary of Group 7B

1. The ions X^- are readily formed. Electron affinity decreases down the group. All the halogens react vigorously with metals. With the s block metals, they form ionic compounds.

2. Covalent compounds and complex ions are formed by the halogens in combination with p block and d block metals. Ionic radii increase down the group, and polarisation of X^- therefore increases down the group. Fluorides are usually ionic.

3. The ions XO^-, XO_2^-, XO_3^- and XO_4^- are formed, except by fluorine.

4. The halogens form molecular compounds with non-metallic elements.

5. The halogens are all oxidising agents. Oxidising power decreases in the order

$$F_2 > Cl_2 > Br_2 > I_2$$

Reducing power of X^- decreases in the order

$$I^- > Br^-$$

Cl^- and F^- are not reductants.

6. Intermolecular forces increase down the group, and lead to a transition in physical state from fluorine (pale yellow gas) and chlorine (green gas), to bromine (brown volatile liquid) and iodine (shiny black solid which sublimes to a violet vapour).

20.13 DIFFERENCES BETWEEN FLUORINE AND OTHER HALOGENS

1. Fluorine is exceptionally reactive because of the low dissociation enthalpy (energy) of the F—F bond.

2. Fluorine compounds are very stable. There are two reasons for this:

Fluorine differs from the rest of the group...

...in its high reactivity...

(*a*) The small size of the fluorine atom allows it to form covalent bonds which are stronger than those of other halogens. Since covalent fluorides have highly negative standard enthalpies of formation, they are stable with respect to dissociation either into their elements or into compounds with lower oxidation numbers. This is why fluorine brings out higher oxidation numbers of other atoms than do the rest of the halogens.

...in the stability of its compounds...

(*b*) The small size of the fluoride ion gives ionic fluorides higher lattice enthalpies than the corresponding compounds of other halogens. Fluorides are often more ionic than other halides. AlF_3 has an ionic structure; $AlCl_3$ has a layer structure, and Al_2Br_6 and Al_2I_6 are molecular compounds. Fluorides sometimes differ in solubility from other halides. The high lattice enthalpy makes calcium fluoride insoluble, in contrast to other calcium halides. Silver fluoride differs from other silver halides in being soluble, owing to the high enthalpy of hydration of the fluoride ion. [p. 207.]

...in its single oxidation state...

3. Fluorine is restricted to an L shell of eight electrons. This means that its only oxidation state is -1, and it explains the inertness of many fluorine compounds, e.g., fluorocarbons.

...and in the strength of the hydrogen bonds which it forms

4. Being the most electronegative of elements, fluorine is able to form strong hydrogen bonds. Hydrogen bonding and the strength of the H—F bond are responsible for the weakly acidic nature of hydrogen fluoride.

5. In its reactions with water and alkalis, fluorine differs from other halogens. It forms no oxo-ions.

20.14 USES OF HALOGENS

Uses of fluorine include the manufacture of 'freon', PTFE

Fluorine is used in the manufacture of fluorohydrocarbons. The commercial refrigerant gas *freon* contains CCl_2F_2, $CClF_3$ and other chlorofluoromethanes. It is extremely unreactive. Tetrafluoroethene, $CF_2{=}CF_2$, can be polymerised to give poly(tetrafluoroethene), ${-}(CF_2{-}CF_2){-}_n$, PTFE. This material is a plastic which resists attack by most chemicals and is used in the chemical industry for the manufacture of corrosion-proof valves and seals. Its insulating property finds it a use for coating electrical wiring, and its low coefficient of friction finds it a use as a non-stick coating for cooking pans and for skis. Because of its volatility and its stability, it is used as the volatile component in aerosol cans.

Uses of chlorine include the manufacture of disinfectants...

Chlorine is used as a domestic bleach, and as a disinfectant in swimming baths. It is used in the manufacture of the gentle antiseptic, sodium chlorate(I), and the powerful weedkiller sodium chlorate(V).

...solvents...
...antiseptics...
...and insecticides

Chlorinated organic compounds find many uses. Tetrachloromethane, CCl_4, and trichloroethene, $CHCl{=}CCl_2$, are used as solvents for grease removal. Many antiseptics are chloro-compounds. TCP contains 2,4,6-trichlorophenol. Chloro-compounds, for example, DDT, have proved to be valuable insecticides [p. 406].

Bromine compounds are used as petrol additives

Bromine compounds, e.g., $C_2H_4Br_2$, are used as petrol additives. When tetraethyl-lead is added to petrol as an antiknock, lead oxides would be deposited in the cylinders in the absence of these organic bromo-compounds. They convert lead into volatile compounds, which are discharged through the vehicle exhaust into the air. Rising concern about the level of lead compounds in the air has led a number of governments, including that of the UK, to ban the use of lead compounds as antiknocks [p. 545]. The UK ban will take effect by 1990.

QUESTIONS ON CHAPTER 20

1. Outline the laboratory preparation of HF, HCl, HBr and HI. Compare these compounds with respect to (a) the strengths of the acids formed in aqueous solution and (b) thermal stability.

2. What trends are observed in Group 7B in the oxidising power of the halogens?

3. Give three general methods for preparing anhydrous halides of elements. Which of the halogens would you expect to stabilise the highest oxidation state of a particular element? Why?

4. 'In some ways, hydrogen resembles the halogens'. Discuss this statement.

5. Describe the manufacture of sodium hydroxide and chlorine from sodium chloride. Explain the principles involved. Why is there a big demand for these two substances?

6. Discuss the chemistry of Group 7B. Suggest reasons for the trends in properties. Include the reactions of the halogens with (a) metals, (b) hydrogen, (c) water, (d) sodium hydroxide solution, (e) solutions of alkali metal halides. Include also the reaction of sodium halides with sulphuric acid and the strengths of the halogen hydracids.

7. Predict the chemistry of astatine, the last member of Group 7B. How would you expect it to react with (a) hydrogen, (b) metals and (c) fluorine?

8. Listed below are some interhalogen compounds:

   ClF

   BrF   BrF$_3$ BrF$_5$ BrCl

       IF$_3$ IF$_5$ IF$_7$ ICl   ICl$_3$ IBr

(a) What is the oxidation state of chlorine in ClF and ICl?

(b) Which diatomic compound would you expect to have the strongest bonds? Explain your answer.

(c) Suggest why iodine does not form a compound ICl$_7$.

(d) Sketch the shapes of the molecules: BrF, BrF$_3$, BrF$_5$. What determines their shapes?

9. (a) Compare the elements chlorine and iodine with respect to:

 (i) the appearance of, and bonding in (1) the elements at room temperature and (2) their vapours;

 (ii) the solubility in water and in aqueous ammonia of their silver compounds;

 (iii) the preparation, ease of oxidation and reaction with water of their compounds with hydrogen.

(b) State how the reaction between iodine and aqueous thiosulphate can be used to determine the concentration of certain aqueous oxidising agents, e.g. iodate(V). (Details of apparatus and manipulations are not required.)

(AEB 82)

10. (a) Reactions of sodium chloride and iodide with concentrated sulphuric acid are represented below:

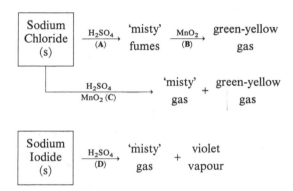

(i) Name the 'misty' gas in reaction **A**.

(ii) Name the 'misty' gas in reaction **D**.
   Complete the equations:

(iii) $Cl^-(s) + H_2SO_4(l) \xrightarrow{\text{(A)}} HSO_4^-(s) + \underline{\quad}(g)$

(iv) $I^-(s) + H_2SO_4(l) \xrightarrow{\text{(D)}} HSO_4^-(s) + \underline{\quad}(g)$.

(v) Name the green-yellow gas in reaction **B**.
Complete the equation:

(vi) $\underline{\quad}(g) + MnO_2(s) \xrightarrow{\text{(B)}} MnCl_2 + \underline{\quad}(g) + \underline{\quad}H_2O$.

(vii) Explain the role of the MnO_2.

(viii) Name the violet vapour in reaction **D**.
Complete the equation representing the formation of the violet vapour from the 'misty' gas:

(ix) $\underline{\quad}(g) + H_2SO_4(l) \xrightarrow{\text{(D)}} \underline{\quad}(g) + SO_2(g) + 2H_2O$.

(x) Explain the role of the H_2SO_4 in (ix).

(xi) Explain, in terms of the bond energies of gases (i) and (ii), why the violet vapour is formed without the use of MnO_2 but the green gas can only be formed if MnO_2 is used.

(*b*) Comment on the nature and extent (i.e. the position of equilibrium) of the reaction:

$$H_2 + X_2 \rightleftharpoons 2HX \ (X = \text{halogen})$$

for the different halogens in terms of the conditions under which they react with hydrogen. In what way does this support your explanation in (*a*) (xi) above?

(*c*) Suggest how the 'misty' gas (*a*) (ii) could be prepared without the violet vapour being produced with it.

(*d*) Outline an experiment, in aqueous solution, which compares the relative reactivities of the halogens.

(*e*) Mention TWO chemical properties of fluorine which are different from those of the other halogens.

(SUJB 82)

20.15 POSTSCRIPT: DDT

A research student called Othmar Geidler made DDT, in 1874. Sixty years later, another chemist called Paul Mueller repeated the synthesis so that he could try out DDT in his work on insecticides. He found that it was extremely poisonous to houseflies and other insects.

Dichlorodiphenyltrichloroethane

DDT, a chloro-compound with insecticidal properties, was used against mosquitoes and lice in the Second World War

This work was done in the 1930s, and when the Second World War started in 1939, chemists had found more uses for DDT. During a war, in addition to those killed in action, many people die of disease through lack of medical supplies, shortage of water, overcrowding and poor sanitation. The use of DDT in the Second World War helped to alleviate some of this misery. When the Allies landed in islands in the Pacific, they faced the danger of malaria as well as enemy forces. By spraying DDT from aeroplanes, they were able to wipe out the mosquito population, and remove the source of malaria. Dusting with DDT kept the troops free from the body lice which had plagued soldiers in earlier wars. After the Allies landed in Italy and occupied Naples, an epidemic of typhus broke out. To kill the lice which carry the disease, the whole area was sprayed with DDT and the population dusted themselves with DDT. Within days, the epidemic was over. In 1948, Mueller was awarded the Nobel prize, for discovering the life-saving properties of DDT.

DDT and related compounds were used as insecticides and herbicides by farmers

After the war, farmers welcomed DDT-related compounds to replace the non-selective insecticides and herbicides which they had been using. DDT killed insects but not

2,4-Dichlorophenoxyethanoic acid
(2,4-D)

2,4,5-Trichlorophenoxyethanoic acid
(2,4,5-T)

farm animals. The herbicides 2,4-D and 2,4,5-T killed weeds and left grass to grow. Dieldrin and aldrin were found to be even more potent than DDT:

Dieldrin

Aldrin

Some insects soon became resistant

As early as 1946, however, it began to appear that the new chloro-compounds were not a perfect solution to the insect problem. Some species of housefly soon became resistant to DDT. In the USA, the cotton farmers had been delighted with the way DDT had attacked the cotton boll weevil, but, by 1960, they were having to spray more and more frequently with higher and higher doses as the weevil became resistant to the insecticide.

These compounds are very stable, and remain in a treated area to be eaten by microorganisms, fish and birds and small animals

In 1962, Rachel Carson, in her book *The Silent Spring*, called for a halt to the widespread and indiscriminate spraying of insecticides and herbicides. These compounds are so stable that they persist for a long time in areas where they have been sprayed. If they are eaten by birds and animals, they cannot be excreted because they are insoluble in water, but they can be stored in the body because they are soluble in fat. In the spring of 1956, large numbers of birds were found dead in cereal-growing regions and were found to have a high content of dieldrin and aldrin, with which the cereals had been treated. Scientists began to investigate the spreading of insecticides and herbicides. They found DDT in penguins in the Antarctic, where the spray had never been used. They reasoned that if DDT is used in a malarial region, it will be taken up by small organisms. If these are eaten by a fish and the fish is eaten by a bird, the bird may carry DDT for hundreds of miles. DDT can concentrate up a food chain. If the content of DDT in seawater is 1×10^{-6} ppm, plankton can have a content of 3×10^{-4} ppm, and fish eating the plankton may contain 0.5 ppm. Birds of prey concentrate DDT further until they may contain 10 ppm. It is possible that fish and birds which have high DDT contents may be eaten by human beings. If DDT is so toxic to insects, can it be completely harmless to human beings? In 1964, the Advisory Committee on Poisonous Substances used in Agriculture and Food Storage recommended that dieldrin and aldrin should be used only when

The use of DDT has declined

drastic measures are needed. Restrictions have been placed on the use of DDT, and consumption of DDT has fallen to about half of what it was in 1964.

In 1967, when the USA forces were fighting in Vietnam, they sprayed 2,4,5-T over the jungle from the air. This compound is a defoliant, and, by destroying the leaves on the trees, it removed the cover of the Viet Cong.

Herbicides can contain dioxin, a deadly poison which causes chloracne and genetic defects

Present as an impurity was another compound, TCDP or dioxin. It is a deadly substance. Minute doses have killed test animals.

2,3,7,8-Tetrachlorodibenzo-para-dioxin
(TCDP or dioxin)

It had two dreadful effects on the Vietnamese and also on American servicemen who came into contact with it. One was a severe form of acne, called chloracne, covering large areas of the body. The other was an effect on the genes. Grossly deformed babies were born to Vietnamese mothers and to the wives of American soldiers. In 1976, an explosion at a chemical factory in Seveso in Italy sent dioxin into the air, and the effect on the population was the same. Children were dreadfully disfigured with chloracne, and mothers gave birth to deformed babies. In 1979, the USA and UK governments suspended the use of 2,4,5-T because of the danger of any dioxin which might be present as impurity.

CHECKPOINT 20G: DDT

†1. Write the formula for 2,4,5-trichlorophenol. Say how you could convert it to 2,4,5-T [p. 406]. Can you explain how this reaction can give rise to dioxin as an impurity?

2. Would you use these insecticides and herbicides if you were a farmer? If your answer is no, what would you do to protect your crops from weeds and pests? (Many people are trying to figure out an answer to this question!)

21
GROUP 6B

21.1 THE MEMBERS OF THE GROUP

The elements of Group 6B are listed in Table 21.1.

Elements		Z	Electron Configuration	AR/nm	IR/nm
Oxygen	O	8	2.6	0.074	0.140
Sulphur	S	16	2.8.6	0.104	0.184
Selenium	Se	34	2.8.18.6	0.117	0.198
Tellurium	Te	52	2.8.18.18.6	0.137	0.221
Polonium	Po	84	2.8.18.32.18.6	0.152	–

TABLE 21.1
The Elements
of Group 6B

There is a transition down the group from non-metallic to metallic properties. All except polonium form E^{2-} ions. All form covalent hydrides, H_2E. Oxygen differs in many respects from the rest of the group [p. 429].

21.2 OXYGEN

Oxygen and respiration

Oxygen constitutes 21% by volume of air. It is the essential part of the air for animal respiration. Oxygen is sufficiently soluble in water to be able to support the respiration of marine animals. The Earth's crust is 89% by mass oxygen.

Its physical properties and chemical reactivity

Oxygen is a colourless diatomic gas, O_2. Properly speaking, we should refer to molecular oxygen as dioxygen. It boils at $-183\,°C$ and freezes at $-219\,°C$. Oxygen reacts directly with most other elements, and forms compounds with all elements except helium, neon, argon and krypton.

21.2.1 MANUFACTURE

Industrial manufacture

Oxygen is obtained industrially by the fractional distillation of liquid air [p. 431]. It is stored under pressure in cylinders.

21.2.2 INDUSTRIAL AND OTHER USES

Oxygen is used in the steel industry...

The steel industry uses oxygen to burn off the carbon which is present as an impurity in cast iron and which makes iron brittle [p. 502]. Since so much oxygen is needed (one tonne of oxygen for every tonne of steel), steel works have their own oxygen

...and for cutting and welding metals...

plants on the same site. Oxygen is also used in 'oxy-acetylene' torches which burn ethyne in oxygen in an extremely exothermic reaction and are used for cutting and welding metals.

...to help patients with breathing difficulties...

Hospitals use oxygen to revive accident victims and patients with breathing difficulties due to complaints such as pneumonia and asthma. Deep-sea divers, mountaineers and high-altitude pilots use oxygen. Space flights consume an enormous amount of oxygen. The Saturn rockets which launched the American astronauts on their journey to the Moon carried 2200 tonnes of liquid oxygen. The first stage burned 450 tonnes of kerosene in 1800 tonnes of oxygen, stages two and three were powered by hydrogen burning in oxygen, and there was also oxygen for the astronauts to breathe.

...and for mountaineers, divers and space travel

21.2.3 LABORATORY PREPARATION

The methods by which oxygen can be obtained in the laboratory are summarised in Figure 21.1.

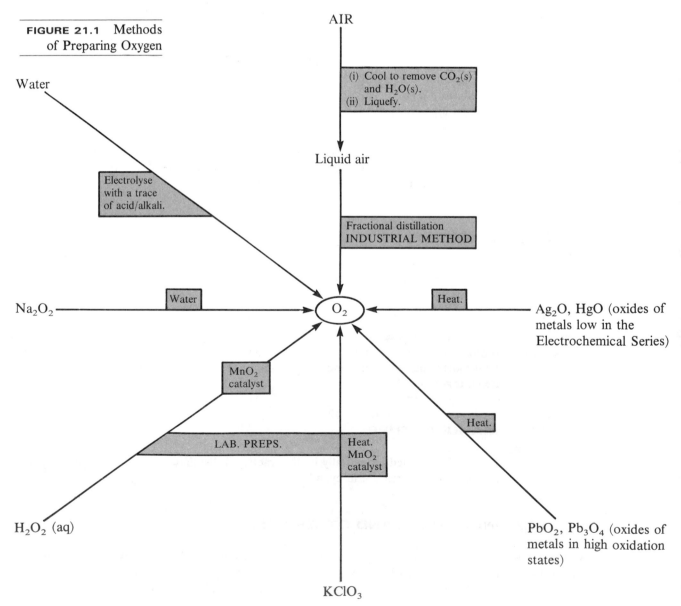

FIGURE 21.1 Methods of Preparing Oxygen

21.3 EXTRACTION OF SULPHUR

Sulphur is found as the element and as metal sulphide ores and as a number of sulphates, e.g., calcium and magnesium sulphates.

H Frasch invented a method of bringing sulphur to the surface from deposits of elemental sulphur below ground level [see Figure 21.2]. It is used in the USA, Poland and Japan.

FIGURE 21.2
The Frasch Process for Mining Sulphur

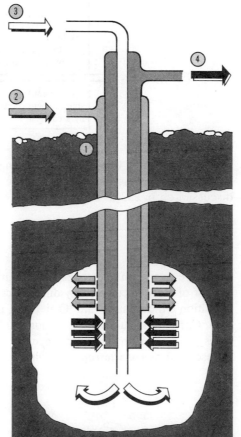

3 Compressed air is sent down the inside pipe.

2 Superheated water at 160 °C and 10 atm is sent down the outside pipe to melt sulphur.

1 A shaft is sunk 200–400 m down to the sulphur-bearing rock. Three concentric pipes are inserted. The outer pipe is 1 m across.

4 A foam of molten sulphur, water and air is forced up through the middle pipe to the surface.

5 The sulphur which solidifies is 99.5% pure.

SULPHUR-BEARING ROCK

CHECKPOINT 21A: OXYGEN AND SULPHUR

1. How is it possible to heat water to 160 °C without turning it into steam? Why is the hot water sent down the outermost pipe in the Frasch process? Why would excavating a mine and digging out sulphur be a dangerous method of obtaining sulphur?

2. Write equations for the thermal decomposition of Ag_2O, HgO, PbO_2, Pb_3O_4, KNO_3, $KClO_3$.

3. Write equations for
(a) the reaction of sodium peroxide with water to give oxygen
(b) the catalytic decomposition of hydrogen peroxide
(c) the electrolysis of acidified water.

21.4 ALLOTROPY

21.4.1 OXYGEN

The element oxygen exists in two forms, dioxygen, O_2, and trioxygen, O_3. The existence of an element in two forms is called **allotropy**. Trioxygen, ozone, is the less stable allotrope:

Allotropy of oxygen...

$$3O_2(g) \rightarrow 2O_3(g); \quad \Delta H^\ominus = 142\,\text{kJ}\,\text{mol}^{-1} \text{ of } O_3$$

...O_2 and O_3

Dioxygen must absorb energy in order to form trioxygen, but trioxygen decomposes spontaneously to form dioxygen.

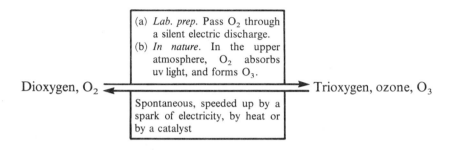

Dioxygen, O_2

(a) *Lab. prep.* Pass O_2 through a silent electric discharge.
(b) *In nature.* In the upper atmosphere, O_2 absorbs uv light, and forms O_3.

Spontaneous, speeded up by a spark of electricity, by heat or by a catalyst

Trioxygen, ozone, O_3

Monotropic allotropy

Since the direction of spontaneous change is always the same under all conditions, the allotropy is described as **monotropic** (moving in one direction). Allotropy is discussed on p. 233.

21.4.2 SULPHUR

Allotropy of sulphur...

...rhombic and monoclinic

Enantiotropic allotropy

There are two main allotropes of sulphur. Both **rhombic** and **monoclinic** sulphur consist of S_8 molecules. They differ in crystal structure [see Figure 21.3]. Above the transition temperature, 95.6 °C at 1 atm, monoclinic sulphur is the stable form; below 95.6 °C, rhombic sulphur is the more stable form. Transition between the two forms takes place in either direction, depending on the temperature. This kind of allotropy is described as **enantiotropic** (moving in both directions) [p. 233].

FIGURE 21.3 Allotropes of Sulphur

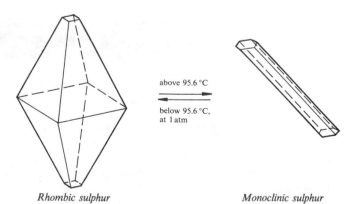

Rhombic sulphur

Yellow transparent crystals, T_m = 113 °C, obtained when sulphur crystallises from solution

above 95.6 °C

below 95.6 °C, at 1 atm

Monoclinic sulphur

Amber coloured needles, T_m = 119 °C, obtained when sulphur solidifies above 95.6 °C

When sulphur is heated, it melts and then undergoes a set of changes, as shown in Figure 21.4.

FIGURE 21.4
The Changes that occur when Sulphur is Heated

Solid sulphur	S_8 rings
↓ Heat.	↓
Transparent yellow liquid	$\bullet S(S)_6 S \bullet$ chains
↓ Heat.	↓
Colour darkens. Liquid reaches maximum viscosity at 200 °C.	Long chains of around 10^5 S atoms become entangled and make the liquid viscous.
↓ Heat.	↓
Liquid becomes mobile at 400 °C.	Chains break up to form smaller units, and the liquid becomes mobile.
↓ Heat.	↓
Liquid sulphur boils at 444 °C.	
↓ Heat.	↓
Vapour	Vapour contains S_8, S_4 and S_2 molecules
↓ Cool.	↓
Solidifies as $S_8(s)$	S_8 molecules

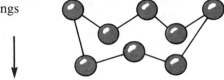

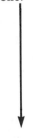

21.5 OZONE, TRIOXYGEN, O_3

There is a layer of ozone 25 km above the Earth

There is a layer of ozone, trioxygen, O_3, surrounding the Earth at a distance of 25 km. It is formed when oxygen absorbs ultraviolet radiation from the Sun. The ozone layer restricts the amount of uv radiation reaching the Earth. If the amount of uv radiation reaching the Earth increased significantly, there might well be an increase in the number of people developing skin cancer. Some people are worried that the exhaust gases from high-flying aircraft are using up the ozone layer. The possible effect on the ozone layer of freons in aerosol sprays is discussed on p. 600.

A silent electric discharge converts O_2 into O_3

In the laboratory, ozone is made by passing a stream of oxygen through a silent electric discharge. (A spark of electricity will cause decomposition of ozone into oxygen.) 'Ozonised oxygen', containing 10% of ozone is formed:

$$3O_2(g) \rightarrow 2O_3(g); \quad \Delta H^{\ominus} = 142\,\text{kJ mol}^{-1} \text{ of } O_3$$

21.5.1 PROPERTIES

Ozone is a stronger oxidising agent than oxygen

Ozone has a characteristic smell. In concentrations above 1000 ppm, it is damaging to health. It is a more powerful oxidising agent than oxygen. It will oxidise lead(II) sulphide, iodide ions and iron(II) salts:

$$O_3(g) + 2H^+(aq) + 2e^- \rightleftharpoons O_2(g) + H_2O(l); \quad E^\ominus = +2.07\,V$$

21.5.2 STRUCTURE

The oxygen molecule $\; :\!\overset{\cdot\cdot}{O}\!=\!\overset{\cdot\cdot}{O}\!: \;$

has a σ bond and a π bond between the two atoms. In the ozone molecule, there is an angle of $117°$ between the bonds. The structure can be represented by the delocalised electron structure:

The structure of ozone

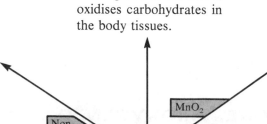

21.6 REACTIONS OF OXYGEN AND SULPHUR

Oxygen combines directly with most elements [see Figure 21.5]. Sulphur combines with most metals when heated and with the non-metallic elements fluorine, chlorine, oxygen and carbon [see Figure 21.6]. It is oxidised by concentrated nitric acid and by concentrated sulphuric acid:

$$S(s) + 6HNO_3(aq) \rightarrow H_2SO_4(l) + 6NO_2(g) + 2H_2O(l)$$

$$S(s) + 2H_2SO_4(l) \rightarrow 3SO_2(g) + 2H_2O(l)$$

FIGURE 21.5 Some Reactions of Oxygen

Animal respiration. Complexes with haemoglobin in the blood, oxidises carbohydrates in the body tissues.

Air oxidises, e.g., alcohol to acid, Fe^{2+} to Fe^{3+}, SO_3^{2-} to SO_4^{2-}.

Acidic: Some of these oxides dissolve in water to form acids; all react with bases to form salts, with the exception of one or two neutral oxides, e.g., NO.

Basic: All these oxides are basic and react with acids to form salts. *Some are amphoteric.

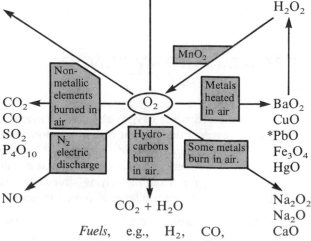

Fuels, e.g., H_2, CO, C_8H_{18}, burn in air; other substances, e.g. NH_3, burn in pure O_2.

FIGURE 21.6 Some
Reactions of Sulphur

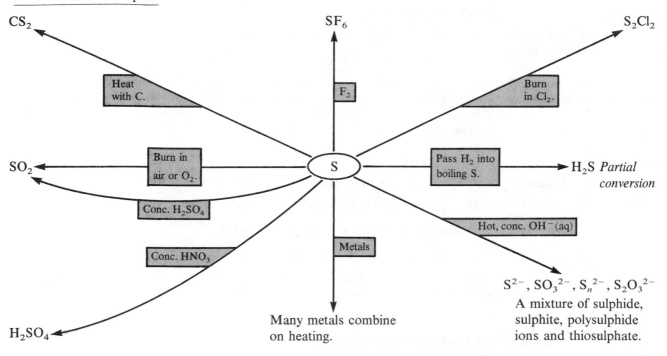

1. Why is oxygen a gas and sulphur a solid?

2. Calculate the standard enthalpy of formation of (a) O^{2-} and (b) S^{2-} by using the following data.

Element	$\Delta^{\ominus}H_{\text{Atomisation}}/$ kJ mol^{-1}	1st electron affinity/kJ mol^{-1}	2nd electron affinity/kJ mol^{-1}
Oxygen	248	−142	+844
Sulphur	223	−200	+532

3. Use the standard bond enthalpies in kJ mol^{-1}: S=S, 431; S—S, 264; O=O, 498 and O—O, 142. Calculate the values of $\Delta H^{\ominus}$ for the changes

(a) $\qquad S_8(g) \rightarrow 4S_2(g)$

(b) $\qquad O_8(g) \rightarrow 4O_2(g)$

Comment on the values you obtain.

21.7 HYDRIDES OF OXYGEN AND SULPHUR

21.7.1 WATER

See p. 347.

21.7.2 HYDROGEN SULPHIDE

Laboratory preparation of hydrogen sulphide . . .

Hydrogen sulphide, H_2S, is an unpleasant-smelling gas which is poisonous at a level of 1000 ppm. It occurs in natural gas and in bad eggs, where it is formed by the decay of proteins. In the laboratory, it is made by the action of dilute hydrochloric acid on iron(II) sulphide:

$$S^{2-}(s) + 2H^+(aq) \rightarrow H_2S(g)$$

Although it is slightly soluble, it can be collected over water.

IONISATION

Hydrogen sulphide is a weak acid, ionising to give hydrogen ions, sulphide ions and hydrogensulphide ions:

...which is a weak, diprotic acid...

$$H_2S(g) + H_2O(l) \rightleftharpoons H_3O^+(aq) + HS^-(aq)$$

$$HS^-(aq) + H_2O(l) \rightleftharpoons H_3O^+(aq) + S^{2-}(aq)$$

...and which will precipitate metal sulphides

The concentration of sulphide ions, although small, is able to precipitate metal ions from solution. Many metal sulphides have very small solubility products [p. 269]. The sulphides of, e.g., Hg^{2+}, Pb^{2+}, Cu^{2+}, and Sn^{2+} are precipitated from solution by hydrogen sulphide.

REDUCING ACTION

Hydrogen sulphide is a reducing agent

Hydrogen sulphide burns to form sulphur dioxide and water:

$$2H_2S(g) + 3O_2(g) \rightarrow 2SO_2(g) + 2H_2O(l)$$

In its reducing reactions, it is oxidised to sulphur:

$$H_2S(aq) \rightleftharpoons S(s) + 2H^+(aq) + 2e^-; \quad E^\ominus = -0.51\,V$$

Hydrogen sulphide reduces acidified manganate(VII), acidified dichromate(VI), iron(III) ions, moist chlorine and moist sulphur dioxide [see Question 2, p. 417].

21.7.3 HYDROGEN PEROXIDE

The arrangement of bonds in H_2O_2

Hydrogen peroxide is a liquid of formula H_2O_2. Its structure is illustrated in Figure 21.7. The molecules are associated by hydrogen bonding, and the liquid is viscous. It boils at 150 °C and freezes at -0.9 °C.

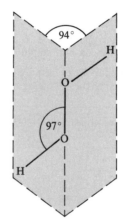

FIGURE 21.7
The Structure of H_2O_2

PREPARATION

The laboratory preparation of hydrogen peroxide

The action of dilute sulphuric acid on barium peroxide, BaO_2, at 0 °C is used for the preparation of hydrogen peroxide:

$$BaO_2(s) + H_2SO_4(aq) \rightarrow BaSO_4(s) + H_2O_2(aq)$$

The insoluble barium sulphate is removed by filtration. The aqueous solution of hydrogen peroxide can be concentrated by distillation under reduced pressure.

Hydrogen peroxide is used as a source of oxygen

Hydrogen peroxide readily decomposes to give oxygen:

$$2H_2O_2(aq) \rightarrow 2H_2O(l) + O_2(g)$$

The decomposition is catalysed by a number of metals and metal oxides, the most popular being manganese(IV) oxide, MnO_2. In the laboratory, hydrogen peroxide is used as aqueous solutions which are designated as, for example, '20 volume', meaning that 1 volume of the solution will provide 20 volumes of oxygen at stp.

OXIDISING AGENT/REDUCING AGENT

Hydrogen peroxide acts as an oxidising agent:

$$H_2O_2(aq) + 2H^+(aq) + 2e^- \rightarrow 2H_2O(l); \quad E^{\ominus} = +1.77\,V$$

and as a reducing agent:

$$H_2O_2(aq) \rightarrow O_2(g) + 2H^+(aq) + 2e^-; \quad E^{\ominus} = -0.68\,V$$

Hydrogen peroxide is an oxidising agent and a reducing agent

Acidic solutions favour the oxidising action, e.g., Fe^{2+} to Fe^{3+}; PbS to $PbSO_4$; I^- to I_2. Alkaline solutions favour the reducing reaction, e.g. ClO^- to Cl^-. Even in acid solution, hydrogen peroxide reduces manganate(VII), MnO_4^-, to Mn^{2+}, with the formation of oxygen [see Question 3]. Titrimetric analysis of hydrogen peroxide is covered on p. 000.

USE

Hydrogen peroxide is used as a bleach for textiles, wood pulp and human hair. The bleaching action depends on its oxidising property. A convenient feature is that the only by-product is water.

========================== **CHECKPOINT 21C: HYDRIDES** ==========================

1. Sketch an apparatus which could be used for the laboratory preparation of hydrogen peroxide.

†2. See p. 416 for the half-reaction equation for the oxidation of hydrogen sulphide. Referring to Chapter 3 if you need to, write half-reaction equations for the reduction of (*a*) acid manganate(VII), MnO_4^-, (*b*) acid dichromate(VI), $Cr_2O_7^{2-}$, (*c*) iron(III) ions, Fe^{3+}, (*d*) moist Cl_2 and (*e*) moist SO_2. Construct equations for the reactions between H_2S and these oxidants.

3. By combining half-reaction equations, construct equations for the following oxidation reactions:

(*a*) I^- to I_2 by acidified H_2O_2
(*b*) PbS to $PbSO_4$ by H_2O_2
(*c*) H_2O_2 to O_2 by Cl_2
(*d*) H_2O_2 to O_2 by acidified $KMnO_4$
(*e*) Fe^{2+} to Fe^{3+} by acidified H_2O_2.

4. Suggest a method for the titrimetric analysis of a solution of hydrogen peroxide. What is the concentration/mol dm^{-3} of '20 volume' hydrogen peroxide?

21.8 OXIDES AND SULPHIDES

Ionic oxides and sulphides containing the ions O^{2-} and S^{2-} are formed

The formation of O^{2-} is an endothermic process. When it forms an ionic structure, however, the small size and double charge on the anion result in a highly negative (exothermic) value for the lattice enthalpy. When the lattice enthalpy compensates for the ionisation enthalpies of the cation and anion, an ionic compound may be formed [see MgO, p. 207]. Many ionic oxides are formed. Since the S^{2-} ion is larger and more easily polarised than the O^{2-} ion, sulphides tend to be more covalent in character than the corresponding oxides. Only the sulphides of Groups 1A and 2A appear to be ionic in character.

The sulphides have more covalent character than the oxides

Oxides and sulphides that dissolve in water give alkaline solutions due to the basic nature of the O^{2-} and S^{2-} ions:

$$O^{2-}(aq) + H_2O(l) \rightarrow 2OH^-(aq)$$

$$S^{2-}(aq) + H_2O(l) \rightarrow HS^-(aq) + OH^-(aq)$$

21.8.1 CLASSIFICATION OF OXIDES

Oxides may be classified according to their formulae [see Table 21.2], or according to their structure [see Figure 21.8], or according to their acid–base properties [see Table 21.3 and Figure 21.9].

Oxide	*Types of bonds in the oxides* E_xO_y
Normal oxides	Bonds are between E and O only. Some are ionic, e.g., $Ca^{2+}O^{2-}$; others are covalent, e.g., CO_2, $(SiO_2)_n$
Peroxides	Bonds are between E and O and also between O atoms. Some are ionic, e.g., $2Na^+ \ O—O^{2-}$; others are covalent, e.g., $H—O—O—H$
Suboxides	Bonds are between E and O and also between E atoms, e.g., C_3O_2, $O{=}C{=}C{=}C{=}O$
Superoxides	Contain the ion O_2^-, e.g., $K^+O_2^-$ (also called *hyperoxides*)
Mixed oxides	e.g., Pb_3O_4, which reacts as a mixture $2PbO \cdot PbO_2$ and Fe_3O_4, which reacts as a mixture $FeO \cdot Fe_2O_3$
Non-stoichiometric oxides	Transition metals form oxides of formula $M_{0-1}O$, e.g., $Fe_{0.9}O$

TABLE 21.2 Classification of Oxides according to Formula

	Oxides of metallic elements	*Oxides of non-metallic elements*
(a)	Oxides of metals in lower oxidation states are basic. Some react with water to give OH^-(aq), e.g., CaO, MgO.	Most are acidic. Some dissolve in water to form solutions with a high concentration of hydrogen ions, e.g., SO_3.
(b)	Others are insoluble in water, but react with acids and acidic oxides, e.g., Fe_2O_3, CuO.	Macromolecular oxides, e.g., $(SiO_2)_n$, $(B_2O_3)_n$, do not dissolve, but react with basic oxides and amphoteric oxides to give salts.
(c)	Strongly basic oxides, e.g., K_2O, CaO react with amphoteric oxides.	A few are neutral, e.g., N_2O, NO, F_2O.
(d)	Some metal oxides are amphoteric, reacting with both basic oxides and acidic oxides, e.g., ZnO, SnO. SnO_2, PbO, PbO_2, Cr_2O_3, Al_2O_3.	

TABLE 21.3 Oxides of Metallic and Non-metallic Elements

FIGURE 21.8
Classification of Oxides
according to Structure

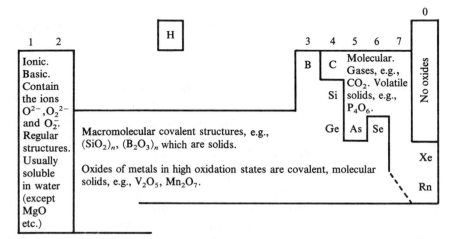

FIGURE 21.9
The Acid–Base
Character of Oxides

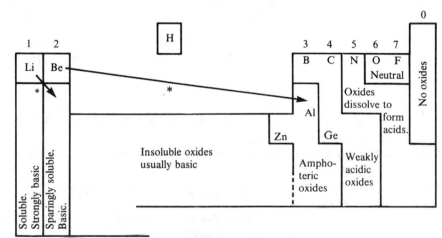

*Diagonal relationship

CHECKPOINT 21D: OXIDES

1. Classify the following (i) according to structure and (ii) according to acid–base character:

(a) Na_2O

(b) CaO

(c) SiO_2

(d) Fe_2O_3

(e) Fe_3O_4

(f) I_2O_5

(g) Cl_2O_7

2. Which member of each of the following pairs is the more acidic oxide?

(a) BaO and CO

(b) MnO and Mn_2O_7

(c) NO and N_2O_5

(d) Cr_2O_3 and CrO_3

(e) CaO and FeO

(f) SO_2 and SeO_2

3. Explain why oxygen and sulphur differ in

(a) the ionic character of MO and MS (where M is a metal)

(b) the boiling temperatures of H_2O and H_2S

(c) the formulae of their fluorides, F_2O and SF_6.

21.9 SULPHUR DIOXIDE

The laboratory preparation from sulphites

The industrial sources of sulphur dioxide, SO_2, are listed on p. 422. A convenient laboratory preparation is the action of a dilute acid on a sulphite. The gas is collected downwards because it is denser than air:

$$SO_3^{2-}(s) + 2H^+(aq) \rightarrow SO_2(g) + H_2O(l)$$

It is also made by the action of hot, concentrated sulphuric acid on copper [p. 425].

21.9.1 PHYSICAL PROPERTIES

Sulphur dioxide is a dense gas with a choking smell. It fumes in air and is extremely soluble in water. It boils at $-10\,°C$, and liquefies at room temperature under a pressure of 3 atm.

The bonding in SO₂ The bonding in SO_2 is shown in Figure 21.10. The $S{=}O$ bonds are double, and the sulphur atom has ten electrons in its valence shell. It has 'expanded its octet'.

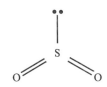

FIGURE 21.10
The Bonding in SO_2

21.9.2 SULPHUROUS ACID AND SULPHITES

Sulphur dioxide is the anhydride of sulphurous acid, which is a weak acid...

Sulphur dioxide dissolves in water to form a solution which contains sulphurous acid:

$$SO_2(aq) + H_2O(l) \rightleftharpoons H_2SO_3(aq)$$

The acid is a weak acid, ionising to give hydrogensulphite ions, HSO_3^-, and sulphite ions, SO_3^{2-}. Attempts to concentrate the solution result in the decomposition of sulphurous acid and the evolution of sulphur dioxide gas.

...It reacts with alkalis to form sulphites or hydrogensulphites

Sulphur dioxide reacts with aqueous sodium hydroxide to form sodium sulphite solution. If more sulphur dioxide is passed into this solution, sodium hydrogensulphite is formed:

$$SO_2(g) + 2NaOH(aq) \rightarrow Na_2SO_3(aq) + H_2O(l)$$

$$Na_2SO_3(aq) + SO_2(g) + H_2O(l) \rightarrow 2NaHSO_3(aq)$$

Moist sulphur dioxide, or sulphurous acid, and sulphites behave as reducing agents:

SO₂ and SO₃²⁻ are reducing agents

$$SO_3^{2-}(aq) + H_2O(l) \rightleftharpoons SO_4^{2-}(aq) + 2H^+(aq) + 2e^-; \quad E^{\ominus} = -0.17\,V$$

Sulphite ions are oxidised by air, chlorine, iron(III) ions, dichromate(VI) ions and manganate(VII) ions. The reaction with dichromate(VI), resulting in a change from orange to blue, and the decolorisation of manganate(VII) are used as tests for sulphur dioxide. Hydrogen sulphide is a stronger reducing agent than sulphur dioxide, and it reduces sulphur dioxide to sulphur.

CHECKPOINT 21E: SULPHUR DIOXIDE

1. Write equations for the ionisation of sulphurous acid.

2. Write half-reaction equations for the oxidants mentioned above.

Combine these with the half-reaction equation for $SO_3^{2-}(aq)$ to obtain equations for the reactions of

$SO_3^{2-}(aq)$ with O_2, Cl_2, $Fe^{3+}(aq)$, $Cr_2O_7^{2-}(aq)$, $MnO_4^-(aq)$.

Obtain an equation for the reaction of H_2S and SO_3^{2-}.

21.10 SULPHUR(VI) OXIDE

Sulphur(VI) oxide exists as molecules of monomer, $SO_3(g)$, trimer, $(SO_3)_3(s)$ and polymer, $(SO_3)_n(s)$ [see Figure 21.11].

The laboratory preparation of sulphur(VI) oxide is illustrated in Figure 21.12. Dry sulphur dioxide and dry oxygen combine on the surface of platinised asbestos at 400 °C. The product solidifies as long needle-shaped crystals in the cooled receiver:

$$2SO_2(g) + O_2(g) \xrightleftharpoons[400\,°C]{Pt} 2SO_3(s)$$

FIGURE 21.11
Molecules of
Sulphur(VI) Oxide

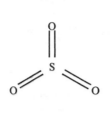

(a) Vapour SO_3

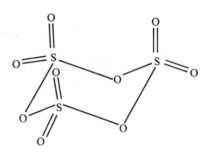

(b) Solid $(SO_3)_3$

(c) Solid $(SO_3)_n$

FIGURE 21.12
Laboratory Preparation
of Sulphur(VI) Oxide

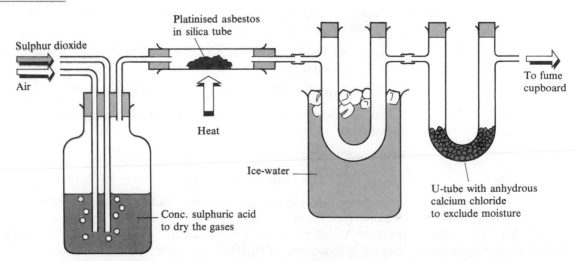

Sulphur dioxide

Air

Platinised asbestos
in silica tube

Heat

To fume
cupboard

Ice-water

Conc. sulphuric acid
to dry the gases

U-tube with anhydrous
calcium chloride
to exclude moisture

21.10.1 PROPERTIES

Sulphur(VI) oxide fumes in moist air, and has acidic reactions and oxidising reactions

Sulphur(VI) oxide is an acid oxide. It fumes in moist air as it reacts with moisture to form a mist of sulphuric acid:

$$SO_3(g) + H_2O(l) \rightarrow H_2SO_4(l)$$

It reacts with basic oxides to form sulphates:

$$CaO(s) + SO_3(g) \rightarrow CaSO_4(s)$$

Sulphur(VI) oxide is an oxidising agent. It takes part in the same oxidising reactions as sulphuric(VI) acid, of which it is the anhydride.

A comparison of SO_2 and SO_3

Sulphur dioxide, SO_2	Sulphur(VI) oxide, SO_3
Colourless gas	Solid. Easily vaporised.
Fumes. Pungent smell	Fumes. Pungent smell
Very soluble	Very soluble
Solution is weakly acidic	Solution is strongly acidic
Reducing agent (e.g., $Fe^{3+} \rightarrow Fe^{2+}$). Combines with $O_2 \rightarrow SO_3$ and with $Cl_2 \rightarrow SO_2Cl_2$	Oxidising agent (e.g., $Br^- \rightarrow Br_2$)

TABLE 21.4
A Comparison of
the Oxides of Sulphur

21.11 SULPHURIC ACID

Manufacture of sulphuric acid needs SO_2

Sulphuric acid is made by the Contact process. The anhydride of sulphuric acid is sulphur(VI) oxide, SO_3. It is made by the oxidation of sulphur dioxide, which is obtained from the following sources:

1. Sulphur is burned in air.

Sources of sulphur dioxide

2. Hydrogen sulphide from crude oil is burned in air.

3. During the extraction of metals from sulphide ores, the ores are roasted in air, giving sulphur dioxide.

4. Anhydrite, $CaSO_4$, is heated with coke and sand in the manufacture of cement, $CaSiO_3$, and sulphur dioxide is a by-product:

$$2CaSO_4(s) + C(s) + 2SiO_2(s) \rightarrow 2CaSiO_3(s) + 2SO_2(g) + CO_2(g)$$

Sulphur dioxide is passed through an electrostatic precipitator to remove dust and impurities.

When sulphur dioxide and oxygen combine

$$2SO_2(g) + O_2(g) \rightleftharpoons 2SO_3(g); \quad \Delta H^\ominus = -197\,kJ\,mol^{-1}$$

$SO_2 + O_2$ combine to form SO_3 in a reaction which reaches equilibrium

the reactants and product reach a state of equilibrium. According to Le Chatelier's Principle [p. 222] the reaction will be favoured by a low temperature and a high pressure. At high pressures, however, sulphur dioxide liquefies, and at low temperatures the rate of attainment of equilibrium is slow. The catalyst vanadium(V) oxide is so effective that a 95% conversion is achieved at 450 °C and 2 atm. The catalyst is easily 'poisoned' by absorbing impurities in the sulphur dioxide. These take up the surface of the catalyst and impair its efficiency. The temperature of the catalyst is maintained at 450 °C in spite of the exothermicity of the reaction by using the heat generated in the catalyst chamber to heat the incoming gases [see Figure 21.13].

The principles which govern the position of equilibrium are discussed in the light of Le Chatelier's Principle

The sulphur(VI) oxide formed is not absorbed in water because the reaction

$$SO_3(g) + H_2O(l) \rightarrow H_2SO_4(l)$$

SO_3 is absorbed in H_2SO_4

is so intensely exothermic as to vaporise the sulphuric acid formed. It can be safely absorbed in 98% sulphuric acid to form fuming sulphuric acid, *oleum*, $H_2S_2O_7$, which reacts with water to form sulphuric acid:

$$SO_3(g) + H_2SO_4(l) \rightarrow H_2S_2O_7(l)$$

$$H_2S_2O_7(l) + H_2O(l) \rightarrow 2H_2SO_4(l)$$

FIGURE 21.13
The Contact Process

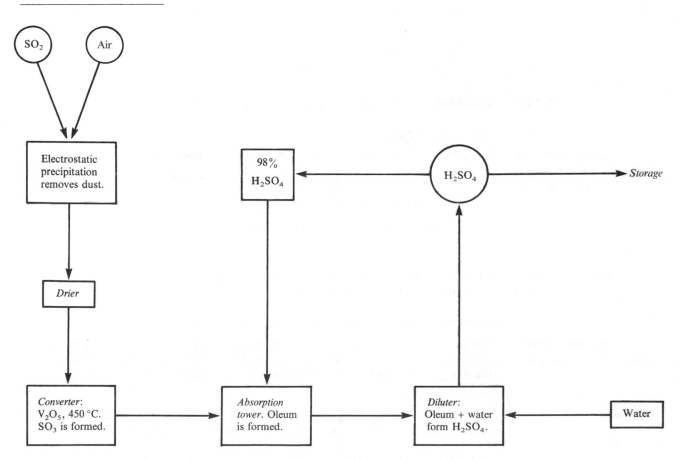

21.11.1 PHYSICAL PROPERTIES

Physical properties of H_2SO_4

Sulphuric acid is a viscous liquid. It is a covalent substance with the structure shown in Figure 21.14(a). The viscosity and the high boiling temperature (270 °C) are due to hydrogen bonding [see Figure 21.14(b)].

FIGURE 21.14
Sulphuric Acid
(*a*) the Structure
(*b*) the Intermolecular
Hydrogen Bonding

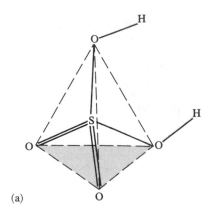

(a)

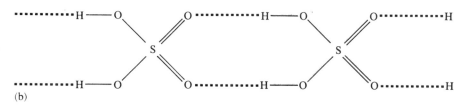

(b)

With water, sulphuric acid forms a constant-boiling mixture [p. 163] containing 98.3% acid.

21.11.2 CHEMICAL PROPERTIES

ACID PROPERTIES

Sulphuric acid is a diprotic acid, the second ionisation being incomplete...

In aqueous solution, sulphuric acid functions as an acid. The ionisation

$$H_2SO_4(aq) + H_2O(l) \rightarrow H_3O^+(aq) + HSO_4^-(aq)$$

is virtually complete. The ionisation

$$HSO_4^-(aq) + H_2O(l) \rightleftharpoons H_3O^+(aq) + SO_4^{2-}(aq)$$

is about 10% complete.

...It reacts as a typical acid

Sulphuric acid in solution reacts with metals and their oxides, hydroxides and carbonates in typical acid fashion.

CONCENTRATED SULPHURIC ACID AS A DRYING AGENT

Concentrated sulphuric acid must be diluted with care

Concentrated sulphuric acid reacts exothermically with water. When a solution is made it is essential to pour the acid into water, stirring to disperse the heat evolved. It is dangerous to add water to concentrated sulphuric acid as small pockets of acid are likely to boil.

It is used to dry gases...

Gases are dried by bubbling them through concentrated sulphuric acid, as shown in Figure 21.12. For basic gases, another drying agent must be used.

CONCENTRATED SULPHURIC ACID AS A DEHYDRATING AGENT

When concentrated sulphuric acid removes water from a mixture, it is termed a **drying agent**. When it removes the elements of water from a compound, with the formation of a new compound, it is described as a **dehydrating agent**. Some dehydrating reactions of concentrated sulphuric acid are:

...and as a dehydrating agent

$$\text{Glucose, } C_6H_{12}O_6(s) \rightarrow \text{Sugar charcoal, } C(s) + H_2O$$

$$\text{Methanoic acid, } HCO_2H(l) \rightarrow CO(g) + H_2O$$

Ethanedioic acid, $HO_2CCO_2H(l) \rightarrow CO(g) + CO_2(g) + H_2O(l)$

Ethanol, $C_2H_5OH(l) \rightarrow$ Ethene, $CH_2{=}CH_2(g)$ or
Ethoxyethane, $C_2H_5OC_2H_5(g)$ depending on the conditions [p. 622]

CONCENTRATED SULPHURIC ACID AS AN OXIDISING AGENT

Hot concentrated sulphuric acid is an oxidising agent. It may be reduced to sulphurous acid, which dissociates to form sulphur dioxide and water:

$$SO_4^{2-}(aq) + 4H^+(aq) + 2e^- \rightleftharpoons SO_2(aq) + 2H_2O(l); \quad E^\ominus = +0.17\,V$$

As an oxidising agent, concentrated sulphuric acid oxidises metals...

With a powerful reducing agent, sulphuric acid may be reduced to hydrogen sulphide.

Some oxidising reactions are:

1. Metals are oxidised to sulphates. Equations represent these reactions only approximately as a mixture of products, SO_2, H_2S and S, is formed:

$$Cu(s) + 2H_2SO_4(l) \rightarrow CuSO_4(aq) + SO_2(g) + 2H_2O(l)$$

...non-metallic elements...

2. Non-metallic elements, e.g., carbon and sulphur, are converted into oxides:

$$C(s) + 2H_2SO_4(l) \rightarrow CO_2(g) + 2SO_2(g) + 2H_2O(l)$$

...and compounds, e.g., HBr, HI, H$_2$S

3. Compounds are oxidised, with the formation of sulphur dioxide or hydrogen sulphide. Hydrogen bromide, iodide and sulphide are oxidised:

$$2HBr(g) + H_2SO_4(l) \rightarrow Br_2(l) + SO_2(g) + 2H_2O(g)$$

$$3H_2S(g) + H_2SO_4(l) \rightarrow 4S(s) + 4H_2O(l)$$

FIGURE 21.15
Reactions of
Concentrated Sulphuric
Acid

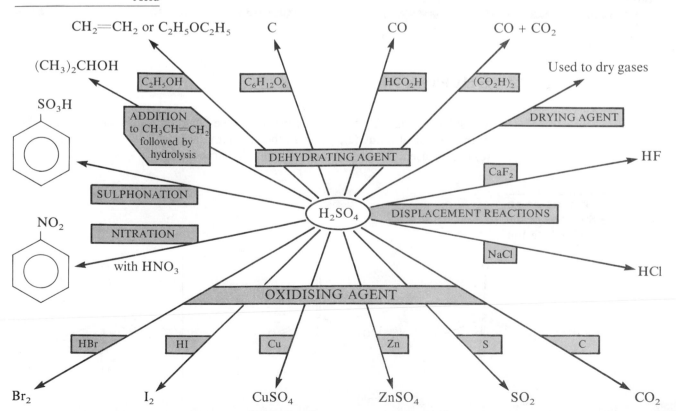

DISPLACEMENT REACTIONS OF CONCENTRATED SULPHURIC ACID

Concentrated sulphuric acid displaces other acids from their salts

When concentrated sulphuric acid is heated with a nitrate, it displaces the other acid from its salt, and nitric acid is formed. In the cold, an equilibrium is set up:

$$NO_3^-(s) + H_2SO_4(l) \rightleftharpoons HNO_3(l) + HSO_4^-(s)$$

If the mixture is heated, nitric acid distils at 120 °C, while sulphuric acid, with T_b 270 °C remains. The removal of nitric acid moves the equilibrium over to the right-hand side. Similarly, the more volatile acids, hydrogen fluoride, hydrogen chloride and phosphoric(V) acid, are displaced from fluorides, chlorides and phosphates(V).

REACTIONS WITH ORGANIC COMPOUNDS

Conc. H_2SO_4 in organic chemistry

Concentrated sulphuric acid is a sulphonating agent, able to introduce the —SO₃H group into an aromatic ring [p. 582]. It is used in nitration [p. 579] and in addition to alkenes [p. 560].

═══════════════════ **CHECKPOINT 21F: SULPHURIC ACID** ═══════════════════

1. Concentrated sulphuric acid reacts with sodium iodide in a two-stage reaction to give iodine. Explain how sulphuric acid is reacting, and write equations for the two steps.

2. Concentrated sulphuric acid reacts with potassium bromide to form a solid and three gases. Name these products, and explain how they come to be formed.

3. A convenient laboratory preparation for carbon monoxide is the action of concentrated sulphuric acid on sodium methanoate, HCO_2Na.
Explain how carbon monoxide is formed. A similar reaction occurs between concentrated sulphuric acid and sodium ethanedioate, $(CO_2Na)_2$. Write equations for the two reactions.

4. What would you expect to see when concentrated sulphuric acid is added to copper(II) sulphate crystals?

5. Phosphorus(V) oxide, P_4O_{10}, is a more powerful dehydrating agent than sulphuric acid. What is the fuming gas that is formed in a reaction between the two chemicals?

6. Give examples, with equations, of reactions in which sulphuric acid acts as (*a*) an acid, (*b*) a dehydrating agent, (*c*) an oxidising agent, (*d*) a sulphonating agent.

21.12 SULPHATES

The sulphate ion, SO_4^{2-}, can be represented [see Figure 21.16] as a resonance hybrid of the canonical forms (a), (b) and others. Alternatively, the delocalised structure (c) can be used. The distribution of bonds is tetrahedral.

FIGURE 21.16
The Structure of the Sulphate Ion

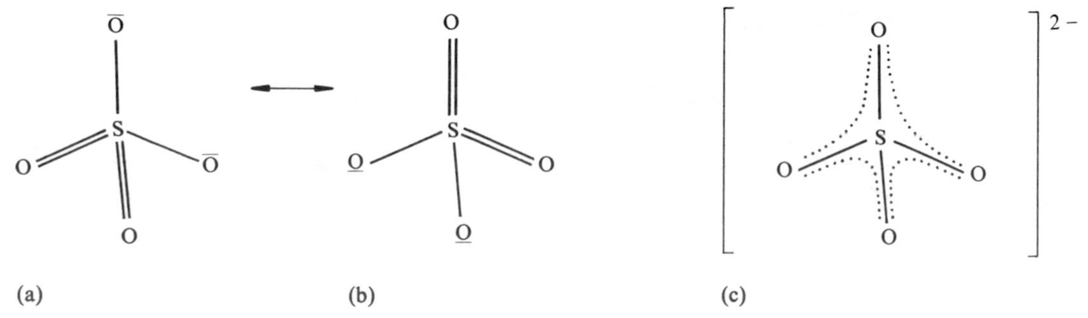

(a)   (b)   (c)

Methods of preparing sulphates

Soluble sulphates are prepared by the action of dilute sulphuric acid on the metal or its oxide, hydroxide or carbonate. Insoluble sulphates are made by precipitation.

The members of Groups 1A and 2A also form hydrogensulphates, which decompose when heated to form sulphates.

Thermal decomposition of $FeSO_4$ and $Fe_2(SO_4)_3$

Some sulphates also decompose on strong heating. Both iron(II) and iron(III) sulphates decompose to form iron(III) oxide:

$$Fe_2(SO_4)_3(s) \rightarrow Fe_2O_3(s) + 3SO_3(g)$$

$$2FeSO_4(s) \rightarrow Fe_2O_3(s) + SO_3(g) + SO_2(g)$$

How to test for sulphates and sulphites

A test for the sulphate ion in solution is the addition of barium chloride solution and dilute hydrochloric acid. A white precipitate of barium sulphate denotes the presence of a sulphate or a hydrogensulphate:

$$SO_4{}^{2-}(aq) + Ba^{2+}(aq) \rightarrow BaSO_4(s)$$

Sulphites give a white precipitate of barium sulphite on the addition of barium chloride solution. When dilute hydrochloric acid is added, the precipitate dissolves with an effervescence of sulphur dioxide.

21.13 THIOSULPHATES

Laboratory preparation of sodium thiosulphate

Thiosulphates contain the ion $S_2O_3{}^{2-}$. The corresponding acid has never been isolated. Sodium thiosulphate is made by boiling a solution of sodium sulphite with sulphur. After filtration and evaporation, crystals of $Na_2S_2O_3 \cdot 5H_2O$ are obtained:

$$SO_3{}^{2-}(aq) + S(s) \rightarrow S_2O_3{}^{2-}(aq)$$

FIGURE 21.17 Some of the Industrial Uses of Sulphuric Acid (The UK consumption totals 130 million tonnes per annum.)

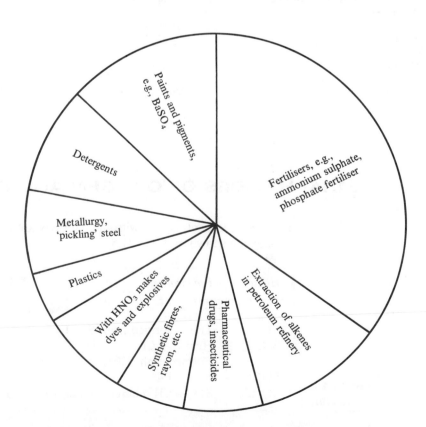

The action of acid on thiosulphate
On the addition of acid to a thiosulphate solution, sulphur is precipitated and sulphur dioxide is evolved:

$$S_2O_3^{2-}(aq) + 2H^+(aq) \rightarrow SO_2(g) + S(s) + H_2O(l)$$

The reaction with I_2 and Cl_2
Sodium thiosulphate is a reducing agent. It is used for the titrimetric analysis of iodine [p. 55] which oxidises thiosulphate to tetrathionate, $S_4O_6^{2-}$. Industrially, thiosulphate is used to remove an excess of chlorine from fabrics after they have been bleached. Chlorine oxidises thiosulphates to sulphates.

CHECKPOINT 21G: SULPHATES

1. The substance $FeSO_4 \cdot 7H_2O$ occurs naturally and has been known for centuries as *green vitriol*. On heating, green vitriol gives a reddish-brown pigment called *jewellers' rouge* and some gaseous products. When these are condensed, another valuable chemical is formed. What is it? What is the formula of jewellers' rouge? Write the equation for the thermal decomposition.

2. Outline the steps you would take in the laboratory preparation of
(a) $MgSO_4 \cdot 7H_2O$ crystals
(b) $Na_2SO_4 \cdot 5H_2O$ crystals
(c) $NaHSO_4 \cdot 6H_2O$ crystals.

3. Copy the following table. In each box, write what you would observe, and explain why.

Solution	Reagents added separately to solutions		
	$BaCl_2(aq)$	$HCl(aq)$	$BaCl_2(aq) + HCl(aq)$
$Na_2SO_4(aq)$ $Na_2SO_3(aq)$ $NaHSO_4(aq)$ $Na_2S_2O_3(aq)$			

4. Draw up a table comparing sulphurous acid, H_2SO_3, and sulphuric acid, H_2SO_4, with respect to
(a) the ease of isolation of the pure acids
(b) the extent of ionisation in aqueous solution
(c) their reactions as typical acids
(d) their power as oxidising agents
(e) their power as reducing agents
(f) any other points.

5. Draw the arrangement of bonds in the species:

$$O_3, S_8, SF_6, SO_2, SO_3, SO_4^{2-}, S_2O_3^{2-}$$

6. Write half-reaction equations for (a) the oxidation of thiosulphate to tetrathionate, (b) the oxidation of thiosulphate to sulphate, (c) the reduction of I_2 to I^-, (d) the reduction of Cl_2 to Cl^-. Combine equations (a) and (c). Combine equations (b) and (d).

21.14 HALIDES OF OXYGEN AND SULPHUR

The compounds of oxygen with the halogens are listed on p. 401. The halides of sulphur are shown in Table 21.5.

Halide	Preparation	Reaction with water
SF_6	$S + F_2$	No reaction
SF_4	Traces formed from $S + F_2$	Hydrolysed $\rightarrow SO_2 + HF$
S_2Cl_2	Heat $S + Cl_2$.	Hydrolysed $\rightarrow SO_2 + HCl$
SCl_2	Formed with S_2Cl_2	Hydrolysed $\rightarrow SO_2 + HCl$
$SOCl_2$	$SO_2 + PCl_5$	Hydrolysed $\rightarrow SO_2 + HCl$
SO_2Cl_2	$SO_2 + Cl_2$ (charcoal catalyst)	Hydrolysed $\rightarrow H_2SO_4 + HCl$

TABLE 21.5
Halides of Sulphur

21.15 A COMPARISON OF OXYGEN AND SULPHUR

21.15.1 SIMILARITIES

1. Both have the electron configuration $(\text{core})ns^2np^4$. They form O^{2-} and S^{2-} and also two covalent bonds.

2. They are non-metallic elements.

3. They are very reactive elements, combining directly with metals and with non-metallic elements.

Oxygen and sulphur are similar in many respects

4. Both show allotropy: oxygen as O_2 and O_3; sulphur as the rhombic and monoclinic forms.

5. The hydrides H_2O and H_2S are stable, while H_2O_2 and H_2S_2 are unstable.

6. The oxides and sulphides of Groups 1A and 2A are ionic.

7. The oxides and sulphides of Group 1A undergo similar hydrolysis reactions:

$$O_2{}^{2-}(aq) + H_2O(l) \rightarrow 2OH^-(aq)$$

$$S^{2-}(aq) + H_2O(l) \rightarrow HS^-(aq) + OH^-(aq)$$

21.15.2 DIFFERENCES

1. Oxygen exists as O_2 molecules (and as O_3 [p. 413]).
Sulphur is usually in the form of S_8 molecules. The reason for the difference is the standard bond enthalpies for the $O{=}O$, $O{-}O$, $S{=}S$ and $S{-}S$ bonds [p. 415].

2. Oxygen is restricted to an L shell of 8 electrons and an oxidation state of -2 (except in peroxides, superoxides and fluorides). Sulphur can use its d orbitals and employ oxidation states of -2, $+2$, $+4$ and $+6$.

3. The diameters of the ions are O^{2-}, 0.280 nm and S^{2-}, 0.368 nm. The S^{2-} ion is therefore more polarisable than the O^{2-} ion. Oxides are more ionic in character than the corresponding sulphides. Ionic sulphides are formed by large cations, e.g., those of Group 1A.

4. Oxygen is more electronegative than sulphur. The difference leads to hydrogen bonding in water but not in hydrogen sulphide.

5. Hydrogen sulphide is a weak acid. Water can act both as a proton-donor and as a proton-acceptor.

6. The allotropy of oxygen is monotropic; that of sulphur is enantiotropic.

QUESTIONS ON CHAPTER 21

1. Describe the Contact process for the manufacture of sulphuric acid. Explain the choice of conditions used for the process. Mention three uses for sulphuric acid.

2. How and under what conditions does sulphuric acid react with (*a*) potassium bromide, (*b*) ethanedioic acid, (*c*) sodium methanoate and (*d*) sucrose?

3. Illustrate the use of sulphuric acid (*a*) as an acid, (*b*) as an oxidising agent, (*c*) as a dehydrating agent, (*d*) in the displacement of other acids from their salts and (*e*) as a sulphonating agent.

4. Sulphur dioxide and chlorine are both used as bleaches. They act on different dyes. Why can sulphur dioxide bleach materials that are not bleached by chlorine?

5. Review the oxidation states of sulphur, giving examples of compounds of sulphur in its different oxidation states. Give examples, with equations, of redox reactions in which sulphur changes its oxidation number.

6. Describe the changes in appearance that occur when sulphur is heated to its boiling temperature. How can the changes be explained in terms of the molecular structure of sulphur?

7. Write the formulae of the two compounds of sulphur with chlorine and oxygen. Say how they can be made from sulphur dioxide. Say how they react with propanoic acid.

8. (*a*) A fluoride of sulphur has the empirical formula SF_4.

0.100 g of the gaseous fluoride occupies a volume of 22.10 cm^3 at 20 °C and 102.1 kPa.

 (i) Find the molecular formula of the gas.

 (ii) What type of bonding is expected to be present in this compound? Give reasons.

(iii) Apart from its physical state at room temperature, give *two* physical properties associated with this compound.

(iv) Suggest a structural formula for the compound.

(*b*) In the manufacture of sulphuric acid the two last stages involve the oxidation of sulphur dioxide followed by the absorption of the oxidised product to form the acid.

Give details of the chemistry and conditions involved in these two stages.

(*c*) Give equations and state the conditions for reactions in which sulphuric acid acts as

 (i) an oxidising agent

 (ii) a dehydrating agent

(iii) a sulphonating agent

(iv) a dibasic acid.

(AEB 81)

22

GROUP 5B

22.1 THE MEMBERS OF THE GROUP

The elements in Group 5B are listed in Table 22.1.

Element	Symbol	Z	Outer shell of electrons	Oxidation states	$T_m/°C$	$T_b/°C$
Nitrogen	N	7	$2s^22p^3$	3,5	-210	-196
Phosphorus	P	15	$3s^23p^3$	3,5	44(white) 590(red)	280
Arsenic	As	33	$4s^24p^3$	3,5	–	613
Antimony	Sb	51	$5s^25p^3$	3,5	630	1380
Bismuth	Bi	83	$6s^26p^3$	3(5)	271	1560

TABLE 22.1
The Members of
Group 5B

22.2 OCCURRENCE, EXTRACTION AND USES

22.2.1 NITROGEN

Nitrogen in nature

Nitrogen constitutes 78% by volume of dry air. It is an essential element in all living things, being present in proteins and nucleic acids. Figure 22.1 illustrates the **nitrogen cycle**. This is the balance between the reactions which take nitrogen out of the air and out of the soil and the reactions which put nitrogen into the air and into the soil. Nitrogen is mined as nitrate ores, for example, *Chile nitre*, $NaNO_3$.

Nitrogen is obtained from liquid air and used in the manufacture of ammonia

For industrial use, nitrogen is obtained by the fractional distillation of liquid air. Nitrogen distils at $-196°C$, while the liquid becomes richer in oxygen, which has a boiling temperature of $-183°C$.

Nitrogen is used for the manufacture of ammonia and thence nitric acid. Gaseous nitrogen is used to provide an inert atmosphere for reactions which cannot be carried out in the presence of oxygen. It is used as a carrier gas in gas–liquid chromatography.

The laboratory preparation

In the laboratory, nitrogen can be made by heating an ammonium salt with a nitrite (not by heating ammonium nitrite because this is an explosive):

$$NH_4^+(s) + NO_2^-(s) \rightarrow N_2(g) + 2H_2O(g)$$

FIGURE 22.1
The Nitrogen Cycle

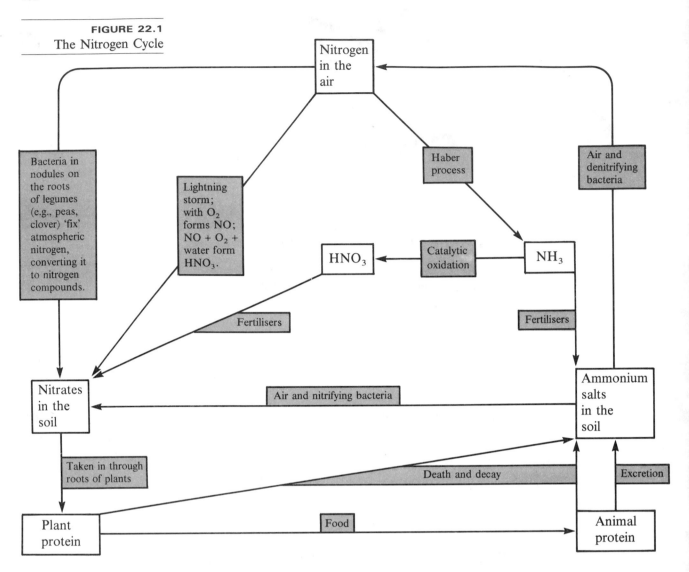

22.2.2 PHOSPHORUS

Phosphorus is extracted from calcium phosphate

Phosphorus is too reactive an element to occur free. It is mined as phosphate ores, e.g., calcium phosphate, $Ca_3(PO_4)_2$. When this is heated with silica and coke in an electric furnace, phosphorus sublimes over.

There are red and white allotropes

The two chief allotropes of phosphorus are the white and red forms. The allotropy is monotropic, red phosphorus being more stable than white under all conditions [see Figures 22.2, 22.3, Table 22.2 and p. 233].

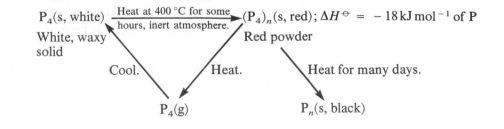

FIGURE 22.2
Allotropy of Phosphorus

Phosphorus is used in match-heads and on the sides of boxes for safety matches.

22.2.3 ARSENIC, ANTIMONY AND BISMUTH

Arsenic, antimony and bismuth are somewhat metallic in nature. They are mined as sulphides.

22.3 STRUCTURE

Dinitrogen consists of diatomic molecules

$$N{\equiv}N$$

The low reactivity of nitrogen can be attributed largely to the strength of the triple bond:

$$N_2(g) \rightarrow 2N(g); \quad \Delta H^{\ominus} = +940\,kJ\,mol^{-1}$$

Figure 22.3 shows the bonding in white and red phosphorus.

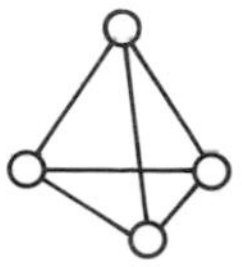

FIGURE 22.3 Bonding in (a) White, (b) Red Phosphorus

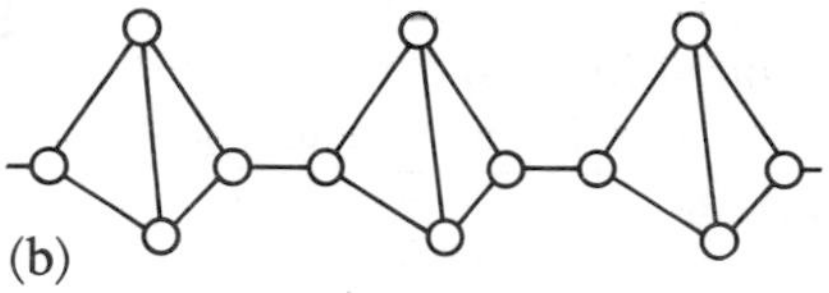

TABLE 22.2 The Allotropes of Phosphorus

	(a)	(b)
Structure	Solid white phosphorus and phosphorus vapour. In P_4 molecules, P uses three p orbitals. The bond angle = 60°, not the usual 90° between p bonds.	Solid red phosphorus. Polymers of P_4 units are arranged in a layer structure.
Reactivity	The strain on the bonds makes them easy to break, and this is why white phosphorus is reactive. It ignites at 35 °C.	The bonds are under less strain, and red phosphorus is less reactive than white. It ignites at 260 °C.
T_m *and solubility*	Intermolecular forces are weak. White $P_4(s)$ has a low T_m of 44 °C, and dissolves in solvents such as carbon disulphide.	The layer structure makes for a higher T_m of 590 °C and a lower solubility. Red phosphorus is insoluble in carbon disulphide.

Arsenic and antimony have layer structures in the solid state. Bismuth has a metallic structure.

22.4 BOND FORMATION

Nitrogen forms N^{3-} ions; other members of the group are 3-covalent...

The members of Group 5B all accept three electrons to attain a full outer octet. The formation of N^{3-} anions is highly endothermic. Ionic nitrides are formed only if their lattice enthalpies are highly exothermic, i.e., by Groups 1A and 2A. The reason the rest of the Group do not form A^{3-} anions is that larger ions, e.g., P^{3-}, cannot approach cations sufficiently closely to form a lattice of high enthalpy. Although Li_3P and Na_3P exist, they do not have ionic structures.

...and show oxidation states of +3 and +5

All the elements of Group 5B show oxidation states of +3 and +5 in their oxo-anions:

$$+3 \text{ in } NO_2^-, PO_2^-, AsO_3^{3-}$$

$$+5 \text{ in } NO_3^-, PO_3^-, PO_4^{3-}, AsO_4^{3-}, SbO_3^- \text{ and } BiO_3^-$$

Some cations are formed, e.g. Sb^{3+}, Bi^{3+}, SbO^+, BiO^+

Antimony and bismuth form the cations $Sb^{3+}(aq)$ and $Bi^{3+}(aq)$, which are readily hydrolysed to the oxo-cations, $SbO^+(aq)$ and $BiO^+(aq)$. They form complex ions, e.g., $BiCl_4^-(aq)$.

All except nitrogen show a covalency of 5 as well as 3

All the elements in this group show a covalency of 3, using the three unpaired p electrons. All except nitrogen show a covalency of 5 by promoting one of the s electrons to a d orbital. This cannot happen with nitrogen because it has no d orbitals at energy levels comparable with the occupied L shell orbitals [see Figure 22.4].

FIGURE 22.4 Bonding Orbitals in N, P in its Ground State and P* in its Excited State

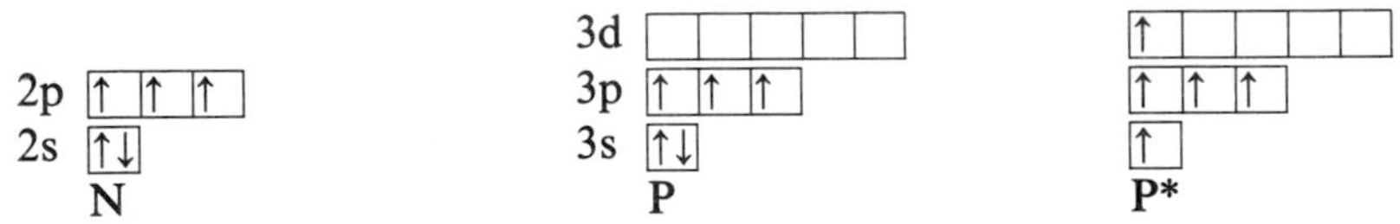

When nitrogen uses its lone pair of electrons to form a coordinate bond, as in NH_4^+ and NO_3^-, it has a covalency of 4.

TABLE 22.3 Reactions of Nitrogen and Phosphorus

Reactant	Nitrogen	Phosphorus
Metals	Ionic or interstitial nitrides formed	Phosphides formed
Oxygen	Some NO formed at high temperatures and in an electric discharge	Ignites, white at 35 °C, red at 260 °C, to form $P_4O_6 + P_4O_{10}$
Sulphur	No reaction	Mixture of sulphides formed
Halogens	No reaction	$PX_3 + PX_5$ formed
Hydrogen	Some NH_3 formed at high pressure	No reaction
Alkali	No reaction	White P (not red) reacts to form $PH_3 + H_2PO_2^-$
Conc. HNO_3	No reaction	H_3PO_4 formed

22.5 THE HYDRIDES

22.5.1 AMMONIA

The Haber process for ammonia

Ammonia is made by the Haber process from nitrogen and hydrogen:

$$N_2(g) + 3H_2(g) \rightleftharpoons 2NH_3(g); \quad \Delta H^{\ominus} = -92\,kJ\,mol^{-1}$$

The reaction is exothermic, and involves a decrease in the number of moles of gas. An application of Le Chatelier's Principle [p. 222] shows that the forward reaction should be assisted by a low temperature and a high pressure. At low temperature, the rate of attainment of equilibrium is low. At high temperature, the position of equilibrium is over to the left. A compromise temperature is adopted, and a catalyst is employed to speed up the attainment of equilibrium concentrations. The conditions employed in industrial plants are 200–1000 atm, 500 °C and, as catalyst, iron with aluminium oxide as a promoter. The yield is about 10%, and unreacted gases are recycled [see Figure 22.5].

FIGURE 22.5(a)
A Flow Diagram of the Haber Process

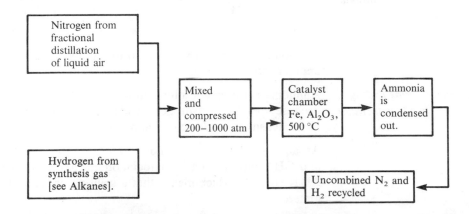

FIGURE 22.5(b)
ICI Billingham. The Haber Process

The laboratory preparation of ammonia from an ammonium salt and a base

In the laboratory, ammonia is made by the action of a strong base on an ammonium salt:

$$NH_4^+(s) + OH^-(s) \rightarrow NH_3(g) + H_2O(g)$$

The reaction is accelerated by warming, and the gas is collected upwards after passing through a drying tower containing calcium oxide. Acidic drying agents cannot be used, and ammonia reacts with calcium chloride to form complex ions.

The bonds in NH_3 are tetrahedrally arranged

The arrangement of bonds in ammonia is tetrahedral, with the lone pair on the nitrogen atom occupying the fourth apex [see Figure 5.5, p. 92]. The lone pair enables ammonia to act as a base, accepting a proton to form the ammonium ion, NH_4^+. This is what happens when ammonia and hydrogen chloride come into contact and form ammonium chloride. This reaction is used as a test for ammonia:

The reaction with hydrogen chloride is a test for ammonia

$$\begin{array}{c} H \\ | \\ H-N:(g) \\ | \\ H \end{array} + H-Cl\,(g) \longrightarrow \left[\begin{array}{c} H \\ | \\ H-N-H \\ | \\ H \end{array}\right]^+ Cl^-(s)$$

Ammonia is a Lewis base, able to use its lone pair to form a coordinate bond to another molecule [p. 251].

Ammonia forms hydrogen bonds with water

Ammonia is extremely soluble in water (1300 volumes per unit volume of water) because it can form hydrogen bonds with water. A saturated solution of ammonia has a density of $0.880\,g\,cm^{-3}$ and is often referred to as 880 ammonia.

An aqueous solution of ammonia is ionised to a small extent, to form NH_4^+ ions and OH^- ions. The basic dissociation constant, K_b, is $1.8 \times 10^{-5}\,mol\,dm^{-3}$ at 298 K [p. 254], which means that a solution of concentration $1.0\,mol\,dm^{-3}$ is 0.4% ionised.

REACTIONS OF AQUEOUS AMMONIA

Basic reactions of NH_3...

1. Ammonia neutralises acids to form ammonium salts.

...precipitation reactions...

2. It precipitates insoluble metal hydroxides from solutions of metal salts. Ammonia is useful for precipitating amphoteric metal hydroxides which dissolve in an excess of a strong alkali (e.g., $Al(OH)_3$, $Pb(OH)_2$).

...complex ion formation

3. Some metal hydroxides dissolve in an excess of ammonia to form soluble ammine complex ions, e.g., $[Ag(NH_3)_2]^+(aq)$ and $[Cu(NH_3)_4]^{2+}(aq)$.

AMMONIA AS A REDUCING AGENT

1. Ammonia is oxidised to nitrogen in reactions with chlorine, chlorate(I) ions and some heated metal oxides:

Ammonia is a reducing agent...

$$2NH_3(g) + 3Cl_2(g) \rightarrow N_2(g) + 6HCl(g)$$

$$2NH_3(aq) + 3ClO^-(aq) \rightarrow N_2(g) + 3Cl^-(aq) + 3H_2O(l)$$

$$2NH_3(g) + 3CuO(s) \rightarrow N_2(g) + 3Cu(s) + 3H_2O(g)$$

...and burns in oxygen to form N_2 or NO

2. Ammonia burns in oxygen to form nitrogen, but, in the presence of platinum, nitrogen monoxide is formed:

$$4NH_3(g) + 3O_2(g) \rightarrow 2N_2(g) + 6H_2O(g)$$

$$4NH_3(g) + 5O_2(g) \xrightarrow{Pt} 4NO(g) + 6H_2O(g)$$

FIGURE 22.6 Some
Reactions of Ammonia

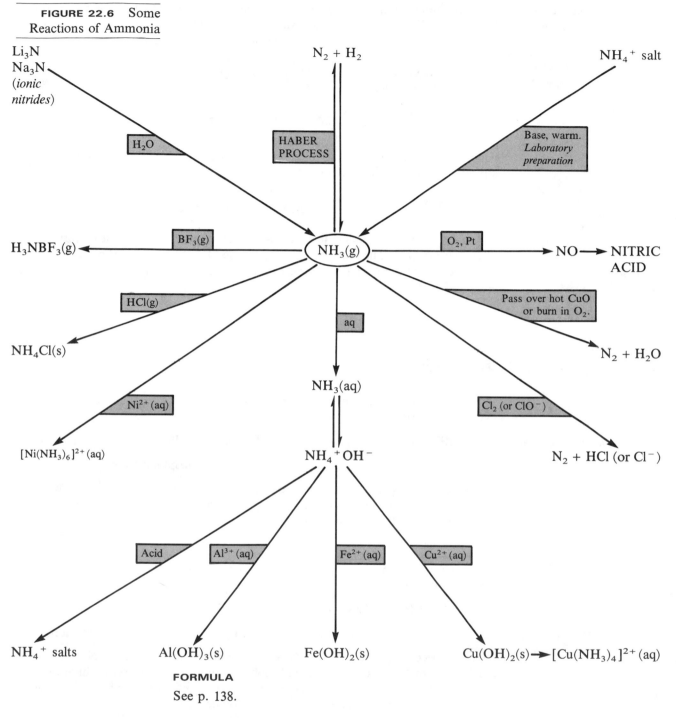

Li₃N
Na₃N
(*ionic nitrides*)

N₂ + H₂

NH₄⁺ salt

H₂O

HABER
PROCESS

Base, warm.
Laboratory preparation

H₃NBF₃(g)   BF₃(g)   NH₃(g)   O₂, Pt   NO → NITRIC ACID

HCl(g)

aq

Pass over hot CuO
or burn in O₂.

NH₄Cl(s)

N₂ + H₂O

Ni²⁺ (aq)

NH₃(aq)

Cl₂ (or ClO⁻)

[Ni(NH₃)₆]²⁺ (aq)

NH₄⁺OH⁻

N₂ + HCl (or Cl⁻)

Acid   Al³⁺ (aq)   Fe²⁺ (aq)   Cu²⁺ (aq)

NH₄⁺ salts   Al(OH)₃(s)   Fe(OH)₂(s)   Cu(OH)₂(s) → [Cu(NH₃)₄]²⁺ (aq)

FORMULA

See p. 138.

LIQUID AMMONIA

Ammonia liquefies at −33 °C at 1 atm. Acid–base reactions in liquid ammonia are described on p. 258.

AMMONIUM SALTS

Ammonium salts are easily decomposed by heat. Ammonium carbonate and ammonium chloride dissociate:

$$(NH_4)_2CO_3(s) \rightleftharpoons 2NH_3(g) + CO_2(g) + H_2O(g)$$

$$NH_4Cl(s) \rightleftharpoons NH_3(g) + HCl(g)$$

Some ammonium salts undergo thermal dissociation . . .

When the ammonium salts of oxidising acids are decomposed on heating, the ammonia formed is oxidised to nitrogen or an oxide of nitrogen. The decomposition of ammonium nitrate(V) and ammonium nitrate(III) (nitrite) can be explosive:

. . . Some decompose explosively

$$NH_4NO_2(s) \xrightarrow[\text{decomposition}]{\text{explosive}} N_2(g) + 2H_2O(g)$$

$$NH_4NO_3(s) \xrightarrow[\text{decomposition}]{\text{explosive}} N_2O(g) + 2H_2O(g)$$

Ammonium dichromate decomposes spectacularly, orange crystals erupting to form a mound of dark green chromium(III) oxide.

$$(NH_4)_2Cr_2O_7(s) \xrightarrow{\text{heat}} N_2(g) + Cr_2O_3(s) + 4H_2O(g)$$

A test for ammonium salts

A test for ammonium salts is the evolution of ammonia when the salt is warmed with a strong base. The reaction can also be used for the estimation of ammonium salts since the amount of base required to decompose the salt can be found by titrimetric analysis [p. 61].

22.5.2 PHOSPHINE

Phosphine, PH_3, is made by heating white phosphorus in concentrated sodium hydroxide solution:

Phosphorus reacts with alkali to form phosphine, PH_3

$$P_4(s) + 3OH^-(aq) + 3H_2O(l) \rightarrow PH_3(g) + 3H_2PO_2^-(aq)$$

$$\text{Phosphinate ion}$$

It is a poisonous gas with an unpleasant smell. When pure, it is stable, but, when prepared by this method, it contains phosphorus vapour and it is spontaneously flammable in air. It burns to form phosphorus(V) oxide:

Phosphides are hydrolysed to give phosphine

$$4PH_3(g) + 8O_2(g) \rightarrow P_4O_{10}(s) + 6H_2O(g)$$

Phosphine is also made by the hydrolysis of ionic phosphides, just as ammonia is made from nitrides:

$$Ca_3P_2(s) + 6H_2O(l) \rightarrow 2PH_3(g) + 3Ca(OH)_2(s)$$

Phosphine is a weak base . . .

Phosphine is a weaker base than ammonia. Since phosphorus and hydrogen have the same electronegativity, the P—H bonds are not polarised like the $\overset{\delta-}{N}$—$\overset{\delta+}{H}$ bonds in ammonia. As a result, the phosphorus atom has a less powerful attraction for a proton. When phosphine does accept a proton, it forms the phosphonium ion, PH_4^+. For example, it reacts with hydrogen iodide:

$$PH_3(g) + HI(g) \rightarrow PH_4I(s)$$

. . . and its salts are easily decomposed

The product, phosphonium iodide, decomposes at 60 °C and is hydrolysed by water to give phosphine:

$$PH_4^+I^-(s) + H_2O(l) \rightarrow PH_3(g) + H_3O^+(aq) + I^-(aq)$$

Phosphine is only slightly soluble in water as it does not take part in hydrogen bonding with water molecules.

. . . it is also a reducing agent

Phosphine is a reducing agent. It reduces silver and copper(II) salts to the free metals, while being oxidised to phosphonic acid.

1. Draw an apparatus which could be used for the laboratory preparation and collection of dry ammonia from ammonium sulphate and calcium hydroxide. Explain the choice of drying agent and method of collection.

2. Describe two methods of preparing phosphine in the laboratory. Why are the methods used different from those for the preparation of ammonia?

3. Compare ammonia and phosphine with regard to (*a*) solubility in water, (*b*) stability and (*c*) basicity. Explain the differences between the two hydrides.

4. Give two examples of reactions in which ammonia acts as a reducing agent.

5. Describe the industrial manufacture of ammonia, explaining the chemical principles involved.

22.6 HALIDES

22.6.1 HALIDES OF NITROGEN

NCl₃ is hydrolysed

NF₃ is stable to hydrolysis

Nitrogen forms halides of formula NX_3. The rest of the group form halides EX_3, and some form EX_5. Nitrogen trichloride is a covalent, highly explosive oil, which is made by a highly dangerous reaction between ammonia and chlorine. It is readily decomposed by water to form ammonia and chloric(I) acid:

$$NCl_3(l) + 3H_2O(l) \rightarrow NH_3(g) + 3HClO(aq)$$

followed by

$$NH_3(aq) + HClO(aq) \rightleftharpoons NH_4^+(aq) + ClO^-(aq)$$

The hydrolysis of a chloride of the first member of a group is unusual and contrasts with the stability to hydrolysis of tetrachloromethane, CCl_4. Nitrogen trifluoride is not hydrolysed.

22.6.2 HALIDES OF PHOSPHORUS

Halide	Preparation	Reaction with water	Other reactions
PCl_3 (PBr₃ and PI₃ are similar.)	Chlorine is passed over white phosphorus. P burns, and PCl₃ distils over.	Easily hydrolysed to H_3PO_3, phosphonic acid.	With $O_2 \rightarrow POCl_3$ With $Cl_2 \rightarrow PCl_5$
PCl_5 (PBr₅ is similar. No PI₅ exists.)	Chlorine is passed over PCl₃.	With cold water, gives POCl₃, phosphorus trichloride oxide. Boiling water gives H_3PO_4, phosphoric(V) acid.	PCl₅, PBr₃, PI₃ react with alcohols ROH → RX and with carboxylic acids $RCO_2H \rightarrow RCOX$

TABLE 22.4 Halides of Phosphorus

The structure of phosphorus pentachloride is the PCl_5 monomer in the vapour state and a structure of $PCl_4^+ PCl_6^-$ ions in the solid state [see Figure 22.7]. Phosphorus pentabromide is also an ionic structure in the solid state, but the ions are $PBr_4^+ Br^-$.

FIGURE 22.7
Structures of (a) $PCl_5(g)$
and (b) $PCl_5(s)$

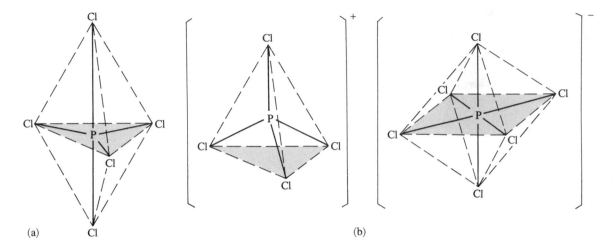

(a)   (b)

CHECKPOINT 22B: HALIDES

1. Draw 'electrons-in-boxes' diagrams to explain why phosphorus can form PCl_3 and PCl_5, but nitrogen can form only NCl_3.

2. Explain why NCl_3 is hydrolysed more slowly than PCl_3, to give different products. Why is NF_3 stable to hydrolysis? How does the ease of hydrolysis change from PCl_3 through $AsCl_3$ and $SbCl_3$ to $BiCl_3$?

22.7   OXIDES

The oxides formed by Group 5B elements are listed in Table 22.5.

	OXIDATION STATE				
Element	*+1*	*+2*	*+3*	*+4*	*+5*
Nitrogen	$N_2O(g)$	$NO(g)$	$N_2O_3(g)$	$NO_2(g)$, $N_2O_4(g)$	$N_2O_5(s)$
Phosphorus			$P_4O_6(s)$	$PO_2(s)$*	$P_4O_{10}(s)$
Arsenic			$As_4O_6(s)$		$As_4O_{10}(s)$
Antimony			$Sb_2O_3(s)$*	$SbO_2(s)$*	$Sb_2O_5(s)$*
Bismuth			$Bi_2O_3(s)$*		

TABLE 22.5
The Oxides of Group 5B

*Macromolecular structure

The gradation in acidic character of the oxides can be represented

Acidic character of oxides

E_2O_3, E_2O_5:	N	P	As	Sb	Bi
Character	Strongly acidic $\longrightarrow$		Weakly acidic $\rightarrow$	Amphoteric (Sb_2O_3 acidic) $\rightarrow$	Weakly basic

Dinitrogen oxide, N_2O, and nitrogen monoxide, NO, are neutral, and the other oxides increase in acidity

$$N_2O_3 < NO_2 < N_2O_5$$

22.7.1 OXIDES OF NITROGEN

DINITROGEN OXIDE

Dinitrogen oxide is made from an ammonium salt and a nitrate

Dinitrogen oxide (nitrogen(I) oxide), N_2O, is made by heating an ammonium salt with a nitrate:

$$NH_4^+(s) + NO_3^-(s) \rightarrow N_2O(g) + 2H_2O(g)$$

It rekindles a glowing splint...

...and is neutral...

The structure

The alternative method of heating ammonium nitrate is not employed because it is explosive. The gas is an anaesthetic called 'laughing gas' with a sickly-sweet smell. It rekindles a glowing splint because this is hot enough to decompose the gas into nitrogen and oxygen. Dinitrogen oxide is insoluble and neutral. Its structure is represented as

N≡≡N⸺O

NITROGEN MONOXIDE

Nitrogen monoxide is formed in lightning flashes and by the oxidation of ammonia

Nitrogen monoxide or nitrogen oxide, NO, is an insoluble, neutral oxide. On meeting air, it reacts rapidly with oxygen to form nitrogen dioxide, NO_2. Nitrogen monoxide is formed during lightning flashes [see Figure 22.1, p. 432] and in the catalytic oxidation of ammonia [p. 436].

Nitrogen monoxide can be made in the laboratory by the following methods:

It is prepared from copper and nitric acid...

1. The reaction of copper with a 50 : 50 solution of nitric acid and water [p. 445].

...and from a nitrite and an acid

2. The reaction of sodium nitrite with a 50 : 50 solution of sulphuric acid in water. Nitrous acid is formed, and disproportionates with the formation of nitrogen monoxide [p. 443].
The gas is purified by absorption in iron(II) sulphate solution to form the compound $FeSO_4 \cdot NO$, which is decomposed by heating it in the absence of air.

Nitrogen monoxide extinguishes a burning splint, but allows magnesium to burn in it.

The structure

The structure can be represented as a resonance hybrid of two odd-electron structures:

$$: \overset{\bullet}{N} = \overset{\bullet\bullet}{O} : \quad \longleftrightarrow \quad : \overset{\bullet\bullet}{N} = \overset{\bullet}{O} :$$

NITROGEN(III) OXIDE

Nitrogen(III) oxide, N_2O_3, is an unstable oxide. It is the anhydride of nitrous acid, HNO_2.

NITROGEN DIOXIDE

Nitrogen dioxide dimerises

Nitrogen has an oxidation state of $+4$ in nitrogen dioxide, NO_2, and in its dimer, dinitrogen tetraoxide, N_2O_4. Nitrogen dioxide is a brown gas which dimerises on cooling to form a colourless liquid:

$$N_2O_4(l) \underset{\text{cool}}{\overset{\text{heat}}{\rightleftharpoons}} N_2O_4(g) \underset{\text{cool}}{\overset{\text{heat}}{\rightleftharpoons}} 2NO_2(g)$$

Colourless   *Brown*

Nitrogen dioxide can be made in the laboratory by the following methods:

It is prepared from nitrates...

1. The action of heat on an anhydrous nitrate (other than those of Group 1A):

$$2Pb(NO_3)_2(s) \rightarrow 2PbO(s) + 4NO_2(g) + O_2(g)$$

The gas can be collected by condensing it as liquid N_2O_4.

...and from nitric acid

2. The action of a metal on concentrated nitric acid (e.g., Cu):

$$Cu(s) + 4HNO_3(conc) \rightarrow Cu(NO_3)_2(aq) + 2NO_2(g) + O_2(g)$$

Reactions of nitrogen dioxide

The decomposition of NO_2...

1. Burning magnesium will continue to burn in nitrogen dioxide. The flame is hot enough to decompose the gas into nitrogen monoxide and oxygen.

...and its oxidising reactions

2. Nitrogen dioxide oxidises reducing agents, e.g., H_2S, SO_2, with the formation of nitrogen monoxide.

NO_2 is the anhydride of nitrous and nitric acids

3. Nitrogen dioxide is a mixed anhydride. It dissolves in water to form a mixture of nitric and nitrous acids:

$$2NO_2(g) + H_2O(l) \rightarrow HNO_2(aq) + HNO_3(aq)$$

The nitrous acid in the solution is unstable, and disproportionates into nitric acid and nitrogen monoxide [p. 444].

A molecule of NO_2 possesses an unpaired electron. [See Figure 22.8.] The structure cannot be represented by any one valence bond structure, such as (a). It can be represented by the delocalised π bonding in structure (b). The unpaired electrons in two NO_2 molecules can pair up to form a N—N bond in N_2O_4, as shown in structure (c).

FIGURE 22.8
The Structure of NO_2
and N_2O_4

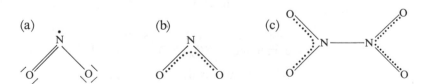

(a)   (b)   (c)

NITROGEN(V) OXIDE

N_2O_5 is the anhydride of nitric acid

Nitrogen(V) oxide, N_2O_5, is a colourless solid. It is the anhydride of nitric acid, HNO_3, and is made by the action of the dehydrating agent phosphorus(V) oxide on nitric acid:

$$4HNO_3(l) + P_4O_{10}(s) \longrightarrow 2N_2O_5(s) + 4HPO_3(l)$$

| Nitric(V) acid | Phosphorus(V) oxide | Nitrogen(V) oxide | Phosphoric(V) acid |

22.7.2 OXIDES OF PHOSPHORUS

Oxide	Preparation	Reactions
Phosphorus(III) oxide, P_4O_6	Burn phosphorus in a limited supply of air.	Is the anhydride of phosphonic acid, H_3PO_3. With $O_2 \rightarrow P_4O_{10}$; with $Cl_2 \rightarrow POCl_3$.
Phosphorus(V) oxide, P_4O_{10}	Burn phosphorus in an excess of air.	With a limited amount of water forms trioxophosphoric(V) acid, HPO_3; with more water forms tetraoxophosphoric(V) acid, H_3PO_4. Used as a drying agent and as a dehydrating agent, e.g., $HNO_3 \rightarrow N_2O_5$; Amide, $RCONH_2 \rightarrow$ nitrile, RCN [p. 710]

TABLE 22.6
The Oxides of Phosphorus

The structures of the molecules P_4O_6 and P_4O_{10} are shown in Figure 22.9.

FIGURE 22.9
Structure of P_4, P_4O_6
and P_4O_{10}

 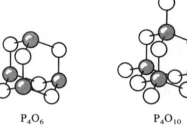

P₄   P₄O₆   P₄O₁₀

=== CHECKPOINT 22C: OXIDES ===

1. Describe the part played by nitrogen monoxide in the nitrogen cycle.

2. Give equations for the reactions of nitrogen monoxide and (a) burning magnesium, (b) acidic manganate(VII), which oxidises NO to NO_3^-.

3. State the products of the reactions of phosphorus(III) oxide with (a) water, (b) oxygen, (c) chlorine. Give equations.

4. State two reactions in which phosphorus(V) oxide reacts as a dehydrating agent. What is formed when P_4O_{10} reacts with (a) cold water, (b) hot water?

5. How can you distinguish by a chemical test between (a) NO_2 and N_2O, (b) N_2O and O_2?

6. Discuss the gradation in acidic character of the oxides of Group 5B.

22.8 THE OXO-ACIDS OF NITROGEN

22.8.1 NITROUS ACID (NITRIC(III) ACID, HNO_2)

Nitrous acid, HNO_2, is a weak acid of pK_a 3.34 at 25 °C. A solution of nitrous acid can be made by adding a mineral acid to a cold, dilute solution of a nitrite:

Nitrites react with mineral acid to give nitrous acid, which disproportionates at room temperature to form NO_2

$$NaNO_2(aq) + HCl(aq) \xrightarrow{5\,°C} HNO_2(aq) + NaCl(aq)$$

The solution is fairly stable below 5 °C and is used in the preparation of diazonium compounds [p. 678]. At room temperature, nitrous acid disproportionates to give nitric acid and nitrogen monoxide, which immediately reacts with air to form nitrogen dioxide:

$$3HNO_2(aq) \rightarrow HNO_3(aq) + 2NO(g) + H_2O(l)$$

$$2NO(g) + O_2(g) \rightarrow 2NO_2(g)$$

This reaction is used to distinguish between nitrites and nitrates.

The nitrite ion is oxidised to the nitrate ion by acidified manganate(VII):

Reducing reactions of nitrites

$$NO_2{}^-(aq) + H_2O(l) \rightleftharpoons NO_3{}^-(aq) + 2H^+(aq) + 2e^-; \quad E^\ominus = -0.94\,V$$

Nitrous acid and nitrites oxidise reducing agents such as iodides and iron(II) salts, with the formation of nitrogen oxide:

$$NO_2{}^-(aq) + 2H^+(aq) + e^- \rightleftharpoons NO(g) + H_2O(l); \quad E^\ominus = +0.99\,V$$

22.8.2 NITRITES (NITRATES(III))

Nitrates of Na and K can be made by heating the nitrates

The nitrites commonly used in the laboratory are those of sodium and potassium. They are made by heating the nitrates:

$$2NaNO_3(s) \rightarrow 2NaNO_2(s) + O_2(g)$$

The structure of $NO_2{}^-$

The structure of the nitrite ion can be represented as

$$\left[\begin{array}{c} N \\ O^{\diagdown}{}^{\diagup}O \end{array} \right]^-$$

22.8.3 NITRIC ACID (NITRIC(V) ACID, HNO₃)

The industrial method of making nitric acid is the catalytic oxidation of ammonia. A flow diagram for the process, which was invented by Ostwald, is shown in Figure 22.10. The reactions which take place are

The catalytic oxidation of ammonia gives nitric acid

$$4NH_3(g) + 5O_2(g) \xrightarrow[900\,°C]{Pt,\,Rh} 4NO(g) + 6H_2O(l) \qquad [1]$$

$$2NO(g) + O_2(g) \rightarrow 2NO_2(g) \qquad [2]$$

$$4NO_2(g) + O_2(g) + 2H_2O(l) \rightarrow 4HNO_3(l) \qquad [3]$$

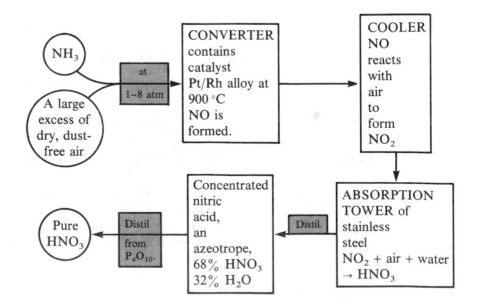

FIGURE 22.10 A Flow Diagram for the Manufacture of Nitric Acid

The azeotrope of
HNO$_3$/H$_2$O

The laboratory acid called 'concentrated nitric acid' is a constant-boiling mixture of 68% nitric acid with water. It can be distilled from phosphorus(V) oxide to give pure nitric acid.

The laboratory
preparation from a nitrate
and conc. sulphuric acid

In the laboratory, nitric acid is made by distilling sodium nitrate and concentrated sulphuric acid:

$$NaNO_3(s) + H_2SO_4(l) \rightarrow HNO_3(l) + NaHSO_4(s)$$

Concentrated nitric acid is usually yellow. In the light, it decomposes to form nitrogen dioxide (the source of the yellow colour) and oxygen:

$$4HNO_3(aq) \rightarrow 4NO_2(g) + O_2(g) + 2H_2O(l)$$

The decomposition is accelerated by heat.

PROPERTIES OF NITRIC ACID

Pure nitric acid is a colourless liquid which boils at 86 °C. The vapour consists of molecules which have the structure

The structure of HNO$_3$

The anhydrous acid is conducting because it is to some extent ionised:

$$2HNO_3 \rightleftharpoons NO_2^+ + NO_3^- + H_2O$$

Ionisation in the absence
of water and in the
presence of water

The degree of ionisation increases in the presence of a proton-acceptor, and in aqueous solution nitric acid is a strong acid:

$$HNO_3(l) + H_2O(l) \rightleftharpoons H_3O^+(aq) + NO_3^-(aq)$$

REACTIONS OF NITRIC ACID

Aqueous nitric acid gives the reactions typical of mineral acids [p. 250]. Metals react to form nitrates. Since nitric acid is an oxidising agent, hydrogen is rarely formed. Only magnesium and calcium react with cold, dilute nitric acid to give hydrogen:

$$Mg(s) + 2HNO_3(aq) \xrightarrow[\text{dilute}]{\text{cold}} Mg(NO_3)_2(aq) + H_2(g)$$

Other metals reduce the nitrate ion in preference to the hydrogen ion. The reduction products formed are governed by the redox equilibria:

Nitric acid is an oxidising
agent, being reduced to a
number of different
products

(a) $\quad NO_3^-(aq) + 2H^+(aq) + e^- \rightleftharpoons H_2O + NO_2$

(b) $\quad NO_3^-(aq) + 4H^+(aq) + 3e^- \rightleftharpoons 2H_2O + NO$

(c) $\quad NO_3^-(aq) + 5H^+(aq) + 4e^- \rightleftharpoons 2\frac{1}{2}H_2O + \frac{1}{2}N_2O$

(d) $\quad NO_3^-(aq) + 8H^+(aq) + 6e^- \rightleftharpoons 2H_2O + H_3\overset{+}{N}OH$

(e) $\quad NO_3^-(aq) + 10H^+(aq) + 8e^- \rightleftharpoons 3H_2O + NH_4^+$

In these reactions, nitrogen is reduced from an oxidation state of +5 in NO_3^- to (a) +4, (b) +2, (c) +1, (d) −1 or (e) −3.

The reaction of conc.
HNO$_3$ with metals . . .

Metals low in the Electrochemical Series (e.g., Cu, Pb) react with concentrated nitric acid according to equation (a) to give NO_2, and with dilute nitric acid according to equation (b) to give NO. Metals high in the Electrochemical Series (e.g., Mg, Zn)

...e.g., Cu, Pb... react with dilute nitric acid according to equation (c), (d) or (e), depending on the concentration and the temperature.

...e.g., Mg, Zn...

...and cations... Cations of metals with variable oxidation states (e.g., Fe^{2+}, Sn^{2+}) are oxidised by concentrated nitric acid.

...non-metals... Non-metallic elements (e.g., S, P) are oxidised by concentrated nitric acid to the acids corresponding to their highest oxidation states (e.g., H_2SO_4, H_3PO_4).

and anions Anions oxidised by concentrated nitric acid include Cl^-, Br^-, I^-, S^{2-}.

Weakly metallic elements are oxidised to hydrated oxides:

$$Sn(s) + 4HNO_3(conc.) \rightarrow SnO_2(aq) + 2H_2O(l) + 4NO_2(g)$$

In reaction with conc. HNO_3 some metals are rendered 'passive' With some metals (e.g., Al, Cr, Fe) concentrated nitric acid reacts to cover the surface with a film of oxide which will not react with any acid. The metal is then described as 'passive'.

Noble metals are attacked by 'aqua regia' The few metals that are not attacked by concentrated nitric acid (e.g., Pt, Au) are called **noble metals**. They are attacked by *aqua regia*, a mixture of concentrated nitric acid and concentrated hydrochloric acid. Concentrated nitric acid reacts with gold to a slight extent to form Au^{3+} ions. Chloride ions remove these Au^{3+} ions to form soluble complex ions:

$$Au^{3+}(aq) + 4Cl^-(aq) \rightleftharpoons AuCl_4^-(aq)$$

Devarda's alloy (Al, Cu, Zn) in alkaline solution reduces nitric acid and nitrates and nitrites to ammonia:

$$4Zn(s) + NO_3^-(aq) + 7OH^-(aq) + 6H_2O(l) \rightarrow 4[Zn(OH)_4]^{2-}(aq) + NH_3(g)$$

A test for a nitrate is its reduction to ammonia This reaction is used as a test for nitrates. Since the test will only be meaningful in the absence of ammonium salts, these are removed by heating the suspected nitrate with alkali before the alloy is added. A separate test must be used to distinguish a nitrate from a nitrite.

CHECKPOINT 22D: OXIDATION REACTIONS

The half-reaction equations for the oxidation reactions of nitric acid are given on p. 445. Write equations for the redox reactions mentioned above by combining the following half-reaction equations:

1. $Cu \rightarrow Cu^{2+} + 2e^-$ and equation (a)

2. $Cu \rightarrow Cu^{2+} + 2e^-$ and equation (b)

3. $Zn \rightarrow Zn^{2+} + 2e^-$ and equation (c)

4. $Fe^{2+} \rightarrow Fe^{3+} + e^-$ and equation (b)

5. $Au \rightarrow Au^{3+} + 3e^-$ and equation (a)

6. $S + 4H_2O \rightarrow SO_4^{2-} + 8H^+(aq) + 6e^-$ and equation (a)

7. $2I^- \rightarrow I_2 + 2e^-$ and equation (a)

8. $P + 4H_2O \rightarrow PO_4^{3-} + 8H^+(aq) + 5e^-$ and equation (a)

9. $2Cl^- \rightarrow Cl_2 + 2e^-$ and equation (a)

10. $Sn^{2+} \rightarrow Sn^{4+} + 2e^-$ and equation (b)

USES OF NITRIC ACID

Nitric acid is used in the manufacture of some explosives and dyes Nitric acid is used in the manufacture of organic nitro-compounds [p. 579]. Some of these are useful explosives (e.g., dynamite and TNT). Others are useful intermediates in the dye industry. A large amount of nitric acid is used in the manufacture of nitrate fertilisers.

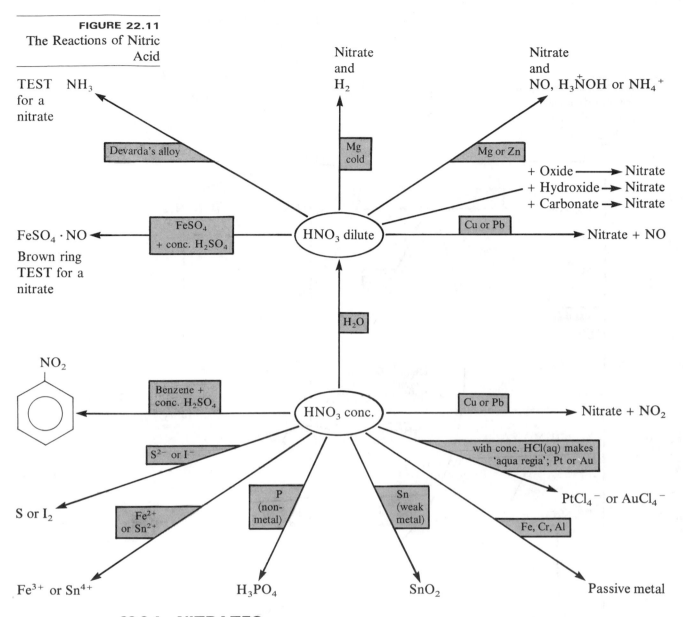

FIGURE 22.11
The Reactions of Nitric Acid

22.8.4 NITRATES

Nitrates are soluble and are decomposed by heat

Nitrates are prepared by the action of nitric acid on metals, metal oxides, hydroxides and carbonates. They all dissolve in water and decompose when heated [see Table 22.7].

Nitrate	Products of thermal decomposition
Group 1A (except Li)	Nitrite + Oxygen: $MNO_2 + O_2$
Most metals	Oxide + Oxygen + Nitrogen dioxide: $MO + O_2 + NO_2$
Unreactive metals (Ag, Hg)	Metal + Oxygen + Nitrogen dioxide: $M + O_2 + NO_2$
Ammonium	Explosive → Dinitrogen oxide, $N_2O + H_2O$

TABLE 22.7 Thermal Decomposition of Nitrates

The brown ring test for a nitrate or a nitrite

One test for a nitrate, the reduction to ammonia by Devarda's alloy, has been described. An alternative, the **brown ring test**, requires the addition to a solution of the suspected nitrate of iron(II) sulphate solution and a few drops of concentrated sulphuric acid. The formation of a ring of the brown complex, $FeSO_4 \cdot NO$, proves the presence of a nitrate or a nitrite.

The nitrate ion has the planar structure represented:

The structure of NO_3^-

FIGURE 22.12 Nitrogen Compounds illustrating the different Oxidation States of Nitrogen

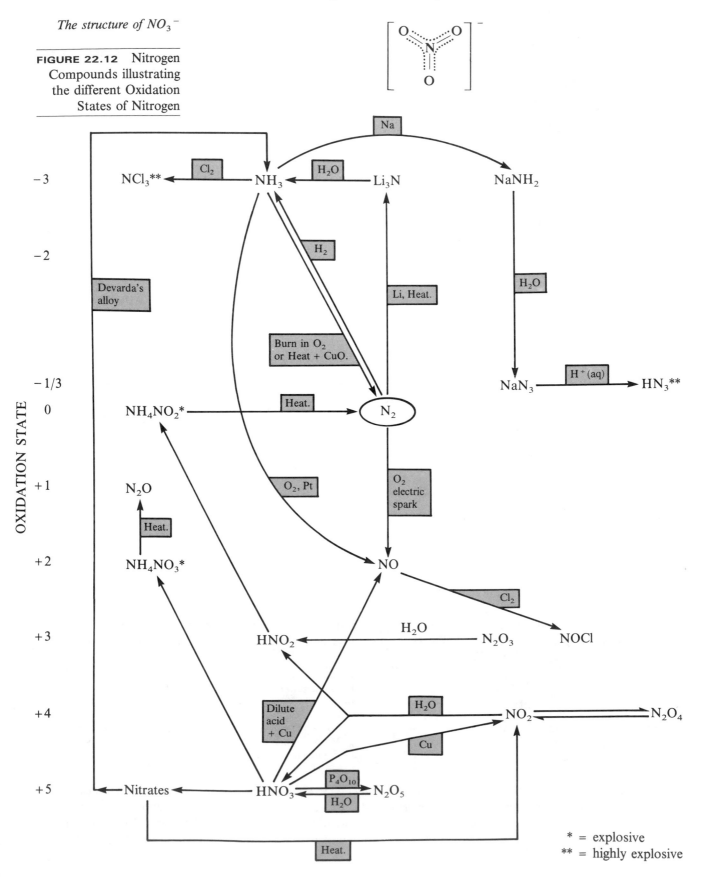

* = explosive
** = highly explosive

22.9 THE OXO-ACIDS OF PHOSPHORUS

The oxo-acids of phosphorus are

H_3PO_2 Phosphinic acid

H_3PO_3 Phosphonic acid

H_3PO_4 Phosphoric(V) acid (tetraoxophosphoric(V) acid in full, or orthophosphoric(V) acid)

$H_4P_2O_7$ Heptaoxodiphosphoric(V) acid (or pyrophosphoric(V) acid)

HPO_3 Trioxophosphoric(V) acid (or metaphosphoric(V) acid)

22.9.1 PHOSPHONIC ACID, H_3PO_3

Phosphonic acid is made by adding water to phosphorus(III) oxide or by hydrolysing phosphorus trichloride:

H_3PO_3 is made from P_4O_6 or from PCl_3

$$P_4O_6(s) + 6H_2O(l) \rightarrow 4H_3PO_3(l)$$

$$PCl_3(l) + 3H_2O(l) \rightarrow H_3PO_3(l) + 3HCl(g)$$

It is a moderately strong diprotic acid, with the structure

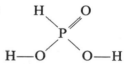

It disproportionates on heating to phosphine and phosphoric(V) acid:

Disproportionation on heating

$$4H_3PO_3(l) \rightarrow PH_3(g) + 3H_3PO_4(l)$$

The acid and its salts are reducing agents. Phosphonates are insoluble, except for those of Group 1A.

22.9.2 PHOSPHORIC(V) ACID, H_3PO_4

Phosphoric(V) acid can be made by boiling phosphorus(V) oxide with water:

H_3PO_4 is made from P_4O_{10} and from $Ca_3(PO_4)_2$

$$P_4O_{10}(s) + 6H_2O(l) \rightarrow 4H_3PO_4(l)$$

It can also be obtained from the calcium phosphate ore, *rock phosphate*, by treatment with concentrated sulphuric acid:

$$Ca_3(PO_4)_2(s) + 6H_2SO_4(l) \rightarrow 3Ca(HSO_4)_2(s) + 2H_3PO_4(l)$$

Phosphoric(V) acid is a deliquescent crystalline solid. It is often sold as an 85% solution in water called 'syrupy phosphoric acid'. The high viscosity of this solution arises from intermolecular hydrogen-bonding.

The salts of H_3PO_4 are Na_2HPO_4 and NaH_2PO_4

Phosphoric(V) acid is a relatively weak triprotic acid. From it can be prepared by titration sodium dihydrogenphosphate(V), NaH_2PO_4, and disodium hydrogen-phosphate(V), Na_2HPO_4. Trisodium phosphate(V), Na_3PO_4, cannot be made in this way because of the extensive hydrolysis of salts of PO_4^{3-}, which is a strong base:

$$PO_4^{3-}(aq) + H_2O(l) \rightleftharpoons HPO_4^{2-}(aq) + OH^-(aq)$$

Most phosphates are insoluble in water. The structure of H_3PO_4 is shown in Figure 22.13. The anions are resonance hybrids, and one canonical structure is shown for each.

FIGURE 22.13
Structure of H_3PO_4 and
the Anions

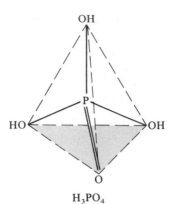

H_3PO_4

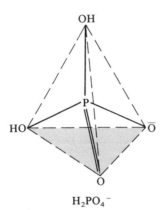

$H_2PO_4^-$

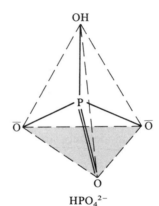

HPO_4^{2-}

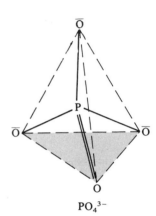

PO_4^{3-}

The action of heat on phosphoric(V) acid is

$H_4P_2O_7$ and HPO_3

$$2H_3PO_4(l) \xrightarrow{270\,°C} H_4P_2O_7(s) + H_2O(g) \xrightarrow{315\,°C} 2HPO_3(s) + 2H_2O(g)$$

Phosphoric(V)   Heptaoxodiphosphoric(V)   Trioxophosphoric(V)
acid   acid   acid

22.9.3 PHOSPHATES

Test for a phosphate(V)   Phosphates(V), PO_4^{3-}, give a yellow precipitate of ammonium phosphomolybdate when warmed with nitric acid and ammonium molybdate solution. Heptaoxo-diphosphates(V), $P_2O_7^{4-}$, and trioxophosphates(V), PO_3^-, also give positive results, but the precipitates are slower to appear.

22.9.4 USES OF PHOSPHORIC ACID AND PHOSPHATES

Uses of phosphates in fertilisers

...the food industry...

...and water softeners

Calcium phosphate occurs naturally and is used in the manufacture of fertilisers. Reaction with concentrated sulphuric or phosphoric or nitric acid yields the fertilisers *superphosphate of lime*, $Ca(H_2PO_4)_2 + CaSO_4$, or *triple superphosphate*, $Ca(H_2PO_4)_2$, or *nitrophos*, $Ca(H_2PO_4)_2 + Ca(NO_3)_2$. Sodium phosphate is used in the food industry. Calgon®, a water softener, is a polymeric sodium phosphate.

Phosphoric acid is used in rust-inhibition

Phosphoric(V) acid is used in the rustproofing of steel. If Fe^{3+} ions are formed by rusting, they react to form insoluble iron(III) phosphate(V) which protects the steel beneath from further attack.

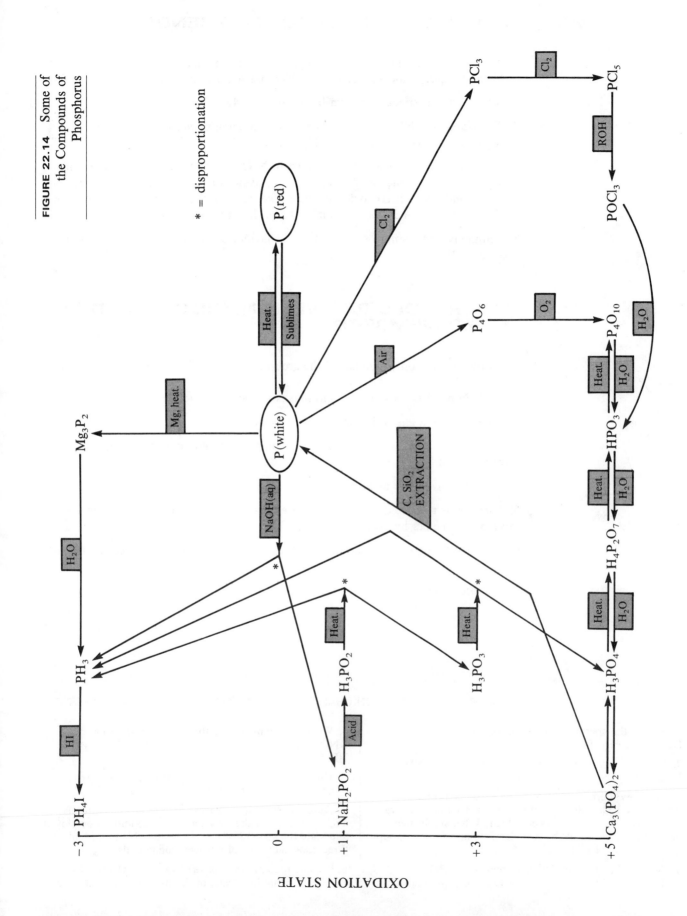

FIGURE 22.14 Some of the Compounds of Phosphorus

* = disproportionation

22.10   A SUMMARY OF GROUP 5B TRENDS

1. The character of the elements changes from non-metallic nitrogen, which consists of N_2 molecules, to metallic bismuth, which has a close-packed metallic structure.

2. The 'inert pair effect' is shown by bismuth [p. 454].

3. The stability to heat of the hydrides, their reducing power and their basic strength all decrease in the order $NH_3 \gg PH_3 > AsH_3 > SbH_3 > BiH_3$.

4. The oxides of nitrogen are acidic (except N_2O and NO), as are the oxides of phosphorus and arsenic. While antimony(III) oxide is amphoteric, antimony(V) oxide is acidic, and bismuth(III) oxide is basic. The basicity of the oxides increases down the group, and the $+3$ oxides are more basic than the $+5$ oxides.

5. Antimony and bismuth show metallic behaviour in the formation of the cations, Sb^{3+} and Bi^{3+}.

22.11   DIFFERENCES BETWEEN NITROGEN AND OTHER MEMBERS

1. Nitrogen is the only member of the group that readily forms multiple covalent bonds.
The $N\equiv N$ bond is very strong, and makes nitrogen unreactive.

2. Dinitrogen is a gas; other members are solids.

3. Nitrogen is the most electronegative member of the group and the only one to form hydrogen bonds.

4. It is the only group member that forms an ion E^{3-}.

5. Nitrogen is restricted to an L shell of electrons. Other members of the group can expand their octets by using their d orbitals.

6. Nitrogen forms oxides other than E_2O_3 and E_2O_5, which are formed by the rest of the group.

7. The properties of nitric acid and nitrous acid are not shared by the corresponding acids of the rest of the group.

QUESTIONS ON CHAPTER 22

1. Sketch the spatial arrangement of bonds in NH_3, NH_4^+, P_4, PF_5, PCl_5, PBr_5.

2. Show the arrangement of valence electrons in NH_3, NH_4^+, NO_2, N_2O_4, NO_3^- and NO.

3. List the differences between the chemical properties of nitrogen and phosphorus. What is the reason for the large difference in reactivity between them? What similarities justify the inclusion of the two elements in the same group of the Periodic Table?

4. Outline the laboratory preparations of (a) PH_3, (b) PH_4I, (c) PCl_3, (d) PCl_5, (e) P_4O_{10}. Give equations.

5. Write equations for the thermal decomposition of KNO_3, $Zn(NO_3)_2$, $Hg(NO_3)_2$, NH_4NO_3, NH_4NO_2 and NH_4Cl.

6. How could you find the concentration of (a) a solution of nitrate ions and (b) a solution of ammonium ions?

7. Describe the Haber process for the manufacture of ammonia. Explain the reasons for the conditions employed, and state the sources of the raw materials used. Mention three important uses of nitrogen compounds.

8. Outline preparations for (a) NO, (b) N_2O_4, (c) N_2O, (d) P_4O_{10}. State how each of the oxides reacts with water.

9. Compare the chemistry of nitrogen and phosphorus with respect to (a) the structure of the elements, (b) the hydrides (bonding, basicity, thermal stability and reaction with oxygen), and (c) the chlorides (preparation, bonding and reaction with water).

10. Draw a phase diagram for phosphorus. Indicate what the lines and areas represent, and state what type of allotropy is shown [p. 232].

11. Discuss the application of the principles of kinetics and equilibrium to the manufacture of (a) ammonia, (b) nitric acid [see Chapters 11 and 14].

†12. Describe and explain what happens when aqueous ammonia is added slowly, until it is in excess, to solutions of (a) copper(II) sulphate, (b) copper(I) chloride and (c) silver nitrate. What use is made of the final products of (b) and (c) in organic chemistry?

13. Explain why the inclusion of nitrogen and phosphorus in the same group of the periodic table is justified. Consider the elements, their hydrides, chlorides and oxides.

14. Explain why nitrogen compounds are added to farmland. Find out, by calculation, which is the best source of nitrogen: sodium nitrate; ammonium nitrate; ammonium sulphate; or urea, CH_4N_2O.

15. (a) Write electron configurations for nitrogen, phosphorus and arsenic in their ground states.
(b) Explain why nitrogen can form N^{3-} ions, but arsenic cannot form As^{3-} ions. Name two elements which react with nitrogen to form ionic nitrides.
(c) Explain why phosphorus and arsenic can show a covalency of 5, but nitrogen cannot. Give examples of compounds in which phosphorus shows a covalency of 5.
(d) Which of the hydrides, NH_3 and PH_3, is the stronger base? Explain the reason for the differences between them.
(e) The boiling temperatures of the hydrides are: $NH_3 = -33\,°C$, $PH_3 = -87\,°C$. Offer an explanation of the difference in terms of structure and bonding.

16. Nitrogen, like boron, aluminium and phosphorus, forms trivalent compounds. Like phosphorus but unlike boron and aluminium it can form compounds with oxidation states of $+5$. Explain these observations and give one suitable illustrative example for each element.
Explain, giving one example for each element, why all the elements can form compounds in which the element has a coordination number of four but in some of their compounds aluminium and phosphorus have higher coordination numbers.

Explain the differences in the acidic behaviour of (a) boric(III) acid and hydrated Al^{3+} and (b) nitric acid and phosphoric(V) acid.

(WJEC 81)

7. (a) For the elements nitrogen and phosphorus compare the trihydrides and pentoxides as follows:
Trihydrides
(i) Shapes of the molecules including the bond angles;
(ii) boiling points (account for the difference);
(iii) stabilities to heat, giving a reason for the difference;
(iv) basicities, illustrated by their reactions with water and acids.

Pentoxides
(i) Physical states at ordinary temperature and pressure;
(ii) structures of the molecules;
(iii) action of water (give equations) on the oxides;
(iv) strengths of the oxoacids formed and the reactions of these with aqueous sodium hydroxide.

(b) In terms of electronic structure and atomic size, account for the similarities and gradual changes in properties of the Group V elements on passing down the group.

(AEB 82)

8. This question concerns the chemistry of *either* sulphur *or* nitrogen. Write your answer in terms of one of these elements only.
(a) Review the principal oxidation states of the element, giving an example of a compound for each oxidation number. Describe and explain *one* 'redox' reaction, showing how the element changes its oxidation number in the course of the reaction described.
(b) Give the name of a hydride of the element and comment on the shape of the molecule. Describe the action of the hydride with water, commenting on points of physico-chemical interest. How does an aqueous solution of the hydride react with a solution of a named metal ion?
(c) Give the names or formulae of
(i) a simple salt,
(ii) a 'large' organic molecule,
containing the element, which are manufactured in the chemical industry. Comment on the properties of each compound, and the reasons for its use; give an outline of its manufacture.

(L 81)

23

GROUP 4B

23.1 THE BONDS FORMED IN GROUP 4B

There is a big gradation in properties down Group 4B

Some of the physical properties of the members of Group 4B are listed in Table 23.1. The gradation in properties from carbon (a non-metallic element) to lead (a metallic element) is much more marked than that in groups on the extreme left or right of the Periodic Table.

Valencies 2 and 4 are met in the group

The elements show valencies of 2 and 4. All the Group 4B elements have the electron configuration (core)ns^2np^2:

	1s	2s	2p
C (ground state)	↑↓	↑↓	↑ ↑ ☐
C* (excited state)	↑↓	↑	↑ ↑ ↑

If an electron is promoted from a 2s orbital to a vacant p orbital, the element can show a covalency of 4. If the energy required to promote the s electron is compensated by the energy released when two additional covalent bonds are formed, the element will show a covalency of 4. Passing down the group, the strength of the covalent bonds formed with other elements decreases and there is an increasing tendency to show a covalency of 2. The behaviour of later members of a group, showing a valency of 2 less than the group valency through a failure to use their s electrons, is called the **'inert pair effect'**. Tin(II) compounds are reducing agents because tin(IV) is the more stable oxidation state of tin, but lead(IV) compounds are oxidising agents because lead(II) is the stable oxidation state for lead.

The 'inert pair effect' favours a valency of 2

Both covalent and ionic compounds are met in Group 4B

Carbon, silicon and germanium form covalent compounds, with a valency of 4. A few compounds of tin and lead are sometimes described as containing the E^{4+} ion, but such compounds are predominantly covalent in character. Tin(II) compounds also are predominantly covalent, but lead(II) compounds are predominantly ionic.

Carbon is restricted to an octet of valence electrons

The coordination number of carbon never exceeds 4 since the number of electrons in the valence shell cannot exceed 8. Later elements in the group can expand their octets. Carbon forms no complex ions as the later elements do, e.g., SiF_6^{2-}, $SnCl_6^{2-}$, $GeCl_6^{2-}$, $PbCl_6^{2-}$. The stability of these complex ions increases as the central atom becomes more electropositive.

Carbon is the only element in the group which forms strong π bonds [p. 103].

The behaviour of lead(II) compounds and some of the properties of tin(II) compounds are attributed to the ions $Pb^{2+}(aq)$ and $Sn^{2+}(aq)$.

454

Element, E	Carbon	Silicon	Germanium	Tin	Lead
Atomic number	6	14	32	50	82
Electron configuration	$(core)2s^22p^2$	$(core)3s^23p^2$	$(core)4s^24p^2$	$(core)5s^25p^2$	$(core)6s^26p^2$
Ionic radius of E^{2+}/nm	–	–	–	0.112	0.120
Covalent radius/nm	0.077	0.117	0.122	0.140	0.154
Standard enthalpy of vaporisation/kJ mol^{-1}	717	440	380	290	180
Density/g cm^{-3}	2.22 (graphite) 3.51 (diamond)	2.33	5.32	7.3	11.3
Electronegativity	2.50	1.75	2.00	1.70	1.55
Ionisation energy (1) MJ mol^{-1}	1.09	0.79	0.76	0.71	0.72
(2)	2.40	1.60	1.50	1.40	1.50
(3)	4.60	3.20	2.30	2.90	3.10
(4)	6.20	4.40	4.40	3.90	4.10
Standard enthalpy of E—E bond/kJ mol^{-1}	348	176	188	150	–
of E—O bond/kJ mol^{-1}	360	374	360	–	–
of E=E bond/kJ mol^{-1}	612	–	–	–	–
of E=O bond/kJ mol^{-1}	743	640	–	–	251
of E—H bond/kJ mol^{-1}	412	338	285	–	–
Structure of element	Macromolecular (diamond, graphite)	Macromolecular (like diamond)		Metallic (and diamond lattice)	Metallic

TABLE 23.1 Group 4B

The principal oxidation
states

The principal oxidation states of the elements are:

Carbon	+4 (-4 in CH_4)
Silicon	+4
Germanium	+4 and +2 (a reducing state)
Tin	+4 and +2 (a reducing state)
Lead	+2 and +4 (an oxidising state)

23.2 THE STRUCTURES OF GROUP 4B ELEMENTS

The structure of carbon...

Carbon has two allotropes, diamond and graphite. Their structures have been described on p. 120 and p. 122 and shown in Figure 6.14, p. 121 and Figure 6.18, p. 122. The allotropy is monotropic:

$$C(diamond) \rightleftharpoons C(graphite); \quad \Delta H^{\ominus} = -2.1\,kJ\,mol^{-1}$$

...which shows
monotropic allotropy

Industrial diamonds are
made from graphite

Although graphite is the more stable allotrope, the activation energy for the change from diamond to graphite is high. As a result, diamond is stable under normal conditions. Since diamond is the denser of the allotropes, the effect of pressure, in accordance with Le Chatelier's Principle, is to increase the stability of diamond relative to graphite. Industrial diamonds are made by subjecting graphite to high temperatures and pressures (e.g., 2000 °C, 10^5 atm), with a catalyst. Graphite is more reactive than diamond: the closed structure of diamond gives rise to high activation energies.

The structures of
charcoal...

The impure forms of carbon (charcoal, etc.) are microcrystalline forms of graphite. Being more finely divided, they are more reactive than graphite.

...silicon and
germanium...

Silicon and germanium have structures of the diamond type. The decrease in melting temperature from carbon through silicon to germanium reflects the decreasing bond enthalpies.

...and tin which shows
enantiotropic allotropy

Tin exists in allotropic forms, two metallic (cubic close-packed) and one of the diamond type. The allotropy is enantiotropic. At low temperatures, β tin changes to α tin, expanding and crumbling as it does so.

$$\alpha \text{ tin} \xrightleftharpoons{13\,°C} \beta \text{ tin} \xrightleftharpoons{161\,°C} \gamma \text{ tin} \xrightleftharpoons{232\,°C} \text{Liquid}$$

	Grey	*White*	
	Diamond structure	Metallic structure	Metallic structure
Density/$g\,cm^{-3}$	5.8	7.3	

...and lead

Lead exists in one form only, which is metallic and cubic close-packed.

23.3 OCCURRENCE, EXTRACTION AND USES OF GROUP 4B ELEMENTS

23.3.1 CARBON

DIAMONDS

Diamonds are used for
jewellery and in industry

Diamonds are mined in Brazil and South Africa. They are prized as jewellery because of their high reflectivity and high refractive index, which makes them sparkle, especially when they have been expertly cut. Small diamonds are used in industry to tip cutting tools and drills.

GRAPHITE

The manufacture of graphite and its uses

Graphite occurs naturally in Germany, the USA and Sri Lanka. The natural supply is supplemented by the manufacture of graphite from coke and sand in an electric furnace:

$$SiO_2(s) + 3C(s) \rightarrow Si(s) + C(s) + 2CO(g)$$

It is used for making electrodes, as a lubricant, as a mixture with clay in 'lead' pencils and for slowing down neutrons in nuclear reactors [p. 23].

Charcoal

Some of the impure forms of carbon are wood charcoal, animal charcoal, sugar charcoal, carbon black, coal and coke.

23.3.2 SILICON

Sources of silicon and purification

Silicon occurs naturally as silicon(IV) oxide, SiO_2, in sand and quartz and also as a number of silicates. It is extracted as shown in Figure 23.1. When it is required in a high state of purity, for use as a semiconductor in transistors, it is purified by zone refining, as shown in Figure 23.2.

FIGURE 23.1
Extraction of Silicon

| SiO_2 impure | → | Heat with coke. | → | Si + impurities | → | Cl_2, heat. | → | $SiCl_4$ + impurities |

Distil.

| Si of high purity | ← | Zone refining | ← | Si | ← | Reduce with H_2. | ← | $SiCl_4$ pure |

FIGURE 23.2
Zone Refining

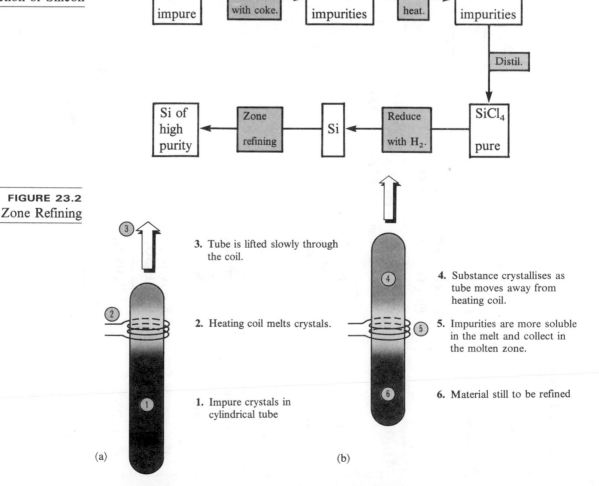

3. Tube is lifted slowly through the coil.

4. Substance crystallises as tube moves away from heating coil.

2. Heating coil melts crystals.

5. Impurities are more soluble in the melt and collect in the molten zone.

1. Impure crystals in cylindrical tube

6. Material still to be refined

(a)   (b)

TABLE 23.2 Reactions of Group 4B

Reagent	Reaction	Comment
O_2 or air, heat.	$E + O_2 \rightarrow EO_2$ $Pb + O_2 \rightarrow PbO \rightarrow Pb_3O_4$	E = C, Si, Ge, Sn Pb only (C → CO in limited air supply)
H_2O, 20 °C	Pb + soft water → $Pb(OH)_2$(aq) + hard water → $PbSO_4$(s), $PbCO_3$(s)	Pb only
Steam, heat.	$E + 2H_2O \rightarrow EO_2 + 2H_2$ $C + H_2O \rightarrow CO + H_2$	E = Sn, Si
S, heat.	$E + 2S \rightarrow ES_2$	All except Pb; Pb → PbS
Cl_2, heat.	$E + 2Cl_2 \rightarrow ECl_4$	All except Pb; Pb → $PbCl_2$
Metals, heat.	Carbides, silicides, alloys of Sn, Pb	
Hot, conc. HCl	$E + 2H^+$(aq) $\rightarrow E^{2+}$(aq) $+ H_2$	Sn and (slowly) Pb
Hot, conc. H_2SO_4	$E + H_2SO_4 \rightarrow E^{n+} + SO_2$ $C + 2H_2SO_4 \rightarrow CO_2 + 2SO_2 + 2H_2O$	E^{n+} = Sn^{4+}, Pb^{2+} C only
Conc. HNO_3	$3E + 4HNO_3 \rightarrow 3EO_2 + 4NO + 2H_2O$ $3Pb + 8HNO_3 \rightarrow 3Pb(NO_3)_2 + 2NO + 4H_2O$	C; Ge and Sn → hydrated oxide. Not Si, Pb Pb only
Aqueous alkali	$Si + 2OH^- + H_2O \rightarrow SiO_3^{2-} + 2H_2$	Not C, Ge, Pb; Sn slowly → $Sn(OH)_6^{2-}$
Molten base	Form SiO_4^{4-}, GeO_4^{4-}, $Sn(OH)_6^{2-}$, $Pb(OH)_4^{2-}$	Not C

23.3.3 GERMANIUM

Semiconductor   Germanium is mined as the sulphide. The element is a semiconductor, and for use in transistors it is, like silicon, purified by zone refining.

23.3.4 TIN

Extraction of tin...   Tin is mined as *tinstone*, SnO_2. The ore is pulverised, washed by flotation, roasted to remove sulphur and reduced to the metal by heating with coke. Tin is used to
...uses   plate iron to prevent it from rusting [p. 503] and as the alloys *type metal* (Sn, Sb, Pb), *bronze* (Sn, Cu) and *solder* (Pb, Sn).

23.3.5 LEAD

Lead is mined as *galena*, PbS. The ore is roasted to form lead(II) oxide, which can be reduced by coke or by lead(II) sulphide:

The extraction of lead...

$$2PbS(s) + 3O_2(g) \rightarrow 2PbO(s) + 2SO_2(g)$$

$$2PbO(s) + PbS(s) \rightarrow 3Pb(s) + SO_2(g)$$

$$PbO(s) + C(s) \rightarrow Pb(s) + CO(g)$$

...and the uses of lead   Traditionally, lead has been used in plumbing, but for many purposes it has been replaced by copper (e.g., water pipes) and by plastics (e.g., gas pipes, guttering and sheathing electrical cables). Alloys of lead are *type metal* (Sn, Sb, Pb) and *solder* (Sn, Pb). Lead is used as a protective shield from radioactivity [p. 19], in accumulators [p. 287] and in the manufacture of the antiknock, tetraethyl-lead, $Pb(C_2H_5)_4$ [p. 545]. Lead compounds are used as pigments, e.g., white basic lead carbonate, $Pb(OH)_2 \cdot PbCO_3$, and the orange pigment 'red lead', Pb_3O_4.

23.4 CHEMICAL REACTIONS OF GROUP 4B

Reactions of Group 4B   Although in Group 4B the difference between the first and last members of the group has reached a maximum, the elements do form a set. The reactions which they have in common are summarised in Table 23.2. As in the other B groups, there are differences between the first member and the second. The demarcation in properties between the Period 2 element and the Period 3 element is maximal in Group 4B. The special features of carbon chemistry are discussed on p. 460.

CHECKPOINT 23A: STRUCTURE AND REACTIONS

1. Describe preparations of (*a*) diamond and (*b*) graphite. List the uses of the two allotropes, and explain the difference in properties between them.

*2. Describe how extremely pure germanium is obtained.

3. Carbon and lead have proton (atomic) numbers 6 and 82. Write the electronic structure of each atom. Explain,

on the basis of their electronic structures, why tin is metallic and carbon is non-metallic.

4. How does the structure of graphite explain (*a*) its ability to mark paper, (*b*) its use as a lubricant and (*c*) its use as an electrical conductor?

23.5 SPECIAL FEATURES OF CARBON CHEMISTRY

23.5.1 CATENATION

Carbon catenates to form —C—C—C— chains

Catenation is the ability of an element to form bonds between atoms of the same element. Carbon catenates to form chains and rings, with single, double and triple bonds. To be able to catenate, an element **E** must have a valency $\geqslant 2$ and must form **E—E** bonds which are similar in strength to those of **E** to other elements, particularly **E—O** bonds. Since elements are exposed to the possibility of reaction with air and water, if **E—O** bonds are stronger than **E—E** bonds, then energy considerations will favour the formation of compounds containing **E—O** bonds rather than **E—E—E** chains [see Question 2, p. 462]. For silicon, it is energetically

Silicon forms —Si—O—Si— bonds in preference

more favourable to form a chain

$$-\overset{|}{\underset{|}{Si}}-O-\overset{|}{\underset{|}{Si}}-O-\overset{|}{\underset{|}{Si}}-$$

while for carbon, energy considerations favour a chain

$$-\overset{|}{\underset{|}{C}}-\overset{|}{\underset{|}{C}}-\overset{|}{\underset{|}{C}}-\overset{|}{\underset{|}{C}}-$$

While no compounds containing Si—Si bonds are found in nature, silica and a variety of silicates contain chains and networks of bonds:

$$-O-\overset{|}{\underset{|}{Si}}-O-\overset{|}{\underset{|}{Si}}-O-$$

SiH$_4$ is unstable

Alkanes are stable at room temperature

The silanes (SiH_4, etc.) and hydrides of other members of the group are thermodynamically unstable (i.e., $\Delta G^{\ominus}$ is negative) with respect to dissociation into the elements, oxidation and hydrolysis. The alkanes (CH_4, etc.) are thermodynamically stable relative to dissociation into the elements and relative to hydrolysis. Although thermodynamically unstable relative to oxidation to carbon dioxide and water, alkanes are inert at room temperature because the activation energy for oxidation is high.

23.5.2 MULTIPLE BONDS

Carbon forms $>C=C<$ bonds and $-C\equiv C-$ bonds

Carbon forms double and triple bonds between carbon atoms (in alkenes, alkynes and arenes); with nitrogen atoms (in nitriles) and with oxygen atoms (in aldehydes, ketones, carboxylic acids and their derivatives). The rest of Group 4B form no corresponding compounds. These compounds involve π bonds, and only elements in the first short period form strong π bonds [p. 104].

23.5.3 GASEOUS OXIDES

The oxides of carbon are gases

Carbon forms gaseous oxides, CO and CO_2, in contrast to the other members. In this group, it is only for carbon that the E=O bond is more than twice as strong as the E—O bond. Carbon forms two double bonds in O=C=O, whereas for silicon the formation of four Si—O bonds is preferred on energy grounds [see Question 2, p. 462].

23.5.4 L SHELL RESTRICTION

Carbon forms no complexes

When carbon forms four covalent bonds, it has completed its octet. This explains the lack of reactivity of many carbon compounds. With no lone pairs of electrons and a complete octet, it cannot give or accept electrons and therefore forms no complexes.

Carbon compounds are not hydrolysed as are those of later members of the group. Comparing the hydrolysis of tetrachloromethane, CCl_4, and silicon tetrachloride, $SiCl_4$, it can be calculated that the hydrolysis

$$ECl_4(l) + 2H_2O(l) \rightarrow EO_2(g \text{ or } s) + 4HCl(g)$$

is thermodynamically feasible for both compounds (i.e., $\Delta G^{\ominus}$ values are negative).

CCl_4 is stable to hydrolysis

Although silicon tetrachloride is hydrolysed with ease, tetrachloromethane is inert to hydrolysis. The reason for the difference in reactivity is that a different mechanism of hydrolysis operates in the two cases. It is believed that, in the hydrolysis of silicon tetrachloride, the first step is the attack by the negatively charged oxygen atom in a water molecule or hydroxide ion on the positively charged silicon atom. Silicon can use an unoccupied 3d orbital to accommodate the lone pair from the oxygen atom, forming a short-lived intermediate which dissociates to form hydrogen chloride and silicon trichloride hydroxide, $SiCl_3OH$. Repetition of these steps gives hydrated silicon(IV) oxide:

The hydrolysis of $SiCl_4$ employs 3d orbitals

$$Si(OH)_4 \text{ or } SiO_2 \cdot xH_2O \longleftarrow Si(OH)Cl_3 + HCl \text{ (g)}$$

Since carbon cannot expand its octet because its empty 3d orbitals are too different in energy from the 2p orbitals to come into play, a similar mechanism cannot operate for the hydrolysis of tetrachloromethane. This must proceed by a mechanism with a much higher activation energy and takes place extremely slowly.

23.5.5 ELECTRONEGATIVITY

Carbon is the most electronegative element in the group

Carbon is the most electronegative member of the group. Whereas the Si—H bond is polarised

$$\overset{\delta+}{Si}—\overset{\delta-}{H}$$

and the silanes behave similarly to ionic hydrides, the C—H bond is polarised

$$\overset{\delta-}{C}—\overset{\delta+}{H}$$

In the dicarbides (e.g., Cu_2C_2), carbon is present as the anion

$$^-C\equiv C^-$$

and the methanides (e.g., Al_4C_3) contain highly polar covalent bonds.

23.5.6 CARBIDES

Binary compounds of carbon and another element of lower or comparable electronegativity are called **carbides**.

Carbon anion in carbides **Ionic carbides** are formed by the most electropositive elements, e.g., CaC_2.

Carbon atoms can be incorporated in metal structures singly... **Interstitial carbides** are formed by the larger transition metals, e.g., Ti, V, W, Mo. The carbon atoms are accommodated in the octahedral holes in the metallic structure to form an alloy of non-stoichiometric composition. The carbon content makes the alloy hard and chemically stable. An example is tungsten carbide, from which cutting tools are made. The smaller transition metal atoms cannot accommodate carbon

...or in chains... atoms within the structure. The carbides of Cr, Mn, Fe, Co, Ni contain chains of carbon atoms and are less hard than interstitial carbides.

...and in macromolecular structures **Covalent carbides** are formed with elements of similar electronegativity, e.g., B, Si. These carbides have a covalent, macromolecular structure, which makes them hard.

CHECKPOINT 23B: CARBON

1. Explain the following statements:

(*a*) Carbon forms a vast number of compounds.

(*b*) Methane boils at a lower temperature than silane, SiH_4, but in Group 6B, H_2O boils at a higher temperature than H_2S.

2. To explain why CO_2 is molecular while SiO_2 exists as a macromolecular structure, calculate the standard enthalpy change for the polymerisation of one EO_2 unit.

(*a*) $O{=}Si{=}O \longrightarrow$

(*b*) $O{=}C{=}O \longrightarrow$

Use the data in Table 23.1, p. 455.

3. Explain the nature of the bonding in

(*a*) $\diagup C{=}C\diagdown$

(*b*) $-C{\equiv}C-$

(*c*) $-C{\equiv}N$

23.6 THE COMPOUNDS OF GROUP 4B

23.6.1 THE HYDRIDES

Reactivity of the hydrides increases down Group 4B The hydrides of Group 4B are summarised in Table 23.3. The reactivity of the hydrides increases down the group. Silanes are spontaneously flammable in air, are readily hydrolysed by bases to hydrated silicon(IV) oxide, are strongly reducing and dissociate into the elements above 400 °C.

Element	Hydrides	Preparation
Carbon	Alkanes, alkenes, alkynes, arenes	Petroleum industry
Silicon	Silanes, Si_nH_{2n+2}; $n = 1$–10	Mg_2Si + Acid
Germanium	Ge_nH_{2n+2}; $n = 1$–6	Mg_2Ge + Acid
Tin	Stannane, SnH_4, only	$Sn \cdot Hg$ + Acid
Lead	Plumbane, PbH_4, only	$Pb \cdot Hg$ + Acid
		ECl_4 + $LiAlH_4$
		(E = Si, Ge, Sn, Pb

TABLE 23.3 Hydrides of Group 4B

23.6.2 THE HALIDES

Stability of EX₄ decreases down the group and decreases from F to I

All the elements of Group 4B form tetrahalides of formula EX_4, where X = F, Cl, Br or I, except for $PbBr_4$ and PbI_4. All the tetrahalides are volatile covalent compounds, except for the tetrafluorides of tin and lead, SnF_4 and PbF_4, which have some ionic character and form macromolecular structures. The tetrahalides are stable with respect to dissociation into the elements, except for lead(IV) chloride which dissociates at room temperature into lead(II) chloride and chlorine. Stability decreases down the group and decreases from fluorides to iodides. Except for tetrachloromethane and tetrafluoromethane, the tetrahalides are hydrolysed in solution:

$$SiCl_4(l) + 2H_2O(l) \rightarrow SiO_2(s) + 4HCl(g)$$

When silicon tetrafluoride is hydrolysed, the products react to form the complex hexafluorosilicate ion, SiF_6^{2-}.

Table 23.4 summarises the preparations of the halides.

Halide	Preparation	Properties
CCl_4	$Cl_2 + CS_2$, Fe catalyst $\rightarrow CCl_4 + S_2Cl_2$	Uses [p. 600]
ECl_4 (except CCl_4)	$E + X_2$, heat $EO_2 + HX(g)$ $EO_2 + HX(aq, conc.)$	$PbCl_4$ must be kept below 5 °C.
$SnCl_2$	$Sn + HCl(g)$, heat $Sn + HCl(aq, conc.)$	Covalent when anhydrous; in solution, Sn^{2+} (aq) is formed; partially hydrolysed to $Sn(OH)Cl(s)$. Reducing agent [p. 675]
$PbCl_2$	$Pb + Cl_2(g)$, heat $Pb + HCl(aq, conc.)$	Ionic [see Figure 23.11, p. 472] The reaction is possible because the insoluble $PbCl_2$ formed is converted by conc. HCl into soluble $PbCl_4^{2-}$ (aq).
PbX_2	Precipitation	Ionic [see Figure 23.11, p. 472]

TABLE 23.4
Halides of Group 4B

═══════════════════════ CHECKPOINT 23C: HALIDES ═══════════════════════

1. (*a*) Why is an aqueous solution of tin(II) chloride cloudy? How can a clear solution be made? (*b*) What happens when this solution reacts with (i) NaOH(aq) and (ii) FeCl₃(aq)?

2. Outline the methods used in the preparation of Group 4B chlorides. Discuss the trends in valency and bond type shown by the chlorides.

3. Suggest methods for the preparation of (*a*) $PbCl_4$, (*b*) $PbCl_2$. Why is $PbCl_2$ very soluble in concentrated hydrochloric acid?

4. How does SiF_4 differ from $SiCl_4$ in its hydrolysis?

23.6.3 THE OXIDES

The basic character of the oxides increases as the group is descended. The $+4$ oxides are more acidic than the $+2$ oxides.

CARBON MONOXIDE, CO

CO is produced by petrol engines

Carbon monoxide is formed when carbon and hydrocarbons burn incompletely. It is present in the exhaust fumes of petrol-driven vehicles [p. 544]. At concentrations above 0.1% carbon monoxide is poisonous [p. 495]. It is the more dangerous for being colourless and odourless.

It is a poisonous gas...

Carbon monoxide is an important reducing agent. It is used for the reduction of iron ores [p. 498].

...and a reducing agent

In the laboratory, carbon monoxide can be made by the action of concentrated sulphuric acid on sodium methanoate or sodium ethanedioate:

The laboratory preparation...

$$HCO_2Na(s) + H_2SO_4(l) \rightarrow NaHSO_4(s) + CO(g) + H_2O(l)$$

$$C_2O_4Na_2(s) + 2H_2SO_4(l) \rightarrow 2NaHSO_4(s) + CO(g) + CO_2(g) + H_2O(l)$$

In the second case, carbon dioxide is removed by bubbling the mixture of gases through alkali.

...electronic structure...

The electronic structure of carbon monoxide can be represented as

...and complex formation

$$\text{:}C\equiv O\text{:}.$$

The lone pair of electrons on the carbon atom enable it to act as a ligand in the formation of carbonyl complexes, e.g., $[Cr(CO)_6]^{3+}$ and $Ni(CO)_4$ [p. 488].

FIGURE 23.3 Some Reactions of Carbon Monoxide

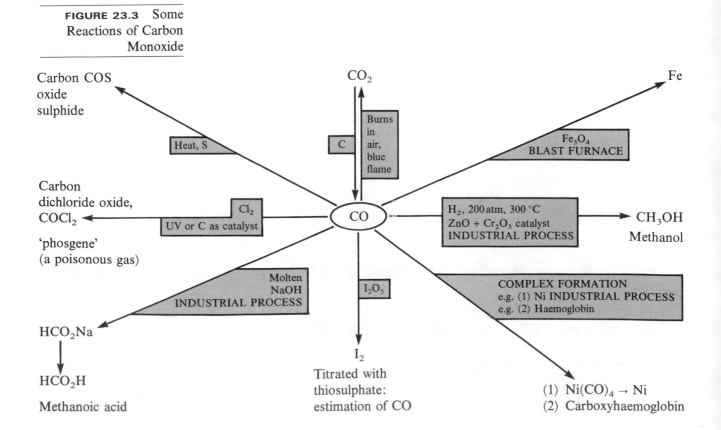

CARBON SUBOXIDE, C$_3$O$_2$

Carbon suboxide has the structure

$$O=C=C=C=O$$

It decomposes at 200 °C into carbon and carbon dioxide

CARBON DIOXIDE, CO$_2$

Photosynthesis

Carbon dioxide is present in air at a level of 0.03% by volume. Plants use carbon dioxide in the process of photosynthesis [p. 192] and both plants and animals evolve carbon dioxide in respiration. The balance of processes which give out carbon dioxide and those which use carbon dioxide is called the carbon cycle and is illustrated in Figure 23.4.

FIGURE 23.4 The Carbon Cycle

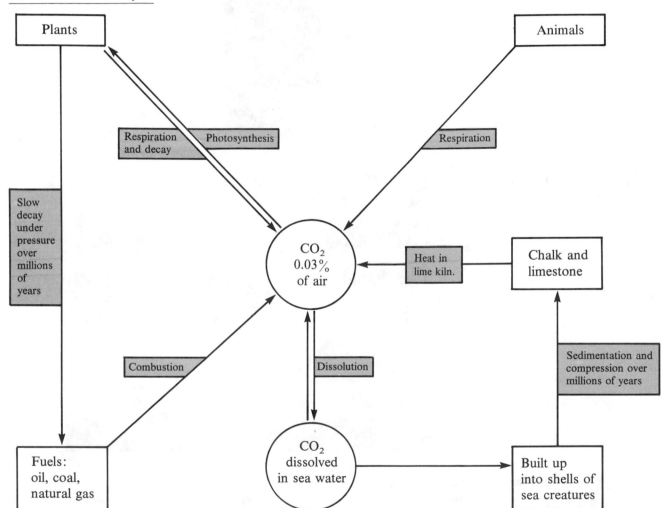

CO$_2$ is used in soft drinks

Carbon dioxide is colourless and odourless, and is slightly soluble in water. Flavoured solutions of carbon dioxide under pressure are the basis of the soft drinks industry.

CO$_2$ is used in fire-extinguishers

Carbon dioxide is non-poisonous, denser than air and does not support combustion. These three factors find carbon dioxide a use in fire-extinguishers. It is stored in cylinders under pressure and released by the opening of a valve [see Figure 23.5]. There are substances which burn with such a hot flame that they can decompose carbon dioxide and continue to burn in the oxygen formed. A magnesium fire could not be extinguished by carbon dioxide.

FIGURE 23.5   A Carbon
Dioxide Fire-extinguisher

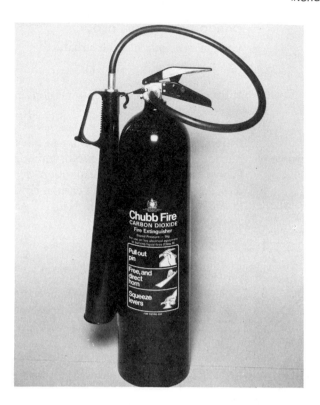

FIGURE 23.6   Drikold,
ICI's Solid CO_2

CO_2 is easily liquefied   The gas can be liquefied at room temperature by a pressure of 60 atm [p. 149]. If
                                the compressed gas is allowed to expand rapidly, solid carbon dioxide is obtained.
                                It is a white solid, with the structure described on p. 120. Its high standard enthalpy
Solid CO_2 is called 'dry   of vaporisation makes it a useful refrigerant, and, since it sublimes to form gaseous
ice'   carbon dioxide, rather than melting, it is known as 'dry ice' or 'Drikold'.

                                Industrially, carbon dioxide is formed as a by-product during the manufacture of
                                quicklime from limestone [p. 363] and in the fermentation of sugars to ethanol
The laboratory   [p. 617]. In the laboratory, it can be made by the action of dilute hydrochloric acid
preparation of CO_2   or dilute nitric acid on marble chips. The gas can be collected over water. If required
                                pure and dry, it is bubbled through water (to remove any hydrogen chloride),
                                through concentrated sulphuric acid and then collected downwards [see Figure 23.7].

FIGURE 23.7
Preparation of Dry
Carbon Dioxide

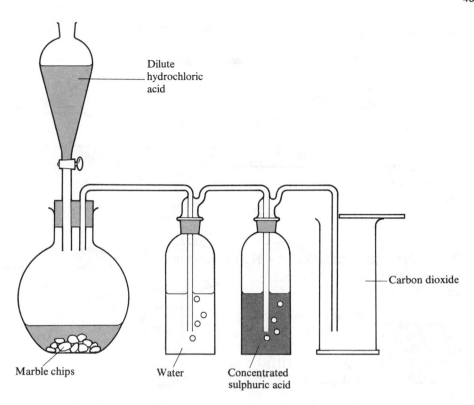

CARBONIC ACID AND CARBONATES

The solution of CO₂ forms carbonic acid

Carbon dioxide is an acidic gas. It is the anhydride of carbonic acid, H_2CO_3, but only about 0.4% of dissolved carbon dioxide is converted into carbonic acid, and when a solution is boiled nearly all the dissolved carbon dioxide is expelled:

$$CO_2(g) + H_2O(l) \rightleftharpoons CO_2(aq) + H_2O(l) \rightleftharpoons H_2CO_3(aq)$$

Carbonic acid is a weak diprotic acid.

Carbon dioxide reacts with bases. When passed into sodium hydroxide solution, it forms first sodium carbonate and then sodium hydrogencarbonate:

$$CO_2(g) + 2NaOH(aq) \rightarrow Na_2CO_3(aq) + H_2O(l) \xrightarrow{CO_2(g)} 2NaHCO_3(aq)$$

Carbonates and hydrogencarbonates

A similar reaction with calcium hydroxide solution (limewater) is used as a test for carbon dioxide. A precipitate of calcium carbonate appears and then dissolves to form soluble calcium hydrogencarbonate:

The limewater test for CO₂

$$CO_2(g) + Ca(OH)_2(aq) \rightarrow CaCO_3(s) + H_2O(l) \xrightarrow{CO_2(g)} Ca(HCO_3)_2(aq)$$

The reaction of CO₂ + NH₃

Carbon dioxide reacts with ammonia to form a white solid, ammonium aminomethanoate, which gives urea when heated:

$$CO_2(g) + 2NH_3(g) \rightarrow H_2NCO_2NH_4(s) \xrightarrow{heat} H_2NCONH_2(s) + H_2O(g)$$

Ammonium aminomethanoate   Urea

The structure of CO₃²⁻

The structure of the planar carbonate ion can be represented as a resonance hybrid of three forms, or by a molecular orbital structure:

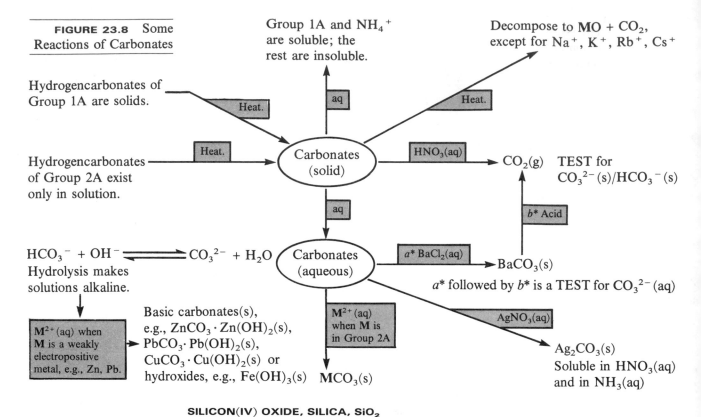

FIGURE 23.8 Some Reactions of Carbonates

Group 1A and NH_4^+ are soluble; the rest are insoluble.

Decompose to **MO** + CO_2, except for Na^+, K^+, Rb^+, Cs^+

Hydrogencarbonates of Group 1A are solids.

Hydrogencarbonates of Group 2A exist only in solution.

Heat.

Heat.

aq

Carbonates (solid)

$HNO_3(aq)$

Heat.

$CO_2(g)$ TEST for CO_3^{2-}(s)/HCO_3^-(s)

aq

b * Acid

$HCO_3^- + OH^- \rightleftharpoons CO_3^{2-} + H_2O$
Hydrolysis makes solutions alkaline.

Carbonates (aqueous)

a * $BaCl_2(aq)$

$BaCO_3(s)$

a * followed by *b* * is a TEST for CO_3^{2-}(aq)

M^{2+}(aq) when **M** is a weakly electropositive metal, e.g., Zn, Pb.

Basic carbonates(s), e.g., $ZnCO_3 \cdot Zn(OH)_2$(s), $PbCO_3 \cdot Pb(OH)_2$(s), $CuCO_3 \cdot Cu(OH)_2$(s) or hydroxides, e.g., $Fe(OH)_3$(s)

M^{2+}(aq) when **M** is in Group 2A

$AgNO_3(aq)$

MCO_3(s)

Ag_2CO_3(s) Soluble in HNO_3(aq) and in NH_3(aq)

SILICON(IV) OXIDE, SILICA, SiO₂

Several forms of silicon(IV) oxide or *silica*, SiO_2, are known. The structure of quartz and other crystalline forms of silica is shown in Figure 6.17, p. 122. Silica melts at 1710 °C to form a viscous liquid. When this liquid is cooled, it forms a glass [p. 123]. Silica glass is used for the manufacture of specialised laboratory glassware. Silica glass transmits infrared and ultraviolet light. It is chemically inert, being attacked only by hydrogen fluoride, damp fluorine and fused bases.

Silica glass is chemically inert and transmits infrared and ultraviolet light

Silica is used in the manufacture of concrete [p. 363], the anhydrite process for making sulphur dioxide [p. 422], the extraction of phosphorus [p. 432] and the manufacture of the abrasive, silicon carbide, SiC.

Uses for silica

Quartz watches

Quartz is used in the manufacture of electronic equipment and timing devices. Quartz watches have recently taken a large slice of the market away from watches with moving parts. The timing mechanism is controlled by a quartz crystal which is induced to vibrate at 32 768 oscillations per second by the application of a small electric field [see Postscript to Chapter 6].

SILICATES

When silicates are formed by the reaction of silicon(IV) oxide and a molten base, the silicate ion that is formed depends on the amounts of base and silica present. The silicate ion may be (a) SiO_4^{4-}, a tetrahedral structure, (b) $Si_2O_7^{6-}$, (c) $Si_3O_9^{6-}$ and (d) chains or sheets of extended $(SiO_3)_n^{2n-}$ and $(SiO_2)_n^{n-}$ anions [see Figure 23.9].

FIGURE 23.9
Silicate Ions

(a) SiO_4^{4-} (b) $Si_2O_7^{6-}$ (d) $(SiO_3)_n^{2n-}$

FIGURE 23.10
A Continuous Sheet of Glass leaving the Pilkington Float Glass Plant

Some aluminosilicates are zeolites, and act as molecular sieves

If some Si^{4+} ions are replaced by Al^{3+} ions, aluminosilicates are formed. To preserve electrical neutrality, another cation, e.g., Na^+ or Ca^{2+}, must be incorporated. Some aluminosilicates lose water on heating to form an open structure which has a large surface area and is porous. These aluminosilicates are called *zeolites* and are used as cation exchangers [p. 174] and as 'molecular sieves'. They retain molecules which are of a size to fit into the cavities in the structure and allow larger and smaller molecules to pass through.

Soda glass is made from silica...

Silica is used in the manufacture of glass. *Soda glass* is a mixture of sodium silicate and calcium silicate, which is made by melting the carbonates with silica at 1500 °C:

$$CaCO_3(s) + SiO_2(s) \rightarrow CaSiO_3(s) + CO_2(g)$$

$$Na_2CO_3(s) + SiO_2(s) \rightarrow Na_2SiO_3(s) + CO_2(g)$$

...as are cobalt glass...

Coloured glasses are made by adding metal oxides to the melt, e.g., cobalt(II) oxide for *blue glass*. Pyrex® is a borosilicate glass, made by adding boron oxide to the melt. It can withstand higher temperatures than soda glass. Sodium silicate, Na_4SiO_4, is a water-soluble glass called *water glass*.

...and water glass

SILICONES

Silicones are polymers

Silicones are polymers formed by the hydrolysis and subsequent dehydration of alkylchlorosilanes. There are three types of structure:

Trialkylmonochlorosilanes, $R_3SiCl \rightarrow R_3Si\!-\!O\!-\!SiR_3$

Dialkyldichlorosilanes, $R_2SiCl_2 \rightarrow$

$$-O-\underset{\underset{R}{|}}{\overset{\overset{R}{|}}{Si}}-O-\underset{\underset{R}{|}}{\overset{\overset{R}{|}}{Si}}-O-\underset{\underset{R}{|}}{\overset{\overset{R}{|}}{Si}}-O-\underset{\underset{R}{|}}{\overset{\overset{R}{|}}{Si}}-$$

$$\begin{array}{ccccccc}
 & | & & | & & | & & | \\
 & R & & O & & R & & O \\
 & | & & | & & | & & | \\
-O-Si-O-Si-O-Si-O-Si- \\
 & | & & | & & | & & | \\
 & O & & R & & O & & R \\
 & & & R & & & & R \\
 & | & & | & & | & & | \\
-O-Si-O-Si-O-Si-O-Si- \\
 & | & & | & & | & & | \\
 & R & & O & & R & & O \\
 & | & & | & & | & & |
\end{array}$$

Monoalkyltrichlorosilanes, $RSiCl_3 \rightarrow$

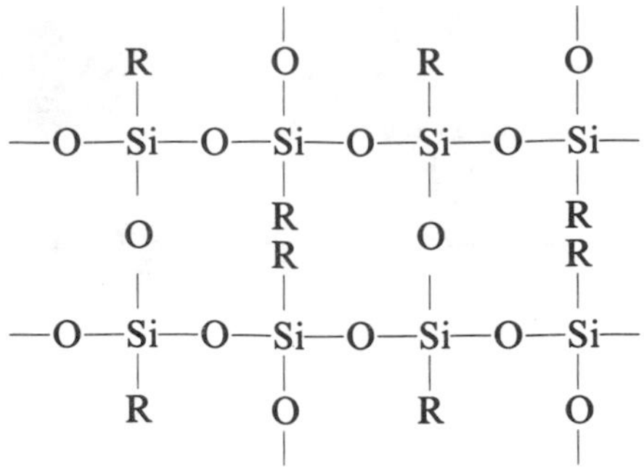

Their resistance to chemical attack finds them many uses

Silicones are resistant to chemical attack and are water-repellent. They are used in paints, varnishes, lubricants and in waterproofing fabrics.

OXIDES OF TIN AND LEAD

The oxides of tin and lead are amphoteric. Their preparation and properties are summarised in Table 23.5.

Oxide	Preparation	Properties
Tin(IV) oxide, SnO_2, the more stable oxide	(a) Heat Sn in air. (b) Sn + conc. HNO_3	Amphoteric; with conc. $H_2SO_4 \rightarrow Sn(SO_4)_2$; with conc. alkali or fused base $\rightarrow$ stannate(IV), $Sn(OH)_6{}^{2-}$
Tin(II) oxide, SnO	Heat SnC_2O_4, tin(II) ethanedioate. (CO and CO_2 are formed and prevent oxidation of SnO.)	Amphoteric, but more basic than SnO_2; with dilute acid $\rightarrow Sn^{2+}$ salt; with alkali $\rightarrow$ stannate(II) $Sn(OH)_4{}^{2-}$
Lead(IV) oxide, PbO_2, brown	Warm Pb^{2+} salt with oxidising agent, e.g., ClO^- (aq).	Lead–acid accumulator [p. 287] Powerful oxidising agent; Heat $\rightarrow PbO + O_2$ with $SO_2 \rightarrow PbSO_4$ with warm HCl $\rightarrow Cl_2$ Amphoteric; with HCl(aq) below $20\,°C \rightarrow PbCl_4(l)$; with alkali or fused base $\rightarrow$ plumbate(IV), $Pb(OH)_6{}^{2-}$
Lead(II) oxide, PbO, yellow and red polymorphs	(a) Heat $Pb(NO_3)_2$. (b) Heat $PbCO_3$.	Amphoteric; with HCl(aq) $\rightarrow PbCl_2$; with alkali or fused base $\rightarrow$ plumbate(II), $Pb(OH)_4{}^{2-}$
Dilead(II) lead(IV) oxide, Pb_3O_4, 'red lead'	Heat PbO at 400 °C in air.	Behaves as a mixed oxide, $2PbO \cdot PbO_2$ With HNO_3(aq) $\rightarrow Pb(NO_3)_2 + PbO_2$ With conc. HCl(aq) $\rightarrow PbCl_2 + Cl_2$

TABLE 23.5   Oxides of Tin and Lead

23.6.4 SULPHIDES

A summary of the sulphides of Group 4B is given in Table 23.6.

Sulphide	Preparation	Properties
Carbon disulphide, CS$_2$	Direct synthesis	Molecular
Silicon(IV) sulphide, SiS$_2$	Direct synthesis	Macromolecular
Tin(II) sulphide, SnS, brown	(a) Direct synthesis (b) Precipitation Sn^{2+}(aq) + H$_2$S(aq)	Dissolves in ammonium polysulphide, (NH$_4$)$_2$S$_2$(aq). Addition of acid precipitates yellow SnS$_2$(s).
Lead(II) sulphide, PbS, black	(a) Direct synthesis (b) Precipitation Pb^{2+}(aq) + H$_2$S(aq)	

TABLE 23.6 Sulphides of Group 4B

23.6.5 SALTS

Tin(II) and lead(II) compounds are ionic. Their reactions are shown in Figure 23.11 and Figure 23.12.

FIGURE 23.11 Some Reactions of Sn^{2+}(aq)

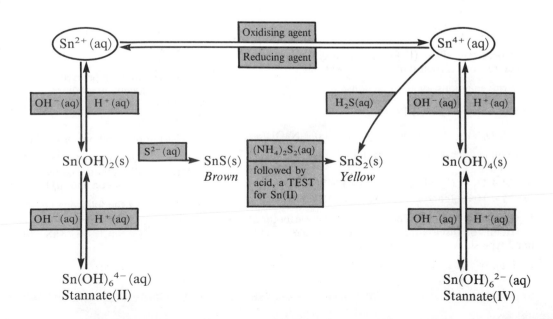

FIGURE 23.12 Some
Reactions of Pb^{2+} (aq)

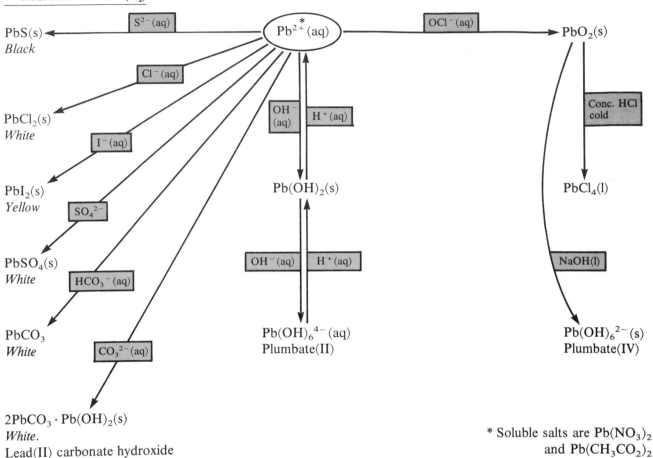

2PbCO$_3$ · Pb(OH)$_2$(s)
White.
Lead(II) carbonate hydroxide

* Soluble salts are Pb(NO$_3$)$_2$
and Pb(CH$_3$CO$_2$)$_2$

CHECKPOINT 23D: OXIDES

1. Describe a method for preparing and collecting carbon monoxide in the laboratory. State the conditions needed and the products formed in the reactions of CO with
(a) NaOH   (b) H$_2$ (c) Cl$_2$ (d) Ni   (e) I$_2$O$_5$
Give equations.

2. Explain why silicon(IV) oxide has a different type of structure from carbon dioxide.

3. Write equations for the reactions:

(a) Pb$_3$O$_4$(s) $\xrightarrow{\text{heat}}$ (b) PbO$_2$(s) $\xrightarrow{\text{heat}}$

(c) Pb$_3$O$_4$(s) + 4HNO$_3$(aq) → (d) PbO(s) + NaOH(aq) →

4. Complete these equations (i) for **M** = Sn, (ii) for **M** = Pb:
(a) **M** + O$_2$ → (b) **M** + Cl$_2$ → (c) **M** + S →
(d) **M** + HNO$_3$(conc.) → (e) **M** + NaOH(conc.) →

5. Outline the methods of preparation available for the oxides of Group 4B. Discuss the group trends in valency and bond type shown by the oxides.

6. State two characteristics of
(a) metallic oxides

(b) non-metallic oxides.
Illustrate the gradation from non-metallic to metallic properties in Group 4B by considering the oxides.

7. Explain the following statements:

(a) It is not necessary to cool carbon dioxide in order to liquefy it.

(b) When cold compressed carbon dioxide is allowed to expand suddenly, it solidifies.

(c) Carbon dioxide is more soluble under pressure [p. 141].

8. List the preparations of (a) CO, (b) CO$_2$ and (c) CS$_2$.

9. Explain why limestone slowly dissolves in water that is saturated with carbon dioxide. What is the connection between this reaction and (a) hard water and (b) stalactites?

10. State how the following compounds can be obtained from tin.
(a) SnCl$_2$ (b) SnCl$_4$ (c) SnS   (d) SnS$_2$ (e) Na$_2$Sn(OH)$_6$

23.7 A COMPARATIVE LOOK AT GROUP 4B

There is a metal to non-metal transition down Group 4B...

In Group 4B, the differences between the first member of the group and the last have reached a maximum. There is a transition from carbon and silicon, which are non-metallic elements, through germanium, which is intermediate in character, a **metalloid**, to tin and lead which are metals. Carbon and silicon have covalent, macromolecular structures. Carbon (except for graphite), is a non-conductor. Silicon and germanium are semiconductors. Tin and lead have metallic structures and are conductors (except for α tin with its diamond-type structure). The reactions with dilute acids and with concentrated nitric acid illustrate the non-metal to metal transition:

...This is illustrated by the reactions with dilute acid and with conc. nitric acid

> *Dilute acid* + C, Si, Ge – no reaction
>
> *Dilute acid* + Sn – reacts
>
> *Dilute acid* + Pb – too low in the Electrochemical Series to react
>
> *Concentrated nitric acid* + C, Si, Ge, Sn → hydrated oxides
>
> *Concentrated nitric acid* + Pb → metal nitrate

Valencies of 2 and 4 are met in the group

The members of the group show valencies of 4 or 2. Passing down the group, there is an increase in the tendency to use a valency of 2 and an increase in the electrovalent character of the bonds.

There is a transition from covalency to electrovalency down the group

Sn and Pb show a valency of 2

C, Si, Ge: covalent compounds, almost exclusively 4-valent

Sn: covalent +4 and ionic +2 states are formed with almost equal ease, but Sn^{2+} is a reducing agent.

Pb: mainly ionic +2 state; also covalent +4 state. Pb(IV) is an oxidising state.

Carbon differs in many respects from the rest of the group...

In addition to the trends down the group, there is a sharp line of demarcation between carbon and silicon. In every group, the first member differs from the rest in having no d orbitals of energy comparable with the occupied p orbitals and in being therefore unable to 'expand its octet'. The main differences between carbon and silicon are:

...Hydrocarbons...

1. Carbon forms a huge number of hydrocarbons. Silicon forms only a few silanes, Si_nH_{2n+2}, which are spontaneously flammable in air [p. 462].

...Oxides...

2. Carbon forms monomeric oxides, CO and CO_2, Silicon forms the polymeric $(SiO_2)_n$ [p. 468].

...Halides...

3. Carbon halides, CX_4, are not hydrolysed, whereas those of silicon, SiX_4, are readily hydrolysed [p. 463].

Carbon forms no complexes

4. Carbon forms no complexes. Silicon and other members of the group can expand their octets to form, e.g., SiF_6^{2-}, $SnCl_6^{2-}$, $PbCl_4^{2-}$.

Halides, EX_4

The +4 halides, EX_4, are, with the exception of the carbon tetrahalides, CX_4, all hydrolysed readily. Of the halides excluding those of carbon, the silicon halides are the most covalent in character and the most readily hydrolysed, to give silicate ions. The halides SnX_4 and PbX_4 are hydrolysed to give basic salts or stannates(IV) or plumbates(IV).

SnX_2 PbX_2

Tin and lead form +2 halides, those of lead being more electrovalent than those of tin.

Oxides increase in basicity down the group   The oxides increase in basicity down the group, and the $+2$ oxides are more basic than the $+4$ oxides. The oxides of carbon and silicon are acidic; those of germanium, tin and lead are amphoteric:

CO_2 dissolves in water to form a weak acid

CO reacts with fused NaOH $\rightarrow$ sodium methanoate

SiO_2 reacts with a fused base $\rightarrow$ a silicate

$\left.\begin{array}{l} SnO_2 \\ SnO \end{array}\right\}$ react with dilute acid $\rightarrow$ Sn^{4+} and Sn^{2+} salts and react with a fused base or concentrated aqueous alkali to form a stannate(IV) or a stannate(II).

$\left.\begin{array}{l} PbO_2 \\ PbO \end{array}\right\}$ react with dilute acids $\rightarrow$ Pb(IV) compounds and Pb^{2+} salts and with a fused base or concentrated aqueous alkali to form a plumbate(IV) or a plumbate(II). PbO is more basic than PbO_2.

QUESTIONS ON CHAPTER 23

1. Outline the preparations of CO_2, SiO_2, CCl_4 and $SiCl_4$. Explain why CO_2 is a gas, whereas SiO_2 is a solid of high melting temperature.

2. Give two properties of (*a*) metallic oxides, (*b*) non-metallic oxides, (*c*) metallic chlorides and (*d*) non-metallic chlorides. Illustrate the transition in properties down Group 4B by considering the properties of (i) their oxides and (ii) their chlorides.

3. Compare the elements of Group 4B with respect to

(*a*) the crystal structure of the element

(*b*) the thermal stability of the hydrides

(*c*) the stability to hydrolysis of the halides

(*d*) the oxidation states of the elements

(*e*) the basicity of the oxides EO_2 and EO.

4. Some chemists view Group 4B as a pair of similar elements, C and Si, followed by a trio of similar elements, Ge, Sn and Pb. What evidence can you offer to support this view?

5. Do you agree with these statements? Explain your answers.

(*a*) The chemistry of carbon is dominated by the ability of the carbon atom to bond to itself and to form multiple bonds to itself and to other atoms.

(*b*) The chemistry of silicon is dominated by the ability of silicon to form strong bonds to oxygen and the reluctance of silicon to bond to itself.

6. What are allotropes?

Sketch the structures of the two allotropes of carbon. How does the structure explain the physical properties of each allotrope?

How could you show that the allotropes are both pure carbon?

7. Outline how you could prepare (*a*) CO from C, (*b*) PbO from Pb, (*c*) SnI_4 from Sn and (*d*) PbI_2 from Pb.

8. Considering Group 4B, discuss

(*a*) the variation of acid/base character of the oxides **MO** and MO_2 down the group

(*b*) the reactivity of the tetrahalides with water

(*c*) the thermal stability of the dichlorides and tetrachlorides. In what ways is carbon atypical of the group?

9. Write a comparative account of the chemistry of the elements of group IV of the Periodic Classification (C, Si, Ge, Sn, Pb) and of their principal compounds, drawing attention to similarities within the group and to broad trends characteristic of this and other groups. Give examples to illustrate what you have to say.

(NI 82)

10. (*a*) Describe the crystal structure of, and the bonding in, both graphite and a typical metal.

(*b*) Compare and contrast the physical properties of graphite and a typical transition metal.

(*c*) Describe briefly and suggest reasons for the differences between the chemistry of carbon and the chemistry of a typical metal.

(JMB 82)

11. This question concerns the group of elements carbon, silicon, tin and lead.

(*a*) Discuss:

(i) the variation of the acid/base character of the oxides **MO** and MO_2 with increase in atomic number

(ii) the difference in acid/base character of the oxides **MO** and MO_2 for any one element

(iii) the reactivity of the tetrachlorides with water

(iv) the relative thermal stability of the dichlorides and tetrachlorides.

(*b*) How could:

(i) carbon monoxide be made from carbon

(ii) lead monoxide be made from lead?

(L 80)

24

THE TRANSITION METALS

24.1 INTRODUCTION

The first transition series in Period 4 are a very similar set of elements

Across the second and third periods of the Periodic Table, there is a gradation in properties, from the alkali metals to the halogens. The fourth period begins in the same way, with an alkali metal (potassium) and an alkaline earth (calcium). The next ten elements do not show the gradation in properties of previous periods: they are remarkably similar to one another in their properties and are all metals. They are called the **first transition series**. Periods five and six also contain transition series. The reason for the similarity of the first transition series is that, considering the series from left to right, while each additional electron is entering the 3d shell, the chemistry of the elements continues to be determined largely by the 4s electrons. From one transition element to the next, the nuclear charge increases by 1 unit, and the number of electrons also increases by 1. Since each additional electron enters the 3d shell, it helps to shield the 4s electrons from the increased nuclear charge, with the result that the effective nuclear charge remains fairly constant across the series of transition elements. The sizes of the atoms and the magnitudes of the first ionisation energies are therefore very similar and the elements have comparable electropositivities. The electron configurations of the first transition series are shown in Table 24.1, in which $(Ar) = 1s^2 2s^2 2p^6 3s^2 3p^6$. You will notice that chromium completes occupying its d orbitals with unpaired electrons at the expense of its 4s electrons, and copper completes its full d^{10} shell at the expense of its 4s electrons. There appears to be a certain measure of stability associated with a full d^{10} shell and with a half-filled d^5 shell.

The difference between transition metals is the number of d electrons. This affects their chemistry less than a difference in s or p electrons

		3d	4s
Sc	(Ar)	↑ _ _ _ _	↑↓
Ti	(Ar)	↑ ↑ _ _ _	↑↓
V	(Ar)	↑ ↑ ↑ _ _	↑↓
Cr	(Ar)	↑ ↑ ↑ ↑ ↑	↑
Mn	(Ar)	↑ ↑ ↑ ↑ ↑	↑↓
Fe	(Ar)	↑↓ ↑ ↑ ↑ ↑	↑↓
Co	(Ar)	↑↓ ↑↓ ↑ ↑ ↑	↑↓
Ni	(Ar)	↑↓ ↑↓ ↑↓ ↑ ↑	↑↓
Cu	(Ar)	↑↓ ↑↓ ↑↓ ↑↓ ↑↓	↑
Zn	(Ar)	↑↓ ↑↓ ↑↓ ↑↓ ↑↓	↑↓

TABLE 24.1 Electron Configurations of the First Transition Series

Transition metals are called d block metals

Scandium, zinc and copper

Transition metals are often referred to as d block metals. They are defined as elements which form some compounds in which there is an incomplete subshell of d electrons. Scandium ($3d^0$ in compounds) and zinc ($3d^{10}$ in compounds) are excluded by this definition, and copper is included only in copper(II) ($3d^9$) compounds. It is convenient to include these metals in a treatment of transition metals, however, on account of the chemical resemblance of their compounds to transition metal compounds.

24.2 PHYSICAL PROPERTIES OF TRANSITION METALS

The metallic bond is stronger than in s block metals

There is a measure of stability in a half-full d subshell

The first transition metals are all hard and dense, good conductors of heat and electricity and possessing useful mechanical properties. Their melting and boiling temperatures [see Figure 24.1] and standard enthalpies of melting are higher than those of s block metals. All these properties are a measure of the strength of the metallic bond. With d electrons as well as s electrons available to take part in delocalisation, the metallic bond is strong in transition metals. Figure 24.1 shows how both melting and boiling temperatures drop at manganese. In manganese, the d subshell is half full, with 5 electrons in 5 orbitals. This electron configuration appears to make the d electrons less available for delocalisation.

FIGURE 24.1 Boiling and Melting Temperatures of Transition Elements

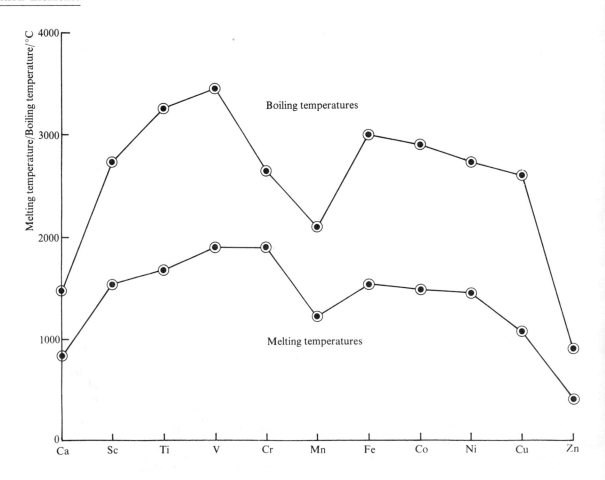

24.2.1 DENSITY

d block metals are denser than s block metals

While the size of the atom, measured by the metallic radius, increases slightly from scandium to zinc, the relative atomic mass increases considerably. The result is an increase in density from scandium to zinc. The d block metals are, in general, denser than the s block metals.

24.2.2 IONISATION ENERGY

Figure 24.2 shows how the first and second ionisation energies increase only slightly from scandium to zinc. The reason is that, as discussed in 24.1, the effective nuclear charge increases only slightly across the series. The increases in the third and fourth ionisation energies across the series are more rapid as d electrons are being removed, and the effective nuclear charge therefore increases by a significant amount from one element to the next.

FIGURE 24.2 Ionisation Energies of Transition Metals

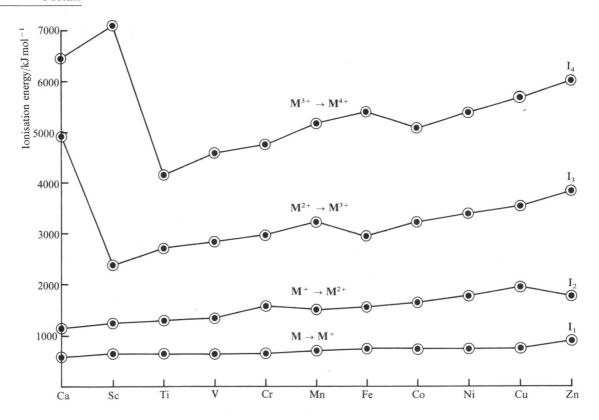

Ionisation energies increase across the period

Each of the curves shows a maximum; these are at Cr^{2+}, at Mn^{3+} and at Fe^{4+}. In each case, the peak in ionisation energy occurs when the removal of an electron disturbs a half-full d^5 shell.

24.3 CHEMICAL PROPERTIES

Features of transition metals

A summary of the differences between d block metals and s block and p block metals is given in Table 24.2. Outstanding features of transition metals are:

1. The variable oxidation state [p. 479].

2. Their use as catalysts [p. 480].

3. Paramagnetism [p. 480].

4. The formation of complex ions [p. 488].

5. The formation of coloured ions [p. 489].

TABLE 24.2
A Comparison of Metals

	s block	*p block*	*d block*
Physical properties	Soft, low melting temperature	Harder, with higher T_m than s block	Harder still, with higher T_m than p block
Reaction with water	React, often vigorously	React only slowly with cold water	
Reaction with non-metallic elements	React vigorously	React less vigorously than s block metals	
Reaction with hydrogen	Form ionic hydrides	Form no hydrides	Some form interstitial hydrides
Bonding	Usually ionic	Usually covalent or complex ions	
Properties of ions	Form simple ions with noble gas configuration	Simple ions have a completed d shell. Easily form complex ions	Some simple ions are formed. Many complex ions are formed.
Complex ions	Simple ions can be loosely hydrated to form colourless complex ions.	Colourless complex ions are formed, rather than simple ions.	Complex ions are formed readily. Usually coloured
Oxidation numbers	Oxidation No. = Group No.	Employ Ox. No. = Gp. No. and also Ox. No. = Gp. No. −2	Ox. No. varies usually by 1 unit at a time, +2, +3 are common.

24.4 METHODS OF EXTRACTION

Extraction by reduction of the oxide

Transition metals occur mainly as sulphides and oxides. The least reactive (e.g., Cu, Au, Pt) are also found **native** (i.e., uncombined). Reduction of the oxide by carbon or carbon monoxide is the usual method of extraction. The theory is covered on pp. 212–4. The following steps may be involved in the extraction:

1. The ore is concentrated. Sometimes flotation is employed: a stream of water carries away debris and leaves the denser ore behind.

2. Sulphide ores are then roasted to convert them to oxides.

3. Heating with coke reduces the oxide to the metal.

4. Carbon is present as a major impurity in the metal. It is removed by heating the metal in a stream of air. Further purification may be achieved by electrolysis (e.g., Cu, Ag, Cr).

24.5 USES OF TRANSITION METALS

Uses of iron, steel...

...and titanium

Iron is our most important metal. Steels are made by alloying iron with carbon and with other transition metals, such as vanadium, manganese, cobalt and nickel [p. 500]. Titanium is a metal with the same kind of mechanical strength as steel and two big advantages: it is less dense than steel, and it does not rust. It is stronger than aluminium. The high cost of titanium has limited its use to applications where no expense is spared. It is used in the construction of space capsules. It is better able than steel to withstand the high temperatures that are experienced when a space capsule re-enters the Earth's atmosphere.

Titanium is being used as a twentieth-century remedy for a nineteenth-century mistake. When repairs were made to the Acropolis in 1896–1933, steel bolts were used to join broken pieces of marble, and steel girders were used to reinforce ancient architraves and porches. The Ancient Greeks, twenty-four centuries ago, had used iron bolts coated with lead. By 1970, the newer steel pins had rusted and, in doing so, had expanded, causing cracks in the marble. The only solution to this serious problem is to remove all the steel pins and replace them with bolts of titanium, which does not rust under any natural conditions.

24.6 OXIDATION STATES

The electron configuration is $(Ar) 4s^2 3d^n$. Once the 4s electrons have been removed, the 3d electrons may also be removed. The difference in energy between the 3d and the 4s electrons is much smaller than the difference between the 3s and the 3p electrons. The oxidation states employed by the elements are shown in Table 24.3.

Some of the oxidation states are uncommon and unstable. The important ones are underlined. The stability of the +2 oxidation state relative to +3 and higher oxidation states increases from left to right across the series. It reflects the increasing difficulty of removing a 3d electron as the nuclear charge increases.

Figure 24.3 shows both the complete range of oxidation states shown by each element and also the important oxidation states in the chlorides and oxides of transition elements. In their lower oxidation states, elements form ionic compounds, but in their higher oxidation states they form covalent compounds [p. 335]. Also included in Figure 24.3 are the carbonyl compounds, in which the metal has an oxidation state of zero.

Sc	Ti	V	Cr	Mn	Fe	Co	Ni	Cu	Zn
								+1	
	+2	+2	+2	+2	+2	+2	+2	+2	+2
+3	+3	+3	+3	+3	+3	+3	+3	+3	
	+4	+4	+4	+4	+4	+4	+4		
		+5	+5	+5					
			+6	+6	+6				
				+7					

TABLE 24.3
Oxidation States of
Transition Metals

24.7 CATALYSIS BY TRANSITION METALS

Many important reactions are catalysed by transition metals...

Transition metals and their compounds are important catalysts. Some industrial reactions which are catalysed by transition metals are: the Contact process (vanadium(V) oxide), the Haber process (iron or iron(III) oxide), the hydrogenation of unsaturated oils (finely divided nickel) and the oxidation of ammonia (platinum or platinum–rhodium alloy). These examples all involve heterogeneous catalysis [p. 314], in which the reactant molecules are adsorbed on the surface of the catalyst. It is likely that the 3d electrons enable the transition metal catalyst to form temporary bonds with reactant molecules. In homogeneous catalysis [p. 313], which is usually found in reactions in solution, the variable oxidation number of the transition metal may enable it to take part in a sequence of reaction stages and emerge unchanged at the end. An example is the oxidation of iodide ions by peroxodisulphate ions, $S_2O_8^{2-}$, according to the equation

$$S_2O_8^{2-}(aq) + 2I^-(aq) \rightarrow 2SO_4^{2-}(aq) + I_2(aq) \qquad [1]$$

...The oxidation of iodide ions is an example of homogeneous catalysis

Iron(II) ions catalyse the reaction, and it is thought that they may provide an alternative route for the reaction via steps [2] and [3]:

$$Fe^{2+}(aq) + \tfrac{1}{2}S_2O_8^{2-}(aq) \rightarrow Fe^{3+}(aq) + SO_4^{2-}(aq) \qquad [2]$$

$$Fe^{3+}(aq) + I^-(aq) \rightarrow Fe^{2+}(aq) + \tfrac{1}{2}I_2(aq) \qquad [3]$$

The alternative route involves two reactions between oppositely charged ions and therefore has a lower activation energy [p. 313] than reaction [1] between ions of the same charge. The Fe^{2+} catalyst is regenerated in step [3].

24.8 PARAMAGNETISM

Transition metal ions with unpaired electron spins are paramagnetic

Paramagnetic substances are weakly attracted by a magnetic field. Any species with an unpaired electron is paramagnetic because there is a magnetic moment associated with the spinning electron. Transition metal ions that have unpaired d electrons are paramagnetic. The greater the number of unpaired electrons, the more paramagnetic is the ion. The metals, iron, cobalt and nickel are **ferromagnetic**, that is, they are strongly attracted to a magnetic field.

FIGURE 24.3 Important
Oxides and Halides of
the Transition Metals

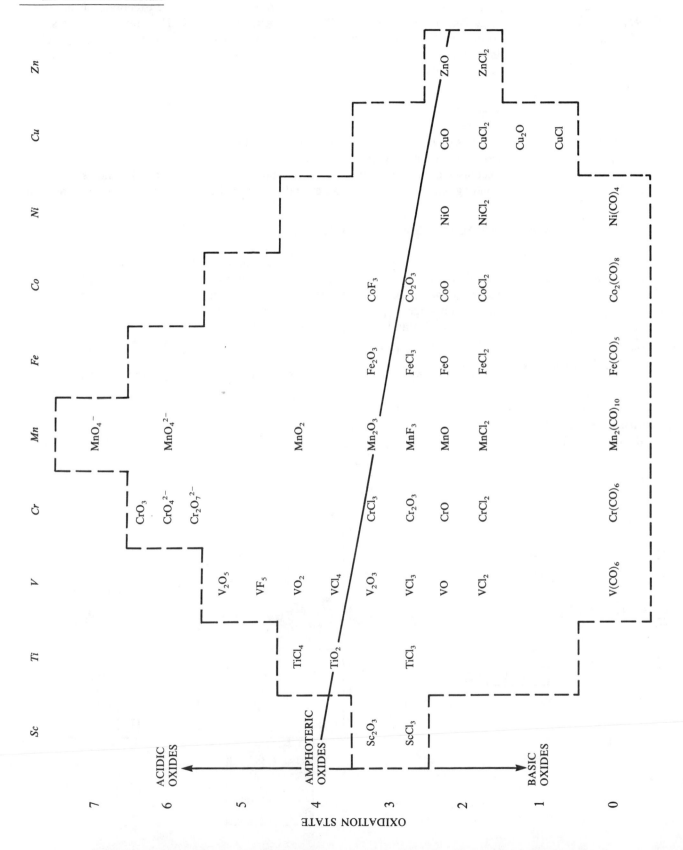

24.9 OXIDES AND HYDROXIDES OF TRANSITION METALS

The oxides are insoluble, black or coloured, with covalent character...

Transition metals react with oxygen to form oxides, with the exception of those (e.g., Ag, Au) that are low in the Electrochemical Series. The important oxides formed by the first transition series are shown in Figure 24.3. They are nearly all insoluble in water and either black or coloured. The covalent character of the bonds is appreciable.

24.9.1 BASICITY

...some are acidic, some basic, some amphoteric

The basicity of the oxides of the transition metals in an oxidation state of +2 increases from left to right across the series. For any one metal, the basicity of the oxides decreases as the oxidation state of the metal increases. Figure 24.3 shows the rough dividing line between basic oxides and acidic oxides. Oxides below the line are basic.

24.9.2 REDUCTION OF OXIDES

The metals can be extracted from oxides

Transition metal oxides can be reduced to the metal. For the less electropositive metals (excluding Ti and V), carbon and carbon monoxide are often used as the reducing agents. In the blast furnace for the extraction of iron, carbon monoxide is the reducing agent [p. 214 and p. 498]. The ore *chromite*, $FeO \cdot Cr_2O_3$, is reduced to an alloy of iron and chromium by heating it with carbon:

Iron is extracted from a ferrochrome alloy

$$FeO \cdot Cr_2O_3(s) + 4C(s) \rightarrow Fe(s) + 2Cr(s) + 4CO(g)$$

The ferrochrome alloy produced is used in the production of stainless steel.

FIGURE 24.4(a) Some Reactions of Titanium

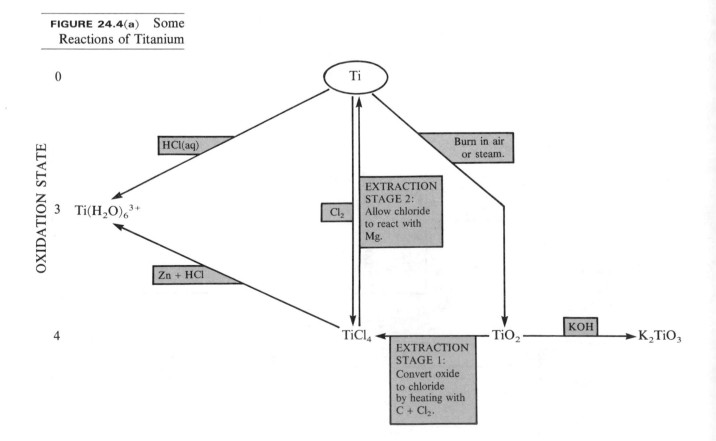

Nickel is purified by the Mond process

Nickel is obtained by reducing nickel(II) oxide with carbon and purified by the Mond process [see Figure 24.17, p. 496].

Thermit reactions use Al to reduce metal oxides

The more electropositive elements cannot be obtained from their oxides by reduction with carbon. Aluminium is used. It reduces metal oxides with the evolution of heat in a set of reactions called thermit reactions. Chromium(III) oxide, Cr_2O_3, vanadium(V) oxide, V_2O_5 and cobalt(II) dicobalt(III) oxide, Co_3O_4, are reduced in this way:

$$Cr_2O_3(s) + 2Al(s) \rightarrow 2Cr(s) + Al_2O_3(s)$$

Titanium is obtained from titanium(IV) oxide as shown in Figure 24.4(a).

FIGURE 24.4(b)
Transistor Package Centres are loaded into an Evaporator, ready to be coated with Titanium

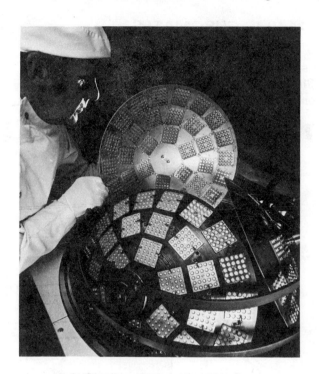

24.9.3 HYDROXIDES

The hydroxides can be made by precipitation

The precipitates are gelatinous and often coloured...

...Some dissolve in ammonia

The hydroxides of transition metals are precipitated from solutions of the metal ions by the addition of hydroxide ions. The colour of the precipitate can often be used to identify the metal present. All the precipitates are gelatinous, owing to hydration, and all are basic. Some are amphoteric, and some form soluble complex ions with ammonia [see Table 24.4].

TABLE 24.4
Transition Metal Hydroxides

Cation	Precipitate	Colour	Reaction with NaOH(aq)	Reaction with NH_3(aq)
Cr^{3+}(aq)	$Cr(OH)_3$	Green	Chromate(III) ion, CrO_3^{3-}(aq)	–
Mn^{2+}(aq)	$Mn(OH)_2$	Beige	–	–
Fe^{2+}(aq)	$Fe(OH)_2$	Green	–	–
Fe^{3+}(aq)	$Fe(OH)_3$	Rust	–	–
Co^{2+}(aq)	$Co(OH)_2$	Pink	Cobaltate(II) ion, $Co(OH)_4^{2-}$(aq)	$Co(NH_3)_6^{2+}$(aq)
Ni^{2+}(aq)	$Ni(OH)_2$	Green	–	$Ni(NH_3)_6^{2+}$(aq)
Cu^{2+}(aq)	$Cu(OH)_2$	Blue	–	$Cu(NH_3)_4^{2+}$(aq)
Zn^{2+}(aq)	$Zn(OH)_2$	Colourless	Zincate ion, $Zn(OH)_4^{2-}$(aq)	$Zn(NH_3)_4^{2+}$(aq)

24.10 OXO–IONS OF TRANSITION METALS

Dichromate(VI) and manganate(VII) are powerful oxidising agents

In their higher oxidation states, the transition metals occur, not as cations M^{2+}, but combined with oxygen as anions MO_4^{3-} and MO_4^{2-}. The most important of these are the vanadate(V) ion, $V_3O_9^{3-}$; chromate(VI), CrO_4^{2-}; dichromate(VI), $Cr_2O_7^{2-}$; manganate(VI), MnO_4^{2-}; and manganate(VII), MnO_4^{-}. The sodium and potassium salts of these anions are soluble in water. Their power as oxidising agents enables them to be used in titrimetric analysis.

24.10.1 CHROMATES

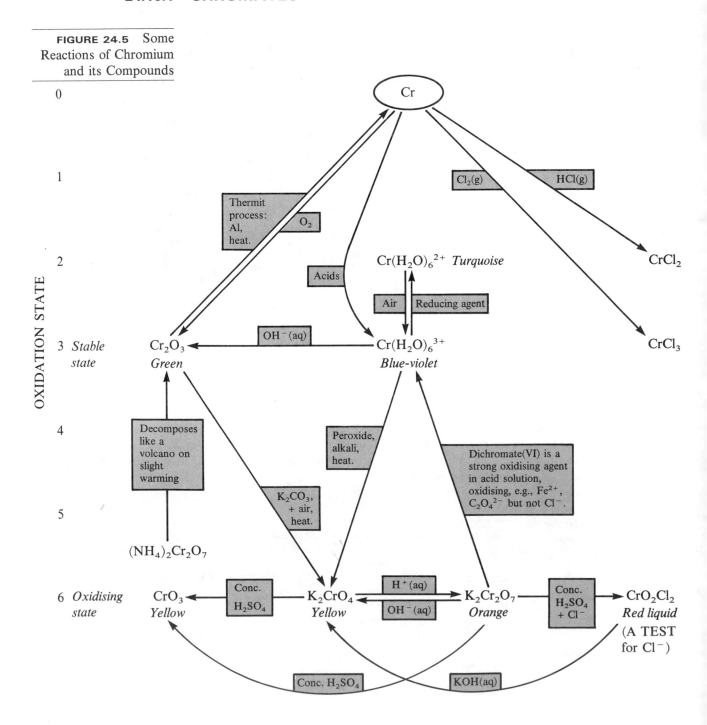

FIGURE 24.5 Some Reactions of Chromium and its Compounds

When a chromium(III) salt is heated in alkaline solution with a peroxide [see Figure 24.5], a chromate(VI) ion is formed:

The preparation of chromate(VI) and of dichromate(VI)

$$2Cr^{3+}(aq) + 4OH^-(aq) + 3O_2{}^{2-}(aq) \rightarrow 2CrO_4{}^{2-}(aq) + 2H_2O(l)$$

The $Cr^{3+}(aq)$ ion is blue-violet; the $CrO_4{}^{2-}(aq)$ ion is yellow. In acid solution, yellow chromate(VI) ions condense to form orange dichromate(VI) ions:

$$2CrO_4{}^{2-}(aq) + 2H^+(aq) \rightleftharpoons Cr_2O_7{}^{2-}(aq) + H_2O(l)$$

Yellow chromate(VI) ion   Orange dichromate(VI) ion

The oxidation state of chromium is +6 in both ions, as can be seen from the structures:

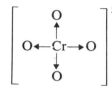

Potassium chromate(VI) is used as an indicator in the titration of silver salts [p. 271].

Potassium dichromate(VI) is an oxidising agent used in titrimetric analysis

Potassium dichromate(VI) is used as an oxidising agent in titrimetric analysis. Since it can be obtained in a high degree of purity and is not deliquescent, it can be used as a primary standard. In the half-reaction [see p. 54 for half-reactions, and p. 279 for $E^\ominus$]

$$Cr_2O_7{}^{2-}(aq) + 14H^+(aq) + 6e^- \rightarrow 2Cr^{3+}(aq) + 7H_2O(l); \quad E^\ominus = +1.33\,V$$

there is a colour change from orange to blue-violet. To give a sharper end-point, a redox indicator such as barium diphenylamine sulphonate is added. Potassium dichromate(VI) in acid solution can be used to estimate iron(II) salts, ethanedioates, iodides and other reducing agents [pp. 55–6]. It can be used in the presence of chloride ions since the standard reduction potential for the system

$$Cl_2(g) + 2e^- \rightleftharpoons 2Cl^-(aq)$$

is +1.36V, and chloride ions are not oxidised by potassium dichromate(VI).

24.10.2 MANGANATES

The preparation of manganate(VI)

Potassium manganate(VI), K_2MnO_4, is a dark green solid formed when manganese(IV) oxide is melted with potassium hydroxide and an oxidising agent such as potassium chlorate(V) or potassium nitrate:

$$MnO_2(s) + 2KOH(s) + [O] \rightarrow K_2MnO_4(s) + H_2O(l)$$

Manganate(VI) disproportionates to form manganate(VII)...

The manganate(VI) ion disproportionates in acid solution into the manganate(VII) ion and manganese(IV) oxide:

$$3MnO_4{}^{2-}(aq) + 4H^+(aq) \rightarrow 2MnO_4{}^-(aq) + MnO_2(s) + 2H_2O(l)$$

Dark green   *Purple*   *Black*

Potassium manganate(VII), often called potassium permanganate, is widely used in acid solution as an oxidising agent in titrimetric analysis:

...this is a powerful oxidising agent used in titrimetric analysis...

$$MnO_4{}^-(aq) + 8H^+(aq) + 5e^- \rightleftharpoons Mn^{2+}(aq) + 4H_2O(l); \quad E^\ominus = +1.51\,V$$

...it oxidises Cl^- ions

With a standard reduction potential of +1.51V, it is a more powerful oxidising agent than potassium dichromate(VI), for which $E^\ominus = +1.33$V. Potassium manganate(VII) cannot be used in solutions containing chloride ions, as it oxidises them to chlorine. Solutions of potassium manganate(VII) are kept in brown bottles because in the presence of light they slowly oxidise water to oxygen. Substances which can be estimated by titration against acidified potassium manganate(VII) are iron(II) salts, hydrogen peroxide and ethanedioates [p. 55, p. 56].

FIGURE 24.6 Some Reactions of Manganese and its Compounds

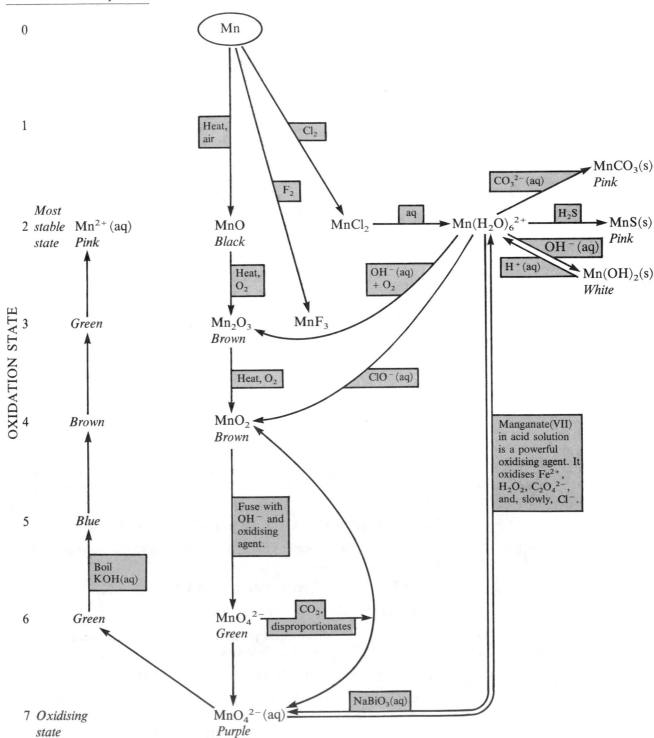

24.11 HALIDES

The reactions of transition metals with halogens...

...fluorine brings out the highest oxidation state

Transition metals react with the halogens. Fluorine brings out high oxidation states in metals, and the fluorides are ionic in character. Chlorine attacks all the metals in the first transition series, but does not always bring out the highest oxidation state: manganese reacts with chlorine to form manganese(II) chloride, $MnCl_2$, whereas with fluorine it forms manganese(III) fluoride, MnF_3. Bromine and iodine also react with metals of the first transition series, although the metals may need to be heated to speed up their reactions with iodine.

24.11.1 CHLORIDES

The anhydrous chloride is made by heating crystals of the hydrate in a stream of dry HCl(g)

Hydrated chlorides can be made by reacting the metal or the metal oxide with hydrochloric acid and allowing the solution to crystallise. An attempt to obtain an anhydrous chloride by heating a hydrate results in hydrolysis and the formation of a basic chloride. If iron(II) chloride crystals are heated, iron(II) chloride hydroxide is formed:

$$FeCl_2 \cdot 6H_2O(s) \rightleftharpoons Fe(OH)Cl(s) + HCl(g) + 5H_2O(g)$$

Anhydrous iron(II) chloride can be obtained, however, if a stream of dry hydrogen chloride is passed over the heated crystals. Equilibrium is driven over to the left.

Anhydrous chlorides are made by synthesis

Anhydrous chlorides are usually made by reacting the metal with a stream of dry hydrogen chloride or chlorine. The apparatus shown in Figure 19.9, p. 381, can be used for the preparation of anhydrous iron(III) chloride. The product sublimes over as molecules of Fe_2Cl_6. If water is added, $Fe^{3+}(aq)$ and $Cl^-(aq)$ ions are formed. If dry hydrogen chloride is passed over heated iron, anhydrous iron(II) chloride is formed.

Many chlorides of the transition metals are soluble

Some transition metal chlorides are molecular (e.g., $TiCl_4$); the rest are macromolecular. Ions are formed when the macromolecular chlorides are either melted or dissolved in water. Most transition metal chlorides are soluble; exceptions are copper(I) chloride, silver chloride and mercury(I) chloride. The ions are stabilised in solution by hydration with the formation of complex aqua ions, e.g., $Fe(H_2O)_6^{3+}$. In the presence of chloride ions, many transition metal chlorides form soluble chloride complex ions, e.g., $CuCl_4^{2-}$.

24.12 SULPHIDES

Sulphides are black or coloured and macromolecular

Many transition metals are found as sulphide ores, for example, FeS_2, $CuFeS_2$, CuS, MnS, NiS, Ag_2S and HgS. The sulphides are black or coloured and are macromolecular. When sulphide ores are roasted in air, the metal oxide and sulphur dioxide are formed. The oxide can be reduced to yield the metal.

Sulphides are made by precipitation or by synthesis

The sulphides can be made by precipitation as they are all insoluble. They can also be made by heating the metals with sulphur. Sulphur does not bring out the highest oxidation state of the metals: iron reacts with sulphur to form iron(II) sulphide, FeS, whereas with oxygen it forms iron(III) oxide, Fe_2O_3.

24.13 COMPLEX COMPOUNDS

Transition metal ions form complex ions by coordination

Transition metals form complexes or coordination compounds. Such complexes are formed by the coordination of lone pairs of electrons from a donor (called a **ligand**) to an atom or cation (called an **acceptor**) which has empty orbitals to accommodate them. A cation may form a complex with a neutral molecule, e.g.

$$[Cu(NH_3)_4]^{2+}$$

or with an oppositely charged ion, e.g.

$$[CuCl_4]^{2-}$$

An atom may form a complex, e.g., $Ni(CO)_4$. The charge remaining on the central atom or ion when the ligands are removed together with their lone pairs is the **oxidation number** of the metal in the complex. The **coordination number** is the number of atoms forming coordinate bonds with the central atom or ion: 2, 4 and 6 are common.

A ligand shares a lone pair of electrons with a transition metal ion

Ligands must possess one or more unshared pairs of electrons. A ligand which can only form one bond to a central atom or ion is called a **monodentate** (literally, 'one tooth') ligand, e.g., NH_3, CN^-. A **polydentate** ('many teeth') ligand can form more than one bond. Examples are the ethanedioate ion

Polydentate ligands form chelates...

$$\begin{array}{c} CO_2{}^- \\ | \\ CO_2{}^- \end{array}$$

and   1,2-diaminoethane $H_2\overset{\bullet\bullet}{N}CH_2CH_2\overset{\bullet\bullet}{N}H_2$

in which the lone pairs on the two nitrogen atoms can form coordinate bonds. The two coordinate bonds formed by each ligand are thought to resemble the claws of a crab (Greek: *chele*), and such compounds are named **chelate compounds** or **chelates** [see Figure 24.12, p. 494]. In the polydentate ligand, bis(1,1-dicarboxymethyl)-1,2-diaminoethane

...edta is an important ligand

$$\begin{array}{ccc} {}^-O_2C-H_2C & & CH_2-CO_2{}^- \\ \diagdown & & \diagup \\ & N-CH_2-CH_2-N \\ \diagup & & \diagdown \\ {}^-O_2C-H_2C & & CH_2-CO_2{}^- \end{array}$$

which is called **edta** (short for its old name), there are unshared electron pairs on four oxygen atoms and on the two nitrogen atoms. This ligand forms six coordinate bonds, and its complex ions are very stable. Zinc ions and most other metal ions can be estimated by complexometric titration against a solution of edta [p. 273].

24.13.1 NAMING

Formula gives central atom followed by ligands

In the formula of a complex ion, the symbol for the central atom appears first and is followed by the anionic ligands and then by neutral ligands, e.g.

$$[CoCl_2(NH_3)_4]^+$$

The formula for the complex ion may be enclosed in square brackets.

The name of the complex gives the name and oxidation state of the central metal cation, e.g., cobalt(III), preceded by the name and number of ligands attached to it, e.g., hexaamminecobalt(III) ion

$$[Co(NH_3)_6]^{3+}$$

The prefixes

di tri tetra penta hexa

are used to show the number of ligands. If several ligands are present, they are listed in alphabetical order, and the prefixes, di, tri, etc., are not allowed to alter this order, e.g.

$$[CrCl_2(H_2O)_4]^+$$

Tetraaquadichlorochromium(III) ion

The system for naming complex ions

If the complex is an anion, the suffix -ate follows the name of the metal, e.g., zincate and chromate. If the metal has a Latin name, then in the complex anion the Latin name of the metal is used, followed by the suffix -ate, e.g.

$$[Fe(CN)_6]^{4-}$$

Hexacyanoferrate(II)

Examples are given in Table 24.5.

TABLE 24.5
Complex Ions

Ligand	Type of complex	Example	Name of Complex
Water	*Aqua-*	$[Cr(H_2O)_6]^{3+}$	Hexaaquachromium(III) ion
Ammonia	*Ammine-*	$[Ag(NH_3)_2]^+$	Diamminesilver(I) ion
Hydroxide ion	*Hydroxo-*	$[Zn(OH)_4]^{2-}$	Tetrahydroxozincate(II) ion
Chloride ion	*Chloro-*	$[CuCl_4]^{2-}$	Tetrachlorocuprate(II) ion
Cyanide ion	*Cyano-*	$[Fe(CN)_6]^{3-}$	Hexacyanoferrate(III) ion
Nitrite ion	*Nitro-*	$[Co(NO_2)_6]^{3-}$	Hexanitrocobaltate(III) ion
Carbon monoxide	*Carbonyl-*	$Ni(CO)_4$	Tetracarbonylnickel(0)
Ethane-1,2-diamine	*Ethane-1,2-diamine-*	$[Cr(en)_3]^{3+}$	Tri(ethane-1,2-diamine) chromium(III) ion
edta	*edta-*	$[Zn(edta)]^{2-}$	edtazincate(II) ion
Cl^-, NH_3	*Mixed*	$[CoCl_2(NH_3)_4]^+$	Tetraamminedichloro-cobalt(III) ion
OH^-, H_2O	*Mixed*	$[Fe(OH)_2(H_2O)_4]^+$	Tetraaquadi-hydroxoiron(III) ion

24.13.2 COLOUR

Transition metal ions are often coloured because electrons move between non-degenerate d orbitals

Transition metal ions are often coloured [see vanadium, Figure 24.7]. In an isolated transition metal atom, the five d orbitals are **degenerate**, that is, they are all at the same energy level. In a complex ion, the d orbitals differ slightly in energy as a result of overlapping differently with the ligands: they are **non-degenerate**. Electrons can jump from one d orbital to another if they absorb energy. For most transition metal complexes, the frequency of light absorbed in these energy transitions is in the visible region of the spectrum, and the ion appears coloured. The colour of the ion is complementary to the colours absorbed. The Sc^{3+} (aq) ion has no d electrons, and is colourless. In the ions Cu^+ and Zn^{2+}, with a d^{10} configuration, no d–d transition is possible, and these ions are colourless. Different ligands affect the energy levels of the d orbitals: $[Cu(H_2O)_4]^{2+}$ is blue, whereas $[Cu(NH_3)_4]^{2+}$ is a very intense deep blue.

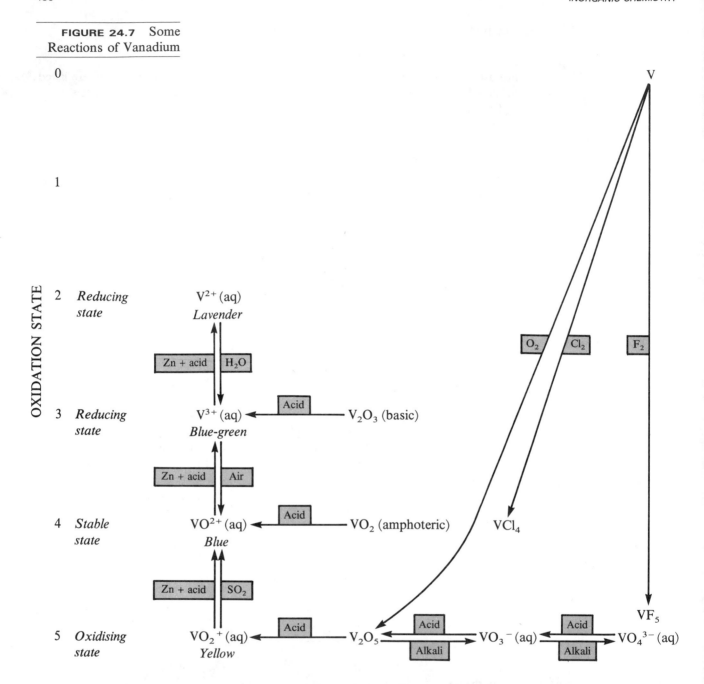

FIGURE 24.7 Some
Reactions of Vanadium

24.13.3 STOICHIOMETRY OF COMPLEX IONS

The formula of the $[Cu(NH_3)_4]^{2+}$ ion has been found by the **partition method** described on p. 168.

Partition methods can be used to find the formula of a complex ion

Colorimetric methods are also used for finding the composition of complex ions. When solutions of nickel(II) sulphate and edta are mixed, coloured complex ions are formed. If the formula of the ion is

$$Ni(edta)_n{}^{x-}$$

then the maximum intensity of colour will be obtained when the two solutions are mixed to give a molar ratio of

$$Ni : edta = 1 : n$$

EXPERIMENTAL DETERMINATION OF THE FORMULA OF Ni(edta)$_n{}^{x-}$

Colorimetric methods can be used...

...e.g., for the Ni^{2+} edta complex

Solutions are made up as shown in Table 24.6. They are put into a colorimeter, as shown in Figure 24.8(a). The instrument gives readings of *absorbance*. This quantity is proportional to the concentration of the species that absorbs light, provided a suitable filter is used. The absorbance is found for each solution.

Solution	Volume of NiSO$_4$/cm^3	Volume of edta/cm^3
	(both 0.05 mol dm^{-3})	
1	0	10
2	1	9
3	2	8
4	3	7
5	4	6
6	5	5
7	6	4
8	7	3
9	8	2
10	9	1
11	10	0

TABLE 24.6

FIGURE 24.8(a) Using a Colorimeter

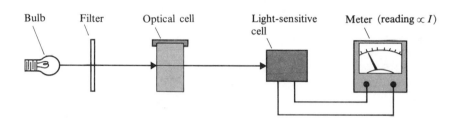

Figure 24.8(b) shows a plot of absorbance against the number of the solution. You can see that the maximum colour intensity corresponds to the formula [Ni(edta)]$^{2-}$.

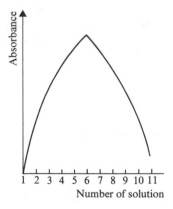

FIGURE 24.8(b) Results of Colorimetry on Ni(edta)$_n{}^{x-}$

24.13.4 STABILITY OF COMPLEX IONS

The dissociation constants of complex ions are covered on p. 272.

24.13.5 STEREOCHEMISTRY

The spatial arrangement of the bonds from the ligands to the central atom or ion depends on the identity of the atomic orbitals used by the central atom or ion. The orientation of bond orbitals has been covered on p. 90–5, summarised on p. 95 and extended on pp. 96–106. It is interesting to look at a selection of the many complex ions formed by transition metals.

24.13.6 COMPLEXES OF CHROMIUM(III)

Cr^{3+}(aq)...

...and complex formation

In aqueous solution, chromium(III) ions exist as blue-violet hexaaquachromium(III) ions, $Cr(H_2O)_6^{3+}$. These ions are present in *chrome alum*, $KCr(SO_4)_2 \cdot 12H_2O$. Water molecules can be replaced by other ligands if these form more stable complexes. In ammonia solution, ammonia molecules replace water molecules to form hexaamminechromium(III) ions, $Cr(NH_3)_6^{3+}$. The octahedral arrangement of ligands is shown in Figure 24.9.

FIGURE 24.9
Hexaammine-
chromium(III) Ion

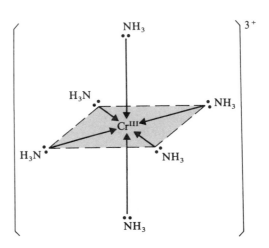

If the six ligands of the octahedron are not identical, isomerism occurs

If the six ligands are not identical, *cis-trans* geometrical isomerism will occur. In the tetraamminedichlorochromium(III) ions, shown in Figure 24.10, the *cis*-form (a) has two chlorine ligands adjacent, whereas in the *trans*-form (b), the chlorine atoms are diagonally opposite.

FIGURE 24.10 *Cis*- and
Trans-tetraammine-
dichlorochromium(III)
ions

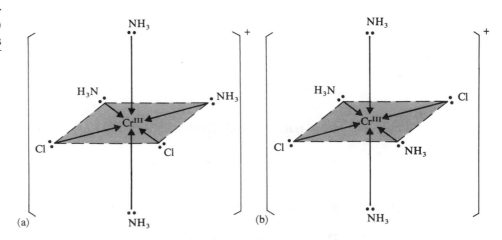

The complex salt $Cr(H_2O)_6Cl_3$ has four isomers. There are three structural isomers, which differ in electrical conductivity and in the fraction of the chloride content that can be precipitated by silver nitrate solution. These three salts are

(a) $[Cr(H_2O)_6]^{3+} \cdot 3Cl^-$ 　　　　　　　*Grey-blue*

(b) $[Cr(H_2O)_5Cl]^{2+} \cdot 2Cl^- \cdot H_2O$ 　　　*Pale green*

(c) $[Cr(H_2O)_4Cl_2]^+ \cdot Cl^- \cdot 2H_2O$ 　　　*Green*

Isomer (c) exists as *cis-* and *trans-* geometrical isomers. See Figure 24.11 and Questions 1, 4 and 10 p. 498.

Bidentate ligands, such as the ethanedioate ion

$$^-O_2C—CO_2^-$$

and 1,2-diaminoethane

$$H_2\overset{\bullet\bullet}{N}CH_2CH_2\overset{\bullet\bullet}{N}H_2$$

Optical isomerism occurs in complexes with bidentate ligands

form complexes which show optical isomerism. Figure 24.12 shows how the two chelate complexes, triethanedioatechromium(III) and tri-1,2-diamino-ethanechromium(III) ions, exist as mirror image forms or **enantiomers**.

FIGURE 24.11　Isomers of $Cr(H_2O)_6Cl_3$

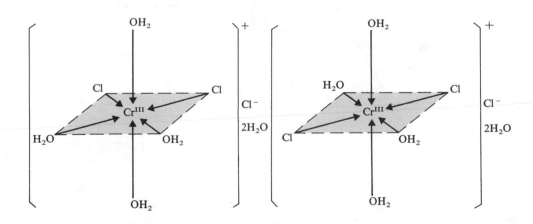

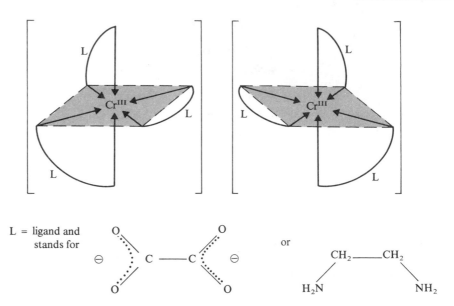

FIGURE 24.12
Enantiomers of
$[Cr(C_2O_4{}^{2-})_3]^{3+}$ and
$[Cr(C_2N_2H_8)_3]^3$

L = ligand and
stands for

or

24.13.7 COMPLEXES OF COBALT(II) AND COBALT(III)

*Cobalt(III) ions form
complexes with various
ligands*

Cobalt(III) ions form complex ions with the ligands H_2O, NH_3, Cl^-, $C_2O_4{}^{2-}$, $H_2NCH_2CH_2NH_2$, CN^-, $NO_2{}^-$ and many others. The complex ions are similar to those of chromium(III). Cobalt(II) ions are more stable than cobalt(III) ions, with the result that Co^{3+} ions are powerful oxidising agents.

FIGURE 24.13 Some
Reactions of Cobalt
Compounds

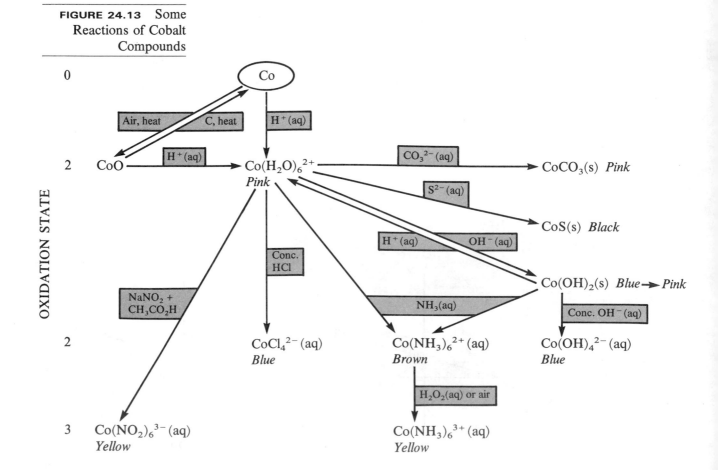

24.13.8 IRON COMPLEXES

Complexes are formed by Fe²⁺ and Fe³⁺ with CN⁻ and H₂O

Iron(II) and iron(III) ions employ d^2sp^3 hybrid ions [p. 105] to form the complexes hexacyanoferrate(II), $[Fe(CN)_6]^{4-}$, and hexacyanoferrate(III), $[Fe(CN)_6]^{3-}$ [see Figure 24.14]. So stable are these complexes that the hydroxides $Fe(OH)_2$ and $Fe(OH)_3$ are not precipitated when hydroxide ions are added to solutions of the complex ions. Both iron(II) and iron(III) ions are hydrated in solution as $[Fe(H_2O)_6]^{2+}$ and $[Fe(H_2O)_6]^{3+}$ [see Figure 24.14].

FIGURE 24.14(a) Hexacyanoferrate(II) Ion (b) Hexaaquairon(III) Ion

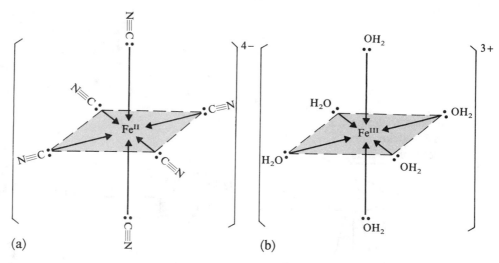

(a) (b)

Thiocyanate ions are used in a test for Fe³⁺ (aq)

Iron(III) ions form a blood-red complex with thiocyanate ions, $[Fe(SCN)(H_2O)_5]^{2+}$ (aq). No such complex is formed by iron(II) ions. The formation of this blood-red complex is therefore used to distinguish iron(III) from iron(II).

Complexes are formed by haemoglobin

Haemoglobin contains iron in the oxidation state +2, with coordination number 6. Figure 24.15 shows how oxygen molecules can form coordinate bonds to the iron atoms. The bonding is reversible, and enables haemoglobin to carry oxygen around the body and release it where it is needed. Carbon monoxide and cyanide ions coordinate more strongly than oxygen to form very stable complexes. They prevent haemoglobin from taking up oxygen, thus acting as poisons.

FIGURE 24.15 Part of the Haemoglobin Molecule

1 Four N atoms in a ring structure occupy four of the Fe orbitals.

2 A nitrogen atom from a protein molecule occupies a fifth orbital.

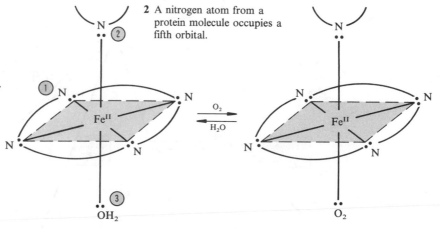

3 The sixth orbital enables O_2 to be bonded reversibly. CO and CN⁻ can form stable complexes by coordinating into this orbital. They then prevent the uptake of O_2.

24.13.9 NICKEL COMPLEXES

Nickel(II) ions form the orange complex tetracyanonickelate(II) ions, $[Ni(CN)_4]^{2-}$. The arrangement of bonds is square planar [see Figure 24.16]. Nickel forms the octahedral complexes $[Ni(H_2O)_6]^{2+}$ and $[Ni(NH_3)_6]^{2+}$ and a tetrahedral complex, $[NiCl_4]^{2-}$.

FIGURE 24.16
Tetracyanonickelate(II)
Ion

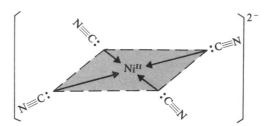

FIGURE 24.17 Some
Reactions of Nickel and
its Compounds

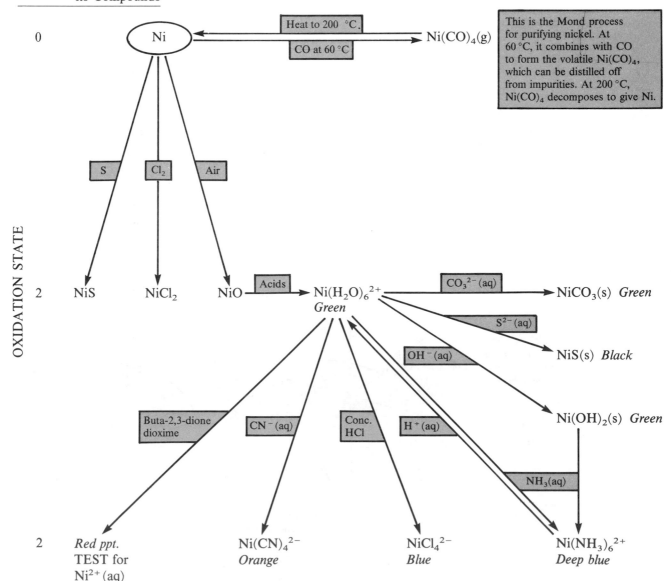

24.13.10 COMPLEXES OF COPPER(I) AND COPPER(II) IONS

The electron configurations of copper and its ions are shown below:

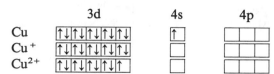

From the electron configurations, it might seem likely that Cu^+ would be more stable than Cu^{2+}. In fact, when standard reduction potentials for the possible reactions are considered

$$Cu^{2+}(aq) + e^- \rightleftharpoons Cu^+(aq); \quad E^\ominus = +0.15\,V$$

$$Cu^+(aq) + e^- \rightleftharpoons Cu(s); \quad E^\ominus = +0.52\,V$$

Cu⁺ is unstable with respect to disproportionation to Cu + Cu²⁺

it follows [from p. 000] that the reaction that takes place is

$$2Cu^+(aq) \rightleftharpoons Cu(s) + Cu^{2+}(aq); \quad E^\ominus = +0.37\,V$$

The copper(I) ion is unstable with respect to disproportionation to copper and the copper(II) ion. Copper(I) ions can be stabilised by the formation of complexes such as

$$[CuCl_2]^- \quad [Cu(CN)_4]^{3-} \quad [Cu(NH_3)_2]^+$$

Cu²⁺ uses four orbitals for complex formation

Copper(II) ions use four bonds in complex formation. The spatial arrangement of the bonds in the hydrated $[Cu(H_2O)_4]^{2+}$ ion is square planar. In solution, two water molecules are loosely coordinated at right angles [see Figure 24.18(a)] in a distorted octahedral configuration. Ammonia displaces water molecules from the pale blue tetraaquacopper(II) ions, converting them into the deep blue tetraamminecopper(II) ions, $[Cu(NH_3)_4]^{2+}$ [see Figure 24.18(b)].

FIGURE 24.18(a)
Tetraaquacopper(II) Ion
(b) Tetraammine-
copper (II) Ion

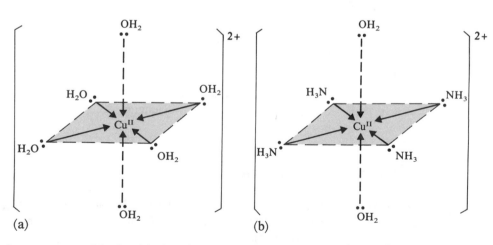

The six ligands take up an octahedral arrangement

In concentrated hydrochloric acid, water molecules in $[Cu(H_2O)_4]^{2+}$ are replaced by chloride ions, with the formation of yellow tetrachlorocuprate(II) ions, $[CuCl_4]^{2-}$:

$$[Cu(H_2O)_4]^{2+}(aq) + 4Cl^-(aq) \rightleftharpoons [CuCl_4]^{2-}(aq) + 4H_2O(l)$$

Blue   Yellow

Cl⁻ and edta complex with Cu²⁺

In more dilute hydrochloric acid, the solution is green due to the presence of both blue $[Cu(H_2O)_4]^{2+}$ ions and yellow $[CuCl_4]^{2-}$ ions. Copper(II) ions also form complexes with ethane-1,2-diamine and with edta. Copper(II) is the most strongly complexing of all the first long period dipositive cations.

1. Name the four complex ions shown in Figure 24.11.

2. Name the complex ion $[CrCl_2(H_2NCH_2CH_2NH_2)_2]^+$. Sketch the two geometrical isomers which have this formula. Which of the two is optically active?

3. Sketch the arrangement of bonds in the complex ions

(a) hexaaquacobalt(III)

(b) hexaamminecobalt(III)

(c) hexafluorocobalt(III)

(d) tri(1,2-diaminoethane)cobalt(III)

(e) tri(ethanedioate)cobalt(III).

4. Explain why the ion $[Co(H_2NCH_2CH_2NH_2)_3]^{3+}$ exists in enantiomeric forms.

5. How many isomers exist of the complex ion $[Co(H_2NCH_2CH_2NH_2)_2Cl_2]^+$?

6. The addition of ammonia solution to an aqueous solution of copper(II) sulphate resulted in the formation of a pale blue precipitate. This dissolved on the addition of more ammonia solution to give a deep blue solution. The precipitate also dissolved in dilute hydrochloric acid to give a pale blue solution and in concentrated hydrochloric acid to give a green solution. Explain these colour changes, giving the name and formula of each coloured species formed.

7. Two isomers exist of formula $CrCl_2(NO_2)(NH_3)_4$. One gives a precipitate with silver nitrate solution; the other does not. Suggest how this difference in behaviour can arise.

8. Name the complex ion $[PtCl_2(NH_3)_4]^{2+}$. Sketch and name the two isomeric forms of this ion.

9. There are two isomers of $Pt(NH_3)_2Cl_2$. What does this tell you about the geometry of the compound?

10. How many isomers exist of the following compounds?

(a) $[Co(NH_3)_6]^{3+}$

(b) $[Co(NH_3)_5Cl]^{2+}$

(c) $[Co(NH_3)_4Cl_2]^+$

(d) $Co(NH_3)_3Cl_3$

11. There are five compounds containing platinum(IV) chloride and ammonia. Werner deduced their formulae from the following information. See whether you can do so. The five compounds are:

Compound	Empirical formula	Number of ions	Number of Cl^- ions
1	$PtCl_4 \cdot 6NH_3$	5	4
2	$PtCl_4 \cdot 5NH_3$	4	3
3	$PtCl_4 \cdot 4NH_3$	3	2
4	$PtCl_4 \cdot 3NH_3$	2	1
5	$PtCl_4 \cdot 2NH_3$	0	0

24.14 IRON

Iron is the most important of metals. The whole of our twentieth-century way of life is based on the use of iron machinery. We shall look in more detail at iron than the rest of the transition metals.

24.14.1 EXTRACTION

Iron is mined as its oxides and sulphides

Iron is mined as its oxides, Fe_2O_3, *haematite*, and Fe_3O_4, *magnetite*, and FeS_2, *iron pyrites*. It is obtained by the reduction of the oxides by carbon monoxide [for the theory, see pp. 212–4]:

$$Fe_2O_3(s) + 3CO(g) \rightarrow 2Fe(s) + 3CO_2(g); \quad \Delta H^\ominus = -27\,kJ\,mol^{-1}$$

The **blast furnace**, in which this process takes place, is illustrated in Figure 24.19. At the bottom of the furnace, coke is oxidised exothermically to carbon dioxide:

$$C(s) + O_2(g) \rightarrow CO_2(g); \quad \Delta H^\ominus = -392\,kJ\,mol^{-1}$$

and the temperature of the furnace is about 1900 °C in this region. Higher up the furnace, carbon dioxide reacts with coke to form carbon monoxide. This reaction is endothermic, and the furnace in this region has a temperature of 1100 °C:

$$CO_2(g) + C(s) \rightarrow 2CO(g); \quad \Delta H^\ominus = 172\,kJ\,mol^{-1}$$

Carbon monoxide reduces iron oxides in the blast furnace

Iron oxides are reduced exothermically, and the iron produced falls to the bottom of the furnace, where the temperature is high enough to melt it, and a layer of molten iron lies on the bottom of the furnace. At the same time, the limestone in the charge dissociates to form calcium oxide and carbon dioxide:

$$CaCO_3(s) \rightleftharpoons CaO(s) + CO_2(g); \quad \Delta H^{\ominus} = 178 \, kJ \, mol^{-1}$$

Iron oxide + coke + limestone produce iron + 'slag'

Calcium oxide combines with silicon(IV) oxide and aluminium oxide, the impurities in the ore, to form a molten 'slag' of calcium silicate(IV) and calcium aluminate(III), which trickles down the stack:

$$CaO(s) + SiO_2(s) \rightarrow CaSiO_3(l)$$

$$CaO(s) + Al_2O_3(s) \rightarrow CaAl_2O_4(l)$$

At the bottom of the furnace, iron and slag are tapped off every few hours. A modern furnace makes 3000 tonnes of iron in a day, using 3000 tonnes of coke and 4000 tonnes of air. If natural gas is injected with the hot air, the consumption of coke can be halved. The quantity of slag produced is about 1 tonne for every tonne of iron. It is used for road making and in the manufacture of cement. A furnace can operate continuously for several years before it needs relining.

FIGURE 24.19(a) The Blast Furnace

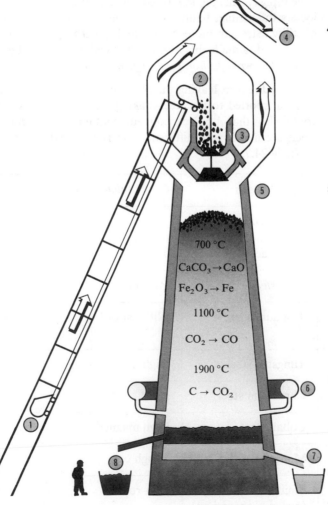

2 The skip discharges its load.

3 The *double-bell* charging system prevents the escape of gases from the furnace. The small bell is lowered to let the charge fall on to the large bell, and then raised. The large bell is lowered to allow the charge to fall into the furnace.

1 A *skip* is loaded with ore, coke and limestone.

4 The *downcomer* takes away exhaust gases to heat the air in 6.

5 The *stack*, a tower of steel plates, lined with heat-resistant bricks, is 30 m high.

6 Hot blasts of air enter through narrow pipes called *tuyeres*, leading from this circular pipe.

7 Molten iron is tapped off into a *ladle*.

8 Slag is run off.

700 °C
CaCO$_3 \rightarrow$ CaO
Fe$_2$O$_3 \rightarrow$ Fe
1100 °C
CO$_2 \rightarrow$ CO
1900 °C
C $\rightarrow$ CO$_2$

FIGURE 24.19(b)
Blast Furnace, Llanwern
Works, Newport, Gwent

24.14.2   CAST IRON

Cast iron or pig iron
contains carbon, which
decreases its ductility

The iron leaving the blast furnace is run into moulds, where it forms solid blocks called 'pigs'. This *pig iron* or *cast iron* contains about 4% carbon. The carbon present lowers the melting temperature of the iron, increases its hardness and decreases its ductility. Strength increases up to 1% carbon and then decreases. If the carbon is removed, a malleable iron is produced, which can be readily worked by ironsmiths. It is called *wrought iron* [see Figure 24.20].

Wrought iron is the purest
form of iron

Steels contain carbon and
other metals alloyed with
iron

Some of the iron leaving the blast furnace is not cast; it is run into giant crucibles and transported to the steel-making section of the iron and steel works. Steels contain less than 1.5% carbon, and contain added metals. Many thousands of different steels are made, with different properties suited to different uses. Some are shown in Table 24.7.

Carbon steels	Composition/% carbon	Uses
Low carbon Medium carbon High carbon	<0.3 0.3–0.7 0.7–1.5	Boiler plates Axles Cutting tools
Alloy steels	*Properties*	*Uses*
Titanium steel	Withstands high temperatures	Gas turbines, spacecraft, Comet aircraft
Tungsten steel	High melting, tough	Tools
Chromium steel	Hard	Ball bearings
Cobalt steel	High magnetic permeability	Magnets
Manganese steel	Tough	Railway points
Stainless steel (nickel + chromium)	Non-rusting	Cutlery, car accessories

TABLE 24.7   Some
Different Steels

FIGURE 24.20 Wrought
Iron

FIGURE 24.21(a) The
LD Process of Steel
Making

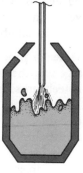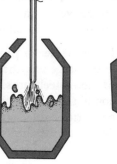

1 Charging.
The *converter* tips to receive
a charge of 300 tonnes of
molten iron.

2 The first blow.
The water-cooled *lance*
directs oxygen and powdered
calcium oxide on to the
surface at a pressure of
10–15 atm.

3 Slagging.
The converter tips
backwards to pour out the
primary slag.

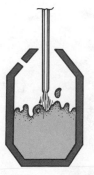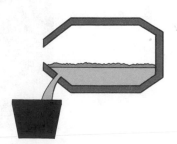

4 The second blow.
The *lance* directs oxygen and
lime into the converter.

5 Pouring.
The converter tilts forward
to pour the steel into a *ladle*,
while the slag remains on
top.

6 Slag remains in the
converter, which tilts to
receive the next charge.

FIGURE 24.21(b)
Charging the Converter

FIGURE 24.21(b)
Charging the Converter

24.14.3   STEEL

Making steel requires the removal of C, S, P

In making steel from iron, carbon and other impurities such as sulphur and phosphorus are converted into their oxides. Gaseous oxides remove themselves; other oxides are removed by combination with a base such as calcium oxide to form a slag.

The Bessemer process

Oxygen-blown converters are used in modern steelworks

The LD process...

...and the Kaldo process

In the Bessemer process, invented in Britain 125 years ago, molten iron was poured into a large tub, the *converter*, and air was blown on to it to oxidise the impurities. Modern steelworks use oxygen-blown converters. The steel is better because it contains no nitrogen, which makes steel brittle. Shown in Figure 24.21 is the LD converter, named after the Austrian towns of Linz and Donawitz, where it was invented. One converter makes about 500 tonnes of steel in an hour. The Kaldo process used in Sweden is similar, but employs a converter which can be rotated. Oxygen is blown in at a pressure of only 2 atm; the rotation assists oxidation.

[See Postscript, p. 516, for modern steelmaking.]

FIGURE 24.21(c)
Slagging

24.14.4   RUSTING

Rusting is an electrochemical process

The mechanism of rusting

Rusting is a serious problem. A large fraction ($\frac{1}{8}$) of the annual UK production of steel (20 million tonnes) is needed simply to replace iron lost through rusting. Rust is hydrated iron(III) oxide, $Fe_2O_3 \cdot xH_2O$. Both water and air are needed for rusting to occur. It is an electrochemical process, with different parts of an iron structure acting as cathodes and anodes. At an anodic region, the process which occurs is

$$Fe(s) \rightarrow Fe^{2+}(aq) + 2e^-$$

At a cathodic region, the process is

$$O_2(aq) + 2H_2O(l) + 4e^- \rightarrow 4OH^-(aq)$$

The presence of dissolved acids and salts in water increases its conductivity and speeds up the process of rusting. If cathodic and anodic areas are close together, precipitation of iron(II) hydroxide, $Fe(OH)_2(s)$, occurs. Air oxidises this to rust, hydrated iron(III) oxide:

$$2Fe(OH)_2(s) + \tfrac{1}{2}O_2(aq) + H_2O(l) \rightarrow Fe_2O_3 \cdot xH_2O(s)$$

24.14.5   PREVENTION OF RUSTING

COATING

Rusting can be prevented by...

Various methods are used to provide a protective coat to exclude water and oxygen.

...paint...

(a) Paint is used for many large objects, e.g., ships and bridges [see Figure 24.22]. A paint containing phosphoric(V) acid is effective as it forms a layer of insoluble iron(III) phosphate(V) with any rust present on the surface.

...oil, grease...

(b) A coat of grease or oil is used for moving parts of machinery.

...zinc coating...
...tin plating...

(c) A coat of another metal may be used; zinc coating, i.e., *galvanising*, and tin plating are used. Since zinc is higher than iron in the Electrochemical Series, even if the coating of zinc is scratched, it will continue to protect the iron underneath from rusting. Tin cans rust if they are scratched because, being higher in the Electrochemical Series, iron is corroded in preference to tin.

...chromium plating...

(d) Chromium plating is used for many car accessories because it is decorative as well as protective. An electrolytic method is used for plating [see Figure 24.3].

FIGURE 24.22
Galvanised Iron Girders
at the Holyhead Ferry
Terminal

FIGURE 24.23
Decorative Chromium
Plating

ALLOYING

...forming alloys, e.g.,
stainless steel...

Stainless steel is an alloy of iron with nickel and chromium (e.g., 18% Ni, 8% Cr). The added metals produce a surface film of metal oxides which is impervious to water.

CATHODIC PROTECTION

...or by cathodic
protection

Sacrificial protection...

If a block of a metal higher in the Electrochemical Series is connected to iron, then that metal acts as the anode, and it is corroded while iron remains intact. Zinc and magnesium are often used. Underground pipes are protected by attaching bags of magnesium scraps at intervals and replacing these from time to time when they have been corroded. The hulls of ships are protected by attaching blocks of zinc, which are sacrificed to protect the iron. This technique is called **sacrificial protection** [see Figure 24.24].

FIGURE 24.24
Sacrificial Protection of
Iron by Zinc

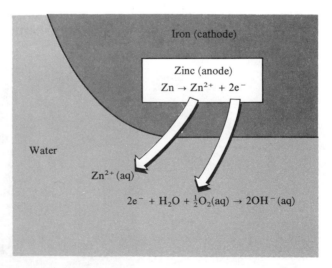

...or the application of a
negative potential, are the
techniques used in
cathodic protection

Another method of cathodic protection is to make iron the cathode by connecting it to the negative side of a battery while a conductor such as graphite is connected to the positive end. The negative potential on the iron structure inhibits the formation of Fe^{2+} (aq) ions.

24.14.6 THE CHEMISTRY OF IRON

FIGURE 24.25 Some
Reactions of Iron

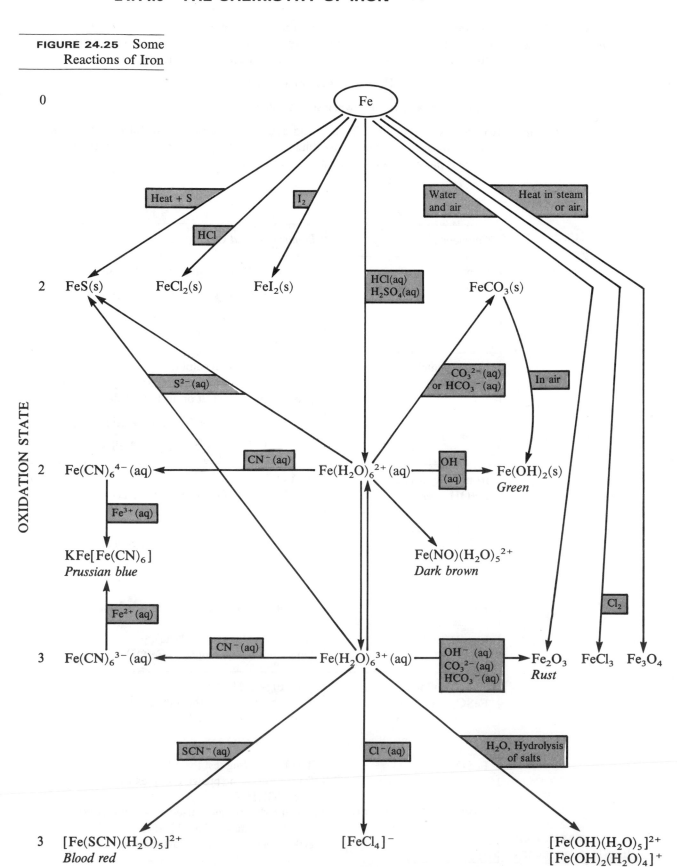

24.14.6   THE CHEMISTRY OF IRON

Iron combines with O_2, S, N_2, C and the halogens

Iron combines on heating with the non-metallic elements oxygen, nitrogen, the halogens, sulphur and carbon. It reacts with water and air to form rust, $Fe_2O_3 \cdot xH_2O$, and with steam to form iron(II) iron(III) oxide, Fe_3O_4, which is magnetic:

$$3Fe(s) + 4H_2O(g) \rightarrow 4H_2(g) + Fe_3O_4(s)$$

This oxide is also formed when iron is heated in air.

It reacts with dilute acids

Iron reacts with dilute sulphuric acid and hydrochloric acid to form iron(II) salts and hydrogen. It is attacked by dilute nitric acid to form iron(III) nitrate and is rendered 'passive' by concentrated nitric acid.

Compound	Preparation and Comments
Iron(II) oxide, FeO	Heat iron(II) ethanedioate in the absence of air: $$FeC_2O_4(s) \rightarrow FeO(s) + CO(g) + CO_2(g)$$ The CO and CO_2 produced prevent oxidation. It is *pyrophoric*: may burst into flame in air.
Iron(II) sulphide, FeS, black	(a) Heat iron with sulphur. (b) Pass $H_2S(g)$ into or add $S^{2-}(aq)$ to a solution of either an Fe^{3+} salt or an Fe^{2+} salt: $$2Fe^{3+}(aq) + S^{2-}(aq) \rightarrow 2Fe^{2+}(aq) + S(s)$$ $$Fe^{2+}(aq) + S^{2-}(aq) \rightarrow FeS(s)$$
Iron(II) hydroxide, $Fe(OH)_2$, green	Add $OH^-(aq)$ to $Fe^{2+}(aq)$. Amphoteric $\rightarrow Fe^{2+}(aq)$ or $Fe(OH)_6^{4-}(aq)$
Iron(II) halides FeF_2, $FeCl_2$ $FeCl_2 \cdot 6H_2O$ $FeBr_2$ FeI_2	Pass $HF(g)$ or $HCl(g)$ over heated iron. See Figure 24.26. Heat iron with bromine vapour. An excess of iron prevents the formation of $FeBr_3$. Heat iron with iodine. (No FeI_3 exists.)
Iron(II) carbonate, $FeCO_3$	Add $CO_3^{2-}(aq)$ to $Fe^{2+}(aq)$. On standing, it changes to $Fe(OH)_2(s)$.
Iron(II) sulphate $FeSO_4 \cdot 7H_2O$, green	[See Figure 24.26.] When heated, gives the anhydrous salt. On further heating, forms iron(III) oxide: $$2FeSO_4(s) \rightarrow Fe_2O_3(s) + SO_2(g) + SO_3(g)$$
$FeSO_4 \cdot NO$	The brown ring test for nitrates involves the formation of the brown complex ion, $[Fe(NO)(H_2O)_5]^{2+}$
$FeSO_4 \cdot (NH_4)_2SO_4 \cdot 6H_2O$	It is a reducing agent used as a primary standard in titrimetric analysis [p. 61]. It is not efflorescent, and not oxidised by air.

TABLE 24.8
Iron(II) Compounds

FIGURE 24.26
Preparation of Iron(II)
Sulphate-7-water,
$FeSO_4 \cdot 7H_2O$
(Hydrochloric acid gives
$FeCl_2 \cdot 6H_2O$.)

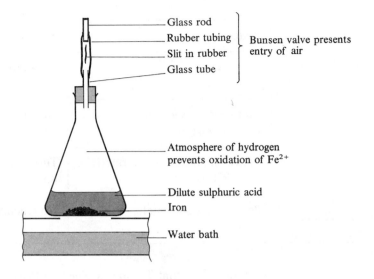

Glass rod
Rubber tubing
Slit in rubber
Glass tube

Bunsen valve presents entry of air

Atmosphere of hydrogen prevents oxidation of Fe^{2+}

Dilute sulphuric acid
Iron
Water bath

TABLE 24.9
Iron(III) Compounds

Compound	*Preparation and Comments*
Iron(III) oxide, Fe_2O_3, rust	(a) Mined as *haematite*. (b) Heat $FeSO_4 \cdot 7H_2O$. (c) Add OH^-(aq) to Fe^{3+}(aq). No $Fe(OH)_3$ exists; hydrated Fe_2O_3 is precipitated. Amphoteric $\rightarrow Fe^{3+}$(aq) or $[Fe(OH)_6]^{3-}$(aq).
Iron(II) iron(III) oxide, Fe_3O_4 blue-black	(a) Mined as *magnetite*. (b) Heat iron in air or steam: $$3Fe(s) + 4H_2O(g) \rightarrow Fe_3O_4(s) + 4H_2(g)$$ It reacts as a mixture of $FeO \cdot Fe_2O_3$: $$Fe_3O_4(s) + 8H^+(aq) \rightarrow Fe^{2+}(aq) + 2Fe^{3+}(aq) + 4H_2O(l)$$
Iron(III) halides $FeCl_3$, $FeBr_3$	Pass Cl_2 or Br_2(g) over heated iron. The apparatus in Figure 19.9, p. 381 can be used. Fe_2Cl_6 or Fe_2Br_6 sublimes over into the receiver. (The reaction between Fe and I_2 gives FeI_2 because Fe^{3+} oxidises I^- to I_2.)

HYDROLYSIS OF IRON SALTS

Fe^{3+} salts are acidic in solution due to hydrolysis

Iron(II) salts are slightly acidic in solution; iron(III) salts are more acidic as the degree of hydrolysis is greater. The small, highly charged Fe^{3+} ion is hydrated, and, as in the case of aluminium salts [p. 379], this leads to hydrolysis:

$$[Fe(H_2O)_6]^{3+} + H_2O \rightleftharpoons [Fe(OH)(H_2O)_5]^{2+} + H_3O^+$$

$$[Fe(OH)(H_2O)_5]^{2+} + H_2O \rightleftharpoons [Fe(OH)_2(H_2O)_4]^+ + H_3O^+$$

Reagent	Reaction of Fe²⁺ (aq)	Reaction of Fe³⁺ (aq)
OH^- (aq)	Gelatinous green ppt. of $Fe(OH)_2$(s)	Gelatinous rust ppt. of Fe_2O_3(s)
Potassium hexacyanoferrate(II), $K_4Fe(CN)_6$(aq)	Green or brown colour	Prussian-blue $KFe[Fe(CN)_6]$(s)
Potassium hexacyanoferrate(III), $K_3Fe(CN)_6$(aq)	Turnbull's blue, $KFe[Fe(CN)_6]$(s)	Green or brown colour
Potassium thiocyanate, KCNS	No reaction	Blood-red colour, $Fe(CNS)^{2+}$ (aq)

TABLE 24.10 Tests for Fe²⁺ and Fe³⁺ Ions

Analysis shows that Turnbull's blue and Prussian blue are both $K^+Fe^{3+}[Fe^{II}(CN)_6]^{4-}$. This bright blue compound is used as a pigment.

24.15 COPPER

Copper is low in the Electrochemical Series

Copper is low in the Electrochemical Series and is found 'native', i.e., uncombined. The chief ores are *copper pyrites*, $CuFeS_2$, and *copper glance*, CuS. In the extraction of copper from copper pyrites, the ore is concentrated by froth flotation. When the ore is roasted with silica and air in a furnace, iron silicate and copper(I) sulphide are formed. Copper(I) sulphide formed in this way and copper(II) sulphide from copper glance can both be converted to copper by roasting in a furnace:

The method of extraction

$$2CuFeS_2(s) + 4O_2(g) + 2SiO_2(s) \rightarrow Cu_2S(s) + 2FeSiO_3(s) + 3SO_2(g)$$

$$Cu_2S(s) + O_2(g) \rightarrow 2Cu(s) + SO_2(g)$$

It is purified by electrolysis

Impure copper is formed. The electrolytic method used to purify copper is illustrated in Figure 24.27.

FIGURE 24.27 Purification of Copper

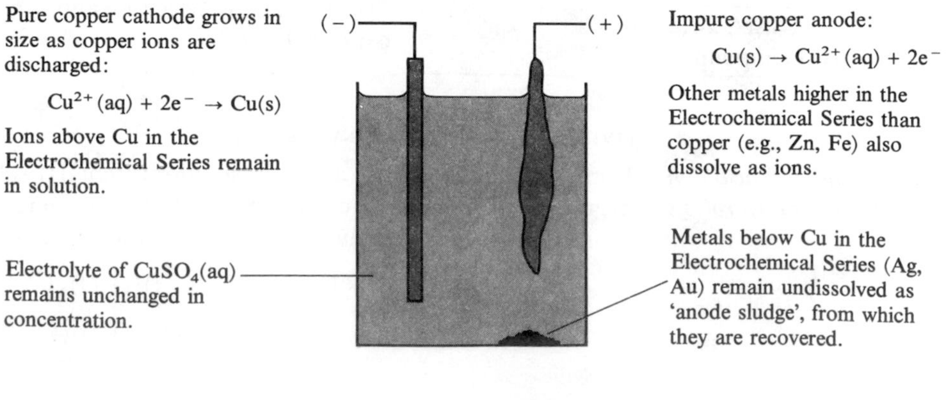

Pure copper cathode grows in size as copper ions are discharged:

$$Cu^{2+}(aq) + 2e^- \rightarrow Cu(s)$$

Ions above Cu in the Electrochemical Series remain in solution.

Electrolyte of $CuSO_4$(aq) remains unchanged in concentration.

Impure copper anode:

$$Cu(s) \rightarrow Cu^{2+}(aq) + 2e^-$$

Other metals higher in the Electrochemical Series than copper (e.g., Zn, Fe) also dissolve as ions.

Metals below Cu in the Electrochemical Series (Ag, Au) remain undissolved as 'anode sludge', from which they are recovered.

FIGURE 24.28
Electrolytic Cells at
Capper Pass

24.15.1 USES OF COPPER

The uses of copper include cooking ware...

...electrical cables...

...and roofing

The high thermal conductivity of copper leads to its use for cooking ware. The high electrical conductivity makes copper wire admirably suitable for electrical circuits and cables. The resistance to corrosion makes copper useful for water pipes. Copper is used as a roofing material because it weathers to acquire a coating of green basic copper carbonate, $CuCO_3 \cdot Cu(OH)_2 \cdot nH_2O$, which lends a colourful touch to a building. Alloys of copper are *coinage metal* (Cu, Ni), *brass* (Cu, Zn) and *bronze* (Cu, Sn).

24.15.2 REACTIONS OF COPPER

Copper reacts with O_2, S, the halogens, and oxidising acids

Copper is low in the Electrochemical Series. It reacts with the oxidising acids, dilute nitric acid, concentrated nitric acid and concentrated sulphuric acid to form copper(II) salts. It combines directly with oxygen, sulphur and the halogens to form copper(II) compounds, except in the case of iodine, with which it forms copper(I) iodide. At temperatures of 800–1000 °C, it combines with oxygen to form copper(I) oxide.

24.15.3   COPPER(II) COMPOUNDS

Compound	Preparation and Comments
Copper(II) oxide, CuO, black	Heat copper(II) nitrate, carbonate or hydroxide. Basic
Copper(II) hydroxide, $Cu(OH)_2$, blue	Add OH^- (aq) to Cu^{2+} (aq). A gelatinous blue precipitate forms. It dissolves in NH_3(aq) to form $Cu(NH_3)_4{}^{2+}$ (aq). Basic
Copper(II) salts	Warm CuO with a dilute acid.
Anhydrous CuX_2	Heat Cu with halogen (except I_2).
Copper(II) sulphide, CuS, black	Pass H_2S(g) into or add S^{2-} (aq) to a solution of a Cu^{2+} salt.

TABLE 24.11

24.15.4   COPPER(I) COMPOUNDS

The stability of copper(I) compounds and complex ions has been discussed on p. 497.

Compound	Preparation and Comments
Copper(I) oxide, Cu_2O, reddish solid	Reduce Cu^{2+} (aq) in alkaline solution with an aldehyde or by SO_2: $$2Cu^{2+}(aq) + 2OH^-(aq) + 2e^- \rightarrow Cu_2O(s) + H_2O(l)$$ The formation of Cu_2O is a test for an aldehyde [p. 652]. With NH_3(aq), Cu_2O forms $[Cu(NH_3)_2]^+$ (aq). With conc. HCl, Cu_2O forms $[CuCl_2]^-$ (aq). With dilute acids, Cu(s) and a Cu^{2+} salt are formed.
Copper(I) chloride, CuCl, white solid	(a) Reduce $CuCl_2$(aq) by Cu(s) or by SO_2(g) to $[CuCl_2]^-$ (aq). (b) Add oxygen-free water. White CuCl(s) precipitates: $$Cu^{2+}(aq) + 4Cl^-(aq) + Cu(s) \rightarrow 2[CuCl_2]^-(aq)$$ $$[CuCl_2]^-(aq) \rightleftharpoons CuCl(s) + Cl^-(aq)$$ Reacts with Cl^-(aq), CN^-(aq) and NH_3(aq) to form complexes $[CuCl_2]^-$ (aq), $[Cu(CN)_4]^{3-}$ (aq), $[Cu(NH_3)_2]^+$ (aq).
Copper(I) bromide	Resembles copper(I) chloride
Copper(I) iodide	(a) Heat copper with iodine. (b) Add I^- (aq) to Cu^{2+} (aq): $$Cu^{2+}(aq) + 5I^-(aq) \rightarrow 2CuI(s) + I_3{}^-(aq)$$

TABLE 24.12

FIGURE 24.29 Some
Reactions of Copper and
its Compounds

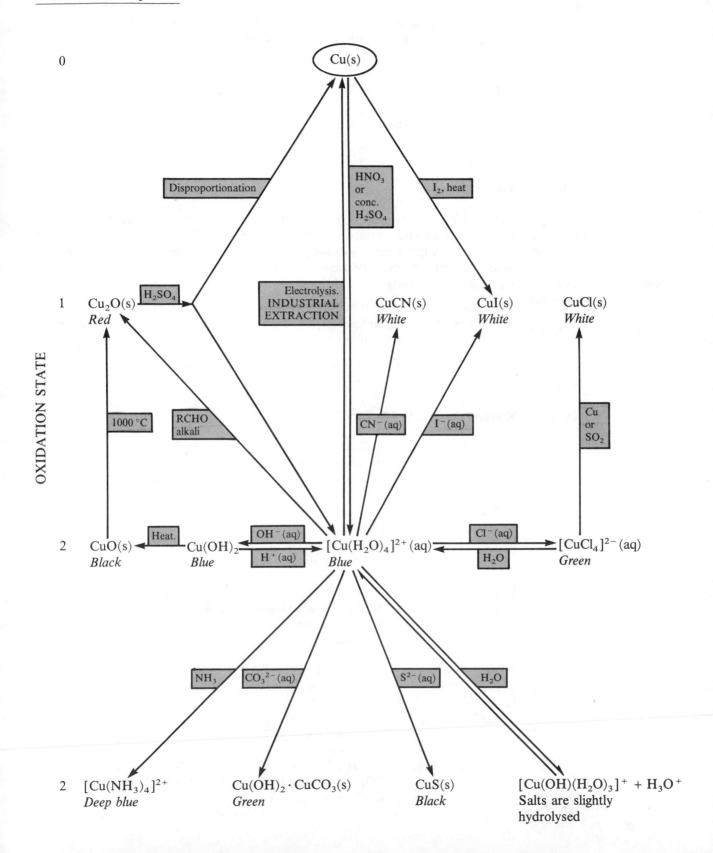

24.16 ZINC

Zinc is extracted from zinc blende or zinc carbonate

Zinc is mined as *zinc blende*, ZnS, and as zinc carbonate, $ZnCO_3$. The sulphide ore is roasted to form the oxide and sulphur dioxide, which is used in the Contact process. Zinc oxide is reduced by coke, and, being a volatile metal, zinc can be distilled from the furnace, leaving less volatile impurities behind.

24.16.1 USES

Its uses include galvanising steel

Zinc is used in galvanising steel and in the production of *brass* (Cu, Zn).

24.16.2 REACTIONS

Zn and Zn^{2+} both have a full d subshell

Zinc resembles transition metals

It reacts with acids and alkalis

Zinc is not a transition metal. It has a full d subshell of electrons, and its ions, Zn^{2+}, which also have a full d subshell, are colourless. The inability of the d electrons to take part in the metallic bonding gives zinc a lower melting temperature and boiling temperature than the transition metals. Zinc resembles transition metals in forming some complex ions and does not resemble the metals of Group 2.

Zinc reacts with acids to form salts of the hydrated zinc(II) ion, $Zn(H_2O)_6^{2+}$. It also reacts with alkalis to form salts of the tetrahydroxozincate ion (or zincate ion), $Zn(OH)_4^{2-}$, with the evolution of hydrogen:

$$Zn(s) + 2OH^-(aq) + 2H_2O(l) \rightarrow [Zn(OH)_4]^{2-}(aq) + H_2(g)$$

24.16.3 COMPOUNDS OF ZINC

Compound	Preparation and Comments
Zinc oxide, ZnO, *white*	Heat zinc carbonate, nitrate or hydroxide. Amphoteric. It becomes yellow when hot. It loses some of its oxygen content, which recombines on cooling. The absence of atoms from the crystal structure leads, here and in other cases, to colour.
Zinc hydroxide, $Zn(OH)_2$	Add OH^- (aq) to a solution of a Zn^{2+} salt. A gelatinous precipitate appears. Amphoteric.
Zinc chloride. $ZnCl_2 \cdot 2H_2O$, $ZnCl_2$	Allow hydrochloric acid to react with Zn, ZnO or $ZnCO_3$. Attempts to make the anhydrous salt from the hydrate lead to Zn(OH)Cl. (a) Heat $ZnCl_2 \cdot 2H_2O$ in dry $Cl_2(g)$ or dry HCl(g). (b) Heat Zn in dry $Cl_2(g)$ or dry HCl(g).
Zinc carbonate, $ZnCO_3$	Add HCO_3^- (aq) to a solution of a Zn^{2+} salt (addition of CO_3^{2-} (aq) leads to the precipitation of the basic carbonate, $ZnCO_3 \cdot 2Zn(OH)_2$).

TABLE 24.13

FIGURE 24.30 RTZ
Electrolytic Zinc Plant at
Budel, Holland

FIGURE 24.31 Some
Reactions of Zinc and
its Compounds

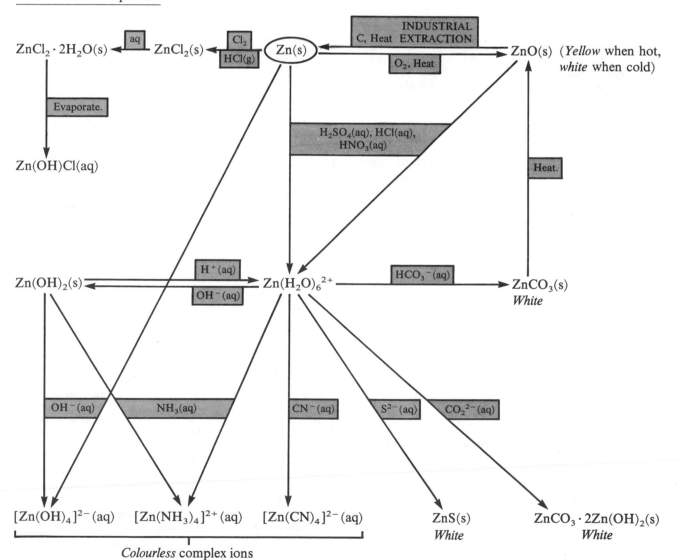

QUESTIONS ON CHAPTER 24

1. Describe the manufacture of steel. What are the advantages and disadvantages of steel, compared with aluminium? Why is the production of steel greater than the total production of all other metals?

2. Compare the chlorides $FeCl_3$ and $AlCl_3$ with respect to (*a*) bond type and (*b*) the behaviour of aqueous solutions. What is the reason for the resemblance?

3. Copper forms a complex

$$[Cu(H_2NCH_2CH_2NH_2)_n]^{2+}$$

which is highly coloured. Explain how the formula of the ion could be found by a colorimetric method.

4. State three characteristics of a transition element, illustrating them by reference to the chemistry of iron. In your answer, refer to the electron configurations of Fe, Fe^{2+} and Fe^{3+}.

5. Give examples of the chemistry of two transition elements other than iron to illustrate
(*a*) the formation of coloured ions in solution
(*b*) the formation of complex ions
(*c*) catalytic activity.

6. Transition elements form compounds in which the metal has variable oxidation states. By referring to *two* of the three metals, vanadium ($Z = 23$), manganese ($Z = 25$) and iron ($Z = 26$), explain how the oxidation states employed by the elements can be explained in terms of their electron structures.

7. (*a*) Outline a laboratory preparation for copper(I) chloride.
(*b*) Explain the reactions which occur when (i) $KCN(aq)$ and (ii) $NH_3(aq)$ is added to an aqueous solution of copper(II) sulphate until no further change results.

8. Outline the similarities and differences in the chemistry of zinc ($Z = 30$) and a characteristic member of the d block. Explain how the similarities and differences arise from the electron configurations.

9. $CuSO_4(s) \xrightarrow{\text{water}} A(aq) \xrightarrow{\text{excess of NaCl(aq)}} B(aq)$

(i) excess SO_2
(ii) water

$E(aq)$ (*Blue-violet*) $\leftarrow$ $D(aq)$ (*Colourless*) $\xleftarrow{\text{NH}_3\text{(aq)}}$ $C(s)$

State the formulae of **A, B, C, D** and **E** and the colours of **A, B** and **C**.

10. Outline the preparation from copper(II) sulphate of (*a*) $Cu(NH_3)_4SO_4 \cdot H_2O$ and (*b*) $CuSO_4 \cdot (NH_4)_2SO_4 \cdot 6H_2O$. Which is the complex salt and which the double salt? How would solutions of (*a*) and (*b*) of the same molar concentrations differ in (i) pH, (ii) colour, (iii) freezing temperature and (iv) reaction with hydrogen sulphide?

11. Explain why potassium manganate(VII) is a useful reagent in titrimetric analysis. Why is it not used as a primary standard? When is potassium dichromate(VI) preferred?
Construct equations for (*a*) the oxidation by acidified potassium manganate(VII) of (i) $Fe^{2+}(aq)$ to $Fe^{3+}(aq)$ and

(ii) $H_2O_2(aq)$ to $O_2(g) + 2H^+(aq)$ and (*b*) the reduction by acidified potassium manganate(VII) of $BiO_3^-(aq)$ to $Bi^{3+}(aq)$.

12. State the reagents and conditions which are needed to bring about the following conversions:
(*a*) $Cr^{3+}(aq) \rightarrow CrO_4^{2-}(aq)$
(*b*) $CrO_4^{2-}(aq) \rightarrow Cr_2O_7^{2-}(aq)$
(*c*) $Cr_2O_7^{2-}(aq) \rightarrow Cr^{3+}(aq)$
(*d*) $Mn^{2+}(aq) \rightarrow MnO_2(s)$
(*e*) $MnO_2(s) \rightarrow MnO_4^{2-}(aq)$
(*f*) $MnO_4^{2-}(aq) \rightarrow MnO_4^-(aq)$

13. Explain the following statements:
(*a*) The compound $CrCl_3 \cdot 6H_2O$ exists in three forms. All produce a precipitate with silver nitrate solution, but in different molar proportions.
(*b*) Addition of alkali to potassium dichromate(VI) solution produces a yellow solution.
(*c*) The addition of barium chloride solution to a solution of potassium dichromate(VI) gives a yellow precipitate.
(*d*) When an excess of potassium iodide solution is added to an aqueous solution of a copper(II) salt, a white precipitate and a brown solution are formed.
(*e*) When dilute sulphuric acid is added to copper(I) oxide, a reddish brown solid and a blue solution are formed.

14. A $25.0\,cm^3$ portion of iron(II) chloride solution required, after acidification, $15.0\,cm^3$ of a $0.0100\,mol\,dm^{-3}$ solution of potassium dichromate(VI). Calculate the concentration of the solution. Explain why potassium manganate(VII) would give an inaccurate result if it were used for this estimation.

15. Write the electronic configurations of Cr, Cr^{2+}, Mn, Mn^{2+} (for Cr, $Z = 24$; for Mn, $Z = 25$). Standard reduction potentials are:

$$Cr^{3+}(aq) + e^- \rightleftharpoons Cr^{2+}(aq); \quad E^\ominus = -0.41\,V$$
$$Mn^{3+}(aq) + e^- \rightleftharpoons Mn^{2+}(aq); \quad E^\ominus = +1.51\,V$$

Deduce from these values which is the more powerful reducing agent, $Cr^{2+}(aq)$ or $Mn^{2+}(aq)$. How is the reducing power of the ions related to their electronic configurations?

16. $Cr^{3+}(aq) \xrightarrow{\text{OH}^-\text{(aq)}} A(s) \xrightarrow{\text{Heat}} B(s) \xrightarrow{\text{Fuse with KOH.}} C(s)$

Fuse with the oxidising agent, KNO_3.

$D(aq) + K_2CrO_4(s)$

Identify **A, B, C** and **D**. What is the oxidation number of chromium in compound **C** and in K_2CrO_4?

17. Explain the following observations:
(*a*) When a solution of iron(II) sulphate is added to a solution of a nitrate in concentrated sulphuric acid, a brown ring forms at the junction of the two layers.
(*b*) When copper(I) sulphate is dissolved in water, a solution of copper(II) sulphate is obtained.
(*c*) Addition of water to copper(II) chloride gives first a green solution and, on dilution, a blue solution.

(d) When aqueous sodium hydroxide is added to a solution of tetraamminecopper(II) sulphate, only a small quantity of copper(II) hydroxide is precipitated. When hydrogen sulphide is passed through the solution, the copper ions are precipitated as copper(II) sulphide.

18. (a) List the chemical reactions that take place in the extraction of iron in the blast furnace. Discuss the physico-chemical principles involved.

(b) Discuss the electrochemical methods used to prevent the rusting of iron and steel.

19. Explain the meaning of each of the following terms. Illustrate your answers with examples from the chemistry of named elements:

(a) transition metal

(b) oxidation number

(c) complex ion

(d) coordination number.

20. This question refers to the reactions

$$MnO_2 \xrightarrow{\text{A}} K_2MnO_4 \xrightarrow{\text{B}} KMnO_4 + MnO_2$$

(a) Give the reagents and conditions employed in reactions **A** and **B**.

(b) Construct an equation for reaction **B**.

(c) State the oxidation number of manganese in each species.

(d) What name is given to reactions of type **B**?

(e) Why is reaction **B** carried out under acidic conditions?

21. Explain the following statements:

(a) $[Fe(CN)_6]^{3-}$ is used to test for iron ions.

(b) $[Ag(NH_3)_2]^+$ is used to test for chloride ions.

(c) Copper(I) chloride is used to distinguish between alkynes.

(d) Copper(I) chloride is used in the preparation of chlorobenzene.

(e) Copper(II) sulphate is used in a test for aldehydes.

22. Sketch the arrangement of bonds in the ions

$$CrO_4^{2-} \quad Cr_2O_7^{2-} \quad [FeCl_4]^- \quad [Ni(CN)_4]^{2-}$$
$$[Ni(NH_3)_6]^{2+} \quad [Fe(CN)_6]^{3-}$$

23. Write an account of the first transition metal series. Consider (a) physical properties, (b) the oxidation states shown by each element, (c) complex formation. Explain how these properties are related to electronic structure.

24. Explain the terms *complex ion* and *ligand*.
Draw the structure of (a) NH_3 and (b) Cl^-.
Explain why ammonia molecules and chloride ions are able to act as ligands. What type of bonds do they employ in complex formation? Give examples of complex ions containing (a) NH_3 and (b) Cl^- as ligands. Sketch the arrangement of ligands in the ions you mention.

25. A compound **X** is made by a reaction between cobalt(II) chloride and ammonia and ammonium chloride and an oxidising agent. Analysis of **X** gives the percentage by mass composition

$$Co = 25.3\%, \ N = 24.0\%,$$
$$H = 5.14\%, \ Cl = 45.6\%$$

Treatment of 1.000 g of **X** with silver nitrate solution gives 0.614 g of silver chloride.

(a) Deduce the formula of **X**, and name it.

(b) Sketch a structure or structures for **X**.

(c) Sketch the structures of the compounds that would be formed if $H_2NCH_2CH_2NH_2$ were used instead of ammonia.

26. The information below refers to various aqueous ions of vanadium in acid solution.

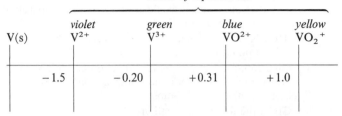

Colour of aqueous ion

	violet V^{2+}	green V^{3+}	blue VO^{2+}	yellow VO_2^+
V(s)				
	-1.5	-0.20	$+0.31$	$+1.0$

Standard redox potential, in V, between adjacent pairs

FIGURE 24.32

(a) Write the oxidation state of vanadium in (i) VO_2^+ (aq), (ii) VO^{2+} (aq)

(b) Write ion/electron half-equations for the redox equilibria in acid solution between the following aqueous ions (i) V^{2+} and V^{3+}, (ii) V^{3+} and VO^{2+}, (iii) VO^{2+} and VO_2^+

(c) State the final colour of the solutions, and write overall ionic equations for the reactions between equal volumes of aqueous solutions containing equal numbers of moles of (i) V^{2+} and VO^{2+}, (ii) V^{3+} and VO_2^+

(d) The standard redox potential for

$$I_2(aq) + 2e^- \rightleftharpoons 2I^-(aq) \text{ is } +0.54 \text{ V}.$$

(i) State which of the vanadium ions represented in Figure 24.32 should oxidise iodide to iodine in acid solution under standard conditions. Give your reasoning.

(ii) What compound could be added to remove any iodine and reveal the colour of the vanadium ion present?

(iii) Write an ionic equation for the reaction in (d) (ii).

(AEB 82)

27. (a) Give *three* examples of ways in which chromium exhibits the properties associated with transition metals.

(b) Comment on the fact that it is possible to isolate three different forms of the hexahydrate of chromium(III) chloride. How would you distinguish them chemically?

(c) Describe briefly, but including all essential reagents and conditions, how you would prepare aqueous solutions of Cr^{2+} (aq) ions and of $Cr_2O_7^{2-}$ (aq) ions, starting from hydrated chromium(III) chloride.

(d) Equal volumes of acidic solutions containing, respectively, 1.00 mol dm^{-3} of Cr^{2+} (aq) and 1.00 mol dm^{-3} of $Cr_2O_7^{2-}$ (aq) were mixed. Calculate the concentration(s) of any chromium-containing ions present in the solution after reaction.

(O & C 82)

28. (a) The preparation of potassium manganate(VII) can be represented by the following equations.

(1) $3MnO_2 + 6OH^- + XO_3^- \rightarrow 3MnO_4^{2-} + 3H_2O + X^-$

(2) $3MnO_4^{2-} + 2H_2O \rightleftharpoons 2MnO_4^- + MnO_2 + 4OH^-$

(i) What change in the oxidation state of manganese is represented by equation (1)?

(ii) Suggest a name for the element represented by **X**.

(iii) During the reaction represented by equation (2) a stream of carbon dioxide is bubbled through the reaction mixture. What is the purpose of this?

(iv) What is meant by the term *disproportionation*? How does equation (2) represent this phenomenon?

(v) What colour changes would accompany this preparation of potassium manganate(VII)?

(vi) How is the potassium manganate(VII) finally isolated from the product mixture?

(b) Give equations and necessary reaction conditions for the preparation, using potassium manganate(VII), of
(i) chlorine, (ii) ethane-1,2-diol.

(c) Given the following equations

$$MnO_4^- + 8H^+ + 5e^- \rightarrow Mn^{2+} + 4H_2O$$

$$H_2O_2 \rightarrow 2H^+ + O_2 + 2e^-$$

$10.0\,cm^3$ of a solution of hydrogen peroxide was diluted to $500\,cm^3$ of aqueous solution. $25.0\,cm^3$ of this diluted solution required $30.0\,cm^3$ of potassium manganate(VII) of concentration $0.02\,mol\,dm^{-3}$ for reduction purposes.

(i) Calculate the concentration, in $g\,dm^{-3}$, of the original solution of hydrogen peroxide.

(ii) What is the maximum volume of oxygen that could be obtained at S.T.P. from $10\,cm^3$ of the original hydrogen peroxide solution if it was treated with manganese(IV) oxide as a catalyst?

(AEB 81)

29. A modern periodic table is divided into 'blocks' of elements, s-block, p-block, etc.

(a) Explain the basis on which elements are assigned to their respective blocks.

(b) For three of the blocks, give an account of the chemical features which are peculiar to the elements in the block, and to their compounds.

Metals can conveniently be classified in accordance with their position in the electrochemical series. Discuss the relation, if any, between the position of an element in that series and the 'block' it is in.

(L 80, S)

24.17 POSTSCRIPT: STEEL

Metallurgists match the composition of a steel with the job that it has to do [see Table 24.7, p. 500]. They are constantly producing new alloys for particular purposes. For making alloy steels of accurately known composition, the electric arc furnace is used [see Figure 24.33]. It takes only 2 or 3 minutes to observe the emission spectrum of a sample of metal [p. 27]. Once the required composition is reached, the alloy can be tapped off.

FIGURE 24.33(a) An Electric Arc Furnace for Making Steel

1 The roof is swung to the side. Steel scrap is tipped in. Then the roof is swung back.

3 Limestone is added through the furnace door. It combines with impurities to form a liquid slag. Measured quantities of alloy metals are added.

4 Samples of steel are removed and analysed. Eventually, the correct composition is reached.

5 The furnace is tilted to this side so that slag can be poured off.

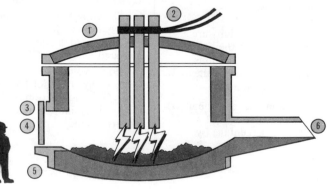

2 Power cables and carbon electrodes. The current is switched on. As each electrode is lowered, electric current jumps the gap, forming an arc between the electrode and the metal. The arc makes a crackling noise like gunfire. The intense heat melts the metal.

6 The furnace is tilted to this side so that molten steel can be poured out through the *tapping spout*.

FIGURE 24.33(b)
Control Panel of an
Electric Arc Furnace

Research workers at the British Iron and Steel Research Association laboratories in
Sheffield developed a new way of making steel. Instead of making steel in batches,
it works continuously [see Figure 24.34]. The process is fast. The tiny droplets of
metal present a vastly increased surface area over which oxygen can reach the impurities,
and oxidation occurs rapidly.

FIGURE 24.34 A Spray
Steelmaker

1 Molten iron from blast
furnace.

2 Jet of oxygen breaks up the
stream of molten iron into
droplets of spray. Impurities
burn to form oxides.

7 The bottom part of the
container is detachable.
Scrap iron or steel can be
added before or during the
process.

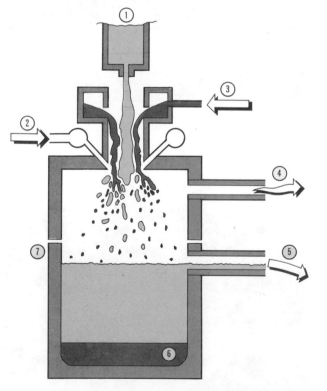

3 Lime (calcium oxide) is fed
in. It forms slag with the
oxides.

4 Waste gases leave.

5 Slag is run off.

6 Steel collects.

A pilot plant in Millom in Cumbria ran successfully from May, 1966 to December,
1966. The Millom Company (a private company) needed to raise capital in order to
expand into an industrial scale plant. The Government and the British Steel
Corporation had to decide whether to invest in a proven batch process or in the
less well tried continuous process. They decided against investing in spray steelmaking,
and in 1968 the Millom Steelworks closed.

APPENDIX 2: TOPICS WHICH SPAN GROUPS OF THE PERIODIC TABLE

A2.1: SOME DETECTIVE WORK

1. Spot the gas. Identify the following gases, giving your reasons, and explaining all the reactions described.

(*a*) **A** is a colourless, poisonous gas. It burns readily in air and reduces heated copper(II) oxide to copper. **A** reacts with chlorine to form another poisonous gas. When **A** is heated with aqueous sodium hydroxide under pressure, a salt is formed. This salt gives **A** when treated with concentrated sulphuric acid.

(*b*) **B** is a gas with a distinctive smell. It is neutral and turns starch-iodide paper blue. When **B** is heated and then cooled back to room temperature, the volume is found to have increased by 50% (measured at the same pressure).

(*c*) **C** is a colourless gas with an unpleasant smell. It dissolves sparingly in water to give a neutral solution. It burns in air, 1 mole of **C** requiring 2 moles of oxygen for complete combustion. Two products are formed, one of which reacts with water to form an acid. **C** combines with an equal volume of hydrogen iodide to form a crystalline salt.

2. Identify these solids, giving reasons and explaining all the reactions described.

(*a*) **A** is a shiny metal, which is resistant to corrosion. When heated, **A** reacts with chlorine to form a compound which dissolves in water to form a green solution. When treated with sodium peroxide, the green solution turns yellow.

(*b*) **B** is a brown solid. When heated, it gives a colourless, odourless gas and a yellow residue. When heated with concentrated hydrochloric acid, **B** gives a gas which bleaches moist litmus paper and a solution which reacts with potassium iodide solution to give a yellow precipitate.

(*c*) **C** is a white solid which melts at 44 °C. It burns readily in air, producing a white solid that reacts with water to form an acid. When **C** is heated with sodium hydroxide solution, it gives a gas which is spontaneously flammable in air.

(*d*) **D** is a solid which dissolves in water to give a pale green acidic solution. Addition of silver nitrate solution precipitates a white solid, and addition of sodium hydroxide solution precipitates a green solid. When **D** is heated in a stream of chlorine, it forms a dark brown sublimate.

3. Identify the liquid. Giving reasons, and explaining the reactions involved, identify the following liquids.

E is a colourless liquid which boils close to room temperature. It is not oxidised by chlorine or by potassium manganate(VII). **E** forms two salts in reactions with potassium hydroxide, and its vapour reacts with ammonia to form a white solid.

F is a colourless, neutral liquid. It reacts as a base towards hydrogen chloride and as an acid towards ammonia. It has a very low electrolytic conductivity and high boiling and freezing temperatures compared with analogous compounds of other elements in the same group of the Periodic Table. **F** reacts reversibly with red hot iron.

G is a colourless aqueous solution. It reacts with potassium manganate(VII) to give oxygen. **G** reacts with lead(II)

sulphide to form a white solid and with iron(II) salts to form iron(III) salts.

H is a volatile red liquid formed when an orange crystalline solid reacts with concentrated hydrochloric acid. **H** reacts with sodium hydroxide solution to form a yellow solution.

4. Some more solids. Identify these solids, with explanations of the chemical reactions described.

J is a solid which colours a flame lilac. Its aqueous solution is yellow, changing to orange when dilute sulphuric acid is added. This orange solution reacts with ethanol when warmed, turning green and evolving a vapour with a distinctive smell.

K is a black solid. When **K** is melted with potassium hydroxide and potassium nitrate, a green solid results. This dissolves in water to form a purple solution and a suspension of **K**.

L is a colourless solid. Its aqueous solution is neutral. When copper(II) sulphate solution is added, a white precipitate and a brown solution are formed. When lead(II) nitrate solution is added, a yellow precipitate appears. When **H** is heated with concentrated sulphuric acid, it evolves a violet vapour.

M is a colourless solid. It combines with chlorine to form a covalent, colourless liquid. **M** dissolves in water to give an acidic solution, which reduces Fe^{3+} salts to Fe^{2+} salts and gives a dark brown precipitate with hydrogen sulphide.

5. Half a dozen solids for you to identify. Give your reasons.

P is an orange-red crystalline solid. When heated, it forms three products, including a green solid and a colourless gas. When **P** is heated with sodium hydroxide solution, an alkaline gas is evolved, and a yellow solution is formed.

Q is a white solid which gives a yellow flame test. When boiled with an excess of ammonium chloride, **Q** evolved a gas which was not identified by simple tests. When added to a solution of iron(II) sulphate in dilute sulphuric acid, **Q** gave a brown colour. The brown colour disappeared on heating, with the evolution of a colourless gas which turned brown on meeting the air.

R is a white solid which gives a lilac flame test. The addition of aqueous barium chloride to a solution of **R** gives a white precipitate, which dissolves in dilute hydrochloric acid with the evolution of a gas which decolorises acidified potassium manganate(VII).

S is a white solid. On warming it with a dilute mineral acid, a pungent-smelling vapour is formed. On strong heating, **S** gave a flammable gas. It gave a yellow flame test.

T is a colourless solid. When it is heated with concentrated sulphuric acid, it gives a gas which burns with a blue flame and is absorbed by ammoniacal copper(I) chloride solution. An aqueous solution of **T** gives a yellow precipitate with sodium hexanitrocobaltate(III) solution. With silver nitrate solution, it gives a white precipitate which dissolves in an excess of **T** solution. When **T** solution is added in excess to a solution of iron(II) sulphate and iron(III) ions, a dark blue colour results.

U is a crystalline green solid, containing water of crystallisation. With silver nitrate solution, aqueous U gives a white precipitate which dissolves in aqueous ammonia. U reacts with hydrogen peroxide in alkaline solution to form a yellow solution, which turns orange when acidified. The yellow solution gives a yellow precipitate with lead(II) nitrate solution.

6. Identify **A**, **B**, **C**, **D** and **E**, giving your reasons.

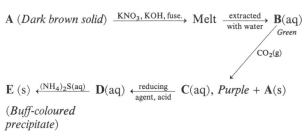

A (*Dark brown solid*) $\xrightarrow{\text{KNO}_3,\ \text{KOH, fuse.}}$ Melt $\xrightarrow[\text{with water}]{\text{extracted}}$ **B**(aq) *Green*

B(aq) ... $\searrow$ $CO_2(g)$

E (s) $\xleftarrow{(\text{NH}_4)_2\text{S(aq)}}$ **D**(aq) $\xleftarrow[\text{agent, acid}]{\text{reducing}}$ **C**(aq), *Purple* + **A**(s)

(*Buff-coloured precipitate*)

7. Identify **P**, **Q**, **R**, **S** and **T**, and give the colours of **Q**, **R**, **S** and **T**.

P(aq) $\xleftarrow{\text{conc. HCl(aq)}}$ $\text{Fe(NO}_3)_3\text{(aq)}$ $\xrightarrow{\text{Na}_2\text{CO}_3\text{(aq)}}$ **Q**(s)

Pale yellow solution

Orange-yellow solution

$\downarrow$ $\text{Zn} + \text{H}_2\text{SO}_4\text{(aq)}$

S(s) $\xleftarrow{\text{NaOH(aq)}}$ **R**(aq) $\xrightarrow{\text{Na}_2\text{CO}_3\text{(aq)}}$ **T**(s)

8. Describe how you could distinguish between the members of the following pairs:

(*a*) CO_3^{2-}(aq) and HCO_3^-(aq)

(*b*) SO_4^{2-}(aq) and SO_3^{2-}(aq)

(*c*) NO_3^-(aq) and NO_2^-(aq)

(*d*) $\text{S}_2\text{O}_3^{2-}$(aq) and SO_4^{2-}(aq)

(*e*) FeCl_3(aq) and $\text{K}_2\text{Cr}_2\text{O}_7$(aq)

(*f*) Ag^+(aq) and Zn^{2+}(aq).

9. (*a*) You are provided with a large number of test-tubes and with nine unlabelled bottles and it is known that these contain approximately $2\,\text{mol dm}^{-3}$ solutions of copper nitrate, copper sulphate, aluminium sulphate, sodium hydroxide, sodium carbonate, barium chloride, sulphuric acid, potassium iodate, potassium iodide. Using no other apparatus or chemicals, devise a scheme whereby you could positively identify the contents of all the bottles.

(*b*) Give examples from the chemistry of copper to illustrate and explain the following concepts: variable oxidation states; non-stoichiometry; formation of complex ions and the dependence of their colour on the surrounding ligands.

(WJEC 81)

A2.2: YOU HAVE SOME EXPLAINING TO DO

1. Explain the following statements:

(*a*) Aqueous solutions of sodium nitrate, aluminium sulphate and potassium cyanide have different pH values.

(*b*) Although silicon tetrachloride is readily hydrolysed by water, carbon tetrachloride is not.

(*c*) Solid beryllium chloride is soluble in ethoxyethane.

(*d*) Calcium oxide melts at a higher temperature than sodium chloride.

(*e*) Ammonium nitrate dissolves in water, although the process is endothermic.

(*f*) Potassium thiocyanate will detect iron(III) ions in the presence of iron(II) ions in aqueous solution.

(*g*) The standard enthalpy of neutralisation of hydrochloric acid by aqueous potassium hydroxide and sodium hydroxide is the same but is different from that obtained when aqueous ammonia is used.

(*h*) Although methane is unaffected by aqueous alkali, silane is violently hydrolysed.

(*i*) When potassium manganate(VII) is used in redox titrations, acidic conditions are employed. Sulphuric acid is often used.

(*j*) The shapes of the molecules BF_3 and PCl_3 are different.

(*k*) In XeF_4, the arrangement of bonds is square planar.

(*l*) There are two isomers with the formula $[\text{Co(NH}_3)_4\text{Cl}_2]^+$.

(*m*) The standard enthalpy of solution of alkali metal fluorides changes from endothermic (e.g., LiF, NaF) to exothermic (e.g., KF, RbF, CsF).

(*n*) The decrease in atomic radius from scandium to zinc is very small.

(*o*) Potassium chromate(VI) is used as an indicator in the titration of a solution of chloride ions with silver nitrate solution.

(*p*) The boiling temperature of ammonia ($-33\,°\text{C}$) is higher than that of phosphine ($-88\,°\text{C}$).

(*q*) Tin(IV) chloride is a fuming liquid; tin(II) chloride is a solid.

(*r*) Silver nitrate solution is acidified with dilute nitric acid before being used to test for the presence of halide ions in solution.

2. Explain the following statements:

(*a*) Although water and ammonia react with anhydrous copper(II) sulphate, methane does not.

(*b*) Dilution of a concentrated solution of copper(II) chloride results in a change of colour from green to blue.

(*c*) Barium carbonate is more stable to heat than magnesium carbonate.

(*d*) The addition of potassium iodide solution to aqueous copper(II) sulphate produces a deep brown solution.

(*e*) Although sulphur forms a hexafluoride, the highest chloride is sulphur tetrachloride.

(*f*) Sulphurous acid is a weaker acid than sulphuric acid. Hydrogen sulphide does not precipitate copper(II) sulphide from a solution of copper(II) sulphate and potassium cyanide.

(g) There are many alkanes but few silanes; there are many giant silicate structures, but no analogous carbonates.

(*h*) The properties of beryllium resemble those of aluminium.

(i)   The dissolution of sodium hydroxide in water is exothermic, whereas the dissolution of sodium nitrite is endothermic.

(j)   When the alkali metals are heated with oxygen, lithium forms the monoxide, sodium forms both the monoxide and peroxide, and the remainder of the metals form superoxides.

(k)   Silver iodide, unlike silver chloride, is insoluble in aqueous ammonia.

(l)   The position of equilibrium in the reaction

$$Mg(s) + Cu^{2+}(aq) \rightleftharpoons Cu(s) + Mg^{2+}(aq)$$

can be predicted from the values of the standard electrode potentials:

$$E^{\ominus}(Mg^{2+}/Mg) = -1.96\,V;$$
$$E^{\ominus}(Cu^{2+}/Cu) = +0.44\,V$$

(m)   A solution of aluminium nitrate boils at a higher temperature than a solution of sodium nitrate of the same molar concentration.

(n)   A solution of aluminium nitrate is more effective in coagulating a sol than an equimolar solution of sodium nitrate.

(o)   Aluminium chloride is a catalyst for a number of organic reactions.

A2.3: PATTERNS IN THE PERIODIC TABLE

1. Metallic character increases from right to left and from top to bottom in the Periodic Table. Illustrate this statement by referring to the period from sodium to chlorine and Group 4B.

2. The first element in a group of the Periodic Table often has properties which differ from those of later members. Illustrate this statement by referring to lithium, beryllium and fluorine.

3. Illustrate one of the trends in the Periodic Table by considering the properties of the oxides: Na_2O, CaO, Al_2O_3, SiO_2 and P_4O_{10}.

4. Shown below are the first ionisation energies $(E/kJ\,mol^{-1})$ of the elements of the second short period $(Z = \text{atomic number})$:

Element	Na	Mg	Al	Si	P	S	Cl	Ar
Z	11	12	13	14	15	16	17	18
E/kJ mol^{-1}	500	740	580	790	1010	1000	1260	1520

(a)   State the meaning of *first ionisation energy*.

(b)   Plot the ionisation energies against atomic number.

(c)   Describe the shape of the plot you obtain, and explain why it has this form.

(d)   Mention one way in which the behaviour of the elements mirrors the form of the graph you have plotted.

5. Distinguish between electronegativity and electron affinity.
How does electronegativity vary (a) from carbon through nitrogen and oxygen to fluorine and (b) from fluorine to iodine? Explain the variations.
Which of these elements is the most powerful oxidising agent? Why?

6. What is meant by the term *diagonal relationship*? Give two examples of pairs of elements which show a diagonal relationship to one another, and quote two chemical reactions for each pair to illustrate the relationship.

7. Give an account of the hydrides of the elements. Relate the nature of the bonding in the hydrides to the position of the element in the Periodic Table and to the properties of the compounds.

8. Compare the following with respect to (w.r.t.) the features cited:

(a)   Ammonia and phosphine w.r.t. stability and basicity.

(b)   Water and hydrogen sulphide w.r.t. boiling temperature and ionisation.

(c)   The halogens w.r.t. oxidising power.

(d)   Magnesium and calcium w.r.t. reactivity with water.

(e)   Carbon dioxide and silicon(IV) oxide w.r.t. structure and acidity.

9. Discuss the elements of the second short period (Na to Ar) with reference to the following features:

(a)   their action on water

(b)   their reaction with dilute sulphuric acid

(c)   methods for the preparation of their chlorides

(d)   the physical and chemical properties of their chlorides.

10. Explain the following statements:

(a)   The atomic radii of the first series of transition elements are similar.

(b)   The ionic radius of sodium is smaller than its atomic radius, but chloride ions have a larger radius than chlorine atoms.

(c)   The first ionisation energy of potassium is smaller than that of sodium.

(d)   The first electron affinity of chlorine is greater (more exothermic) than that of iodine.

(e)   The ion Pb^{2+} is stable, but C^{2+} does not exist.

(f)   The salts NaCl and NaF have similar crystal structures, but CsCl and CsF have different crystal structures.

(g)   Beryllium chloride shows evidence of covalent bonding, but calcium chloride is ionic.

11. Explain the trends in the atomic radii of the elements (a) along Period 3 (Na–Cl) and (b) down Group 7B (F–I). Show how these trends help to explain the changes in the chemistry of the elements.

12. The bonding in sodium chloride is different from that in aluminium chloride. Discuss this statement, referring to the action of (a) heat and (b) cold water on the two compounds.

13. Why are elements described as s block, p block or d block?

Which features are characteristic of the elements in each of these three blocks?

14. Discuss the oxides formed by (*a*) the first period (Li–F) and (*b*) Group 4B (C–Pb). Mention (i) the structure and bonding and (ii) the acid–base properties of the oxides.

15. Write an account of the elements in the second short period (Na–Cl). Include a discussion of their structures and bonding and the chemistry of their oxides and hydrides.

16. Give an account of the chlorides of the elements in the second short period from sodium to sulphur. Mention the methods of preparation, the type of bonding and the reaction of the chlorides with water.

17. Plot a graph of ionic radius (vertical axis) against atomic number (horizontal axis) for the elements sodium to chlorine. Explain the variation in atomic radius in terms of electron configurations.

Ion	Na^+	Mg^{2+}	Al^{3+}	Si^{4+}	P^{3-}	S^{2-}	Cl^-
Radius/nm	0.095	0.065	0.050	0.041	0.212	0.184	0.181

18. Beryllium and aluminium are positioned diagonally to one another in the Periodic Table. Why would this relationship lead you to expect similar properties for the two elements? Point out ways in which the chemistry of the elements bears out your prediction.

19. This question is about trends in properties and the reactions of elements in that part of the Periodic Table shown.

Groups

		I	II	III	IV	V	VI	VII
	2	Li	Be	B	C	N	O	F
	3	Na	Mg	Al	Si	P	S	Cl
Periods	*4*	K	Ca	Ga	Ge	As	Se	Br
	5	Rb	Sr	In	Sn	Sb	Te	I

(*a*) List, as in the printed table, the symbols for those elements which react readily with cold water to form alkaline solutions; and give details, with a balanced equation, for the reaction which you consider to be the most vigorous.

(*b*) List, as in the printed table, the symbols for those elements which form hydrides characterised by their having comparatively low boiling points. For the most stable hydride of each group, draw the electron bond diagram and sketch the shape of its molecule.

(*c*) For the elements of Period 3, in their highest oxidation states where there are several, state the formulae of the oxides and the corresponding acids or hydroxides which they form. Specify the oxidation number in each case.

(*d*) List, as in the printed table, the symbols for those elements of Period 3 and Group IV which form nitrates. Explain what can be deduced about the nature of an element from the results of its reactions (if any) with concentrated nitric acid and with dilute nitric acid.

(*e*) Write a concise comparative account of the chemistry of fluorine and iodine to illustrate the trends in properties of the elements in Group VII.

(SUJB 81)

20. (*a*) (i) Define the term *ionisation energy* and explain the difference between first and second ionisation energies.

(ii) Describe the measurement of the first ionisation energy of an atom, for example sodium, by an electron bombardment technique.

(*b*) Discuss, and account for the general relationship between the ionisation energies of the elements and their electronic configurations, and in particular comment on the ionisation energies listed below.

Ionisation Energies in kJ mol^{-1}

Element	*1st*	*2nd*	*3rd*	*4th*
Sodium	496	4561	6913	–
Potassium	419	3069	4400	–
Caesium	376	2420	3300	–
Magnesium	738	1450	7731	10 540
Calcium	590	1146	4942	–
Boron	801	2428	3660	25 020
Aluminium	578	1817	2745	11 580

(O 83, S)

Part 4

ORGANIC CHEMISTRY

25

ORGANIC CHEMISTRY

25.1 ORGANIC CHEMISTRY

Organic chemistry was originally described as the chemistry of compounds found in living things; in plants and animals. All such naturally occurring compounds contain carbon, and it was thought that some 'vital force' was needed for their formation. Sucrose, $C_{12}H_{22}O_{11}$, was obtained from sugar cane; urea, $CO(NH_2)_2$, was obtained from urine; and glycerol, $CH_2OHCHOHCH_2OH$, was obtained from

Organic compounds were originally obtained from natural sources

the saponification of mutton fat. Until 150 years ago, the only carbon compounds that had been made in the laboratory were very simple compounds such as carbon dioxide, calcium carbide, CaC_2, and potassium cyanide, KCN. When F Wöhler, in 1828, made urea from an inorganic salt, ammonium cyanate, NH_4CNO, he changed

...until Wöhler synthesised urea from an inorganic salt

the definition of organic chemistry. Today, the term organic chemistry refers to the chemistry of millions of carbon compounds. Some of them have been extracted from plant or animal sources, but many more have been made by organic chemists in their laboratories.

Carbon forms more compounds than any other element. With the ground state electron configuration $1s^2 2s^2 2p^2$ [p. 42], it has very little tendency to form positive or negative ions. In order to achieve a stable outer octet of electrons, it forms four covalent

Carbon forms four covalent bonds

bonds. The electron pairs in these bonds occupy sp^3 hybrid orbitals. They have axes directed to the corners of a regular tetrahedron with the carbon nucleus at the centre.

When a carbon atom combines with four hydrogen atoms, it forms a molecule of methane, CH_4 [see Figure 25.1(a)]. If two carbon atoms join, each can still combine with three hydrogen atoms to form a molecule of ethane, C_2H_6 [see Figure 25.1(b)]. A model of a molecule of propane, C_3H_8, is shown in Figure 25.1(c).

FIGURE 25.1(a)
Models of CH_4

FIGURE 25.1(b)
Models of C_2H_6

FIGURE 25.1(c)
Models of C_3H_8

25.2 HYDROCARBONS

Compounds of carbon and hydrogen are called **hydrocarbons**. The compounds shown above, methane, ethane and propane, are **alkanes**. They possess only single bonds and have the general formula

$$C_nH_{2n+2}$$

The alkanes are a homologous series

A series of compounds with similar chemical properties, in which members differ from one another by the possession of an additional CH_2 group, is called a **homologous series**. The first ten members of the unbranched-chain alkane series are:

The names of alkanes and alkyl groups

CH_4	Methane	C_6H_{14}	Hexane
C_2H_6	Ethane	C_7H_{16}	Heptane
C_3H_8	Propane	C_8H_{18}	Octane
C_4H_{10}	Butane	C_9H_{20}	Nonane
C_5H_{12}	Pentane	$C_{10}H_{22}$	Decane

The groups of atoms CH_3—, C_2H_5— and C_nH_{2n+1}— are called *methyl*, *ethyl* and *alkyl* groups.

25.3 ISOMERISM AMONG ALKANES

Figure 25.1 shows models of methane, ethane and propane. The next member of the series, butane, C_4H_{10}, can be modelled in two ways:

Isomeric compounds have the same molecular formulae but different structural formulae

Molecule (a) is a continuous-chain or unbranched-chain molecule, and (b) is a branched-chain molecule. The two formulae correspond to different compounds. Compound (a) is called butane (or sometimes, *normal*-butane or *n*-butane) and boils at $-0.5\,°C$. Compound (b) is called 2-methylpropane, and boils at $-12\,°C$.

The existence of different compounds with the same molecular formulae but different structural formulae is called **isomerism**. There are five isomeric hexanes of formula C_6H_{14}. They are:

$CH_3CH_2CH_2CH_2CH_2CH_3$ Hexane (sometimes referred to as *normal--hexane* or *n*-hexane)

The isomers of hexane

$CH_3CH_2CH_2CHCH_3$ 2-Methylpentane
　　　　　　|
　　　　　CH_3

$CH_3CH_2CHCH_2CH_3$ 3-Methylpentane
　　　　|
　　　CH_3

　　　　CH_3
　　　　|
$CH_3CH_2C{-}CH_3$ 2,2-Dimethylbutane
　　　　|
　　　　CH_3

$CH_3CH{-}CHCH_3$ 2,3-Dimethylbutane
　　|　　|
　CH_3　CH_3

25.4 SYSTEM OF NAMING HYDROCARBONS

The IUPAC system of nomenclature

The names of these isomers are given in accordance with the International Union of Pure and Applied Chemistry (IUPAC) system of nomenclature. The procedure followed is:

1. Name the longest unbranched carbon chain.

2. Name the substituent groups.

3. Give the positions of the substituent groups.

$CH_3{-}CH_2{-}CH_2{-}CH{-}CH_3$ 2-Methylpentane
　　　　　　　　　|
　　　　　　　　CH_3

One could count from the other end and call it 4-methylpentane, but the IUPAC system is to count from the end which will give the lower locant (number) for the position of a substituent group.

$$CH_3—CH_2—CH—CH—CH_3$$
$$H_3CCH_3$$

2,3-Dimethylpentane

$$CH_3—CH_2—CH_2—CH—CH—CH_3$$
$$H_3CC_2H_5$$

3,4-Dimethylheptane

In spite of the way it is written, you should be able to see that the longest unbranched chain has seven carbon atoms.

$$CH_3—CH—CH_2—CH—CH_3$$
$$ClBr$$

2-Bromo-4-chloropentane

The substituent groups are named in alphabetical order.

$$CH_3—CH_2—CH_2—CH—CH—CH_3$$
$$BrCl$$

3-Bromo-2-chlorohexane

The substituents are named in alphabetical order, not in the numerical order of the locants.

To construct the formula of a compound from its name, e.g., 2,3-dichloro-4-methylhexane, first write the carbon atoms of the hexane part of the molecule:

$$C—C—C—C—C—C$$

Then put in Cl atoms on carbons 2 and 3, and a CH_3 group on carbon 4:

$$C—C—C—C—C—C$$
$$ClClCH_3$$

Fill in the hydrogen atoms to give each carbon atom a valency of four:

$$CH_3—CH—CH—CH—CH_2—CH_3$$
$$ClClCH_3$$

25.4.1 UNSATURATED HYDROCARBONS

Alkanes are saturated hydrocarbons

The alkanes are not the only hydrocarbons. There are also alkenes and alkynes. Alkanes are said to be **saturated** hydrocarbons as they contain only single bonds between carbon atoms. Alkenes and alkynes are **unsaturated** hydrocarbons: they contain multiple bonds between carbon atoms. The simplest alkene is ethene

Alkenes and alkynes are unsaturated hydrocarbons: they contain multiple bonds

$$H_2C{=}CH_2$$

formerly called ethylene. It is the first member of the homologous series of alkenes, which have the general formula C_nH_{2n}. Alkynes contain one or more carbon–carbon triple bonds. Ethyne

Alkanes, alkenes and alkynes are aliphatic hydrocarbons

$$HC{\equiv}CH$$

is the first member of the homologous series of alkynes, which have the general formula C_nH_{2n-2}.

The names of the other members of the series are given in Table 25.1. All these hydrocarbons are classified as **aliphatic** hydrocarbons. Aliphatic means 'fatty' in Greek, the connection being that fats contain large alkyl groups, e.g. $C_{15}H_{31}—$.

No. of C atoms	Alkane		Alkene		Alkyne		Alkyl group	
1	CH_4	Methane					CH_3-	Methyl
2	C_2H_6	Ethane	C_2H_4	Ethene	C_2H_2	Ethyne	C_2H_5-	Ethyl
3	C_3H_8	Propane	C_3H_6	Propene	C_3H_4	Propyne	C_3H_7-	Propyl
4	C_4H_{10}	Butane	C_4H_8	Butene	C_4H_6	Butyne	C_4H_9-	Butyl
5	C_5H_{12}	Pentane	C_5H_{10}	Pentene	C_5H_8	Pentyne	$C_5H_{11}-$	Pentyl
n	C_nH_{2n+2}		C_nH_{2n}		C_nH_{2n-2}		C_nH_{2n+1}	

TABLE 25.1 Names of Aliphatic Hydrocarbons

In naming alkenes and alkynes, the positions of the multiple bonds must be stated:

$$CH_2=CH-CH_2-CH_3 \qquad \text{But-1-ene}$$

How to state the position of the double bond in an alkene

The *but*-part of the name shows that there are 4 carbon atoms. The *-ene* suffix shows that there is a C=C double bond. The number 1 indicates that the double bond is between carbon atoms 1 and 2. Count from the end that will give the lowest numbers, not 3 and 4.

$$CH_3-CH=CH-CH_3 \qquad \text{But-2-ene}$$

$$CH_2=CH-CH=CH_2 \qquad \text{Buta-1,3-diene}$$

$$CH_2=\underset{\underset{CH_3}{|}}{C}-CH_2CH_3 \qquad \text{2-Methylbut-1-ene}$$

$$CH_3-CH=CH-\underset{\underset{CH_3}{|}}{CH}-CH_3 \qquad \text{4-Methylpent-2-ene}$$

The double bond is numbered first and then the methyl group.

How to state the position of the triple bond

$$CH_3-\underset{\underset{CH_3}{|}}{CH}-C\equiv C-CH_3 \qquad \text{4-Methylpent-2-yne}$$

$$CH_3-CH=CH-C\equiv CH \qquad \text{Pent-2-en-4-yne}$$

The double bond is numbered first, then the triple bond.

25.5 ALICYCLIC HYDROCARBONS

A second set of hydrocarbons is the alicyclic hydrocarbons. They contain rings of carbon atoms. Examples are

Alicyclic hydrocarbons

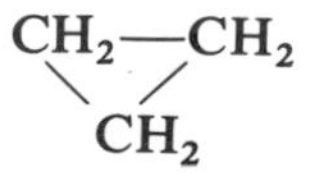

Cyclopropane 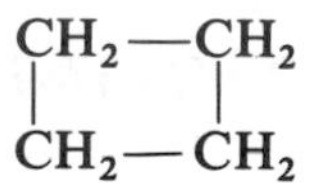   Cyclobutane 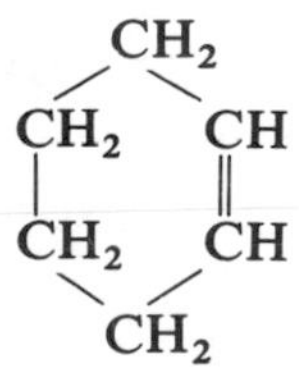   Cyclohexene

[See Figure 25.2.]

FIGURE 25.2
Cyclohexane and
Cyclopropane

FIGURE 25.2
Cyclohexane and
Cyclopropane

25.6   AROMATIC HYDROCARBONS

*Aromatic hydrocarbons
are related to benzene;
they are called arenes*

A third group of hydrocarbons is the **aromatic** hydrocarbons. They are related to benzene. The first benzene compounds to be isolated had pleasant aromas, and gave this group of hydrocarbons their name. Members of the group of aromatic hydrocarbons, the **arenes**, are: benzene, C_6H_6; methylbenzene, $C_6H_5CH_3$; and naphthalene, $C_{10}H_8$. The structure of these compounds is discussed on p. 107 and in Chapter 28. The group C_6H_5— is called a **phenyl** group, and the group $C_{10}H_7$— is called a **naphthyl** group. Both are **aryl** groups.

*The phenyl and naphthyl
groups are aryl groups*

The subsets of hydrocarbons are summarised in Figure 25.3.

FIGURE 25.3   Classes
of Hydrocarbons

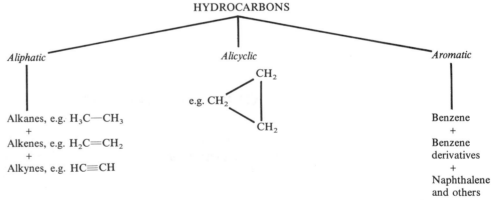

25.7   FUNCTIONAL GROUPS

The double bond in the alkenes is responsible for most of the chemical reactions of these compounds. The group of atoms

$$\diagdown C = C \diagup$$

is called the **functional group** of the alkenes.

Similarly, alkynes possess the functional group

$$-C\equiv C-$$

The reactions of a homologous series depend on the functional group

In your study of the reactions of the hydrocarbons, you will come across some other classes of compounds with different functional groups, and it will help to deal with their names in advance.

The halogenoalkanes (or haloalkanes) have the formula RX where R is an alkyl group and X is a halogen:

CH_3Br	Bromomethane
C_2H_5Cl	Chloroethane
$C_2H_4Cl_2$	Dichloroethane

How to name halogenoalkanes

There are two isomers. Figure 25.4(a) shows 1,2-dichloroethane, $ClCH_2CH_2Cl$, with the chlorine atoms attached to different carbon atoms. Figure 25.4(b) shows 1,1-dichloroethane, CH_3CHCl_2, with the chlorine atoms on the same carbon atom.

FIGURE 25.4
(a) 1,2-Dichloroethane,
(b) 1,1-Dichloroethane

(a)   (b)

How to name alcohols...

The **alcohols** or **alkanols** are compounds with the functional group —OH, a hydroxyl group. They are named by taking the name of the alkane with the same number of carbon atoms and changing the ending from *-ane* to *-anol*:

CH_3OH	Methanol
C_2H_5OH	Ethanol
C_3H_7OH	Propanol. There are two isomers

$CH_3CH_2CH_2OH$ Propan-1-ol

CH_3CHOH Propan-2-ol
|
CH_3

A number must be used to show the position of the hydroxyl group in the carbon chain.

Carboxylic acids or **alkanoic acids** have the functional group

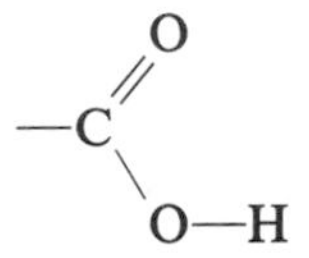

$$
\begin{array}{c}
\ \ \ \ \ \ \ O \\
\ \ \ \ \ \ \ \parallel \\
-C \\
\ \ \ \ \ \ \ \backslash \\
\ \ \ \ \ \ \ O-H
\end{array}
$$

...and carboxylic acids They are named by taking the alkane with the same number of carbon atoms and changing the ending from *-ane* to *-anoic acid*.

CH_3CO_2H Ethanoic acid
(There are *two* carbon atoms, that in CH_3 and that in CO_2H.)

$CH_3CH_2CH{=}CHCO_2H$ Pent-2-enoic acid
(The C of the CO_2H group is C-1, and the double bond lies between C-2 and C-3.)

CHECKPOINT 25A: NOMENCLATURE

1. Name the following compounds:

(a) $CH_3{-}CH_2{-}CH{-}CH_3$
$\phantom{(a) CH_3{-}CH_2{-}CH}{|}$
$\phantom{(a) CH_3{-}CH_2{-}CH}CH_3$

(b) $CH_3CH{=}CHCH_2CH_3$

(c) CH_3CHCH_3
${|}$
$CH_2CH{=}CH_2$

(d) $CH_3CH_2CHCHCH_3$
${|}\ \ {|}$
$Cl\ \ Br$

(e) $CH_3CH_2CH_2OH$

(f) $CH_3CH_2CH_2CH_2CO_2H$

2. Write structural formulae for:
(a) Heptane
(b) 2-Chloro-3-methylhexane
(c) 3-Bromo-2-chloroheptane
(d) Pentan-2-ol
(e) Hex-2-ene
(f) Butanoic acid

25.8 REACTIONS OF ORGANIC COMPOUNDS

25.8.1 TYPES OF REACTIONS

The reactions of organic compounds fall into four classes. These are listed below.

SUBSTITUTION

There are four types of organic reactions An atom or group of atoms replaces another:

$C_2H_5Cl + OH^- \rightarrow C_2H_5OH + Cl^-$

Chloroethane + Hydroxide ion → Ethanol + Chloride ion

ADDITION

Two molecules react to form one:

$Br_2 + CH_2{=}CH_2 \rightarrow BrCH_2CH_2Br$

Bromine + Ethene → 1,2,-Dibromoethane

ELIMINATION

One molecule reacts to form more than one:

$$C_2H_5OH \rightarrow C_2H_4 + H_2O$$

Ethanol $\rightarrow$ Ethene + Water

REARRANGEMENT

One molecule reacts to give a different molecule:

$$CH_3\!-\!CH\!-\!CH\!=\!CH_2 \rightleftharpoons CH_3\!-\!CH\!=\!CH\!-\!CH_2\!-\!Cl$$
$$\quad\quad\;\; |$$
$$\quad\quad\;\; Cl$$

3-Chlorobut-1-ene $\quad\rightleftharpoons\quad$ 1-Chlorobut-2-ene

25.8.2   TYPES OF BOND FISSION

There are two types of bond fission...   When the bond breaks, each of the bonded atoms takes one of the pair of electrons. **Free radicals** are formed. These are atoms or groups of atoms with unpaired electrons:

HOMOLYTIC FISSION

...homolysis...   When the bond breaks, each of the bonded atoms takes one of the pair of electrons. **Free radicals** are formed. These are atoms or groups of atoms with unpaired electrons.

$$CH_3CH_2CH_3 \xrightarrow{\text{heat}} CH_3CH_2\!\cdot + \cdot CH_3$$

Propane   Ethyl radical   Methyl radical

$$Cl_2 \xrightarrow{\text{sunlight}} 2Cl\cdot$$

Chlorine molecule $\rightarrow$ Chlorine atoms or free radicals

Energy must be supplied, either as heat or light, to break the bond. The free radicals formed possess this energy, and are very reactive.

HETEROLYTIC FISSION

...and heterolysis   When the bond breaks, one of the bonded atoms takes both of the bonding electrons to form an anion. The rest of the molecule becomes a cation. An ion with a positively charged carbon atom is called a **carbocation**. An ion with a negatively charged carbon atom is called a **carbanion**:

$$(CH_3)_3C\!-\!Cl \rightarrow (CH_3)_3C^+ + Cl^-$$

1-Chloro-1,1-dimethylethane $\rightarrow$ A carbocation + Chloride ion

$$CH_3COCH_2CO_2C_2H_5 + OH^- \rightleftharpoons CH_3CO\overline{C}HCO_2C_2H_5 + H_2O$$

Ethyl 3-oxobutanoate   A carbanion + Water

25.8.3   TYPES OF REAGENT

In a covalent bond between **A** and **B**, if **A** is more **electronegative** [p. 74] than **B**, the distribution of bonding electrons can be represented as

$$\overset{\delta-}{A}\!-\!\overset{\delta+}{B}$$

Covalent bonds can be polar

The bond is described as **polar**. The reagents which attack organic compounds seek out either the slightly positive ($\delta +$) end of the bond or the slightly negative ($\delta -$) end of the bond. There are two main classes of reagent:

NUCLEOPHILIC REAGENTS

Nucleophiles attack centres of positive charge

Negative ions, e.g., OH^-, CN^-, and compounds in which an atom has an unshared pair of electrons, e.g., NH_3, are **nucleophilic** (nucleus-seeking). They attack the positive end of a polar bond.

ELECTROPHILIC REAGENTS

Electrophiles attack centres of negative charge

A reagent which attacks a region where the electron density is high is called an **electrophile**. Examples of electrophilic reagents are the nitryl cation, NO_2^+, and sulphur(VI) oxide, SO_3 [see Figure 28.5, p. 582].

25.8.4 REACTION MECHANISM

The mechanism of a reaction is the sequence of steps from start to finish

The stoichiometric equation for an organic reaction does not tell you how the reaction takes place. There may be a series of reactions in between the mixing of the reactants and the formation of the products. The sequence of steps by which the reaction takes place is called the **reaction mechanism**. The mechanism is worked out from a study of the kinetics of the reaction [see Chapter 14]. Other techniques, such as spectroscopy [see Chapter 35] and the incorporation of radioisotopes into a reactant, are also used.

25.9 ISOMERISM

Isomerism is the existence of different compounds with the same molecular formulae but different structural formulae. There are various types of isomerism.

25.9.1 STRUCTURAL ISOMERISM

A structural formula shows the order in which atoms are bonded together

The structural formula shows the sequence in which the atoms in a molecule are bonded. A structural formula can be written in full, with every bond drawn, or it can be written by joining groups of atoms in sequence, provided the formula for each of the groups is unambiguous. The structural formula for 2-methylpropane is shown in Figure 25.5.

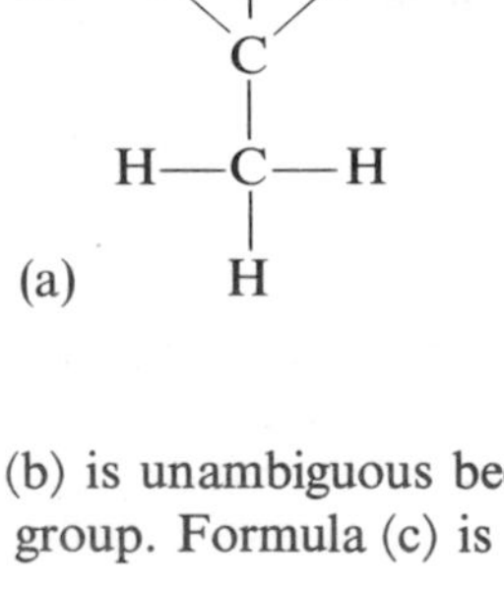

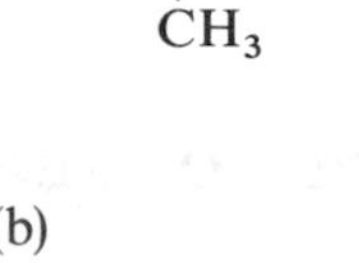

FIGURE 25.5
2-Methylpropane

Formula (b) is unambiguous because there is only one way of writing the bonds in a —CH_3 group. Formula (c) is a condensed way of writing formula (b).

CHAIN ISOMERISM

In chain isomers, the carbon 'skeletons' differ

The isomers have different carbon chains. They possess the same functional group, and belong to the same homologous series, e.g.

$$CH_3CH_2CH_2CH_3 \qquad CH_3CHCH_3$$
$$\qquad\qquad\qquad\qquad\qquad\quad |$$
$$\qquad\qquad\qquad\qquad\qquad\ CH_3$$

Butane   2-Methylpropane

POSITIONAL ISOMERISM

The position of a functional group in the carbon skeleton differs between positional isomers

These isomers have a substituent group in different positions in the same carbon 'skeleton'. The isomers are chemically similar because they possess the same functional group, e.g.

(*a*) Propan-1-ol, $CH_3CH_2CH_2OH$ and propan-2-ol, $CH_3CH(OH)CH_3$.

(*b*) Pent-1-ene, $CH_3CH_2CH_2CH{=}CH_2$ and pent-2-ene, $CH_3CH_2CH{=}CHCH_3$.

(*c*) This type of isomerism is found in benzene derivatives. The structural formula of benzene, C_6H_6, was discussed on p. 107. If two chlorine atoms replace two hydrogen atoms to form $C_6H_4Cl_2$, three different compounds can be formed. Their names are given below:

1,2-Dichlorobenzene   1,3-Dichlorobenzene   1,4-Dichlorobenzene

(*ortho*-dichlorobenzene)   (*meta*-dichlorobenzene)   (*para*-dichlorobenzene)

FUNCTIONAL GROUP ISOMERISM

These isomers have different functional groups and belong to different homologous series. Some examples follow:

In another type of isomerism, the functional group is different in the isomers

(*a*) An alcohol and an ether, e.g.,

ethanol, C_2H_5OH, and methoxymethane (dimethyl ether) CH_3OCH_3.

(*b*) An aldehyde and a ketone, e.g.,

propanal, CH_3CH_2CHO, and propanone, CH_3COCH_3.

(*c*) A carboxylic acid and one or more esters, e.g.,

butanoic acid, $CH_3CH_2CH_2CO_2H$ and methyl propanoate, $C_2H_5CO_2CH_3$, ethyl ethanoate, $CH_3CO_2C_2H_5$, propyl methanoate, $HCO_2CH_2CH_2CH_3$, and (methylethyl) methanoate, $HCO_2CH(CH_3)_2$.

(*d*) A nitrile and an isonitrile, e.g.,

ethanonitrile, $CH_3C{\equiv}N$, and ethanoisonitrile, $CH_3\overset{+}{N}{\equiv}\overset{-}{C}$.

CHECKPOINT 25B: STRUCTURAL ISOMERISM

1. Draw structural formulae for three compounds of formula C_5H_{12} and for five compounds of formula C_6H_{14}.

2. Draw the formulae of the three structural isomers of formula C_3H_8O. How many can you find for $C_4H_{10}O$?

3. There are ten compounds with the formula $C_4H_8Cl_2$. Draw their structural formulae.

4. The molecular formulae C_4H_8O and $C_5H_{10}O$ each correspond to a number of carbonyl compounds, which contain the group

$$\text{\Large$\diagdown$}\atop\text{\Large$\diagup$}\hspace{-0.3em}C\!=\!O$$

Sketch their structural formulae.

5. Supply structural formulae for the acids and esters which have the molecular formulae (*a*) $C_5H_{10}O_2$ and (*b*) $C_6H_{12}O_2$.

TAUTOMERISM

Two isomeric forms of a compound may exist in dynamic equilibrium. Aldehydes and ketones exhibit **keto-enol** tautomerism. The equilibrium between the forms is represented

The position of a hydrogen atom differs between tautomers

$$CH_3\!-\!\underset{\underset{O}{\|}}{C}\!-\!CH_3 \quad \rightleftharpoons \quad CH_2\!=\!\underset{\underset{O\!-\!H}{|}}{C}\!-\!CH_3$$

the *keto* form of propanone   the *enol* form of propanone

The **tautomers** differ in the position of a hydrogen atom. They are called the **keto** form (for the $\text{\Large$\diagdown$}\atop\text{\Large$\diagup$}\hspace{-0.3em}C\!=\!O$ group in ketones) and the **enol** form (*ene* for $-C\!=\!C-$ and *ol* for the $-OH$ group). The percentage of the enol tautomer is small, but it is invoked to explain some of the reactions of carbonyl compounds [p. 646 and p. 317].

25.9.2 STEREOISOMERISM

There are various types of stereoisomerism

Stereoisomers have the same molecular formula and also the same structural formula. The difference between them is the arrangement of the bonds in space. Stereoisomerism can be (*1*) *cis-trans* isomerism, (*2*) conformational isomerism or (*3*) optical isomerism.

CIS-TRANS ISOMERISM

Restriction of rotation about a C=C double bond gives rise to cis-trans isomerism

The planar arrangement of bonds in $R_2C\!=\!CR_2$ was discussed on p. 103. The CR_2 groups are not free to rotate about the double bond. In a compound

$$R_1R_2C\!=\!CR_3R_4$$

R_1 and R_3 may be on the same side of the double bond (the *cis*-isomer) or on opposite sides (the *trans*-isomer). *Cis*- and *trans*-butenedioic acid are shown in Figure 25.6.

FIGURE 25.6
Models of Butenedioic
Acids

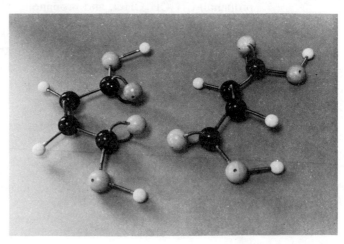

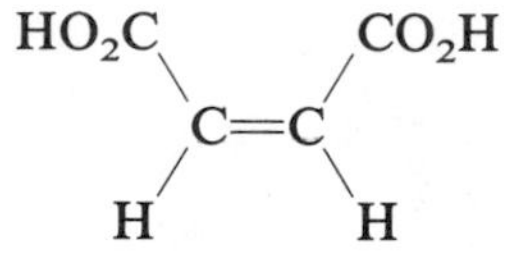

cis-Butenedioic acid 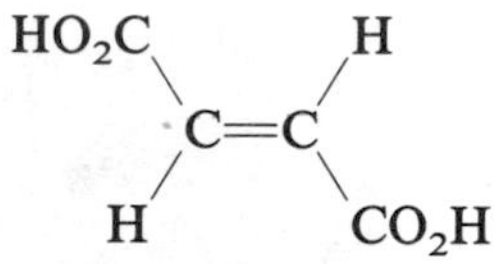   *trans*-Butenedioic acid

The physical and chemical properties of cis-trans *isomers differ*

They do not have the same physical and chemical properties. They differ in melting temperature (*cis* 135 °C, *trans* 287 °C), in solubility (*cis* being 100 times more soluble than *trans*) and in dipole moments. The *cis* acid (formerly called maleic acid) forms an anhydride on gentle heating, but the *trans* isomer (fumaric acid) does not:

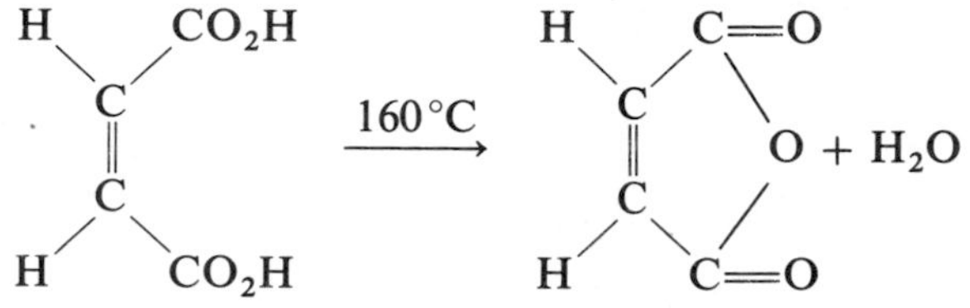

cis-Butenedioic acid   *cis*-Butenedioic anhydride

Examples are butenedioic acid...

On strong heating, *trans*-butenedioic acid forms the anhydride of *cis*-butenedioic acid, showing that rotation about the double bond is possible at higher temperatures.

Ethanal oxime, $CH_3CH{=}N{-}OH$, exists in *cis* and *trans* forms. The lone pair of electrons on the nitrogen atom occupies one bonding position:

...ethanal oxime...
$CH_3CH{=}NOH$

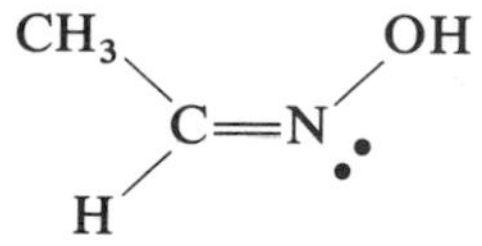

cis-Ethanal oxime 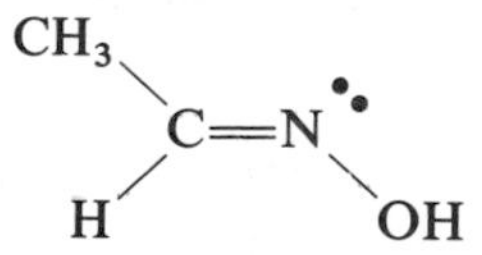   *trans*-Ethanal oxime

...and some inorganic compounds...

For *cis-trans* isomerism in inorganic compounds, see p. 492.

CONFORMATIONAL ISOMERISM

Cyclohexane shows isomerism

Cyclohexane, C_6H_{12}, exists in two forms which differ in conformation and are described as the **boat** and **chair** forms [see Figure 25.7].

FIGURE 25.7(a) The Isomers of Cyclohexane

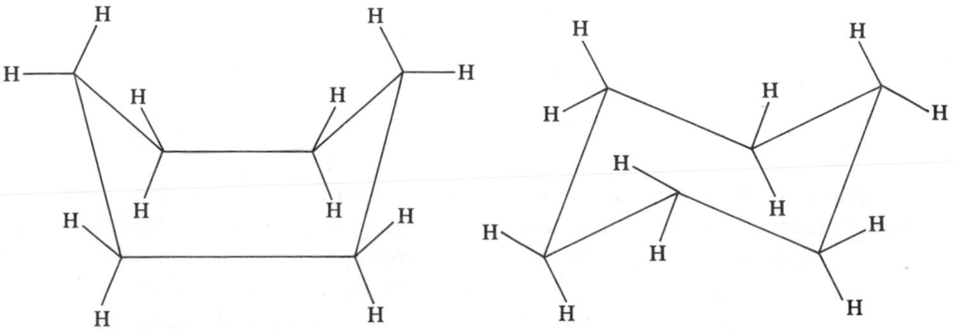

The boat form   The chair form

FIGURE 25.7(b) Models
of Cyclohexane, in its
Boat and Chair Forms

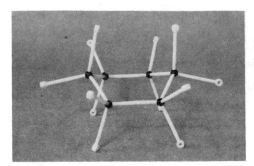 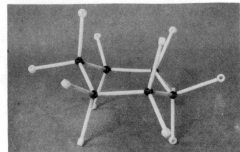

CHECKPOINT 25C: STEREOISOMERISM

1. Write structural formulae for (*a*) $CH_3CH{=}CHCl$,
(*b*) $CH_3CH{=}CHC_2H_5$, (*c*) $ClCH{=}C(CH_3)CO_2H$,
(*d*) $C_2H_5C(CH_3){=}CHCH_3$.

2. Make a model of cyclopentane, C_5H_{10}. How many ways
can you find of replacing 2H by 2Cl to give the structures
of the isomers of $C_5H_8Cl_2$?

3. Dinitrogen difluoride, F—N=N—F, exists in isomeric
forms. Draw their formulae.

OPTICAL ISOMERISM

If a beam of light is passed through a Nicol prism (of *calcite*, $CaCO_3$) or a piece of
polaroid, the emergent light vibrates in a single plane. It is said to be **plane-polarised**
[see Figure 25.8].

FIGURE 25.8
Plane-polarisation of
Light

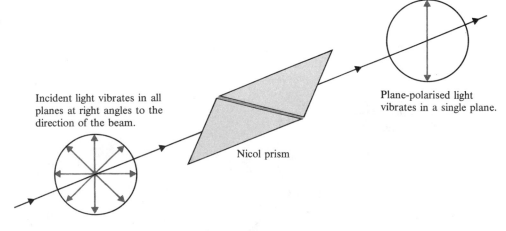

Incident light vibrates in all
planes at right angles to the
direction of the beam.

Nicol prism

Plane-polarised light
vibrates in a single plane.

*Optical activity is the
ability to rotate the plane
of polarisation of
plane-polarised light*

Certain substances, either in crystalline form or in solution, have the ability to rotate
the plane of polarisation of plane-polarised light. They are said to be **optically active**.
The effect is measured quantitatively in a **polarimeter** [see Figure 14.4, p. 295]. If
the plane of polarisation rotates in a clockwise direction, viewed from the direction
of the emergent beam, the rotation is designated (+); an anticlockwise rotation is
designated (−).

Optically active compounds are chiral...

Compounds which are optically active are **chiral**. Their molecules have no plane or axis or centre of symmetry [p. 99]. Chiral compounds have two different types of molecules. They are **enantiomers**, i.e., mirror images of one another [see Figure 5.18, p. 100 and Question 1 below]. The $(+)$-enantiomer rotates the plane of polarisation of plane-polarised light clockwise. The $(-)$-enantiomer produces an anticlockwise rotation. An equimolar mixture of $(+)$ and $(-)$-enantiomers is optically inactive and is called a **racemic mixture** or **racemate**.

...Their molecules have no plane of symmetry...

...They exist in $(+)$- and $(-)$-forms and as optically inactive mixtures called racemates...

There are three forms of 2-hydroxypropanoic acid (lactic acid), $CH_3CH(OH)CO_2H$. The racemic mixture is found in sour milk; the $(+)$-enantiomer is present in muscle. The $(-)$-enantiomer does not occur naturally and has to be obtained from the racemic mixture. All three forms have the same chemical properties. The physical properties of the $(+)$ and $(-)$-enantiomers are the same. They differ from those of the racemate because it is a mixture, not a pure compound.

...$(+)$- and $(-)$-forms are called enantiomers

Enantiomers react in the same way in chemical reactions with achiral reagents. They may differ in biochemical reactions. Enzymes are catalysts found in plants and animals. An enzyme and its substrate (the substance which requires the enzyme in order to react) fit together 'like a key in a lock'. The geometry of the substrate is important, and enzymes can distinguish between enantiomers. *Penicillium glaucum* (a mould) feeds on $(+)$-2-hydroxypropanoic acid (lactic acid) but not on its $(-)$-enantiomer.

*RESOLUTION OF RACEMATES

Chiral reagents can be used in the resolution of a racemic mixture...

Racemic mixtures can be separated into $(+)$- and $(-)$-enantiomers. The separation is referred to as **optical resolution**. One method, due to L Pasteur, is to treat the racemate with an optically active compound. An acid, $(+)(-)$-**A**, will react with a base, $(+)$-**B**, to give two salts, $(+)$-**A**$(+)$-**B** and $(-)$-**A**$(+)$-**B**. The salts are not enantiomers: they are **diastereoisomers** and have different physical properties. If they differ in solubility, they can be separated by fractional crystallisation.

...Enzymes can sometimes be used

Biochemical methods of separation can sometimes be used. The mould *Penicillium glaucum* feeds on $(+)$-2,3-dihydroxybutanedioic acid (tartaric acid) and leaves $(-)$-tartaric acid behind. It is therefore possible to obtain the $(-)$-isomer from a racemic mixture. The method has the disadvantage that only one of the enantiomers is obtained, and this is not the one that is of interest to biochemists.

INORGANIC COMPOUNDS

Optical isomerism among inorganic compounds is met in Chapter 24, p. 493.

CHECKPOINT 25D: OPTICAL ACTIVITY

1. (*a*) Construct models of the enantiomers of 2-hydroxypropanoic acid, $CH_3CH(OH)CO_2H$, and draw them.

(*b*) Find out what happens when you substitute a hydrogen atom for the —OH group.

2. Alanine, an amino acid, has the formula $H_2NCH(CH_3)CO_2H$. Construct a model of the molecule. Do you think that alanine should be optically active? Replace —CH_3 by —H. This is the formula of the amino acid, glycine. Do you think that glycine should be optically active?

3. What is *optical activity*? How could you demonstrate that a compound is optically active?

Draw structural formulae for the optical isomers of molecular formula (*a*) $C_3H_6O_3$, (*b*) $C_4H_{10}O$.

*Outline a chemical method for separating a racemic base into its enantiomers.

4. Discuss the isomerism shown by the following compounds. State the differences (if any) in physical and chemical properties that exist between the isomers of each formula.

(a) C_4H_{10}

(b) C_4H_8

(c) $CH_3CH(OH)CO_2H$

(d) $C_6H_5CH=CH-CO_2H$

(e) $C_6H_5-\underset{\underset{OH}{|}}{CH}-CN$

(f) $\underset{\underset{CHCO_2H}{\diagdown}}{CH_2}-CHCl$

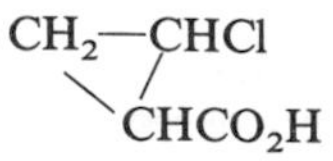

(g) $C_3H_6Br_2$

(h) C_3H_7Cl

(i) $C_6H_5CH_2CH(NH_2)CO_2H$

=== QUESTIONS ON CHAPTER 25 ===

1. (a) What is meant by the term *optical activity*?

(b) What conditions must be fulfilled in order that a compound may exhibit optical activity?

(c) What are the essential requirements for a structure to exhibit geometrical (*cis/trans*) isomerism?

(d) Indicate which of the following structures may exhibit stereoisomerism, and, where they do, draw a diagram of each stereoisomer. You should indicate clearly what type(s) of stereoisomerism is/are involved.

$$CH_3CH_2CH_2CH(OH)CH_3$$

$$CH_3CH_2CH(OH)CH_2CH_3$$

$$CH_3CH=CHCOOCH(CH_3)C_6H_5$$

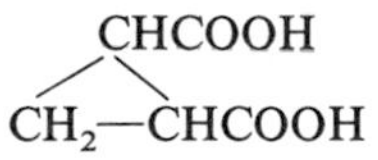

$$\underset{CH_2-CHCOOH}{\overset{CHCOOH}{\diagup}}$$

(O & C 82)

2. Write an essay on the *Stereochemistry of Covalent Molecules* in which you discuss:

(a) the shapes adopted by simple inorganic molecules in terms of the electron-pair repulsion theory,

(b) *cis/trans* isomerism in *both* organic and inorganic molecules,

(c) optical isomerism in organic molecules.

(NI 82)

[For (a) and (c), see Chapter 5; for (b), see also Chapter 24.]

26

THE ALKANES

26.1 PETROLEUM OIL

Alkanes are fuels...

The most important feature of the alkanes is their use as fuels. A huge fraction of the energy we use comes from the combustion of alkanes. The gas in our cookers, the petrol in our cars, aviation fuel and diesel oil for powering ships and electric generators – all these fuels are mixtures of alkanes. The source of these fuels is either crude petroleum oil or natural gas. Deposits of crude oil and natural gas usually occur together as they are formed by the same slow decay of marine animals and plants. Crude oil is found in many parts of the world. Figure 26.1 shows an oil rig in the North Sea.

26.1.1 FRACTIONAL DISTILLATION OF CRUDE OIL

...They are obtained from crude oil by fractional distillation

Crude oil is a mixture of about 150 compounds. It is difficult to ignite. To yield volatile substances which can be used as fuels, crude oil is fractionally distilled. The theory is covered on p. 160. Figure 26.2 represents a fractionating column for use in separating crude oil into fractions. Each fraction is a mixture of hydrocarbons which boil over a limited range of temperature [see Table 26.1].

FIGURE 26.1
An Oil Rig

FIGURE 26.2 An Industrial Fractionating Column

Fraction	Boiling temperature/°C	Length of carbon chain	Use
Refinery gas	20	C_1–C_4	Fuel: domestic heating, gas cookers
Light petroleum	20–60	C_5–C_6	Solvent
Light naphtha	60–100	C_6–C_7	Solvent
Gasoline (petrol)	40–205	C_5–C_{12}	Fuel for the internal combustion engine (cars etc.)
Kerosene (paraffin)	175–325	C_{12}–C_{18}	Fuel for jet engines
Gas oil	275–400	C_{18}–C_{25}	Does not vaporise easily. Used in diesel engines, where it is injected into compressed air to make it ignite. Used in industrial furnaces, being introduced as a fine mist to help the oil to burn
Lubricating oil	Non-volatile	C_{20}–C_{34}	Lubrication
Paraffin wax	Solidifies from lubricating oil fraction	C_{25}–C_{40}	Polishing waxes, petroleum jelly
Bitumen (asphalt)	Residue	$>C_{30}$	Road surfacing, roofing

TABLE 26.1
The Fractions obtained from Crude Oil

26.1.2 PETROCHEMICALS

Besides being used as fuels, all these fractions have another, very important use. They are the foundation of the petrochemicals industry. From them are manufactured thousands of compounds: plastics, paints, solvents, rubbers, detergents and many medicines are petrochemicals.

26.2 PHYSICAL PROPERTIES

The C—H bond has a weak dipole...

Since the electronegativities of carbon and hydrogen are 2.5 and 2.1 respectively

$$\overset{\delta-}{C}\!-\!\overset{\delta+}{H} \text{ bonds}$$

...and the intermolecular forces are weak

have only very weak dipole moments. Weak attractive forces exist between dipoles in neighbouring molecules [p. 80], and van der Waals forces [p. 81] also come into play. The attractive forces are so weak that the lower alkanes, from methane to butane, are gases at room temperature and pressure. Linear molecules of higher homologues can align themselves in a parallel arrangement so that dipole–dipole interactions and van der Waals forces can operate along the whole length of the molecule. The alkanes from C_5 to C_{17} are liquids, while those with larger molecules are solids. Since branched-chain molecules are more spherical in shape than unbranched-chain hydrocarbons, the attractive forces between molecules are more restricted. The boiling temperatures of branched alkanes are therefore lower than those of their straight-chain isomers. The boiling temperatures of unbranched-chain alkanes are plotted against molar mass in Figure 26.3.

Branched-chain alkanes are more volatile than the unbranched-chain isomers

FIGURE 26.3 Boiling Temperatures of *n*-Alkanes

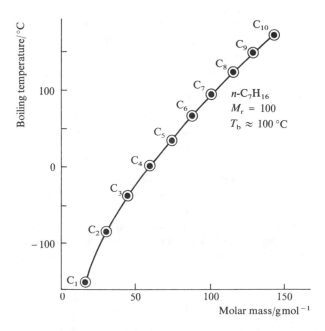

The difference in boiling temperatures between the C_1 and C_2 alkanes is 73 °C, while the C_9 and C_{10} alkanes differ by only 25 °C. It is therefore more difficult to separate the higher members by fractional distillation.

Liquid alkanes float on water...

The liquid alkanes are less dense than water: oil floats on water. The higher members are viscous liquids, the viscosity increasing with increasing molecular mass as the attractive forces between molecules increase.

Alkanes are only slightly soluble in water. Water molecules interact because of the strong dipoles in the

...Alkanes cannot form hydrogen bonds with water and are therefore insoluble

$$\overset{\delta-}{O} \!\!-\!\! \overset{\delta+}{H} \text{ bonds}$$

[See Figure 4.21, p. 84.] The hydrogen bonds formed are stronger than any interaction which can occur between water molecules and the non-polar alkane molecules. Dissolution is therefore not favoured by energy considerations.

26.3 REACTIONS OF ALKANES

26.3.1 COMBUSTION

Alkanes burn with the release of energy

The most important reaction of the alkanes is combustion. They burn to form the harmless products, carbon dioxide and water, in an exothermic reaction [for $\Delta H^{\ominus}$, see Chapter 10]:

$$CH_4(g) + 2O_2(g) \rightarrow CO_2(g) + 2H_2O(l); \quad \Delta H^{\ominus} = -890 \, \text{kJ mol}^{-1}$$

Liquid alkanes, such as octane, must be vaporised before they will burn:

$$C_8H_{18}(g) + 12\tfrac{1}{2}O_2(g) \rightarrow 8CO_2(g) + 9H_2O(l); \quad \Delta H^{\ominus} = -5510 \, \text{kJ mol}^{-1}$$

Incomplete combustion gives the poisonous gas, carbon monoxide

This is the reaction which takes place in the internal combustion engine. The hydrocarbons in gasoline have boiling temperatures of around 150 °C and will vaporise in the internal combustion engine. If the supply of oxygen is insufficient, incomplete combustion will take place, and the poisonous gas carbon monoxide will be formed. It is the more dangerous in that, being odourless, it gives no warning of its presence. It combines with haemoglobin, the red pigment in the blood, to form a very stable complex, carboxyhaemoglobin. This cannot combine with oxygen, and haemoglobin is unable to do its job of transporting oxygen around the body [see Figure 24.15, p. 495]. There is always a certain percentage of carbon monoxide, around 5%, in the exhaust gases of motor vehicles.

In the internal combustion engine, compression of petrol vapour and air can lead to auto-ignition...

When combustion of petrol vapour occurs inside the cylinders in a car engine, a large volume of hot gases is formed. The gases force the piston down the cylinder, and the power generated is transmitted to the wheels. For smooth running it is essential that the ignition of gasoline vapour and air takes place when the piston is at the right point in the cylinder. To obtain the maximum energy from the fuel, the engine design must ensure that all the petrol vapour is burnt. For this reason the mixture of petrol vapour and air is compressed, and this compression can lead to **auto-ignition**. The ignition of gases takes place before the spark. It results in a sudden rise in pressure, which delivers a blow to the piston. The engine makes a metallic sound called **knocking**. The best fuels for resistance to knocking are branched-chain hydrocarbons, such as 2,2,4-trimethylpentane:

...which causes 'knocking'

$$H_3C\!-\!\underset{\underset{\displaystyle CH_3}{|}}{CH}\!-\!CH_2\!-\!\underset{\underset{\displaystyle CH_3}{|}}{\overset{\overset{\displaystyle CH_3}{|}}{C}}\!-\!CH_3$$

The octane number of a fuel measures its resistance to knocking

This compound, which used to be called *iso*-octane, has been assigned an *octane number* of 100. Heptane, $CH_3CH_2CH_2CH_2CH_2CH_3$, has very bad knocking properties and has been assigned an *octane number* of 0. The **octane number** of a petrol is found by comparing its performance with a mixture of heptane and 2,2,4-trimethylpentane. If it has the same performance as a mixture of 25% heptane and 75% 2,2,4-trimethylpentane, then its octane number is 75.

TEL is an antiknock

After a great deal of research, it was found that tetraethyl lead, $Pb(C_2H_5)_4$, TEL, reduced the knocking properties of unbranched hydrocarbons. It is called an **antiknock**. To prevent lead(IV) oxide from coating the cylinders when TEL is added to petrol, 1,2-dibromoethane, $BrCH_2CH_2Br$, must also be added. Waste lead is converted to volatile lead(IV) bromide, which passes out with the exhaust gases. Four-star petrol contains 6 g of TEL per gallon, and the number of vehicles on the road is so large that the emission of lead is causing concern. When ingested in only moderate quantities, it makes people depressed, and increases their reaction times. A growing body of people think that city dwellers are being affected by the lead content of the air. The UK Government has decided that by 1990 all new petrol-driven vehicles will run on lead-free petrol. Another approach to the problem of knocking does exist: this is to convert fuels with poor knocking properties into branched-chain compounds. This is done by *cracking*, which is discussed below.

Lead compounds in car exhaust gases are causing concern

Lubricating oil reduces engine wear

Lubricating oil does not vaporise at the engine temperature. It is used to ease the movement of the pistons in the cylinders by reducing friction, thus reducing wear and prolonging the life of the engine.

26.3.2 CRACKING

'Cracking' is used to obtain more of lower molecular mass alkanes, which are more volatile

The petroleum fractions with 1 to 12 carbon atoms in the molecule are in demand in larger quantities than the fractions with bigger molecules. The petroleum industry uses **pyrolysis** (splitting by heat) of high molar mass alkanes to give hydrocarbons with smaller molecules, which are more easily vaporised and are therefore more useful fuels:

Alkane with large molecules $\xrightarrow[\text{at 450 °C over catalyst of Al}_2\text{O}_3/\text{SiO}_2]{\text{Vapour passed}}$ Alkane with smaller molecules + Alkene + Hydrogen

e.g. $2CH_3CH_2CH_3(g) \rightarrow CH_4(g) + CH_3CH{=}CH_2(g) + CH_2{=}CH_2(g) + H_2(g)$
Propane   Methane   Propene   Ethene   Hydrogen

The industry calls this type of reaction **cracking**.

26.3.3 ALKYLATION

'Alkylation' is used to make branched-chain alkanes

Since branched-chain compounds have higher octane numbers than straight-chain compounds, a good deal of research has gone into the synthesis of branched-chain compounds. They are made by the **alkylation** of alkenes by alkanes in the presence of a catalyst.
In alkylation reactions

Tertiary alkane + Alkene $\xrightarrow[\text{conc. H}_2\text{SO}_4 \text{ as catalyst}]{20\,°C}$ Branched-chain alkane

e.g. $(CH_3)_3CH + CH_3CH{=}CHCH_2CH_3 \rightarrow CH_3CH_2CHCH_2CH_3$
                                                                         $C(CH_3)_3$

2-Methyl-   Pent-2-ene   3-(Dimethylethyl)pentane
propane

26.3.4 REFORMING

'Reforming' converts alkanes into aromatic compounds, e.g., benzene

A huge number of important chemicals are derived from benzene. One source of benzene is petroleum. Unbranched-chain alkanes are converted into aromatic compounds by the process of **reforming**. For example

$$C_6H_{14}(l) \xrightarrow[\substack{\text{compressed to 40 atm} \\ \text{passed over Al}_2\text{O}_3 \text{ as} \\ \text{catalyst (or Pt, 10 atm)}}]{\text{Vaporised at 500 °C}} C_6H_6(l) + 4H_2(g)$$

Hexane   Benzene   Hydrogen

'Platforming' uses a platinum catalyst

Much work has gone into the effectiveness of different catalysts. If a platinum catalyst is used, the process is called **platforming**.

26.3.5 THE 'CUMENE' PROCESS FOR MAKING PHENOL

See Chapter 30, p. 630.

26.3.6 OTHER REACTIONS OF ALKANES

The alkanes used to be called the **paraffins**, a name derived from the Latin for *little liking*, implying that this class of compounds had little liking for the usual chemical

Alkanes are not reactive

reagents. Alkanes do not react with dilute acids or alkalis or with oxidising agents. At high temperatures they react with nitric acid vapour. All the reactions of alkanes,

They undergo some substitution reactions

apart from combustion, are **substitution** reactions, of the form

$$\text{RH} + \text{XY} \rightarrow \text{RX} + \text{HY}$$

26.3.7 HALOGENATION

Substitution by halogens occurs in sunlight

Under certain conditions, alkanes react with halogens. If a stoppered test-tube containing hexane and a drop of liquid bromine is left to stand at room temperature in the dark, nothing happens. The colour of the bromine is still as intense after three or four days. If the solution is exposed to sunlight, the colour fades in a few minutes, and the acidic, fuming gas hydrogen bromide can be detected. The reaction that has occurred is

$$C_6H_{14}(l) + Br_2(l) \rightarrow C_6H_{13}Br(l) + HBr(g)$$

with the formation of bromohexanes. Since it takes place in the presence of light, it is called a **photochemical reaction**. Alkanes can be chlorinated and brominated photochemically.

Produced by chlorination of CH₄ are CH₃Cl, CH₂Cl₂, CHCl₃, CCl₄

When methane reacts with chlorine in sunlight, one or more chlorine atoms may replace hydrogen atoms, depending on the amounts of halogen and alkane present. The formation of chloromethane

$$CH_4(g) + Cl_2(g) \rightarrow CH_3Cl(g) + HCl(g)$$

...They are useful solvents

may be followed by the formation of dichloromethane, CH_2Cl_2, trichloromethane (chloroform), $CHCl_3$, and tetrachloromethane, CCl_4. Chloroalkanes are useful solvents, and the mixture of products formed in the chlorination of an alkane may find use as a solvent, without the need for isolation of individual compounds.

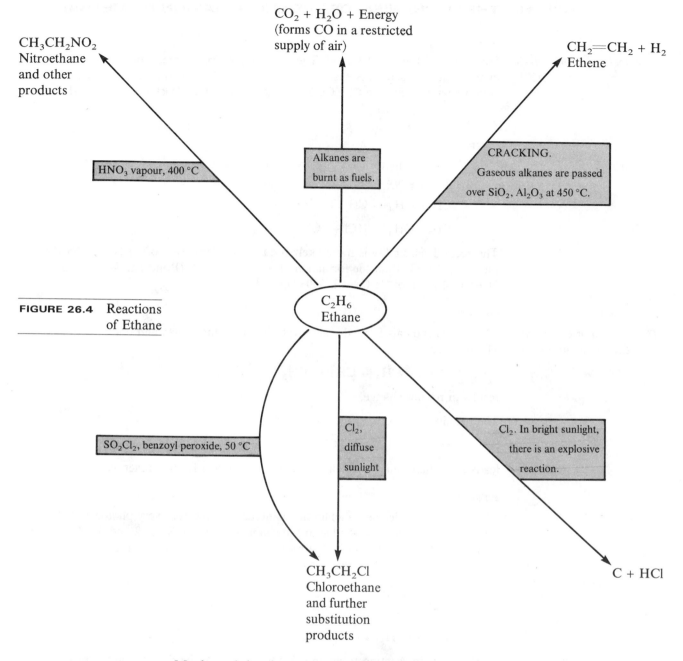

CO$_2$ + H$_2$O + Energy
(forms CO in a restricted
supply of air)

CH$_3$CH$_2$NO$_2$
Nitroethane
and other
products

CH$_2$=CH$_2$ + H$_2$
Ethene

HNO$_3$ vapour, 400 °C

Alkanes are
burnt as fuels.

CRACKING.
Gaseous alkanes are passed
over SiO$_2$, Al$_2$O$_3$ at 450 °C.

C$_2$H$_6$
Ethane

FIGURE 26.4 Reactions
of Ethane

SO$_2$Cl$_2$, benzoyl peroxide, 50 °C

Cl$_2$,
diffuse
sunlight

Cl$_2$. In bright sunlight,
there is an explosive
reaction.

CH$_3$CH$_2$Cl
Chloroethane
and further
substitution
products

C + HCl

Much work has been done on the subject of how this reaction takes place, when
methane is so unreactive towards other reagents. The mechanism proposed for the
chlorination of methane takes into account the following experimental observations.

Experimental observations
on the reaction,
CH$_4$ + Cl$_2$

1. Reaction takes place rapidly in sunlight or above 300 °C but not in the dark at
room temperature.

2. Thousands of molecules of chloromethane are formed for each photon of light
absorbed.

3. A little ethane is formed.

4. If a trace of tetramethyl lead, Pb(CH$_3$)$_4$, is added, the reaction will take place in
the dark or at room temperature. This substance is known to dissociate into methyl
radicals, ·CH$_3$.

A mechanism which fits these observations is as follows.

26.3.8 THE MECHANISM OF THE CHLORINATION OF METHANE

STEP 1

*Photochemical homolysis
of Cl_2*

Homolysis of the Cl—Cl bond. The necessary energy comes from the light absorbed or the heat supplied. It is easier to split the Cl—Cl bond than the C—H bond. (Bond energy terms are Cl—Cl, $242 \, kJ \, mol^{-1}$; C—H, $435 \, kJ \, mol^{-1}$ [p. 203].)

$$Cl_2 \xrightarrow[\text{or heat}]{\text{light}} 2Cl\cdot$$

STEP 2

*There are two possible
reactions of $Cl\cdot$ with CH_4*

The chlorine atoms formed are very reactive. Since they are surrounded by methane molecules, there are two possible reactions:

$$Cl\cdot + CH_4 \rightarrow CH_3Cl + H\cdot$$

$$Cl\cdot + CH_4 \rightarrow HCl + \cdot CH_3$$

The second possibility is more likely because the formation of an H—Cl bond is more exothermic than the formation of a C—Cl bond. (Bond energy terms are H—Cl, $431 \, kJ \, mol^{-1}$; C—Cl, $350 \, kJ \, mol^{-1}$.)

STEP 3

*The reaction of $CH_3\cdot$ with
Cl_2 results in a chain
reaction*

The methyl radicals formed collide with methane molecules and chlorine molecules. The reaction

$$\cdot CH_3 + CH_4 \rightarrow CH_4 + \cdot CH_3$$

results in no net change.

The reaction

$$\cdot CH_3 + Cl_2 \rightarrow CH_3Cl + \cdot Cl$$

leads to a chain reaction because the chlorine atoms formed react as in Step 2.

STEP 4

Thousands of molecules of chloromethane are formed for every photon of light absorbed. The high yield is due to the chain reaction — Steps 2 and 3. The reason why the yield is not higher is that radicals can combine with each other and bring the chain to an end. The reactions

*There are three
chain-terminating
reactions*

$$2Cl\cdot \rightarrow Cl_2$$

$$2CH_3\cdot \rightarrow C_2H_6$$

$$Cl\cdot + \cdot CH_3 \rightarrow CH_3Cl$$

bring the chain reaction to an end. Some ethane can be detected in the product.

*Summary of the
mechanism*

> To summarise, the steps in the chlorination of methane to chloromethane are:
>
> *Chain initiation* $Cl_2 \xrightarrow[\text{or heat}]{\text{light}} 2Cl\cdot$
>
> *Chain propagation* $Cl\cdot + CH_4 \rightarrow HCl + \cdot CH_3$
> $\cdot CH_3 + Cl_2 \rightarrow CH_3Cl + Cl\cdot$
>
> *Chain termination* $2Cl\cdot \rightarrow Cl_2$
> $2 \cdot CH_3 \rightarrow C_2H_6$
> $Cl\cdot + \cdot CH_3 \rightarrow CH_3Cl$

FORMATION OF CH₂Cl₂

Further Cl atoms can be introduced

Step 3 can give rise to the chain:

$$CH_3Cl + Cl\cdot \rightarrow HCl + \cdot CH_2Cl$$

$$\cdot CH_2Cl + Cl_2 \rightarrow CH_2Cl_2 + Cl\cdot$$

CH_2Cl_2 can undergo further chlorination to $CHCl_3$ and CCl_4.

26.3.9 BROMINATION

Methane reacts with the other halogens

The yield per photon is less for bromination than for chlorination because the step

$$Br\cdot + CH_4 \rightarrow HBr + \cdot CH_3$$

is more endothermic than the corresponding step in chlorination.

26.3.10 IODINATION

Iodination is slow and reversible:

$$RH + I_2 \rightleftharpoons RI + HI$$

Iodoalkanes can be reduced to alkanes by HI.

26.3.11 FLUORINATION

Fluorination is dangerously exothermic because of the low bond dissociation enthalpy (energy) of fluorine and the high enthalpy (energy) of C—F bonds.

26.3.12 SULPHUR DICHLORIDE DIOXIDE

Chlorination can also be effected by SO₂Cl₂

Sulphur dichloride dioxide, SO_2Cl_2, is another chlorinating agent. In the presence of a source of free radicals (such as benzoyl peroxide) it will chlorinate alkanes at 50 °C.

26.3.13 NITRATION OF ALKANES

Nitration gives a mixture of products

Alkanes can be nitrated by nitric acid vapour at 400 °C. A mixture of products is formed: for example, ethane gives nitromethane as well as nitroethane.

26.4 PREPARATION OF ALKANES

Since alkanes are obtained from the fractional distillation of crude petroleum oil, the laboratory methods for their preparation are not much used [see Figure 26.5].

FIGURE 26.5 Methods of Preparing Ethane and Methane and other Reactions which result in the Formation of Alkanes

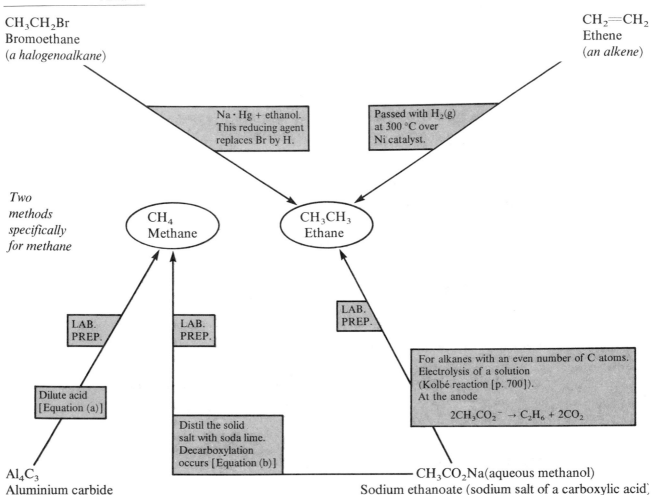

CH_3CH_2Br
Bromoethane
(*a halogenoalkane*)

$CH_2{=}CH_2$
Ethene
(*an alkene*)

Na · Hg + ethanol.
This reducing agent
replaces Br by H.

Passed with $H_2(g)$
at 300 °C over
Ni catalyst.

Two methods specifically for methane

CH_4
Methane

CH_3CH_3
Ethane

LAB. PREP.

LAB. PREP.

LAB. PREP.

Dilute acid
[Equation (a)]

For alkanes with an even number of C atoms.
Electrolysis of a solution
(Kolbé reaction [p. 700]).
At the anode

$$2CH_3CO_2{}^- \rightarrow C_2H_6 + 2CO_2$$

Distil the solid
salt with soda lime.
Decarboxylation
occurs [Equation (b)]

Al_4C_3
Aluminium carbide

CH_3CO_2Na(aqueous methanol)
Sodium ethanoate (sodium salt of a carboxylic acid)

(a): $Al_4C_3(s) + 12H_2O(l) \rightarrow 3CH_4(g) + 4Al(OH)_3(s)$

(b): $CH_3CO_2Na(s) + NaOH(s) \rightarrow CH_4(g) + Na_2CO_3(s)$

QUESTIONS ON CHAPTER 26

1. Give the names of the following compounds:

(*a*) $CH_3CH_2CH_2CH_3$

(*b*) $CH_3CH_2CH_2CH_2CH_2CH_3$

(*c*) $CH_3CH_2CH(CH_3)CH_3$

(*d*) $CH_3C(CH_3)_2CH_2CH_3$

(*e*) $CH_3CH_2CHCH_3$
             $|$
             CH_2CH_3

(*f*) CH_3CHCH_2Cl
            $|$
            $CH_2CH_2CH_3$

(*g*)   Cl
                 $|$
      $CH_3CH_2CCH_2CH_2CH_3$
                 $|$
                 CH_3

(*h*)   Cl   OH
               $|$ $|$
      $CH_3CH{-}CH{-}CHCH_3$
               $|$
               CH_2CH_3

(*i*) $(CH_2)_6$

(*j*) $C_2H_5CH_2CHCH_2CHCH_3$
                   $|$ $|$
                   Br   I

2. Write structural formulae for the following:

(*a*) 2,3-dimethylpentane,

(*b*) 2,4,5-trimethylheptane,

(*c*) 3-ethyl-2,4-dimethylheptane,

(*d*) 2,2,4-trimethylhexane,

(*e*) 3-bromo-2-chloropentane.

3. Which member of the following pairs of alkanes has the higher boiling point?

(*a*) (i) butane and (ii) heptane,

(*b*) (i) 2-methylbutane and (ii) pentane,

(*c*) (i) hexane and (ii) 2,3-dimethylbutane,

(*d*) (i) hexane and (ii) cyclohexane.

4. Name all the products formed in the following reactions:

(*a*) the combustion of octane,

(*b*) the bromination of methane,

(*c*) the reaction between sulphur dichloride dioxide and methane,

(*d*) the reactions between $CH_3\cdot$ and $H\cdot$.

5. How can butane be prepared from (*a*) pentanoic acid, (*b*) propanoic acid and (*c*) bromobutane?

6. Write structural formulae for the compounds with the formulae

(*a*) C_4H_{10} (*b*) $C_2H_4Cl_2$ (*c*) C_5H_{12} (*d*) $C_4H_8Cl_2$

7. Write equations for all the steps that occur in the reaction

$$CH_3Cl + Cl_2 \rightarrow CH_2Cl_2 + HCl$$

What names are given to these steps?

8. Explain briefly what is meant by the following terms:

(*a*) fractional distillation

(*b*) pyrolysis

(*c*) catalytic cracking

(*d*) reforming

(*e*) platforming

(*f*) knocking

(*g*) octane number

9. Name the substitution products formed when propane reacts with (*a*) concentrated nitric acid, (*b*) bromine, (*c*) sulphur dichloride dioxide. State the conditions under which these reactions will occur.

10. A car drives 100 000 miles in its lifetime. It does 35 miles to the gallon, and runs on three-star petrol, which contains 3.5 g of TEL per gallon. How much lead does this car use in its lifetime? What happens to this lead?

11. The following reactions take place during the 'free radical' chlorination of methane.

Initiation

$$Cl_2(g) \xrightarrow{hv} 2Cl\cdot(g) \qquad [1]$$

Propagation

$$Cl\cdot(g) + CH_4(g) \rightarrow CH_3\cdot(g) + HCl(g) \qquad [2]$$

$$Cl_2(g) + CH_3\cdot(g) \rightarrow CH_3Cl(g) + Cl\cdot(g) \qquad [3]$$

The bromination of methane proceeds in a similar way using bromine instead of chlorine.

You are given the following bond enthalpies, D, in $kJ\,mol^{-1}$:

$$D(Cl—Cl) = +242, D(C—Cl) = +349,$$
$$D(H—Cl) = +431, D(Br—Br) = +193,$$
$$D(C—Br) = +293, D(H—Br) = +366,$$
$$D(C—H) = +412,$$

and also that the energy associated with blue light is about $-270\,kJ\,mol^{-1}$.

(*a*) Calculate the molar enthalpy changes for reactions [1], [2] and [3], and the overall enthalpy change for the three reactions for

 (i) the chlorination of methane and

 (ii) the bromination of methane.

(*b*) State with a reason in each case

 (i) whether there is any overall change in disorder during the chlorination of methane and the bromination of methane,

 (ii) whether both the chlorination and the bromination of methane would be expected to be feasible.

(*c*) (i) Which of the three reaction steps [1], [2] and [3] is energetically the least favourable? State why this step determines the rate of the overall reaction.

 (ii) Why is the bromination of methane much slower than the chlorination of methane? State how the rate can be increased.

(*d*) Draw a continuous reaction profile (energy against progress of reaction) for reactions [2] and [3] for the bromination of methane.

[See also Chapter 10.]

(AEB 82)

27

ALKENES AND ALKYNES

27.1 ALKENES

The names of alkenes

Alkenes are a homologous series of aliphatic hydrocarbons with the general formula C_nH_{2n}. Typical members of the series are listed below:

$CH_2{=}CH_2$	Ethene
$CH_3CH{=}CH_2$	Propene
$CH_3CH_2CH{=}CH_2$	But-1-ene
$CH_3CH{=}CHCH_3$	But-2-ene

They are called **unsaturated** hydrocarbons because the double bond can open to allow them to take up more hydrogen atoms or other species.

27.2 SOURCE OF ALKENES

Alkenes are obtained in the petroleum industry by the process of cracking [p. 545]. The laboratory preparation is usually the dehydration of an alcohol. Figure 27.1 summarises the methods employed.

27.3 PHYSICAL PROPERTIES OF ALKENES

The volatility and solubility of alkenes are similar to those of the corresponding alkanes.

27.4 STRUCTURAL ISOMERISM

27.4.1 CARBON CHAIN

Both branched-chain and unbranched-chain isomers of alkenes exist, as with alkanes.

27.4.2 POSITIONAL ISOMERISM

Isomerism can involve the carbon chain or the position of the double bond

The position of the double bond must be given in the name of an alkene. Butene has two isomers, but-1-ene, with the double bond between carbon atoms 1 and 2; and but-2-ene, with the double bond between carbon atoms 2 and 3.

27.4.3 *CIS-TRANS* ISMERISM OF ALKENES

Inhibition of rotation about the C=C bond gives rise to cis-trans isomerism

In ethene [see Figure 27.2(a)], the four hydrogen atoms all lie in the same plane. Rotation of the CH_2 groups about the C=C bond is inhibited by the requirements for the formation of a strong π bond [p. 103]. The restriction of rotation is the reason why but-2-ene has two isomers, which are shown in Figure 27.2(b). The geometry of the molecules is different. The isomer with both CH_3— groups on the same side of the double bond is called the *cis*-isomer, and the isomer with the CH_3— groups on opposite sides of the double bond is called the *trans*-isomer.

The cis-trans *isomers of but-2-ene*

cis-But-2-ene

$trans$-But-2-ene

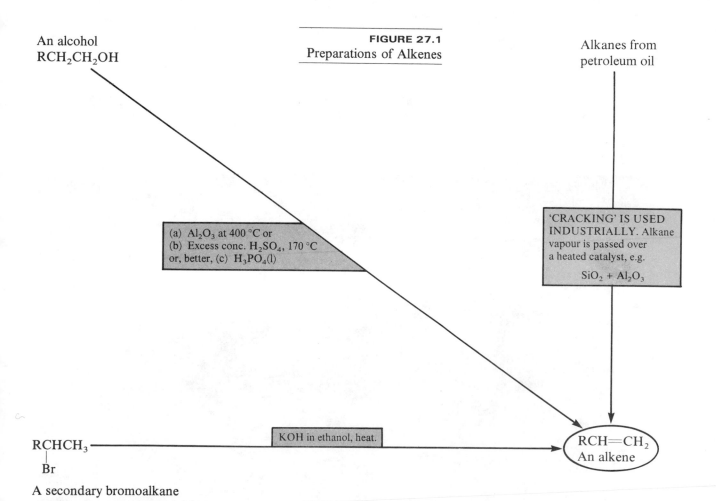

An alcohol
RCH_2CH_2OH

FIGURE 27.1
Preparations of Alkenes

Alkanes from petroleum oil

(a) Al_2O_3 at 400 °C or
(b) Excess conc. H_2SO_4, 170 °C
or, better, (c) $H_3PO_4(l)$

'CRACKING' IS USED INDUSTRIALLY. Alkane vapour is passed over a heated catalyst, e.g.

$SiO_2 + Al_2O_3$

KOH in ethanol, heat.

$RCHCH_3$
|
Br

$RCH=CH_2$
An alkene

A secondary bromoalkane

(This reaction gives a fair yield with a 2° halogenoalkane, a good yield with a 3° halogenoalkane and a very poor yield with a 1° halogenoalkane.)

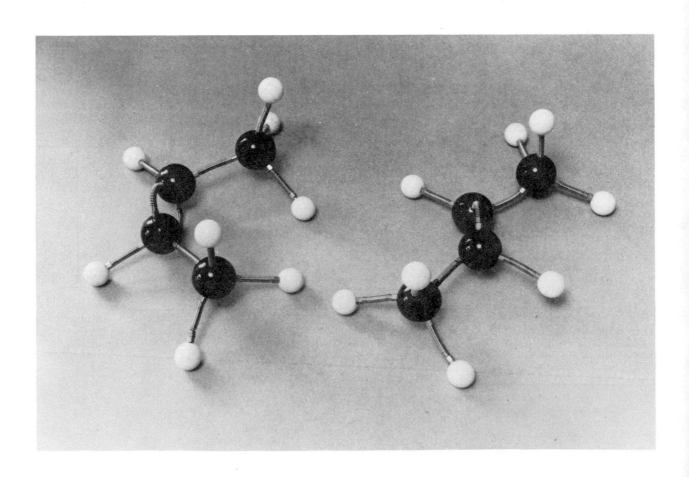

27.5 REACTIVITY OF ALKENES

The two unsaturated carbon atoms in the

$$\ce{\overset{\diagup}{\underset{\diagdown}{C}}=\overset{\diagdown}{\underset{\diagup}{C}}} \text{ bond}$$

Alkenes are more reactive than alkanes...

are joined by a σ bond and also by a π bond [pp. 102–3]. The cloud of electrons which forms the π bond lies above and below the plane of the three sp^2 hybrid bonds formed by each of the unsaturated carbon atoms. In this position, the π electrons are more susceptible than the σ electrons to attack by an electrophilic reagent [see Figure 27.3]. This is why alkenes are so much more reactive than alkanes.

FIGURE 27.3
The Reactivity of Ethene

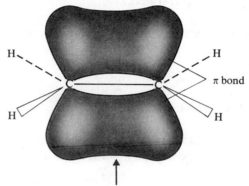

X^+ electrophile attacks cloud of π electrons

27.6 COMBUSTION

...and are not used as fuels, because they can be used instead for the manufacture of important chemicals

Alkenes burn to form carbon dioxide and water. They show a greater tendency than alkanes to undergo incomplete combustion to form carbon monoxide and carbon. In any case, alkenes are not used as fuels. The double bond makes them reactive compounds, and they are the starting point in the manufacture of many important chemicals.

27.7 ADDITION REACTIONS

Although alkenes can form substitution products, with chlorine in sunlight for example, they more frequently take part in addition reactions.

27.7.1 HYDROGEN

Hydrogen adds to alkenes with Ni powder as catalyst

In the presence of a catalyst, hydrogen adds across a

$$\ce{\overset{\diagup}{\underset{\diagdown}{C}}=\overset{\diagdown}{\underset{\diagup}{C}}} \text{ bond}$$

to form a saturated compound. Palladium is an extremely effective catalyst, but nickel is used industrially because it is less expensive:

$$R_2C{=}CR_2'(g) + H_2(g) \xrightarrow[\text{100 °C, 4 atm}]{\text{Ni powder}} R_2CH{-}CHR_2'$$

Catalytic hydrogenation, using nickel powder, is an important process in the food industry. Plant oils, such as sunflower seed oil and peanut oil, are 'polyunsaturates': they are esters of carboxylic acids which contain more than one carbon–carbon double bond. Although they are useful for cooking, these oils are less valuable commercially than expensive animal fats such as butter. Animal fats are esters of saturated carboxylic acids. Hydrogenation is used to convert unsaturated edible oils into edible fats. In soft margarine some of the double bonds remain: the degree of softness can be controlled by regulating the amount of hydrogenation.

Hydrogenation converts unsaturated edible oils into edible fats

...e.g., margarine

27.7.2 ALKANES

Tertiary alkanes add to alkenes

Tertiary alkanes will form adducts with alkenes [p. 545].

27.7.3 HALOGENS

Chlorine adds across a double bond...

Chlorine gas will react with ethene at room temperature, without catalysis by light or peroxides, to form liquid 1,2-dichloroethane:

$$H_2C{=}CH_2(g) + Cl_2(g) \rightarrow ClCH_2CH_2Cl(l)$$

...as do bromine and iodine

Bromine and iodine will also add across the double bond. When an alkene is bubbled through bromine in a solvent, such as tetrachloromethane, the bromine colour disappears as bromine adds across the double bond:

$$CH_2{=}CH_2(g) + Br_2(CCl_4) \rightarrow BrCH_2CH_2Br(CCl_4)$$

1,2-Dibromoethane

A test for a multiple bond is the decolorisation of a bromine solution without gas evolution

The decolorisation of bromine solutions without the evolution of hydrogen bromide (a fuming, pungent gas) is a test for a multiple bond. Bromine water is also decolorised by alkenes to give a different product, a **bromoalcohol**:

$$CH_2{=}CH_2(g) + Br_2(aq) + H_2O(l) \rightarrow BrCH_2CH_2OH(aq) + HBr(aq)$$

2-Bromoethanol

27.7.4 HYDROGEN HALIDES

Hydrogen halides add across the double bond

The hydrogen halides, HCl, HBr and HI, all add across the double bond. When hydrogen bromide adds to ethene, bromoethane is formed:

$$CH_2{=}CH_2(g) + HBr(g) \rightarrow CH_3CH_2Br(g)$$

Addition of hydrogen halide

When hydrogen halides add to an unsymmetrical alkene, two products are possible. The addition of hydrogen bromide to propene could give either

$$CH_3CH{=}CH_2(g) + HBr(g) \rightarrow CH_3\overset{|}{C}HCH_3(l)$$

Br

2-Bromopropane

or $CH_3CH_2CH_2Br(l)$

1-Bromopropane

Markovnikov's Rule predicts the product of the addition of HX to an alkene

In fact, the product is almost entirely 2-bromopropane. The Russian chemist, Markovnikov, formulated a rule for predicting which addition product would be formed. **Markovnikov's Rule** can be stated in this form: in addition of a compound HX to an unsaturated compound, hydrogen becomes attached to the unsaturated carbon atom which carries the larger number of hydrogen atoms. Thus, in addition of hydrogen iodide to 2-methylbut-2-ene, the Markovnikov Rule predicts the formation of 2-iodo-2-methylbutane:

$$\underset{\substack{\text{CH}_3}}{\overset{\substack{\text{CH}_3}}{}} \underset{\substack{\text{CH}_3}}{\overset{\substack{\text{CH}_3}}{}} \text{C}=\text{C} \quad \text{(g)} + \text{HI(g)} \rightarrow \text{CH}_3-\underset{\substack{|\\ \text{I}}}{\overset{\substack{\text{H}_3\text{C}\\ |}}{\text{C}}}-\underset{\substack{|\\ \text{H}}}{\overset{\substack{\text{CH}_3\\ |}}{\text{C}}}-\text{H(l)}$$

with hydrogen adding to the carbon atom which already has one hydrogen atom bonded to it. This is in fact the major product of the reaction, with a trace of 2-iodo-3-methylbutane also being formed:

$$\text{CH}_3-\underset{\substack{|\\ \text{H}}}{\overset{\substack{\text{H}_3\text{C}\\ |}}{\text{C}}}-\underset{\substack{|\\ \text{I}}}{\overset{\substack{\text{CH}_3\\ |}}{\text{C}}}-\text{H(l)}$$

27.7.5 MECHANISM OF ADDITION OF BROMINE TO ALKENES

At first sight it might seem that the reaction could take place in one step if the bromine molecule approaches the alkene from the right direction:

A simple one step mechanism for the addition of bromine to alkenes does not fit the facts...

$$\underset{\substack{\text{H}\quad\text{H}}}{\overset{\substack{\text{H}\quad\text{H}}}{\overset{\displaystyle\text{C}}{\underset{\displaystyle\text{C}}{\|}}}} \quad \underset{\substack{\text{Br}\\ \text{Br}}}{} \rightarrow \underset{\substack{\text{H}-\text{C}-\text{Br}\\ \text{H}-\text{C}-\text{Br}}}{\overset{\substack{\text{H}\\ |}}{}}$$

A number of workers have found evidence that this is not the course of the reaction. Only two of these pieces of evidence are given here.

EVIDENCE

...For example, these two experimental observations on the addition of bromine to alkenes

(a)   Ethene reacts with bromine in a solvent such as tetrachloromethane to give 1,2-dibromoethane. With bromine water, it gives a mixture of 2-bromoethanol, 1,2-dibromoethane and hydrogen bromide:

$$\text{CH}_2{=}\text{CH}_2 + \text{Br}_2 \xrightarrow{\text{in CCl}_4} \text{BrCH}_2\text{CH}_2\text{Br}$$
$$\xrightarrow[\text{in H}_2\text{O}]{} \text{BrCH}_2\text{CH}_2\text{OH} + \text{BrCH}_2\text{CH}_2\text{Br} + \text{HBr}$$

(The yield of $\text{BrCH}_2\text{CH}_2\text{OH}$ is too great to be explained by hydrolysis of $\text{BrCH}_2\text{CH}_2\text{Br}$).

(b)   Ethene and bromine water containing sodium chloride give three products, 2-bromoethanol, 1-bromo-2-chloroethane and 1,2-dibromoethane:

$$\text{CH}_2{=}\text{CH}_2 + \text{Br}_2 + \text{H}_2\text{O} + \text{NaCl} \rightarrow \text{HOCH}_2\text{CH}_2\text{Br} \;(\textit{major product})$$
$$+ \text{ClCH}_2\text{CH}_2\text{Br} \;(\textit{smaller amount})$$
$$+ \text{BrCH}_2\text{CH}_2\text{Br} \;(\textit{a trace})$$

THE PROPOSED MECHANISM

The formation of the chloro-compound in (b) needs some explaining. Since sodium chloride does not react with ethene, there must be formed from bromine and ethene a positively charged species which will react with chloride ions:

$$CH_2{=}CH_2 \xrightarrow{Br_2} CH_2{-}CH_2 \xrightarrow{Cl^-} ClCH_2CH_2Br$$
$$\underset{\overset{+}{Br}}{\diagdown\diagup}$$

The following mechanism is proposed for the formation of the positive ion.

1. Ethene has both a σ bond and a π bond between the two carbon atoms. Bromine is an electrophile. When a bromine molecule approaches an ethene molecule, the π electron cloud interacts with the approaching bromine molecule, causing a polarisation of the Br—Br bond:

When the π electron cloud of ethene reacts with bromine...

2. The π electrons become gradually more attached to the $\delta+$ Br atom, and the electrons of the Br—Br bond become gradually more polarised until the association of ethene and bromine is transformed into a positive ion, called a *bromonium ion*, and a bromide ion:

...the bromonium ion intermediate which is formed...

The positive charge is stabilised by delocalisation. The negative charge resides on Br$^-$, which is a good leaving group. These factors stabilise the transition state that precedes the bromonium ion intermediate [pp. 310–3].

3. The positive bromonium ion is immediately attacked by a bromide ion to form the product, 1,2-dibromoethane:

...quickly reacts to form the product

This reaction step is very fast compared with the other steps.

4. In reaction (b), i.e., with Br$_2$(aq) + NaCl(aq), the cation reacts in three ways:

$$CH_2{-}CH_2 + Br^- \longrightarrow BrCH_2CH_2Br \qquad\qquad [1]$$
$$\underset{\overset{+}{Br}}{\diagdown\diagup}$$

$$CH_2{-}CH_2 + H_2O \longrightarrow \underset{H}{\overset{H}{\diagdown}}\overset{+}{O}{-}CH_2CH_2Br \xrightarrow{H_2O} HOCH_2CH_2Br + H_3O^+$$
$$\underset{\overset{+}{Br}}{\diagdown\diagup} \qquad\qquad\qquad\qquad\qquad\qquad\qquad\qquad\qquad [2]$$

$$CH_2 \overset{+}{\underset{Br}{—}} CH_2 + Cl^- \longrightarrow ClCH_2CH_2Br \qquad [3]$$

A summary

> To summarise, the addition of halogens to alkenes can be formulated as
>
> $$CH_2{=}CH_2 \quad X{-}X \rightarrow CH_2 \overset{X}{\underset{+}{—}} CH_2 + X^- \rightarrow X{-}CH_2CH_2{-}X$$

A curly arrow represents the movement of a pair of electrons from the tail of the arrow to its tip. The addition of an electrophile to an alkene results in the formation of a cation which can accept electrons from a nucleophile, such as Br^-, Cl^-, NO_3^-, HSO_3^- or H_2O.

CONFIRMATION OF THE MECHANISM

Methyl groups stabilise the carbocation intermediate and the transition state that precedes it

Confirmation of this mechanism is seen in the effect of alkyl groups on the rate. Propene, $CH_3CH{=}CH_2$, reacts twice as fast as ethene does, and 2,3-dimethylbut-2-ene, $(CH_3)_2C{=}C(CH_3)_2$, reacts 14 times faster than ethene does. The reason is that methyl groups are electron-releasing [p. 257]. They tend to decrease the positive charge on a carbocation and thus stabilise it:

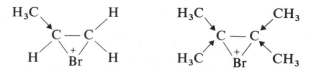

The more stable a carbocation is, the lower is the energy of the transition state that precedes it [pp. 310–3], assuming that the transition state is similar to the carbocation intermediate. The rate of formation of the transition state is therefore increased by methyl groups. Propenoic acid, $CH_2{=}CHCO_2H$, adds bromine slowly. The reason is that the carboxyl group, by withdrawing electrons, decreases the stability of the carbocation.

27.7.6 MECHANISM OF ADDITION OF HYDROGEN HALIDE TO ALKENES

Addition of HX to an alkene involves interaction of the electrophile with the double bond

A mechanism similar to that for the addition of bromine is proposed. A molecule of hydrogen halide, $H{-}X$ is permanently polarised as $\overset{\delta+}{H}{-}\overset{\delta-}{X}$. The π electrons of the alkene bond to the electrophilic H atom:

$$\begin{array}{c} CH_2 \\ \| \\ CH_2 \end{array} H \rightarrow X \rightarrow \overset{+}{C}H_2{-}CH_3 + X^-$$

This is the slow step in the reaction. Once formed, the carbocation reacts rapidly with halide ions:

$$\overset{+}{C}H_2CH_3 + X^- \rightarrow XCH_2CH_3$$

EVIDENCE

The rate of addition increases with increasing acid strength of HX

(a)   The rate of addition increases in the order

$$HF < HCl < HBr < HI$$

Markovnikov's Rule is explained by the carbocation theory

This is the order of increasing acid strengths, i.e., the order of readiness to release a proton. The reaction of the carbocation with X^- is rapid, and its speed is much the same for all X^- ions.

(b)   Addition follows Markovnikov's Rule:

$$CH_3CH{=}CH_2 + HCl \rightarrow CH_3CHClCH_3(90\%) + CH_3CH_2CH_2Cl(10\%)$$

Carbocations are produced during the formation of the two products:

$$CH_3{-}CH{=}CH_2 + HCl \rightarrow CH_3{-}\overset{+}{C}H{-}CH_3 + Cl^- \qquad [1]$$

$$CH_3{-}CH{=}CH_2 + HCl \rightarrow CH_3{-}CH_2{-}\overset{+}{C}H_2 + Cl^- \qquad [2]$$

Since alkyl groups are electron-releasing [p. 257], they tend to decrease the charge on a cation, and thus stabilise it. The carbocation

$$CH_3 \rightarrow \overset{+}{C}H \leftarrow CH_3 \qquad\qquad [1]$$

is therefore more stable than

$$CH_3 \rightarrow CH_2 \rightarrow \overset{+}{C}H_2 \qquad\qquad [2]$$

The more stable the carbocation, the lower is the energy of the transition state that precedes it. The rate of formation of carbocation [1] is therefore the greater, and reaction [1] predominates.

Alkyl groups increase the rate of addition

(c)   The rate of reaction of alkenes increases in the order

$$CH_2{=}CH_2 < RCH{=}CH_2 < R_2C{=}CH_2$$

(where R is an alkyl group).
This is in accordance with the carbocation theory since the alkyl groups confer stability on the carbocations in the order

$$R_2\overset{+}{C}CH_3 > R\overset{+}{C}HCH_3 > \overset{+}{C}H_2CH_3$$

This leads to reaction rates in the same order (as discussed in (b) above).

27.7.7 THE PEROXIDE EFFECT

Free radical addition does not obey Markovnikov's Rule

Some alkenes react with hydrogen bromide more quickly and give a different product from that predicted by Markovnikov's Rule. This type of addition happens in the presence of peroxides, which are known to initiate the formation of bromine atoms by homolysis of the H—Br bond. The mechanism of addition is different. Hydrogen chloride and hydrogen iodide do not share this behaviour.

27.7.8 CONCENTRATED SULPHURIC ACID

Addition of conc. H_2SO_4 yields the hydrogensulphate...

The addition of sulphuric acid across a double bond is similar to that of a hydrogen halide. Markovnikov's Rule is followed. When ethene is bubbled into concentrated sulphuric acid at room temperature, ethyl hydrogensulphate is formed:

$$CH_2{=}CH_2(g) + H_2SO_4(l) \xrightarrow{\text{cold}} H_2C{-}CH_2(l)$$
$$\quad\quad\quad\quad\quad\quad\quad\quad\quad\quad\quad\quad\,\, H \quad OSO_2OH$$

Ethyl hydrogensulphate

...which can be hydrolysed to the alcohol

The product, ethyl hydrogensulphate, when added to water and warmed, is hydrolysed to ethanol:

$$CH_2{=}CH_2 + H_2SO_4 \rightarrow C_2H_5SO_4H \xrightarrow[\text{warm}]{H_2O} C_2H_5OH + H_2SO_4$$

The net result is the addition of H·OH across the double bond. The industrial method of accomplishing this is the catalytic hydration of ethene. Ethene and steam are passed at 300 °C and 60 atm over phosphoric acid absorbed on silica pellets:

$$CH_2{=}CH_2(g) + H_2O(g) \xrightarrow[H_3PO_4\text{ catalyst}]{300\,°C,\,60\,atm} C_2H_5OH(g)$$

Ethene   Ethanol

The reaction makes possible the manufacture of the important solvent ethanol from ethene, which is a product of the petroleum industry.

27.7.9   ALKALINE POTASSIUM MANGANATE(VII)

Oxidation of C═C by alkaline KMnO₄...

Potassium manganate(VII) in alkaline solution is a weak oxidising agent [p. 282]. When ethene is bubbled into alkaline potassium manganate(VII) solution, the purple colour fades as ethene is oxidised to ethane-1,2-diol:

...forming a diol...

$$H_2C{=}CH_2 \xrightarrow[KMnO_4,\,OH^-]{} HOCH_2{-}CH_2OH$$

Ethene   Ethane-1,2-diol

...with decolorisation is a test for C═C

A diol is a compound containing two hydroxyl groups [p. 615]. A brown suspension of manganese(IV) oxide, MnO_2, appears. The decolorisation of ice-cold, dilute alkaline manganate(VII) solution is a test for a carbon–carbon multiple bond as few organic compounds are oxidised by this weak oxidising agent.

27.7.10   OZONE

Ozone adds to alkenes to form ozonides

Ozone, or trioxygen, O_3, adds across the double bond to form an **ozonide**. Since the ozonide formed is an explosive compound, it is prepared below 20 °C in a non-aqueous solvent:

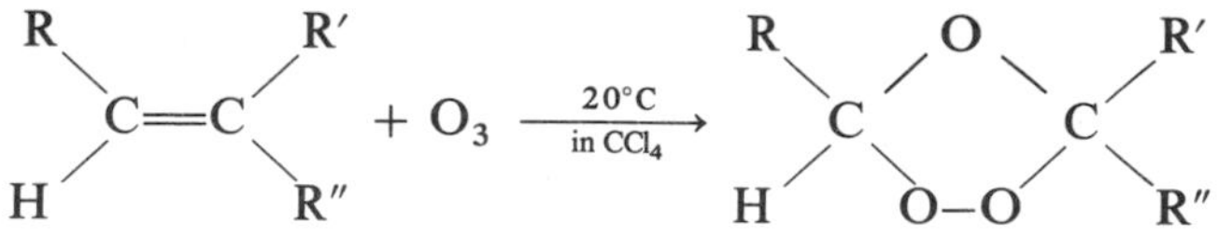

On hydrolysis, it splits up into two carbonyl compounds [see Chapter 31], and hydrogen peroxide, which may oxidise the other products:

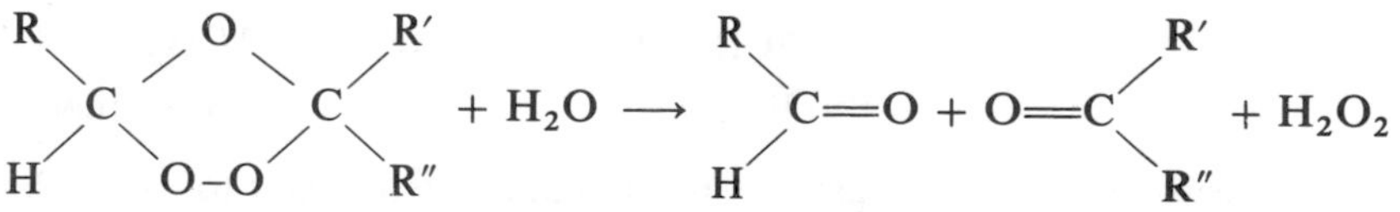

Ozonolysis can be used to locate the position of the C═C bond...

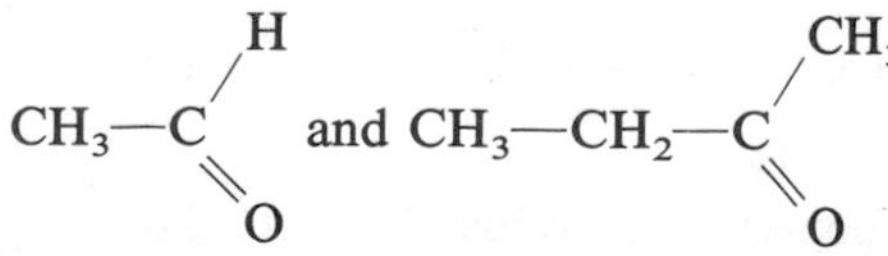

When zinc and ethanoic acid are used as a combined hydrolysing and reducing agent, the possibility of oxidation is avoided. The carbonyl compounds formed in the hydrolysis of the ozonide can be identified. The usefulness of **reductive ozonolysis**, i.e., the formation of an ozonide followed by its hydrolysis under mildly reducing conditions, is that it can be used to locate the position of the double bond in an alkene. If the ozonolysis of hexene gives the carbonyl compounds then putting these compounds together gives the formula of the original hexene:

$$CH_3-\underset{\underset{H}{|}}{C}=O \quad \text{and} \quad O=\underset{\underset{CH_3}{|}}{C}-CH_2CH_3 \quad \text{come from} \quad CH_3-\underset{\underset{H}{|}}{C}=\underset{\underset{CH_3}{|}}{C}-CH_2CH_3$$

...The amount of ozone absorbed gives the number of C=C bonds per molecule

Ozone reacts quantitatively with alkenes: 1 mole of ozone is absorbed by 1 mole of an alkene which possesses one double bond per molecule. This reaction can be used to find the number of double bonds in a compound.

Example How many double bonds are present in a molecule of C_6H_{10}, given that 0.082 g of the compound absorbs 48 cm^3 of ozone at room temperature?

Method

Gas molar volume = 24 dm^3 at rtp

Moles of ozone = 48/24000 = 2.0×10^{-3} mol

Moles of hydrocarbon = 0.082/82 = 1.0×10^{-3} mol

2 moles of ozone react with 1 mole of hydrocarbon: C_6H_{10} must contain 2 double bonds per molecule.

27.7.11 OSMIUM(VIII) OXIDE OsO$_4$

OsO$_4$

When osmium(VIII) oxide adds to an alkene in ethereal solution, a precipitate of an osmate ester appears. This is hydrolysed by heating for 2–3 hours with aqueous-alcoholic sodium sulphite to give a diol:

Diols

$$RCH=CHR' + OsO_4 \xrightarrow[\text{in ether}]{\text{dissolved}} \underset{\underset{Os}{\overset{O \quad O}{\diagdown \diagup}}}{RCH-CHR'} \xrightarrow{H^+(aq)} \underset{OH \quad OH}{RCH-CHR'} + H_2OsO_4$$

27.7.12 OXYGEN

Ethene reacts with oxygen...

In the presence of finely divided silver as catalyst, oxygen in the air combines with ethene to form epoxyethane. For safety, the reaction is carried out under carbon dioxide:

$$2CH_2=CH_2 + O_2 \xrightarrow[\text{(CO}_2)]{\text{Ag, 300 °C}} 2\underset{\text{Epoxyethane}}{CH_2-CH_2}$$
Ethene

...to form epoxyethane...

The product reacts readily with water to form ethane-1,2-diol:

$$\underset{\text{Epoxyethane}}{H_2C-CH_2} + H_2O \rightarrow \underset{\text{Ethane-1,2-diol}}{HOCH_2CH_2OH}$$

...which reacts with water to form ethane-1,2-diol

This compound is sold under the name *antifreeze*. Added to the water in a car radiator it lowers the freezing temperature [p. 181]. Alcoholysis of epoxyethane yields compounds which are useful solvents:

$$\underset{\text{Epoxyethane}}{H_2C-CH_2} + \underset{\text{Ethanol}}{C_2H_5OH} \xrightarrow{\text{acid}} \underset{\text{2-Ethoxyethanol}}{HOCH_2CH_2OC_2H_5}$$

FIGURE 27.4 Reactions of Alkenes

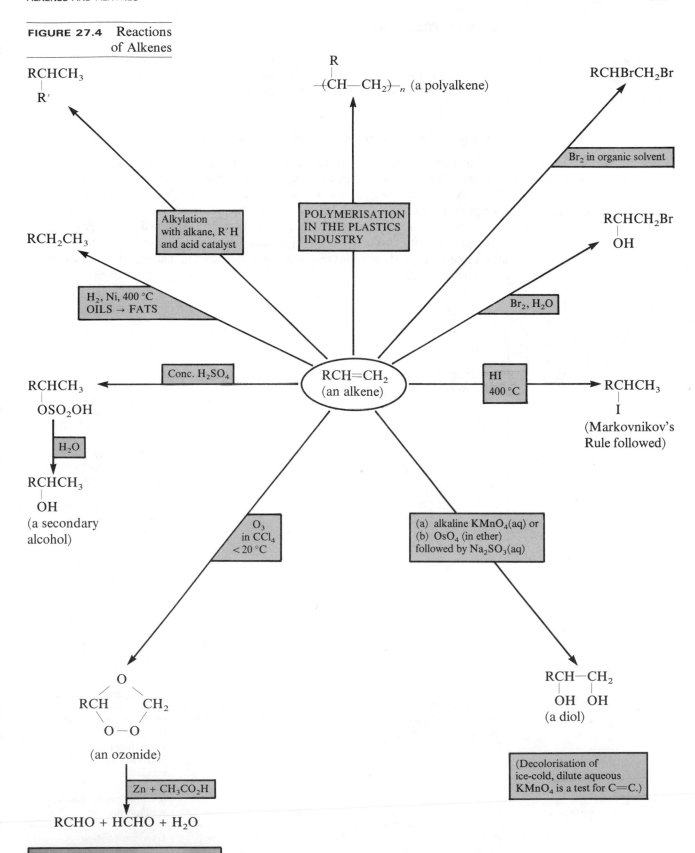

27.7.13 POLYMERISATION

Monomers and polymers

Another addition reaction of alkenes is **polymerisation**. Many molecules of **monomer** add together to form one large molecule of **polymer**. The polymer may have a molecular mass of several thousand. In the case of ethene

$$n\text{CH}_2{=}\text{CH}_2(g) \rightarrow {-}(\text{CH}_2{-}\text{CH}_2){-}_n(s)$$

The polymer formed is named *poly(ethene)*, often called *polythene* for short.

Ethene polymerises to form polyethene...

...which is an important plastics material

The first polymerisation of ethene was accomplished in 1933 by the use of very high pressure (1000 atm) and oxygen as the catalyst. A free radical mechanism operates. Research has developed the use of powerful catalysts, which enable the addition to take place at atmospheric pressure. The polyethene formed under high pressure is a low density, extremely pliable material, while the polymer formed at low pressure is of a higher density and is tougher. Polyethene is a **plastic**: it can be moulded easily into a multitude of shapes. Some polyalkenes are listed in Table 27.1. Plastics are discussed on p. 735.

TABLE 27.1 Polymers of Ethene and other Alkenes

Name	Monomer	Polymer	Uses
Poly(ethene) (polythene)	$\text{CH}_2{=}\text{CH}_2$	${-}(\text{CH}_2{-}\text{CH}_2){-}_n$	Polyethene has 40% of the polyalkene market. Low density polyethene is made at high pressure (1000–2000 atm) at 100–300 °C. It has a low melting temperature. It is used in packaging and for plastic bags and toy making. High density polyethene is made at low pressure (5–25 atm) at 20–50 °C, with a catalyst. It has a higher T_m and is used for kitchenware, food boxes, bowls, buckets, etc.
Poly(propene)	$\text{CH}_3\text{CH}{=}\text{CH}_2$	$\left(\text{CH}_2{-}\underset{\overset{\displaystyle |}{\text{CH}_3}}{\text{CH}}\right)_n$	Poly(propene) is tougher than poly(ethene). Used to make ropes and for packaging. Its high T_m enables it to resist boiling water.
Poly(chloroethene) (polyvinylchloride, PVC)	$\text{CH}_2{=}\text{CHCl}$	${-}(\text{CH}_2{-}\text{CHCl}){-}_n$	PVC is more rigid than polyethene. Used as a building material, e.g., guttering and electrical insulation. With added plasticisers, PVC is used for macintoshes, wellingtons, etc.
Poly(tetrafluoro-ethene) (PTFE, Teflon)	$\text{CF}_2{=}\text{CF}_2$	${-}(\text{CF}_2{-}\text{CF}_2){-}_n$	Used for coating surfaces to reduce friction, e.g., non-stick frying pans

Name	Monomer	Polymer	Uses
Poly(phenylethene) (polystyrene)	$CH_2{=}CH$ ⬡	$-(CH_2{-}CH)_n$ ⬡	Made into polystyrene foam by dissolving the polymer in a solvent and then vaporising the solvent. Used as insulation and for packaging. Polystyrene is flammable.
Poly(2-chloro-butadiene) (Neoprene)	$CH_2{=}CH$ | $ClC{=}CH_2$	$\left(\begin{array}{c}CH_2{-}CH{-} \\ \| \\ ClC{=}CH_2\end{array}\right)_n$	Used in the manufacture of synthetic rubber
Poly(ethenyl-ethanoate) (polyvinylacetate, PVA)	CH_3CO_2 | $CH{=}CH_2$	CH_3CO_2 | $-(CH{-}CH_2)_n$	Records. More flexible than PVC. Used in emulsion paints
Poly(methyl 2-methyl-propenoate) (Perspex®, Plexiglas®)	CH_3 | $C{=}CH_2$ | CO_2CH_3	$\left(\begin{array}{c}CH_3 \\ \| \\ {-}C{-}CH_2{-} \\ \| \\ CO_2CH_3\end{array}\right)_n$	Transparent. Used as a substitute for glass
Poly(propenonitrile) (polyacrylonitrile) (Acrilan®, Orlon®, Courtelle®)	$CH_2{=}CHCN$	$\left(\begin{array}{c}CH_2{-}CH \\ \| \\ CN\end{array}\right)_n$	Used as *acrylic* fibres for making clothing, blankets and carpets

CHECKPOINT 27A: ALKENES

1. Name the following compounds:

(a) $CH_3CH{=}CHCH_2CH_2CH_3$

(b) $CH_3C{=}CHCH_2CHCH_3$ with CH_3 and CH_3 substituents

(c) CH_3CH_2 \ C=C / with H and CH_2CH_2Cl

(d) $CH_3CHClCH{=}CH_2$

(e) CH_3 \ C=C / $CH_2CH_2CH_3$, with H and H

2. Write structural formulae for the following:

(a) Pent-1-ene

(b) 3-Chlorohexa-2,4-diene

(c) Buta-1,3-diene

(d) 4,4-Dimethylpent-2-ene

3. A compound C_6H_{12}, after ozonolysis, gives two products, one of which is propanone, $(CH_3)_2CO$. Which of the following is its formula?

(a) $CH_3CH{=}CHCH(CH_3)_2$

(b) $CH_3CH_2CH{=}C(CH_3)_2$

(c) $CH_3C{=}CHCH_3$ with C_2H_5 substituent

(d) $(CH_3)_2C{=}C(CH_3)_2$ or

(e) $CH_3CH{=}C(CH_3)C_2H_5$

4. Give the name and formula of the product from each of the reactions

(a) $CH_3CH{=}CH_2 + HBr \rightarrow$

(b) $(CH_3)_2C{=}CH_2 + Br_2 + NaOH \rightarrow$

(c) $(CH_3)_2C{=}CH_2 + H_2SO_4 \rightarrow$

5. Give the names and formulae of the products formed when the following reagents add to propene:

(a) chlorine in tetrachloromethane

(b) chlorine water.

Write the mechanism for each reaction.

6. Complete the following equations. Indicate the conditions needed for reaction.

(a) $CH_2{=}CH_2 + \qquad \rightarrow CH_3CH_2OH$

(b) $CH_3CH{=}CH_2 + HBr + Peroxide \rightarrow$

(c) $(CH_3)_2C{=}CH_2 + Br_2 + H_2O + HNO_3 \rightarrow$

(d) $CH_3CH{=}CH_2 + H_3PO_4 \rightarrow$

(e) $CH_3CH_2CH{=}CH_2 + \qquad \rightarrow CH_3CH_2CHOHCH_2OH$

(f) $(CH_3)_2CHCH{=}CHCH_3 + \qquad \rightarrow (CH_3)_2CHCHO + CH_3CHO$

7. When propene is bubbled through chlorine water containing nitrate ions, three products are formed. Give the names and formulae of the three products, and explain how they come to be formed. What product do you think would be formed in the reaction between propene and nitrogen chloride oxide, NOCl, which reacts as NO^+Cl^-?

8. Why do alkenes show geometrical isomerism, whereas alkanes do not? Draw and name the isomers of $CH_3CH{=}CHCl$, $CH_3CH{=}CHC_2H_5$, $ClCH{=}CHBr$ and $ClCH{=}CBrCH_3$.

9. Describe the principal types of reaction present in a homolytic chain reaction with reference to the photo-chlorination of methane, mentioning the importance of bond energies. Give examples of the energy profiles for the principal types of reactions.

Extend the concepts to hydrocarbon cracking reactions by deducing the probable reactions and products for the cracking of propane, the principal product being ethene.

Suggest an explanation for the observation that treatment of one of the products of the cracking of propane with bromine water gives 1-bromopropan-2-ol and not 2-bromo-propan-1-ol.

For the cracking of propane the activation energy is $220 \, kJ \, mol^{-1}$. Calculate the ratio of the rates of reaction at 850 K and 750 K. ($R = 8.31 \, J \, mol^{-1} K^{-1}$.)

[See also Chapter 14.]   (WJEC 81)

27.8 ALKYNES

Alkynes are aliphatic hydrocarbons with a $C{\equiv}C$ triple bond.

The general formula of this homologous series is C_nH_{2n-2}. They are named by changing the name of the corresponding alk*ane* to end in *-yne*. Some members are:

The names of alkynes

$HC{\equiv}CH$	Ethyne
$CH_3{-}C{\equiv}C{-}CH_3$	But-2-yne
$CH_3{-}CH{=}CH{-}C{\equiv}CH$	Pent-2-en-4-yne

The model of ethyne in Figure 27.5 shows that the four atoms lie in a straight line.

FIGURE 27.5 Ethyne

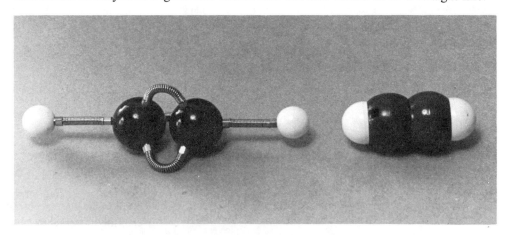

27.9 SOURCE OF ETHYNE

The combustion of ethyne...

...is used in welding

Ethene is used in some syntheses

Ethyne (formerly called acetylene) is a valuable compound. It burns in oxygen in an extremely exothermic reaction.

$$CH{\equiv}CH(g) + 2\tfrac{1}{2}O_2(g) \rightarrow 2CO_2(g) + H_2O(l); \quad \Delta H^{\ominus} = -1300 \, kJ \, mol^{-1}$$

Oxy-acetylene torches are used for cutting and welding metals. Ethyne is also valued because its unsaturated nature makes it a starting-point for the manufacture of a variety of organic compounds. It is, however, being superseded by ethene as a starting-point in organic syntheses as ethene is a cheaper starting-material.

It is made from natural gas

Ethyne is manufactured industrially from natural gas. The partial combustion at high temperature of methane [see Figure 27.6], the main component of natural gas, gives ethyne:

$$2CH_4(g) + 1\tfrac{1}{2}O_2(g) \rightarrow C_2H_2(g) + 3H_2O(l)$$

Ethyne may undergo an explosive reaction

Work with ethyne is difficult as it tends to undergo an explosive reaction to form carbon and hydrogen. It is stored in solution in propanone under pressure.

FIGURE 27.6
Preparations of Alkynes

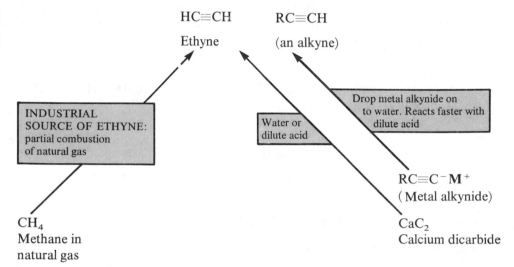

27.10 REACTIVITY OF ALKYNES

Accessibility of π electrons explains the reactivity of alkynes

The unsaturated carbon atoms in the —C≡C— bond are joined by a σ bond and two π bonds [p. 104]. The two π bonds lie in mutually perpendicular planes. The exposed position of the π electron clouds makes them accessible to attack by electrophilic reagents [see Figure 27.7].

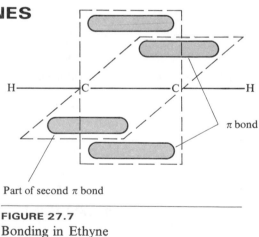

Part of second π bond

FIGURE 27.7
Bonding in Ethyne

27.11 PHYSICAL PROPERTIES

In physical properties alkynes resemble alkanes and alkenes.

27.12 ADDITION REACTIONS WITH ELECTROPHILES

Alkynes react with electrophilic reagents such as hydrogen, halogens and hydrogen halides, to form adducts. The reactions are similar to those of alkenes.

27.12.1 HYDROGEN

Addition of hydrogen to alkynes yields alkanes

Hydrogen adds across the triple bond in the presence of platinum, palladium or nickel as catalyst, to form an alkane:

$$CH_3C{\equiv}CH + 2H_2 \xrightarrow{\text{Ni as catalyst}} CH_3CH_2CH_3$$

27.12.2 CHLORINE AND BROMINE

Addition of Cl_2, Br_2 gives halogenoalkenes and halogenoalkanes

Chlorine and bromine add to alkynes in solution in the presence of metal halides as catalysts to form dihalogenoalkenes and then tetrahalogenoalkanes. The intermediate compound can be isolated:

$$CH{\equiv}CH + X_2 \xrightarrow[\text{as catalyst}]{\text{Metal halide}} \underset{\underset{X}{|}}{CH}{=}\underset{\underset{X}{|}}{CH} \xrightarrow{X_2} \underset{\underset{X}{|}}{\overset{\overset{X}{|}}{CH}}{-}\underset{\underset{X}{|}}{\overset{\overset{X}{|}}{CH}}$$

In the absence of a solvent, chlorine reacts in an explosive reaction with ethyne to form carbon and hydrogen çhloride:

$$CH{\equiv}CH + Cl_2 \rightarrow 2C + 2HCl$$

27.12.3 HYDROGEN HALIDES

Addition of HCl, HBr

Hydrogen halides add in the presence of light or a metal halide catalyst to form a halogenoalkene, which can be isolated and, on further reaction, a dihalogenoalkane. Markovnikov's Rule is obeyed [p. 557]. When hydrogen chloride adds to ethyne in the presence of mercury(II) chloride as catalyst, chloroethene can be isolated. This intermediate will react with more hydrogen chloride to form 1,1-dichloroethane:

$$CH{\equiv}CH(g) + HCl(g) \xrightarrow[\text{or light}]{HgCl_2} \underset{\text{Chloroethene}}{CH_2{=}CHCl(g)} \xrightarrow{HCl} \underset{\text{1,1-Dichloroethane}}{CH_3CHCl_2(g)}$$

PVC can be made from ethyne...

Chloroethene is the monomer from which poly(chloroethene), the important plastic known as *polyvinyl chloride* or *PVC*, is obtained. The industrial manufacture of chloroethene is now based on the reaction of ethene with oxygen and hydrogen chloride:

...or from ethene

$$CH_2{=}CH_2(g) + HCl(g) + \tfrac{1}{2}O_2(g) \xrightarrow[\text{CuCl}_2\text{ catalyst}]{250\,°C} CH_2{=}CHCl(g) + H_2O(g)$$

27.13 ADDITION REACTIONS WITH NUCLEOPHILES

Alkynes react with some nucleophilic reagents which do not attack alkenes.

27.13.1 HYDROGEN CYANIDE

HCN reacts with ethyne to form the monomer from which acrylic fibre is made

Hydrogen cyanide, HCN, reacts with ethyne when the two gases are brought into contact in the presence of the catalyst copper(I) chloride in solution in hydrochloric acid at 80 °C:

$$CH{\equiv}CH(g) + HCN(g) \xrightarrow[80\,°C]{CuCl \text{ in HCl}} \underset{\text{Propenonitrile}}{CH_2{=}CHCN(l)}$$

Propenonitrile is formed, the monomer from which poly(propenonitrile) or *polyacrylonitrile* is made.

FIGURE 27.8
Reactions of Ethyne

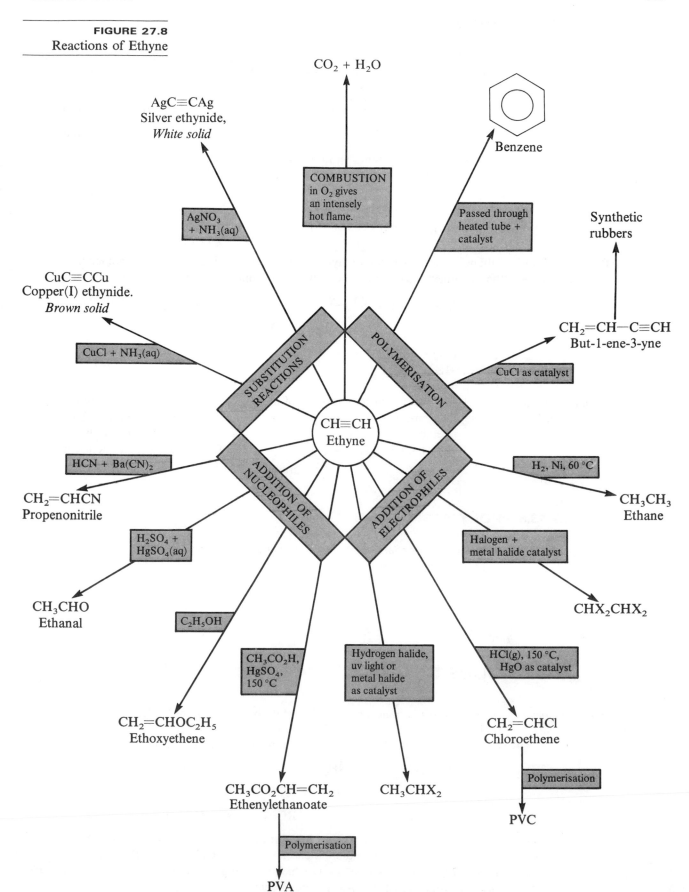

27.13.2   WATER

Water adds across the triple bond to give ethanal

Water adds across the triple bond in the presence of a catalyst, mercury(II) sulphate in 40% sulphuric acid. Ethyne gives the unstable intermediate, ethenol, which rearranges to give ethanal:

$$CH{\equiv}CH(g) + H_2O(l) \xrightarrow[\text{in 40\% } H_2SO_4, \, 90\,°C]{HgSO_4} [CH_2{=}CHOH(l)] \rightarrow CH_3CHO(l)$$
$$\text{Ethanal}$$

This process is used industrially for the manufacture of ethanal, but it is being replaced by methods based on ethanol and ethene.

27.13.3   ETHANOIC ACID

CH_3CO_2H yields the monomer of PVA

Ethanoic acid reacts with ethyne to give ethenyl ethanoate, the monomer used for the manufacture of the plastic poly(ethenylethanoate) or 'polyvinylacetate', PVA:

$$CH{\equiv}CH(g) + CH_3CO_2H(l) \xrightarrow[\text{at 200 °C}]{\text{Cd or Zn salt}} CH_3CO_2CH{=}CH_2(l)$$
$$\text{Ethenyl ethanoate}$$

27.13.4   DIMERISATION

Dimerisation leads to synthetic rubber

Dimerisation occurs in the presence of copper(I) chloride and ammonium chloride to give but-1-en-3-yne, an intermediate in the manufacture of synthetic rubbers:

$$CH{\equiv}CH(g) + CH{\equiv}CH(g) \xrightarrow[\text{NH}_4\text{Cl}]{\text{CuCl}} CH_2{=}CH{-}C{\equiv}CH(l) \rightarrow \begin{array}{l}\text{Used in the} \\ \text{manufacture of} \\ \text{synthetic rubbers}\end{array}$$
$$\text{But-1-en-3-yne}$$

27.13.5   POLYMERISATION

Polymerisation gives benzene

Polymerisation occurs when ethyne is passed under pressure over an organo-nickel catalyst at 70 °C. Benzene is formed:

$$3C_2H_2(g) \xrightarrow[\text{catalyst at 70 °C}]{\text{organo-nickel}} C_6H_6(l)$$
$$\text{Benzene}$$

27.14   SUBSTITUTION REACTIONS

Substitution of the acidic H by some metals . . .

A terminal triple bond reacts with certain bases, the terminal hydrogen atom being replaced by a metal ion. These bases are: sodium amide in liquid ammonia, copper(I) ions in aqueous ammoniacal solution, and silver ions in aqueous ammoniacal solution:

$$RC{\equiv}CH(g) + NaNH_2(NH_3) \rightarrow RC{\equiv}CNa(NH_3) + NH_3(NH_3)$$

$$RC{\equiv}CH(g) + Cu^+(aq) \rightarrow RC{\equiv}CCu(s) + H^+(aq)$$

$$RC{\equiv}CH(g) + Ag^+(aq) \rightarrow RC{\equiv}CAg(s) + H^+(aq)$$

. . . a test for a terminal triple bond

These reactions are a test for a terminal triple bond. In the case of ethyne, the products are sodium ethynide (or sodium dicarbide), $NaC{\equiv}CNa$, copper(I) ethynide (or copper(I) dicarbide), $CuC{\equiv}CCu$, which is a red precipitate, and silver ethynide (or silver dicarbide), $AgC{\equiv}CAg$, a white precipitate.

The acidity of H in RC≡CH The reason for the acidity of the hydrogen atom attached to a terminal triple bond is that the negative charge on the anion, RC≡C⁻, can be delocalised by interaction with the cloud of π electrons in the C≡C bond. This reduces the charge on the terminal carbon atom, and reduces its attraction for a proton.

FIGURE 27.9
Relationships between
Alkanes, Alkenes,
Alkynes and
Halogeno-compounds

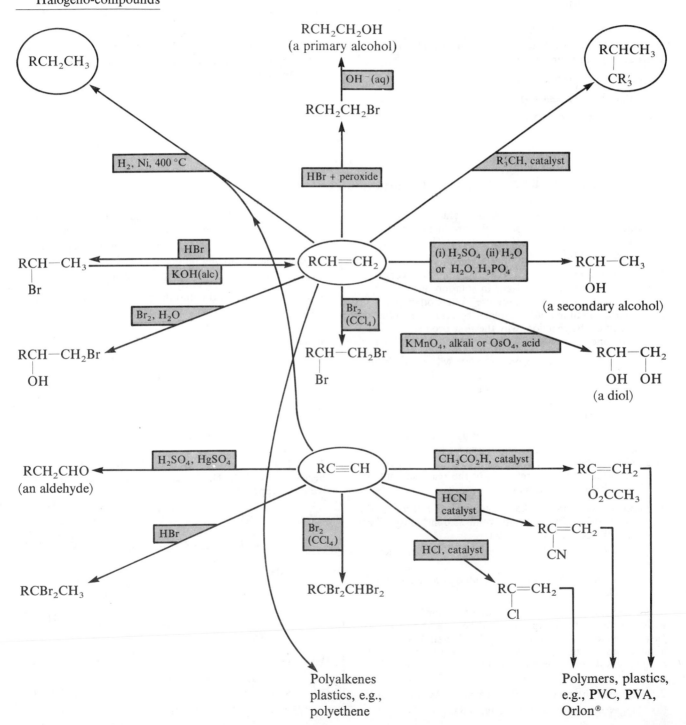

================ CHECKPOINT 27B: ALKYNES ================

1. Propyne reacts differently from propene with one of the following reagents. Which is it? (*a*) bromine water, (*b*) hydrogen chloride, (*c*) acidified potassium manganate(VII), (*d*) alkaline potassium manganate(VII), or (*e*) ammoniacal copper(I) chloride.

2. Name the following compounds, and say which one will form a precipitate with an aqueous solution of silver nitrate:

(*a*) $CH_3CHClCH_2CH_3$

(*b*) $CH_3CH=CHCH_3$

(*c*) $CH_2=CHCH=CH_2$

(*d*) $CH\equiv CCH_2CH_3$

(*e*) $CH_3C\equiv CCH_3$

3. Which is more likely to be used as a fuel, ethene or ethyne? Explain your answer.

4. Name the following compounds:

(*a*) $CH_3CH=CHC\equiv CCH_3$

(*b*) $CH_2=CHCH_2C\equiv CH$

(*c*) $CH\equiv CCHBrCH=CH_2$

(*d*) $CH_3C\equiv CCH=CHCH_3$

(*e*) $C_2H_5C\equiv CCHBrCH_2CH=CH_2$

5. What is the product of catalytic hydration of (*a*) $CH_3CH=CH_2$ and (*b*) $CH_3CH_2C\equiv CH$?

6. Suggest a laboratory preparation for

(*a*) $CH_2=CHCN$

(*b*) CH_3CHO

(*c*) $C_2H_5OCH=CH_2$

(*d*) $CH_3CO_2CH=CH_2$

7. What is the major product formed when propyne reacts with (*a*) hydrogen chloride with a mercury(II) oxide catalyst, (*b*) hydrogen chloride in ultraviolet light, (*c*) ethanol and (*d*) ethanoic acid?

================ QUESTIONS ON CHAPTER 27 ================

1. When ethene is bubbled through a solution containing bromine and potassium chloride, the products are CH_2BrCH_2Br and CH_2BrCH_2Cl, but no CH_2ClCH_2Cl is formed. Explain this behaviour.

2. What is the major product of the reaction between hydrogen bromide and but-1-ene in solution in tetra-chloromethane at room temperature? What other product is formed in small amount? Explain, by referring to the mechanism of the reaction, why the first product is the major one.

3. What is formed when propene reacts with (*a*) bromine water and (*b*) concentrated sulphuric acid? What is the product of hydrolysis of the compound formed in reaction (*b*)? Discuss the mechanisms of these two reactions, pointing out the similarity between them.

4. Some hydrocarbons are described as 'saturated'; others as 'unsaturated'. What are the differences between these classes of hydrocarbons, in structure and in reactivity? Illustrate your answer by referring to the reactions of two named aliphatic hydrocarbons.

5. Measurements on an alkene showed that $100\,cm^3$ of the gas weighed $0.231\,g$ at $25\,°C$ and $1\,atm$. $25.0\,cm^3$ of the alkene reacted with $25.0\,cm^3$ of hydrogen.
Find the molar mass of the alkene, and give its molecular formula.
Give the names and structural formulae of alkenes with this formula.

6. (*a*) $195\,cm^3$ of oxygen were added to $20\,cm^3$ of hydrocarbon **A**. After explosion and cooling, the residual gases occupied a volume of $145\,cm^3$. After shaking with aqueous potassium hydroxide, the volume of the residual oxygen was $45\,cm^3$ (temperature and pressure remaining constant).

 (i) Calculate the molecular formula of the hydrocarbon **A**.

 (ii) Suggest a suitable structural formula.

(*b*) Compare **A** with a named hydrocarbon **B**, containing two more hydrogen atoms per molecule by commenting on

 (i) the relative number of non cyclic isomers they possess,

 (ii) their relative boiling points,

 (iii) their relative enthalpy changes of formation.

(*c*) Give equations and name the products formed when **A** reacts with

 (i) hydrogen iodide,

 (ii) oleum followed by hydrolysis with water,

 (iii) a saturated solution of bromine in water.

(*d*) Outline *two* methods for the preparation of hydrocarbon **A** involving the use of

 (i) sodium hydroxide,

 (ii) hydrogen.

(AEB 81)

†7. (*a*) For each of parts (i) to (ix) below *only one* of the alternatives **A, B, C, D, E** is correct. Answer each part by giving the appropriate letter.

$$CH_3-\underset{\underset{CH_3}{|}}{\overset{\overset{CH_3}{|}}{CH}}-CH-CH_3 \qquad CH_3-CH_2-\overset{\overset{CH_3}{|}}{CH}-\overset{\overset{CH_3}{|}}{CH}-CH_3$$

$$\textit{A} \qquad\qquad\qquad\qquad \textit{B}$$

$$\underset{H_2C-CH_2}{\overset{H_2C-\overset{H}{\underset{|}{C}}}{|\qquad\qquad |\ CH_3}} \qquad CH_3-\overset{\overset{CH_3}{|}}{\underset{\underset{CH_3}{|}}{C}}-CH_3 \qquad \underset{H_2C-CH_2}{\overset{\overset{CH_3}{\underset{|}{CH}}}{H_2C\qquad CH_2}}$$

$$\textit{C} \qquad\qquad\qquad \textit{D} \qquad\qquad\qquad \textit{E}$$

(i) Which structure is under the greatest strain?

(ii) Which structure will exist as optical isomers?

(iii) Which is isomeric with pentane?

(iv) Which has a systematic name ending in -propane?

(v) Which structure could form *two and only two* different monochloro-compounds when hydrogen atoms are replaced by chlorine?

(vi) Which structure has the most carbon atoms in *one* plane?

(vii) Which structure requires the least amount of oxygen per mole for complete combustion?

(viii) Which is the least volatile?

(ix) Which could *not* be produced by addition of hydrogen to an alkene?

(*b*) Write the systematic names for compounds **A, B, C, D** and **E** in part (*a*).

(*c*) Using the structures **A, B, C, D** and **E**, and others as required, explain and illustrate what you understand by the terms

(i) 'structural isomerism';

(ii) 'optical isomerism'.

(SUJB 82)

[See Chapter 25 for (*a*) i–vi, Chapter 7 for vii, Chapter 26 for viii and Chapter 25 for (*b*) and (*c*).

28

BENZENE AND OTHER ARENES

28.1 BENZENE

Benzene is an arene

Benzene, C_6H_6, is the simplest member of the class of hydrocarbons called **aromatic hydrocarbons** or **arenes**. It is a colourless liquid with a characteristic smell.

For many years there was speculation over the structure of benzene. After 1834, when the molecular formula was established as C_6H_6, people put forward unsaturated formulae, such as

From the formula, C_6H_6, one might expect it to be an unsaturated compound...

$$CH_2=C=CH-CH=C=CH_2$$

In fact, benzene appeared to be strangely unreactive in comparison with the alkenes. In 1865, A Kekulé suggested the cyclic hexatriene structure, written in full in (a) and schematically in (b).

...but in fact it does not show much resemblance to the alkenes

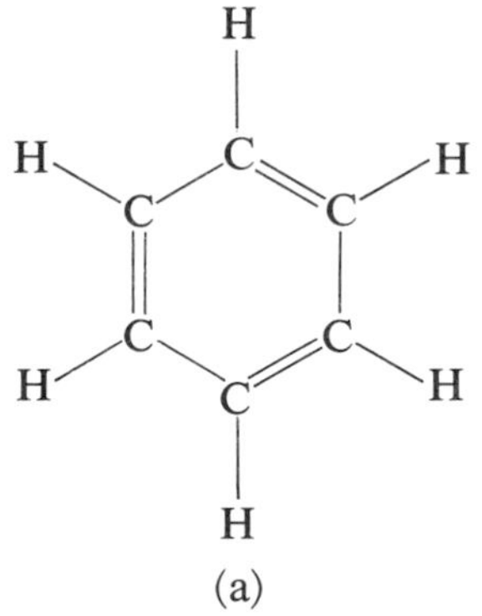

(a)    (b)

The Kekulé formula explains the number of isomeric substitution products formed by benzene...

Kekulé wanted to explain why benzene gave only one monosubstitution product, C_6H_5X, and three isomeric disubstitution products, $C_6H_4X_2$. No linear formula would correctly account for the number of substitution products, but the cyclic structure solved the problem [see Question 11, p. 589].

...but does not explain its lack of reactivity

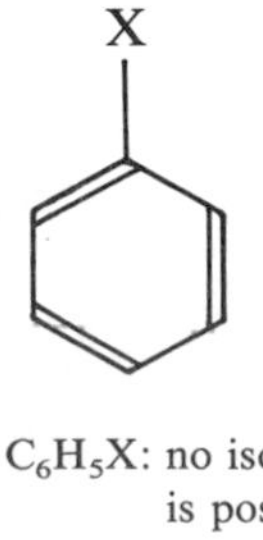

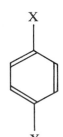

C_6H_5X: no isomerism is possible

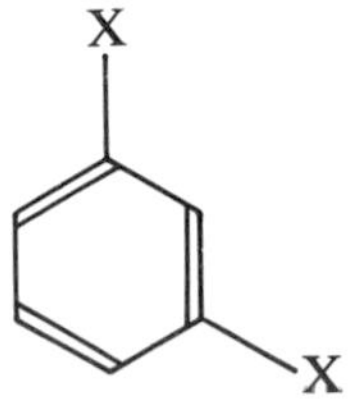

$C_6H_4X_2$: the three isomers

The Kekulé formula was not accepted by all chemists because it implies that benzene should show the same addition reactions as alkenes.

The Kekulé formula has now been superseded by the delocalised formula

Years later, X ray work showed that all the C—C bonds in benzene have the same length, 0.139 nm. This is intermediate between the C—C single bond length of 0.154 nm and the C═C double bond length of 0.133 nm. The Kekulé structure was superseded by the delocalised structure, which has been described on p. 107 and shown in Figure 5.28, p. 107. The delocalised formulae for benzene and other arenes are shown below:

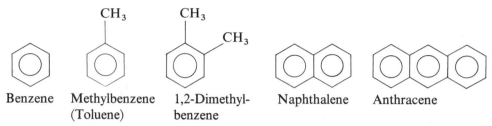

Benzene   Methylbenzene   1,2-Dimethyl-   Naphthalene   Anthracene
           (Toluene)   benzene

Other arenes

The group C_6H_5- is called a *phenyl* group. Phenyl and substituted phenyl groups are called *aryl* groups. Some derivatives of benzene are:

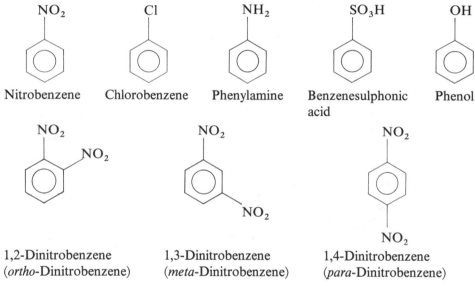

Nitrobenzene   Chlorobenzene   Phenylamine   Benzenesulphonic   Phenol
                                                acid

1,2-Dinitrobenzene   1,3-Dinitrobenzene   1,4-Dinitrobenzene
(*ortho*-Dinitrobenzene)   (*meta*-Dinitrobenzene)   (*para*-Dinitrobenzene)

Naming substituted benzenes

Isomers with two substituents in the 1,2-, 1,3-, and 1,4-positions are often called *ortho*-, *meta*- and *para*-isomers or *o*-, *m*- and *p*- for short.

28.2 SOME MORE NAMES OF AROMATIC COMPOUNDS

Derivatives of benzene are:

$C_6H_5NH_3{}^+Cl^-$	Phenylammonium chloride	(Anilinium chloride)
$C_6H_5CO_2H$	Benzenecarboxylic acid	(Benzoic acid)
$C_6H_5CO_2C_2H_5$	Ethyl benzenecarboxylate	(Ethyl benzoate)
C_6H_5COCl	Benzenecarbonyl chloride	(Benzoyl chloride)
$C_6H_5CONH_2$	Benzenecarboxamide	(Benzamide)
C_6H_5CN	Benzenecarbonitrile	(Benzonitrile)
C_6H_5CHO	Benzenecarbaldehyde	(Benzaldehyde)
$C_6H_5COCH_3$	Phenylethanone	
C_6H_5OH	Phenol	
$C_6H_5NH_2$	Phenylamine	(Aniline)
$C_6H_5OCH_3$	Methoxybenzene	

The names given are the IUPAC names. The names in parentheses are traditional names which are still in widespread use, and are acceptable. The naming of benzene derivatives containing two or more substituent groups is done by giving the group which is nearest to the top of the above list the number 1 position in the benzene ring. Other groups are numbered by counting from position 1 in the manner which gives them the lowest **locants** (numbers). Some examples are given below:

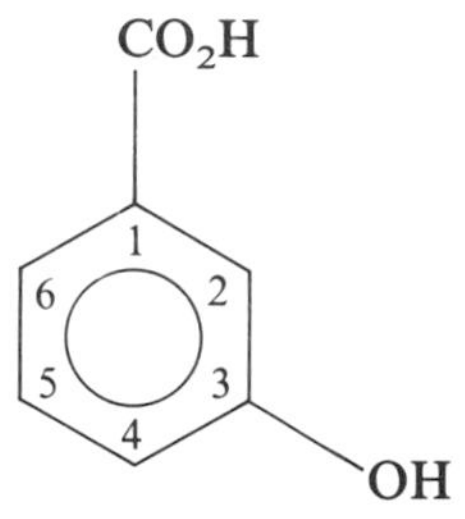

The —CO_2H group is higher up the list than —OH, and is regarded as the principal group, occupying position 1. The —OH group is in position 3, not 5, as the lower locant (number) is used.

3-Hydroxybenzenecarboxylic acid (3-Hydroxybenzoic acid)

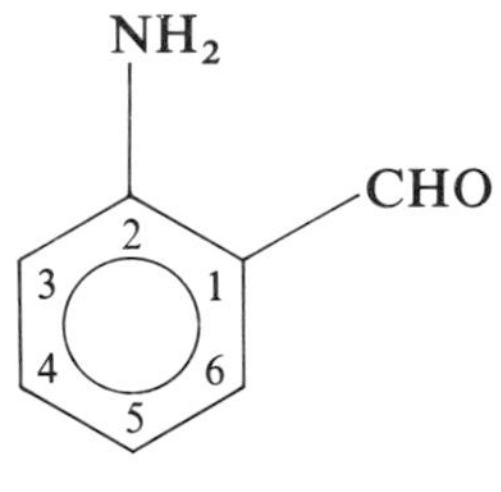

The —CHO group, being higher up the list, is regarded as the principal group and given position 1.

The —NH_2 group is in position 2, not 6. The name *ortho*-aminobenzaldehyde could also be used.

2-Aminobenzenecarbaldehyde (2-Aminobenzaldehyde)

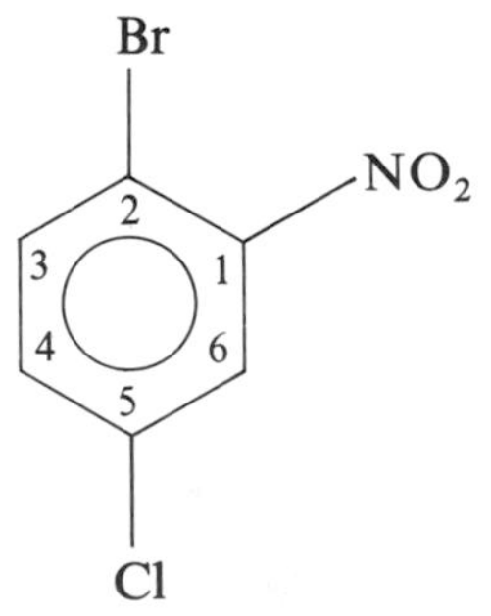

These groups are not on the list. They are listed alphabetically.

2-Bromo-5-chloronitrobenzene

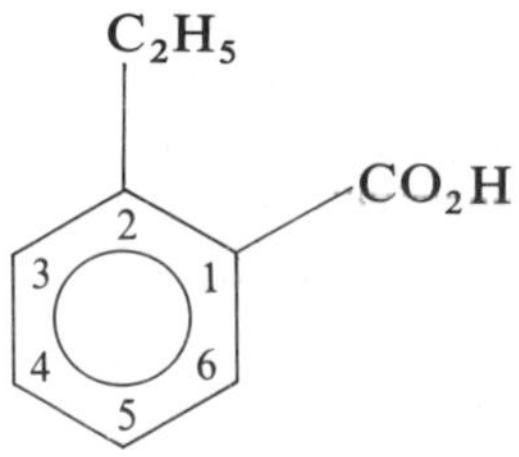

The functional group —CO_2H, is given the locant 1.

2-Ethylbenzenecarboxylic acid (2-Ethylbenzoic acid)

FIGURE 28.1 Methods of Preparing Benzene

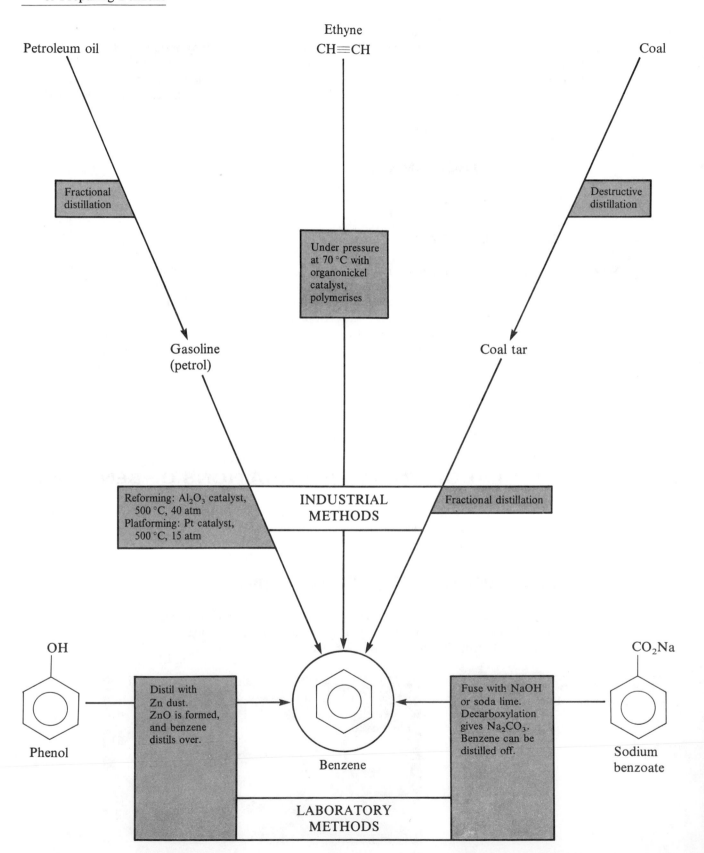

28.3 PHYSICAL PROPERTIES OF BENZENE

Benzene burns, and has a toxic vapour

Benzene is a colourless liquid of boiling temperature 80 °C and melting temperature 5.5 °C. It is immiscible with water. It dissolves many substances and is soluble in other organic solvents. When benzene and other arenes burn, particles of carbon are produced in addition to carbon dioxide and water, and these particles make the flame smoky and luminous. Benzene is toxic; inhalation over a period of time leads to anaemia and leukaemia. Other arenes, e.g., methylbenzene, are much less toxic.

28.4 SOURCES OF BENZENE

28.4.1 PETROLEUM OIL

See p. 546.

28.4.2 COAL TAR

Sources of benzene

The coal gas industry heats coal in retorts in the absence of air to 1000 °C. *Destructive distillation* takes place. Coal gas (a fuel) and ammonia (a source of fertilisers) are formed. A black viscous liquid distillate of coal tar is obtained. The residue in the retorts is coke. Besides being a valuable smokeless fuel, it is a reducing agent which is used in industrial processes such as the extraction of metals from oxide ores. Fractional distillation of coal tar gives methylbenzene and other aromatic compounds.

28.4.3 ETHYNE

See p. 570.

28.5 LABORATORY PREPARATIONS OF BENZENE

There is no need to prepare benzene in the laboratory, but sometimes it is useful to be able to replace a substituent group in a benzene derivative by hydrogen. Two methods of doing this are shown in Figure 28.1.

28.6 REACTIVITY OF BENZENE

The benzene ring is a planar hexagon with a cloud of delocalised π electrons lying above and below the ring. The reactions of benzene usually involve the attack of an electrophile on the cloud of π electrons.

FIGURE 28.2 A Model of the Benzene Molecule

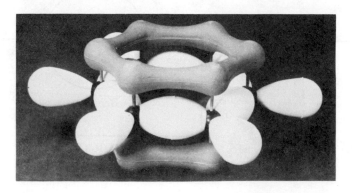

28.7 ADDITION REACTIONS OF BENZENE

The unsaturated character of benzene allows it to undergo addition reactions. Figure 28.3 shows the addition of hydrogen, ozone and halogen.

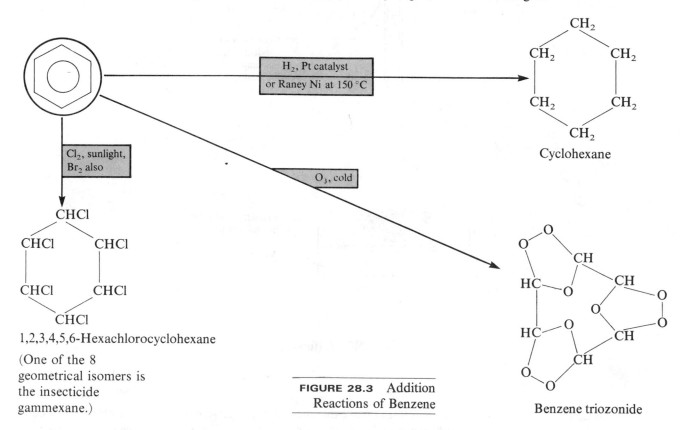

Cyclohexane

1,2,3,4,5,6-Hexachlorocyclohexane

(One of the 8 geometrical isomers is the insecticide gammexane.)

FIGURE 28.3 Addition Reactions of Benzene

Benzene triozonide

28.8 SUBSTITUTION REACTIONS

28.8.1 NITRATION

Benzene reacts with a 'nitrating mixture' to give nitrobenzene...

The substitution of a —H atom by a —NO_2 group is called **nitration**. To obtain nitrobenzene, $C_6H_5NO_2$, benzene is refluxed on a water bath at 60 °C with a 'nitrating mixture', as shown in Figure 28.4. Although concentrated nitric acid will slowly nitrate benzene, a mixture of concentrated nitric and sulphuric acids is much more effective:

Benzene (l) + HNO_3(l) $\xrightarrow[\substack{\text{conc. } HNO_3 + \\ \text{conc. } H_2SO_4}]{\text{Reflux, 60 °C}}$ Nitrobenzene (l) + H_2O(l)

...which is an intermediate in the preparation of many derivatives of benzene

Nitrobenzene is a pale yellow liquid which can be separated from benzene by distillation under reduced pressure. Its preparation from benzene is an important reaction as it is the first step in the introduction of many different substituent groups into the benzene ring. The —NO_2 group can be reduced to —NH_2, to form phenylamine, $C_6H_5NH_2$, which gives rise to all the compounds described on p. 684.

If the temperature is raised to 95 °C, and fuming nitric acid is used, 1,3-dinitrobenzene is formed.

FIGURE 28.4
Apparatus for Refluxing
Benzene and Nitrating
Mixture

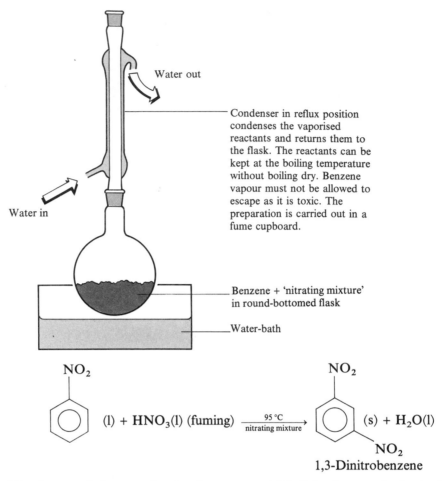

Water out

Condenser in reflux position
condenses the vaporised
reactants and returns them to
the flask. The reactants can be
kept at the boiling temperature
without boiling dry. Benzene
vapour must not be allowed to
escape as it is toxic. The
preparation is carried out in a
fume cupboard.

Water in

Benzene + 'nitrating mixture'
in round-bottomed flask

Water-bath

1,3-Dinitrobenzene

*Dinitrobenzene can also
be formed by nitrating
benzene*

Nitrobenzene is less reactive than benzene, and 1,3-dinitrobenzene is obtained together
with nitrobenzene and benzene. Since it is a solid, 1,3-dinitrobenzene crystallises
out when the mixture is poured into cold water. No trinitrobenzene is formed.

*Nitration of methylbenzene
produces TNT*

When methylbenzene (toluene), $C_6H_5CH_3$, is nitrated, 2,4,6-trinitromethylbenzene
(2,4,6-trinitrotoluene) is formed.

This shows that the methyl group makes
methylbenzene more reactive than benzene.
The product, which is called TNT for
short, is a powerful explosive.

Nitration of aromatic compounds is important for the production of this and other
explosives.

28.8.2 MECHANISM OF NITRATION

*What is the nitrating
agent?*

Chemists were intrigued by the puzzle that, although concentrated nitric acid alone
is a poor nitrating agent, a mixture of concentrated nitric and sulphuric acids is a
powerful reagent. It seemed that the two acids must react to produce a species
which can attack benzene. Much research work has gone into the unravelling of
this mystery, and only a glimpse of this exciting area of study can be given here.
Two of the pieces of evidence which help to show what this nitrating agent is are
discussed here.

(a) The freezing temperature depression of concentrated sulphuric acid produced by dissolving concentrated nitric acid in it is four times that expected [p. 185]. It is suggested that sulphuric acid, being a stronger acid than nitric acid, can to some extent donate a proton to nitric acid:

$$H_2SO_4 + HNO_3 \rightleftharpoons HSO_4^- + \overset{H}{\underset{H}{>}}\overset{+}{O}-NO_2$$

Evidence from freezing temperature depression

Nitric acid, in accepting a proton, is acting as a base. The protonated nitric acid splits up to form a molecule of water and the cation, NO_2^+, which is called a *nitryl cation*.

Formation of the nitryl cation...

$$\overset{H}{\underset{H}{>}}\overset{+}{O}-NO_2 \rightleftharpoons H_2O + NO_2^+$$

The molecule of water is protonated by a second molecule of sulphuric acid:

$$H_2O + H_2SO_4 \rightleftharpoons H_3O^+ + HSO_4^-$$

Adding up the three steps gives

$$HNO_3 + 2H_2SO_4 \rightleftharpoons NO_2^+ + H_3O^+ + 2HSO_4^-$$

A molecule of nitric acid thus produces four particles, and the freezing temperature depression is four times that expected.

...The nitryl cation as a nitrating agent

(b) The salt nitryl chlorate(VII), $NO_2^+ClO_4^-$, has been prepared and shown to nitrate benzene. The nitryl cation must be a nitrating agent.

NO_2^+ is an electrophile...

Once formed, the nitryl cation is attracted by the cloud of delocalised π electrons which lies above and below the plane of the benzene ring [see Figure 28.2, p. 578]. The NO_2^+ cation adds to the ring to form a short-lived intermediate, in which both

...which adds to benzene to form a reactive intermediate...

the entering —NO_2 group and the leaving —H atom are bonded to the ring. The symmetry of the annular cloud of delocalised π electrons is disrupted. The positive charge contributed by the NO_2^+ ion is distributed over the ring. The intermediate rapidly loses a proton, restoring the symmetry and stability of the benzene ring. The proton is immediately picked up by a hydrogensulphate ion:

...this intermediate rapidly loses a proton to form the product

Reactive intermediate

$$H^+ + HSO_4^- \xrightarrow{fast} H_2SO_4$$

The overall rate of the reaction is determined by the slowest step in the sequence. This is the rate at which the C—NO_2 bond is formed in the intermediate.

28.8.3 SULPHONATION

Benzene can be sulphonated by fuming H_2SO_4

Sulphonation is the substitution of a —H atom by an —SO_3H group. Benzenesulphonic acid, $C_6H_5SO_3H$, is obtained by refluxing benzene with concentrated sulphuric acid for many hours, or warming with fuming sulphuric acid (which contains SO_3) at 40 °C for 20–30 minutes:

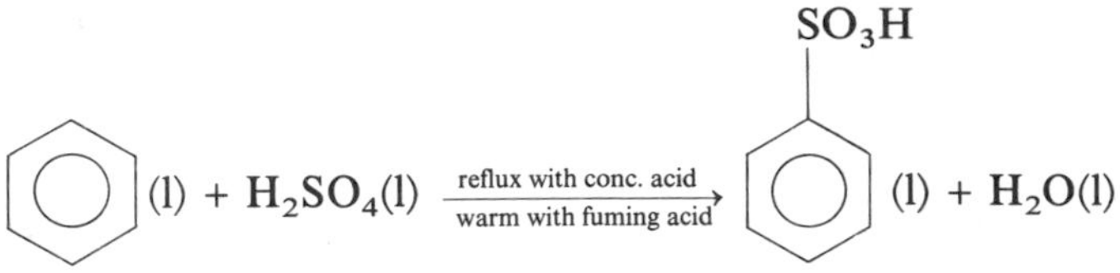

$$\text{(l)} + H_2SO_4\text{(l)} \xrightarrow[\text{warm with fuming acid}]{\text{reflux with conc. acid}} \overset{\displaystyle SO_3H}{\text{(l)}} + H_2O\text{(l)}$$

Benzenesulphonic acid

Sulphonic acids are important in the preparation of phenols and in the manufacture of detergents

The importance of this reaction is that the —SO_3H group can be replaced by hydrolysis to give an —OH group. Benzenesulphonic acid is converted into phenol, C_6H_5OH. This is the easiest way of preparing phenol in the laboratory. Sulphonation is also important in the manufacture of detergents. It is because detergents contain —SO_3H groups that they lather even in hard water, as the calcium and magnesium salts of sulphonic acids are soluble.

28.8.4 MECHANISM OF SULPHONATION

SO_3 is a sulphonating agent

A more powerful sulphonating agent than concentrated sulphuric acid is fuming sulphuric acid, which contains sulphur(VI) oxide, SO_3. It is therefore thought that SO_3 is the electrophile which attacks the benzene ring.
The three S=O bonds are all polarised, each O atom being $\delta-$. The cumulative $\delta+$ charges on the S atom make it highly positively polarised and thus a powerful electrophile.

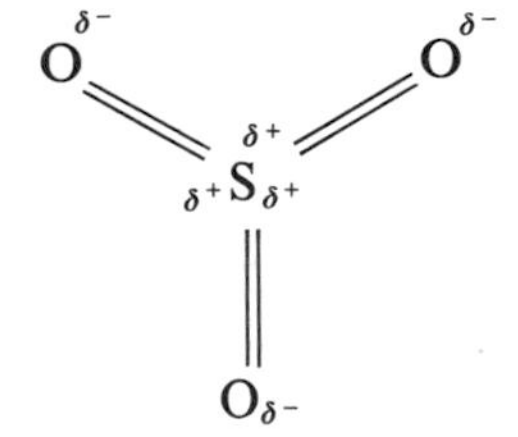

FIGURE 28.5
The SO_3 molecule

28.8.5 ALKYLATION

An alkyl group can be introduced into the benzene ring by the reaction of a halogenoalkane with benzene:

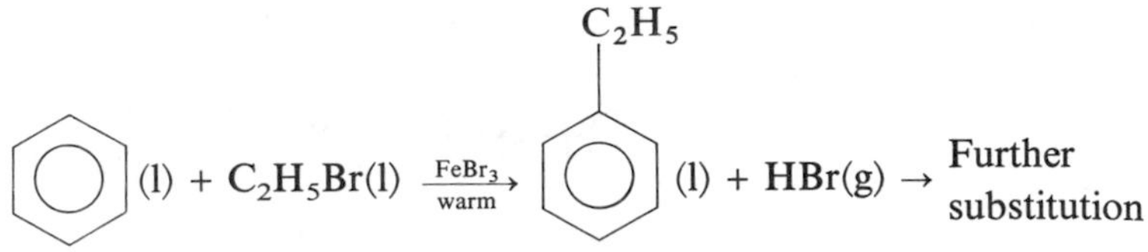

$$\text{(l)} + C_2H_5Br\text{(l)} \xrightarrow[\text{warm}]{FeBr_3} \overset{\displaystyle C_2H_5}{\text{(l)}} + HBr\text{(g)} \rightarrow \text{Further substitution}$$

Benzene   Bromoethane   Ethylbenzene

Alkylation needs a catalyst + RX

The reaction takes place under the influence of a catalyst, such as aluminium chloride or iron(III) bromide. More than one alkyl group enters the ring as the alkylbenzene formed is more reactive than benzene. The effectiveness of the catalysts was discovered by C Friedel and J M Crafts, and reactions of this type are called **Friedel–Crafts reactions**. The mechanism is discussed on p. 584.

Alkenes are alkylating agents... Alkylation can also be effected by the use of an alkene and a Friedel–Crafts catalyst together with an acid such as HCl or H_3PO_4:

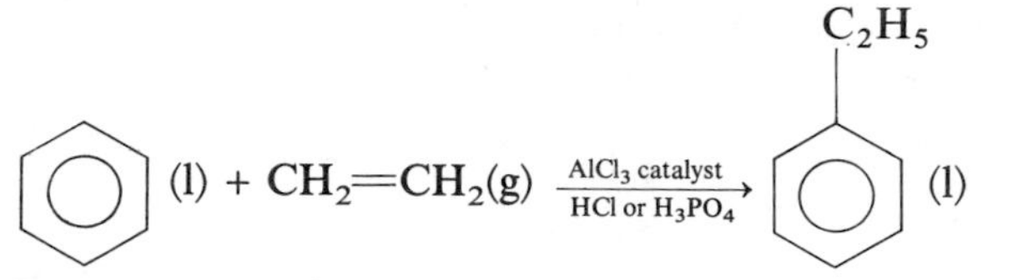

Benzene (l) + $CH_2{=}CH_2$(g) $\xrightarrow[\text{HCl or }H_3PO_4]{\text{AlCl}_3\text{ catalyst}}$ Ethylbenzene (C_2H_5) (l)

Benzene   Ethene   Ethylbenzene

...used to make styrene and cumene This is the method used industrially to make ethylbenzene. The product is dehydrogenated to give phenylethene (styrene), from which polystyrene is made [see Table 27.1, p. 565]. Another Friedel–Crafts reaction used in industry is the reaction between benzene and propene to make cumene and thence phenol [p. 630].

Dimethyl sulphate, $(CH_3O)_2SO_2$, is a popular reagent for the introduction of methyl groups. It is toxic, and must be used in a fume cupboard and prevented from touching the skin.

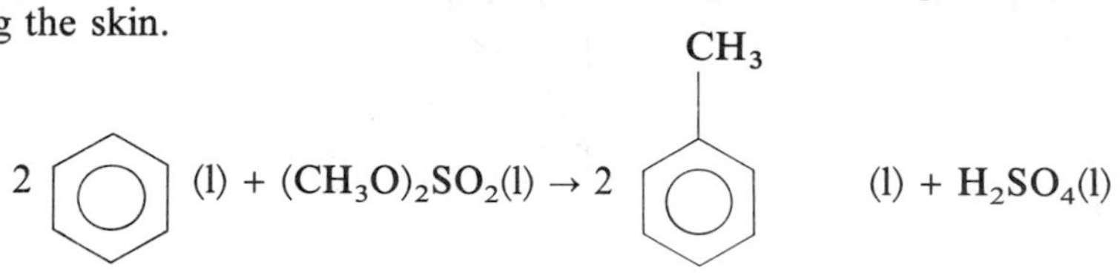

2 Benzene (l) + $(CH_3O)_2SO_2$(l) $\rightarrow$ 2 Methylbenzene (CH_3) (l) + H_2SO_4(l)

Benzene   Dimethyl sulphate   Methylbenzene

28.8.6 ACYLATION

Acylation, the introduction of an acyl group

$$-\overset{\displaystyle O}{\underset{\displaystyle R}{\overset{\|}{C}}}$$

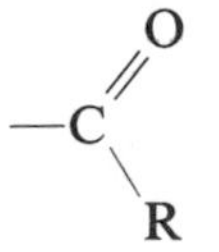

is another Friedel–Crafts reaction:

Acylation needs RCOCl + catalyst

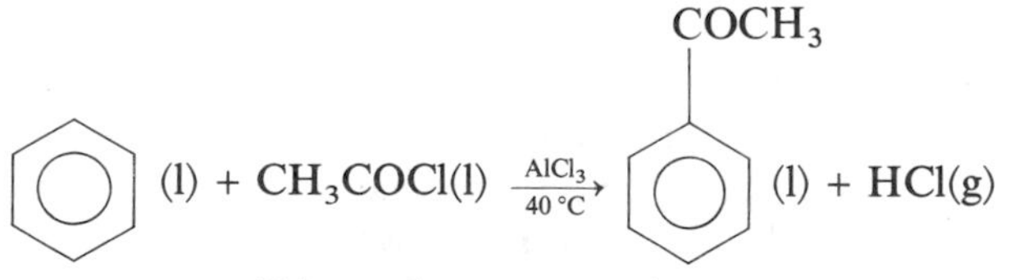

Benzene (l) + CH_3COCl(l) $\xrightarrow[40\,°C]{\text{AlCl}_3}$ Phenylethanone ($COCH_3$) (l) + HCl(g)

Ethanoyl chloride   Phenylethanone

An acid chloride or acid anhydride, in the presence of a Friedel–Crafts catalyst, will acylate benzene and its derivatives [see Figure 28.6]. Only one acyl group enters the ring as the acyl compound formed is less reactive than benzene. This gives acylation an advantage over alkylation, which yields polyalkyl derivatives. A method of preparing pure ethylbenzene is to introduce the group —$COCH_3$ by acylation and then reduce it to —CH_2CH_3:

The reduction of an acyl group gives an alkyl group

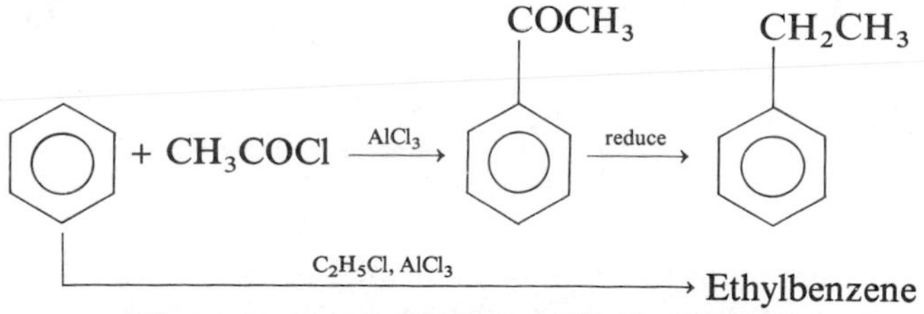

Benzene + CH_3COCl $\xrightarrow{\text{AlCl}_3}$ ($COCH_3$) $\xrightarrow{\text{reduce}}$ (CH_2CH_3)

Benzene $\xrightarrow{C_2H_5Cl,\ \text{AlCl}_3}$ Ethylbenzene

Gives some polyethylbenzenes in addition.

FIGURE 28.6
Friedel–Crafts Acylation

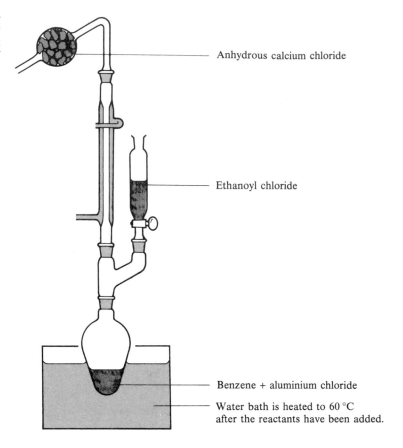

Anhydrous calcium chloride

Ethanoyl chloride

Benzene + aluminium chloride

Water bath is heated to 60 °C
after the reactants have been added.

28.8.7 HALOGENATION

The addition of halogens to the benzene ring [p. 579] takes place under conditions which favour the formation of free radicals, i.e., sunlight or high temperature. The first step in the addition is the homolysis of the halogen molecule, e.g.

$$Cl_2 \xrightarrow{hv} 2Cl$$

Chlorine and bromine can be substituted in the ring in the presence of a Friedel–Crafts catalyst...

A different type of reaction takes place at room temperature in the presence of a Friedel–Crafts catalyst. Substitution then occurs. In recognition of the way in which they ease the path of halogens into the benzene ring the Friedel–Crafts catalysts are referred to as **halogen-carriers**. Aluminium and iron act as catalysts because during the reaction they are converted into their chlorides or bromides, which act as halogen-carriers:

(l) + Cl_2(g) $\xrightarrow[\text{room temperature}]{\text{Al or AlCl}_3}$ (l) + HCl(g)

Cl

Chlorobenzene

(l) + Br_2(l) $\xrightarrow[\text{room temperature}]{\text{Fe or FeBr}_3}$ (l) + HBr(g)

Br

Bromobenzene

Iodine cannot be introduced in this way.

...which is called a
halogen-carrier

Why does the presence of a halogen-carrier result in a completely different reaction—substitution instead of addition? Much work has been done on the mode of action of Friedel–Crafts catalysts, and only an outline can be given here.

28.8.8 MECHANISM OF ALKYLATION, ACYLATION AND HALOGENATION

*The mechanism:
Friedel–Crafts catalysts
act as Lewis acids...*

Friedel–Crafts catalysts are halides like $AlCl_3$, $AlBr_3$, BF_3 and $FeBr_3$. They function as Lewis acids [p. 251] since the central atom can accept a pair of electrons. Aluminium chloride can accept a pair of electrons from a chloride ion. In the complex ion formed, $AlCl_4^-$, aluminium has eight electrons in its outer shell:

$$Cl-Al-Cl + Cl^- \rightarrow \left[Cl-Al-Cl \right]^-$$

Let us see how the ability to form complex ions by acting as Lewis acids can explain the catalytic effect of these halides.

ALKYLATION

When a Friedel–Crafts catalyst is dissolved in a halogenoalkane, the solution formed is a weak electrical conductor. This could be explained by the formation of an ionic complex. In the case of bromoethane and aluminium bromide

*...They form complexes
with halogenoalkanes*

$$C_2H_5Br + AlBr_3 \rightleftharpoons C_2H_5^+ AlBr_4^-$$

*A carbocation complex
attacks benzene to form
an intermediate...*

The attack on the π electron cloud of the benzene ring could be by the carbocation, $C_2H_5^+$ or by the complex, $C_2H_5^+ AlBr_4^-$. In the case of primary and secondary carbocations, it is thought that the complex is the attacking electrophile. Tertiary carbocations are more stable, and in the case of $(CH_3)_3C^+$, for example, it may be the carbocation that attacks benzene:

*...which loses a proton to
form the product*

The reactive intermediate that is formed quickly loses a proton to form the product, ethylbenzene. The catalyst, aluminium bromide, is regenerated, and hydrogen bromide is evolved:

Reactive intermediate Ethylbenzene

When alkenes are used for alkylation a proton acid must be present, in addition to a Lewis acid. The first step in the reaction is protonation of the alkene to form a carbocation which attacks the benzene ring.

ACYLATION

The mechanism of acylation: the F-C catalyst + RCOCl form a complex...

Again, the Friedel–Crafts catalyst acts as a Lewis acid. In the reaction with an acyl chloride, a complex containing an acyl cation, $\overset{+}{R}CO$, is formed:

$$RCOCl + AlCl_3 \rightleftharpoons \overset{+}{R}CO\ AlCl_4^{\ -}$$

The electrophile which attacks the benzene ring could be either the acyl cation, $\overset{+}{R}CO$, or the complex, $\overset{+}{R}CO\ AlCl_4^{\ -}$. There is evidence that the attacking reagent is the whole complex:

...the electrophilic complex attacks benzene...

Reactive intermediate   Phenylketone

...to form a reactive intermediate which is rapidly converted into the product

The rate-determining step is the formation of the RCO—ring bond. The intermediate that is formed rapidly loses a proton to yield the product.

HALOGENATION WITH THE AID OF A HALOGEN-CARRIER

In the bromination of benzene a bromine molecule approaches a benzene molecule and encounters the annular cloud of delocalised π electrons above and below the plane of the ring [see Figure 28.2, p. 558]. As the π electron cloud interacts with the bromine molecule, the Br—Br bond becomes polarised:

The role of a halogen-carrier is explained in terms of polarising the X—X bond

This will remind you of the interaction between Br_2 and the π electrons of the C=C bond in alkenes [p. 558]. Benzene is less reactive than an alkene, and the reaction proceeds only if there is a Lewis acid (e.g., $FeBr_3$) present. This accepts a pair of electrons from the $\delta-$ Br atom in the polarised Br_2 molecule, and enables the Br—Br bond to split. A bromonium ion and a $FeBr_4^{\ -}$ complex ion are formed. The bromonium ion rapidly loses a proton to form bromobenzene, with the regeneration of the catalyst, $FeBr_3$:

Iodination of benzene does not occur, but some benzene derivatives, containing activating substituents, are iodinated.

A SUMMARY

Alkylation, first step

$$RX + AlX_3 \rightleftharpoons R^+\ AlX_4^{\ -}$$

followed by

Summary of the role of Friedel–Crafts catalysts

Acylation, first step

$$RCOX + AlX_3 \rightleftharpoons R\overset{+}{C}O \; AlX_4{}^-$$

followed by

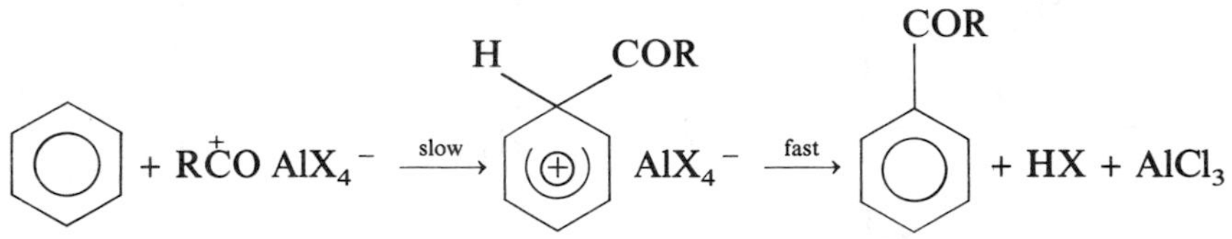

Halogenation, first step

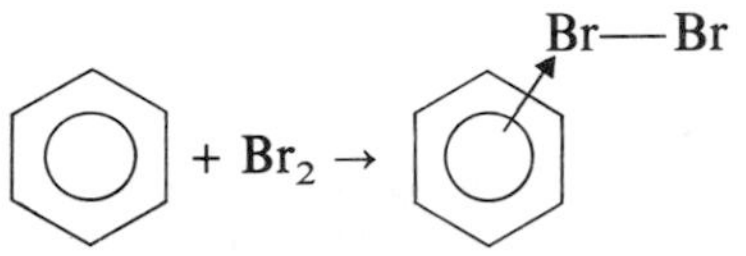

followed by

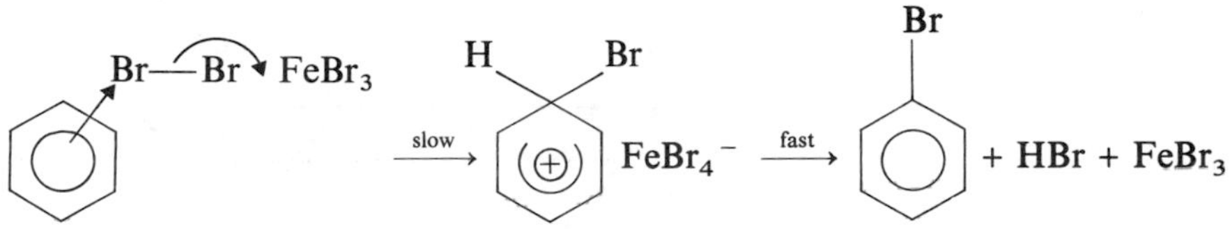

Reagent	Benzene 	Cyclohexene 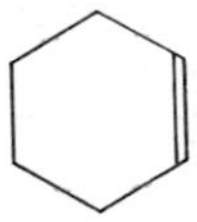
Br$_2$(CCl$_4$)	No reaction	Rapid decolorisation
KMnO$_4$, alkaline, cold	No reaction	Rapid decolorisation
HBr(g)	No reaction	Rapid addition
Air, catalyst, 150 °C	No reaction	Oxidised to hexane-1,6-dioic acid
H$_2$, catalyst, 200 °C	Slowly adds to give cyclohexane	Adds rapidly to give cyclohexane
Halogen, sunlight, boil	Slow addition gives C$_6$H$_6$X$_6$	Rapid addition gives C$_6$H$_{10}$X$_2$
Conc. HNO$_3$, H$_2$SO$_4$	Substitution gives nitrobenzene	Oxidation gives hexane-1,6-dioic acid
Conc. H$_2$SO$_4$	No reaction when cold. Slow sulphonation when heated	Absorbed by H$_2$SO$_4$
Friedel–Crafts reactions	Alkylation or acylation	No reaction

TABLE 28.1
A Comparison of an
Arene and a Cycloalkene

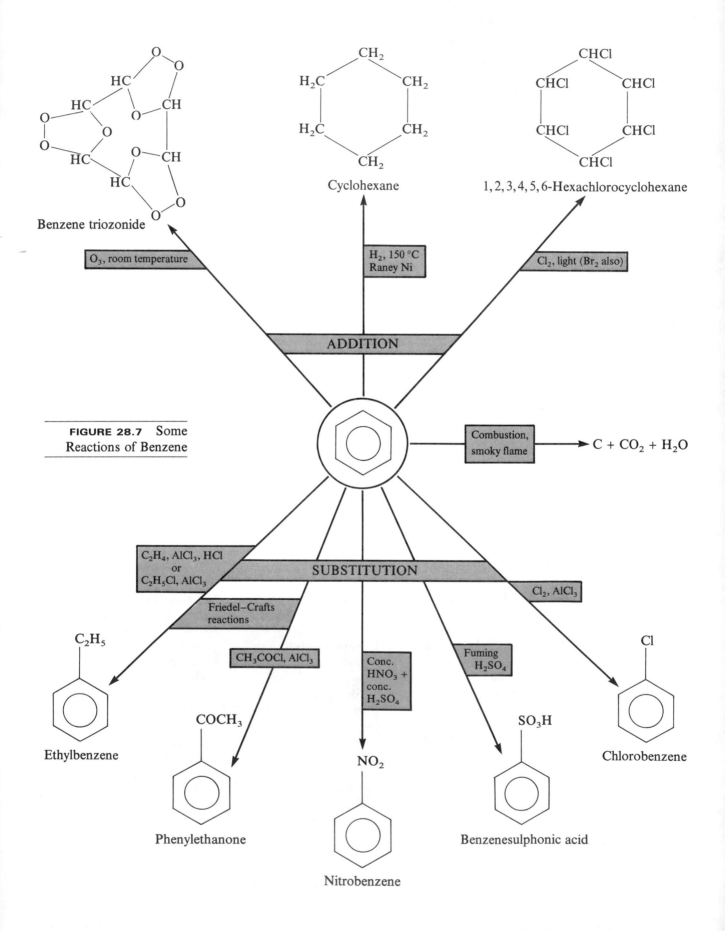

FIGURE 28.7 Some Reactions of Benzene

======== CHECKPOINT 28A: BENZENE ========

1. Explain why:

(*a*) 1 mole of benzene reacts with 3 moles of chlorine in the presence of ultraviolet light, without the formation of hydrogen chloride, and

(*b*) 1 mole of benzene reacts with 1 mole of chlorine in the presence of iron(III) chloride with the formation of hydrogen chloride.

2. Gammexane is an insecticide, with the formula

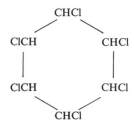

Refer to the boat and chair forms of cyclohexane [p. 537]. How many isomers are there of hexachlorocyclohexane?

3. Compare the reactions of benzene with those of cyclohexene.

4. What type of reagent attacks the benzene ring to form substitution compounds? State the attacking reagents which are involved in (*a*) the nitration of benzene, (*b*) the sulphonation of benzene, (*c*) the alkylation of benzene, and (*d*) the acylation of benzene.

5. What are the components of the 'nitrating mixture' used to make nitrobenzene from benzene? What is the nitrating agent that they produce? Give an equation for the formation of this nitrating agent from the nitrating mixture, and explain how benzene reacts with it.

6. Two substitution reactions of benzene require the use of concentrated sulphuric acid. What are these reactions? How does the reagent act in each case?

7. Name two reagents that react with both ethene and benzene. Give the structural formulae and names of the products.

8. Name two reactions in which benzene differs from ethene. Point out the structural resemblance between benzene and ethene and the reason for the difference in reactivity towards electrophiles.

9. Benzene reacts with propene in the presence of hydrogen chloride to form (1-methylethyl)benzene. Write the equation for the reaction. Write the formula of the carbocation formed when hydrogen chloride protonates propene. Write the formula of the transition state formed when benzene adds to the carbocation. This transition state loses a proton to become (1-methylethyl)benzene. What happens to the proton?

10. Before Kekulé proposed his formula for benzene, people tried out formulae such as

$$CH{\equiv}C{-}CH{=}CH{-}CH{=}CH_2$$

State two reactions which you would expect of a compound with this formula and which are not typical of benzene.

11. Long ago, the formula

$$CH_2{=}C{=}CH{-}CH{=}C{=}CH_2$$

was suggested for benzene. On the basis of this formula, how many isomers would you expect of the substitution products (*a*) C_6H_5X, (*b*) $C_6H_4X_2$, (*c*) $C_6H_3X_3$? How many isomeric substitution products with each of these formulae are formed by benzene?

28.9 METHYLBENZENE (TOLUENE): PHYSICAL PROPERTIES

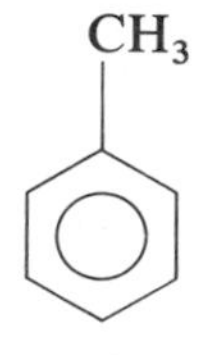

Methylbenzene, or toluene, $C_6H_5CH_3$, resembles benzene. Its physical properties are similar: the boiling temperature is higher (111 °C) and the melting temperature lower (−95 °C).

28.10 INDUSTRIAL SOURCE AND USES

Methylbenzene is used as a solvent and a petrol additive

Like benzene, methylbenzene is obtained from petroleum oil and from coal tar [see Figure 28.1, p. 577].

Methylbenzene is used as a solvent, as a source of the explosive trinitrotoluene (TNT), and as an additive to petrol, in which it improves the antiknock quality.

28.11 LABORATORY PREPARATIONS

Friedel–Crafts alkylation and a Fittig reaction

Methylbenzene is not often made in the laboratory as it is readily available. It can be made from benzene by a Friedel–Crafts alkylation [p. 582]. Another method is from a halogenobenzene by a Fittig reaction:

Bromo-   Iodomethane   Methylbenzene
benzene

28.12 REACTIONS OF THE RING

There are two sets of reactions, one set involving the aromatic ring in substitution or addition and the other involving reactions of the methyl group or 'side chain', as it is called.

28.12.1 SUBSTITUTION

Methylbenzene is more reactive than benzene towards electrophiles...

Methylbenzene is more reactive than benzene towards the electrophilic reagents which substitute in the benzene ring. This is because the methyl group pushes electrons into the ring [p. 593]. Milder conditions are employed than in the reactions of benzene. They are:

...Conditions for substitution are milder...

Chlorination/bromination: Cl_2/Br_2 with halogen carrier at 20 °C

Nitration: conc. HNO_3 + conc. H_2SO_4 at 30 °C

Sulphonation: fuming H_2SO_4 (containing SO_3) at 0 °C

Alkylation: RCl, $AlCl_3$ at 20 °C

Acylation: RCOCl or $(RCO)_2O$, $AlCl_3$, warm

A mixture of 1,2- and 1,4-substituted methylbenzenes is obtained in each case, e.g.

2-Nitromethylbenzene and 4-Nitromethylbenzene

If the temperature is raised, two groups or three groups are introduced, e.g.

...More than one group is substituted

2,4-Dinitromethylbenzene and 2,4,6-Trinitromethylbenzene (TNT)

28.12.2 ADDITION

Addition reactions are the same as those of benzene [see Figure 28.3, p. 579].

28.13 REACTIONS OF THE SIDE CHAIN

28.13.1 HALOGENATION

In sunlight, Cl_2 and Br_2 substitute the CH_3— group, not the ring

When chlorine is bubbled through boiling methylbenzene in strong sunlight or ultraviolet light, substitution occurs in the side chain:

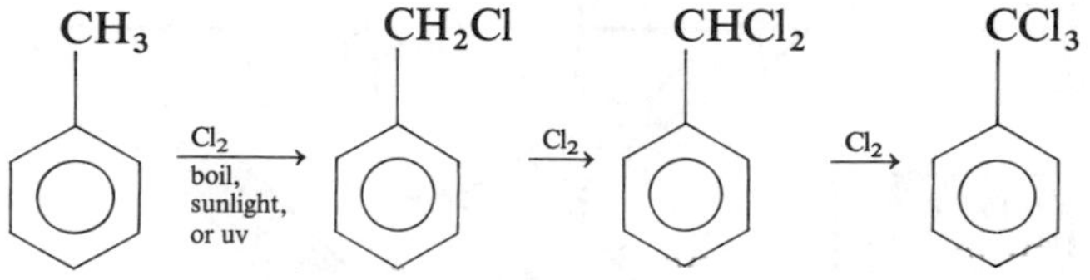

(Chloromethyl)benzene

(Dichloromethyl)benzene

(Trichloromethyl)benzene

Reaction involves free radicals, as does the halogenation of methane

Bromination occurs under the same conditions to give similar products. The conditions employed here favour the formation of free radicals, and the reaction proceeds in a similar way to the halogenation of alkanes [p. 546]. Note the difference between the product of free radical halogenation in the side chain and halogenation in the presence of a halogen carrier, when the halogen substitutes in the aromatic ring.

28.13.2 OXIDATION

Oxidation of —CH_3 gives —CHO or —CO_2H

The powerful oxidising agents acidified potassium manganate(VII) and acidified potassium dichromate(VI) will oxidise the side chain, —CH_3, to the carboxylic acid group, —CO_2H. The reaction mixture must be refluxed for several hours. A milder oxidising agent, manganese(IV) oxide, MnO_2, or chromium dichloride oxide, $CrOCl_2$, oxidises —CH_3 to the aldehyde group, —CHO:

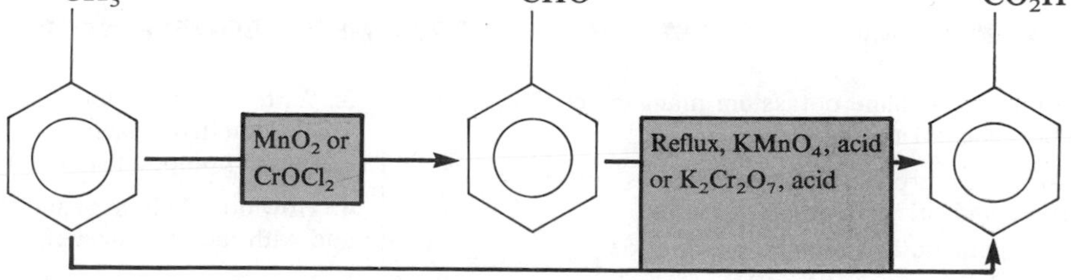

Benzenecarbaldehyde
(benzaldehyde)

Benzenecarboxylic acid
(benzoic acid)

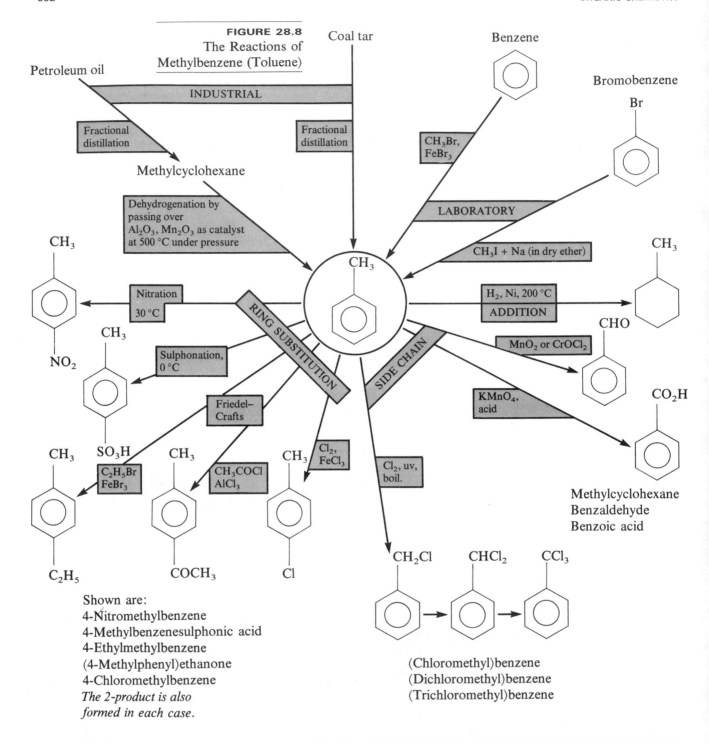

FIGURE 28.8
The Reactions of
Methylbenzene (Toluene)

Shown are:
4-Nitromethylbenzene
4-Methylbenzenesulphonic acid
4-Ethylmethylbenzene
(4-Methylphenyl)ethanone
4-Chloromethylbenzene
*The 2-product is also
formed in each case.*

Methylcyclohexane
Benzaldehyde
Benzoic acid

(Chloromethyl)benzene
(Dichloromethyl)benzene
(Trichloromethyl)benzene

CHECKPOINT 28B: METHYLBENZENE

1. Explain why alkaline potassium manganate(VII) attacks methylbenzene but not benzene.

2. What are the products of the reactions between
(i) benzene and (ii) methylbenzene with
(a) chloromethane and aluminium chloride
(b) ethanoyl chloride and iron(III) chloride
(c) concentrated nitric and concentrated sulphuric acid.

3. Write structural formulae for 2-chloromethylbenzene, 4-chloromethylbenzene and (chloromethyl)benzene. How can these compounds be made from methylbenzene?

4. How do (a) chlorine and (b) nitric acid react with benzene and with methylbenzene?

5. How is methylbenzene obtained (a) industrially and (b) in the laboratory, starting from benzene?

28.14 THE EFFECT OF SUBSTITUENT GROUPS ON FURTHER SUBSTITUTION IN THE BENZENE RING

28.14.1 THE INDUCTIVE EFFECT

Polarisation of a σ bond gives rise to an inductive effect

In a bond between carbon and a more electronegative element X the electron cloud will be denser at the X end of the bond than at the C end. The bond is polarised. This polarising effect of X is called an **inductive effect**. If X is more electronegative than carbon (e.g., F, Cl, Br), X is said to have a negative inductive ($-I$) effect. Alkyl groups are electron-donating, and are said to have a positive inductive ($+I$) effect [p. 258]:

$$C \rightarrow X \qquad\qquad C \leftarrow R$$

$$-I \text{ effect} \qquad\qquad +I \text{ effect}$$

28.14.2 THE MESOMERIC EFFECT

Polarisation of a π bond gives rise to a mesomeric effect

In a multiple bond, the π electrons shift towards the more electronegative of the bonded atoms. In a carbonyl group, the arrangement of electrons is in between the two structures

$$\diagdown C = O \qquad \text{and} \qquad \diagdown \overset{+}{C} - O^-$$

The actual structure is represented as

$$\diagdown C \overset{\frown}{=} O \qquad \text{or} \qquad \diagdown \overset{\delta+}{C} = \overset{\delta-}{O}$$

The shift in π electrons is called a **mesomeric effect**.

28.14.3 SUBSTITUENTS IN THE BENZENE RING

The reagents which attack the benzene ring are mainly electrophilic reagents (such as NO_2^+). If one of the hydrogen atoms in the benzene ring is replaced by a substituent group, the compound formed will differ in reactivity from benzene. If the substituent withdraws electrons, the compound will be less reactive than benzene.

A substituent with a $+I$ effect, such as CH_3—, donates electrons to the benzene ring. It is said to **activate** the ring. A substituent with a $-I$ effect, such as —NO_2, withdraws electrons and **deactivates** the ring:

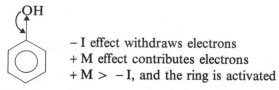

CH₃ — Electrons supplied to the ring. More easily attacked by NO_2^+.

NO₂ — Electrons withdrawn from the ring. Less easily attacked by NO_2^+.

The substituent —OH is more complicated. In addition to its $-I$ effect, withdrawing electrons from the ring, another effect operates. The oxygen atom has two lone pairs of electrons. It can feed these electrons into the π orbitals of the benzene ring by means of a mesomeric effect, which is shown by a curved arrow representing the movement of a pair of electrons:

OH

$-I$ effect withdraws electrons
$+M$ effect contributes electrons
$+M > -I$, and the ring is activated

The +M effect of the hydroxyl group is greater than the −I effect, and a hydroxyl group therefore activates the benzene ring.

28.14.4 FURTHER SUBSTITUTION IN THE BENZENE RING

An activating substituent
is ortho/para-*directing*

An activating substituent in the benzene ring increases the availability of electrons more in the *ortho-* and *para*-positions than in the *meta*-positions. A second substituent therefore enters in the *ortho-* or *para*-positions.

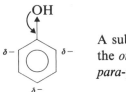

The *ortho-* and *para*-positions are activated more than the *meta*-positions.

A substituent enters in the *ortho-* or *para*-position.

...A deactivating group is
meta-directing

If a deactivating group is present in the ring, it withdraws electrons more from the *ortho-* and *para*-positions than from the *meta*-positions, and is therefore *meta*-directing.

The *ortho-* and *para*-positions are deactivated more than the *meta*-positions.

A substituent enters in the *meta*-position.

Halogens are exceptional. They are deactivating and *ortho/para*-directing.

Activating groups		*Deactivating groups*	
—R —OH —OR —NH₂ —NHR —NR₂	*Ortho/para*-directing. Note that only single bonds are present.	![structures]	*Meta*-directing. Note that these groups contain multiple bonds.
		—Halogen	*Ortho/para*-directing. Single bonds only

QUESTIONS ON CHAPTER 28

1. Describe a laboratory method for the preparation of nitrobenzene from benzene. Include: the reagents, the necessary conditions, the equation for the reaction and the method of purifying the product.

2. Under what conditions does benzene react with (*a*) chlorine, (*b*) chloroethane, (*c*) ethanoyl chloride and (*d*) sulphuric acid?

3. Discuss the mechanisms of the Friedel–Crafts acylation and alkylation of benzene.

4. Describe the reactions of (*a*) propene and (*b*) benzene with (i) H_2, (ii) Cl_2 and (iii) H_2SO_4. State the conditions required for reaction, and give equations.

Discuss the mechanisms of the reactions of chlorine with propene and with benzene.

5. List the reactions of methylbenzene which are (*a*) typical of benzene and (*b*) not shared by benzene.

*6. Explain the reason for the difference in the reactivity towards electrophiles of benzene and methylbenzene.

7. What type of reaction is most characteristic of benzene? Give, in the form of equations, *three* different examples of this type of reaction.

What type of catalyst is frequently employed in these reactions? Outline the mechanism of any *one* such catalysed reaction, indicating clearly how the catalyst functions.

Predict a possible mechanism for the reaction

$$C_6H_5Si(CH_3)_3 + Br_2 \rightarrow C_6H_5Br + (CH_3)_3SiBr.$$

(O & C, 82)

8. Give the name and formula of one example of an alkane, an alkene, an alkyne and an aromatic hydrocarbon.

Compare and contrast the reactions of the named compounds with (*a*) bromine, (*b*) potassium manganate(VII) (potassium permanganate), and (*c*) sulphuric acid.

Give a mechanism for (i) the reaction of the alkane with bromine, (ii) the reaction of the alkene with sulphuric acid, (iii) the reaction of the aromatic hydrocarbon with a nitrating mixture.

(L 81)

29

HALOGENOALKANES AND HALOGENOARENES

29.1 HALOGENOALKANES

The homologous series called **halogenoalkanes** (or haloalkanes) have the functional group

$$-C-X$$

where X = F, Cl, Br or I. Some members of the series are listed:

Names of some halogenoalkanes...

C_2H_5Cl — Chloroethane.
The name is derived from ethane, as the compound can be thought of as ethane, C_2H_6, with a Cl atom substituted for an H atom.

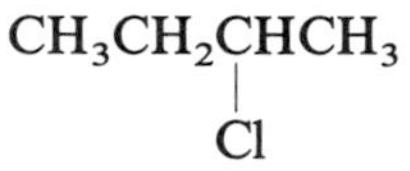

$CH_3CH_2CHCH_3$
|
Cl — 2-Chlorobutane.
The name is taken from that of the longest unbranched chain. The **locant**, 2, gives the position of the chlorine atom, counting from the end which gives the lower number for the locant.

...and halogenoalkenes

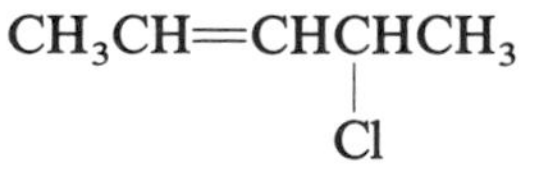

$CH_3CH{=}CHCHCH_3$
|
Cl — 4-Chloropent-2-ene.
The double bond is numbered first; then the chlorine atom.

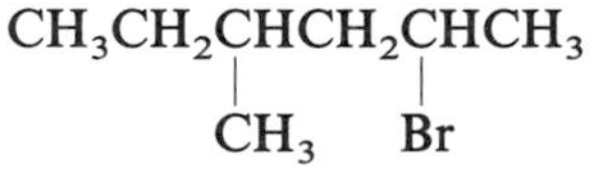

$CH_3CH_2CHCH_2CHCH_3$
| |
CH_3 Br — 2-Bromo-4-methylhexane.
The longest chain is 6C: the compound is a derivative of hexane. The substituents are placed in alphabetical order and then numbered.

CH_3CHCl_2 — 1,1-Dichloroethane

CH_2ClCH_2Cl — 1,2-Dichloroethane

$CHCl_3$ — Trichloromethane (chloroform)

CCl_4 — Tetrachloromethane (carbon tetrachloride)

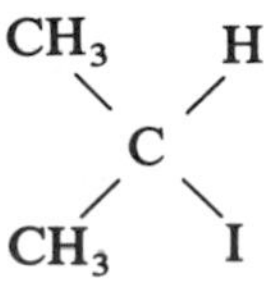

CH_3 H
  \   /
   C   — 2-Iodopropane
  /   \
CH_3 I

        CH_3
         \
CH_3— C —I   — 2-Iodo-2-methylpropane
         /
       CH_3

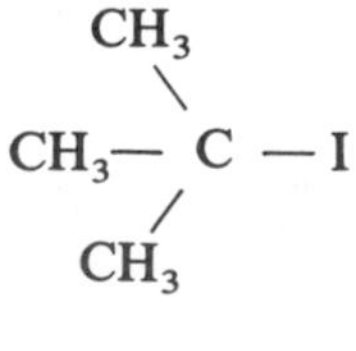

29.1.1 PRIMARY, SECONDARY AND TERTIARY HALOGENOALKANES

Primary, secondary and tertiary

There are three types of halogenoalkanes:

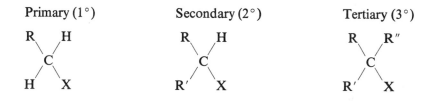

Primary (1°)   Secondary (2°)   Tertiary (3°)

29.2 PHYSICAL PROPERTIES

29.2.1 VOLATILITY

Fluoroalkanes and chloroalkanes have polar molecules...

Electronegativities are F = 4.0, Cl = 3.0, Br = 2.8, I = 2.5, C = 2.5. The strong polarity of C—F and C—Cl bonds gives rise to attraction between the dipoles in neighbouring molecules:

$$\overset{\delta+}{CH_3}—\overset{\delta-}{Cl} \cdots \overset{\delta+}{CH_3}—\overset{\delta-}{Cl} \cdots \overset{\delta+}{CH_3}—\overset{\delta-}{Cl}$$

...As a result, their boiling temperatures are higher than those of the corresponding alkanes...

Energy must be supplied to separate the molecules, and the boiling temperatures of fluoroalkanes and chloroalkanes are therefore higher than those of alkanes of similar molecular mass.

...This is not the case for bromoalkanes and iodoalkanes

One bromine atom has the same mass as six CH_2 groups. A bromoalkane therefore has a smaller volume than an alkane of the same molecular mass. A smaller volume means less interaction between molecules and a lower boiling temperature. The boiling temperatures of bromoalkanes and iodoalkanes are less than those of alkanes of the same molecular mass.

29.2.2 SOLUBILITY

Halogenoalkanes have a low solubility in water

The polar molecules can interact with water molecules, but the attractive forces set up are not as strong as the hydrogen bonds present in water. Halogenoalkanes therefore, although they dissolve more than alkanes, are only slightly soluble in water.

29.2.3 SMELL

These compounds have a sweet, slightly sickly smell.

29.2.4 DENSITY

Chloroalkanes are less dense than water; bromoalkanes and iodoalkanes are denser than water.

29.3 LABORATORY METHODS OF PREPARING HALOGENOALKANES

Preparative methods are summarised in Figures 29.1 and 29.2.

FIGURE 29.1
Laboratory Preparations
of Halogenoalkanes

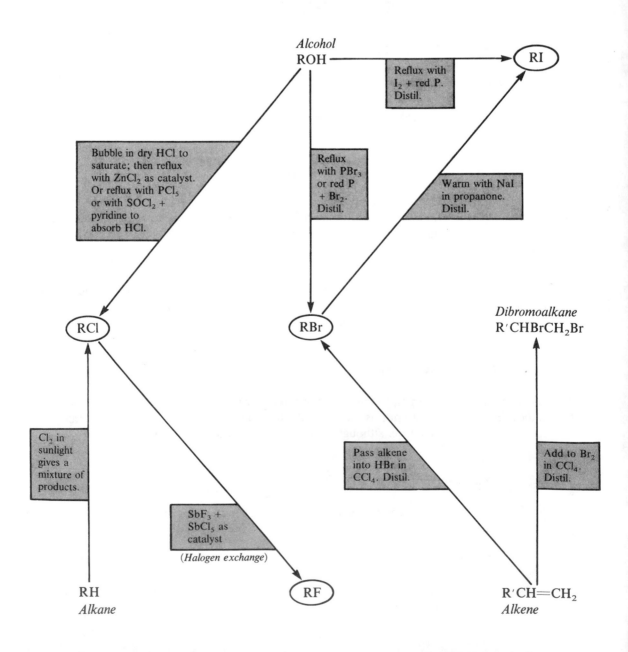

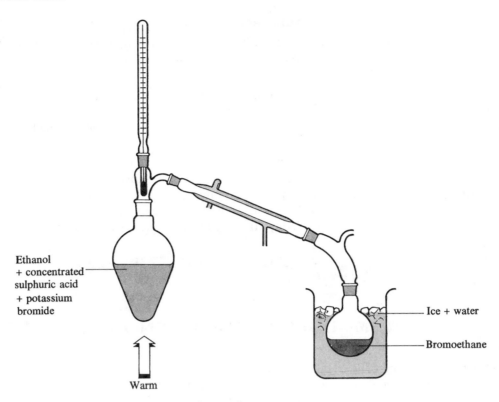

FIGURE 29.2
Preparation of
Bromoethane from
Ethanol

Ethanol
+ concentrated
sulphuric acid
+ potassium
bromide

Ice + water

Bromoethane

Warm

29.4 INDUSTRIAL MANUFACTURE

*Industrial manufacture
from Cl₂ + alkane...*

Industrially, chloroalkanes are the most readily manufactured as chlorine is available from the electrolysis of brine. The chlorination of alkanes yields a mixture of monochloroalkanes and polychloroalkanes, from which the components can be obtained by distillation. If the product is to be used as a solvent, the fact that it consists of a mixture of chloroalkanes does not matter:

$$C_2H_6(g) + Cl_2(g) \xrightarrow{light} C_2H_5Cl(g) + HCl(g) \xrightarrow{Cl_2} C_2H_4Cl_2(g) \xrightarrow{Cl_2}$$

Ethane   Chloroethane   Dichloroethane

The addition of hydrogen chloride to an alkene is the method used when a mono-chloroalkane is required:

*...and from
HCl + alkene...*

$$CH_3CH{=}CH_2(g) + HCl(g) \rightarrow CH_3CHCH_3(l)$$
$$\qquad\qquad\qquad\qquad\qquad\qquad\quad |$$
$$\qquad\qquad\qquad\qquad\qquad\qquad\ Cl$$

Propene   2-Chloropropane

Alkenes are obtained from the petrochemicals industry.

Alternatively, an alcohol can be vaporised and passed with hydrogen chloride over a heated catalyst:

*...and from
alcohol + HCl*

$$CH_3OH(g) + HCl(g) \xrightarrow[catalyst]{heat} CH_3Cl(g) + H_2O(g)$$

Methanol   Chloromethane

TETRACHLOROMETHANE

Manufacture of CCl₄   Tetrachloromethane, CCl_4, is made by passing chlorine into carbon disulphide in the presence of aluminium chloride as catalyst:

$$CS_2(l) + 3Cl_2(g) \xrightarrow{\text{AlCl}_3 \text{ catalyst}} CCl_4(l) + S_2Cl_2(l)$$

Tetrachloromethane

FLUORINATION

Fluorination is too difficult: RF is made from RCl   Fluorination of alkanes is not employed because fluorine and hydrogen fluoride, which is produced in the reaction, both attack metals and glass and are therefore difficult to work with. Fluoroalkanes are obtained from chloroalkanes by the action of antimony(III) fluoride in the presence of antimony(V) chloride:

$$3RCl + SbF_3 \xrightarrow{\text{SbCl}_5 \text{ catalyst}} 3RF + SbCl_3$$

29.5 USES OF HALOGENOALKANES

Halogenoalkanes are used as grease solvents...   Halogenoalkanes dissolve oil and grease. They are used in the dry-cleaning industry and also for cleaning articles which carry a film of oil or grease from the machinery used in their manufacture. Tetrachloromethane was once the prime halogenoalkane

...CCl₄ is toxic...   in this area. It is toxic as, when inhaled in quantity, it dissolves fat from the liver and kidneys. Less toxic are dichloromethane, CH_2Cl_2, trichloroethene, $CCl_2{=}CHCl$, called 'trichlor' in industry, and tetrachloroethene, $CCl_2{=}CCl_2$.

...Trichlor is safer   Halogenoalkanes which have boiling temperatures just below room temperature can easily be liquefied by a slight increase in pressure. This property makes them suitable

They are also used as refrigerant liquids   for use as the liquid in refrigerators. The most suitable refrigerator liquids are fluoro-compounds, called fluorocarbons for short, or 'freons'. Dichlorodifluoromethane, CCl_2F_2, and 1,2-dichloro-1,1,2,2,-tetrafluoroethane, $CClF_2CClF_2$, are used.

Being easily liquefied by pressure, chlorofluorohydrocarbons vaporise when the pressure is reduced and are suitable for use as propellant liquids in aerosol sprays. They are very unreactive towards most reagents. They reach the upper atmosphere before they find anything with which they will react. Here, they undergo a photochemical

Freons are stable and used in aerosol cans...   reaction with ozone, converting it to oxygen. The ozone layer performs a useful function in absorbing ultraviolet rays from the Sun. Without this absorption, we should receive too much ultraviolet radiation, and this would cause skin cancer.

...they react with ozone in the upper atmosphere   Some people believe that chlorofluorohydrocarbons in aerosol sprays are using up too much of the ozone layer, and endangering us from ultraviolet radiation. There is, as yet, no evidence of a reduction in the ozone layer.

Halogenoalkanes are also used as fire-extinguishers   Fully halogenated alkanes, being non-flammable, volatile and dense, are used in fire extinguishers. Tetrachloromethane, CCl_4, was used as a fire extinguisher, but, at high temperatures, there is a danger of its being oxidised to the poisonous gas phosgene, $COCl_2$. It also has a toxic effect on the liver and kidneys and has been replaced by the safer gas dibromochlorofluoromethane, CBr_2ClF, called 'BCF'.

29.6 REACTIONS

The reactions of halogenoalkanes are shown in Figure 29.3.

FIGURE 29.3 Reactions
of Halogenoalkanes

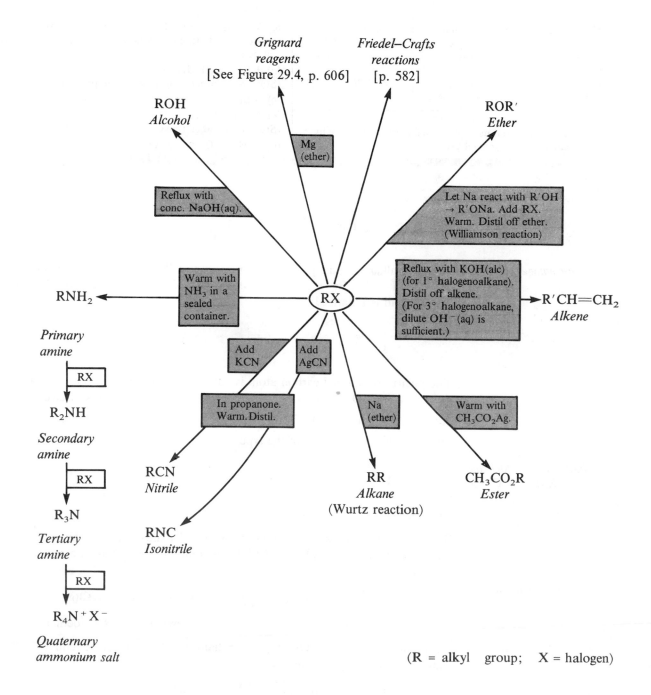

Grignard
reagents
[See Figure 29.4, p. 606]

Friedel–Crafts
reactions
[p. 582]

ROH
Alcohol

ROR′
Ether

Mg
(ether)

Reflux with
conc. NaOH(aq).

Let Na react with R′OH
→ R′ONa. Add RX.
Warm. Distil off ether.
(Williamson reaction)

Reflux with KOH(alc)
(for 1° halogenoalkane).
Distil off alkene.
(For 3° halogenoalkane,
dilute OH⁻(aq) is
sufficient.)

RNH$_2$

Warm with
NH$_3$ in a
sealed
container.

RX

R′CH=CH$_2$
Alkene

*Primary
amine*

RX

Add
KCN

Add
AgCN

R$_2$NH

In propanone.
Warm. Distil.

Na
(ether)

Warm with
CH$_3$CO$_2$Ag.

*Secondary
amine*

RX

R$_3$N

RCN
Nitrile

RR
Alkane
(Wurtz reaction)

CH$_3$CO$_2$R
Ester

*Tertiary
amine*

RX

RNC
Isonitrile

R$_4$N$^+$X$^-$

*Quaternary
ammonium salt*

(R = alkyl   group;   X = halogen)

====== CHECKPOINT 29A: HALOGENOALKANES ======

1. Name the following:

(a) $CH_3CH_2CHC(CH_3)_3$
$\quad\quad\quad\quad\;\; |$
$\quad\quad\quad\quad\;\; I$

(b) $(CH_3)_3CBr$

(c) $CH_3CH{=}CHC(Cl)(CH_3)_2$

2. Explain the differences between the boiling temperatures of the following compounds:

Compound	Molar mass/g mol^{-1}	Boiling temperature/°C
C_5H_{12}	72	36
C_3H_7Cl	79	46
C_3H_8	44	-42

3. State four industrial uses of the halogenoalkanes. Why do fluoroalkanes find special uses?

4. What reagents and what conditions are needed to bring about these reactions?
(a) $C_2H_5OH \rightarrow C_2H_5Cl$
(b) $C_3H_7OH \rightarrow C_3H_7I$
(c) $(CH_3)_2CHOH \rightarrow (CH_3)_2CHBr$
(d) $C_3H_7Br \rightarrow C_3H_7I$

5. Under what conditions does HCl(g) react with
(a) propanol and (b) propene? Give the names and formulae of the products.

6. Give the name of the product formed in the following reactions, and state the conditions under which reaction will occur:
(a) $KBr + C_3H_7OH \rightarrow$
(State what other reagent must be present.)
(b) $KBr + C_3H_7Cl \rightarrow$
(c) $I_2 + C_3H_7OH \rightarrow$
(State what other reagent must be present.)
(d) $C_3H_7I + C_2H_5CO_2Ag \rightarrow$

29.7 REACTIVITY

Why are halogenoalkanes reactive?

Halogenoalkanes undergo substitution reactions. The electronegativity values $C = 2.5$, $I = 2.5$, $Br = 2.8$, $Cl = 3.0$, $F = 4.0$ show that, except for C—I, the C—X bond is polarised:

The C—X bond is polarised . . .

$$\overset{\delta_+}{\underset{/}{\overset{\backslash}{C}}}{-}\overset{\delta_-}{X}$$

The positively charged carbon atom is susceptible to attack by nucleophiles, anions such as OH^- and CN^- and compounds with lone pairs of electrons such as $\text{:}NH_3$ and $H_2\overset{\cdot\cdot}{\underset{\cdot\cdot}{O}}$. The halide ion, X^-, makes a stable leaving group, and substitution reactions of the type

. . . Nucleophiles attack the carbon in C—X

$$Y\text{:} + {-}\overset{\delta_+}{\underset{/}{\overset{\backslash}{C}}}{-}\overset{\delta_-}{X} \rightarrow Y{-}\overset{/}{\underset{\backslash}{C}}{-} + X^-$$

therefore occur.

These substitution reactions are not very fast. To substitute —X by —NH_2, the halogenoalkane must be heated with 0.880 ammonia in a 'bomb', a sealed metal container:

Conditions needed for reaction

$$C_2H_5Cl(g) + NH_3(aq) \xrightarrow[\text{pressure develops}]{\text{heat in a 'bomb'}} C_2H_5NH_2(g) + HCl(g)$$

Chloroethane   Ethylamine or Aminoethane

To substitute —X by —OH, a primary halogenoalkane must be refluxed with aqueous alkali for about an hour:

$$C_2H_5Cl(g) + OH^-(aq) \xrightarrow{\text{reflux}} C_2H_5OH(g) + Cl^-(aq)$$

Chloroethane   Ethanol

A suitable apparatus is shown in Figure 28.4.

The ease of reaction depends on the ease of breaking the C—X bond. Average standard bond enthalpies in $kJ\,mol^{-1}$ are C—F = 484, C—Cl = 338, C—Br = 276, and C—I = 238. Thus, the ease of bond-breaking is

$$C—I > C—Br > C—Cl > C—F$$

The ease of reaction also depends on the stability of the *leaving group*. The ease of formation of X^- (aq) is

$$F^-(aq) > Cl^-(aq) > Br^-(aq) > I^-(aq)$$

The two factors are in opposition. The C—F bond is so strong that fluoroalkanes are extremely unreactive. The order of reactivity of the other halogenoalkanes towards nucleophiles is usually

$$RI > RBr > RCl$$

Elimination of HX with the formation of an alkene takes place in reactions with strong bases

Halogenoalkanes also undergo reactions in which hydrogen halide is eliminated and an alkene is formed. The nucleophiles which take part in substitution reactions (e.g., OH^-, CN^-) have basic properties. They are able to abstract a hydrogen atom from a halogenoalkane, while the halogen is expelled as halide ion. Thus, two reactions, substitution and elimination, are in competition:

$$RCH_2CH_2X + OH^-(aq) \rightarrow RCH_2CH_2OH + X^-(aq);\ Substitution$$

$$RCH_2CH_2X + OH^-(aq) \rightarrow RCH{=}CH_2 + H_2O + X^-(aq);\ Elimination$$

Substitution reactions are favoured by weakly basic nucleophiles, e.g., CN^-; elimination by strongly basic reagents, e.g., OH^-. A high concentration of base in a non-aqueous solvent (e.g., potassium hydroxide in ethanol) at a reflux temperature favours elimination.

The relative extents to which substitution and elimination take place depend on the structure of the molecule, as well as on the basic strength of the nucleophile. Elimination becomes progressively more important in the order

$$Primary < Secondary < Tertiary\ halogenoalkane$$

When a tertiary halogenoalkane reacts with a base, the main product is an alkene:

$$(CH_3)_3CCl(l) + OH^-(aq) \rightarrow (CH_3)_2C{=}CH_2(l) + H_2O(l) + Cl^-(aq)$$

Tertiary halogenoalkanes tend to undergo elimination reactions

To replace —X by —OH in a tertiary halogenoalkane, no base is needed:

$$(CH_3)_3CCl(l) + H_2O(l) \xrightarrow[\text{at 25 °C}]{\text{80% aqueous ethanol}} (CH_3)_3COH(l) + HCl(aq)$$

For a given aqueous base, the reactions which take place are
Primary halogenoalkane: Substitution
Secondary halogenoalkane: Substitution/Elimination
Tertiary halogenoalkane: Elimination.

1. How could the following reactions be carried out? State the necessary reagents and conditions:

(a) $C_4H_9OH \rightarrow C_4H_9Cl$

(b) $(C_2H_5)_3COH \rightarrow (C_2H_5)_3CCl$

(c) $C_3H_7I \rightarrow C_3H_7OH$

(d) $C_2H_5I \rightarrow C_2H_5OCH_3$

(e) $C_3H_7Br \rightarrow CH_3CO_2C_3H_7$

2. Explain why halogenoalkanes undergo substitution reactions (a) more readily than alkanes and (b) fairly slowly.

3. How can the following compounds be made from bromopropane?

(a) propylamine   (c) hexane

(b) propene   (d) butanonitrile

4. Give named examples of primary, secondary and tertiary halogenoalkanes.

29.8 THE MECHANISM OF HYDROLYSIS OF HALOGENOALKANES

Hydrolyses of halogenoalkanes are S_N reactions

The hydrolysis of a halogenoalkane is a **substitution** reaction. The attacking species is a **nucleophile**. Hydrolyses are therefore described as S_N **reactions**. A great deal of work has been done on the mechanisms of hydrolysis reactions.

29.8.1 HYDROLYSIS OF PRIMARY HALOGENOALKANES

EXPERIMENTAL EVIDENCE

The alkaline hydrolysis of a primary halogenoalkane

$$RX(l) + OH^-(aq) \rightarrow ROH(aq) + X^-(aq)$$

is second-order, following the rate expression

$$Rate = [RX][OH^-]$$

where $[RX]$ = concentration of RX.

Experimental evidence on the hydrolysis of primary halogenoalkanes

One can deduce from this evidence that a molecule of RX and an OH^- ion must collide before reaction will occur. Calculation shows that the rate of reaction is much less than the rate at which RX and OH^- collide. Only a small fraction of the collisions result in reaction. It must be necessary for the two species not only to collide but also to collide with enough energy to overcome the repulsion between the hydroxide ion and the halogenoalkane molecule.

MECHANISM

The mechanism...

...The making of a new bond eases the breaking of the old bond...

...A transition state is formed

It is known that, in order to minimise repulsion between RX and OH^-, OH^- approaches the carbon atom attached to X on the opposite side of the molecule from X:

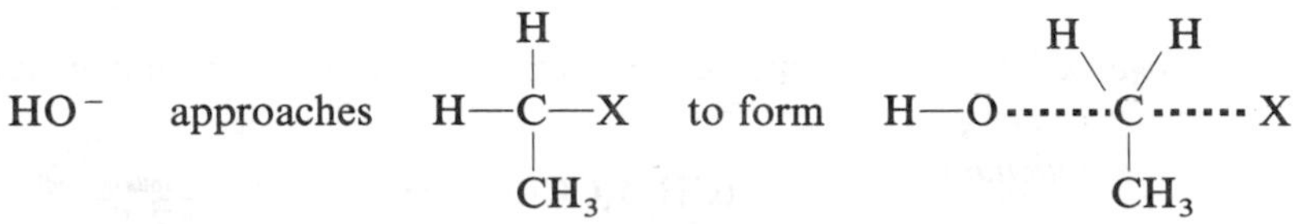

As a bond forms between O and C, the C—X bond weakens, and a transition state is reached in which C is partially bonded both to O and X. Once the transition state has been formed, it is rapidly converted into the products. A transition state has a momentary existence [p. 311]. The reactive intermediates that have been postulated in other reactions are longer-lived.

A summary of the mechanism

The proposed mechanism for the alkaline hydrolysis of primary halogenoalkanes is

$$HO^- + H-\underset{\underset{CH_3}{|}}{\overset{\overset{H}{|}}{C}}-X \xrightarrow[\text{step}]{\text{rate-determining}} \left[H-O\cdots\cdots\underset{\underset{CH_3}{|}}{\overset{\overset{H\quad H}{\diagdown\diagup}}{C}}\cdots\cdots X \right]^- \xrightarrow{\text{fast}} H-O-\underset{\underset{CH_3}{|}}{\overset{\overset{H}{|}}{C}}-H + X^-$$

Transition state

The movement of the bonding electrons is often shown by a curved arrow

$$HO^- \quad CH_2-I \rightarrow HO-CH_2 + I^-$$
$$\quad\quad\quad |\qquad\qquad\qquad |$$
$$\quad\quad CH_3\qquad\qquad CH_3$$

29.8.2 HYDROLYSIS OF TERTIARY HALOGENOALKANES

The hydrolysis of tertiary halogenoalkanes is first-order

An example of the hydrolysis of a tertiary halogenoalkane is the hydrolysis of 1,1-dimethylchloroethane in aqueous ethanol:

$$(CH_3)_3CCl(l) + H_2O(l) \rightarrow (CH_3)_3COH(l) + HCl(aq)$$

Experimental results show that the reaction is first order with respect to the halogenoalkane:

$$Rate = k[(CH_3)_3CCl]$$

It is inferred that the slow, rate-determining step must involve the halogenoalkane alone. It is suggested that this step is the dissociation of the halogenoalkane into a carbocation and a halide ion:

A carbocation intermediate is postulated

$$(CH_3)_3CCl \xrightarrow{\text{slow, rate-determining step}} (CH_3)_3C^+ + Cl^-$$

The carbocation formed reacts rapidly with water molecules:

$$(CH_3)_3C^+ + H_2O \xrightarrow{\text{fast step}} (CH_3)_3COH + H^+(aq)$$

This step is fast, and the rate of the overall reaction is determined by the rate at which the carbocations are formed.

A summary of the mechanism

The proposed mechanism for the hydrolysis of tertiary halogenoalkanes is

$$R_3CX \xrightarrow{\text{rate-determining step}} R_3C^+ + X^-$$

followed by $\quad R_3C^+ + H_2O \xrightarrow{\text{fast}} R_3COH + H^+(aq)$

29.8.3 HYDROLYSIS OF SECONDARY HALOGENOALKANES

Secondary halogenoalkanes

The behaviour of secondary halogenoalkanes is intermediate between primary and tertiary halogenoalkanes, depending on the nature of the halogenoalkane and the solvent.

29.8.4   *DESCRIPTION OF MECHANISM

S_N1 and S_N2 reactions

The hydrolyses of halogenoalkanes are S_N reactions, nucleophilic substitution reactions. In the hydrolysis of primary (and some secondary) halogenoalkanes, two species, RX and OH$^-$, are involved in the formation of the transition state: the rate-determining step is bimolecular. The reaction is described as an **S_N2 reaction**. In the hydrolysis of tertiary (and some secondary) halogenoalkanes, only one species is required for the formation of the transition state: the rate-determining step is unimolecular. Such reactions are designated as **S_N1 reactions**.

29.8.5   **RELATED REACTIONS**

Other reactions of halogenoalkanes involve S_N mechanisms

The same S_N mechanisms are found to operate in the attack of other nucleophiles on halogenoalkanes. Such reactions are those of (*a*) alkoxide ions, e.g., $C_2H_5O^-$, (*b*) cyanide ions, CN^-, (*c*) carboxylate ions, e.g., $CH_3CO_2^-$, (*d*) ammonia, $:NH_3$, and amines, e.g., $C_2H_5\overset{..}{N}H_2$.

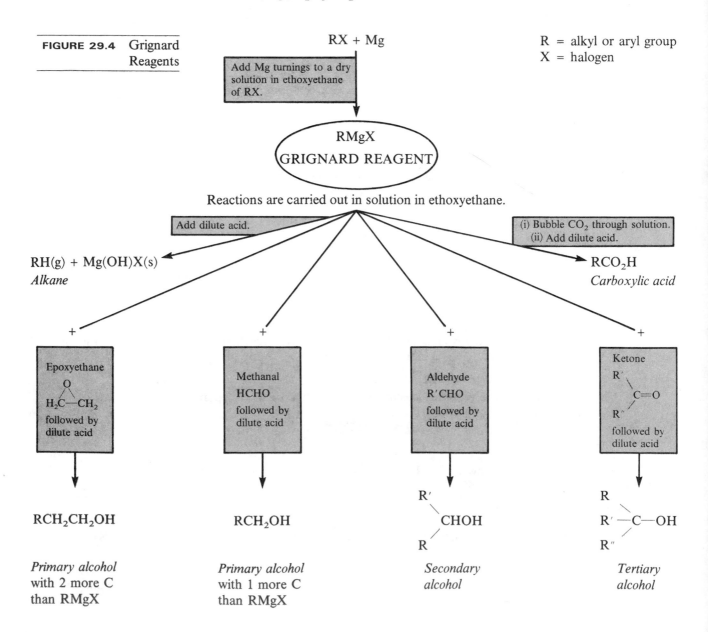

FIGURE 29.4   Grignard Reagents

29.9 *GRIGNARD REAGENTS

Grignard reagents, RMgX

When a dry ethereal solution of a halogenoalkane is added to magnesium turnings, an exothermic reaction occurs. The cloudy liquid formed is a solution of a Grignard reagent. These compounds have the formula R—Mg—X, where R is an alkyl group or an aryl group and X is a halogen. They undergo many reactions which enable them to be used as the starting point in the synthesis of a variety of compounds. Some of the reactions of Grignard reagents are shown in Figure 29.4.

*CHECKPOINT 29C: HALOGENOALKANES

1. Give the names and formulae of all the different alcohols that can be made from the Grignard reagent C_2H_5MgBr, provided that the following reagents are also available: epoxyethane, methanal, ethanal, propanone.

2. Sketch the transition state which is thought to be formed in the alkaline hydrolysis of 1-bromopropane. What leads a molecule of bromopropane to form this transition state? Why do hydroxide ions not attack alkanes?

What is the difference between an S_N1 reaction and an S_N2 reaction? Explain how the hydrolysis of 1-bromo-1,1-dimethylethane differs from that of 1-bromopropane in the manner in which the rate of the reaction depends on the concentrations of the reactants. What reason has been suggested to account for the difference?

29.10 HALOGENOALKENES

Halogenoalkenes are unreactive

The C—Cl bond is strong as it interacts with π electrons of C=C

Halogenoalkenes, e.g., chloroethene, $CH_2{=}CHCl$, are very unreactive. They are not hydrolysed by aqueous alkali or attacked by other nucleophiles. The reason is that the p orbitals of the doubly bonded carbon atoms overlap with the p orbitals of the chlorine atom to form a π bond [see Figure 29.5]. This makes the C—Cl bond much stronger than that in chloroalkanes, where there is only a σ bond between carbon and chlorine.

FIGURE 29.5
Bonding in Chloroethene

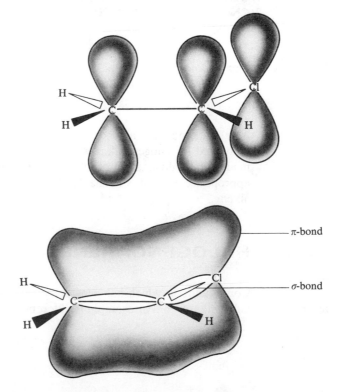

π-bond

σ-bond

The chemical stability of chloroalkenes and their solvent properties explain their use as degreasing agents. 'Trichlor', trichloroethene, $CHCl=CCl_2$, is much used as a degreasing agent.

Halogenoalkenes polymerise to give valuable plastics, for example, PVC and neoprene [p. 564].

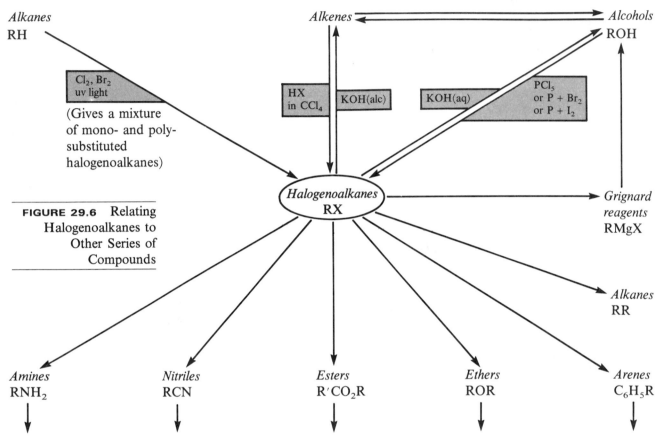

FIGURE 29.6 Relating Halogenoalkanes to Other Series of Compounds

All these compounds contain functional groups which enable them to react to form a variety of other compounds.

Note   Alkanes and alkenes are obtained from the petroleum industry. They can be converted to halogenoalkanes, from which a variety of different compounds can be made. The halogenation of alkanes gives a mixture of halogenoalkanes. If the appropriate alcohol is available, it is a more convenient source of a monohalogeno-alkane.

29.11 HALOGENOARENES

Halogenoarenes

Compounds in which halogens are substituents in the benzene ring are called halogenoarenes. An example is chlorobenzene:

29.12 PREPARATION

The methods of preparation of halogenoarenes are shown in Figure 29.8.

29.13 REACTIVITY OF HALOGENOARENES

The halogen atom in a halogenoarene is very much less reactive than that in a halogenoalkane. The reason is similar to the reason for the lack of reactivity of the halogenoalkenes. The p orbitals of the halogen atom interact with the p orbitals of the six carbon atoms in the benzene ring to form a delocalised cloud of π electrons [see Figure 29.7]. This π bond adds to the strength of the σ bond between the halogen atom and the ring, and makes the halogen atom very difficult to displace.

The C—X bond interacts with the ring...

...π bonding adds to the strength of the σ bond...

FIGURE 29.7 Electron Distribution in Halogenoarenes

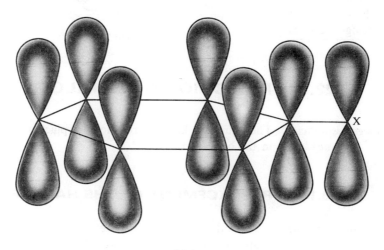

p orbital

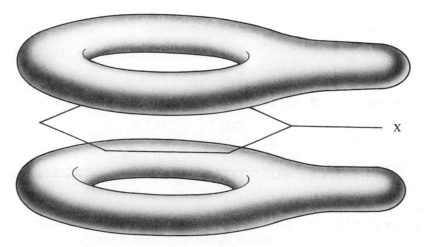

Delocalised cloud of π electrons

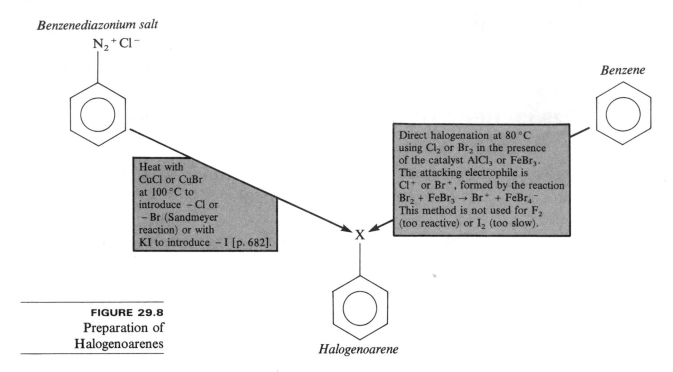

Benzenediazonium salt

$$N_2{}^+ Cl^-$$

Heat with CuCl or CuBr at 100 °C to introduce – Cl or – Br (Sandmeyer reaction) or with KI to introduce – I [p. 682].

Benzene

Direct halogenation at 80 °C using Cl_2 or Br_2 in the presence of the catalyst $AlCl_3$ or $FeBr_3$. The attacking electrophile is Cl^+ or Br^+, formed by the reaction $Br_2 + FeBr_3 \rightarrow Br^+ + FeBr_4{}^-$ This method is not used for F_2 (too reactive) or I_2 (too slow).

X

Halogenoarene

FIGURE 29.8
Preparation of
Halogenoarenes

29.14 REACTIONS OF HALOGENOARENES

...The halogen is difficult to displace

There are two kinds of reactions: replacement of the halogen atom and substitution in the benzene ring.

29.14.1 REPLACEMENT OF THE HALOGEN ATOM

Stringent conditions are needed for hydrolysis of halogenoarenes

The conditions needed for the replacement of a halogen atom are exemplified in the manufacture of phenol, C_6H_5OH. One industrial method is to heat chlorobenzene and aqueous sodium hydroxide together at 150 atm and 350 °C. The sodium phenoxide formed is converted into phenol by the action of dilute acid:

$$C_6H_5Cl(l) + 2NaOH(aq) \xrightarrow[350\,°C]{150\,atm} C_6H_5ONa(aq) + NaCl(aq) + H_2O(l)$$

Chlorobenzene   Sodium phenoxide

$$C_6H_5ONa(aq) + HCl(aq) \rightarrow C_6H_5OH(l) + NaOH(aq)$$

Sodium phenoxide   Phenol

X can be replaced by R

In the Fittig reaction [p. 590], a halogen is replaced by an alkyl group: e.g.

$$C_6H_5Cl \rightarrow C_6H_5CH_3$$

29.14.2 SUBSTITUTION IN THE BENZENE RING

Halogenoarenes are less reactive than benzene

The benzene ring undergoes the usual reactions [see Figure 29.9, p. 611]. Halogen substituents withdraw electrons, thus deactivating the benzene ring [p. 594] and are *ortho/para*-directing [p. 594]. A comparison of a halogenoalkane and a halogenoarene is made in Table 29.1, p. 612.

FIGURE 29.9 The Reactions of Chlorobenzene, a Typical Halogenoarene

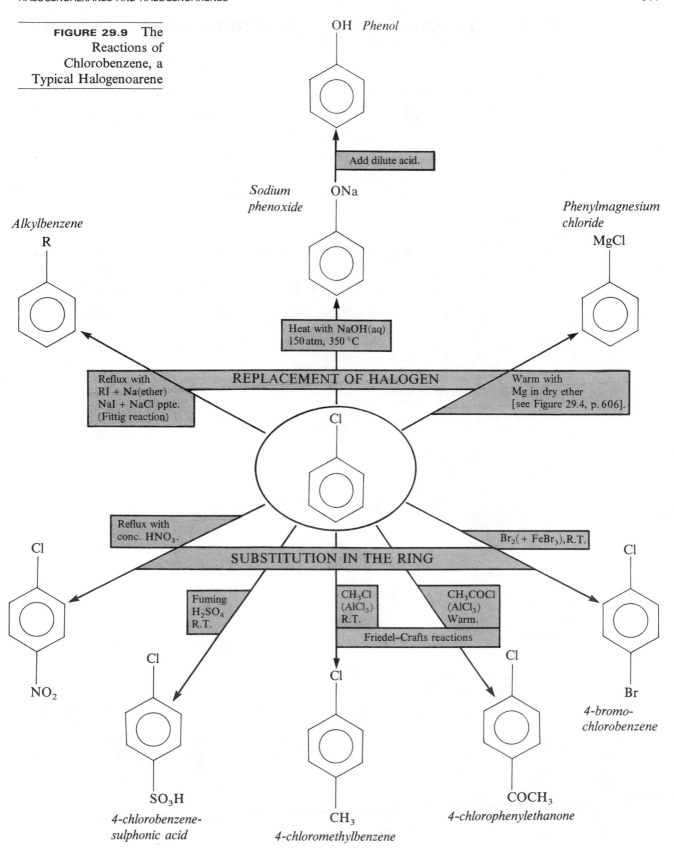

In each substitution reaction, a mixture of the 2- and 4-isomers (the *ortho-* and *para-*isomers) is formed.

29.15 COMPARISON OF HALOGENOALKANES AND HALOGENOARENES

	Halogenoalkanes, RX	Halogenoarenes, ArX
(1)	—OH replaces —X → Alcohol RX + NaOH(aq) → ROH + NaX	—ONa replaces —X : NaOH(conc.) 150 atm, 400 °C → ArONa
(2)	—OR′ replaces X → Ether (Williamson synthesis) RX + R′ONa → ROR′ + NaX	No such reaction
(3)	R′ replaces X → Alkane (Wurtz reaction) R′X + RX + 2Na $\xrightarrow{\text{dry ether}}$ RR′ + 2NaX	R replaces X → Arene (Fittig reaction) ArX + RI + 2Na → ArR + NaX + NaI
(4)	—MgX replaces X → Grignard reagent RX + Mg $\xrightarrow{\text{dry ether}}$ RMgX	—MgX replaces X → Grignard reagent ArX + Mg $\xrightarrow{\text{dry ether}}$ ArMgX
(5)	—CN replaces X → Nitrile RX + NaCN → RCN + NaX	
(6)	—NC replaces X → Isonitrile RX + AgCN → RNC + AgX	
(7)	—CO$_2$R′ replaces X → Ester RX + R′CO$_2$Ag → R′CO$_2$R + AgX	No comparable reactions
(8)	—NH$_2$ replaces X → Amine RX + NH$_3$(alc) → RNH$_3{}^+$ X$^-$	
(9)	Conversion to alkenes RCH$_2$CH$_2$X + KOH(alc) → RCH=CH$_2$	
(10)		Substitution in the aromatic ring Nitration: conc. HNO$_3$/H$_2$SO$_4$, R.T.
(11)		Sulphonation: fuming H$_2$SO$_4$, R.T.
(12)	No comparable reactions	Halogenation: X$_2$/FeX$_3$, R.T.
(13)		Alkylation: RCl/AlCl$_3$, R.T.
(14)		Acylation: RCOCl/AlCl$_3$, R.T.

TABLE 29.1
Comparison of
Halogenoalkanes
and Halogenoarenes

═══ CHECKPOINT 29D: HALOGENOARENES ═══

1. How does chlorobenzene differ in reactivity from benzene? Give two examples of reactions of the two compounds which illustrate the difference in reactivity. What explanation can you give for the difference?

2. What examples can you find of compounds C$_6$H$_5$X which can be made by the route

$$C_6H_6 \rightarrow C_6H_5Cl \rightarrow C_6H_5X$$

and which cannot be made by the direct route

$$C_6H_6 \rightarrow C_6H_5X?$$

3. Compare the rates of reaction of aqueous alkali with C$_6$H$_5$Br, with C$_3$H$_7$Br and with CH$_3$CH=CHBr. Give a reason for the differences.

1. State the conditions needed for bromine to react with

(a) ethane to form bromoethane

(b) benzene to form bromobenzene

(c) benzene to form hexabromocyclohexane

(d) methylbenzene to form (bromomethyl)benzene

(e) methylbenzene to form 2-bromomethylbenzene

Compare the reactions of (i) silver nitrate solution and (ii) aqueous alkali with the products in (a), (b) and (d).

2. Describe a laboratory preparation for bromoethane. Write an equation for the reaction. State what impurities may be present, and how you would purify a sample of bromoethane made in this way.

3. (a) Describe how you could make 2-bromobutane from a named alcohol.

(b) Give the structures and names of the isomers of 2-bromobutane. Compare the rates at which they react with aqueous sodium hydroxide.

(c) State the conditions under which 2-bromobutane reacts with the following reagents. Name the products, and write equations for the reactions.

(i) OH^-(aq)   (iii) KCN

(ii) KOH(ethanol)   (iv) NH_3

4. What chemical tests could you do to distinguish between the members of the following pairs of compounds?

(a) $CH_3CH_2CH_2CH_2Cl$ and $CH_3CH_2CH=CHCl$

(b)

5. (a) Show, by giving equations, conditions and reaction types, how the following conversions may be effected:

(i) iodoethane into ethanol,

(ii) 2-bromopropane into propene,

(iii) iodoethane into ethyl propanoate.

(b) Give *one* chemical test which would distinguish between the members of each of the following pairs of compounds:

(i) 1-chloropropane and ethanoyl (*acetyl*) chloride, [p. 702]

(ii) 2-chloropropane and 2-bromopropane,

(iii) 1-chloropropane and chlorobenzene.

In each case give the reagents and conditions for the reactions you describe.

See also Chapter 33.9, p. 702, for (b)(i).   (JMB 82)

6. (a) (i) What is the chemical nature of the principal constituents of a petroleum fraction boiling in the range 300–400 K and by what large scale industrial process is ethene obtained from such a fraction?

(ii) The step actually leading to the formation of ethene involves a reaction of the type:

$$CH_3CH_2\cdot + CH_3CH_2CH_2\cdot$$
$$\rightarrow CH_2=CH_2 + CH_3CH_2CH_3$$

By what type of bond cleavage will the ethyl and propyl radicals have been produced? Formulate a reaction mechanism for the interaction of these two radicals in the step shown above which accounts for the observed products and for the formation of two new covalent bonds and the fission of a third. How would an ethyl radical react with a chlorine molecule?

(b) How is chloroethene (vinyl chloride) prepared from ethene, and converted into poly(chloroethene), PVC. Poly(chloroethene), PVC, is an amorphous material, but poly(1,1-dichloroethene), prepared by an exactly analogous polymerisation method, is a crystalline material. What structural features in the two polymers might account for this difference?

(c) Chloroethene is now recognised as a cancer-producing agent, but poly(chloroethene) is harmless. What are the differences between the polymer and its monomer which could be important in explaining this difference?

(O 83, S)

*7. The hydrolysis of chloroethane C_2H_5Cl is a *nucleophilic, bimolecular* substitution reaction, whereas the hydrolysis of 2-chloro-2-methylpropane

is a *nucleophilic, unimolecular* reaction.

(a) Explain the terms in italics.

(b) Discuss the mechanisms of the two reactions.

(c) Describe in outline one experiment to show that the tertiary halogen compound hydrolyses at a faster rate than the primary.

(d) Give, and explain the results of, the reactions of each of the halogen compounds with sodium ethoxide in ethanol solution.

(L 81, S)

30

ALCOHOLS, PHENOLS AND ETHERS

30.1 ALCOHOLS

30.1.1 NOMENCLATURE

Aliphatic alcohols have the general formula $C_nH_{2n+1}OH$

Aliphatic alcohols are a homologous series. They all possess the same functional group, a hydroxyl group attached to a saturated carbon atom

$$\begin{array}{c} \diagdown \\ -C-OH \\ \diagup \end{array}$$

They all have the formula ROH, where R is an alkyl group

$$C_nH_{2n+1}-$$

They can be regarded as alkanes in which one H atom is replaced by a —OH group, and are often called the **alkanols**. The names are arrived at by adapting the name of the parent alkane by changing the terminal *-ane* to *-anol*. The first members of the series are:

Names

CH_3OH	Methanol	C_4H_9OH	Butanol
C_2H_5OH	Ethanol	$C_5H_{11}OH$	Pentanol
C_3H_7OH	Propanol	$C_6H_{13}OH$	Hexanol

FIGURE 30.1 Ethanol

'Alcohol'

Ethanol is by far the most important member of the series [see Figure 30.1], and is often referred to simply as 'alcohol'.

614

30.1.2 ISOMERISM

Primary, secondary and tertiary alcohols

Isomerism occurs. Primary (1°), secondary (2°) and tertiary (3°) alcohols exist. Their formulae are:

Primary (1°)   Secondary (2°)   Tertiary (3°) alcohols

The three isomeric butanols, of formula C_4H_9OH, are:

$$CH_3CH_2CH_2CH_2OH \qquad CH_3CH_2CHCH_3 \qquad (CH_3)_3COH$$
$$\qquad\qquad\qquad\qquad\qquad | $$
$$\qquad\qquad\qquad\qquad\qquad OH$$

Butan-1-ol (1°)   Butan-2-ol (2°)   2-Methylpropan-2-ol (3°)

30.1.3 POLYHYDRIC ALCOHOLS

Polyhydric alcohols contain more than one —OH group. Diols contain two, and triols contain three hydroxyl groups:

$$CH_2-CH_2 \qquad\qquad CH_2-CH-CH_2$$
$$|\quad\ \ | \qquad\qquad\qquad |\quad\ \ |\quad\ \ |$$
$$OH\quad OH \qquad\qquad OH\quad OH\quad OH$$

Ethane-1,2-diol   Propane-1,2,3-triol (glycerol)

30.1.4 ARYL ALCOHOLS

Aromatic alcohols or aryl alcohols contain a benzene ring. The simplest are:

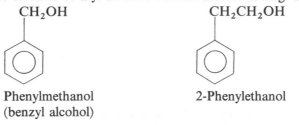

Phenylmethanol
(benzyl alcohol)

2-Phenylethanol

Aryl alcohols differ from phenols

The hydroxyl group is separated from the benzene ring by a saturated carbon atom and behaves as it does in aliphatic alcohols. If the hydroxyl group is in the ring, as in C_6H_5OH, the compound is not an alcohol: it is a **phenol**, and its reactions are quite different.

30.2 PHYSICAL PROPERTIES

30.2.1 VOLATILITY

Hydrogen bonding raises the boiling temperatures of alcohols...

Aliphatic alcohols with less than 12 carbon atoms and the lower aryl alcohols are liquids at room temperature. The boiling temperatures are higher than those of alkanes of comparable molecular mass (e.g., values of $T_b/°C$ are: $C_2H_5OH = 78°$ and $C_3H_8 = -42°$; $C_7H_{15}OH = 180°$ and $C_8H_{18} = 126°$. The reason is that the highly polar nature of the $\overset{\delta-}{—O}—\overset{\delta+}{H}$ bond leads to hydrogen bonding between molecules of alcohols [see Figure 30.2].

FIGURE 30.2 Hydrogen bonding between Alcohol Molecules

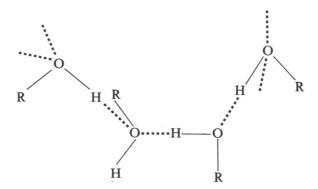

When the liquid is vaporised, energy must be supplied to break the hydrogen bonds and to convert the association of molecules into monomers in the vapour phase.

...as do an increasing molecular mass and branching

The boiling temperatures increase with increasing molecular mass, and branched-chain isomers have lower boiling temperatures than unbranched-chain isomers.

30.2.2 SOLUBILITY

Lower alcohols are soluble in water. Molecules of alcohol can displace molecules of water in the hydrogen-bonded association of water molecules.

FIGURE 30.3 Hydrogen Bonding in Aqueous Ethanol

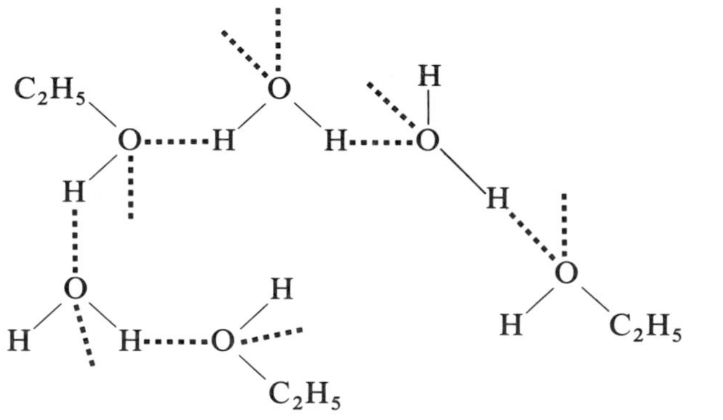

Alcohols of high molecular mass form fewer hydrogen bonds than the numerous water molecules they displace. Energy considerations therefore make the higher alcohols less soluble than the lower members.

30.2.3 AZEOTROPES

An azeotrope is formed by ethanol and water

Ethanol forms a constant-boiling azeotropic mixture with water [p. 163]. It contains 95.6% ethanol and boils at 78.1 °C. Pure ethanol can be obtained by distilling the azeotrope from a drying agent such as calcium oxide. Other alcohols also form azeotropes with water.

30.2.4 SOLVENT POWER

Alcohols dissolve polar solutes and non-polar solutes

Alcohols are good solvents. The polar —OH group enables them to dissolve sodium hydroxide and potassium hydroxide. The non-polar hydrocarbon part of the molecule enables alcohols to dissolve substances like hexane.

30.2.5 DENSITY

Aliphatic alcohols are less dense than water; aryl alcohols are slightly denser.

30.2.6 VISCOSITY

Ethanol is a low viscosity 'runny' liquid like water. The extensive hydrogen bonding increases the viscosity in polyhydric alcohols; glycerol, $CH_2OHCH(OH)CH_2OH$, is a very viscous liquid.

30.3 INDUSTRIAL SOURCES OF ALCOHOLS

30.3.1 HYDRATION OF ALKENES

Catalytic hydration of alkenes gives alcohols

Ethanol is manufactured by the hydration of ethene [p. 561]. Hydration of alkenes in the presence of a catalyst is used for the preparation of other alcohols too.

30.3.2 NATURAL GAS: A SOURCE OF METHANOL

Methanol is obtained from synthesis gas

Methanol can be manufactured from natural gas. First, methane from natural gas is passed with steam over a catalyst to convert it into a mixture of carbon monoxide and hydrogen known as *synthesis gas*:

$$CH_4(g) + H_2O(g) \xrightarrow[\text{Ni as catalyst}]{\text{900 °C, high pressure}} CO(g) + 3H_2(g)$$

Then synthesis gas is passed over another catalyst at high pressure and moderate temperature to give methanol:

$$2H_2(g) + CO(g) \xrightarrow[\text{Cr}_2\text{O}_3 + \text{ZnO as catalyst}]{\text{400 °C, high pressure}} CH_3OH(l)$$

30.3.3 FERMENTATION: A METHOD USED FOR ETHANOL

Ethanol is made by the fermentation of sugars

The fermentation method is used to make alcoholic drinks. Fruit juices such as grape juice contain the sugar glucose, $C_6H_{12}O_6$. When yeast is added, the sugar 'ferments' to form wine (a solution of ethanol) and carbon dioxide:

$$C_6H_{12}O_6(aq) \xrightarrow{\text{yeast}} 2C_2H_5OH(aq) + 2CO_2(g)$$

Glucose   Ethanol

Wines contain about 12% ethanol

Yeast is a living plant, containing the enzyme zymase, which catalyses the reaction. Juices of other fruits (plums, apples, pears, etc.) can be fermented to give wine. Wines contain about 12% ethanol. When the ethanol content reaches this level it kills the yeast, and fermentation stops.

Fractional distillation of fermented liquors gives 'spirits'

The starch present in potatoes and grain is also used as a source of ethanol. It is first hydrolysed enzymically to give glucose. Ethanol produced from fermentation has a concentration of 7–14% of ethanol by volume. Beer and cider are 7–8% ethanol, and wines are about 12% ethanol. Some people like drinks with a higher ethanol content. Whisky, gin and brandy have 40% ethanol. These 'spirits' are made by fractional distillation of fermented liquors [see Figure 8.10, p. 161 for fractional distillation].

Ethanol is the only alcohol which people drink, and it is referred to as 'alcohol'. It is a toxic substance if taken in large quantities. Methanol is much more toxic than ethanol: drinking methanol leads to blindness and then to death. Methanol is added to ethanol to give 'industrial methylated spirit', which is unfit to drink and therefore carries tax at a lower rate than ethanol. For domestic sale, industrial methylated spirit has some unpleasant-tasting substance added to it and also a purple dye as a warning. This liquid is referred to as 'meths'. Desperate alcoholics can be reduced to drinking meths; the effects prove fatal. The higher alcohols are unpleasant-tasting liquids of moderate toxicity.

'Meths' is a mixture of methanol and ethanol

People enjoy alcohol for its taste and for its relaxing effect. It is not a stimulant. Medically, ethanol is classified as a central depressant. It dulls the senses, and impairs the functions of the mind and body. This is why an activity such as driving a car, which requires rapid responses from the mind and the body, is very dangerous for someone who is intoxicated or 'under the influence of alcohol'.

Alcohol is a depressant

An old-fashioned way of measuring the ethanol concentration of spirits is to quote the 'degrees proof' (° proof). The measurement was made by pouring the spirit over gunpowder. If the gunpowder would not burn, the spirit was 'under-proof': it contained too much water. The spirit that would just allow gunpowder to burn was called 100° proof spirit. It is 60% ethanol. Whisky is 70° proof, i.e., 42% ethanol.

'Degrees proof' and the gunpowder test

30.4 LABORATORY METHODS OF PREPARATION

30.4.1 FROM HALOGENOALKANES

Alcohols can be prepared by hydrolysis of RX...

(a) Hydrolysis of a halogenoalkane, RX, gives the alcohol ROH [p. 601].

(b) *Grignard reagents, RMgX, afford a means of preparing alcohols [see Figure 29.4, p. 606].

30.4.2 FROM ESTERS

...and of esters, RCO_2R'...

When an ester, RCO_2R', is refluxed with dilute acid or alkali [p. 706], a carboxylic acid or its salt and an alcohol are formed:

$$CH_3CO_2C_2H_5(l) + NaOH(aq) \rightarrow CH_3CO_2Na(aq) + C_2H_5OH(l)$$

Ethyl ethanoate   Sodium   Ethanol
                                   ethanoate

30.4.3 HYDRATION OF ALKENES

...by the hydration of alkenes...

The alkene is bubbled into concentrated sulphuric acid to form an alkyl hydrogensulphate. When this is diluted with water and distilled, an alcohol is obtained:

$$CH_2{=}CH_2(g) + H_2SO_4(l) \rightarrow CH_3CH_2HSO_4(l)$$

$$CH_3CH_2HSO_4(l) + H_2O(l) \rightarrow CH_3CH_2OH(l) + H_2SO_4(l)$$

Ethyl hydrogensulphate   Ethanol

30.4.4 REDUCTION OF ALDEHYDES, KETONES AND CARBOXYLIC ACIDS

...by reduction of carbonyl compounds

The reduction of aldehydes and ketones is covered on p. 651. Carboxylic acids and acid chlorides and esters can be reduced, as described on p. 699, p. 702 and p. 705.

30.4.5 ARYL ALCOHOLS

The Cannizzaro reaction gives aryl alcohols

Aryl alcohols can be made from aryl aldehydes by the Cannizzaro reaction [p. 658].

FIGURE 30.4 Methods of Preparing Ethanol, a Typical Alcohol

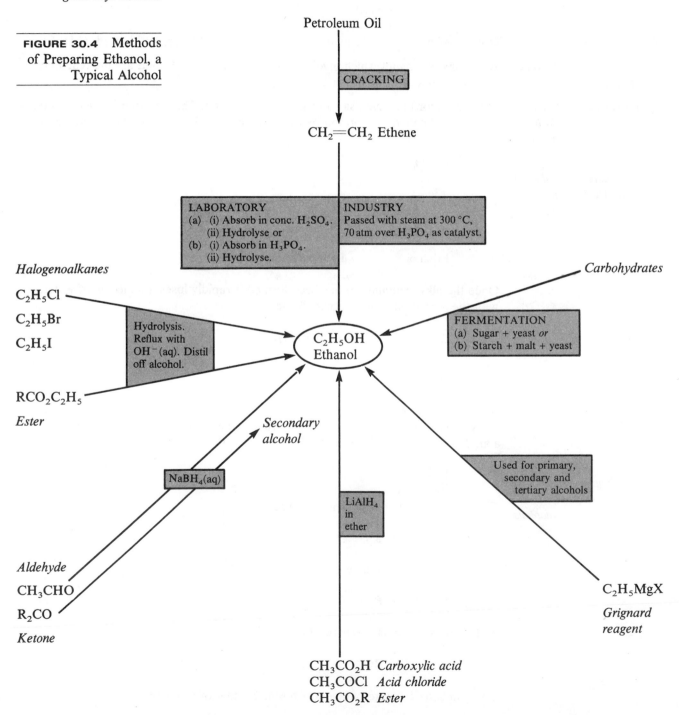

30.5 REACTIVITY OF ALCOHOLS

Alcohols are weak acids with very low values of K_a

In alcohols, the —O—H bond is polarised as

$$\overset{\delta-}{-O}\!-\!\overset{\delta+}{H}$$

There is a tendency for H^+ to dissociate in the presence of a base [p. 251]. An example of ethanol acting as an acid is its slight dissociation in water to form ethoxide ions and hydrogen ions:

$$C_2H_5OH + H_2O \rightleftharpoons C_2H_5O^- + H_3O^+$$

It is an even weaker acid than water, with $K_a = 10^{-16}\,mol\,dm^{-3}$ at 25 °C [p. 254].

RO—H bond fission is involved in some reactions

Reactions of ethanol which involve fission of the RO—H bond are the reactions with sodium and with carboxylic acids [p. 621].

R—OH bond fission is involved in other reactions . . .

Other reactions involve fission of the R—OH bond. The first step in such reactions is the coordination of a proton by the lone pairs of electrons on the oxygen atom in

$$R\!-\!\overset{\bullet\bullet}{\underset{\bullet\bullet}{O}}\!-\!H$$

. . . The bond is weakened by ROH acting as a base . . .

For example, in the reaction of hydrogen iodide with ethanol

$$C_2H_5OH + HI \rightleftharpoons C_2H_5\overset{+}{O}H_2 + I^-$$

Ethanol   Ethyloxonium ion

Once the alkyloxonium ion has been formed it rapidly loses a molecule of water, with the formation of a halogenoalkane:

$$C_2H_5\overset{+}{O}H_2 + I^- \rightarrow C_2H_5I + H_2O$$

Ethyloxonium   Iodoethane
iodide

The ability of the oxygen atom to accept a proton is greatest in tertiary alcohols. Alkyl groups are electron-donating, and reduce the positive charge on the alkyloxonium ion. The basic strengths of alcohols increase in the order:

$$1° < 2° < 3°$$

. . . The R—OH bond splits most readily in 3° alcohols

The acid strengths decrease in the order:

$$1° > 2° > 3°$$

Tertiary alcohols favour reactions of the type

$$R\!-\!OH \rightarrow R^+$$

and primary alcohols prefer reactions of the type

$$RO\!-\!H \rightarrow RO^-$$

The amphoteric nature of alcohols resembles that of water [p. 251].

30.6 REACTIONS OF ALCOHOLS

30.6.1 FISSION OF RO—H

REACTION WITH SODIUM

Alcohols react with sodium

Alcohols react with sodium to give hydrogen and a sodium alkoxide. The reaction is much slower than the reaction of water and sodium:

$$2C_2H_5OH(l) + 2Na(s) \rightarrow 2C_2H_5O^-Na^+(s) + H_2(g)$$

Ethanol   Sodium ethoxide

ESTERIFICATION

They react with carboxylic acids to give esters

Alcohols and carboxylic acids react to give esters. The functional groups of acids and esters are (where R is an alkyl group)

Carboxylic acid   Carboxylic ester

Esterification takes place much faster in the presence of a catalyst, such as hydrogen chloride or concentrated sulphuric acid. Acid anhydrides and acid chlorides also react with alcohols to give esters:

$$CH_3CO_2H(l) + C_2H_5OH(l) \underset{\text{conc. } H_2SO_4}{\overset{HCl(g) \text{ or}}{\rightleftharpoons}} CH_3CO_2C_2H_5(l) + H_2O(l)$$

Ethanoic   Ethanol   Ethyl ethanoate
acid

30.6.2 FISSION OF R—OH BOND

HALOGENATION

(a) Chlorination

Dry HCl(g) or conc. HCl replaces OH by Cl

Dry hydrogen chloride is bubbled through the anhydrous alcohol in the presence of anhydrous zinc chloride as catalyst. When the solution is saturated, it is refluxed on a water bath. For secondary and tertiary alcohols concentrated hydrochloric acid can be used. For tertiary alcohols no zinc chloride catalyst is needed:

$$C_2H_5OH(l) + HCl(g) \xrightarrow[\text{ZnCl}_2 \text{ catalyst}]{\text{reflux}} C_2H_5Cl(l) + H_2O(l)$$

The order of rates of reaction is

tertiary > secondary > primary alcohol

(b) Iodination

Red P + I$_2$ replace OH by I

Red phosphorus and iodine are used as a source of hydrogen iodide (concentrated sulphuric acid cannot be used because it oxidises HI to I$_2$):

$$C_2H_5OH(l) + HI(g) \xrightarrow{\text{distil}} C_2H_5I(l) + H_2O(l)$$

Ethanol   Iodoethane

(c) Halogenation using Phosphorus Halides

RBr and RI are formed by reaction with PBr$_3$ and PI$_3$

If it is available, phosphorus(III) bromide is used. Alternatively, red phosphorus and bromine are used to generate PBr$_3$ in the reaction mixture. Phosphorus(III) iodide is generated by the addition of iodine to red phosphorus and the alcohol. The mixture is refluxed on a water bath, and then the halogenoalkane is distilled over:

$$3C_2H_5OH(l) + PBr_3(l) \xrightarrow[\text{distil}]{\text{reflux}} 3C_2H_5Br(l) + H_3PO_3(l)$$

Ethanol   Bromo-   Phosphonic acid
                                 ethane

PCl$_5$ replaces OH by Cl

Phosphorus(V) chloride reacts in the cold with alcohols. Phosphorus(III) chloride cannot be used:

$$C_2H_5OH(l) + PCl_5(l) \xrightarrow{\text{room temperature}} C_2H_5Cl(l) + POCl_3(l) + HCl(g)$$

Ethanol   Chloroethane   Phosphorus
                                                            trichloride oxide

A test for the —OH group

The formation of hydrogen chloride on reaction with phosphorus(V) chloride is a test for the presence of a hydroxyl group in a compound.

(d) Chlorination with Sulphur Dichloride Oxide

SOCl$_2$ is a convenient chlorinating agent

A halogenoalkane is formed when an alcohol is refluxed with sulphur dichloride oxide, SOCl$_2$, in the presence of pyridine. This organic base assists the reaction by absorbing the hydrogen chloride as it is formed:

$$C_6H_5CH_2OH(l) + SOCl_2(l) \xrightarrow[\text{pyridine}]{\text{reflux with}} C_6H_5CH_2Cl(l) + HCl(g) + SO_2(g)$$

The convenient feature of this method is that the by-products, being gases, do not contaminate the product.

DEHYDRATION

Reaction with conc. H$_2$SO$_4$...

A primary alcohol reacts with cold concentrated sulphuric acid to form an alkyl hydrogensulphate:

$$C_2H_5OH(l) + HO{-}S(=O)(=O){-}OH(l) \rightarrow C_2H_5O{-}S(=O)(=O){-}OH(l) + H_2O(l)$$

Ethanol   Sulphuric acid   Ethyl hydrogensulphate

...results in dehydration to give an ether...

If an excess of alcohol is used and the reaction mixture is warmed to 140 °C, an ether is formed:

(excess alcohol, <170 °C)

$$C_2H_5OH(l) + C_2H_5HSO_4 \xrightarrow[\text{C}_2\text{H}_5\text{OH (in excess)}]{140\,°C} C_2H_5OC_2H_5(g) + H_2SO_4(l)$$

Ethoxyethane (or diethyl ether)

FIGURE 30.5
Dehydration of Ethanol
to Ethene

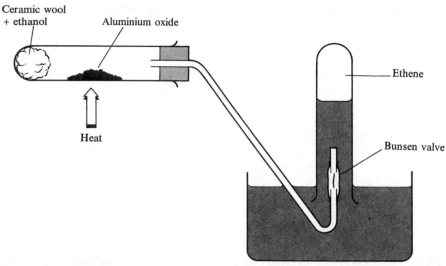

Ceramic wool + ethanol
Aluminium oxide
Ethene
Heat
Bunsen valve

Ethers possess the group

$$\equiv\!C\!-\!O\!-\!C\!\equiv$$

...or to give an alkene (excess conc H_2SO_4, >170 °C)

If an excess of concentrated sulphuric acid is used, and the temperature is raised to 170 °C, water is eliminated, with the formation of an alkene:

$$C_2H_5HSO_4(l) \xrightarrow[\text{conc. } H_2SO_4 \text{ (in excess)}]{170\,°C} C_2H_4(g) + H_2SO_4(l)$$

Ethyl hydrogensulphate   Ethene

Other dehydrating agents are Al_2O_3 and H_3PO_4

Phosphoric(V) acid and aluminium oxide at 300 °C [see Figure 30.5] also act as dehydrating agents. Secondary alcohols give a mixture of products. For example, butan-2-ol gives but-1-ene and but-2-ene:

$$2CH_3CH_2CHCH_3(l) \xrightarrow[300\,°C]{Al_2O_3} CH_3CH_2CH\!=\!CH_2 + CH_3CH\!=\!CHCH_3 + 2H_2O$$
$$\hspace{3.2em}|$$
$$\hspace{3.2em}OH$$

Butan-2-ol   But-1-ene   But-2-ene

The major product formed is the alkene with the highest number of alkyl groups, in this case, but-2-ene.

OXIDATION

Primary alcohols are oxidised to aldehydes and to acids...

Primary alcohols are oxidised to **aldehydes**, a series of compounds which have the functional group

$$R\!-\!C\overset{\displaystyle H}{\underset{\displaystyle O}{\big\langle}}$$

They can be further oxidised to carboxylic acids:

$$R\!-\!CH_2OH \xrightarrow{[-2H]} R\!-\!C\overset{\displaystyle H}{\underset{\displaystyle O}{\big\langle}} \xrightarrow{[+O]} R\!-\!C\overset{\displaystyle O\!-\!H}{\underset{\displaystyle O}{\big\langle}}$$

Primary   Aldehyde   Carboxylic acid
alcohol

...Secondary alcohols are oxidised to ketones...

Secondary alcohols are oxidised to ketones, a series of compounds possessing the functional group

$$\begin{array}{c} R \\ \diagdown \\ \diagup \\ R' \end{array} C{=}O$$

$$\begin{array}{c} R \quad H \\ \diagdown \diagup \\ C \\ \diagup \diagdown \\ R' \quad OH \end{array} \xrightarrow{[-2H]} \begin{array}{c} R \\ \diagdown \\ C{=}O \\ \diagup \\ R' \end{array}$$

Secondary   Ketone
alcohol

...Tertiary alcohols resist oxidation

Tertiary alcohols are resistant to oxidation. A powerful acidic oxidising agent converts them into a mixture of carboxylic acids.

A number of oxidising agents can be used. Acidified sodium dichromate solution at room temperature will oxidise primary alcohols to aldehydes and secondary alcohols to ketones. At higher temperatures primary alcohols are oxidised further to acids:

$$RCH_2OH \xrightarrow[\text{room temperature}]{Na_2Cr_2O_7,\, H^+\,(aq)} RC\!\!\begin{array}{c}H\\ \diagup \\ \diagdown\\ O\end{array} \xrightarrow[\text{50\,°C}]{Na_2Cr_2O_7,\, H^+\,(aq)} RC\!\!\begin{array}{c}OH\\ \diagup \\ \diagdown\\ O\end{array}$$

$$R_2CHOH \xrightarrow{Na_2Cr_2O_7,\, H^+\,(aq)} R_2C{=}O$$

Acid + dichromate are used in a breathalyser test

The dichromate solution turns from the orange colour of $Cr_2O_7{}^{2-}$ (aq) to the blue colour of Cr^{3+} (aq). This colour change is the basis for the 'breathalyser test'. The police can ask a motorist to exhale through a tube containing some orange crystals. If the crystals turn blue, it shows that the breath contains a considerable amount of ethanol vapour.

Other oxidising agents

Acid + KMnO₄...   (a) Acidified potassium manganate(VII) solution.

This is too powerful an oxidising agent to stop at the aldehyde: it oxidises primary alcohols to acids. It oxidises secondary alcohols to ketones.

...oxygen...   (b) Oxygen in the presence of silver as a catalyst [p. 648].

...copper catalyst...   (c) Copper as a catalyst for dehydrogenation [p. 648].

...and halogens   (d) Chlorine, bromine and iodine in the presence of alkali.

All secondary alcohols and one primary alcohol, ethanol, give a positive result in the haloform reaction [p. 657].
The reactions of alcohols are summarised in Figure 30.6 and Table 30.1.

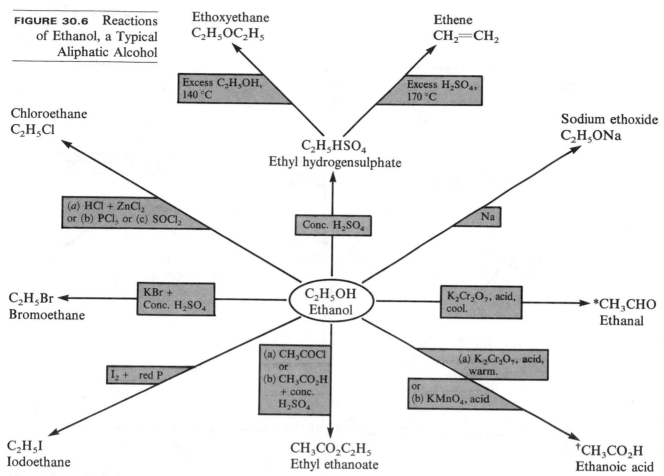

FIGURE 30.6 Reactions of Ethanol, a Typical Aliphatic Alcohol

*Secondary alcohols are oxidised to ketones.   †Tertiary alcohols are more difficult to oxidise. With powerful, acidic oxidising agents, they form a mixture of acids.

Reagent	Primary alcohol	Secondary alcohol	Tertiary alcohol
Acidified $K_2Cr_2O_7$ (orange)	Aldehyde, RCHO formed (blue Cr^{3+} (aq) formed)	Ketone, R_2CO formed (blue Cr^{3+} (aq) formed)	Resists oxidation
Conc. H_2SO_4	Alkene formed slowly	Intermediate in speed	Alkene formed fast
Conc. HCl + $ZnCl_2$. Add to alcohol and place in boiling water bath.	Cloudiness due to formation of RCl is slow to appear. (Anhydrous conditions are needed for primary alcohols.)	Cloudiness appears in 5 minutes	Cloudiness appears in 1 minute owing to the formation of RCl, which is insoluble in water.

TABLE 30.1 Methods of Distinguishing between Primary, Secondary and Tertiary Alcohols

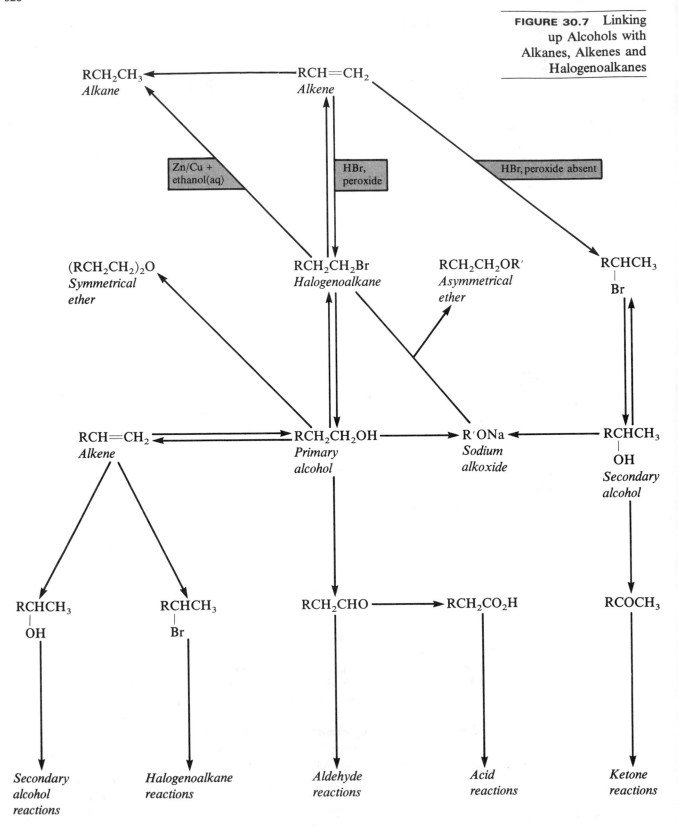

Note that the group —CH₂CH₂OH can be converted into a number of different functional groups. All these groups have characteristic reactions, to be studied in later chapters. By means of these reactions it is possible to make a large number of compounds in two stages, starting from alcohols.

CHECKPOINT 30A: ALCOHOLS

1. Name the reagents and describe the conditions needed to convert C_2H_5OH into

(a) C_2H_4

(d) C_2H_5Cl

(b) C_2H_5I

(e) $C_2H_5OCOCH_3$

(c) C_2H_5Br

(f) $C_2H_5OC_2H_5$

2. Name and give the structural formulae of the products formed when propan-1-ol reacts with

(a) Na

(d) conc. $H_2SO_4 > 170\,°C$

(b) PCl_5

(e) CH_3CO_2H

(c) conc. $H_2SO_4 < 170\,°C$

3. A $CH_3CH_2CH_2CH_2OH$ **C** $CH_3CH(CH_3)CH_2OH$
   B $CH_3CH_2CH(OH)CH_3$ **D** $(CH_3)_3COH$

Name these four alcohols, and explain how you could distinguish between

(a) **A** and **B**

(c) **C** and **D**

(b) **B** and **D**

4. Explain how you could synthesise from propan-1-ol

(a) propan-2-ol

(b) 1,2-dichloropropane

(c) propane-1,2-diol

5. Outline a way in which you could obtain propanonitrile, C_2H_5CN, starting from ethanol.

6. Give the names and structural formulae of a primary, a secondary and a tertiary alcohol, and explain how you could distinguish between them.

7. Compare the reactivity of (a) C_2H_5OH and H_2O, (b) C_2H_5OH and C_2H_5Cl, and (c) C_2H_5OH and $C_2H_5OC_2H_5$. Explain the reasons why the members of each pair of compounds react differently.

8. Compounds **A** and **B** are colourless liquids with the formula $C_4H_{10}O$. **A** reacts with sodium, with the formation of hydrogen. It reacts with conc. HI to form **C**, C_4H_9I, and with conc. H_2SO_4 to form **D**, C_4H_8. **B** does not react with sodium, but reacts with conc. HI to form **E**, C_2H_5I. Identify **A**, **B**, **C**, **D** and **E**, and write equations for all the reactions.

9. Identify the intermediate compound in each of the following syntheses. State the reagents and the conditions which are needed for each step in the synthesis:

$$C_2H_5OH \rightarrow A \rightarrow C_2H_5CN$$

$$C_2H_5OH \rightarrow B \rightarrow C_2H_5NH_2$$

$$CH_3CH(Br)CH_3 \rightarrow C \rightarrow CH_3COCH_3$$

$$CH_3CH{=}CH_2 \rightarrow D \rightarrow CH_3CH(OH)CH_3$$

$$(C_2H_5)_2O \rightarrow E \rightarrow F \rightarrow CH_3CO_2H$$

$$CH_3CH{=}CH_2 \rightarrow G \rightarrow CH_3CH_2CH_2OH$$

30.7 POLYHYDRIC ALCOHOLS

Ethane-1,2-diol is used as antifreeze

Ethane-1,2-diol, $HOCH_2CH_2OH$, is used in car radiators as antifreeze and as a de-icing fluid on aeroplane wings. It is often called by its old name of ethylene glycol. Its reactions are similar to those of monohydric alcohols.

Glycerol is a triol

Propane-1,2,3-triol is usually known by its old name of glycerol. It is a by-product in the manufacture of soap [p. 707]. Glycerol is used for the manufacture of the ester which it forms with nitric acid, propane-1,2,3-trinitrate or glyceryl trinitrate:

$$
\begin{array}{l}
CH_2OH \\
| \\
CHOH + 3HNO_3 \rightarrow \\
| \\
CH_2OH
\end{array}
\quad
\begin{array}{l}
CH_2ONO_2 \\
| \\
CHONO_2 \\
| \\
CH_2ONO_2
\end{array}
$$

Propane-1,2,3-triol   Propane-1,2,3-trinitrate
(*Glycerol*)   (*Glyceryl trinitrate*)

Nitroglycerine is the major component of dynamite

This ester is also called *nitroglycerine*. It is an explosive which is detonated by shock. When nitroglycerine is absorbed on kieselguhr (a type of clay), it is called *dynamite*. It is still a powerful explosive, but it is less sensitive to shock and can be handled with safety. Dynamite was invented by the Swedish chemist, A B Nobel, who invested some of the rewards for his invention to endow the Nobel prizes for great achievements in science and the arts.

30.8 ARYL ALCOHOLS

Benzyl alcohol is an aryl alcohol

The reactions of phenylmethanol (benzyl alcohol) are very similar to those of primary aliphatic alcohols. (See Figure 30.8.)

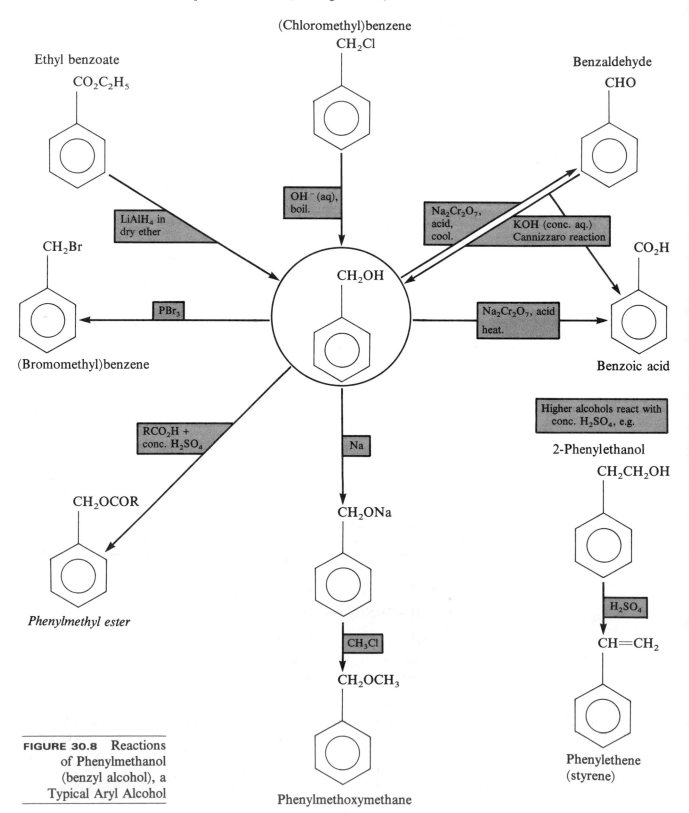

FIGURE 30.8 Reactions of Phenylmethanol (benzyl alcohol), a Typical Aryl Alcohol

CHECKPOINT 30B: ARYL ALCOHOLS

1. How would you distinguish between the following?

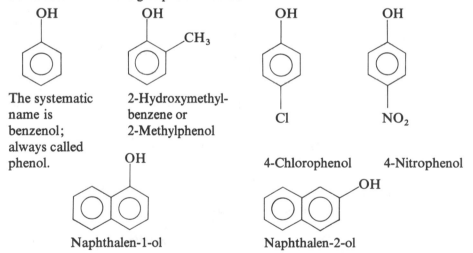

2. A liquid **P**, of formula $C_8H_{10}O$, reacted with sodium to give hydrogen. With concentrated sulphuric acid, **P** gave a liquid, **Q**, which was separated by distillation. When a trace of benzoyl peroxide (a source of free radicals) was added, **Q**, with a molar mass of $104\,g\,mol^{-1}$, was converted into a solid foam, **R**, with a very high molar mass. Identify **P**, **Q** and **R**, and explain the reactions which have taken place, giving equations.

3. In the synthetic route

Benzene → **A** → **B** → **C** → **D** →

identify **A**, **B**, **C** and **D**. State the reagents and conditions needed for each step in the route.

4. Identify **E** and **F** in the synthetic pathway shown, and state how each step may be carried out:

→ **E** → **F** →

30.9 PHENOLS

30.9.1 FUNCTIONAL GROUP

Phenols   Phenols are compounds containing a hydroxyl group attached to an aromatic ring. Some members of the group are listed below:

OH

The systematic name is benzenol; always called phenol.

2-Hydroxymethyl-benzene or 2-Methylphenol

4-Chlorophenol   4-Nitrophenol

Naphthalen-1-ol   Naphthalen-2-ol

30.9.2 PHYSICAL PROPERTIES

Phenol + water = the antiseptic, 'carbolic acid'

The vapour is toxic. Solid or liquid phenol burns the skin

Most phenols are colourless solids. Phenol itself melts at 42 °C. The presence of water lowers its melting temperature, and a mixture of phenol and water is a liquid at room temperature. It is called 'carbolic acid', and is a powerful antiseptic. A century ago, a surgeon called J Lister began to wonder why patients were dying some time after operations when the initial surgery appeared to have been satisfactory. He realised that their wounds were becoming infected both during the operation and afterwards in the ward. By spraying phenol mist around the operating theatre and the wards, he was able to reduce the death rate among his patients. Phenol is not a convenient antiseptic as the vapour is toxic, and if the solid or liquid comes into contact with the skin, it causes burns. Research workers have come up with a number of substituted phenols which are better at killing bacteria and less caustic to the skin. One of these is 2,4,6-trichlorophenol, which is present in TCP® and Dettol®.

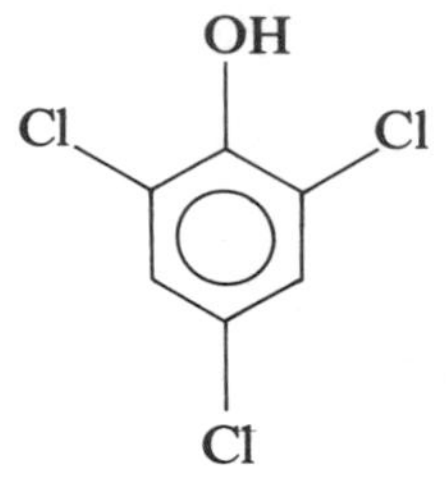

$$\text{(structure of 2,4,6-trichlorophenol: benzene ring with OH at top, Cl at positions 2, 4, 6)}$$

The presence of hydrogen bonds makes the melting temperatures of phenols higher than those of hydrocarbons of comparable molecular mass.

30.10 SOURCES OF PHENOL

30.10.1 PETROLEUM OIL

The major source of phenol is the petrochemical industry. The *cumene* process is used to make phenol. Benzene and propene are obtained from petroleum oil by cracking. They take part in the reactions shown below:

*Phenol is manufactured by
the 'cumene' process...*

$$C_6H_6 + CH_3CH{=}CH_2 \xrightarrow[\substack{\text{at 250 °C, 30 atm,}\\ \text{over } H_3PO_4 \text{ as}\\ \text{catalyst}}]{\text{vapours passed}} C_6H_5CH(CH_3)_2$$

Benzene   Propene   (1-Methylethyl)benzene
                                                                 'Cumene'

Oxidised by air
(+ catalyst)

$$C_6H_5OH + (CH_3)_2CO \xleftarrow{H_2SO_4(aq)} C_6H_5{-}\underset{\underset{CH_3}{|}}{\overset{\overset{CH_3}{|}}{C}}{-}O{-}OH$$

Phenol   Propanone   'Cumene' hydroperoxide

30.10.2 SODIUM BENZENESULPHONATE

*...from sodium
benzenesulphonate...*

Both in industry and in the laboratory, fusion of sodium benzenesulphonate with sodium hydroxide is used to make sodium phenoxide. This is dissolved in water and acidified to give phenol:

$$C_6H_5SO_3Na + 2NaOH(s) \xrightarrow{\text{fuse}} C_6H_5ONa(s) + Na_2SO_3(s) + H_2O(g)$$

Sodium benzenesulphonate   Sodium phenoxide

$$2C_6H_5ONa(aq) + CO_2(g) + H_2O(l) \rightarrow 2C_6H_5OH(l) + Na_2CO_3(aq)$$

Sodium phenoxide   Phenol

30.10.3 CHLOROBENZENE

...or from chlorobenzene

Under drastic conditions, 400 °C and 150 atm, aqueous sodium hydroxide will hydrolyse chlorobenzene. The product, sodium phenoxide, is converted into phenol by acidification:

$$C_6H_5Cl(l) + 2NaOH(aq) \xrightarrow{400\,°C,\ 150\,atm} C_6H_5ONa(aq) + NaCl(aq) + H_2O$$

Chlorobenzene   Sodium phenoxide

This is a method of preparation which is used industrially. The conditions are difficult to obtain in the laboratory. (You will remember from p. 609 why the hydrolysis of halogenoarenes is so difficult.)

30.10.4 DIAZONIUM COMPOUNDS

The laboratory preparation of phenol uses diazonium compounds

The easiest method to use in the laboratory is the hydrolysis of diazonium compounds [p. 682].

30.11 REACTIONS OF PHENOL

The reactions of phenol are of two kinds: (a) the reactions of the hydroxyl group and (b) substitution in the aromatic ring.

30.11.1 REACTIONS OF THE —OH GROUP

DISSOCIATION

Phenol dissociates in water:

$$C_6H_5OH(aq) + H_2O(l) \rightleftharpoons C_6H_5O^-(aq) + H_3O^+(aq)$$

Phenol is a weak acid...

Dissociation occurs to only a slight extent: phenol is a very weak acid, with $pK_a = 10.0$. The fact that it is an acid makes it dissolve more readily in sodium hydroxide solution than in water. It forms a solution of sodium phenoxide, $C_6H_5O^-\ Na^+$:

$$C_6H_5OH(l) + NaOH(aq) \rightarrow C_6H_5ONa(aq) + H_2O(l)$$

...weaker than H_2CO_3...

...but stronger than C_2H_5OH

Since phenol is a weaker acid than carbonic acid, it will not displace carbon dioxide from carbonates as carboxylic acids do. Phenols are stronger acids than alcohols. The ethoxide ion, $C_2H_5O^-$, readily accepts a proton to form C_2H_5OH: ethanol is a very weak acid, with $pK_a = 16.0$. In the phenoxide ion, a p orbital of the oxygen atom overlaps with the π orbital of the ring carbon atoms [see Figure 30.9]:

FIGURE 30.9
Interaction between the Charge on the Oxygen Atom and the Ring in Phenol

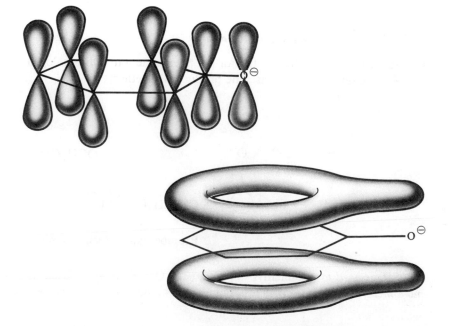

The phenoxide ion is stabilised by charge delocalisation

The negative charge is to some extent spread around the ring, and the —O⁻ group is less likely to accept a proton to form C_6H_5OH. The presence of electron-withdrawing groups in the ring attracts electrons away from the oxygen atom and so stabilises the phenoxide ion. This is why 4-chlorophenol is a stronger acid than phenol and why 2,4,6-trinitrophenol is a very much stronger acid than phenol, with $pK_a = 0.42$. Electron-donating groups, such as CH_3, in the ring decrease the acid strength of phenol:

Substituents affect the acid strength of phenol

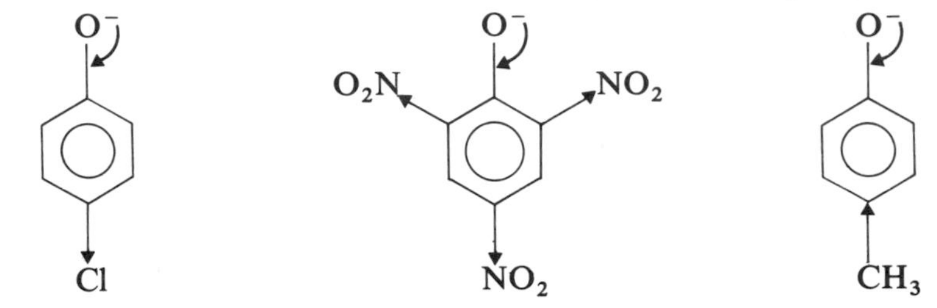

—Cl and —NO_2 assist the interaction of —O⁻ with the ring.

—CH_3 opposes the interaction of —O⁻ with the ring.

ESTERIFICATION

Before esterification, a phenol, ArOH, must be converted into the ion ArO⁻

Esterification occurs less readily than with alcohols. First, the phenol must be converted into the phenoxide ion; then it will react with acid chlorides and anhydrides:

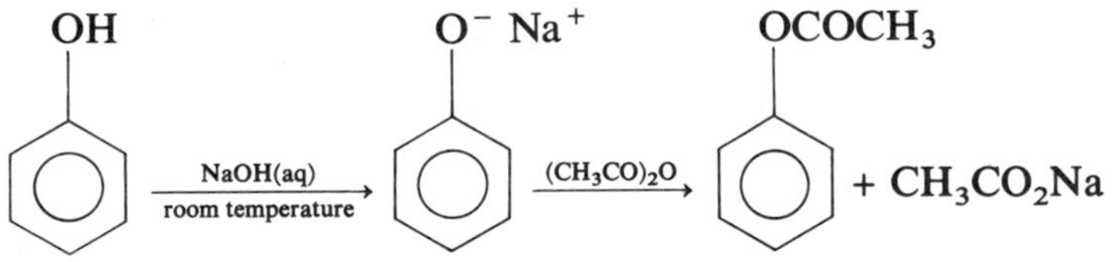

Phenol   Sodium phenoxide   Phenyl ethanoate

ETHER FORMATION

The reaction between ArO⁻ + RX gives an ether

The reaction of the phenoxide ion with a halogenoalkane is known as Williamson's synthesis. The product is an ether:

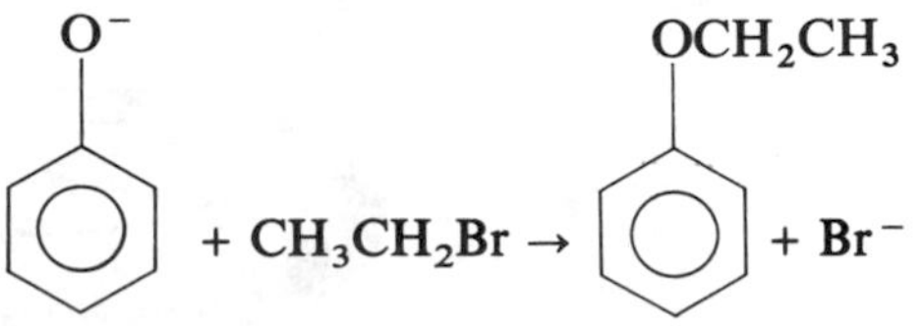

Bromoethane Phenoxyethane

$(CH_3)_2SO_4$ is used to make methyl ethers

Dimethyl sulphate, $(CH_3O)_2SO_2$, can be used as a methylating agent. It costs less than halogenomethanes. It must be used with extreme care because of its toxic nature.

REPLACEMENT BY HALOGENS

Reaction with PCl₅ is
slow

The replacement of —OH by a halogen takes place much less readily than with alcohols. Hydrogen halides do not react, and phosphorus(V) halides react slowly to give a poor yield of the halogenoarene:

$$C_6H_5OH(s) + PCl_5(s) \quad \rightarrow \quad C_6H_5Cl(l) + POCl_3(s) + HCl(g)$$

Phenol   Phosphorus(V)   Chloro-   Phosphorus
           chloride   benzene   trichloride oxide

REPLACEMENT OF —OH BY —H

Phenol can be converted
into benzene

When phenol is distilled with zinc dust, benzene distils over, leaving zinc oxide behind:

$$C_6H_5OH(s) + Zn(s) \xrightarrow{\text{distil}} C_6H_6(l) + ZnO(s)$$

Phenol   Benzene

30.11.2 SUBSTITUTION IN THE RING

Phenol is more reactive
than benzene towards
electrophiles...

Phenol is more reactive than benzene towards electrophilic reagents. The reason is that the interaction between the lone pairs on the oxygen atom in —OH or —O⁻ and the ring increases the availability of electrons in the aromatic ring [p. 593]. Milder conditions are employed in substitution reactions than are needed for benzene. These are:

NITRATION

...Dilute HNO₃ will
nitrate phenol...

Dilute nitric acid at room temperature:
(The reaction proceeds via nitrosation by NO^+ followed by oxidation.) When concentrated nitric acid is used, 2,4,6-trinitrophenol is formed. This is a powerful explosive known as 'picric acid'.

SULPHONATION

Concentrated sulphuric acid at room temperature.

HALOGENATION

...No halogen-carrier is
needed for halogenation
with Cl₂, Br₂...

Cl_2/Br_2, no halogen-carrier, at room temperature:
Chlorine, in the absence of solvent, gives 2- and 4-chlorophenol. Bromine, in a non-polar solvent such as carbon disulphide or tetrachloromethane, gives 2- and 4-bromophenol. Bromine water gives a precipitate of 2,4,6-tribromophenol. The faster reaction in water is due to the presence of phenoxide ions. In non-polar solvents, phenol molecules are the reacting species.

FRIEDEL—CRAFTS REACTIONS

...Alkylation uses an
alcohol or an alkene...

Alkylation: Alcohol or alkene in the presence of sulphuric acid as catalyst, on gentle warming:
A low yield is obtained. Halogenoalkanes give poorer results.

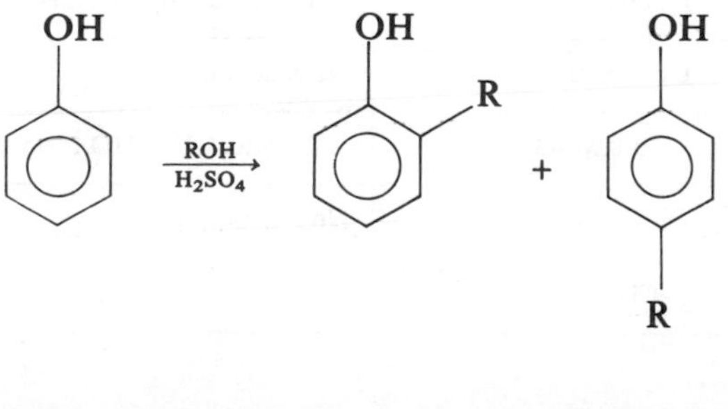

...Acylation uses the
Fries rearrangement

Acylation: Acid chloride or anhydride with Friedel–Crafts catalyst:
Alternatively, esterification followed by the Fries rearrangement can be used:

Phenyl ethanoate   2-Ethanoylphenol

4-Ethanoylphenol

REACTION WITH DIAZONIUM COMPOUNDS

See p. 684.

REACTION WITH METHANAL

See p. 656.

30.11.3   TEST FOR PHENOL

A test for phenol   In solution, phenol reacts with iron(III) chloride to form a violet complex. Other phenols give different colours, and this reaction is used as a test for phenols.

	Reagent	*Ethanol*, C_2H_5OH	*Phenol*, C_6H_5OH
		Liquid at room temperature. Characteristic smell.	Solid at room temperature. Characteristic smell.
(1)	Water	Miscible. Neutral solution	Partially miscible. Weakly acidic solution
(2)	Sodium	$C_2H_5ONa + H_2(g)$ formed	$C_6H_5ONa + H_2(g)$ formed
(3)	PCl_5	Vigorous reaction $\rightarrow$ HCl(g)	Slow reaction. Poor yield
(4)	CH_3COCl, $(CH_3CO)_2O$	Readily $\rightarrow$ ester	Readily $\rightarrow$ ester
(5)	C_6H_5COCl + NaOH	Readily $\rightarrow$ ester	Readily $\rightarrow$ ester
(6)	Organic acid	Ethyl ester formed	No reaction
(7)	HCl, HBr, HI	Halogenoethanes formed	No reaction
(8)	H_2, Ni catalyst	No addition	Addition $\rightarrow C_6H_{11}OH$
(9)	HNO_3(aq)	No reaction	$O_2NC_6H_4OH$ formed
(10)	Conc. H_2SO_4	Dehydration $\rightarrow$ ethene or ether	Sulphonation $\rightarrow HOC_6H_4SO_3H$
(11)	Br_2, H_2O	No reaction	White ppt. of $HOC_6H_2Br_3$
(12)	I_2, OH^-(aq)	CHI_3, yellow ppt. formed	No reaction
(13)	Distil with Zn.	No reaction	Benzene formed
(14)	Acid $KMnO_4$	Aldehyde, CH_3CHO formed	Mixture of oxidation products
(15)	Diazonium salt	No reaction	Yellow dye produced
(16)	$FeCl_3$	No colour produced	Violet colour produced

Table 30.2
Comparison of Ethanol
and Phenol

FIGURE 30.10
Reactions of Phenol

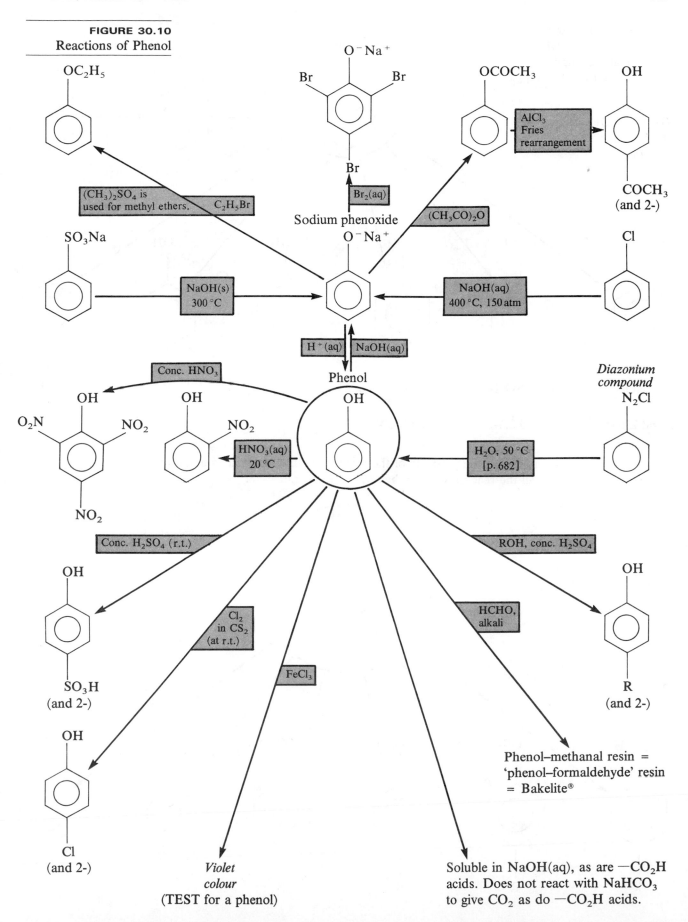

FIGURE 30.11 Some of
the Routes by which
Aromatic Compounds
are made from Phenol

(I) The —OH group
directs other substituents
into the ring.

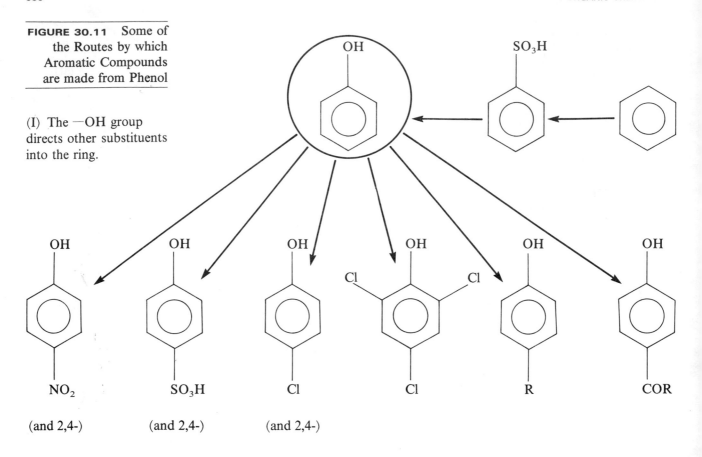

(and 2,4-)   (and 2,4-)   (and 2,4-)

Every one of the products of these reactions may be used as a starting point
for all the reactions shown below.

(II) The —OH group is
replaced by a different
group.

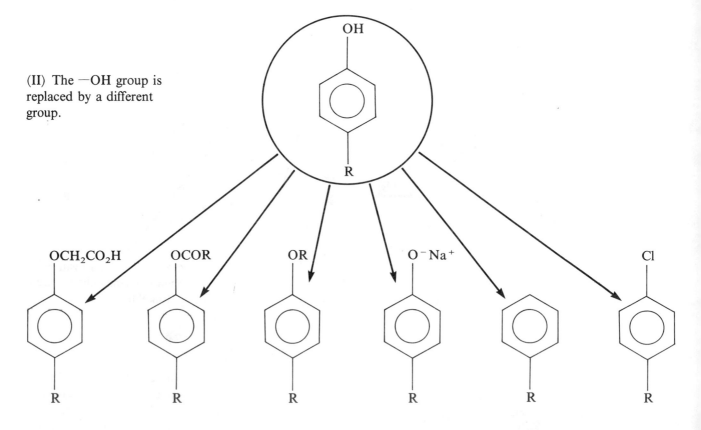

CHECKPOINT 30C: PHENOLS

†1. Copy the scheme shown in Figure 30.12 and fill in the necessary reagents and conditions. After reading Chapter 32, come back and fill in **X** and **Y**.

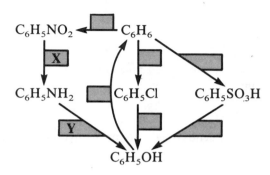

FIGURE 30.12

2. Compare the reactions of phenol and benzyl alcohol.

3. Starting from phenol, how could you make the following compounds?

(*a*)

(*b*)

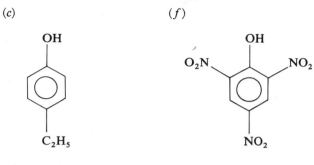

(*c*)

(*d*)

(*e*)

(*f*)

(*g*)

(*h*)

4. Explain why phenol is a stronger acid than ethanol and why 'picric acid', 2,4,6-trinitrophenol, is stronger still.

30.12 ETHERS

Saturated aliphatic ethers form a homologous series of formula

$$R—O—R'$$

where R and R′ are alkyl groups. Examples are:

Names of aliphatic ethers

CH_3OCH_3	Methoxymethane (or dimethyl ether)
$CH_3OC_2H_5$	Methoxyethane (or ethyl methyl ether)
$C_2H_5OC_2H_5$	Ethoxyethane (or diethyl ether)

If the two alkyl groups are the same, the ether is described as **symmetrical**; if R and R′ are different, the ether is said to be **unsymmetrical**. The examples show the method of naming: the ethers are named as alkoxy derivatives of alkanes. Ethoxyethane is well known as a solvent and as an anaesthetic, and it is often referred to as *'Ether'* simply 'ether'. Aliphatic ethers have the general formula

$$C_nH_{2n+2}O$$

They are isomeric with alcohols.

Epoxyethane is important in synthesis

The cyclic ether, epoxyethane

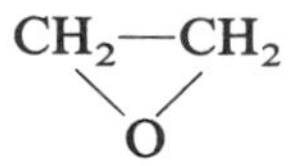

$$CH_2 \!-\! CH_2$$
$$\diagdown \quad \diagup$$
$$O$$

has important applications in synthesis [see Figure 29.4, p. 606].

Aromatic ethers have the formula

ArOR

where Ar is an aryl group, such as phenyl or substituted phenyl or naphthyl, and R is either an alkyl group or an aryl group. Examples are:

Aromatic ethers

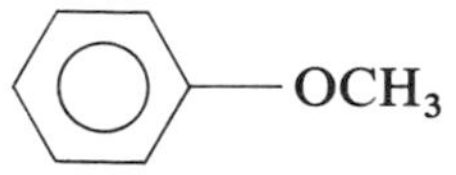

⬡—OCH₃   Phenoxymethane (or methyl phenyl ether)

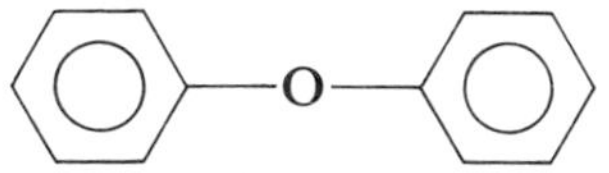

⬡—O—⬡   Phenoxybenzene (or diphenyl ether)

30.12.1 PHYSICAL PROPERTIES

VOLATILITY

Ethers are volatile, highly flammable...

The lower aliphatic ethers are gases or highly volatile liquids. Since ethers contain polar

$$\overset{\delta+}{C} \!-\! \overset{\delta-}{O} \text{ bonds}$$

the dipoles in neighbouring molecules interact, and the resulting attractive forces make the boiling temperatures of ethers a little higher than those of alkanes. Phenoxybenzene is a solid which melts at 28 °C. The vapours of ethers are highly flammable. As there is no possibility of hydrogen bonding, ether molecules are not associated, and the boiling temperatures are much lower than those of corresponding alcohols.

SOLUBILITY

...and only slightly soluble in water

The lower ethers are slightly soluble in water. The hydrogen bonds formed between water molecules and ether molecules are weak, and, as the hydrocarbon part of the molecule increases, solubility decreases. Ethers are important solvents for covalent substances.

DENSITY

There is danger of fire from the dense, flammable vapour

Ether vapours are denser than air. If ether is spilt in a laboratory the vapour, being denser than air, lingers at bench level and is therefore easily ignited. This is why chemists never handle ether anywhere near a flame.

ETHER EXTRACTION

Ether extraction is used to separate a covalent substance from water

The technique of **ether extraction** is used to obtain a covalent substance from solution in water. It depends on two factors:
(a) Covalent substances are more soluble in ethoxyethane than in water.
(b) Ethoxyethane and water are immiscible.

When an aqueous solution of a covalent substance is shaken with ethoxyethane in a separating funnel [see Figure 30.13], the covalent substance passes into the ether layer. This is separated and dried with, e.g., anhydrous sodium sulphate. The ethoxyethane is removed by distillation over a water-bath. Since it boils at 38 °C, this is easy to accomplish [for the theory, see p. 166].

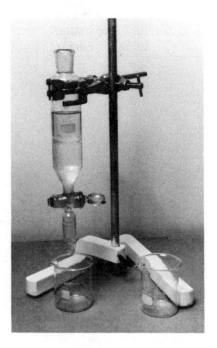

FIGURE 30.13
A Separating Funnel

30.12.2 INDUSTRIAL SOURCE OF ETHOXYETHANE

Ethoxyethane is manufactured from ethanol

The industrial method of manufacture of ethoxyethane is to pass ethanol vapour into a mixture of ethanol and concentrated sulphuric acid at 140 °C:

$$2C_2H_5OH(g) \xrightarrow[140\,°C]{\text{conc. } H_2SO_4} C_2H_5OC_2H_5(g) + H_2O(g)$$

Ethanol must be in excess, and the temperature must be kept below 170 °C; otherwise further dehydration to ethene occurs.

30.12.3 LABORATORY METHODS OF PREPARATION

The industrial method can be adopted in the laboratory [see Figure 30.14].

30.12.4 WILLIAMSON'S SYNTHESIS

RONa + R'X give unsymmetrical ethers

This method will give unsymmetrical ethers, ROR′, as well as symmetrical ethers, ROR. The sodium derivative of an alcohol or phenol is boiled with a halogenoalkane until sodium halide crystals appear. Then the mixture is distilled to give an ether:

$$C_2H_5ONa(alc) + C_3H_7Br(alc) \rightarrow C_2H_5OC_3H_7(g) + NaBr(s)$$

Sodium ethoxide	1-Bromopropane	
		Ethoxypropane

30.12.5 REACTIONS OF ETHERS

Ethers are unreactive...

Ethers are unreactive. They burn readily, on account of their high volatility, to form carbon dioxide and water.

FIGURE 30.14
Laboratory Preparation
of Ethoxyethane

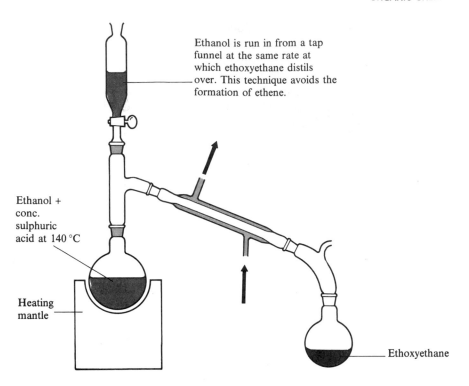

Ethanol is run in from a tap funnel at the same rate at which ethoxyethane distils over. This technique avoids the formation of ethene.

Ethanol + conc. sulphuric acid at 140 °C

Heating mantle

Ethoxyethane

The oxygen atom carries two lone pairs of electrons, which allow it to be protonated to form an oxonium ion:

$$R\text{—}\overset{\cdot\cdot}{\underset{\cdot\cdot}{O}}\text{—}R + HA \longrightarrow R\text{—}\overset{+}{\underset{\underset{\displaystyle H}{|}}{\overset{\cdot\cdot}{O}}}\text{—}R + A^-$$

...but are protonated by strong acids...

The oxonium ion can be attacked by nucleophiles. This is what happens when an aliphatic ether is refluxed with an excess of hydrogen iodide:

$$R\text{—}\overset{\cdot\cdot}{\underset{\cdot\cdot}{O}}\text{—}R' \xrightarrow[\text{Reflux}]{\text{excess HI}} R\text{—}\overset{+}{\underset{\underset{\displaystyle H}{|}}{\overset{\cdot\cdot}{O}}}\text{—}R' \xrightarrow{I^-} RI + R'OH$$

...they are split by the action of HI on refluxing

The alcohol R'OH reacts immediately with more HI:

$$R'OH \xrightarrow{HI} R'\overset{+}{O}H_2 \xrightarrow{I^-} R'I + H_2O$$

The net result is the formation of two iodoalkanes:

$$R\text{—}O\text{—}R' + 2HI \rightarrow RI + R'I + H_2O$$

The other hydrogen halides, HCl and HBr, can take the place of HI, but are less effective.

Aryl alkyl ethers react with boiling hydriodic acid. The products are a phenol and an iodoalkane:

$$ArOR + HI \rightarrow ArOH + RI$$

In the presence of air and sunlight, ethers form peroxides. These compounds are unstable and explode if heated. To avoid their formation, ethers are stored in brown, airtight bottles.

FIGURE 30.15
Reactions and
Preparations of Ethers

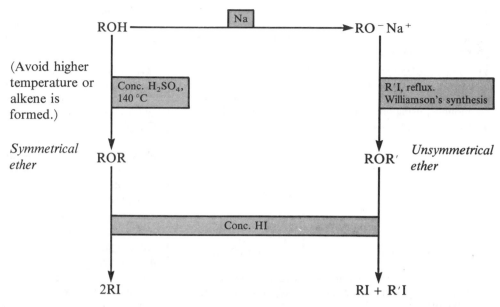

(Avoid higher
temperature or
alkene is
formed.)

*Symmetrical
ether*

*Unsymmetrical
ether*

═══════════════ CHECKPOINT 30D: ETHERS ═══════════════

1. Explain why ethoxyethane is (*a*) more volatile than ethanol, (*b*) less dense than water and (*c*) a better solvent than water for organic compounds.

2. Say how you could prepare the following from propan-1-ol:
(*a*) $CH_3CH_2CH_2OCH_2CH_2CH_3$
(*b*) $CH_3CH_2CH_2OCH_2CH_3$

3. Give the name and formula of the ether formed as a result of the reaction between concentrated sulphuric acid and butan-2-ol. What other substance is formed? What precautions are taken to ensure a good yield of the ether?

4. (*a*) State the reagents you would need and the conditions you would employ to make phenoxyethane, $C_6H_5OC_2H_5$.
(*b*) Give one reaction of phenoxyethane.

5. How could you make the ether $C_6H_5CH_2OCH_3$, starting from benzene and employing common laboratory reagents?

═══════════════ QUESTIONS ON CHAPTER 30 ═══════════════

1. Explain why ethanol dissolves in water, whereas ethane does not.

2. Outline an industrial method for the manufacture of ethanol. What products are formed when ethanol reacts with (*a*) concentrated sulphuric acid and (*b*) ethanoic acid? State the conditions under which the reactions occur.

3. Compare and contrast the reactions of phenol and ethanol with (*a*) concentrated sulphuric acid, (*b*) ethanoic acid, (*c*) ethanoyl chloride, (*d*) iron(III) chloride solution.

4. There are four isomeric butenes. On reaction with concentrated sulphuric acid, followed by hydrolysis, three of the four give the same alcohol, while the fourth butene gives a different alcohol. Deduce the structures of the two alcohols, and show how they are formed from the butenes.

5. Wr   ctural formulae for **A**, **B** and **C**:

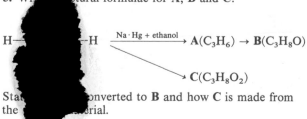

H—   —H $\xrightarrow{\text{Na·Hg + ethanol}}$ $A(C_3H_6) \rightarrow B(C_3H_8O)$

$\rightarrow C(C_3H_8O_2)$

Sta   onverted to **B** and how **C** is made from the   rial.

6. Outline two methods of preparing phenol, starting from benzene. What products are formed when phenol reacts with (*a*) benzoyl chloride, (*b*) ethanoic anhydride, (*c*) concentrated sulphuric acid, (*d*) bromine water?

7. Complete the following reactions. State the necessary conditions for reaction to occur:

(*a*) $CH_3CH(OH)CH_3 + H_2SO_4 \rightarrow$

(*b*) $C_6H_5SO_3Na + NaOH \rightarrow$

(*c*) $CH_3CH_2OH + Na \rightarrow \xrightarrow{CH_3I}$

(*d*) $CH_3CH_2C{=}CH_2 + H_3PO_4 \rightarrow$
                $|$
               CH_3

(*e*) $(CH_3)_3COH + HCl \rightarrow$

8. Give the names and structural formulae of the alcohols with the molecular formula C_3H_8O. Describe how you could distinguish between them by means of chemical tests.

9. How can ethanol be made from (*a*) ethene and (*b*) bromoethane? How and under what conditions does ethanol react with (i) ethanoic acid and (ii) sulphuric acid?

10. How would you prepare the following compounds from phenol?
(a) sodium phenoxide, (b) 2-nitrophenol, (c) 2,4,6-tribromophenol, (d) phenoxyethane.

11. Name the compounds **A** and **B**. Both are soluble in ethoxyethane. Neither is soluble in water. Describe how you could separate **A** and **B** from a mixture of the two compounds, using only sodium hydroxide, hydrochloric acid and ethoxyethane.

A   B

OH ... $CH(CH_3)_2$ OCH_3 ... $CH(CH_3)_2$

12. Compare the reactions of ethanol and phenol with (a) sodium, (b) sodium hydroxide, (c) sulphuric acid and (d) ethanoic acid.

13. Phenol is obtained from the distillation of coal tar. Mixed with it are benzene and phenylamine. Describe how you could (a) obtain phenol from this mixture, (b) test the sample to see whether it is a phenol, (c) purify the sample, and (d) test its purity.

14. How does phenol react with (a) ethanoic anhydride and (b) benzoyl chloride? State the necessary conditions for reaction to occur. Write equations for the reactions.

15. Describe chemical tests which would allow you to distinguish between the following pairs:

(a) Cyclohexanol and phenoxymethane.

(b) Ethane-1,2-diol and propan-2-ol.

(c) Cyclohexanol and cyclohexene.

(d) Cyclohexanol and phenol.

16. Outline the *cumene* process for the manufacture of phenol. State the sources of the raw materials, and the conditions under which the steps in the process are carried out.

†17. (a) (i) In terms of structure, what is the difference between a phenol and an aromatic alcohol?

(ii) Draw the structural formulae of the isomeric phenols and aromatic alcohols having the molecular formula C_7H_8O.

(iii) How could the reagents sodium, iron(III) chloride and phosphorus(V) chloride be used to distinguish between phenols and aromatic alcohols?

(iv) Outline with the necessary reagents and conditions, how phenol (C_6H_6O) could be prepared from benzene.

(b) Explain the following observations concerning phenol C_6H_6O.

(i) Addition of aqueous bromine results in a white precipitate.

(ii) Addition to a cold aqueous nitrogen containing salt produces a yellow dye.

(iii) A solution of phenol in aqueous sodium hydroxide, when shaken with benzoyl chloride, produces a white precipitate.

(c) Outline how a mixture of phenol and benzoic acid could be separated.
See Chapter 32 for (a) (iv) and (b) (ii).

(AEB 81)

31

ALDEHYDES AND KETONES

31.1 THE FUNCTIONAL GROUP

In aldehydes and ketones the functional group is the carbonyl group

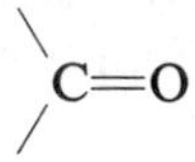

$$\backslash C{=}O$$

The formulae of aldehydes and ketones

Aldehydes have a hydrogen atom attached to the carbonyl carbon atom; ketones have two alkyl or aryl groups:

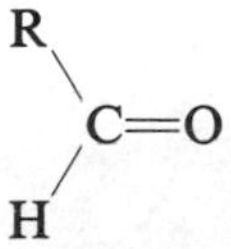

Aldehyde

(In the first member of the series, methanal, R = H.)

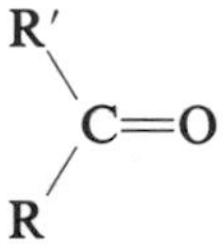

Ketone

(R and R′ may be aliphatic or aromatic.)

The general formula of saturated aliphatic aldehydes and ketones is $C_nH_{2n}O$. In addition, there are cyclic ketones of formula $C_nH_{2n-2}O$, for example, cyclohexanone. The simplest aromatic aldehyde and ketone are benzenecarbaldehyde and phenylethanone:

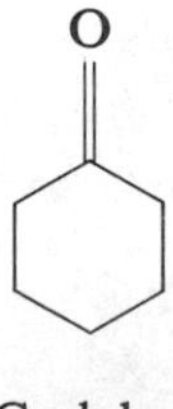

Cyclohexanone

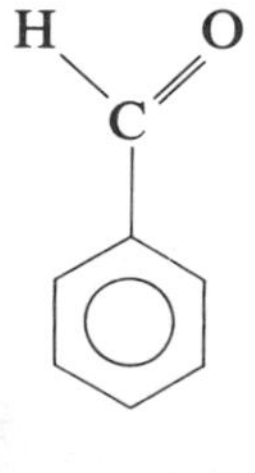

Benzenecarbaldehyde (or benzaldehyde)

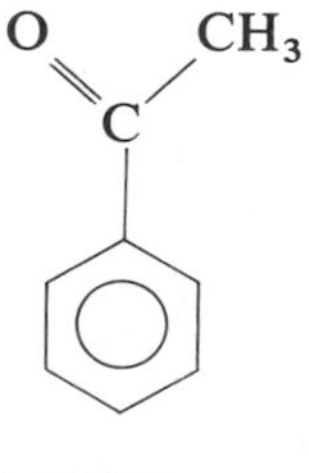

Phenylethanone

31.2 NOMENCLATURE FOR ALDEHYDES AND KETONES

The system of naming carbonyl compounds

Aldehydes and ketones are named after the hydrocarbon with the same number of carbon atoms. The name of the parent alkane is changed to end in *-al* for aldehydes and *-one* for ketones. The position of the carbonyl group in ketones is indicated by a locant preceding the suffix, *-one* [see Table 31.1(b)].

(a) *Aldehydes*

Formula	Name
HCHO	Methanal (formerly formaldehyde)
CH_3CHO	Ethanal (formerly acetaldehyde)
$CH_3(CH_2)_3CHO$	Pentanal
$C_6H_5CH_2CHO$	Phenylethanal
C_6H_5CHO	Benzenecarbaldehyde (or benzaldehyde)

(b) *Ketones*

Formula	Name
CH_3COCH_3	Propanone (formerly acetone)
$CH_3COCH_2CH_3$	Butanone (formerly ethyl methyl ketone)
$CH_3CO(CH_2)_2CH_3$	Pentan-2-one
$CH_3CH_2COCH_2CH_3$	Pentan-3-one
$C_6H_5COCH_3$	Phenylethanone

TABLE 31.1
The Names of some
Aldehydes and Ketones

FIGURE 31.1 Models of
Ethanal, CH_3CHO and
Propanone, $(CH_3)_2CO$

31.3 THE CARBONYL GROUP

In the group $R_2C{=}O$, the carbon atom forms three single σ bonds, which are coplanar.
The unused p orbital of the carbon atom overlaps with one of the unused p orbitals
of the oxygen atom to form a π orbital [see Figure 31.2].

FIGURE 31.2 The

Carbonyl Group, $\diagup\!\!\diagdown C\!=\!O$

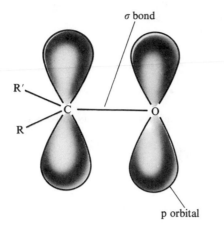
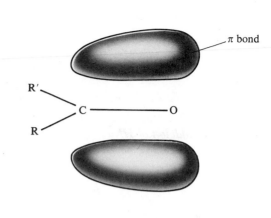

σ bond

π bond

p orbital

The π electrons in the

$\diagup\!\!\diagdown C\!=\!O$ *bond are*

extensively polarised

The electrons in the σ and π bonds are drawn towards the more electronegative oxygen atom. To represent this polarisation the carbonyl group is written as

$$\diagup\!\!\diagdown C\!=\!O \quad \text{or} \quad \overset{\delta+}{\diagup}\!\!\diagdown C\!=\!\overset{\delta-}{O}$$

Since π electrons are polarised much more extensively than σ electrons, the dipole in a

$$\diagup\!\!\diagdown C\!=\!O \text{ bond}$$

is greater than that in a

$$\diagup\!\!\diagdown C \rightarrow Cl \text{ bond}$$

The electronic structure of the carbonyl group is responsible for the reactions of aldehydes and ketones.

31.3.1 ATTACK BY NUCLEOPHILES

Nucleophiles attack

$$\overset{\delta+}{\diagup}\!\!\diagdown C\!=\!\overset{\delta-}{O} \text{ to form an}$$

intermediate . . .

Nucleophilic reagents are attracted by the partial positive charge on the carbon atom of the

$$\diagup\!\!\diagdown C\!=\!O \text{ group}$$

At the approach of a nucleophile (e.g., CN^-), the π electrons are repelled away from the carbon atom towards the oxygen atom. In the reactive intermediate which is formed, oxygen bears a negative charge, and the carbon atom is surrounded by four electron pairs.

$$N\!\equiv\!\overset{\ominus}{C}\!: \overset{R'}{\underset{R}{\diagdown}}C\!=\!\overset{..}{\underset{..}{O}}: \longrightarrow \left[N\!\equiv\!C\!-\!\overset{R'}{\underset{R}{\diagdown}}C\overset{..}{\underset{\underset{\ominus}{..}}{O}}: \right]$$

Nucleophile attacks $\overset{\delta+}{C}$

C is 4-valent; O is negative

...which acts as a base   The intermediate acts as a base:

$$\left[N\equiv C-\underset{R}{\overset{R'}{C}}\!\!-\!\!\overset{..}{\underset{..}{O}}{\!}^{\ominus} \right] + HA \longrightarrow N\equiv C-\underset{R}{\overset{R'}{C}}\!\!-\!\!\overset{..}{\underset{..}{O}}\!\!-\!\!H + A^-$$

The complete reaction is   The overall reaction is the addition of HCN across the double bond. In general, if
the addition of HNu   Nu: is the nucleophile
across the double bond

$$R_2C{=}O + HNu \rightarrow R_2C\overset{O-H}{\underset{Nu}{<}}$$

The ease with which the reaction occurs is in contrast with the attack by a nucleophile on a saturated carbon atom [see the hydrolysis of halogenoalkanes, p. 604].

31.3.2 COMPARISON WITH THE ALKENE GROUP

$\diagdown$
$\diagup C{=}O$ *does not react*   The carbonyl group resembles the alkene group in that both groups contain a
                        σ bond and a π bond between the bonded atoms. One might expect that the π electrons
with electrophiles   in the

$\diagdown$ $\diagup$
as does $C{=}C$
$\diagup$ $\diagdown$

$$\diagdown \atop \diagup C{=}O \text{ bond} \qquad \text{like those in}$$

$$\diagdown \quad \diagup \atop \diagup C{=}C \diagdown$$

would react with electrophiles, such as Br_2 and HBr. This does not happen. The reason is that the oxygen atom in the carbonyl group, being electronegative, is able to keep control of the π electrons and does not make them available for bonding to an electrophile. To summarise:

$$\diagdown{\overset{\delta+}{}}\quad{\overset{\delta-}{}} \atop \diagup C{=}O$$

polar bond

The electron-deficient carbon atom is attacked by nucleophiles.

$$\left[\diagdown \quad \diagup \atop \diagup C{=}C \diagdown \right]$$

non-polar bond

The double bond is attacked by electrophiles.

31.3.3 KETO—ENOL TAUTOMERISM

There is another set of reactions. These originate in the tautomerism shown by aldehydes
The equilibrium mixture   and ketones [p. 536]. In propanone, the tautomers are
of tautomers gives the

$$CH_3-\underset{\underset{O}{\|}}{C}-CH_3 \rightleftharpoons CH_3-\underset{\underset{OH}{|}}{C}{=}CH_2$$

$\diagdown$ $\diagup$
reactions of $C{=}C$,
$\diagup$ $\diagdown$ keto-Propanone   enol-Propanone

$\diagdown$
$\diagup C{=}O$ *and* $-OH$ There is only a fraction of a per cent of the enol present. The enol has a different
                        set of reactions: it has the reactions of the hydroxyl group and the reactions of the
                        alkene group. If the small fraction of enol present is used up in a reaction, more

will be formed from the keto-tautomer to maintain equilibrium. The whole of the equilibrium mixture can therefore undergo the reactions of the enol-tautomer. The iodination of propanone proceeds through the formation of the enol [p. 317].

31.3.4 COMPARISON OF ALDEHYDES AND KETONES

Aldehydes are more reactive than ketones

Aldehydes are more reactive than ketones. The reason is chiefly that the presence of two alkyl groups in ketones hinders the approach of attacking reagents to the carbonyl group. Another factor is that alkyl groups are electron-donating and reduce the partial positive charge on the carbonyl atom [see Figure 31.3].

Approach of nucleophile is hindered.

FIGURE 31.3
Hindrance by Alkyl
Groups to Nucleophilic
Attack

CHECKPOINT 31A: CARBONYL COMPOUNDS

1. Write structural formulae for
(a) butanal
(b) pentan-2-one
(c) pentan-3-one
(d) cyclohexanone
(e) 3-methylhexanal
(f) pentane-2,4-dione
(g) 4-chloro-2-methylhexan-3-one
(h) *trans*-but-2-enal
(i) 4-hydroxybenzaldehyde
(j) 2-chlorobenzaldehyde

2. Name the following:
(a) $CH_3CH_2CH_2CHO$
(b) $CH_3CH_2COCH_3$
(c) $C_2H_5COC_2H_5$
(d) $(CH_3)_2CHCHO$
(e) $(CH_3)_2CHCOCH_3$
(f) $CH_3CH{=}CHCHO$
(g) $CH_3CH{=}CHCOCH_3$

(h) Cl—⟨O⟩—COCH₃

(i) O₂N—⟨O⟩—CHO

3. What feature of the carbonyl group is chiefly responsible for the reactive nature of aldehydes and ketones?

4. Why does bromine attack

$$\overset{\diagdown}{}C{=}C\overset{\diagup}{}$$

but not

$$\overset{\diagdown}{}C{=}O\,?$$

*5. Propanone contains only 0.01% of the enol tautomer, yet pentane-2,4-dione contains 80% enol. Draw the *keto* and *enol* forms of this dione. Can you explain how the stability of the enol tautomer arises?

6. Explain why there is intermolecular hydrogen bonding in $C_2H_5CH_2OH$ but not in C_2H_5CHO.

31.4 INDUSTRIAL MANUFACTURE AND USES

31.4.1 METHANAL

Methanal is made from methanol by aerial oxidation...

Methanal (*formaldehyde*) is obtained by the aerial oxidation of methanol in the presence of a silver catalyst.

...It is used as formalin and in the manufacture of plastics, antiseptics and explosives

Methanal is a gas ($T_b = -21\,°C$). It is used as a 40% aqueous solution, called *formalin*, for the preservation of biological specimens. The solid trimethanal (*paraformaldehyde*), $(CH_2O)_3$, is a convenient source of methanal, which it yields on heating. Methanal is used in the manufacture of the tough plastics, Delrin® [p. 654] and Bakelite® [p. 656], the antiseptic hexamethylene tetramine and the explosive *cyclonite* [p. 654].

31.4.2 ETHANAL

Ethanal is made from ethanol and from ethyne...

Ethanal is made from ethanol (*a*) by aerial oxidation in the presence of silver as catalyst or (*b*) by catalytic dehydrogenation over copper. Ethyne is another source of ethanal [p. 570].

...It is used to make chloral and DDT

Ethanal is used as a starting point in the manufacture of certain organic compounds. Examples are ethanoic acid and chloral ('knockout drops').

31.4.3 PROPANONE

Propanone is made from propan-2-ol by dehydrogenation and is used as a solvent

Propanone is made by the dehydrogenation of propan-2-ol using a copper catalyst. It is a widely used, inexpensive industrial solvent, with $T_b = 56\,°C$. If a solvent with a higher boiling temperature is needed, butanone is used ($T_b = 80\,°C$). Propanone is formed in the fermentation of some sugars and starches and is found in the breath and urine of diabetics.

31.4.4 OCCURRENCE

Some carbonyl compounds are found in natural products

Many aldehydes and ketones occur naturally. A few examples are:

Benzaldehyde (A derivative occurs in almonds.)

3-Phenylpropenal (In cinnamon)

Vanillin (Vanilla flavouring)

Testosterone (a male sex hormone)

31.5 LABORATORY PREPARATIONS

31.5.1 FROM ALCOHOLS

OXIDATION OF ALCOHOLS

A laboratory method of preparation is the oxidation of alcohols to aldehydes and ketones...

Oxidation of primary alcohols gives aldehydes, provided that conditions are controlled to avoid further oxidation to a carboxylic acid [p. 623]. Oxidation of secondary alcohols gives ketones. Acidified sodium dichromate(VI) is often employed for oxidation in the laboratory. The oxidising agent is added slowly to the alcohol [see Figure 31.4]. The temperature is kept below the boiling temperature of the alcohol and above that of the carbonyl compound. (Carbonyl compounds are more volatile than the corresponding alcohols.) Arranging for an excess of alcohol over oxidant and distilling off the aldehyde as it is formed avoid further oxidation. Ketones are in little danger of further oxidation:

$$C_2H_5OH(l) + [O] \xrightarrow[\text{acid, warm}]{Na_2Cr_2O_7} CH_3CHO(l) + H_2O(l)$$

Ethanol   Ethanal

$$(CH_3)_2CHOH(l) + [O] \xrightarrow[\text{acid, warm}]{Na_2Cr_2O_7} (CH_3)_2CO(l) + H_2O(l)$$

Propan-2-ol   Propanone

FIGURE 31.4
Preparation of
Aldehydes and Ketones

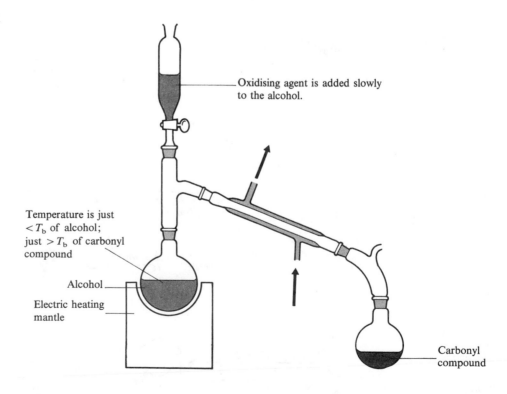

Oxidising agent is added slowly to the alcohol.

Temperature is just $< T_b$ of alcohol; just $> T_b$ of carbonyl compound

Alcohol

Electric heating mantle

Carbonyl compound

DEHYDROGENATION OF ALCOHOLS

Carbonyl compounds are made from alcohols...

Catalytic dehydrogenation of alcohols can be used in the laboratory as well as in industry [p. 648].

31.5.2 FROM ACID CHLORIDES

...Other sources are: the reduction of acid chlorides...

An **acid chloride**

$$RC \overset{\displaystyle O}{\underset{\displaystyle}{\diagup}} Cl$$

may be reduced to an aldehyde by hydrogen in the presence of a catalyst. To avoid further reduction to an alcohol, the catalyst, palladium on barium sulphate, is 'poisoned' by the addition of some deactivating material:

$$C_2H_5COCl(g) + H_2(g) \xrightarrow[\text{'poisoned' Pd, BaSO}_4]{\text{heat with}} C_2H_5CHO(l) + HCl(g)$$

Propanoyl chloride   Propanal

31.5.3 FROM ALKENES

...the ozonolysis of alkenes...

Carbonyl compounds are formed during reductive ozonolysis of alkenes. The reaction is used to locate the positions of double bonds [p. 561].

31.5.4 AROMATIC KETONES BY FRIEDEL—CRAFTS ACYLATION

Friedel–Crafts acylation gives aromatic ketones

Aromatic ketones are obtained by the Friedel–Crafts acylation reactions [p. 583].

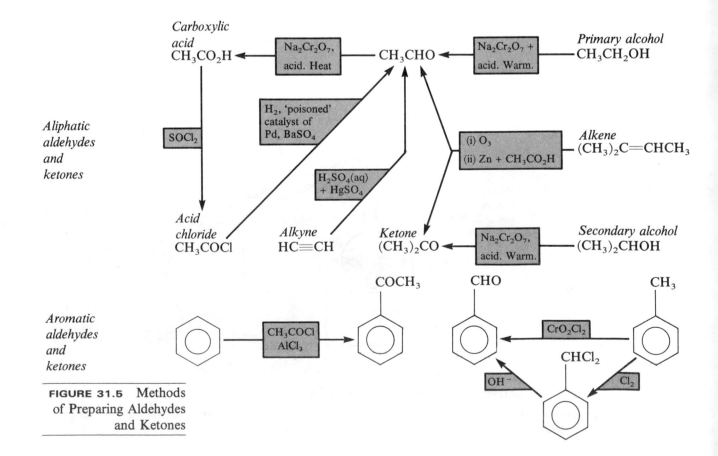

Aliphatic aldehydes and ketones

Aromatic aldehydes and ketones

FIGURE 31.5 Methods of Preparing Aldehydes and Ketones

31.5.5 AROMATIC ALDEHYDES FROM METHYLARENES

Oxidation of —CH₃ attached to an aromatic ring gives an aromatic aldehyde

Aromatic aldehydes can be made by the oxidation of a methyl group attached to the benzene ring. Chromium dichloride dioxide is often employed as the oxidising agent:

$$C_6H_5CH_3 \xrightarrow{CrO_2Cl_2} C_6H_5CHO$$

Methyl-benzene   Benzenecarbaldehyde (benzaldehyde)

Another route is the alkaline hydrolysis of 1,1-dihalogenocompounds:

$$C_6H_5CH_3 \xrightarrow{Cl_2} C_6H_5CHCl_2 \xrightarrow{OH^-(aq)} C_6H_5CHO$$

Aliphatic aldehydes can be made in this way, but the dihalogenocompounds are usually made from the aldehydes.

CHECKPOINT 31B: PREPARATIVE METHODS

1. Aldehydes are often prepared from alcohols. What difficulty arises in this method of preparation? How is it overcome?

2. Draw the structural formulae of the alkenes that yield the following products on ozonolysis:

(a) ethanal and propanone
(b) propanol, propanone and butandial
(c) methanal and pentane-2,4-dione
(d) propandial and methanal.

31.6 REACTIONS OF CARBONYL COMPOUNDS

31.6.1 REDUCTION

Aldehydes are reduced to primary alcohols by H₂ or by LiAlH₄ in ether or by NaBH₄(aq)

Hydrogen adds across the carbonyl group, reducing aldehydes to primary alcohols and ketones to secondary alcohols. The reaction is more difficult than addition to a carbon–carbon double bond, and usually requires heat, pressure and a metal catalyst, (Pt or Ni):

(R′ can be H)

In the laboratory, reduction is usually effected by the use of a metal hydride. Lithium tetrahydridoaluminate, $LiAlH_4$, is used in solution in ethoxyethane, which must be dry because the hydride reacts with water. It releases hydride ions which attack the carbonyl carbon atom. The reaction with aldehydes is very vigorous, and the less powerful reducing agent, sodium tetrahydridoborate, $NaBH_4$, can be used. This is a more convenient reagent as it can be used in aqueous or methanolic solution:

Ketones are reduced by H₂ or by LiAlH₄ in ether to give secondary alcohols

31.6.2 OXIDATION

Aldehydes are oxidised to carboxylic acids...

Aldehydes are readily oxidised to carboxylic acids by acidified potassium dichromate(VI) or acidified potassium manganate(VII). Alternatively, the aldehyde vapour can be passed with air over a heated catalyst or over silver oxide:

$$R-C\overset{H}{=}O(g) + O_2(g) \xrightarrow[\text{Pt or Cu as catalyst}]{\text{heat}} R-C\overset{O}{\underset{OH}{\diagdown}} (g)$$

Aromatic aldehydes are less easily oxidised than aliphatic aldehydes.

...Ketones are more difficult to oxidise

Ketones are more difficult to oxidise than aldehydes. With a powerful oxidising agent, in acid conditions, they are oxidised to a mixture of carboxylic acids.

31.6.3 DISTINGUISHING BETWEEN ALDEHYDES AND KETONES

The difference in the ease of oxidation provides a means of distinguishing between aldehydes and ketones

The relative ease with which aldehydes are oxidised gives a means of distinguishing them from ketones. The gentle oxidising agent, Fehling's solution, contains complex copper(II) ions. Warm Fehling's solution is reduced to a reddish precipitate of copper(I) oxide, Cu_2O, by aldehydes but not by ketones. Tollens' reagent, a solution of complex silver ions, $[Ag(NH_3)_2]^+$, is reduced to a silver mirror when warmed with aldehydes but not by ketones.

31.6.4 ADDITION REACTIONS

HYDROGEN CYANIDE

The carbonyl group,

$$\diagdown\atop{/}C=O$$

is attacked by nucleophiles. As mentioned on p. 645, the cyanide ion, CN^-, is such a nucleophile. When hydrogen cyanide is added to a cold (10–20 °C) solution of a carbonyl compound, it adds across the double bond. The reaction occurs rapidly in the presence of a base and is inhibited by acids:

HCN adds across $\diagdown\atop{/}C=O$

to give a hydroxynitrile (a cyanohydrin)...

$$\begin{array}{c}CH_3 \\ \diagdown \\ \diagup \\ C_2H_5\end{array}C=O(l) + HCN(g) \xrightarrow{\text{base}} \begin{array}{c}CH_3 \quad OH \\ \diagdown \;\; \diagup \\ C \\ \diagup \;\; \diagdown \\ C_2H_5 \quad CN\end{array}(s)$$

Butanone

2-Hydroxy-2-methylbutanonitrile
(the cyanohydrin of butanone)

The product that is formed is described as the **cyanohydrin** of the carbonyl compound. It is a hydroxynitrile. The **nitrile** group

$$-C\equiv N$$

can be hydrolysed to the carboxyl group

$$-CO_2H$$

The hydrolysis of hydroxynitriles is a useful way of making hydroxycarboxylic acids:

... which can be hydrolysed to a hydroxyacid

$$\begin{array}{c} CH_3 \quad OH \\ \diagdown \; / \\ C \\ \diagup \; \diagdown \\ C_2H_5 \quad CN \end{array} \xrightarrow[\text{reflux}]{\text{conc. HCl(aq)}} \begin{array}{c} CH_3 \quad OH \\ \diagdown \; / \\ C \\ \diagup \; \diagdown \\ C_2H_5 \quad CO_2H \end{array}$$

2-Hydroxy-2-methyl-   2-Hydroxy-2-methylbutanoic acid
butanonitrile

To avoid the danger of working with the poisonous gas hydrogen cyanide, it is generated in the reaction mixture by the action of dilute sulphuric acid on potassium cyanide.

SODIUM HYDROGENSULPHITE

NaHSO₃ adds across the C=O bond to give a crystalline hydrogensulphite compound, from which the carbonyl compound can be regenerated

Another reagent which adds across the carbonyl group is sodium hydrogensulphite:

$$\begin{array}{c} C_2H_5 \\ \diagdown \\ \quad C{=}O + NaHSO_3 \\ \diagup \\ C_2H_5 \end{array} \underset{\text{acid or alkali}}{\overset{\text{excess of NaHSO}_3}{\rightleftharpoons}} \begin{array}{c} C_2H_5 \quad OH \\ \diagdown \; / \\ C \\ \diagup \; \diagdown \\ C_2H_5 \quad SO_3Na \end{array}$$

Pentan-3-one   Sodium hydrogensulphite
                              compound of pentan-3-one

These derivatives are used to purify carbonyl compounds ...

Aldehydes react more readily than ketones, possibly on account of the steric hindrance to the approach of the attacking reagent offered by two alkyl groups. The reaction is carried out by shaking together the carbonyl compound and an excess of a saturated sodium hydrogensulphite solution at room temperature. The hydrogensulphite derivative crystallises out of the solution. On treatment with acid or alkali, it liberates the free aldehyde or ketone. The pair of reactions are used to purify carbonyl compounds. The hydrogensulphite compound is precipitated, recrystallised from water and then decomposed.

... and for conversion into hydroxynitriles

With sodium cyanide, hydrogensulphite compounds react to give hydroxynitriles (cyanohydrins):

$$\begin{array}{c} C_2H_5 \quad OH \\ \diagdown \; / \\ C \\ \diagup \; \diagdown \\ CH_3 \quad SO_3{}^-Na^+ \end{array} + Na^+CN^- \rightarrow \begin{array}{c} C_2H_5 \quad OH \\ \diagdown \; / \\ C \\ \diagup \; \diagdown \\ CH_3 \quad CN \end{array} + 2Na^+ + SO_3{}^{2-}$$

Sodium hydrogensulphite   2-Hydroxy-2-methylbutanonitrile
compound of butanone

Hydroxynitriles are often made from carbonyl compounds by these two steps in order to avoid the use of hydrogen cyanide.

WATER AND ALCOHOL

H₂O adds across $\diagdown$C=O$\diagup$

In aqueous solution, carbonyl compounds exist in equilibrium with diols:

$$\begin{array}{c} R' \\ \diagdown \\ \quad C{=}O + H_2O \rightleftharpoons \\ \diagup \\ H \end{array} \begin{array}{c} R' \quad OH \\ \diagdown \; / \\ C \\ \diagup \; \diagdown \\ H \quad OH \end{array}$$

R′OH adds to form acetals

Aldehydes react with alcohols to form addition products called **acetals**:

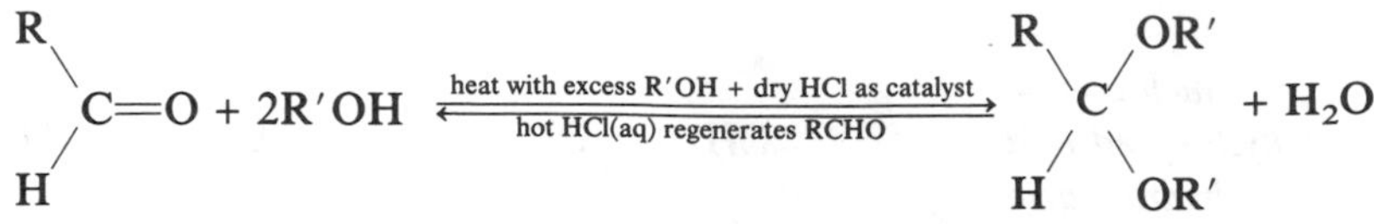

$$\underset{H}{\overset{R}{>}}C=O + 2R'OH \xrightleftharpoons[\text{hot HCl(aq) regenerates RCHO}]{\text{heat with excess R'OH + dry HCl as catalyst}} \underset{H\quad OR'}{\overset{R\quad OR'}{C}} + H_2O$$

An acetal

Ketals exist, but cannot be made in the same way.

AMMONIA

NH₃ adds across the

$\searrow C=O$ *bond to form a hydroxyamine*

Ammonia adds to aliphatic aldehydes (except methanal) in solution in dry ethoxyethane:

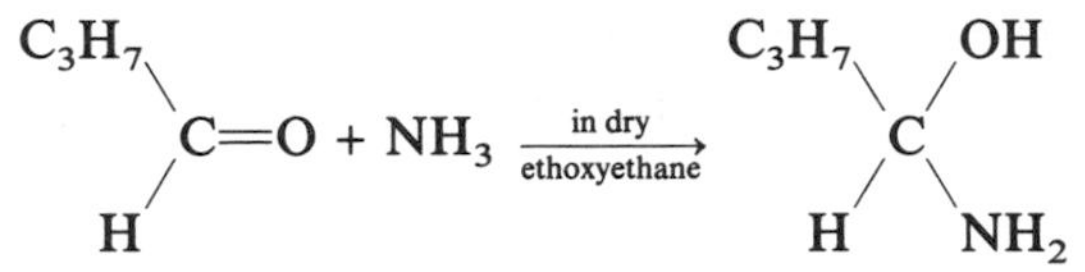

$$\underset{H}{\overset{C_3H_7}{>}}C=O + NH_3 \xrightarrow[\text{ethoxyethane}]{\text{in dry}} \underset{H\quad NH_2}{\overset{C_3H_7\quad OH}{C}}$$

Butanal   1-Aminobutan-1-ol

A white precipitate of a hydroxyamine (also called an *aldehyde-ammonia*) is formed. These compounds are unstable and are hydrolysed by water to the parent aldehydes.

Methanal and NH₃ react in a different manner to form hexamethylene tetramine

Methanal, aromatic aldehydes and some ketones react with ammonia in a different way. Instead of addition compounds they form condensation products through the elimination of water. Methanal and ammonia form hexamethylene tetramine (hexamine):

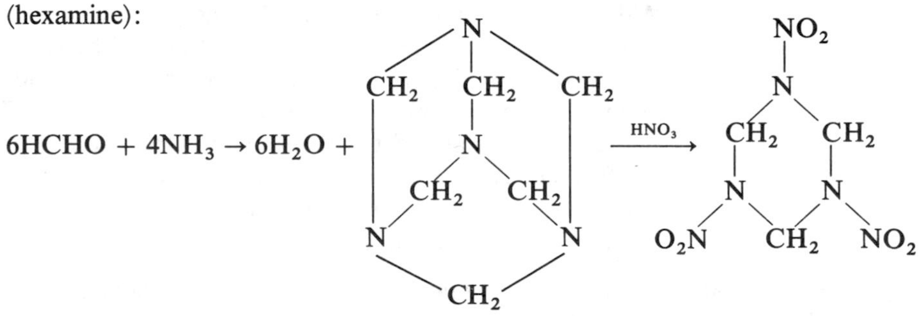

$$6HCHO + 4NH_3 \rightarrow 6H_2O +$$

Hexamethylene tetramine   *Cyclonite* (explosive)

This product is a solid and is a convenient source of methanal, which it liberates on being treated with a dilute acid. It is used medicinally as a urinary antiseptic. On reaction with concentrated nitric acid, it gives the important explosive *cyclonite*.

POLYMERISATION

Methanal polymerises to form poly(methanal)...

...a useful plastics material

If a solution of methanal is evaporated, poly(methanal), $(CH_2O)_n$, is formed. This polymer (Delrin®) is a tough plastics material which is used as a substitute for wood and metal. Both methanal and ethanal form trimers.

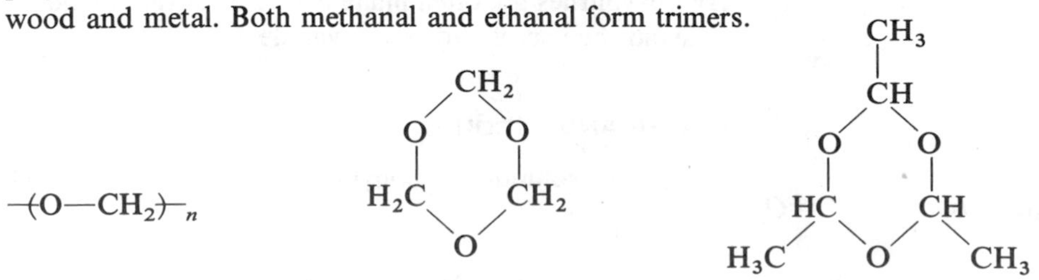

$\{O\!-\!CH_2\}_n$

Poly(methanal)

Methanal trimer
(*trioxane*)

Ethanal trimer
(*paraldehyde*)

Ethanal trimerises

Ethanal trimer (*paraldehyde*) is a liquid which is made by adding a few drops of concentrated sulphuric acid to ethanal. It has been used as a hypnotic (for inducing sleep).

Ketones and aromatic aldehydes do not polymerise.

31.6.5 ADDITION—ELIMINATION (OR CONDENSATION) REACTIONS

Compounds XNH₂ give products as the result of addition followed by elimination

Carbonyl compounds form adducts with many derivatives of ammonia, XNH_2, adducts which immediately eliminate a molecule of water to form a product $R_2C=NX$. In the reactions with hydroxylamine, the products are **oximes**:

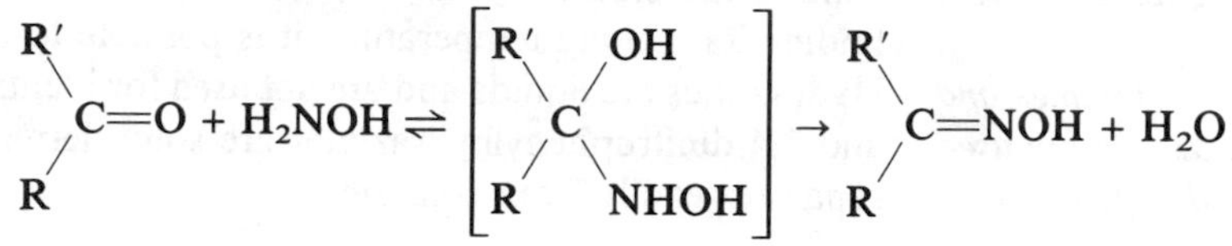

$$\underset{\text{Hydroxylamine}}{\overset{R'}{\underset{R}{>}}C=O + H_2NOH} \rightleftharpoons \underset{\text{Intermediate}}{\left[\overset{R'}{\underset{R}{>}}C\overset{OH}{\underset{NHOH}{<}}\right]} \rightarrow \underset{\text{An oxime}}{\overset{R'}{\underset{R}{>}}C=NOH + H_2O}$$

Since H_2O is eliminated, these **addition—elimination** reactions are described as **condensation** reactions.

Similar reactions are listed in Table 31.2.

Reaction with a weak base	Product
Hydroxylamine H_2NOH	An oxime $\overset{R}{\underset{R'}{>}}C=N-OH + H_2O$
Hydrazine H_2NNH_2	A hydrazone $\overset{R}{\underset{R'}{>}}C=N-NH_2 + H_2O$
Phenylhydrazine 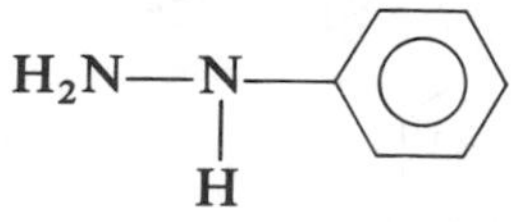 $H_2N-\underset{H}{N}-\bigcirc$	A phenylhydrazone $\overset{R}{\underset{R'}{>}}C=N-\underset{H}{N}-\bigcirc + H_2O$ 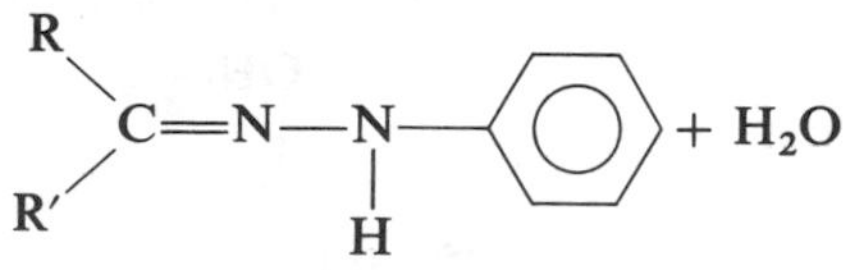
2,4-Dinitrophenylhydrazine 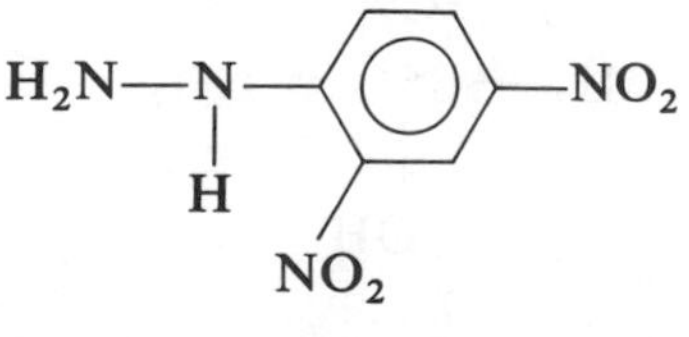 $H_2N-\underset{H}{N}-\underset{NO_2}{\bigcirc}-NO_2$	A 2,4-dinitrophenylhydrazone 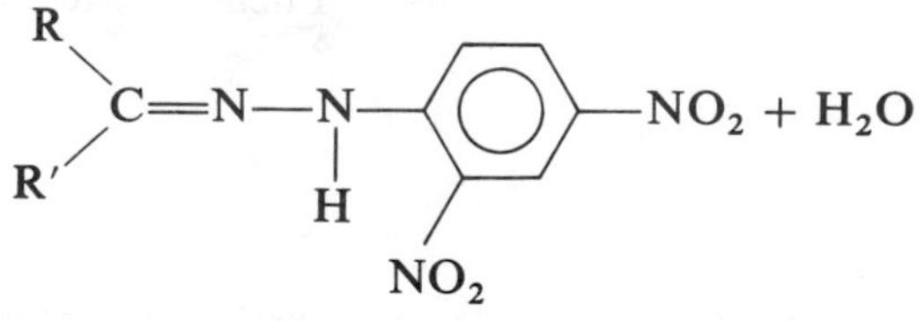 $\overset{R}{\underset{R'}{>}}C=N-\underset{H}{N}-\underset{NO_2}{\bigcirc}-NO_2 + H_2O$

TABLE 31.2
Addition–Elimination Reactions of Carbonyl Compounds (Reactions of RCOR′, where R can be H (in aldehydes) or an alkyl group (in ketones).)

Note

Examples of the names of some of these derivatives are:

$(CH_3)_2C=N-NH_2$ Propanone hydrazone

$CH_3CH=N-OH$ Ethanal oxime (an aldoxime)

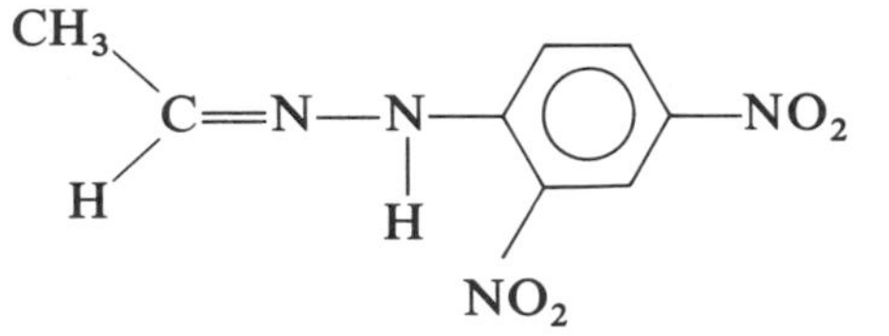

Ethanal 2,4-dinitrophenylhydrazone

H_2NOH gives oximes...

...Oximes and phenylhydrazones are used to identify carbonyl compounds...

...Brady's reagent is used to detect carbonyl compounds

Oximes are stable crystalline solids. By preparing the oxime, recrystallising it and finding its melting temperature, it is possible to identify an aldehyde or ketone. Hydrazones are liquids and are not used for identification purposes. Phenylhydrazones and 2,4-dinitrophenylhydrazones are solid derivatives of limited solubility, which separate readily from solution.

A solution of 2,4-dinitrophenylhydrazine in methanol and sulphuric acid is called Brady's reagent. When it is added to a solution of a carbonyl compound, the 2,4-dinitrophenylhydrazone separates rapidly. Aliphatic carbonyl compounds give orange or yellow derivatives, while those of aromatic carbonyl compounds are darker in colour. Brady's reagent can therefore be used to test for the presence of carbonyl compounds. The reaction is quantitative, and the precipitate can be dried and weighed to reveal the amount of carbonyl compound present.

PLASTICS

Urea-methanal polymers are useful plastics, e.g., Bakelite®

A number of useful plastics are formed by addition–elimination reactions. The *urea-formaldehyde* resins are polyamides made from carbamide (urea) and methanal (formaldehyde):

$$n\text{HCHO} + n\text{H}_2\text{N}-\underset{\underset{\text{O}}{\|}}{\text{C}}-\text{NH}_2 \rightarrow -(\text{CH}_2-\text{NHCONH})_n + n\text{H}_2\text{O}$$

Methanal   Carbamide   'Urea-formaldehyde' resin

Bakelite® is formed by **condensation polymerisation** from phenol and methanal. It is a thermosetting resin [p. 735].

The structural formula of Bakelite®

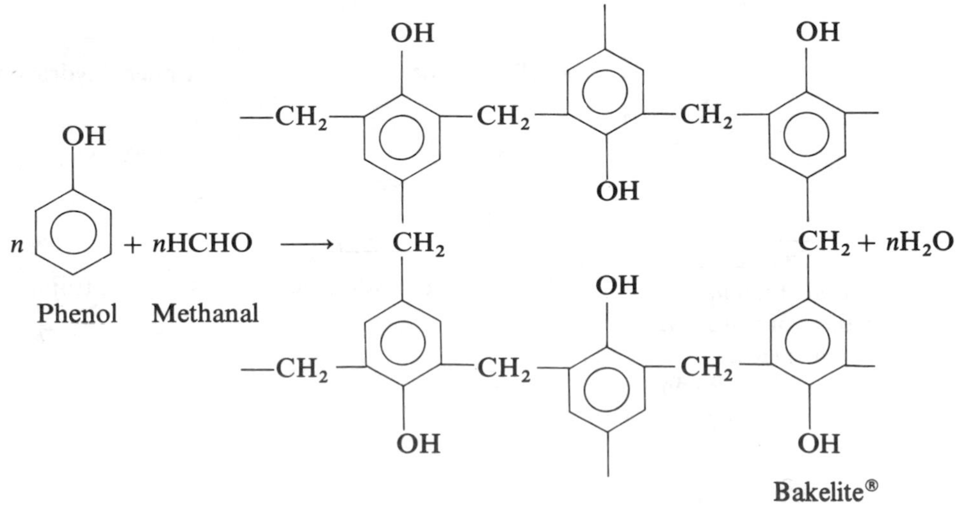

Bakelite®

31.6.6   CHLORINATION

PCl₅ converts $\diagdown C = O \diagup$
to $\diagdown CCl_2 \diagup$

Phosphorus(V) chloride reacts with carbonyl compounds, replacing the oxygen atom by two chlorine atoms:

$$R_2CO + PCl_5 \rightarrow R_2CCl_2 + POCl_3$$

Chlorine replaces the hydrogens on the α carbon atoms...

Chlorine reacts with carbonyl compounds. The hydrogen atoms on the **α carbon atoms** (those adjacent to the carbonyl atoms) are easily substituted by chlorine atoms. This is because the electron-withdrawing effect of the carbonyl group weakens the C—H bonds. When chlorine is bubbled through ethanal, trichloroethanal, CCl_3CHO, is formed. Propanone forms a mixture of products, and benzaldehyde forms benzoyl chloride:

...e.g., in the formation of CCl₃CHO, chloral

$$CH_3\overset{\displaystyle H}{C}{=}O + 3Cl_2 \rightarrow CCl_3\overset{\displaystyle H}{C}{=}O + 3HCl$$

Ethanal   2,2,2-Trichloroethanal

$$CH_3COCH_3 \xrightarrow{Cl_2} CH_3COCH_2Cl \xrightarrow{Cl_2} CH_3COCHCl_2 \rightarrow etc.$$

Propanone   Chloropropanone   1,1-Dichloropropanone

$$C_6H_5\overset{\displaystyle H}{C}{=}O + Cl_2 \rightarrow C_6H_5\overset{\displaystyle Cl}{C}{=}O + HCl$$

Benzaldehyde   Benzoyl chloride

Trichloroethanal is a colourless liquid called *chloral*. It combines with water to form a colourless crystalline solid, *chloral hydrate*, $CCl_3CH(OH)_2$. Chloral hydrate was used to induce sleep.

...Chloral is used in medicine...

Trichloroethanal is used industrially. It condenses with chlorobenzene in the presence of concentrated sulphuric acid:

...and in the manufacture of DDT

The product, 1,1-di(4-chlorophenyl)-2,2,2-trichloroethane (DDT) is a powerful insecticide [p. 406].

31.6.7   THE HALOFORM REACTION

Chloroform from CH₃COR, chlorine and alkali

Methyl carbonyl compounds, i.e., compounds containing a

$$CH_3-\underset{\underset{\displaystyle O}{\|}}{C}-C\diagup \quad group$$

give chloroform, $CHCl_3$, when warmed with an aqueous solution of sodium chlorate(I), NaClO (or with chlorine and alkali):

$$\underset{R}{\overset{CH_3}{\diagdown}}C{=}O + 3Cl_2 \rightarrow \underset{R}{\overset{CCl_3}{\diagdown}}C{=}O + 3HCl$$

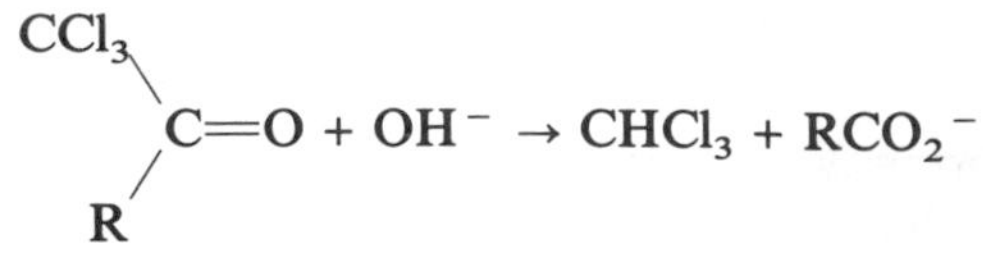

$$\begin{array}{c} CCl_3 \\ \diagdown \\ C{=}O + OH^- \rightarrow CHCl_3 + RCO_2^- \\ \diagup \\ R \end{array}$$

Iodoform from CH₃COR and I₂ and alkali or NaIO…

Iodoform from CH_3COR and I_2 and alkali or NaIO…

If potassium bromide or potassium iodide is dissolved in the reacting mixture, bromoform, $CHBr_3$, or iodoform, CHI_3, is formed. Iodoform, tri-iodomethane, is precipitated as fine yellow crystals with a characteristic smell. The reaction is used as a test for the group

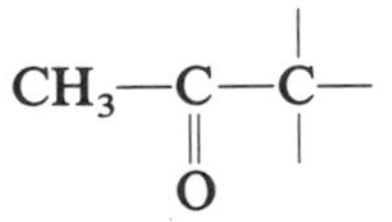

$$\begin{array}{ccc} & & | \\ CH_3{-}C{-}C{-} \\ & \| & | \\ & O & \end{array}$$

…and from $CH_3CH(OH)R$

Alcohols of formula $CH_3CH(OH)R$ are oxidised by sodium iodate(I) to CH_3COR and therefore give a positive iodoform test:

$$CH_3CH_2OH \xrightarrow{\text{NaIO}} CH_3CHO \xrightarrow{\text{NaIO}} CI_3CHO \xrightarrow{\text{OH}^-} CHI_3$$

Ethanol   Ethanal   Tri-iodo-ethanal   Tri-iodomethane

31.6.8 ALDOL REACTION

Aldehydes and ketones with α H atoms dimerise in the aldol reaction…

In the presence of a dilute alkali, some aldehydes and ketones dimerise to form a hydroxycarbonyl compound called an **aldol**. The carbonyl compound must possess an α hydrogen atom (on the carbon atom adjacent to the carbonyl group):

$$2CH_3CHO \xrightarrow{\text{OH}^-\text{(aq)}} \begin{array}{c} CH_3CHCH_2CHO \\ | \\ OH \end{array}$$

Ethanal   3-Hydroxybutanal

Methanal and benzaldehyde do not have α hydrogen atoms and do not dimerise.

31.6.9 CANNIZZARO REACTION

…Those without α H atoms undergo the Cannizzaro reaction

Carbonyl compounds which contain no α hydrogen atoms react differently with alkali. They undergo the Cannizzaro reaction, in which both oxidation to an acid and reduction to an alcohol occur:

$$2C_6H_5CHO + NaOH \rightarrow C_6H_5CO_2Na + C_6H_5CH_2OH$$

Benzaldehyde   Sodium benzoate   Benzyl alcohol

$$2HCHO + NaOH \rightarrow HCO_2Na + CH_3OH$$

Methanal   Sodium methanoate   Methanol

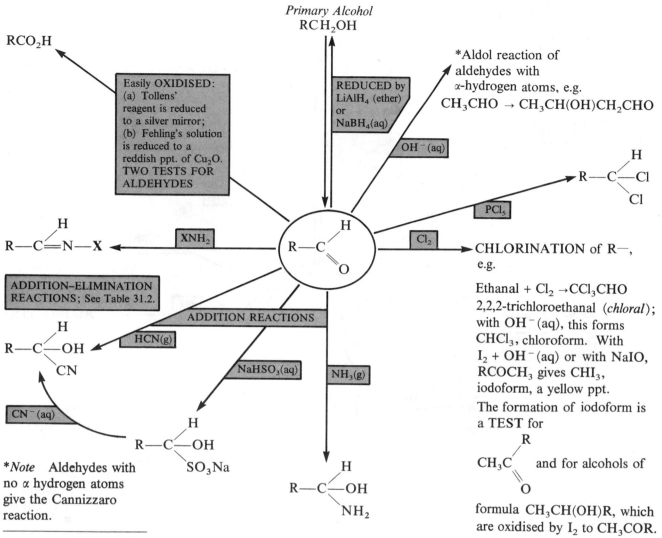

FIGURE 31.6
A Summary of the
Reactions of Aldehydes

*Note Aldehydes with no α hydrogen atoms give the Cannizzaro reaction.

TABLE 31.3
A Comparison of
Aliphatic and Aromatic
Aldehydes

Reagent	Ethanal, CH₃CHO Product	Benzaldehyde, C₆H₅CHO Product
KMnO₄, acid	CH₃CO₂H	C₆H₅CO₂H
LiAlH₄ (ethoxyethane)	CH₃CH₂OH	C₆H₅CH₂OH
PCl₅ or SOCl₂	CH₃CHCl₂	C₆H₅CHCl₂
HCN	CH₃CH(OH)CN	C₆H₅CH(OH)CN
NaHSO₃	CH₃CH(OH)SO₃Na	C₆H₅CH(OH)SO₃Na
C₆H₅NHNH₂	CH₃CH=NNHC₆H₅	C₆H₅CH=NNHC₆H₅
NH₂OH	CH₃CH=NOH	C₆H₅CH=NOH
NH₃	CH₃CH(OH)NH₃	(C₆H₅CH)₆N₄*
NaOH(aq)	CH₃CH(OH)CH₂CHO	C₆H₅CH₂OH + C₆H₅CO₂Na
Cl₂	CCl₃CHO	C₆H₅COCl
Cl₂ + Halogen-carrier	—	3-ClC₆H₄CHO
Conc. HNO₃ + Conc. H₂SO₄	—	3-O₂NC₆H₄CHO
Hot conc. H₂SO₄	—	3-HO₃SC₆H₄CHO
Fehling's solution	Red Cu₂O	—
Tollens' reagent	Ag mirror	—
KI, KI, NaClO	CHI₃, iodoform	— *like methanal
Conc. H₂SO₄	(C₂H₄O)₃, paraldehyde	—

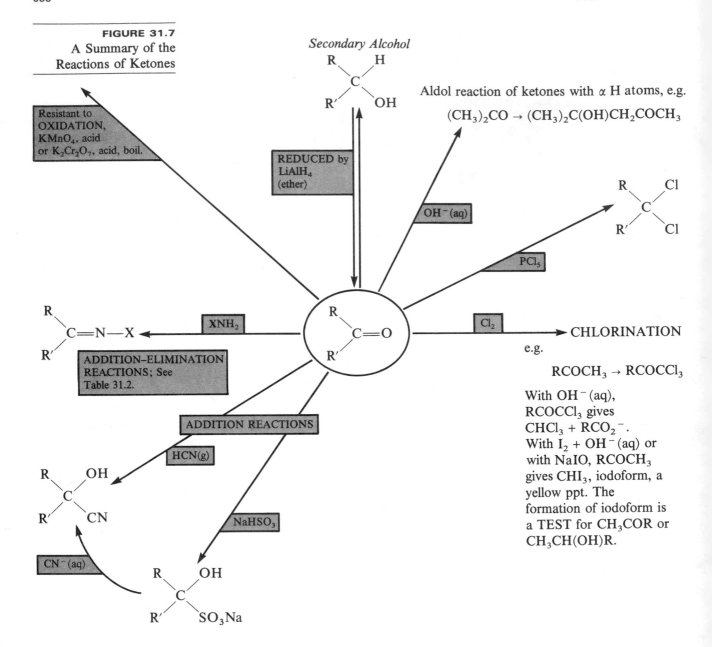

FIGURE 31.7
A Summary of the
Reactions of Ketones

CHECKPOINT 31C: REACTIONS OF CARBONYL COMPOUNDS

1. Draw structural formulae for

(a) butanal oxime
(b) benzaldehyde oxime
(c) propanone hydrazone
(d) propanone phenylhydrazone
(e) pentan-2-one phenylhydrazone
(f) butanone 2,4-dinitrophenylhydrazone
(g) the hydrogensulphite compound of benzaldehyde.

2. Draw structural formulae for

(a) butanone cyanohydrin
(b) benzaldehyde cyanohydrin
(c) propanal diethyl acetal
(d) the product of reductive ozonolysis of cyclohexene.

3. Write the formulae for the organic products formed in these reactions:

(a) $C_2H_5COCH_3 + NH_2OH \rightarrow$
(b) $C_6H_5CHO + LiAlH_4 \rightarrow$
(c) $C_2H_5CHO + [Ag(NH_3)_2]^+ (aq) \rightarrow$
(d) $C_6H_5CH{=}CHCHO + H_2 \xrightarrow{Pd}$
(e) $C_6H_5CH{=}CHCHO + NaBH_4 \rightarrow$
(f) $C_2H_5COCH_3 + LiAlH_4 \rightarrow$

4. State two reactions of aldehydes which are also characteristic of ketones and two which are not shared by ketones. What is the reason for the difference?

5. Give the structural formulae of the products of the reactions of butanone with

(a) HCN

(b) NH_2OH

(c) Br_2

(d) $I_2 + NaOH(aq)$

(e) $LiAlH_4$

(f) 2,4-dinitrophenylhydrazine

6. State the reagents and conditions needed for the preparation of the following from ethanal:

(a) chloroform

(b) ethene

(c) ethyl ethanoate

(d) 3-hydroxybutanal.

7. Draw the structural formulae for the products of the reactions of

(a) butanone and hydrazine

(b) cyclohexanone and hydroxylamine

(c) phenylethanone and 2,4-dinitrophenylhydrazine.

8. Which of the following give a positive iodoform test?

(a) $CH_3COC_2H_5$

(b) CH_3CH_2CHO

(c) $CH_3CH_2CH(OH)CH_3$

(d) ICH_2CHO

9. The addition of alkali to ethanal and to benzaldehyde results in reactions of different types. State the products of the two reactions.

10. Of the compounds **A, B, C** and **D**, which compound (a) undergoes the aldol reaction, (b) gives a positive iodoform test, (c) undergoes the Cannizzaro reaction and (d) is reduced to a secondary alcohol?

A, $CH_3CH_2COCH_2CH_3$ **B,** $CH_3CH_2CH_2CH_2CHO$

C, $C_6H_5COCH_3$ **D,** C_6H_5CHO

11. Hexanol and hexanal are present in solution in ethoxyethane. How can you obtain a pure sample of hexanal without distillation?

*12. Which of the following compounds reacts faster with sodium iodate(I): $CH_3CH(OH)CH_2CH_2OH$ or $CH_3CH_2COCH_3$? Explain your choice and state the products of the reactions.

*13. The compound 2-chloro-1-phenylethanone is a powerful lachrymator. Write its structural formula, and say how it could be made from benzene.

14. From the formula [p. 654] it might appear that the bonds in hexamethylene tetramine are strained. Construct a 3D model of the molecule. What bond angles do you find? What structure does the arrangement of atoms call to mind? (Remember to include the lone pairs on the nitrogen atoms.)

31.7 THE MECHANISMS OF THE REACTIONS OF ALDEHYDES AND KETONES

In the $\overset{\delta+}{\diagdown}C = \overset{\delta-}{O}$ *group, there are two reactive sites...*

Aldehydes and ketones possess the polar carbonyl group

$$\overset{\delta+}{\diagdown}C = \overset{\delta-}{O}$$

Carbonyl compounds undergo addition reactions and condensation reactions. There are two possibilities to be considered: the $\delta-$ oxygen atom of the carbonyl group may react with an electrophile, or the $\delta+$ carbon atom of the carbonyl group may react with a nucleophile.

31.7.1 THE ADDITION OF HYDROGEN CYANIDE TO FORM A CYANOHYDRIN

...In the addition of HCN, it is the $\overset{\delta+}{C}$ that is attacked by CN^- ...

$$\diagdown C = O + HCN \rightarrow \diagdown C \diagup^{OH}_{CN}$$

The rate of reaction is increased by the presence of a base and decreased by hydrogen ions. This is what would happen if the concentration of cyanide ions were a crucial factor. Hydrogen cyanide dissociates:

$$HCN(aq) + H_2O(l) \rightleftharpoons H_3O^+(aq) + CN^-(aq)$$

...The evidence comes from kinetic studies

Addition of hydrogen ions suppresses dissociation, while addition of hydroxide ions removes hydrogen ions and increases the degree of dissociation and the concentration of CN^- ions. It is likely that the rate-determining step (R.D.S.) involves CN^- ions. There are two possibilities:

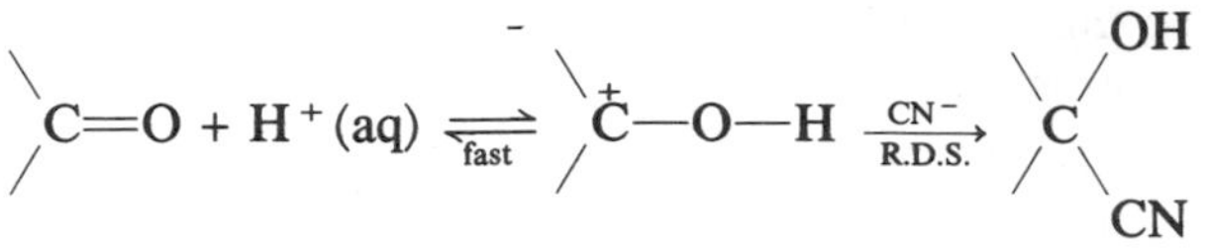

$$\ce{>C=O} + H^+(aq) \underset{\text{fast}}{\rightleftharpoons} \ce{>\overset{+}{C}-O-H} \xrightarrow[\text{R.D.S.}]{CN^-} \ce{>C<^{OH}_{CN}} \qquad [1]$$

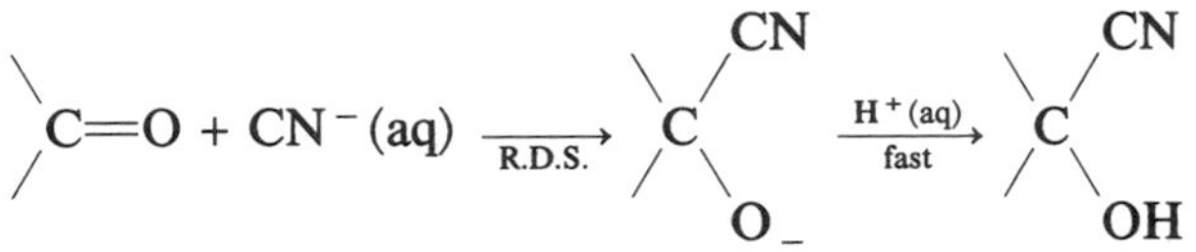

$$\ce{>C=O} + CN^-(aq) \xrightarrow{\text{R.D.S.}} \ce{>C<^{CN}_{O^-}} \xrightarrow[\text{fast}]{H^+(aq)} \ce{>C<^{CN}_{OH}} \qquad [2]$$

In [1], the step involving CN^- is the attack on a positive ion. This type of reaction is fast, and is unlikely to be the rate-determining step. In [2], CN^- is involved in an attack on the polar

$$\overset{\delta+}{C}=\overset{\delta-}{O} \text{ group}$$

The second step is a fast reaction between $H^+(aq)$ and a negatively charged intermediate. Mechanism [2] involves CN^- in the rate-determining step, and is more likely to be correct.

A summary of the mechanism of cyanohydrin formation

The proposed mechanism for the addition reactions of aldehydes and ketones is illustrated by the formation of a cyanohydrin:

$$HCN(aq) \rightleftharpoons H^+(aq) + CN^-(aq)$$

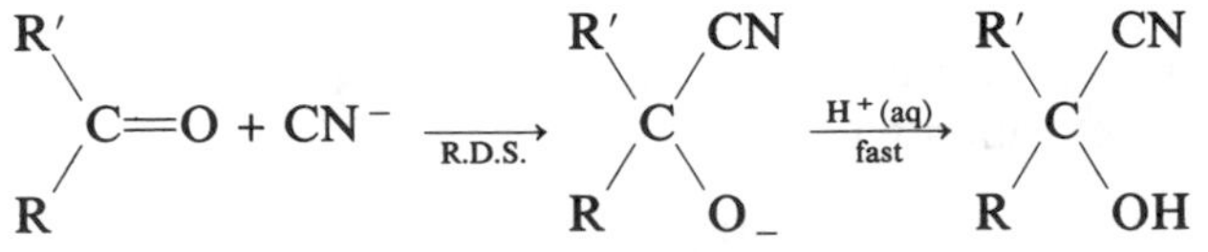

$$\ce{R'R>C=O} + CN^- \xrightarrow{\text{R.D.S.}} \ce{R'R>C<^{CN}_{O^-}} \xrightarrow[\text{fast}]{H^+(aq)} \ce{R'R>C<^{CN}_{OH}}$$

31.7.2 THE MECHANISMS OF ADDITION–ELIMINATION REACTIONS

In the addition–elimination reactions between a carbonyl compound and hydroxylamine...

One example is the reaction of an aldehyde or ketone with hydroxylamine to form an oxime:

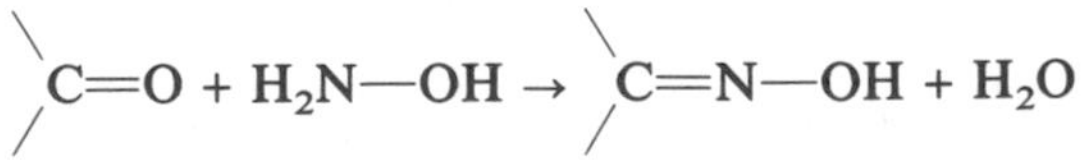

$$\ce{>C=O} + H_2N-OH \rightarrow \ce{>C=N-OH} + H_2O$$

...firstly, the nucleophile
adds...

The spectroscopic studies and kinetic studies that have been made of this reaction have led to the formulation of the following mechanism.

(a) The addition of the nucleophile to form an intermediate is followed by a fast rearrangement:

...then the intermediate
rearranges...

$$
\begin{array}{ccc}
\text{R}' \\
\diagdown \\
\text{C}=\text{O} + \text{H}_2\overset{\bullet\bullet}{\text{N}}-\text{X} \rightarrow \\
\diagup \\
\text{R}
\end{array}
\quad
\begin{array}{c}
\text{R}' \quad \overset{+}{\text{N}}\text{H}_2-\text{X} \\
\diagdown\diagup \\
\text{C} \\
\diagup\diagdown \\
\text{R} \quad \text{O}_-
\end{array}
\quad \xrightarrow{\text{fast}} \quad
\begin{array}{c}
\text{R}' \quad \text{NH}-\text{X} \\
\diagdown\diagup \\
\text{C} \\
\diagup\diagdown \\
\text{R} \quad \text{OH}
\end{array}
$$

...A proton adds...

(b) The addition of a proton is followed by the elimination of H_2O:

...and water is eliminated

$$
\begin{array}{c}
\text{R}' \quad \text{NH}-\text{X} \\
\diagdown\diagup \\
\text{C} \\
\diagup\diagdown \\
\text{R} \quad \text{OH}
\end{array}
\quad \underset{\text{H}^+\text{(aq)}}{\rightleftharpoons} \quad
\begin{array}{c}
\text{R}' \quad \overset{\bullet\bullet}{\text{N}}\text{H}-\text{X} \\
\diagdown\diagup \\
\text{C} \\
\diagup\diagdown \\
\text{R} \quad \overset{+}{\text{O}}\text{H}_2
\end{array}
\quad \rightarrow \quad
\begin{array}{c}
\text{R}' \\
\diagdown \\
\text{C}=\text{N}-\text{X} + \text{H}_3\text{O}^+ \\
\diagup \\
\text{R}
\end{array}
$$

R is an alkyl group; R′ is an alkyl group or hydrogen.

$X = -OH$ or $-NH_2$ or $-NHC_6H_5$ or $-NHC_6H_3(NO_2)_2$.

31.7.3 A COMPARISON OF ALKENES AND CARBONYL COMPOUNDS

*Addition to C=O starts
with attack by a
nucleophile; addition to
C=C starts with attack
by an electrophile*

The addition reactions of carbonyl compounds contrast with those of alkenes. The first step in addition to

$$
\begin{array}{c}
\diagdown \\
\text{C}=\text{O} \\
\diagup
\end{array}
$$

is the addition of a nucleophile, Nu: to form

$$
\begin{array}{c}
\diagdown \\
\text{C}-\text{O}^- \\
\diagup| \\
\text{Nu}
\end{array}
$$

In the addition to

$$
\begin{array}{c}
\diagdown\diagup \\
\text{C}=\text{C} \\
\diagup\diagdown
\end{array}
$$

the first step is the addition of an electrophile. A nucleophile does not add to a carbon–carbon double bond because it is repelled by the unpolarised π electrons of the C=C bond.

The hypothetical adduct

$$
\begin{array}{c}
\diagdown\diagup \\
\text{C}-\text{C} \\
\diagup|^-\diagdown \\
\text{Nu}
\end{array}
$$

is not formed.

Carbon is less electronegative than oxygen, and this type of species is much less stable than

FIGURE 31.8 Aldehydes
in the Preparation of
Other Classes of
Compounds

where the negative charge is carried by an oxygen atom.

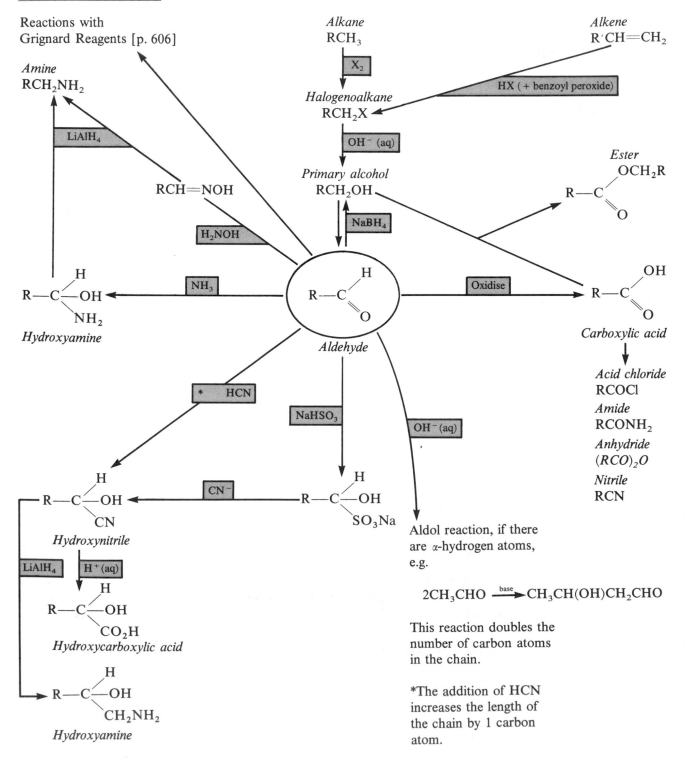

Reactions with
Grignard Reagents [p. 606]

Amine
RCH_2NH_2

$LiAlH_4$

$RCH{=}NOH$

H_2NOH

NH_3

$R{-}\overset{\displaystyle H}{\underset{\displaystyle NH_2}{C}}{-}OH$

Hydroxyamine

Alkane
RCH_3

X_2

Halogenoalkane
RCH_2X

OH^- (aq)

Primary alcohol
RCH_2OH

$NaBH_4$

$R{-}\overset{\displaystyle H}{\underset{\displaystyle O}{C}}$

Aldehyde

Alkene
$R'CH{=}CH_2$

HX (+ benzoyl peroxide)

Ester
$R{-}\overset{\displaystyle OCH_2R}{\underset{\displaystyle O}{C}}$

Oxidise

$R{-}\overset{\displaystyle OH}{\underset{\displaystyle O}{C}}$

Carboxylic acid

Acid chloride
$RCOCl$

Amide
$RCONH_2$

Anhydride
$(RCO)_2O$

Nitrile
RCN

*   HCN

$NaHSO_3$

OH^- (aq)

$R{-}\overset{\displaystyle H}{\underset{\displaystyle CN}{C}}{-}OH$

Hydroxynitrile

CN^-

$R{-}\overset{\displaystyle H}{\underset{\displaystyle SO_3Na}{C}}{-}OH$

$LiAlH_4$ H^+ (aq)

$R{-}\overset{\displaystyle H}{\underset{\displaystyle CO_2H}{C}}{-}OH$

Hydroxycarboxylic acid

$R{-}\overset{\displaystyle H}{\underset{\displaystyle CH_2NH_2}{C}}{-}OH$

Hydroxyamine

Aldol reaction, if there
are α-hydrogen atoms,
e.g.

$$2CH_3CHO \xrightarrow{\text{base}} CH_3CH(OH)CH_2CHO$$

This reaction doubles the
number of carbon atoms
in the chain.

*The addition of HCN
increases the length of
the chain by 1 carbon
atom.

CHECKPOINT 31D: REACTIVITY

1. The alkene group reacts with compounds of formula **HX**. What can **X** be?
The carbonyl group reacts with compounds of formula **HY**. What can **Y** be? Compare the reactivity of the alkene group and the carbonyl group in reactions of this type.

2. The formation of butanone cyanohydrin is speeded up by the addition of cyanide ions and retarded by the addition of hydrogen ions. Explain these observations.

3. (a) Sketch the activated complex that is formed in the reaction

$$OH^- + CH_3CH_2Br \rightarrow CH_3CH_2OH + Br^-$$

How many pairs of bonding electrons surround the carbon atom attached to the halogen? Which atoms carry negative charge?

(b) Sketch the intermediate in the reaction

$$HCN + CH_3CHO \rightarrow CH_3CH(OH)CN$$

How many pairs of bonding electrons surround the central carbon atom? Which atoms carry negative charge?

(c) Which of the two species, the activated complex in (a) or the intermediate in (b), appears to you to be the more stable? Explain your answer.

*4. Why does HBr attack $\diagdown C = C \diagup$ but not $\diagdown C = O$?

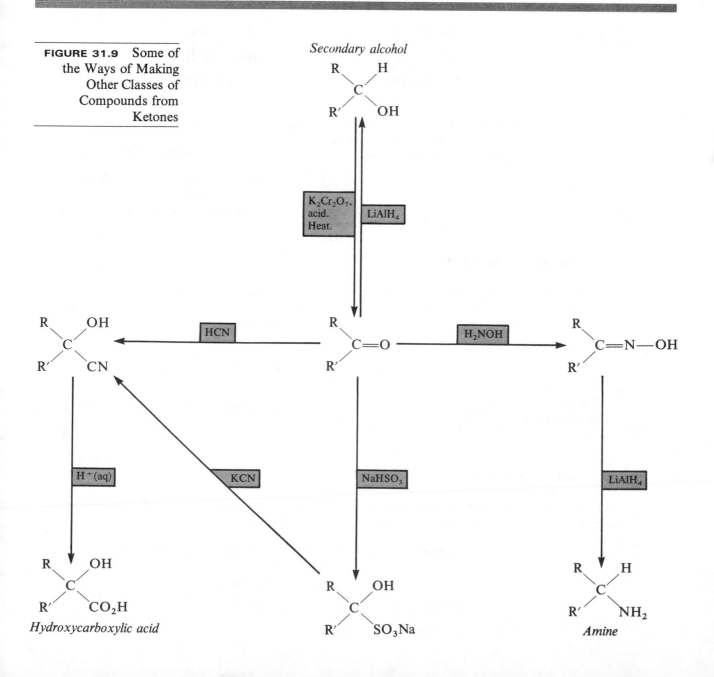

FIGURE 31.9 Some of the Ways of Making Other Classes of Compounds from Ketones

31.8 CARBOHYDRATES

The general formula of carbohydrates is $C_m(H_2O)_n$

Carbohydrates are an important, naturally occurring group of compounds which contain carbonyl groups. This group includes sugars and starch, cellulose and a number of antibiotics. The name carbohydrate is derived from the molecular formulae of these compounds, which in many cases can be written $C_m(H_2O)_n$, e.g., glucose, $C_6H_{12}O_6$.

Sugars contain one or more carbonyl groups and other functional groups

Carbohydrates are polyfunctional. Glucose contains five hydroxyl groups as well as an aldehyde group. Fructose contains a ketone group. Both glucose and fructose have the formula $C_6H_{12}O_6$ and are monosaccharides:

$$
\begin{array}{ccc}
\text{CHO} & & \text{CH}_2\text{OH} \\
\text{H—C—OH} & & \text{C=O} \\
\text{HO—C—H} & & \text{HO—C—H} \\
\text{H—C—OH} & & \text{H—C—OH} \\
\text{H—C—OH} & & \text{H—C—OH} \\
\text{CH}_2\text{OH} & & \text{CH}_2\text{OH}
\end{array}
$$

(+)-Glucose is an **aldose**.
It reduces Fehling's solution and Tollens' reagent and also reduces, and therefore decolorises, bromine water.

(−)-Fructose is a **ketose**.
The —OH groups affect the properties of the $\diagdown$C=O group. Unlike other ketones, fructose reduces Fehling's solution and Tollens' reagent. It does not reduce and decolorise bromine water.

31.8.1 DISACCHARIDES

Sucrose, maltose and lactose are disaccharides

Disaccharides have the formula $C_{12}H_{22}O_{11}$. They are hydrolysed by dilute acids and by enzymes to monosaccharides. Sucrose (cane or beet sugar), maltose (in malt) and lactose (in milk) are the commonest disaccharides.

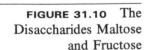

Maltose: 2 glucose ring molecules linked by the elimination of H_2O between 2OH groups

---
FIGURE 31.10 The Disaccharides Maltose and Fructose

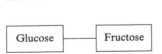

Sucrose: a glucose molecule and a fructose molecule linked as in maltose

31.8.2 POLYSACCHARIDES

Starch is made by photosynthesis...

Starch and cellulose are polysaccharides. They are polymers of glucose. Starch is formed as a result of photosynthesis in green plants:

$$6CO_2(g) + 6H_2O(l) \xrightarrow[\substack{\text{sunlight and chlorophyll} \\ \text{in green plants}}]{\text{photosynthesis}} C_6H_{12}O_6(aq) + 6O_2(g)$$

Glucose

$$nC_6H_{12}O_6(aq) \rightarrow (C_6H_{10}O_5)_n(s) + nH_2O(l)$$

Glucose   Starch

...A test for starch is the formation of a blue colour with iodine

Starch is found in potatoes, cereals and rice. With iodine, it gives an intense blue colour, which is used as a test for iodine or for starch. Starch does not reduce Fehling's solution or Tollens' reagent.

Cellulose is another polymer of glucose

Cellulose is the main structural component of cell walls, and wood contains large quantities of cellulose. It cannot be digested by animals: they do not produce the enzymes which will hydrolyse it. Cows have living in their alimentary canals bacteria which hydrolyse cellulose.

QUESTIONS ON CHAPTER 31

1. Pentan-2-ol yields a ketone when heated with potassium dichromate solution under acid conditions.
(*a*) Give the structural formula and name of the ketone.
(*b*) What is the change in the oxidation number of chromium during the reaction?
(*c*) What other reagents will oxidise the alcohol?
(*d*) How can the product be identified as a ketone?

2. Compare the reactions of ethanal and propanone with
(*a*) a few drops of concentrated hydrochloric acid,
(*b*) aqueous ammoniacal silver nitrate, (*c*) 2,4-dinitrophenyl-hydrazine, (*d*) sodium hydrogensulphite.

3. Compare and contrast the addition reactions of ethene and ethanal. Include a discussion of the mechanisms of the reactions.

4. Suggest a method of synthesis for phenylethanone, $C_6H_5COCH_3$. Summarise the addition and addition–elimination reactions of the compound.

5. State the reactions of propanone with (*a*) hydroxylamine, (*b*) lithium tetrahydridoaluminate, (*c*) iodine and alkali, (*d*) hydrogen cyanide. Why is reaction (*d*) so slow in the absence of a catalyst? What catalyst is used to speed up the reaction? How does it function?

6. Outline a method for the preparation of propanone from propene. What structural isomer of propanone exists? How would you distinguish between the two isomers?

7. Three different compounds of formula C_8H_8O give precipitates with 2,4-dinitrophenylhydrazine and are reduced by $LiAlH_4$ to compounds of formula $C_8H_{10}O$. Suggest structures for the three compounds, and say how you could distinguish between them. The isomerism does not arise from a difference in the orientation of substituents in the benzene ring.

8. A compound, **P**, of formula C_6H_{10}, was subjected to reductive ozonolysis. **Q**, $C_6H_{10}O_2$, was formed. **Q** gave a silver mirror with ammoniacal silver nitrate solution and reacted with 2,4-dinitrophenylhydrazine. One mole of **Q** reacted with 2 moles of DNP to form a yellow, crystalline solid. Deduce the identity of **P** and **Q**.

9. A liquid, **A**, of formula C_7H_6O, was refluxed with concentrated aqueous sodium hydroxide. Ethoxyethane extraction of the resulting solution, followed by drying and distillation, gave as the main product a liquid, **B**, of boiling temperature 205 °C. **B** reacted with phosphorus(V) chloride to give a pungent gas. When the solution from which **B** had been extracted was acidified, a white crystalline solid, **C**, precipitated. On recrystallisation from hot water it appeared as flaky white crystals. A solution of **C** effervesced when sodium hydrogencarbonate was added. Identify **A**, **B** and **C**, and explain the reactions described.

10. (*a*) What type of reaction takes place between benzenecarbaldehyde (benzaldehyde) and 2,4-dinitrophenyl-hydrazine? Under what conditions does reaction occur?
(*b*) Write equations for the reactions of benzenecarbaldehyde with (i) 2,4-dinitrophenylhydrazine, (ii) hydroxylamine, (iii) hydrazine. In reaction (iii), two products may be formed, with molecular formulae $C_7H_8N_2$ and $C_{14}H_{12}N_2$. Write structural formulae for these compounds.

11. **A** has the formula $C_5H_{12}O$. On oxidation it gives **B**, of formula $C_5H_{10}O$. **B** reacts with phenylhydrazine and gives a positive result in an iodoform test. **A** is dehydrated by concentrated sulphuric acid to **C**, C_5H_{10}. Reductive ozonolysis of **C** gives butanal. What is **A**?

12. **P** has the formula $C_5H_8O_2$. It forms a compound by reaction with sodium hydrogensulphite which has the formula $C_5H_8O_2(SO_3HNa)_2$. **P** gives a positive iodoform test, a silver mirror with Tollens' reagent and can be reduced to pentane. What is **P**?

13. **Q**, of formula $C_5H_{12}O_2$, reacts with ethanoic anhydride to form **R**, formula $C_9H_{16}O_4$. **Q** does not react with phenylhydrazine, but on oxidation it forms **S**, $C_5H_8O_2$, which reacts with phenylhydrazine to form $C_5H_8(=NNHC_6H_5)_2$, and also reduces Fehling's solution. **S** reacts with sodium iodate(I) to give iodoform and butane-1,4-dioic acid. Deduce the identity of **Q**, and explain the reactions described.

14. Describe chemical tests which you could do to distinguish between the following compounds:
(*a*) C_2H_5CHO, CH_3COCH_3, $C_2H_5CH_2OH$ and $(CH_3)_2CHOH$
(*b*) C_6H_5CHO, $C_6H_5CH_2OH$, $C_6H_5COCH_3$ and $C_6H_5CH=CH_2$

15. How can the following conversions be brought about?
(*a*) $CH_3COCH_3 \rightarrow CH_3COCH_2OH$
(*b*) $CH_3COCH_3 \rightarrow CH_3CH=CH_2$
(*c*) $C_6H_5CHO \rightarrow C_6H_5CH_2Br$

16. (*a*) Compare the reactions of ethanal and propanone with (i) aqueous ammoniacal silver nitrate, (ii) ammonia, (iii) hydrogen cyanide, (iv) 2,4-dinitrophenylhydrazine.
(*b*) How does the presence of a catalyst assist reaction (iii)?
(*c*) Explain why there is only one oxime of propanone, while there are two isomeric oximes of ethanal.

17. A compound **X** has the formula $C_5H_{10}O$. It gives a 2,4-dinitrophenylhydrazone. Ammoniacal silver nitrate oxidises **X** to **Y**. Sodium tetrahydridoborate reduces **X** to **Z**. With aqueous potassium hydroxide, **X** gives a mixture of **Z** and the potassium salt of **Y**. Deduce the identity of

X, **Y** and **Z**, and explain the reactions described.

*18. The formula of glucose is

CHO
|
(CHOH)$_4$
|
CH$_2$OH

From your knowledge of the reactions of —OH groups and —CHO groups, say how you would prepare the following substances from glucose:

(*a*) hexane-1,2,3,4,5,6-hexol

(*b*) 2,3,4,5,6-pentahydroxyhexanoic acid

(*c*) sugar charcoal

(*d*) pentaethanoylglucose.

19. This question is concerned with the properties of ethanal (acetaldehyde), a simple aldehyde.

(*a*) Give a reagent, or set of reagents, which will convert ethanal into:

 (i) CH$_3$CO$_2$H

 (ii) CHI$_3$ + HCO$_2$Na

 (iii) CH$_3$CH(OH)$_2$

(*b*) Give balanced equations for reactions (*a*) (ii) and (iii) above.

(*c*) Give the structure of the products obtained when ethanal reacts with

 (i) 2,4-dinitrophenylhydrazine

 (ii) hydrogen cyanide

 (iii) dilute, aqueous alkali

(*d*) Calculate the enthalpy of reaction (*a*) (iii) above on the assumption that the reaction takes place between gaseous reagents to give gaseous products. Appropriate bond enthalpies (kJ mol^{-1}) are:

C=O 743; C—O 360; O—H 463.

(*e*) What other data would you require to enable you to convert the enthalpy of the reaction which you obtained in (*d*) into the enthalpy of the more realistic reaction between liquid phase reagents to give dilute, aqueous CH$_3$CH(OH)$_2$? [You are not expected to quote values for these data or to work out the enthalpy of this reaction.]

[See Chapter 10 for (*d*) and (*e*).]   (O & C 82)

20. Compound **A** represents benzaldehyde. The scheme below describes the type of reactions involved and the molecular formula of the products.

Product

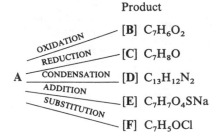

A — OXIDATION → [B] C$_7$H$_6$O$_2$
— REDUCTION → [C] C$_7$H$_8$O
— CONDENSATION → [D] C$_{13}$H$_{12}$N$_2$
— ADDITION → [E] C$_7$H$_7$O$_4$SNa
— SUBSTITUTION → [F] C$_7$H$_5$OCl

(*a*) For each reaction identify the compound by either name or structural formula, naming the reagents responsible and stating the conditions necessary.

(*b*) How does the addition reaction involved in producing **E** differ from addition reactions involving alkenes?

(*c*) How does the reaction of benzaldehyde with concentrated alkali differ from that of ethanal?

(*d*) Give *two* different chemical tests which could be used to distinguish between benzaldehyde and phenylethanone (acetophenone).

(AEB 81)

32

AMINES

32.1 NOMENCLATURE FOR AMINES

Primary, secondary and tertiary (1°, 2°, 3°) amines...

Amines are organic derivatives of ammonia. They are classified as primary, secondary or tertiary amines according to the number of alkyl or aryl groups attached to the nitrogen atom. Examples are:

CH_3NH_2

$$\begin{array}{c} CH_3 \\ \diagdown \\ NH \\ \diagup \\ CH_3 \end{array}$$

$$\begin{array}{c} CH_3 \\ CH_3-N \\ CH_3 \end{array}$$

CH_2NH_2 (on benzene ring)

...some examples...

Methylamine
a primary
(1°) amine

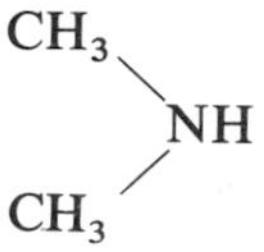

Dimethylamine
a secondary
(2°) amine

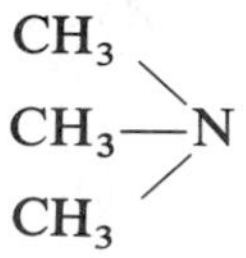

Trimethylamine
a tertiary
(3°) amine

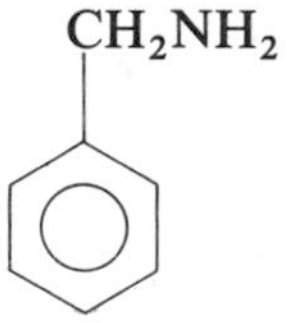

Phenylmethylamine
(or benzylamine)
1° amine

NH_2 (on benzene ring)

NH_2 (on benzene ring with CH_3)

$NHCH_3$ (on benzene ring)

$N(CH_3)_2$ (on benzene ring)

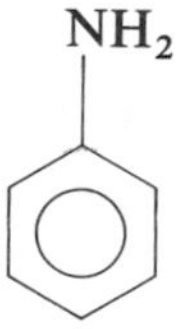

Phenylamine
(or aniline)
1° aromatic
amine

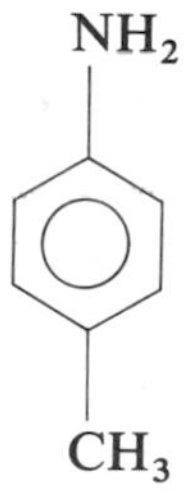

4-Methyl-
phenylamine

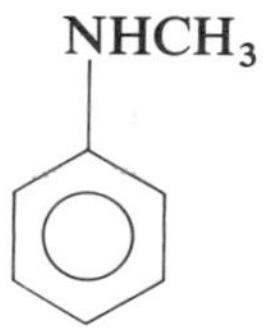

N-Methyl-
phenylamine
2° aromatic
amine

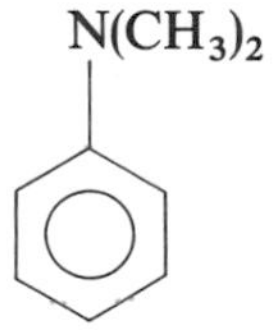

N,N-Dimethyl-
phenylamine
3° aromatic
amine

An aromatic amine has the nitrogen atom directly attached to the aromatic ring. In phenylmethylamine the nitrogen atom is not directly attached to the ring: this is not an aromatic amine; it is a phenyl-substituted alkylamine.

...an alternative name...

...amino-compounds

One system of naming is illustrated by the examples given above. The alkyl groups or aryl groups attached to the nitrogen atom are named, and the ending *-amine* is added. Alternatively, the compounds may be named as amino-substituted alkanes or arenes:

$CH_3CH_2CH_2NH_2$

$$\begin{array}{c} CH_3CHNH_2 \\ | \\ CH_3 \end{array}$$

$H_2NCH_2CO_2H$

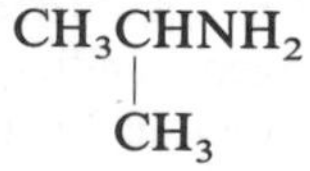

$HO-\langle\bigcirc\rangle-NH_2$

1-Aminopropane

2-Aminopropane

2-Aminoethanoic
acid

4-Aminophenol

Related to the amines are the **quaternary** (4°) ammonium compounds, in which nitrogen is tetravalent. They are named as alkyl-substituted ammonium salts:

Quaternary ammonium compounds

$(CH_3)_4N^+I^-$

$$CH_3\overset{\overset{\displaystyle C_2H_5}{|}}{\underset{\underset{\displaystyle CH_3}{|}}{N^{\pm}}}C_3H_7\ Br^-$$

Tetramethylammonium iodide

Ethyldimethylpropylammonium bromide

Other nitrogen compounds will be covered in Chapter 33. They are the amides, nitriles and isonitriles. They are named as derivatives of acids:

Amides, nitriles, isonitriles

Carboxylic acid $R\overset{\overset{\displaystyle O}{\|}}{-C}-OH$ e.g. $CH_3CH_2\overset{\overset{\displaystyle O}{\|}}{C}-OH$
Propanoic acid

Amide $R\overset{\overset{\displaystyle O}{\|}}{-C}-NH_2$ e.g. $CH_3CH_2\overset{\overset{\displaystyle O}{\|}}{C}-NH_2$
Propanamide

Nitrile $R-C\equiv N$ e.g. $CH_3CH_2C\equiv N$
Propanonitrile

Isonitrile $R-\overset{+}{N}\equiv\bar{C}$ e.g. $CH_3CH_2\overset{+}{N}\equiv\bar{C}$
Isocyanoethane

Other derivatives of acids which will be met in this chapter are:

Acid chlorides and anhydrides

Acid chloride $R\overset{\overset{\displaystyle O}{\|}}{-C}-Cl$ e.g. $CH_3CH_2\overset{\overset{\displaystyle O}{\|}}{C}-Cl$
Propanoyl chloride

Acid anhydride

$$R-C\overset{\displaystyle O}{\diagup}$$
$$\diagdown O$$
$$R-C\diagup$$
$$\diagdown O$$

e.g.

$$CH_3CH_2C\overset{\displaystyle O}{\diagup}$$
$$\diagdown O$$
$$CH_3CH_2C\diagup$$
$$\diagdown O$$

Propanoic anhydride

32.2 NATURAL OCCURRENCE

Proteins...

...amino acids...

...DNA...

Compounds with amino groups are widely distributed in nature. Amino acids, of formula $H_2NCH(R)CO_2H$, are the building blocks from which proteins are made. Their chemistry is described on p. 713. The way in which hydrogen bonding between amino groups and other groups maintains the three-dimensional configuration of proteins is illustrated in Figure 4.25, p. 86. The bases in DNA (deoxyribonucleic acid) are amines. Hydrogen bonding between the bases maintains the double helical structure of DNA, as shown in Figures 4.26, 4.27, 4.28, pp. 87–8.

...drugs

Many powerful drugs and medicines are amines, e.g., morphine (a powerful pain-killer), strychnine (a powerful poison) and LSD (a hallucinogen).

32.3 PHYSICAL PROPERTIES

Intermolecular hydrogen bonding is weak

The N—H bond is polar, more polar than C—H but less polar than O—H. Intermolecular hydrogen bonding in amines is therefore weaker than in alcohols, and the lower molecular mass amines (up to C_3) are gases at room temperature. The lower molecular mass amines dissolve in water as they can form hydrogen bonds with water molecules. Phenylamine and other amines with a large hydrocarbon part of the molecule are only sparingly soluble in water but are soluble in organic solvents. Since phenylamine is soluble in fatty tissues, it can be absorbed through the skin. The ease of absorption, combined with its toxicity, makes phenylamine a somewhat dangerous substance.

Solubility

Toxicity

Odour

Amines have a characteristic smell. That of the lower members of the series resembles ammonia. Higher members have an odour described as 'fishy'. This is because amines are formed when protein material decomposes; dimethylamine and trimethylamine are found in rotting fish. Higher amines are found in decaying animal flesh.

Amines are formed when proteins decay

32.4 INDUSTRIAL MANUFACTURE AND USES

32.4.1 METHYLAMINE AND ETHYLAMINE

Alcohols are a source of amines

Methylamine is made by passing methanol vapour with ammonia under pressure over alumina as catalyst at 400 °C:

$$CH_3OH(g) + NH_3(g) \xrightarrow[400\,°C]{Al_2O_3} CH_3NH_2(g) + H_2O(g)$$

Dimethylamine and trimethylamine are also formed. Dimethylamine is used for the manufacture of solvents, for jet fuel and rocket fuel.

Ethylamine is made in a similar way from ethanol and ammonia.

32.4.2 PHENYLAMINE

Phenylamine is used to make dyes and rubber...

Phenylamine is used for the manufacture of dyes [p. 683]. Its derivatives find use in the rubber industry. There are two chief industrial methods of manufacture:

...It is manufactured industrially from nitrobenzene or chlorobenzene

(a) Reduction of nitrobenzene, $C_6H_5NO_2$, by
 (i) catalytic hydrogenation, or
(ii) iron and hydrochloric acid (similar to the laboratory method [p. 675]).

(b) Reaction between ammonia and chlorobenzene at 200 °C, under high pressure, in the presence of copper(I) oxide as catalyst.

32.5 BASICITY OF AMINES

Amines are weak bases, resembling ammonia

The nitrogen atom in ammonia has a lone pair of electrons, which enable it to act as a base ($pK_b = 4.74$) [p. 250]. In the same way, amines are weak bases. Their solutions are alkaline, and they react with acids to form salts:

$$\underset{\underset{H}{\overset{\displaystyle H}{|}}}{CH_3-\overset{\bullet}{\underset{}{N}}{\bullet}} + H_2O \xrightleftharpoons[pK_b = 3.36]{} \underset{\underset{H}{\overset{\displaystyle H}{|}}}{CH_3-\overset{+}{N}-H} + OH^-$$

$$(CH_3)_3N\!: + H_2O \underset{pK_b = 4.20}{\rightleftharpoons} (CH_3)_3\overset{+}{N}\!-\!H + OH^-$$

The gas methylamine reacts with hydrogen chloride to form a white crystalline solid, methylammonium chloride, $CH_3NH_3^+ Cl^-$. The salt is involatile, odourless and soluble in water. The addition of a strong base, such as sodium hydroxide, liberates the weak base methylamine from its salt:

$$CH_3NH_3^+ Cl^- (s) + OH^- (aq) \rightarrow CH_3NH_2(g) + H_2O(l) + Cl^- (aq)$$

Trimethylamine forms trimethylammonium salts, e.g., $(CH_3)_3\overset{+}{N}H$ Br^-, and the salts of phenylamine are called phenylammonium salts, e.g., $C_6H_5NH_3^+ I^-$.

Aliphatic amines are more basic than ammonia...

Aliphatic amines are stronger bases than ammonia in aqueous solution because the positive charge in the alkylammonium ion can be shared between the nitrogen atom and a carbon atom. This distribution of charge stabilises the cation. In secondary and tertiary amines, the positive charge is spread over two or three carbon atoms in addition to the nitrogen atom. Basicity might be expected to increase in the order

$$NH_3 < 1° \text{ amine} < 2° \text{ amine} < 3° \text{ amine}$$

...The order of basic strengths depends on spreading the positive charge and on hydrogen bonding...

Another factor which comes into play is the stabilisation of the alkylammonium ion through solvation by water molecules:

...These factors combine to give
$$NH_3 < 3° < 1° < 2°$$

The number of hydrogen atoms available for hydrogen bonding increases in the order

$$3° \text{ amine} < 2° \text{ amine} < 1° \text{ amine} < NH_3$$

The combination of the two factors leads to the order of basic strengths

$$NH_3 < 3° \text{ amine} < 1° \text{ amine} < 2° \text{ amine}$$

This order is not always followed: bulky substituents may decrease the basicity of an amine.

Aromatic amines are weaker bases because of the delocalisation of the N lone pair

Aromatic amines are weaker bases. The electron pair on the nitrogen atom is partially delocalised by interaction with the π electron cloud of the benzene ring [see Figure 32.1]. As a result of delocalisation, the lone pair is less available for coordination to a proton.

FIGURE 32.1
Phenylamine

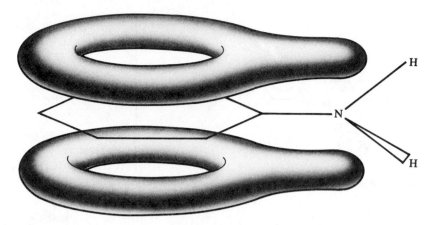

Substituents affect basicity

Substituents in the benzene ring affect the basicity of phenylamine. Electron-donating groups, e.g. CH₃O— and HO—, especially in the *para*-position, increase the basicity; electron-withdrawing groups, e.g. —NO₂ and —Cl, decrease the basicity [see Table 32.1].

Base	pK_b
NH_3	4.75
CH_3NH_2	3.36
$(CH_3)_2NH$	3.28
$(CH_3)_3N$	4.20
$C_6H_5NH_2$	9.38
$4\text{-}HOC_6H_4NH_2$	8.53
$4\text{-}O_2NC_6H_4NH_2$	13.02
$4\text{-}ClC_6H_4NH_2$	10.07

TABLE 32.1 Values of pK_b for Some Amines

32.5.1 NOTE ON AMIDES

Amides, RCONH₂, are not bases

Amides

$$R-\overset{\displaystyle O}{\overset{\|}{C}}-NH_2$$

are not basic. The electron-withdrawing character of the carbonyl group reduces the electron density on the nitrogen atom:

$$-\overset{\displaystyle O}{\overset{\displaystyle \nwarrow}{C}}\overset{H}{\underset{\displaystyle \underset{H}{N:}}{}}$$

━━━━━━━━━━━━━━━ **CHECKPOINT 32A: FORMULAE** ━━━━━━━━━━━━━━━

1. Draw structural formulae for (*a*) ethylmethylpropylamine, (*b*) 2-aminoheptane, (*c*) 3-amino-4-chloro-octane, (*d*) diethyldimethylammonium chloride, (*e*) *N,N*-dimethyl-phenylamine and (*f*) 4-nitrophenylamine.

2. Name the following compounds, and say whether they are 1°, 2°, 3° or 4° amines:

(*a*) CH₃—N—H
        |
        CH₃

(b) C_2H_5—N—C_2H_5
                |
               CH_3

(c)
        (two benzene rings attached to N—H)

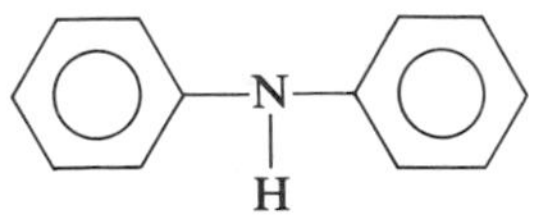

(d) I——⟨ring⟩——NHC_2H_5

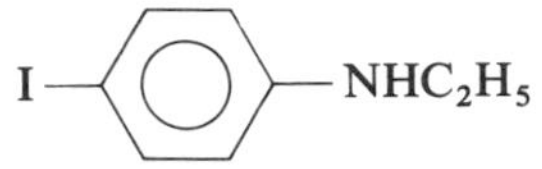

(e)
$$CH_3—\overset{\overset{CH_3}{|}}{\underset{\underset{CH_3}{|}}{N}}—C_2H_5 \; Br^-$$

(f) H_3C——⟨ring⟩——$NHCH_3$

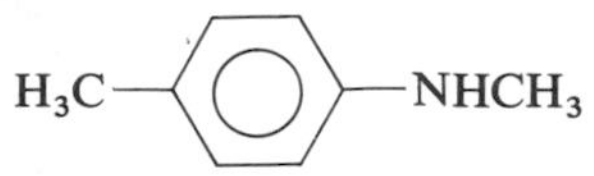

3. Which of the following has the highest boiling temperature?

$(C_2H_5)_3N$ $(CH_3CH_2CH_2)_2NH$ $CH_3(CH_2)_5NH_2$

32.6 LABORATORY PREPARATIONS

32.6.1 THE REACTION OF AMMONIA AND A HALOGENOALKANE

RX + NH₃ form a mixture of 1°, 2°, 3° and 4° ammonium salts

Ammonia and a halogenoalkane (in solution in ethanol) are heated together in a **bomb** (a metal container which will withstand high pressure):

$$RX(alc) + NH_3(alc) \xrightarrow{\text{heat in a bomb}} R\overset{+}{N}H_3 \, X^-(s)$$

A mixture of primary, secondary, tertiary and quaternary ammonium salts is formed:

$$NH_3 \xrightarrow{CH_3I} CH_3\overset{+}{N}H_3 \, I^- \xrightarrow{CH_3I} (CH_3)_2\overset{+}{N}H_2 \, I^- \xrightarrow{CH_3I} (CH_3)_3\overset{+}{N}H \, I^-$$
$$+ \, HI \qquad\qquad + \, HI$$

$$\xrightarrow{CH_3I} (CH_3)_4\overset{+}{N} \, I^-$$
$$+ \, HI$$

The products are separated by the addition of alkali to liberate the free amines, followed by fractional distillation.

32.6.2 REDUCTION OF NITROGEN COMPOUNDS

(a) Reduction of a Hydroxyamine

RCH(OH)NH₂ is reduced by H₂ in the presence of a catalyst

Aldehydes react with ammonia to form hydroxyamines. These adducts can be made and reduced in one combined reaction by allowing the carbonyl compound to react with ammonia and hydrogen at a high temperature in the presence of a nickel and chromium catalyst:

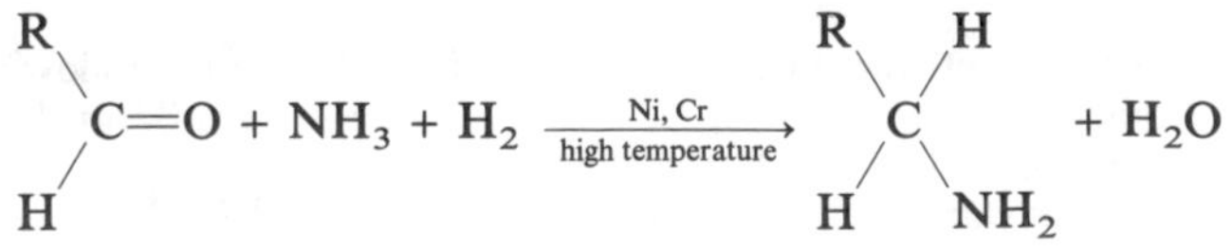

$$\overset{R}{\underset{H}{>}}C=O + NH_3 + H_2 \xrightarrow[\text{high temperature}]{Ni, Cr} \overset{R \quad H}{\underset{H \quad NH_2}{C}} + H_2O$$

Aldehyde   *Primary amine*

(b) Reduction of a Nitrile $R—C{\equiv}N$

(c) Reduction of an Oxime, $R_2C{=}N—OH$

(d) Reduction of an Amide, $RCONH_2$

Lithium tetrahydridoaluminate in ethoxyethane solution is used as the reducing agent for (b), (c) and (d). The use of an acidic reagent which would hydrolyse the starting material must be avoided.

(e) Reduction of a Nitro-compound

ArNH₂ is made by the reduction of ArNO₂

Phenylamine, $C_6H_5NH_2$, is made by the reduction of nitrobenzene, $C_6H_5NO_2$. Tin(II) chloride is made in the reaction mixture from tin and hydrochloric acid [see Figure 32.2]. In reducing the nitro-compound, tin(II) chloride is oxidised to tin(IV) chloride, which reacts with hydrochloric acid to form the complex ion, $[SnCl_6]^{2-}$. The reaction product is therefore $(C_6H_5\overset{+}{N}H_3)_2\,[SnCl_6]^{2-}$. The addition of concentrated alkali liberates phenylamine, and the addition of sodium chloride to the mixture reduces the solubility of phenylamine in water. Steam distillation is employed to separate phenylamine. Extraction with ethoxyethane separates phenylamine from the water in the distillate. Distillation removes ethoxyethane from the dried extract. The product can be purified by distillation under reduced pressure. Phenylamine decomposes when heated to its normal boiling temperature.

Practical details for the preparation of phenylamine

FIGURE 32.2 Reduction of Nitrobenzene to Phenylamine

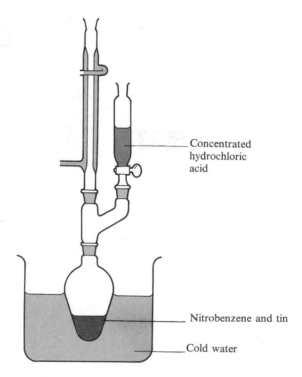

Concentrated hydrochloric acid

Nitrobenzene and tin

Cold water

32.6.3   FROM AN AMIDE BY THE HOFMANN DEGRADATION

RCONH₂ can be converted into RNH₂ by the Hofmann degradation

The action of bromine and concentrated alkali is to convert an amide of formula $RCONH_2$ into an amine of formula RNH_2:

$$CH_3CONH_2 + Br_2 + 4NaOH \rightarrow CH_3NH_2 + 2NaBr + Na_2CO_3 + 2H_2O$$

Ethanamide   Methylamine

This reaction, the Hofmann degradation, reduces the number of atoms in the carbon chain [p. 737].

FIGURE 32.3 Methods of Preparing Amines

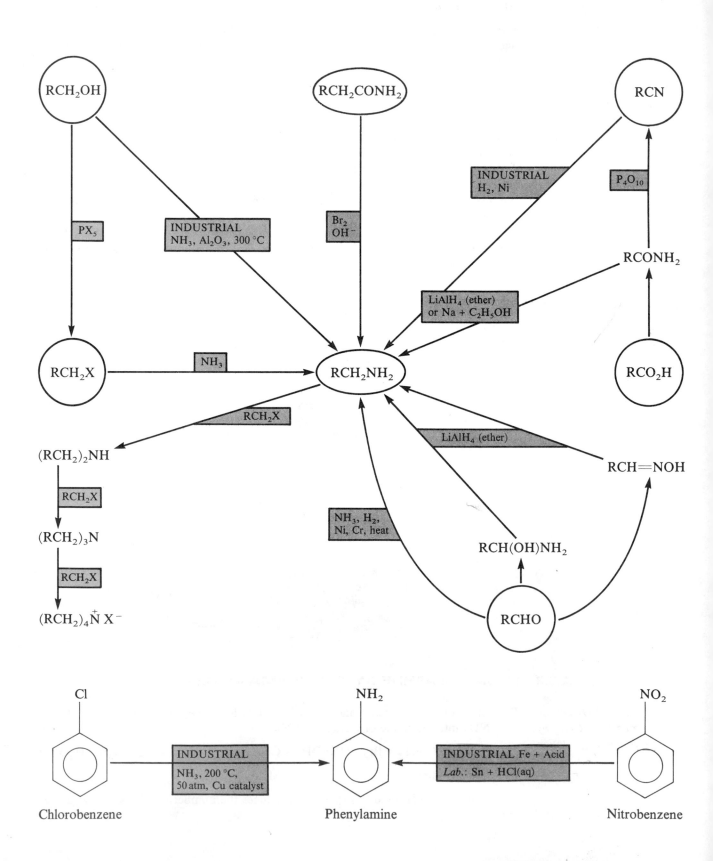

======= **CHECKPOINT 32B: PREPARATIONS OF AMINES** =======

1. Explain how you would carry out the following conversions. Some of them involve more than one step:

(a) $CH_3CH_2Cl \rightarrow CH_3CH_2NH_2$

(b) $CH_3CH_2Cl \rightarrow CH_3CH_2CH_2NH_2$

(c) ⬡—CHO ⟶ ⬡—CH₂CH₂NH₂

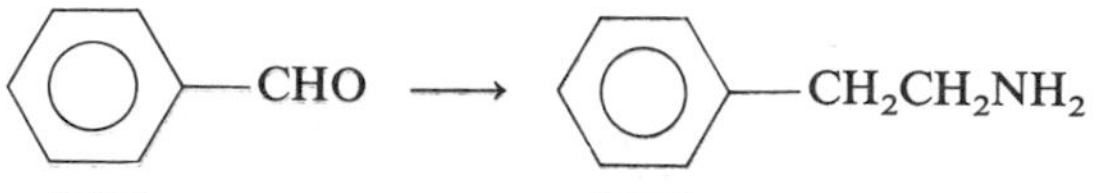

*(d) ⬡—CH₃ ⟶ ⬡(—CH₃)(—NH₂)

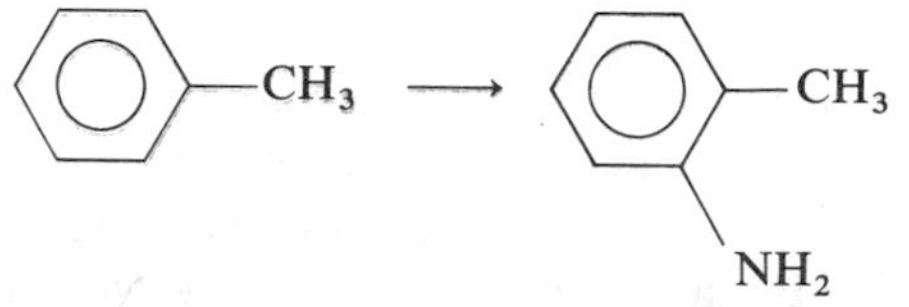

(e) $CH_3CH{=}CH_2 \rightarrow (CH_3)_2CHNH_2$

(f) ⬡—CH₃ ⟶ H₂N—⬡—CH₃

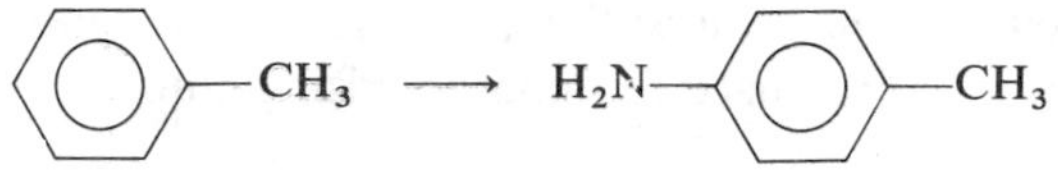

2. Phenylamine can be made from nitrobenzene. Describe how the reduction is carried out. Explain why the product of the reaction is made alkaline before being steam-distilled. Describe how the product is extracted from the steam-distillate, and how it is finally purified.

3. Complete the following equations, indicating reagents and conditions:

(a) $(CH_3)_3N + C_2H_5I \xrightarrow{\text{conditions}}$

(b) ⬡—$NO_2 + Zn + HCl(aq) \rightarrow$

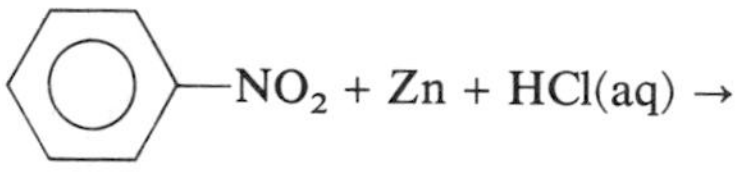

(c) $CH_3COC_2H_5 + NH_3 + H_2 \xrightarrow[\text{conditions}]{\text{catalyst}}$

(d) $C_2H_5CONH_2 \xrightarrow[\text{solvent}]{\text{reducing agent}} C_2H_5CH_2NH_2$

(e) $(C_3H_7)_2C{=}O \xrightarrow{\text{reagent}} (C_3H_7)_2C{=}NOH$

$\xrightarrow[\text{solvent}]{\text{reagent}} (C_3H_7)_2CHNH_2$

4. A is an amine which is insoluble in water. P is a phenol which is insoluble in water. E is a solution of A and P in ethoxyethane. How could you separate A and B from the solution E, using only acid and alkali and ethoxyethane?

32.7 THE REACTIONS OF AMINES

32.7.1 ACYLATION

—NH₂ can be acylated to form —NHCOR...

Acylation is the conversion of the groups

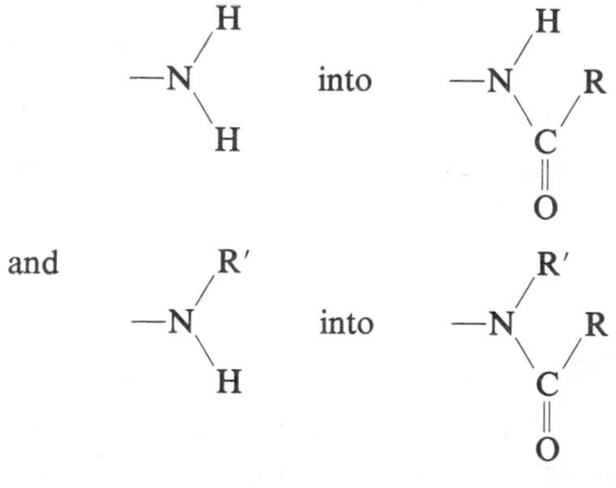

Tertiary amines, having no replaceable hydrogen atoms, are not acylated. The compounds formed are **amides**. The acylating agent can be an acid chloride, RCOCl or, preferably, an acid anhydride, (RCO)₂O:

...by RCOCl or by (RCO)₂O...

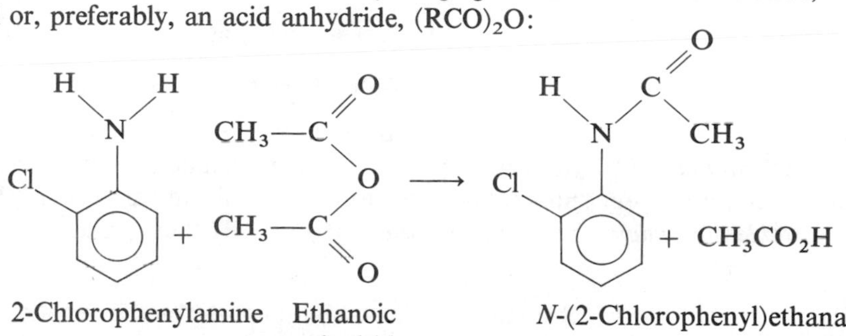

2-Chlorophenylamine   Ethanoic anhydride   *N*-(2-Chlorophenyl)ethanamide

Benzoylation is accomplished by the use of benzoyl chloride and an excess of alkali:

and by
C_6H_5COCl + *NaOH*

Amides are solids. When pure, they have sharp melting temperatures. Chemists often prepare an amide when they want to identify an amine. After purifying the derivative they find its melting temperature. Then they look at tables which list the melting temperatures of many amides. By identifying the derivative, they have succeeded in identifying the parent amine.

Amides are used for identification of amines

Benzoyl derivatives have higher melting temperatures than, for example, ethanoyl and propanoyl derivatives, and they are often used for 'characterising' amines.

32.7.2 ALKYLATION

Alkylation of RNH₂ gives 1°, 2°, 3° amines and quaternary ammonium salts

Halogenoalkanes react with ammonia and with amines, replacing the hydrogen atoms by alkyl groups:

$$NH_3 \xrightarrow{RX} RNH_2 \xrightarrow{RX} R_2NH \xrightarrow{RX} R_3N \xrightarrow{RX} R_4\overset{+}{N}\,X^-$$

Halogenoarenes do not react in this way.

Primary amines are converted into secondary amines and secondary amines into tertiary amines. The final product is a quaternary ammonium salt, e.g., $(CH_3)_4\overset{+}{N}\,I^-$, tetramethylammonium iodide.

32.7.3 REACTIONS WITH NITROUS ACID

Nitrous acid HNO₂ reacts with aliphatic 1° amines to form unstable R—$\overset{+}{N}$≡N ions which decompose to give N₂...

Amines undergo a number of different reactions with nitrous acid. Since it is an unstable compound, nitrous acid is generated in the reaction mixture by the action of a mineral acid on sodium nitrite at 5 °C. [Fig. 32.4, p. 679.]

Aliphatic primary amines and aromatic primary amines react to form a cation

$$R—\overset{+}{N}\!\equiv\!N$$

called a **diazonium ion**. The diazonium compounds from primary aromatic amines are stable in solution at 5 °C. Their reactions are important, and are described on p. 682. The diazonium compounds from primary aliphatic amines decompose to form carbocations, with the elimination of nitrogen:

$$R\overset{+}{N}H_3(aq) + HNO_2(aq) \rightarrow R—\overset{+}{N}\!\equiv\!N(aq) + 2H_2O(l)$$

$$R—\overset{+}{N}\!\equiv\!N(aq) \rightarrow R^+(aq) + N_2(g)$$

...with aromatic 1° amines to form more stable diazonium compounds...

The reaction is quantitative: the volume of nitrogen evolved is a measure of the amount of primary aliphatic amine present. The carbocation, R^+, can react in a number of ways, to form an alkene, an alcohol, ROH, an ether, ROR, and also, by reaction with the $NaNO_2 + HCl(aq)$ present, RCl and RNO_2.

Secondary amines, both aliphatic and aromatic, are nitrosated by nitrous acid to form *N*-nitrosoamines, which are yellow oils of a highly carcinogenic nature.

...with 2° amines to form N-nitrosoamines (yellow oils)...

$$(C_2H_5)CH_3NH(l) + HNO_2(aq) \rightarrow (C_2H_5)CH_3N-N=O(l) + H_2O(l)$$

N-methylethylamine   *N*-methyl-*N*-nitrosoethylamine

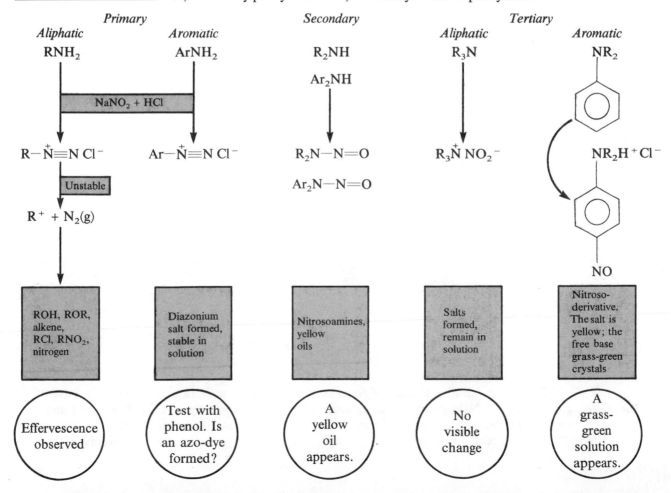

N-methylphenylamine   *N*-methyl-*N*-nitrosophenylamine

...with aliphatic 3° amines to form salts...

Aliphatic tertiary amines, lacking a replaceable hydrogen atom, cannot be diazotised or nitrosated. They simply form salts with nitrous acid:

$$R_3N(l) + HNO_2(aq) \rightarrow R_3\overset{+}{N}H\ NO_2^-(aq)$$

...Aromatic 3° amines are nitrosated in the ring

Aromatic tertiary amines are nitrosated in the ring, in the *para*-position to the activating —NR$_2$ group:

FIGURE 32.4 Reactions of Nitrous Acid with Amines

N,N-Dimethylphenylamine   *N,N*-Dimethyl-4-nitrosophenylamine

Primary		Secondary	Tertiary	
Aliphatic	*Aromatic*		*Aliphatic*	*Aromatic*
RNH$_2$	ArNH$_2$	R$_2$NH	R$_3$N	NR$_2$
		Ar$_2$NH		

NaNO$_2$ + HCl

$R-\overset{+}{N}\equiv N\ Cl^-$	$Ar-\overset{+}{N}\equiv N\ Cl^-$	$R_2N-N=O$	$R_3\overset{+}{N}\ NO_2^-$	$NR_2H^+Cl^-$
		$Ar_2N-N=O$		

Unstable

$R^+ + N_2(g)$

NO

ROH, ROR, alkene, RCl, RNO$_2$, nitrogen	Diazonium salt formed, stable in solution	Nitrosoamines, yellow oils	Salts formed, remain in solution	Nitroso-derivative. The salt is yellow; the free base grass-green crystals

Effervescence observed	Test with phenol. Is an azo-dye formed?	A yellow oil appears.	No visible change	A grass-green solution appears.

32.7.4 SUBSTITUTION IN THE AROMATIC RING

The —NH₂ group
activates the ring

Aromatic amines undergo substitution in the ring. The $-NH_2$ group activates the ring, and is *ortho/para*-directing (like $-OH$ [p. 594]). In many cases, it is difficult to obtain a monosubstituted compound.

HALOGENATION

Phenylamine reacts quantitatively with bromine water to give a white precipitate of 2,4,6-tribromophenylamine:

Halogenation gives a
tri-halogeno-derivative

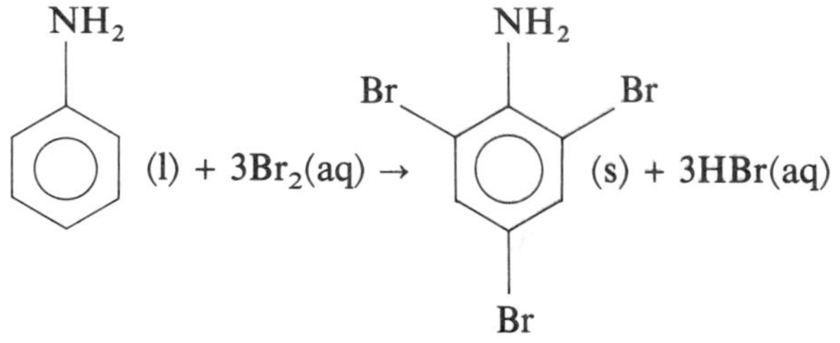

2,4,6-Tribromophenylamine

...unless the —NHCOR
derivative is halogenated

If a monobromo-derivative is required, the $-NH_2$ group is ethanoylated to convert it into the less powerfully activating $-NHCOCH_3$ group. Then the acyl derivative is brominated to give a mixture of the *ortho-* and *para*-bromo-derivatives. Hydrolysis restores the $-NH_2$ group:

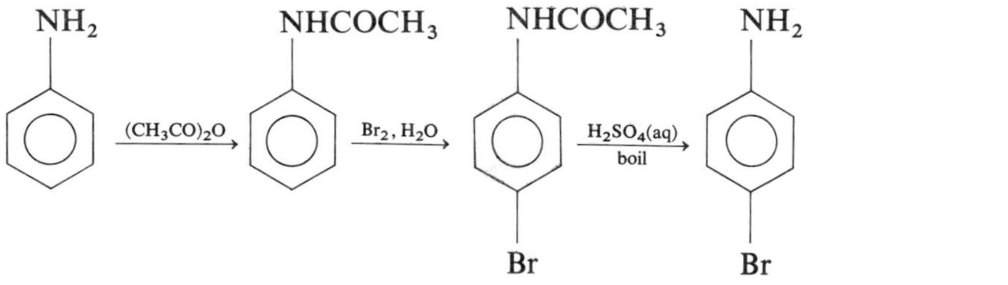

Phenylamine   *N*-Ethanoylphenylamine   4-Bromophenylamine and (2-)
                                                   N-Ethanoyl-4-bromophenylamine

SULPHONATION

When phenylamine reacts with concentrated sulphuric acid at 180 °C, the reaction takes several hours and yields 4-aminobenzenesulphonic acid. With fuming sulphuric acid reaction is faster, but a mixture of products is obtained:

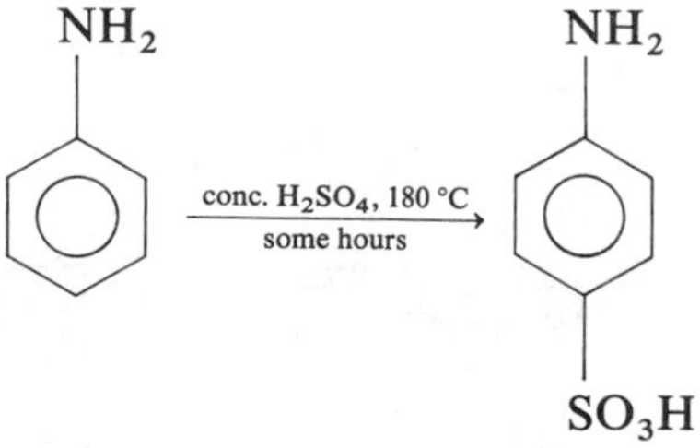

4-Aminobenzenesulphonic acid

NITRATION

The —NH₂ group must
be converted to
—NHCOR before
nitration

A nitrating mixture of concentrated nitric and sulphuric acids may oxidise the $-NH_2$ group as well as nitrating the ring. If the $-NH_2$ group is acylated, the acyl-derivative can be nitrated at room temperature, to give a mixture of *ortho-* and *para*-nitro-compounds. Hydrolysis with sulphuric acid yields the nitrophenylamines:

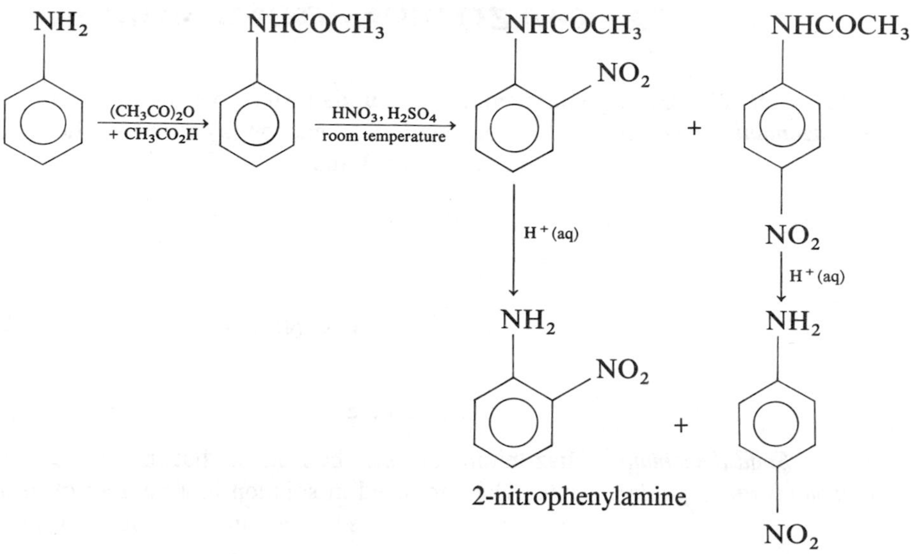

2-nitrophenylamine

4-nitrophenylamine

A TEST FOR A PRIMARY AMINE

A test for a primary amine

When warmed with trichloromethane and a few drops of alkali in ethanolic solution, a primary amine gives a foul smell. This is caused by the formation of an isonitrile, $R-\overset{+}{N}\equiv\overset{-}{C}:$

$$CH_3NH_2 + CHCl_3 + 3KOH \rightarrow CH_3NC + 3KCl + 3H_2O$$

Methylamine $\qquad\qquad\qquad$ Isocyanomethane

CHECKPOINT 32C: REACTIONS

1. Describe the tests you could do to distinguish between the members of the following pairs of compounds:

(a) $CH_3CH_2CH_2NH_2$ and $CH_3CH_2NHCH_3$

(b) $C_4H_9\overset{+}{N}H_3$ Cl^- and $(CH_3)_4\overset{+}{N}$ Cl^-

(c)

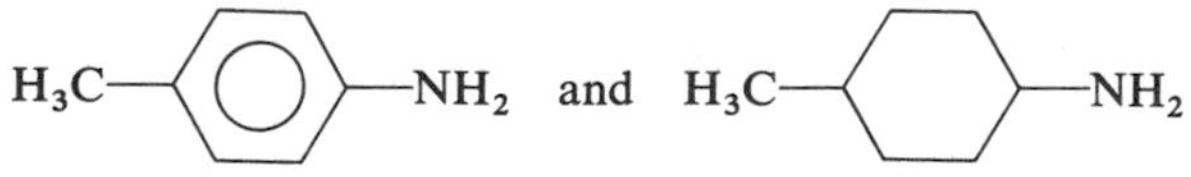

(d)

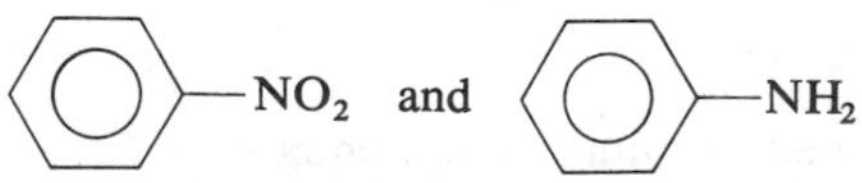

2. (a) How can phenylamine be made from benzene?

(b) How does phenylamine react with the following?

(i) hydrochloric acid

(ii) ethanoyl chloride

(iii) cold sodium nitrite and hydrochloric acid.

3. How could you bring about the conversion of

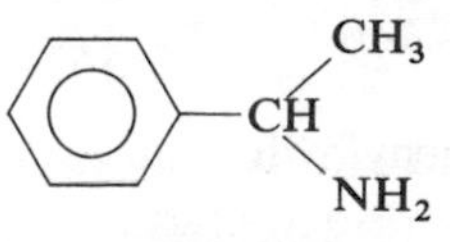

into the following?

(a)

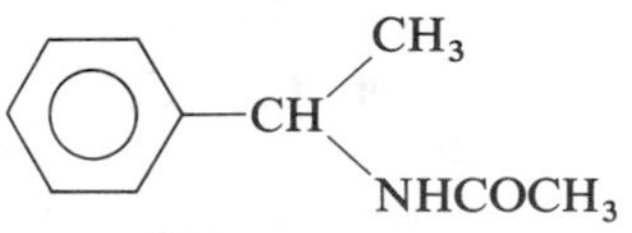

(b)

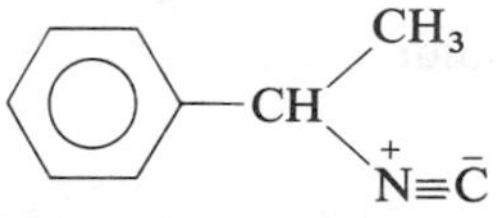

4. Which of the following compounds reacts with nitrous acid to give (a) nitrogen, (b) a yellow oil, (c) no visible change?

(i) $CH_3CH_2N(CH_3)_2$

(ii) $CH_3CH_2CH_2NHCH_3$

(iii) $CH_3CH_2CH_2CH_2NH_2$

(iv) $(CH_3)_3CNH_2$

*5. Write the structural formula for phenylamine in solution in sulphuric acid. Why is a mixture of *ortho-*, *meta-* and *para*-derivatives formed on sulphonation? [p. 594]

32.8 DIAZONIUM COMPOUNDS

ArNH₂ + HNO₂ form a diazonium compound

Diazonium compounds result from the reaction between primary aromatic amines and nitrous acid. The reaction is carried out below 10 °C to avoid the decomposition of both nitrous acid and the product:

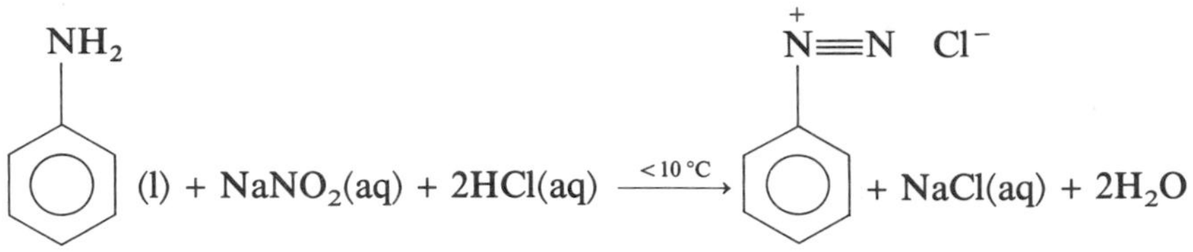

Phenylamine   Benzenediazonium chloride

Solid diazonium compounds are explosive

Diazonium salts can be isolated, but they are unstable and explosive in the solid state. They are used in solution in a number of important preparations. The $-\overset{+}{N}_2$ group can be replaced by a number of different groups.

32.8.1 REPLACEMENT OF $-\overset{+}{N}{\equiv}N$ BY $-$OH

They are used in solution

When warmed in acidic solution, diazonium compounds form phenols, with the evolution of nitrogen:

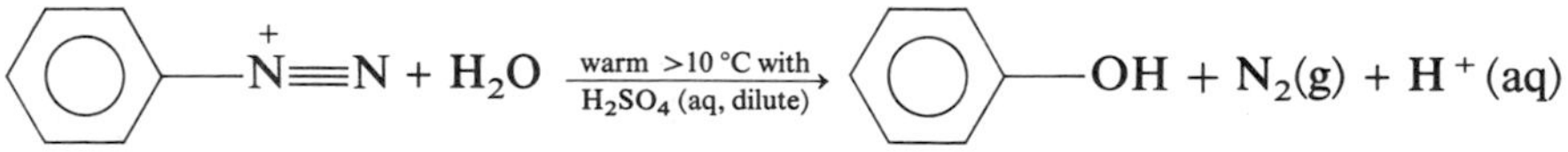

On warming, they give phenols

To avoid a reaction between the phenol formed and the diazonium compound, [p. 683] the diazonium compound is added slowly to a large excess of boiling dilute sulphuric acid. The volatile phenol distils over.

32.8.2 REPLACEMENT OF $-\overset{+}{N}{\equiv}N$ BY $-$HALOGEN OR $-$CN

$-\overset{+}{N}{\equiv}N$ can be replaced by $-$Cl, $-$Br, $-$I or $-$CN...

In a set of reactions named after T Sandmeyer, $-\overset{+}{N}{\equiv}N$ is replaced by $-$Cl, $-$Br or $-$CN. The diazonium compound is warmed to 100 °C with the appropriate reagent and catalyst. These are shown in Figure 32.5. If potassium iodide solution is added, $-\overset{+}{N}{\equiv}N$ is replaced by $-$I.

32.8.3 REDUCTION

...and reduced to $-$NHNH₂ and $-$H

Reduction by tin(II) chloride converts benzenediazonium compounds into phenylhydrazines. When phosphinic acid (H_3PO_2) is used as the reducing agent, $-\overset{+}{N}_2$ is replaced by $-$H:

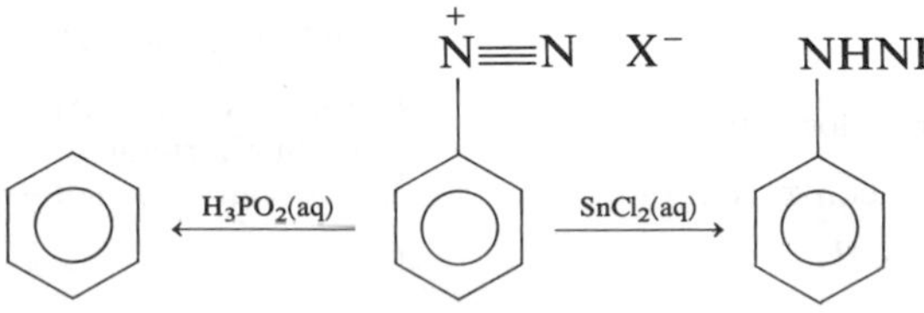

Benzene   Benzene-   Phenylhydrazine
                                 diazonium
                                 halide

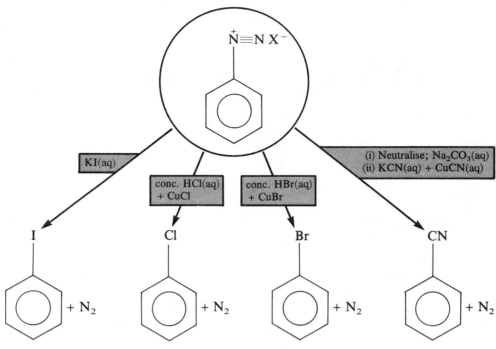

32.8.4 COUPLING REACTIONS

Diazonium compounds will react with phenols and aromatic amines to form azo-dyes

Diazonium ions are electrophiles. The diazonium cation will attack reactive nucleophilic sites, such as the *para*-positions in phenols and aromatic amines. The reaction results in the formation of a $-N=N-$ bond between two aromatic rings, and is called **azo-coupling**. The compounds formed are highly coloured, and many are used as dyes.

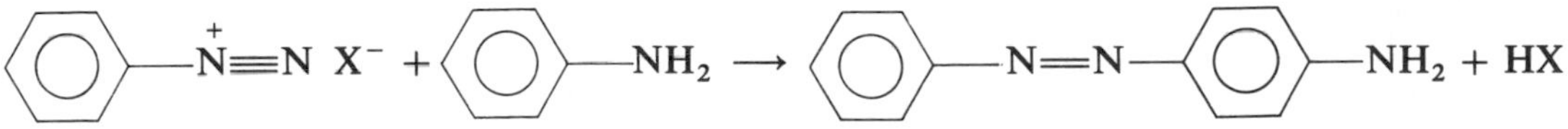

Phenylamine
(aniline)

4-Phenylazophenylamine
(aniline yellow, the first
azo-dye to be made)

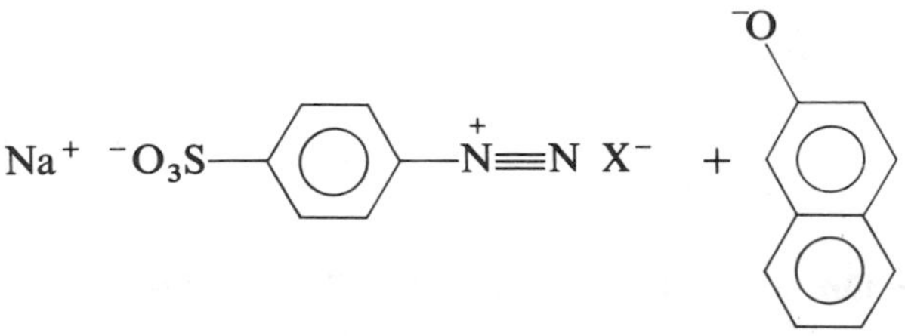

Sodium 4-diazobenzenesulphonate

Naphthalen-2-ol

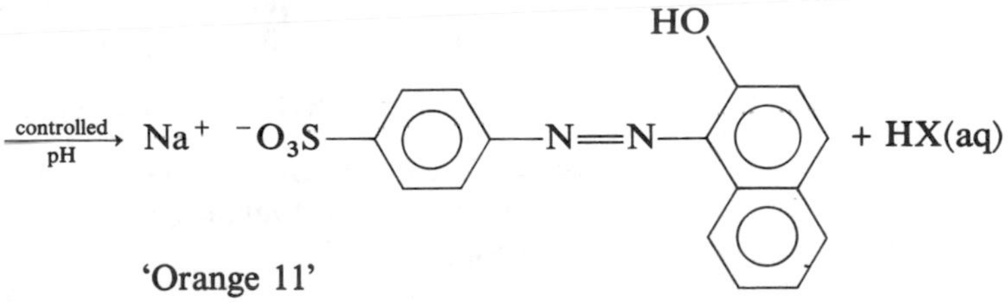

$\xrightarrow[\text{pH}]{\text{controlled}}$

'Orange 11'

(The $-SO_3^- \ Na^+$ substituent makes
this dye soluble in water.)

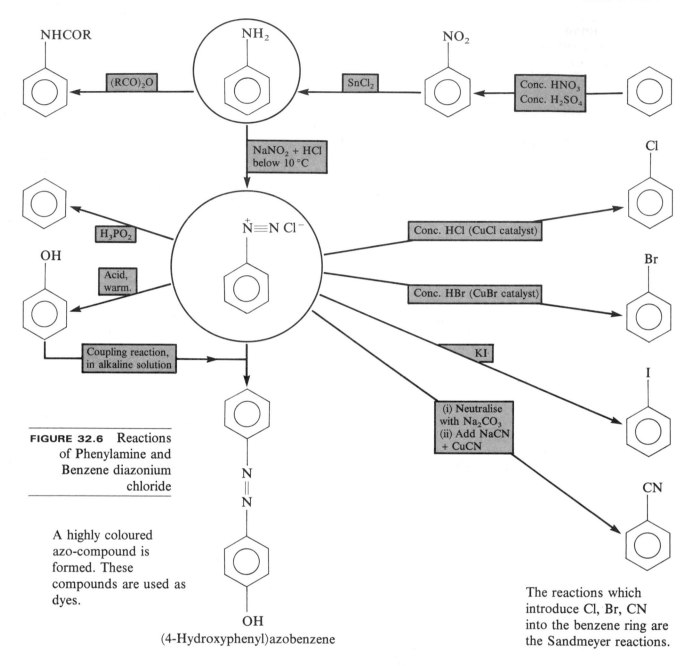

FIGURE 32.6 Reactions of Phenylamine and Benzene diazonium chloride

A highly coloured azo-compound is formed. These compounds are used as dyes.

(4-Hydroxyphenyl)azobenzene

The reactions which introduce Cl, Br, CN into the benzene ring are the Sandmeyer reactions.

32.8.5 THE IMPORTANCE OF DIAZONIUM COMPOUNDS

Diazonium compounds are used in synthesis

Diazonium compounds open up the possibility of making a number of benzene derivatives which cannot be made directly from benzene [See Figure 32.6.] You saw how halogenoalkanes opened up the route

Alkane, $RH \rightarrow$ *Halogenoalkane*, $RX \rightarrow RY$

where Y can be any of a number of groups. Halogenoarenes are not very reactive, and it is often diazonium compounds that open up routes to aromatic compounds [see Figure 32.7]:

Arene → *Nitroarene* → *Arylamine* → *Diazonium compound* → *Substituted arene*

$ArH \rightarrow ArNO_2 \rightarrow ArNH_2 \rightarrow Ar\overset{+}{N}_2 X^- \rightarrow ArY$

FIGURE 32.7 Routes
from Aromatic Amines
to Other Compounds

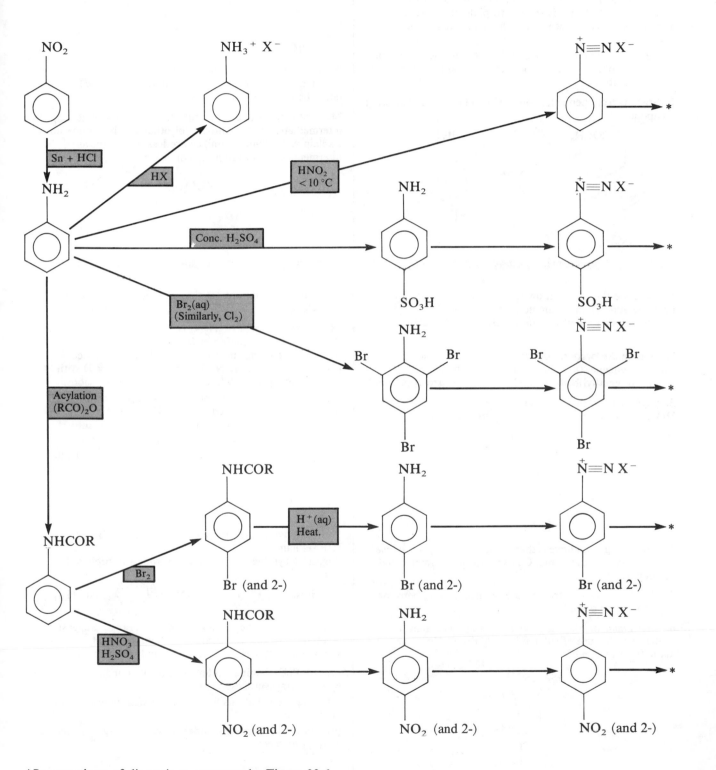

*See reactions of diazonium compounds, Figure 32.6.

CHECKPOINT 32D: DIAZONIUM COMPOUNDS

1. (*a*) Describe how you could prepare a solution of benzenediazonium chloride. How can the compound be converted into phenol? How can phenol be separated from the solution?

(*b*) Under what conditions will a phenol react with a diazonium salt? What is the electrophile in this reaction? Why does benzene not react with benzenediazonium chloride?

(*c*) How could you use the solution of benzenediazonium chloride to obtain (i) chlorobenzene, (ii) iodobenzene, (iii) benzonitrile?

2. Starting from benzene, how could you make the following compounds?

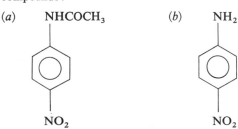

(*a*) NHCOCH$_3$ (*b*) NH$_2$

(*c*) OH   (*d*)

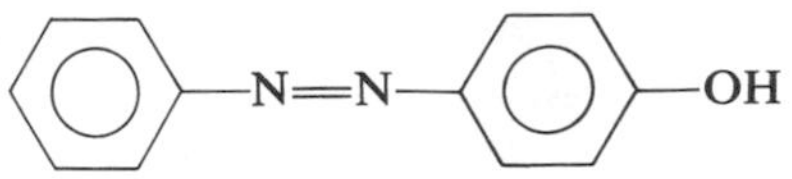

3. How could you convert nitrobenzene into (*a*) phenylamine, (*b*) phenol, (*c*) chlorobenzene, (*d*) 1,3-dichlorobenzene, (*e*) an azo-dye?

*4. Write an essay on the importance of amines as intermediates in the synthesis of other organic compounds. Explain why aliphatic amines are less important as intermediates than aromatic amines.

QUESTIONS ON CHAPTER 32

1. Describe the preparation of (*a*) 2-aminopropane from propane and (*b*) phenylamine from benzene. Give any details you can about the mechanisms of the reactions involved.

2. Outline the preparation, starting from benzene, of (*a*) nitrobenzene, (*b*) phenylamine, (*c*) bromobenzene, (*d*) (bromomethyl)benzene and (*e*) a named azo-compound.

3. Suggest a synthesis for choline chloride, HOCH$_2$CH$_2$N(CH$_3$)$_3$ Cl$^-$, starting from ethene.

*4.

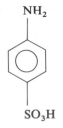

NH$_2$

CH$_2$—CH—CH$_3$

The formula given above is that of the stimulant, *benzedrine*®. Give it a systematic name. Can you suggest how it could be prepared?

5. How can iodo-4-nitrobenzene be made, using benzene as the starting point?

6. '*Sulphanilic acid*', 4-aminobenzenesulphonic acid, is used in the manufacture of sulphanilamide drugs. How can it be made from benzene?

NH$_2$

SO$_3$H

*7. Compare and contrast the reactions of phenol and phenylamine.

*8. A compound, **A** (C$_4$H$_7$N) is reduced by LiAlH$_4$ to **B** (C$_4$H$_{11}$N). Ethanoylation of **B** gives **C** (C$_6$H$_{13}$NO), and treatment of **B** with iodomethane followed by aqueous NaOH gives **D** (C$_5$H$_{13}$N). Further treatment of **D** with iodomethane gives a white solid **E** containing I$^-$ ions.

When **A** is treated with hot aqueous sulphuric acid, **F** (C$_4$H$_8$O$_2$) is formed. Reduction of **F** with LiAlH$_4$ gives **G** (C$_4$H$_{10}$O), and concentrated H$_2$SO$_4$ converts **G** into **H** (C$_4$H$_8$). Treatment of **H** with HBr gives (CH$_3$)$_3$CBr.

Identify the compounds **A** to **H**, giving your reasons, and explaining the reactions mentioned.

*9. C$_2$H$_5$O—⟨◯⟩—NHCOCH$_3$

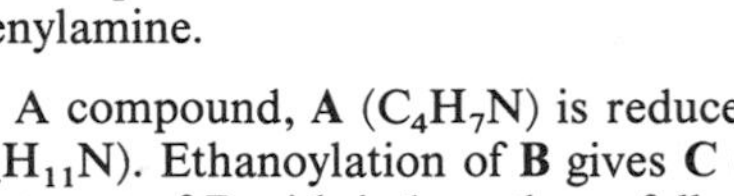

Shown above is the formula of Phenacetin®. Give its IUPAC name.
Suggest a synthesis of phenacetin from 4-nitrophenol.

10. HO$_3$S—⟨◯⟩—N==N—⟨◯⟩—N(CH$_3$)$_2$

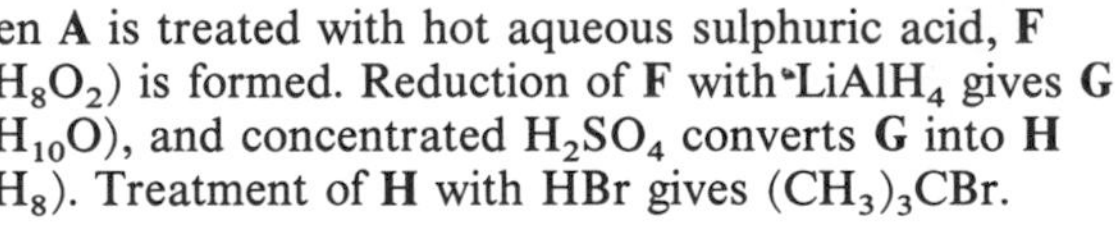

Shown above is the formula of methyl orange. Suggest a method of making the compound from 4-aminobenzenesulphonic acid, 4-HO$_3$SC$_6$H$_4$NH$_2$, and phenylamine.

11. (*a*) Phenylamine (aniline) can be obtained from benzene by a *two*-stage process.

(i) Give the equations for the reactions involved in the *two* stages.

(ii) State the conditions under which the reagents used are allowed to react.

(b) In the isolation and purification of phenylamine, the processes of steam distillation and ether extraction are employed.

(i) Describe these *two* processes.

(ii) Indicate the physical principles on which these processes are based.

(iii) What criteria are used to assess the purity of the phenylamine obtained?

(c) (i) By means of a suitable reaction, explain why phenylamine is classed as both a Bronsted base and a Lewis base.

(ii) How does the strength of phenylamine, as a base, compare to that of ethylamine (aminoethane)? Give a reason for your answer.

(d) Give the name and the structure of the major organic product obtained when phenylamine reacts with

(i) ethanoic anhydride,

(ii) nitrous acid below 10 °C,

(iii) nitrous acid above 10 °C,

(iv) aqueous bromine.

(AEB 81)

12. (a) How and under what conditions does butylamine react with

(i) concentrated hydrobromic acid;

(ii) aqueous copper(II) sulphate;

(iii) ethanoyl chloride (acetyl chloride);

(iv) nitrous acid (nitric(III) acid)?

In each case, indicate the experimental conditions for the reaction, describe what happens, write an equation and give the names and formulae of the reaction products.

†(b) Describe and explain the tests you would perform in order to demonstrate the presence of nitrogen in butylamine.

[See Chapter 34.3 for (b).]   (SUJB 81)

13. Define basicity. List compounds **A**, **B**, **C** and **D** below in order of decreasing basicity, giving reasons for the chosen order.

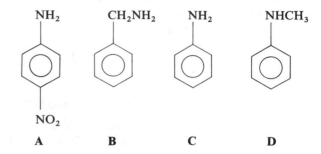

Devise *one* synthesis for each of compounds **A** and **B** starting from benzene. Give all necessary reagents, essential conditions and equations for the reactions you describe.

(JMB 82, S)

32.9 POSTSCRIPT: QUATERNARY AMMONIUM COMPOUNDS

Acetylcholine is a quaternary ammonium salt which is important in the transmission of nerve impulses . . .

A quaternary compound which occurs naturally is acetylcholine chloride. (*Acetyl* = ethanoyl; *choline* is shown below.) When a voluntary nerve cell is stimulated, acetylcholine is released at the nerve endings. After it has been active in transmitting a nerve impulse, acetylcholine is removed. The removal is achieved by the enzyme acetylcholinesterase which catalyses the hydrolysis of acetylcholine to choline and ethanoic (acetic) acid:

$$CH_3CO_2CH_2CH_2{-}\overset{\overset{\displaystyle CH_3}{|}\;+}{\underset{\underset{\displaystyle CH_3}{|}}{N}}{-}CH_3\;Cl^- + H_2O \xrightarrow[\text{acetylcholinesterase}]{\text{catalysed by}} CH_3CO_2H + HOCH_2CH_2{-}\overset{\overset{\displaystyle CH_3}{|}\;+}{\underset{\underset{\displaystyle CH_3}{|}}{N}}{-}CH_3\;Cl^-$$

Acetylcholine chloride

Ethanoic   Choline chloride
(acetic) acid

. . . It is destroyed by enzyme action

Only in the presence of the specific enzyme does the reaction take place rapidly. An enzyme and its **substrate** (e.g., acetylcholinesterase and acetylcholine) are described as fitting together 'like a lock and key'. There are substances which attach themselves to the site on the enzyme that is needed by the substrate. The quaternary ammonium compound, '*decamethonium*' will do this:

$$(CH_3)_3\overset{+}{N}(CH_2)_{10}\overset{+}{N}(CH_3)_3$$

1,10-Di(trimethylammonio)decane (or 'decamethonium')

Enzyme inhibitors interfere with the transmission of nerve impulses

You can see the resemblance between 'decamethonium' and acetylcholine. Once decamethonium is attached to the enzyme, acetylcholinesterase is prevented or **inhibited** from doing its job, and nerve impulses are not transmitted. Decamethonium was used as a muscle relaxant. The naturally occurring base, *curare*, which acts in a similar way, was also used medicinally. Curare has other uses too: some South American Indians use it to tip their poison arrows. The quantities which they use cause paralysis and death in their victims.

Nerve gases, e.g. DFP and insecticides e.g., Parathion® act on acetylcholinesterase

Other compounds which do not resemble acetylcholine in structure also inhibit the enzyme acetylcholinesterase. Nerve gases (e.g., DFP) work in this way. Insecticides like Parathion® have the same effect on insects:

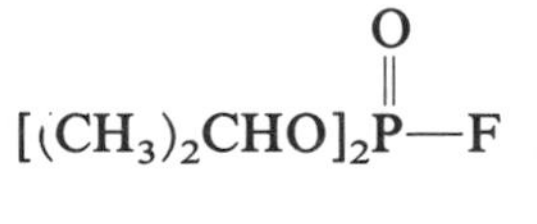

$$[(CH_3)_2CHO]_2\overset{\overset{\displaystyle O}{\|}}{P}—F \qquad\qquad (C_2H_5O)_2\overset{\overset{\displaystyle S}{\|}}{P}—O—\bigcirc—NO_2$$

Di(1-methylethoxy)fluorophosphate (DFP) 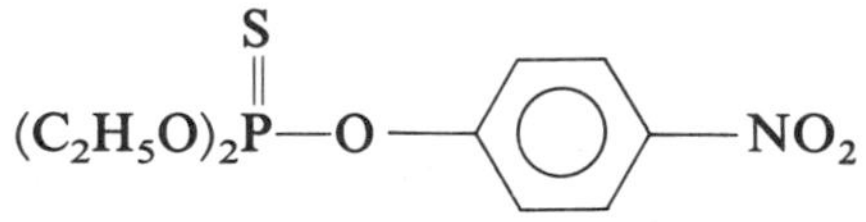   Parathion®

Parathion® and similar insecticides can affect human beings. Agricultural workers who use the compounds regularly and who do not wear protective clothing can absorb sufficient insecticide to damage their health.

33

ORGANIC ACIDS AND THEIR DERIVATIVES

33.1 INTRODUCTION

Most organic acids contain a carboxyl group or a sulphonic acid group:

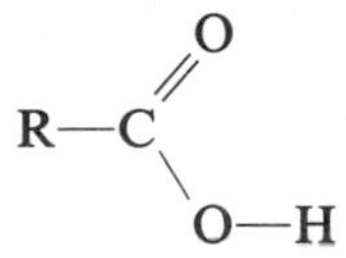

Carboxylic acid

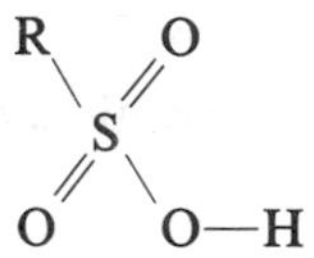

Sulphonic acid

This chapter is devoted to carboxylic acids and their derivatives, followed by a summary of the reactions of sulphonic acids in Figure 33.16, p. 718.

Carboxylic acids are weakly ionised...

Carboxylic acids ionise to some extent to give hydrogen ions, and are neutralised by bases to form salts:

$$RCO_2H(aq) + H_2O(l) \rightleftharpoons RCO_2^-(aq) + H_3O^+(aq)$$

$$RCO_2H(aq) + OH^-(aq) \rightarrow RCO_2^-(aq) + H_2O(l)$$

...They form salts, many of which are soluble

The salts are usually soluble in water. This property makes carboxylic acids easy to extract from natural sources, and is the reason why they were among the first organic compounds to be isolated. Ethanoic (acetic) acid was obtained from sour wine; butanoic (butyric) acid from butter; 2-hydroxypropanoic (lactic) acid from sour milk and benzoic acid from gum benzoin. Aliphatic carboxylic acids were called 'fatty acids' because esters of several of the higher members are fats.

33.2 NOMENCLATURE FOR ORGANIC ACIDS AND THEIR DERIVATIVES

Aliphatic carboxylic acids are named after the alkane with the same number of carbon atoms. The ending *-ane* is changed to *-anoic acid*. Examples are:

The IUPAC names of some acids

HCO_2H	Methanoic acid (formerly called formic acid)
CH_3CO_2H	Ethanoic acid (formerly acetic acid)
$CH_3CH_2CO_2H$	Propanoic acid (formerly propionic acid)

The C in the $-CO_2H$ group is always given the number 1, and substituents are given locants:

$CH_3-CH-CO_2H$ 2-Hydroxypropanoic acid
      |
     OH

$CH_3CH=CHCH_2CO_2H$ Pent-3-enoic acid

HO_2C-CO_2H Ethanedioic acid (formerly oxalic acid)

CH_2CO_2H Butanedioic acid (formerly succinic acid)
|
CH_2CO_2H

Aromatic carboxylic acids are named by adding the suffix *-carboxylic acid* to the name of the parent hydrocarbon. Alternatively, the suffix *-oic acid* can be used:

                $-CO_2H$ Benzenecarboxylic acid (or benzoic acid)

O_2N-⬡$-CO_2H$ 4-Nitrobenzenecarboxylic acid (or 4-nitrobenzoic acid)

The derivatives of carboxylic acids covered in this chapter contain the **acyl** group

$$R-C\underset{\diagdown}{\overset{\diagup O}{}}$$

and are listed in Table 33.1. The nitriles are included because their reactions link up with those of acids.

33.3 PHYSICAL PROPERTIES OF ACIDS AND THEIR DERIVATIVES

33.3.1 ACIDS

The lower aliphatic acids are liquids...

The aliphatic acids C_1-C_{10} are liquids. Anhydrous ethanoic acid freezes at 17 °C, and is often called *glacial* ethanoic acid. The boiling temperatures increase with increasing molecular mass. Aromatic acids are crystalline solids with melting temperatures above those of aliphatic acids of comparable molecular mass.

The lower members of the aliphatic carboxylic acids have penetrating odours. Vinegar is a 3% solution of ethanoic acid. Butanoic acid is the substance you smell in rancid butter.

Hydrogen bonding takes place between molecules of carboxylic acids. In the vapour phase and in solution in organic solvents, dimerisation occurs:

...associated by hydrogen bonds...

$$R-C\underset{\diagdown}{\overset{\diagup O\cdots\cdots H-O}{}}\underset{O-H\cdots\cdots O}{}C-R$$

The facility which acids have for forming two hydrogen bonds per molecule makes their boiling temperatures higher than those of corresponding alcohols.

Formula	Name	Example	
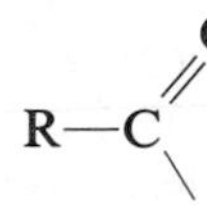 R—C(=O)\ (Acyl group structure)	Acyl group	CH_3CO-	Ethanoyl group
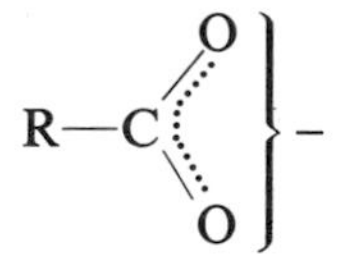 R—C with two O (resonance) }⁻	Carboxylate ion	$CH_3CO_2^-$	Ethanoate ion
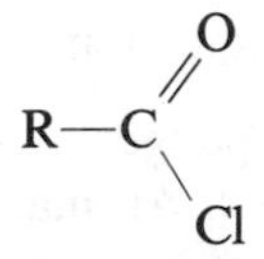 R—C(=O)—Cl	Acid chloride	CH_3COCl	Ethanoyl chloride
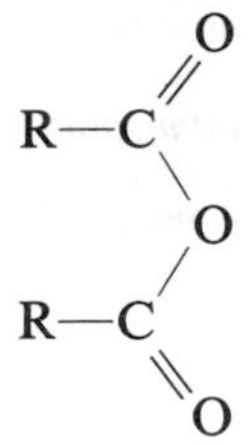 R—C(=O)—O—C(=O)—R	Acid anhydride	$(CH_3CO)_2O$	Ethanoic anhydride
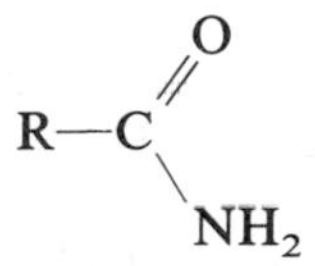 R—C(=O)—NH$_2$	Amide	CH_3CONH_2	Ethanamide
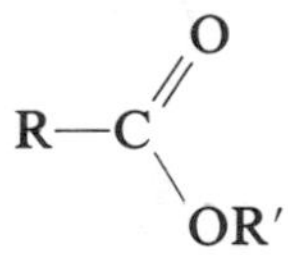 R—C(=O)—OR′	Carboxylic ester	$CH_3CO_2C_2H_5$	Ethyl ethanoate
R—C≡N	Nitrile	CH_3CN	Ethanonitrile

TABLE 33.1 The Names and Formulae of some Derivatives of Acids

...and dissolving in water as they form hydrogen bonds to water molecules

Carboxylic acids of fairly low molecular mass dissolve in water. The dimers dissociate to form monomers in order to form hydrogen bonds to water molecules:

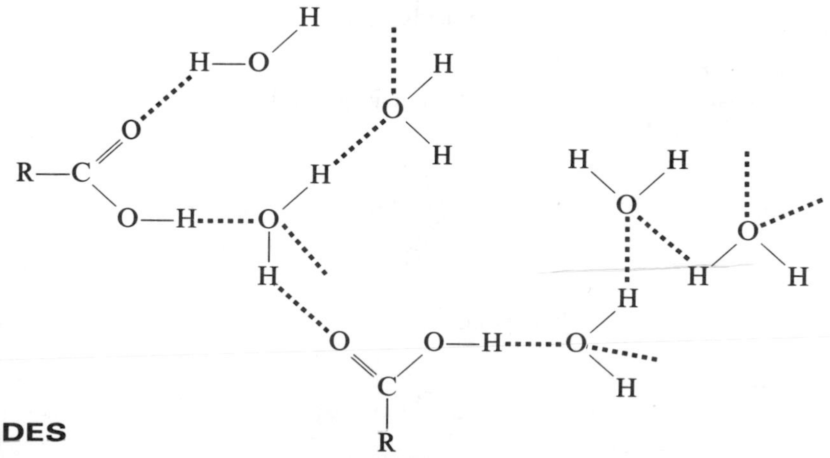

33.3.2 AMIDES

Amides are solids...

...which dissolve in water

Amides, with two hydrogen atoms in each —CONH$_2$ group, can form more hydrogen bonds than acids. Even the lowest members of the series (except HCO_2NH_2) are solids, have higher boiling temperatures than the corresponding acids, and have little odour. The ability to form hydrogen bonds with water molecules makes amides more soluble in water than other acid derivatives.

33.3.3 ESTERS, CHLORIDES, ANHYDRIDES AND NITRILES

Other derivatives are volatile liquids...

Esters, chlorides, anhydrides and nitriles form no hydrogen bonds. Except for anhydrides, their boiling temperatures are lower than those of the corresponding acids. These derivatives are volatile and odorous. There are, however, strong dipole–dipole interactions between the molecules, arising from the polarity of the

$$\diagdown \!\! C \!\!=\!\! O \;\; \text{group}$$

Even the lowest members of the series are liquids at room temperature.

...Chlorides are fuming liquids...

Acid chlorides are colourless liquids with a pungent smell and a lachrymatory action. Anhydrides also have a pungent smell. Aromatic anhydrides are solids.

...Esters have a fruity smell

Esters are colourless liquids with pleasant fruity odours. Examples are:

$CH_3CH_2CH_2CO_2C_2H_5$ Ethyl butanoate   *apple odour*

$CH_3CO_2(CH_2)_7CH_3$ Octyl ethanoate   *orange odour*

33.3.4 SALTS

The salts of carboxylic acids are ionic solids

The salts of carboxylic acids are electrovalent compounds and are therefore involatile crystalline solids. They are often soluble as, on dissolution, the ions are hydrated as the $-CO_2^-$ group forms hydrogen bonds with water molecules. The lower members of the series are soluble, but in higher members the large hydrocarbon part of the molecule makes the compounds insoluble.

33.4 REACTIVITY OF CARBOXYLIC ACIDS

The $\diagdown \!\! C \!\!=\!\! O$ and $-OH$ groups in $-CO_2H$ do not react as they do in carbonyl compounds and alcohols

The carboxyl group is so named because it contains a *carb*onyl group and a hydr*oxyl* group. The two groups influence each other to such an extent that the reactions of carboxylic acids bear little resemblance to those of either carbonyl compounds or alcohols

The carboxylic acid group is

$$-C\!\!\diagup^{\!\!\diagup O}_{\diagdown O\!\leftarrow\! H}$$

The polar

$$\diagdown \!\! C \!\!=\!\! O \;\; \text{group}$$

attracts electrons away from the $-O-H$ bond, and makes it easier for the hydrogen atom to ionise than is the case in the $-O-H$ bond in an alcohol. The flow of electrons from the $-OH$ group towards the carbonyl carbon atom reduces the $\delta+$ charge on the carbonyl carbon atom, with the result that it is not attacked by the nucleophiles that attack carbonyl compounds.

The charge in RCO$_2$$^-$ is delocalised. RCO$_2$$^-$ is a weaker base than RO$^-$

When a carboxylate ion, RCO$_2$$^-$, is formed, the negative charge on the ion is shared equally between two oxygen atoms. The structure of RCO$_2$$^-$ can be represented as a resonance hybrid or by a molecular orbital picture:

$$R-C{\overset{O}{\underset{O-}{}}} \leftrightarrow R-C{\overset{O-}{\underset{O}{}}} \quad or \ R-C{\overset{O}{\underset{O}{}}} \Big\}-$$

The delocalisation of the charge makes the carboxylate ion less ready to accept a proton than is an alkoxide ion, R—O$^-$. The carboxylate ion, RCO$_2$$^-$ is therefore a weaker base than the alkoxide ion, RO$^-$, and RCO$_2$H is a stronger acid than ROH.

Carboxylic acids are weak acids...

...Substituents affect the strength of acids...

Carboxylic acids are much weaker than the common mineral acids. The dissociation constant, K_a, for ethanoic acid is 1.8×10^{-5} mol dm^{-3} [p. 254]. This means that in a 1 mol dm^{-3} solution of the acid, 3 molecules in a thousand are ionised. Substituents affect the strength of acids [see Table 33.2 and p. 257].

...Benzoic acid is stronger than aliphatic acids...

Benzoic acid is stronger than ethanoic acid because the negative charge on the —CO$_2$$^-$ group, being delocalised by interaction with the π electron cloud of the benzene ring, is less available for attaching a proton to form —CO$_2$H [see Figure 33.1].

FIGURE 33.1 The Benzoate Ion

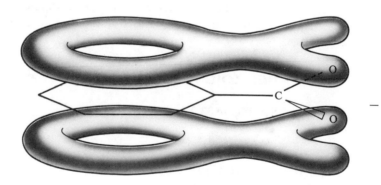

...Carboxylic acids are stronger than carbonic acid

Carboxylic acids are stronger than carbonic acid. The evolution of carbon dioxide that occurs when a carboxylic acid reacts with sodium hydrogencarbonate is used to distinguish carboxylic acids from weaker acids, such as phenols. Some substituted phenols also give a positive result.

TABLE 33.2 pK_a Values for Some Carboxylic Acids and Phenols

Carboxylic acid	pK_a	Benzoic acid	pK_a	Phenol	pK_a
HCO$_2$H	3.75	C$_6$H$_5$CO$_2$H	4.20	C$_6$H$_5$OH	10.00
CH$_3$CO$_2$H	4.76	4-O$_2$NC$_6$H$_4$CO$_2$H	3.43	4-ClC$_6$H$_4$OH	9.38
CH$_3$CH$_2$CO$_2$H	4.87	4-H$_3$CC$_6$H$_4$CO$_2$H	4.34	4-O$_2$NC$_6$H$_4$OH	7.15
(CH$_3$)$_3$CCO$_2$H	5.05			2,4,6-(O$_2$N)$_3$C$_6$H$_2$OH	0.42
ClCH$_2$CO$_2$H	2.86				
CCl$_3$CO$_2$H	0.65				

33.5 INDUSTRIAL SOURCE AND USES

33.5.1 METHANOIC ACID

HCO₂H is obtained from CO

Carbon monoxide and aqueous sodium hydroxide react at 200 °C under pressure to form sodium methanoate. On acidification with mineral acid, this salt gives methanoic acid:

$$CO(g) + NaOH(aq) \xrightarrow[200\,°C]{pressure} HCO_2Na(aq) \xrightarrow{acid} HCO_2H(aq)$$

33.5.2 ETHANOIC ACID

1. The aerial oxidation of alkanes (C_5–C_7) from petroleum can be carried out at high temperature and pressure to yield ethanoic acid.

CH₃CO₂H is made from petroleum and from ethyne

2. Ethanal can be obtained from the hydration of ethyne [p. 570] and then oxidised by air in the presence of a catalyst to give ethanoic acid:

$$HC\equiv CH \xrightarrow[catalyst]{H_2O} CH_3CHO \xrightarrow[catalyst]{air} CH_3CO_2H$$

USE

Much of the ethanoic acid produced is converted into ethanoic anhydride, which is used in the manufacture of the fabric acetate rayon and the drug aspirin [p. 721]. Chloroethanoic acid is used in the manufacture of the weedkiller 2,4-D [p. 699].

33.5.3 BENZOIC ACID

See Figure 33.2.

33.6 LABORATORY PREPARATIONS OF CARBOXYLIC ACIDS

33.6.1 OXIDATION

PRIMARY ALCOHOLS AND ALDEHYDES

Oxidation of alcohols, aldehydes...

Potassium dichromate(VI) and potassium manganate(VII), both in acid solution, are often used as oxidising agents. Primary alcohols are oxidised via aldehydes to carboxylic acids:

$$RCH_2OH \rightarrow RCHO \rightarrow RCO_2H$$

ALKENES

...and alkenes gives acids

Alkenes are oxidised by acidified potassium manganate(VII):

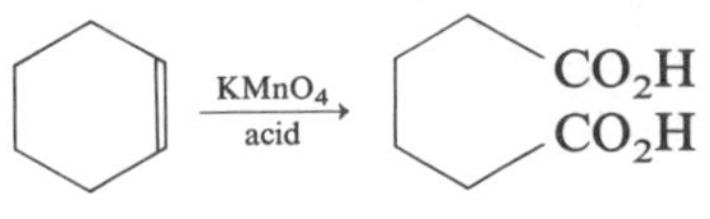

Cyclohexene   Hexane-1,6-dioic acid

METHYLBENZENE

Aromatic side chains are oxidised to CO₂H

Benzoic acid is made by oxidising methylbenzene. Industrially, air is used as the oxidising agent. Even if the group attached to the aromatic ring is larger than —CH₃, benzoic acid is the oxidation product:

FIGURE 33.2

Preparations of Benzoic Acid

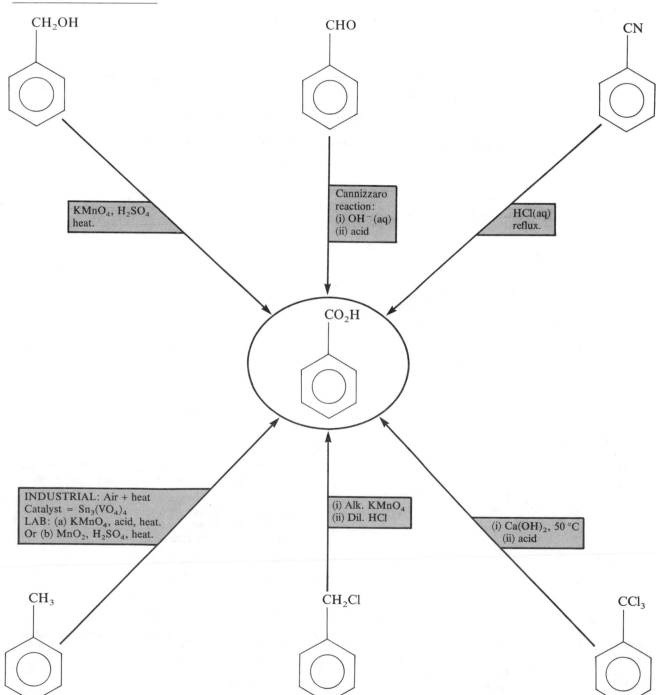

33.6.2 HYDROLYSIS

NITRILES

Hydrolysis of RCN gives RCO_2H

Nitriles are hydrolysed to carboxylic acids by being boiled under reflux with aqueous alkali or mineral acid. The amide, $RCONH_2$, is an intermediate:

$$RCN(l) \xrightarrow[\text{or } H^+(aq)]{NaOH(aq)} RCONH_2(aq) \xrightarrow{H_2O} RCO_2Na(aq) + NH_3(g) \text{ or } RCO_2NH_4(aq)$$

ESTERS

Esters are hydrolysed to acids

Esters

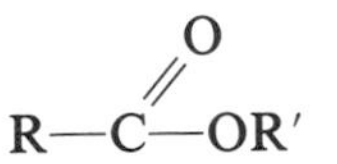

are hydrolysed to alcohols and carboxylic acid salts when they are boiled under reflux with aqueous alkali:

$$RCO_2R'(l) + NaOH(aq) \rightarrow RCO_2Na(aq) + R'OH(aq)$$

The carboxylic acid is obtained by acidifying the salt with a mineral acid:

$$RCO_2Na(aq) + HCl(aq) \rightarrow RCO_2H(aq) + NaCl(aq)$$

33.6.3 GRIGNARD REAGENTS

See Figure 29.4, p. 606.

33.6.4 BENZOIC ACID

See Figure 33.2.

CHECKPOINT 33A: ACIDS

1. Name the following acids:
(*a*) $CH_3(CH_2)_5CO_2H$ (*b*) $(CH_3)_2CHCO_2H$ (*c*) CF_3CO_2H

(*d*)
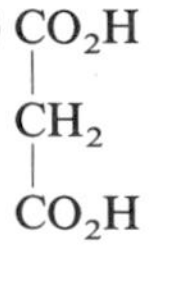
(*e*)
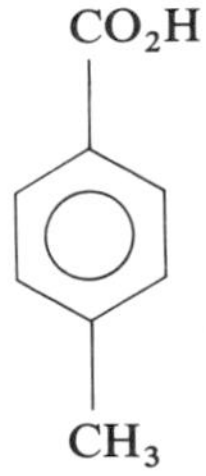
f)
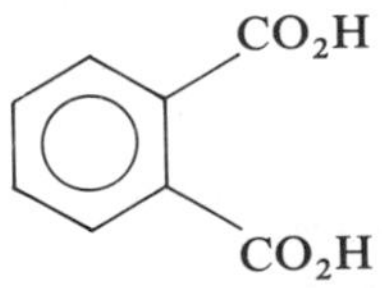
(*g*)
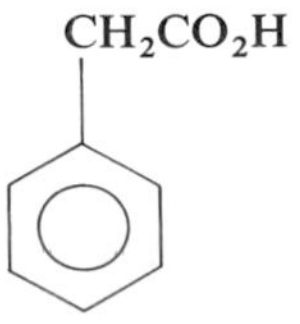
(*h*)
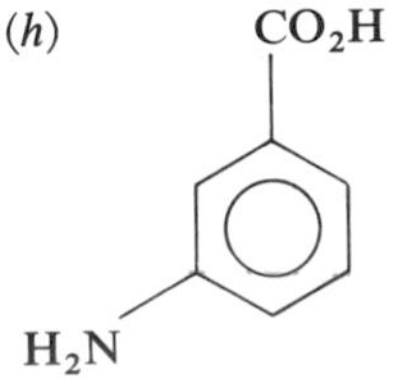

2. Arrange in order of increasing boiling temperature the compounds: (*a*) $CH_3CH_2CO_2H$, (*b*) CH_3CH_2CHO, (*c*) $CH_3CH_2CH_2OH$, (*d*) $CH_3CH_2CH_3$.
Explain the reasons for the order you give.

3. Draw a sketch to show hydrogen bonding in (*a*) $CH_3CONH_2(s)$, (*b*) $CH_3CONH_2(aq)$.

4. Explain the existence of two forms of (*a*) butene-1,4-dioic acid, (*b*) 2-chloropropanoic acid.

***5.** Arrange in order of increasing acid strength: (*a*) phenol, (*b*) benzoic acid, (*c*) carbonic acid, (*d*) trichloroethanoic acid.

6. What products are formed by the reaction of pent-2-ene with (*a*) cold, alkaline potassium manganate(VII), (*b*) hot acidic potassium manganate(VII)?

†**7.** Complete the following equations:

(*a*) $RCH_2Cl + \quad \rightarrow RCH_2CN$

(*b*) $HC{\equiv}CH + \quad \rightarrow H_2C{=}CHCN$

(*c*) $H_2C{=}CHCN + \quad \rightarrow H_2C{=}CHCO_2H$

Indicate the conditions needed for each reaction. What is the IUPAC name for $H_2C{=}CHCO_2H$? The trivial name for this compound is acrylic acid. What is its industrial importance? Why cannot it be made from $H_2C{=}CHCl$? [For help, see p. 607].

8. A liquid **A** of formula C_7H_6O, is boiled with concentrated aqueous sodium hydroxide. Ethoxyethane extraction of the products, followed by drying and distillation, gave as the main product a liquid **B** of boiling temperature 205 °C. **B** reacted with phosphorus(V) chloride to give a pungent gas. When the solution from which **B** had been extracted was acidified, a white crystalline solid **C** was precipitated. On recrystallisation from hot water, it appeared as flaky white crystals. A solution of **C** effervesced when sodium carbonate was added. Identify **A**, **B** and **C**, and explain the reactions described.

9. (a) Outline a method for the conversion

$$CH_3CHO \rightarrow CH_3CH(OH)CO_2H$$

(b) Name the product.

(c) How could you show that the product is (i) a secondary alcohol and (ii) an acid?

(d) What effect would you expect the acid to have on plane-polarised light? Explain your answer.

33.7 REACTIONS OF CARBOXYLIC ACIDS

33.7.1 SALT FORMATION

Carboxylic acids form salts by reaction with metals, carbonates, hydrogencarbonates and alkalis:

A test for —CO_2H is the evolution of CO_2 from $NaHCO_3$

$$2RCO_2H(aq) + Mg(s) \rightarrow (RCO_2)_2Mg(aq) + H_2(g)$$

$$2RCO_2H(aq) + Na_2CO_3(s) \rightarrow 2RCO_2Na(aq) + CO_2(g) + H_2O(l)$$

$$RCO_2H(aq) + NaOH(aq) \rightarrow RCO_2Na(aq) + H_2O(l)$$

The evolution of carbon dioxide from sodium hydrogencarbonate is used as a test to distinguish carboxylic acids from weaker acids, such as phenols. The reaction with alkali can be carried out by titration against a standard alkali to find the concentration of a solution of the acid. A suitable indicator must be used [p. 259].

Stronger acids displace carboxylic acids from their salts

The salts are crystalline solids, most of which are soluble in water. Organic acids can be extracted from a mixture of products by dissolving them in aqueous sodium hydroxide. Acidification with a mineral acid may precipitate the free acid from its salt:

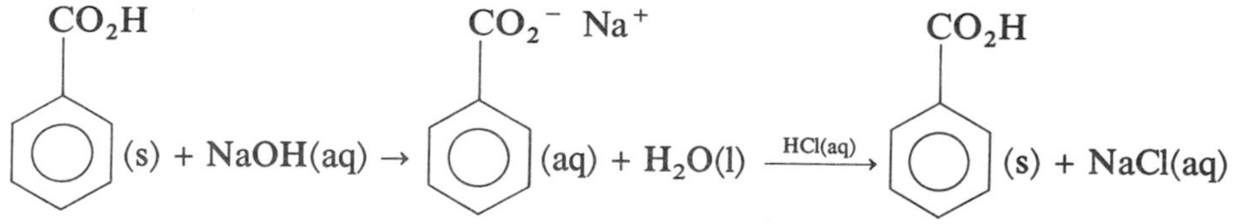

*Carboxylic acids will form salts with amines. This reaction is used in the optical resolution of racemic mixtures [p. 539].

33.7.2 ESTERIFICATION

Reaction with an alcohol gives an ester

Carboxylic acids react with alcohols in the presence of concentrated sulphuric acid to form esters. The acid and alcohol are boiled under reflux with the concentrated acid or, alternatively, hydrogen chloride can be bubbled through the mixture. If the alcohol is labelled with ^{18}O, analysis of the products in a mass spectrometer shows that all the ^{18}O is present in the ester and none is in the water. This proves that the bonds that are broken in the reaction are:

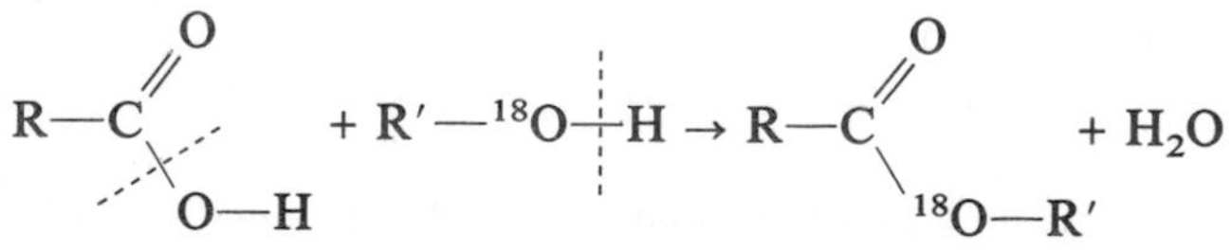

In the esterification of tertiary alcohols and some secondary alcohols, there is fission of the

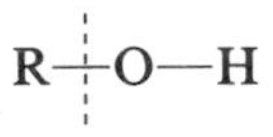

$$R \!\!-\!\!|\!\!-\!\! O \!-\! H$$

bond, and the yield of ester is low.

33.7.3 CONVERSION INTO ACID CHLORIDES

RCOCl is formed by reaction with PCl₅ or SOCl₂

The conversion

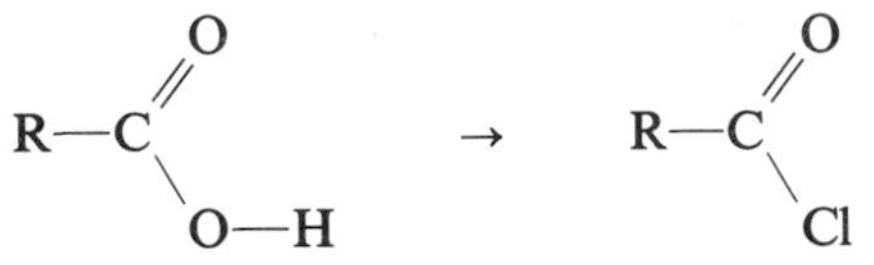

is accomplished by the action of phosphorus(V) chloride, PCl_5, or by sulphur dichloride oxide, $SOCl_2$. The latter is more convenient as the by-products are gaseous:

$$C_2H_5CO_2H(l) + PCl_5(l) \rightarrow C_2H_5COCl(l) + POCl_3(l) + HCl(g)$$
Propanoic acid   Propanoyl chloride

$$C_6H_5CO_2H(s) + SOCl_2(l) \rightarrow C_6H_5COCl(l) + SO_2(g) + HCl(g)$$
Benzoic acid   Benzoyl chloride

33.7.4 CONVERSION INTO AMIDES

The conversion

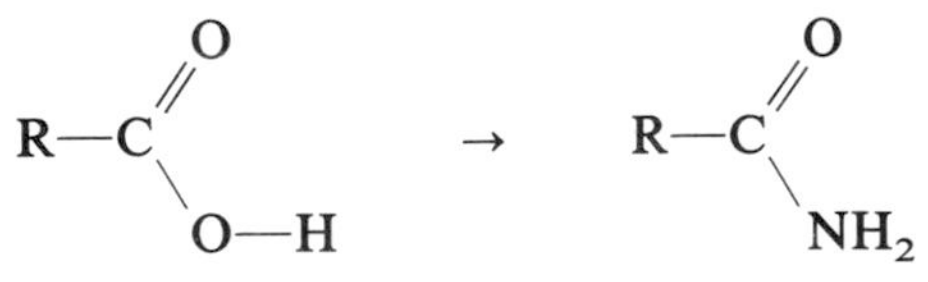

Amides are made from ammonium salts by distillation with acid

is accomplished via the ammonium salt of the acid. The ammonium salt is made by the action of the acid on ammonium carbonate (not ammonia solution because of salt hydrolysis). When heated to 100–200 °C, with an excess of the parent acid, the ammonium salt is dehydrated to give the amide:

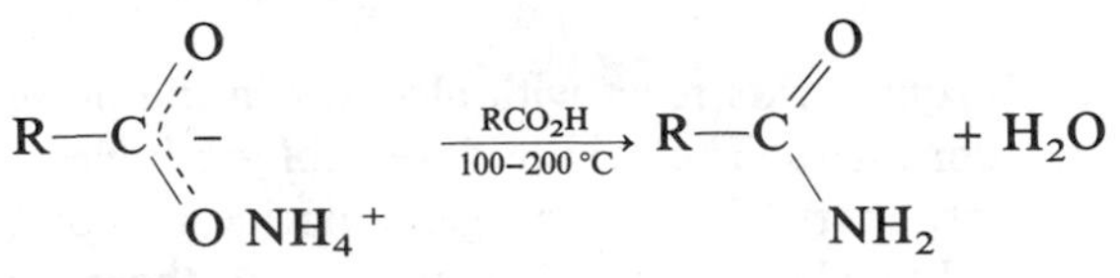

The reason for the presence of an excess of the free acid is that ammonium salts dissociate on being heated:

$$RCO_2{}^- NH_4{}^+ (s) \rightleftharpoons RCO_2H(g) + NH_3(g)$$

The addition of acid displaces the equilibrium to the left.

33.7.5 HALOGENATION

Hydrogen atoms in the α position to —CO$_2$H can be replaced by halogens

When chlorine is bubbled through boiling ethanoic acid, in sunlight and in the presence of iodine or red phosphorus as catalyst, chloroethanoic acid is formed. It is a crystalline solid which melts at 61 °C:

$$CH_3CO_2H(l) + Cl_2(g) \xrightarrow[\text{I}_2 \text{ or red P}]{\text{boil}} CH_2ClCO_2H(s) + HCl(g)$$

If the flow of chlorine is continued and the temperature is raised to the boiling temperature of the product, further substitution occurs to give dichloro- and trichloroethanoic acid:

$$CH_2ClCO_2H(l) \xrightarrow{Cl_2} CHCl_2CO_2H(l) + HCl(g) \xrightarrow{Cl_2} CCl_3CO_2H(s) + HCl(g)$$

Homologues of ethanoic acid are halogenated only in the α position. A phosphorus(III) halide is used as a catalyst:

$$RCH_2CO_2H(l) + Br_2(l) \xrightarrow{PBr_3} RCHBrCO_2H \xrightarrow[\text{PBr}_3]{Br_2} RCBr_2CO_2H + HBr(g)$$

As mentioned previously, halogenoacids are stronger acids than the parent acids.

2,4-D is made from chloroethanoic acid

Chloroethanoic acid is used in the manufacture of the weedkiller, 2,4-dichloro-phenoxyethanoic acid, which is marketed as 2,4-D [p. 406]:

Sodium 2,4-dichloro-
phenoxide

2,4-Dichlorophenoxyethanoic acid

33.7.6 REDUCTION

Acids can be reduced to alcohols...

To convert a carboxylic acid into an alcohol, a powerful reducing agent is needed. Lithium tetrahydridoaluminate dissolved in ethoxyethane will effect reduction:

$$RCO_2H \xrightarrow[\text{in ethoxyethane}]{LiAlH_4} RCH_2OH$$

33.7.7 DECARBOXYLATION OF A SODIUM SALT

...decarboxylated to give a hydrocarbon

When the sodium salt of a carboxylic acid is heated with soda lime, the carboxyl group is removed as a carbonate and a hydrocarbon is formed:

33.7.8 ELECTROLYSIS OF A SODIUM SALT

...and also electrolysed to give alkanes

The Kolbé method of making alkanes is the electrolysis of the sodium salt of a carboxylic acid. An alkane with an even number of carbon atoms is evolved at the anode [p. 550]:

$$2RCO_2Na(aq) + 2H_2O(l) \rightarrow \underbrace{RR(g) + 2CO_2(g)}_{\text{At the anode}} + \underbrace{H_2(g) + 2NaOH(aq)}_{\text{At the cathode}}$$

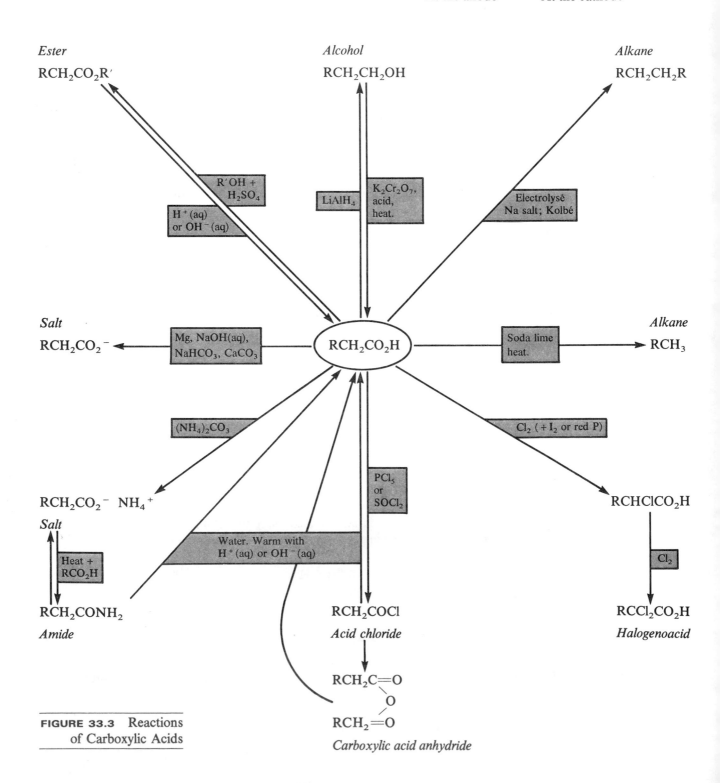

FIGURE 33.3 Reactions of Carboxylic Acids

CHECKPOINT 33B: REACTIONS OF ACIDS

1. A student has read that rancid butter contains butanoic acid. She has supplies of ethoxyethane and laboratory acids and alkalis. How can she obtain a sample of butanoic acid? How can she (*a*) purify her sample and (*b*) test its purity? Butanoic acid is a liquid at room temperature.

2. Sodium propanoate solution is electrolysed. Give the formulae of the ions present in solution. State what products are liberated at the electrodes.

How could propanoic acid be converted into (*a*) a stronger acid, (*b*) a buffer solution? [p. 266]

3. Suggest a method for the conversion

$$C_2H_5OH \rightarrow CH_3CH(OH)CO_2H$$

The product, 2-hydroxypropanoic acid, is a *racemic* mixture, which can be *resolved* by using *optically active* bases. Explain the terms in italics. Outline the practical method for obtaining a resolution.

4. Explain why the α hydrogen atoms in $CH_3CH_2CO_2H$ can be substituted by halogens but not the β hydrogen atoms.

5. Butene-1,4-dioic acid exists in two isomeric forms. Draw their structural formulae. What type of isomerism is involved? Which of these isomers will not readily form an anhydride? [p. 537]

On the addition of hydrogen bromide, both acids give two isomeric compounds. The same two products are obtained from both acids. Draw the structural formulae of the two adducts. Explain the type of isomerism involved.

6. Predict the results of the reactions of the compound shown below with (*a*) bromine water, (*b*) ethanoic anhydride, (*c*) hydrogen and nickel, (*d*) lithium tetrahydridoaluminate, (*e*) ozone.

$$H_2N\text{—}\langle\bigcirc\rangle\text{—}CH{=}CHCO_2H$$

7. Explain why the carbonyl group in a carboxylic acid is less reactive than that in an aldehyde or a ketone.

***8.** Compare the —OH group in ethanol, ethanoic acid and phenol.

33.8 DERIVATIVES OF CARBOXYLIC ACIDS

The derivatives of carboxylic acids that contain an acyl group are shown in Figure 33.4 in order of decreasing reactivity.

FIGURE 33.4 Acid Derivatives in Order of Decreasing Reactivity

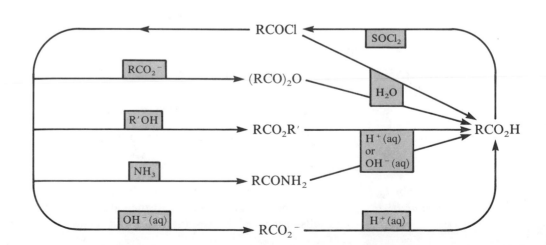

33.9 ACID CHLORIDES

33.9.1 REACTIVITY

With the functional group

Nucleophiles attack the carbonyl C atom in RCOCl... the reactions of acid chlorides depend upon the attack by nucleophiles on the $\delta+$ carbon atoms. In benzoyl chloride and other aromatic acid chlorides, the $\delta+$ charge is spread over the ring, and the carbon atom is less susceptible to nucleophilic attack than it is in aliphatic acid chlorides.

33.9.2 REACTIONS OF ACID CHLORIDES

HYDROLYSIS

...H_2O is an example Acid chlorides are insoluble in water but are hydrolysed quickly to a carboxylic acid and hydrochloric acid. Aliphatic acid chlorides fume in moist air to give hydrogen chloride:

$$CH_3COCl(l) + H_2O(l) \rightarrow CH_3CO_2H(l) + HCl(g)$$

ESTER FORMATION

...ROH and ArO^- ... Acid chlorides, RCOCl, and aroyl chlorides, ArCOCl, react readily with phenols in alkaline solution and with alcohols in the presence of pyridine (a base which removes the HCl produced) to form esters:

$$CH_3COCl(l) + C_2H_5OH(l) \xrightarrow{\text{pyridine}} CH_3CO_2C_2H_5(l) + HCl$$

Ethanoyl   Ethanol   Ethyl ethanoate
chloride

Sodium   Phenyl ethanoate
phenoxide

AMIDE FORMATION

...NH_3, RNH_2, R_2NH... Amides are formed by reactions between acid chlorides and ammonia or a primary or secondary amine:

$$RCOCl + R'_2NH \rightarrow RCONR'_2 + HCl$$

The reaction takes place readily at room temperature.

REDUCTION

...H_2 with Pt or Pd catalyst... Acid chlorides are reduced to alcohols by a stream of hydrogen in the presence of platinum or palladium as catalyst. If the catalyst is 'poisoned' by the addition of some substance which makes it less effective, an aldehyde is formed [see Figure 33.5].

FIGURE 33.5 The
Reduction of Acid
Chlorides

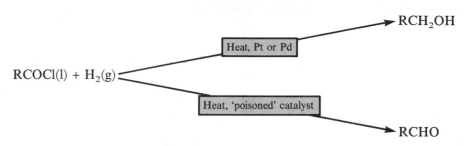

ANHYDRIDE FORMATION

...and RCO₂⁻

When an acid chloride is heated with the sodium salt of the carboxylic acid, the acid anhydride distils over:

$$C_2H_5CO_2Na(s) + C_2H_5COCl(l) \xrightarrow{\text{distil}} (C_2H_5CO)_2O(l) + NaCl(s)$$

Sodium   Propanoyl   Propanoic anhydride
propanoate   chloride

KETONE FORMATION

See Friedel–Crafts acylation, p. 583 and Figure 33.6.

33.10 ACID ANHYDRIDES

The reactions of acid anhydrides are similar to those of acid chlorides

Anhydrides are insoluble in water but react with it to form the parent acid:

$$(RCO)_2O(l) + H_2O(l) \rightarrow 2RCO_2H(aq)$$

Anhydrides react in a similar way to acid chlorides. They form esters with alcohols and phenols [see Figure 33.6]; they form amides with ammonia and primary and secondary amines and are Friedel–Crafts acylating agents:

$$(CH_3CO)_2O + \langle\bigcirc\rangle-O^- Na^+ \rightarrow \langle\bigcirc\rangle-O_2CCH_3 + CH_3CO_2^- Na^+$$

Ethanoic   Sodium   Phenyl ethanoate
anhydride   phenoxide

$$(CH_3CO)_2O + C_2H_5NHCH_3 \rightarrow C_2H_5NCH_3 + CH_3CO_2H$$
$$\underset{\displaystyle COCH_3}{|}$$

N-Methyl-   *N*-Ethyl-*N*-methylethanamide
ethylamine

The industrial importance of ethanoic anhydride is featured in the Postscript, p. 721.

Methanoic acid does not form an anhydride. Dehydration of this acid yields carbon monoxide [p. 464].

FIGURE 33.6 Reactions of Acid Chlorides and Anhydrides

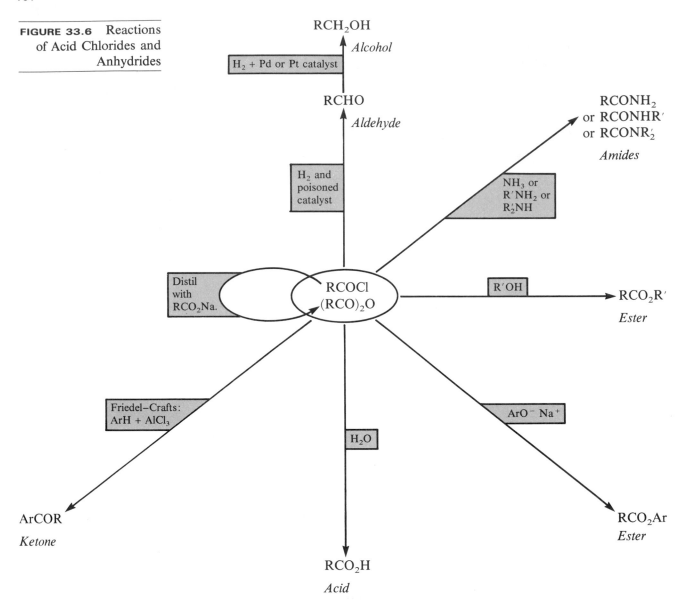

CHECKPOINT 33C: ACID CHLORIDES AND ANHYDRIDES

1. Complete the following equations:

(*a*) $C_2H_5CO_2H + PCl_5 \rightarrow$

(*b*) $C_6H_5CO_2H + SOCl_2 \rightarrow$

(*c*) $(C_6H_5CO)_2O + NH_3 \rightarrow$

(*d*) $(CH_3)_2\overset{*}{C}HCOCl + C_2H_5OH \rightarrow$

2. State how you could prepare ethanoyl chloride and ethanoic anhydride. Compare the reactions of the two compounds with (*a*) water, (*b*) ethanol, (*c*) phenol, (*d*) phenylamine.

3. Compare the reactions of chloroethane and ethanoyl chloride. Explain the reason for the differences in behaviour.

4. Into a large excess of water were put *m* grams of ethanoyl chloride. The products remained in solution and were neutralised by 50.0 cm^3 of a solution of sodium hydroxide of concentration $0.100 \text{ mol dm}^{-3}$. Find *m*.

5. Describe how you could use benzoyl chloride, C_6H_5COCl, to distinguish between phenylamine and phenylmethylamine.

†**6.** Describe how you could distinguish ethanoyl chloride from benzoyl chloride.
[Remember Chapter 32.]

33.11 ESTERS

Esters contain the functional group

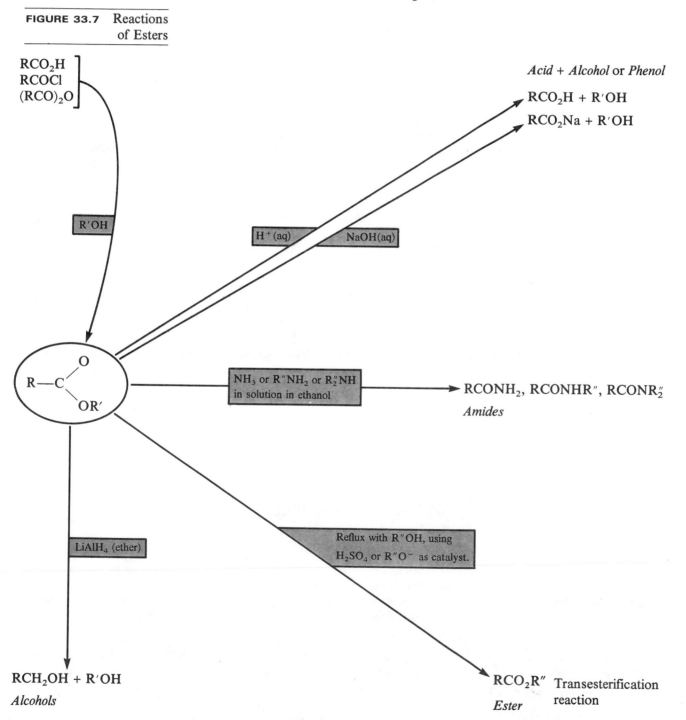

where R and R′ may be alkyl or aryl groups and may be the same or different. The reactions of esters are summarised in Figure 33.7.

FIGURE 33.7 Reactions of Esters

33.11.1 ESTER HYDROLYSIS

The hydrolysis of esters to acid + alcohol is catalysed by H⁺ (aq)

The hydrolysis of an ester gives an equilibrium mixture of carboxylic acid and alcohol or phenol. The attainment of equilibrium is catalysed by the presence of mineral acids:

$$CH_3CO_2C_2H_5(l) + H_2O(l) \underset{}{\overset{H^+(aq)}{\rightleftharpoons}} CH_3CO_2H(aq) + C_2H_5OH(aq)$$

OH⁻ ions catalyse the R.D.S. and also move the position of equilibrium in favour of the products

The reaction also proceeds faster in alkaline conditions. Hydroxide ions catalyse the rate-determining step in the hydrolysis:

$$RCO_2R' + H_2O \underset{}{\overset{OH^-}{\rightleftharpoons}} RCO_2H + R'OH$$

When some RCO_2H molecules have been formed, OH^- ions react with them to form a salt:

$$RCO_2H + OH^- \rightarrow RCO_2^- + H_2O$$

Salt formation removes RCO_2H from the equilibrium mixture. The hydrolysis reaction therefore goes further towards completion in the presence of a base. The base is used up in the reaction:

$$RCO_2R' + NaOH \rightarrow RCO_2Na + R'OH$$

For the alkaline hydrolysis of oils see p. 705.

33.11.2 POLYESTERS

Dicarboxylic acids and diols react to form polyesters...

When esterification takes place between an acid with two carboxyl groups and an alcohol with two hydroxyl groups, a polymer is formed. An example is the reaction between ethane-1,2-diol and 1,4-benzenedicarboxylic acid:

...e.g., Terylene®

The method of manufacture of Terylene®

The polyester formed has the trade names of Terylene® and Dacron®, and is used for making fabrics. Terylene® has about 80 units per chain. It can be melted without decomposition. The molten polymer is forced through tiny holes and cooled to form thin fibres. When the fibres are stretched, polymer molecules become aligned parallel to the axis of the fibre, packing closely together to make a strong fibre. Terylene® clothing is both soft and hard-wearing.

Nucleic acids are polyesters of vital importance

Nucleic acids are polyesters of phosphoric acid. Carbohydrates, proteins and nucleic acids are the three enormous, naturally occurring groups of polymers. Nucleic acids are not covered in this book, except for a reference to the structure of DNA [p. 87].

33.11.3 FATS AND OILS; SOAPS AND DETERGENTS

Fats and oils are esters of propane-1,2,3-triol and acids

Fats and oils are together classified as **lipids**. They are esters of propane-1,2,3-triol (*glycerol*) and a long chain carboxylic acid. Two examples are:

$$CH_2OCOC_{15}H_{31}$$
$$|$$
$$CHOCOC_{15}H_{31}$$
$$|$$
$$CH_2OCOC_{15}H_{31}$$

Propane-1,2,3-trihexadecanoate
(*Glyceryl palmitate*)
is a component of animal fats.

$$CH_2OCO(CH_2)_7CH{=}CH(CH_2)_7CH_3$$
$$|$$
$$CHOCO(CH_2)_7CH{=}CH(CH_2)_7CH_3$$
$$|$$
$$CH_2OCO(CH_2)_7CH{=}CH(CH_2)_7CH_3$$

Propane-1,2,3-trioctadec-9-enoate
(*Glyceryl oleate*)
occurs in olive oil.

Fats are saturated; oils are unsaturated...

Fats occur both in plants and animals, and are a source of energy in our diet. If the carboxylic acid groups are largely saturated, the ester is a solid fat; if the hydrocarbon chains are highly unsaturated, the ester is an oil. The difference in melting temperatures arises from the fact that saturated hydrocarbon chains can pack more closely together than unsaturated chains. In these esters, the configuration at the double bonds is always *cis*, so that the molecules are bent, and cannot pack closely. There is a better sale for solid fats than for oils, and the conversion of oils to fats is a profitable business. It is accomplished by hydrogenation in the presence of nickel as a catalyst. Margarine is made in this way from corn oil and soya bean oil. Recently, it has been suggested that unsaturated esters are metabolised more easily than saturated fats, and lead to less cholesterol in the blood. Sales of unsaturated fats have increased, but the matter is far from settled, and debate continues.

...Hydrogenation converts oils into fats

Oils are used in varnishes and paints

Saturated fats are very stable compounds. Unsaturated oils undergo oxidation by air. This is what happens when fats and oils turn rancid. The oxidation of oils is put to good use in paints and varnishes. When the oil is spread in a thin film, the surface is exposed to oxygen, and a free radical chain reaction is initiated by oxygen. The surface layer of oil is converted into giant polymer molecules to give a hard, tough surface. Catalysts are added to assist polymerisation.

Saponification is the alkaline hydrolysis of fats to give soaps

An important use of fats is soap-making. The alkaline hydrolysis or **saponification** (Latin: soap-making) of fats gives glycerol and the sodium or potassium salt of the carboxylic acid. This is a soap. On addition of salt, the soap separates from solution, and can be skimmed off the surface:

$$CH_2OCOC_{17}H_{35}$$
$$|$$
$$CHOCOC_{17}H_{35}(s) + 3NaOH(aq) \rightarrow$$
$$|$$
$$CH_2OCOC_{17}H_{35}$$

Propane-1,2,3-trioctadecanoate
(*Glyceryl tristearate*)

$$CH_2OH$$
$$|$$
$$CHOH(l) + 3C_{17}H_{35}CO_2Na(s)$$
$$|$$
$$CH_2OH$$

Sodium octadecanoate (*stearate*)
Propane-1,2,3-triol (*glycerol*)

This process has been carried out since the Iron Age. Animal fats were boiled with water and the ashes of a wood fire, which contained potassium carbonate.

Soaps cannot work in hard water

Sodium soaps have limited solubility in water and can be obtained as solid cakes. Potassium soaps are more soluble and are used as gels in shampoos and shaving creams. The calcium and magnesium salts of soaps are insoluble. When a soap, such as sodium hexadecanoate, is added to hard water, an insoluble 'scum' of calcium and magnesium hexadecanoates is formed. To allow soap to lather readily, hard water must be softened by the removal of Ca^{2+} and Mg^{2+} ions [p. 349].

Soaps can emulsify fats and oils because they contain both hydrophilic and lipophilic groups

Soap cleans because it can **emulsify** fats and oils, i.e., convert them into a suspension of tiny droplets in water. Dirt is held to fabrics by a thin film of oil or grease, and this film must be removed before the dirt can be rinsed away. Soap owes its emulsifying action to the combination of polar and non-polar groups in its structure. At one end is a highly polar carboxylate ion, which is **hydrophilic** (attracted to water) and **lipophobic** (repelled by oils and fats). At the other end is a long hydrocarbon chain which is hydrophobic and lipophilic. When soap is added to water, the hydrophilic carboxylate ions dissolve in the water, while the hydrophobic hydrocarbon ends do not. The result is a surface layer one molecule thick [see Figure 33.8] which reduces the surface tension of water.

Soap interacts with an interface between oil and water when clothes are washed. The carboxylate end dissolves in the water, and the hydrocarbon end dissolves in the oil. Soap ions arrange themselves round each droplet of oil to form a **micelle** [see Figure 33.9]. If the water is agitated, the micelle can float free of the fabric. As the surface of each micelle is negatively charged, the drops repel one another and do not coalesce.

FIGURE 33.8
A Monomolecular Layer
of Soap on the Surface
of Water

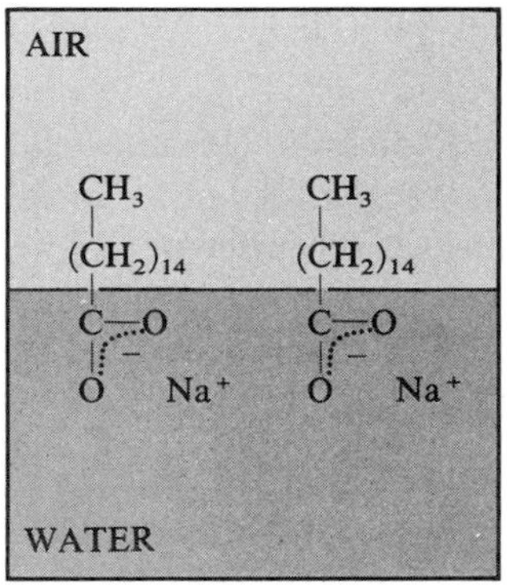

FIGURE 33.9
A Micelle: a Drop of Oil
surrounded by Soap
Ions

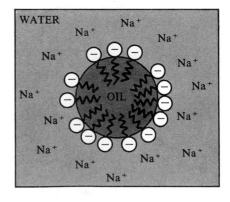

Synthetic detergents are the sodium salts of sulphonic acids . . .

The sodium salts of certain alkyl hydrogensulphates and benzenesulphonic acids also have detergent (cleaning) properties:

$$CH_3(CH_2)_{10}CH_2O-\overset{\displaystyle O}{\underset{\displaystyle O}{\overset{\|}{\underset{\|}{S}}}}-O^- Na^+$$

Sodium dodecanyl sulphate

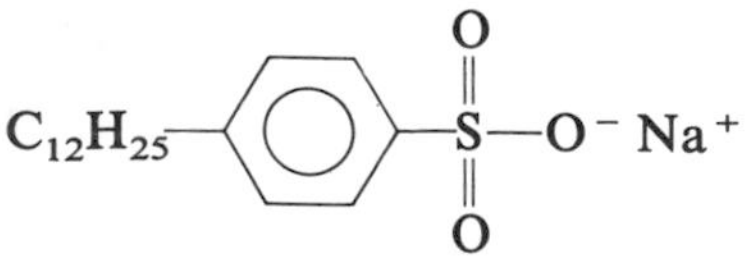

$$C_{12}H_{25}-\overset{\displaystyle O}{\underset{\displaystyle O}{\overset{\|}{\underset{\|}{S}}}}-O^- Na^+$$

Sodium 4-dodecanylbenzene-
sulphonate

. . . They can work in hard water

They have an advantage over soaps in that their calcium and magnesium salts are soluble, allowing them to work in hard water [p. 349]. A disadvantage of soapless detergents is that they are not biodegradable. When these detergents are discharged into rivers, the microorganisms present in the water cannot destroy them. The water becomes polluted, and fish and other aquatic creatures die.

FIGURE 33.10
Detergent Foam

FIGURE 33.10
Detergent Foam

CHECKPOINT 33D: ESTERS

1. Write the names and structural formulae of nine isomeric esters with the formula $C_5H_{10}O_2$.

2. What products are formed in the alkaline hydrolysis of this ester?

$$C_3H_7C\overset{\|}{\underset{O}{}}{}^{18}OC_2H_5$$

3. From which alcohol and acid are the following esters made?

(a) propyl propanoate

(b) methyl benzoate

(c) $HCO_2C_2H_5$

(d) $CH_3(CH_2)_3CO_2CH_3$

(e) $(CH_3)_2CHO_2CCH_3$

(f) $C_6H_5CO_2CH(CH_3)_2$

4. Outline how you would carry out the following conversions:

(a) $C_6H_5NH_2 \rightarrow C_6H_5OH$

(b) $C_6H_5NH_2 \rightarrow C_6H_5COCl$

More than one step is needed in each case.

(i) How could you obtain from the products of (a) and (b) a sample of phenyl benzoate, $C_6H_5CO_2C_6H_5$?

(ii) Starting from 10.0 g of phenylamine, and using no other organic compound, what is the maximum yield of the ester?

5. The complete hydrolysis of 1.76 g of an ester of a monocarboxylic acid and a monohydric alcohol required 2.0×10^{-2} mol of sodium hydroxide. Find the molar mass of the ester. Deduce its molecular formula, and write the names and structural formulae of all the esters with this molecular formula.

6. Compare the reactions of the —C—O—C— group in ethers and esters, explaining the reason for the differences.

*7. Mutton fat is largely an ester of *glycerol* and *stearic acid*

$$CH_3(CH_2)_{16}CO_2H$$

Linseed oil is the *glyceryl* ester of *linolenic* acid

$$CH_3CH_2CH{=}CHCH_2CH{=}CHCH_2CH{=}CH(CH_2)_7CO_2H$$

(a) Give the IUPAC names of the acids.

(b) The configuration at each double bond is *cis*. Write the formula of *linolenic* acid to show this. Can you explain why linseed oil is a liquid at room temperature while mutton fat is solid?

*8. Write the structural formula for the ester of propane-1,2,3-triol with *linolenic* acid [see Question 7]. What are the products of the reactions of this ester with the following reagents?

(a) NaOH, heat

(b) H_2, Ni

(c) Br_2

(d) O_3 followed by $Zn + CH_3CO_2H$

(e) $LiAlH_4$

9. Naturally occurring fats and oils are the esters of acids with an even number of carbon atoms. Acids with an odd number of carbon atoms are rare. Suggest a method of increasing the length of an aliphatic acid chain by one carbon atom:

$$RCH_2CO_2H \rightarrow RCH_2CH_2CO_2H$$

More than one step will be needed.

33.12 AMIDES

The methods of preparation have been described, and are summarised in Figure 33.12, p. 712.

Amides are hydrolysed to ammonium salts or acids...

Amides are hydrolysed in a similar way to esters:

$$RCONH_2(aq) + H_2O(l) \xrightarrow{H^+(aq),\, heat} RCO_2NH_4(aq)$$

$$RCONH_2(aq) + H_2O(l) \xrightarrow{OH^-(aq),\, heat} RCO_2^-(aq) + NH_3(aq)$$

Dehydration of an amide by distillation with phosphorus(V) oxide gives a nitrile:

$$RCONH_2(s) \xrightarrow{P_4O_{10}\, distil} RCN(l) + H_2O(l)$$

...and converted to nitriles, alcohols, amines...

The action of nitrous acid ($NaNO_2$ + HCl) is to replace —NH_2 by —OH, giving a carboxylic acid and nitrogen:

$$RCONH_2(aq) + HNO_2(aq) \rightarrow RCO_2H(aq) + N_2(g) + H_2O(l)$$

Reduction of an amide to an amine can be accomplished by using hydrogen with nickel as a catalyst or an ethoxyethane solution of lithium tetrahydridoaluminate:

$$RCONH_2(s) \xrightarrow[or\ LiAlH_4(ether)]{H_2,\ Ni,\ heat} RCH_2NH_2(l)$$

...and into the amines which have one carbon atom less than the amides

In the Hofmann degradation [see Figure 33.12], reaction with bromine and concentrated alkali gives an amine with one carbon atom less than the parent amide:

$$RCONH_2(aq) + Br_2(aq) + 4OH^-(aq) \rightarrow RNH_2(l) + 2Br^-(aq) + CO_3^{2-}(aq) + 2H_2O(l)$$

33.12.1 POLYAMIDES

Polyamides are made by condensation, e.g. nylon 66

Nylon is a polyamide, made by **condensation polymerisation** from a dicarboxylic acid and a diamine. Nylon 66 is the type that is manufactured in the largest quantities. It is made from a six-carbon dicarboxylic acid and a six-carbon diamine:

$$n\ H_2N(CH_2)_6NH_2 + n\ HOC(CH_2)_4COH \rightarrow \overline{}(HN(CH_2)_6NHC(CH_2)_4C)\overline{}_n + 2n\ H_2O$$
$$\overset{\|}{O}\qquad\overset{\|}{O}\qquad\qquad\qquad\overset{\|}{O}\qquad\overset{\|}{O}$$

Hexane-1,6-diamine   Hexane-1,6-dioic acid
(in water)   (in CCl_4)

Nylon forms at the interface between the aqueous solution of the diamine and the tetrachloromethane solution of the dicarboxylic acid. It is removed as it is formed, to allow further reaction to take place [see Figure 33.11].

33.12.2 UREA

Urea

$$\overset{\textstyle O}{\underset{\textstyle H_2N—C—NH_2}{\|}}$$

is the diamide of carbonic acid, carbamide. It was the first organic compound to be

Urea is a product of protein metabolism...

made from inorganic material. F Wöhler, in 1828, made it by evaporating an aqueous solution of ammonium cyanate:

$$NH_4CNO(aq) \xrightarrow{\text{evaporate}} H_2NCONH_2(s)$$

...and is also made synthetically for use as a fertiliser...

Urea is formed as an end-product of protein metabolism. It is widely used as a fertiliser. It is made commercially by heating carbon dioxide and ammonia under pressure:

$$CO_2(g) + 2NH_3(g) \xrightarrow[\text{pressure}]{\text{heat under}} H_2NCONH_2(s)$$

As the reaction proceeds slowly in the reverse direction, urea slowly releases ammonia into the soil.

...for the manufacture of plastics, and sleeping pills

Urea is also required in quantity for the manufacture of urea-methanal polymers [p. 656].

The barbiturate drugs which are used for inducing sleep are amides made from urea and esters of dicarboxylic acids:

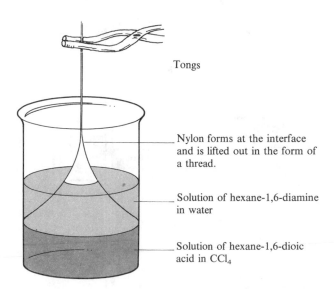

Phenobarbital

These drugs are habit-forming, and are available only on prescription.

FIGURE 33.11
Making Nylon

Tongs

Nylon forms at the interface and is lifted out in the form of a thread.

Solution of hexane-1,6-diamine in water

Solution of hexane-1,6-dioic acid in CCl$_4$

33.13 NITRILES

Aliphatic nitriles, RCN are made from halogenoalkanes, RX...

...Aromatic nitriles, ArCN, are made from diazonium compounds

Nitriles

$$R—C≡N$$

differ from the other acid derivatives in that they do not possess a carbonyl group. They are readily prepared from acids and readily yield acids on hydrolysis. Aliphatic nitriles are prepared from halogenoalkanes; aromatic nitriles from diazonium compounds [see Figure 33.12].

FIGURE 33.12
Reactions of Amides
and Nitriles

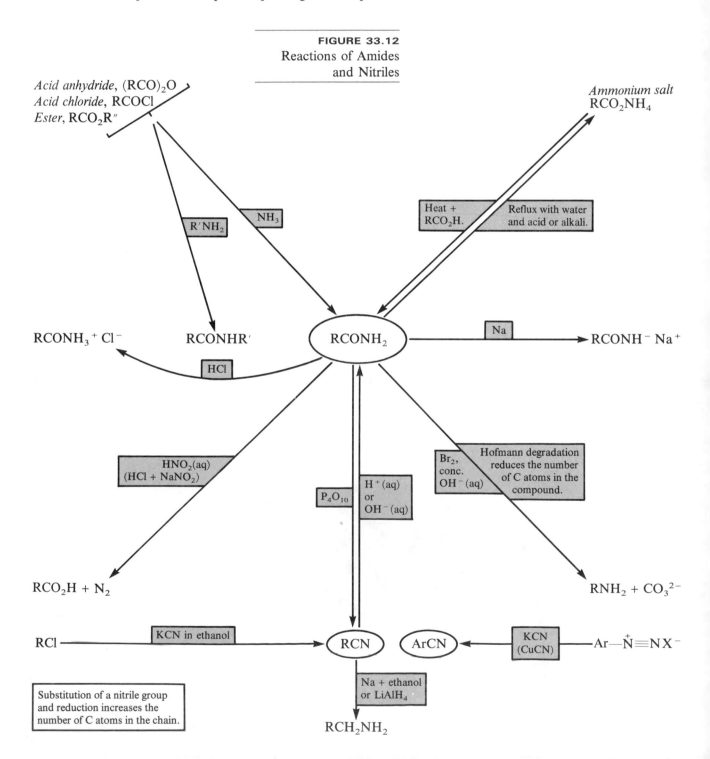

$Acid\ anhydride,\ (RCO)_2O$
$Acid\ chloride,\ RCOCl$
$Ester,\ RCO_2R''$

$Ammonium\ salt$
RCO_2NH_4

R′NH₂

NH₃

Heat + RCO₂H.

Reflux with water and acid or alkali.

$RCONH_3{}^+\ Cl^-$ $RCONHR'$ $RCONH_2$ Na $RCONH^-\ Na^+$

HCl

$HNO_2(aq)$
$(HCl + NaNO_2)$

P_4O_{10}

$H^+(aq)$
or
$OH^-(aq)$

Br_2, conc. OH^- (aq)

Hofmann degradation reduces the number of C atoms in the compound.

$RCO_2H + N_2$

$RNH_2 + CO_3{}^{2-}$

RCl

KCN in ethanol

RCN   ArCN

KCN
(CuCN)

$Ar—\overset{+}{N}≡N\,X^-$

Substitution of a nitrile group and reduction increases the number of C atoms in the chain.

Na + ethanol
or LiAlH₄

RCH_2NH_2

1. Say how the following conversions could be brought about:

(a) $RCONH_2 \rightarrow RCO_2H$

(b) $RCONH_2 \rightarrow RCH_2NH_2$

(c) $RCONH_2 \rightarrow RNH_2$

(d) $RCO_2R' \rightarrow RCONH_2 + R'OH$

(e) $(RCO)_2O \rightarrow RCONHR'$

(f) $RCOCl \rightarrow RCHO$

(g) $RCOCl \rightarrow RCH_2OH$

2. (a) Write the name and structural formula of (i) a carboxylic acid amide and (ii) a primary aliphatic amine.

(b) Describe two ways in which the properties of amides differ from those of amines and two ways in which they show resemblance.

(c) How can amide (i) be converted into an amine?

(d) How can amine (ii) be converted into an amide?

3. State how you could prepare the following compounds, using the organic compound cited as the only organic substance and any inorganic reagents you need:

(a) $CH_3CH_2CH_2CO_2H$ from $CH_3CH_2CH_2OH$

(b) $CH_3CO_2C_2H_5$ from CH_3CH_2OH

(c) CN   CONH$_2$ CO$_2$H   CO$_2$H

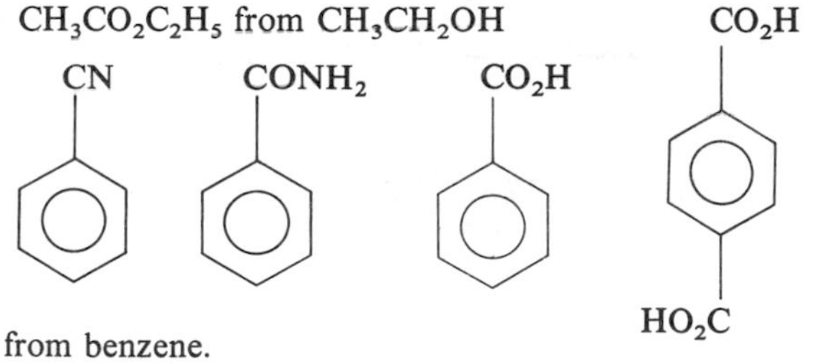

all from benzene. HO$_2$C

4. Write a structural formula for the main product of the reaction

$$ClC(CH_2)_4CCl + H_2N(CH_2)_6NH_2 \rightarrow$$
(with C=O groups shown by ‖ O below each C)

5. Identify compounds **A**, **B**, **C**, **D** and **E**, and reagents x, y and z in the following scheme of reactions. Write equations for the reactions involved:

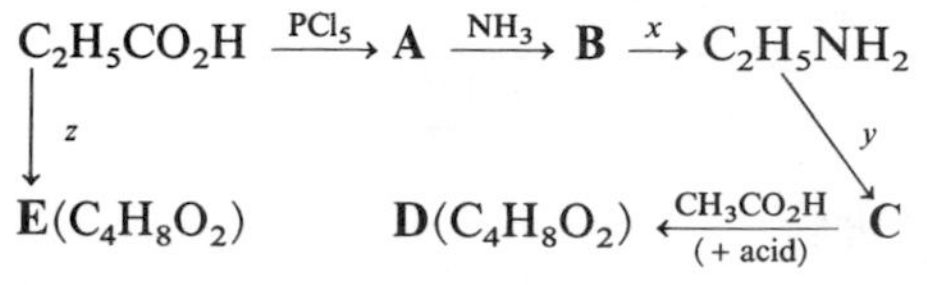

$$C_2H_5CO_2H \xrightarrow{PCl_5} \mathbf{A} \xrightarrow{NH_3} \mathbf{B} \xrightarrow{x} C_2H_5NH_2$$

$$\downarrow z \qquad\qquad\qquad\qquad\qquad\qquad \downarrow y$$

$$\mathbf{E}(C_4H_8O_2) \qquad \mathbf{D}(C_4H_8O_2) \xleftarrow[\text{(+ acid)}]{CH_3CO_2H} \mathbf{C}$$

6. Identify **A**, **B**, **C**, **D**, **E**, **F** and **G** in the scheme of reactions shown below. Write equations for the reactions portrayed:

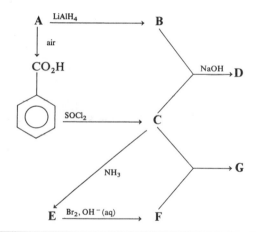

33.14 AMINO ACIDS AND PROTEINS

Proteins are one of the three large classes of natural polymers

There are three large classes of natural polymers: polysaccharides, nucleic acids and proteins. Proteins are polyamides. The units from which they are formed are amino acids, of formula

$$H_2NCHCO_2H$$
$$|$$
$$R$$

Since the —NH$_2$ group is on the α carbon atom (that adjacent to the —CO$_2$H group), they are correctly known as α amino acids. Table 33.3 lists some of the 20 naturally occurring α amino acids.

FIGURE 33.13
Relationships between
Carboxylic Acids and
other Series

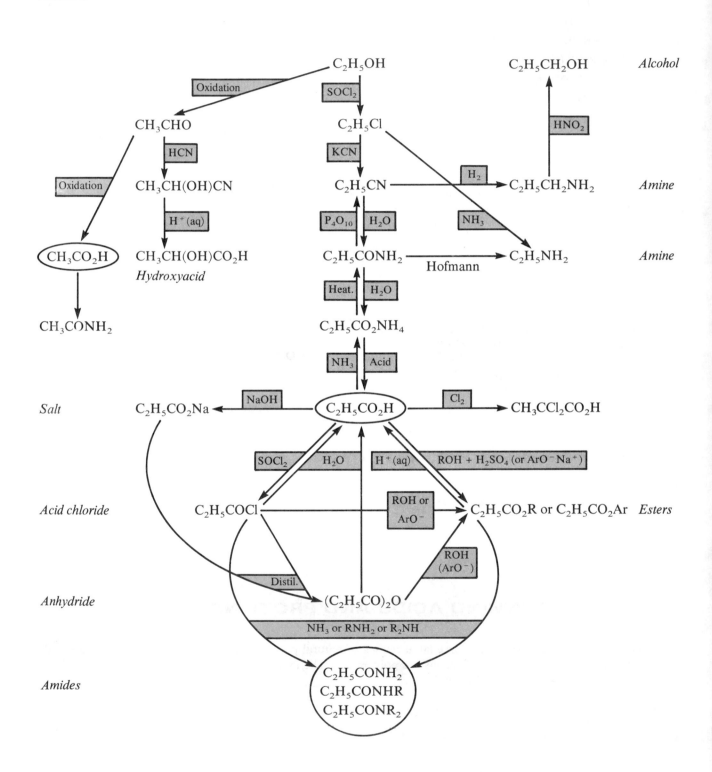

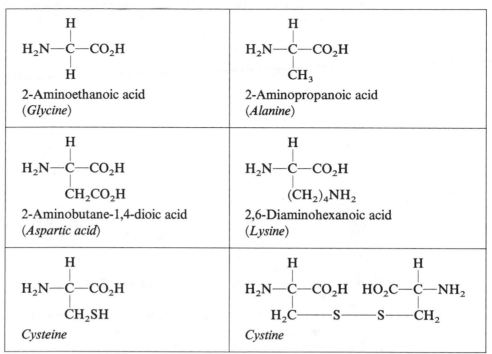

TABLE 33.3 Some Naturally occurring α Amino Acids

$H_2N-\overset{\overset{\displaystyle H}{|}}{\underset{\underset{\displaystyle H}{|}}{C}}-CO_2H$

2-Aminoethanoic acid
(*Glycine*)

$H_2N-\overset{\overset{\displaystyle H}{|}}{\underset{\underset{\displaystyle CH_3}{|}}{C}}-CO_2H$

2-Aminopropanoic acid
(*Alanine*)

$H_2N-\overset{\overset{\displaystyle H}{|}}{\underset{\underset{\displaystyle CH_2CO_2H}{|}}{C}}-CO_2H$

2-Aminobutane-1,4-dioic acid
(*Aspartic acid*)

$H_2N-\overset{\overset{\displaystyle H}{|}}{\underset{\underset{\displaystyle (CH_2)_4NH_2}{|}}{C}}-CO_2H$

2,6-Diaminohexanoic acid
(*Lysine*)

$H_2N-\overset{\overset{\displaystyle H}{|}}{\underset{\underset{\displaystyle CH_2SH}{|}}{C}}-CO_2H$

Cysteine

$H_2N-\overset{\overset{\displaystyle H}{|}}{\underset{\underset{\displaystyle H_2C}{|}}{C}}-CO_2H \quad HO_2C-\overset{\overset{\displaystyle H}{|}}{\underset{\underset{\displaystyle CH_2}{|}}{C}}-NH_2$
$\qquad\qquad S——S$

Cystine

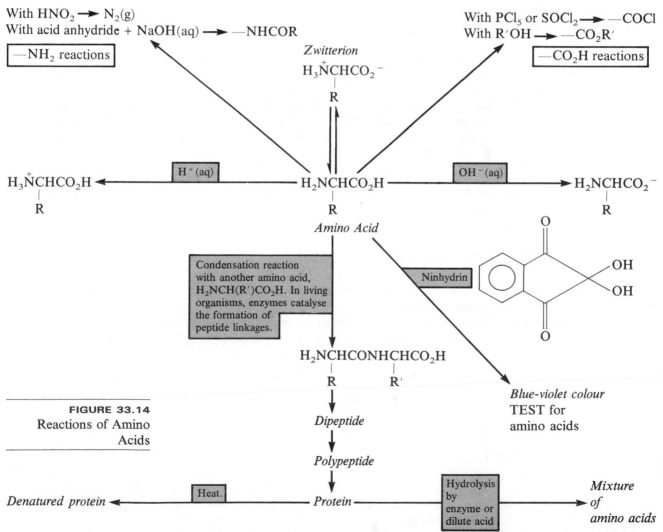

With $HNO_2 \longrightarrow N_2(g)$
With acid anhydride + NaOH(aq) $\longrightarrow$ —NHCOR

—NH₂ reactions

With PCl_5 or $SOCl_2 \longrightarrow$ —COCl
With R′OH $\longrightarrow$ —CO₂R′

—CO₂H reactions

Zwitterion
$H_3\overset{+}{N}CHCO_2^-$
$\quad | $
$\quad R$

$H_3\overset{+}{N}CHCO_2H$ ⟵ H⁺(aq) — H_2NCHCO_2H — OH⁻(aq) ⟶ $H_2NCHCO_2^-$
$\quad |$ $\quad |$ $\quad |$
$\quad R$ $\quad R$ $\quad R$

Amino Acid

Condensation reaction with another amino acid, $H_2NCH(R′)CO_2H$. In living organisms, enzymes catalyse the formation of peptide linkages.

Ninhydrin

Blue-violet colour
TEST for amino acids

$H_2NCHCONHCHCO_2H$
$\quad | \qquad\qquad |$
$\quad R \qquad\qquad R′$

Dipeptide

Polypeptide

Denatured protein ⟵ Heat. ⟵ *Protein* — Hydrolysis by enzyme or dilute acid ⟶ *Mixture of amino acids*

FIGURE 33.14
Reactions of Amino Acids

Glycine can be made from ethanoic acid...

All amino acids, except glycine, have an asymmetric carbon atom and show optical isomerism. Glycine is a white solid of T_m 235 °C. It can be made from ethanoic acid and from methanal:

...or from methanal

$$CH_3CO_2H \xrightarrow[\text{pass Cl}_2]{\text{boil}} ClCH_2CO_2H \xrightarrow[\text{large excess}]{\text{conc. NH}_3\text{(aq)}} H_2NCH_2CO_2H + NH_4Cl$$

Ethanoic   Chloroethanoic   *Glycine*
acid   acid

$$HCHO \xrightarrow[\text{dil. acid}]{\text{KCN +}} H_2C\begin{smallmatrix}OH\\CN\end{smallmatrix} \xrightarrow{\text{conc. NH}_3\text{(aq)}} H_2C\begin{smallmatrix}NH_2\\CN\end{smallmatrix} \xrightarrow[\text{acid}]{\text{dil.}} H_2C\begin{smallmatrix}NH_2\\CO_2H\end{smallmatrix}$$

Methanal   Hydroxy-   Aminoethano- *Glycine*
            ethanonitrile   nitrile

33.14.1 REACTIONS OF AMINO ACIDS

There are two functional groups, $-NH_2$ and $-CO_2H$, each of which is responsible for a typical set of reactions [see Figure 33.12].

Amino acids form zwitterions, containing $-NH_3^+$ and $-CO_2^-$

In acidic solution, the $-NH_2$ group ionises as $-\overset{+}{N}H_3$; in alkaline solution, the $-CO_2H$ group ionises as $-CO_2^-$. At intermediate values of pH, the **zwitterion**

$$H_3\overset{+}{N}CH(R)CO_2^-$$

exists (German: *Zwitter*, mongrel). The reason why amino acids are high-melting solids is that they crystallise as zwitterions. In solution, the pH at which the concentration of the zwitterion is maximal is called the **isoelectric point**. It differs for different amino acids. In **electrophoresis**, an amino acid will move towards the cathode or the anode, depending on the pH. This behaviour is the basis of the electrophoretic method (Greek: *phoresis*, being carried) of separating amino acids according to their isoelectric points.

Amino acids can be separated by electrophoresis or chromatography

Another method of separating amino acids is chromatography [p. 172]. The positions of the colourless amino acids on the chromatogram are revealed by spraying the paper with ninhydrin. This reagent gives a blue-violet colour with amino acids.

33.14.2 PEPTIDES AND PROTEINS

The polyamides formed by amino acids are peptides and proteins

The most important reaction of amino acids is polymerisation. In living organisms, enzymes catalyse the polymerisation of amino acids to form peptides and proteins. Polymerisation occurs through the formation of amide groups by the reaction of an $-NH_2$ group in one molecule with a $-CO_2H$ group in another molecule:

$$H_2N-\overset{\overset{R}{|}}{\underset{\underset{H}{|}}{C}}-\overset{\overset{O}{\|}}{C}-OH + H-\overset{\overset{H}{|}}{N}-\overset{\overset{R}{|}}{\underset{\underset{H}{|}}{C}}-\overset{\overset{O}{\|}}{C}-OH \rightarrow H_2N-\overset{\overset{R}{|}}{\underset{\underset{H}{|}}{C}}-\overset{\overset{O}{\|}}{C}-\overset{\overset{H}{|}}{N}-\overset{\overset{R}{|}}{\underset{\underset{H}{|}}{C}}-\overset{\overset{O}{\|}}{C}-OH$$

A dipeptide   + H_2O

The dipeptide formed can go on polymerising to form a long chain of amino acid residues linked through $-CONH-$ groups. The $-CONH-$ group is called a **peptide linkage**.

Fibrous proteins have linear molecules...

When the number of amino acid residues is 100 or more, the polymer is called a **protein**. Proteins are a diverse group of polymers. Fibrous proteins have linear molecules, and are insoluble in water and resistant to acids and alkalis. Examples are *keratin* (in hair, nails, horn and feathers), *collagen* (in muscles and tendons), *elastin* (in arteries and tendons) and *fibroin* (in silk).

...Globular proteins have a complicated 3D structure...

The molecules of globular proteins, e.g., *albumin* (in egg white) and *casein* (in milk), adopt a complicated three-dimensional structure. The polypeptide chain is held in position by attractions between polar groups, e.g., $-CO_2^-$ in aspartic acid and $-\overset{+}{N}H_3$ in lysine and also by cystine bridges [see Table 33.3, p. 714]. The $-SH$ groups in two cysteine amino acid residues can be oxidised to form an $-S-S-$ bridge between two strands of polypeptide. Anything which interferes with the three-dimensional configuration, such as acids, alkalis and a rise in temperature, is said to **denature** the protein. Globular proteins are soluble in water and are easily denatured.

...e.g., enzymes

Enzymes are globular proteins, and their catalytic activity depends on their three-dimensional structure. If the enzyme is denatured by a rise in temperature or an extreme of pH, or even by violent agitation of the solution, it is no longer able to function as a catalyst.

Hair protein can be permanently waved

Hair protein owes some of its texture to $-S-S-$ bonds in cystine. Permanent waving techniques utilise this fact. When hair is soaked in a gentle reducing agent, the $-S-S-$ bridges are broken by reduction to $-SH$ groups. While the hair proteins are no longer cross-linked, they are arranged into curls. Re-oxidation by soaking the hair in a mild oxidising agent reforms the disulphide bonds and holds the hair in its new configuration [see Figure 33.15].

FIGURE 33.15
Permanent Waving

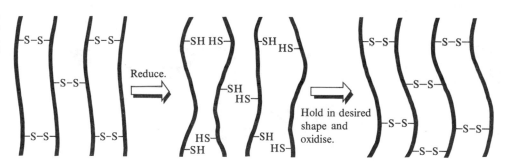

CHECKPOINT 33F: AMINO ACIDS AND PROTEINS

1. Outline how you could make *alanine*, $H_2NCH(CH_3)CO_2H$, from ethanal. Give the IUPAC name for alanine. Predict two reactions which alanine would undergo.

2. (a) Outline the preparation of *glycine*, aminoethanoic acid, starting from ethanoic acid.

(b) Which of these formulae is the better representation of the structure of glycine?

$$H_2NCH_2CO_2H \quad \text{or} \quad H_3\overset{+}{N}CH_2CO_2^-$$

Explain your choice.

(c) Write equations for two reactions of glycine.

3. What class of polymers are made from monomers such as glycine and alanine? Give the names of three of these polymers.

What is the name and the structure of the functional group in these polymers? State one reaction which is typical of this functional group.

4. How could you distinguish between aminoethanoic acid and ethanamide?

5. A 0.1110 g specimen of an amino acid was dissolved in water and treated with nitrous acid. The volume of nitrogen produced was 16.0 cm³ at 1 atm and 293 K. Calculate the molar mass of the amino acid. (GMV = 24.0 dm³ at rtp.)

6. A biochemist is analysing a mixture of amino acids. She adds 1.00×10^{-3} g of pure deuterioalanine, $H_2NCH(CD_3)CO_2H$, to the mixture. After separating the alanine present by chromatography, she finds that it contains $9.5 \times 10^{-3}\%$ deuterium by mass. What mass of alanine was present in the original mixture?

33.15 *SULPHONIC ACIDS

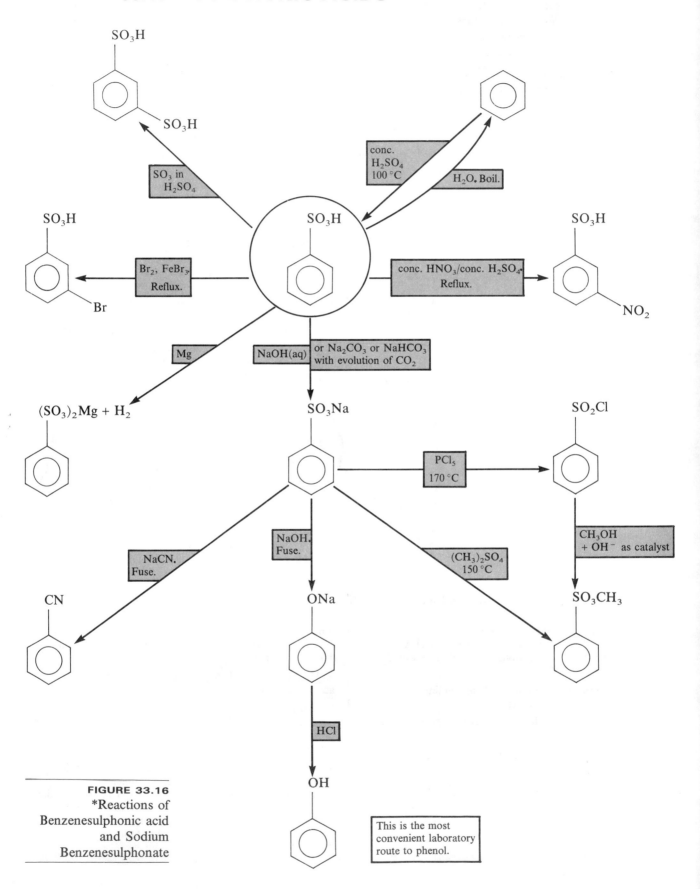

FIGURE 33.16
*Reactions of
Benzenesulphonic acid
and Sodium
Benzenesulphonate

This is the most
convenient laboratory
route to phenol.

QUESTIONS ON CHAPTER 33

1. How can you distinguish between the members of the following pairs of compounds? Suggest two tests, one of which will give a positive result with the first compound, and the other of which will give a positive result with the second compound.

(a) CH_3CH_2CHO and $CH_3CH_2CO_2H$

(b) $CH_3COC_2H_5$ and $CH_3CO_2C_2H_5$

(c)

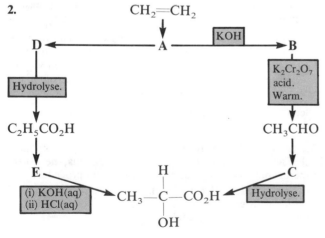

and

(d) and

(e) and

(f) and

(g) HCO_2H and CH_3CO_2H

(h) CH_3CO_2H and Cl_2CHCO_2H

(i) $C_6H_5NH_2$ and $C_6H_5CONH_2$

2.

FIGURE 33.17

Identify the compounds **A**, **B**, **C**, **D** and **E** in Figure 33.17. Name the product. How would you expect this acid to react with (a) Cl_2, (b) $SOCl_2$, (c) Na_2CO_3, (d) plane-polarised light?

3. What is formed when bromine reacts with (a) methane, (b) benzene, (c) propene, (d) phenol, (e) ethanoic acid and (f) ethanamide? State the conditions under which the reactions occur.

4. How could you carry out the following conversions? Use no organic compounds other than the one cited. Use any inorganic substances you need.

(a) $CH_3CH_2CH_2OH \rightarrow CH_3CH_2CO_2CH_2CH_2CH_3$

(b) $CH_3CH_2CH_2OH \rightarrow CH_3CH_2CH_2CO_2CH_2CH_2CH_3$

(c) $(CH_3)_2CHOH \rightarrow (CH_3)_2C(OH)CO_2H$

(d) $CH_3CHO \rightarrow CH_3CO_2CH_2CH_3$

More than one step will be needed.

5. **C** is an organic compound which dissolves in warm water to form an acidic solution. **C** reacts with phenol in the presence of a base to form an ester, and reacts with diethylamine to form an *N,N*-disubstituted amide. Benzene reacts with **C** in the presence of aluminium chloride to form diphenylmethanone.

Deduce the identity of **C**. Explain the reactions described.

6. Consider the three compounds $C_2H_5NH_2$, $C_6H_5NH_2$ and CH_3CONH_2.

(a) Name the compounds, and give the names of the series to which they belong.

(b) Compare the basicity of the $-NH_2$ group in the compounds. Explain how the differences arise from differences in structure.

(c) Describe the behaviour of the compounds with a solution of sodium nitrite in dilute hydrochloric acid.

7. Write the formulae for phenylamine and phenyl-methylamine. Describe how you could use benzoyl chloride to distinguish between them.

8. Starting from an aliphatic carboxylic acid, how could you make (a) an acid chloride, (b) an acid anhydride, (c) an amide, (d) an ester? Give the necessary conditions for reaction, and give equations for the reactions. State the conditions under which each of the products will react with water. Write equations, and name the products of the hydrolyses.

9.

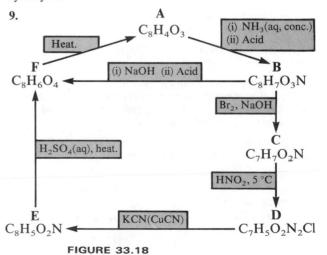

FIGURE 33.18

Write structural formulae for compounds **A**–**F**. Explain the reactions shown in Figure 33.18.

10. Discuss the factors that affect the strength of aliphatic acids.

11. Explain why a carboxylic acid, although it possesses a

$$\diagdown C{=}O \text{ group}$$

does not react with the nucleophiles which attack aldehydes and ketones.

12. When a solution of chlorine in hot ethanoic acid is irradiated with ultraviolet light, 2-chloroethanoic acid is formed. Write equations for all the steps that occur in this photochemical reaction.

13. Two esters have the molecular formula $C_6H_{12}O_2$. Both show optical isomerism. When heated with aqueous sodium hydroxide, **A** gives sodium ethanoate and another product, and **B** gives methanol and another product. Write structural formulae for **A** and **B**.

14. $\mathbf{P} = \begin{array}{c} CHCO_2H \\ \| \\ CHCO_2H \end{array}$ $\mathbf{Q} = \begin{array}{c} CH_2CO_2H \\ | \\ CH_2CO_2H \end{array}$

$$\mathbf{R} = CH_3{-}\underset{\underset{OH}{|}}{\overset{\overset{H}{|}}{C}}{-}CO_2H \qquad \mathbf{S} = CH_3{-}\underset{\underset{CO_2H}{|}}{\overset{\overset{H}{|}}{C}}{-}CO_2H$$

(*a*) Why does **P** exist in isomeric forms, whereas **Q** does not? What name is given to this type of isomerism?

(*b*) Why does **R** display optical isomerism, whereas **S** does not?

15. An optically active ester of molecular formula $C_8H_{16}O_2$ was hydrolysed by aqueous alkali to give as one product a liquid of molecular formula C_3H_8O, which gave a positive iodoform test. Write a structural formula for the ester.

16. Explain these statements:

(*a*) Ethanoyl chloride is more reactive towards water than is chloroethane.

(*b*) Laboratory preparations of 2-hydroxypropanoic acid yield a racemic mixture.

17. Place the following compounds in order of increasing acid dissociation constant. Explain why they follow this order.

CH_3CO_2H, C_2H_5OH, C_6H_5OH, Cl_2CHCO_2H, H_2O, $C_6H_5CO_2H$

18. Write a structural formula for aminoethanoic acid (glycine). How does this structure explain why the amino acid is (*a*) a crystalline solid, (*b*) soluble in water and (*c*) insoluble in ethoxyethane?

What happens if a potential difference is applied across a solution of aminoethanoic acid (*a*) at pH 5, (*b*) at pH 9?

What is the peptide bond? Give the formulae of the amino acids that are formed by hydrolysis of the peptide bonds in the peptide shown below:

$$H_2NCH_2CONHCHCONHCHCONHCHCO_2H$$
$$\underset{\underset{SH}{|}}{CH_2}\underset{\underset{CO_2H}{|}}{CH_2}\underset{}{CH_3}$$

19. Describe how, given ethanol as your only organic compound, you would prepare ethyl ethanoate.

Explain why the hydrolysis of esters is usually carried out in the presence of a base, rather than an acid.

Explain why ketones are not hydrolysed, although, like esters, they possess a carbonyl group.

20. Outline the laboratory preparation of a pure sample of ethyl ethanoate, starting from ethanol and ethanoic acid. Write an expression for the equilibrium constant for the reaction involved and briefly discuss the factors that influence the choice of reaction conditions.

An ester of molecular formula $C_9H_{10}O_2$ was heated under reflux with aqueous sodium hydroxide. On distilling the product, a colourless liquid was obtained which, on warming with iodine and aqueous sodium hydroxide, gave a pale yellow precipitate. The residue in the distilling flask gave a white precipitate on acidification with hydrochloric acid. Identify the ester, explaining your reasoning, and write a balanced equation for its reaction with sodium hydroxide.

(C 82)

21. (*a*) For the preparation of an alkanoic acid from the corresponding aldehyde

 (i) name the oxidising agent you would use,

(ii) give the conditions, and write an equation for the reaction.

(*b*) For the alkanoic acid chosen in (a)

 (i) write the structural formulae of its anhydride and its acid amide,

(ii) compare the conditions required to convert each of them to the acid.

(*c*) 'An acid is a proton-donor, a base is a proton-acceptor. Therefore, the addition of hydrogen bromide to propene is an acid–base reaction.'

Discuss this statement by considering the mechanism of the addition of hydrogen bromide to propene stating which part of the mechanism could be considered an acid–base reaction.

(*d*) Write *two* general equations, one to show neutralisation in water as the solvent, the other to show neutralisation in liquid ammonia as the solvent. Hence give a reason why it is important to refer to the solvent before we classify a reaction as an acid–base reaction.

[See Chapter 12 for (c) and (d).]

(AEB 82)

22. (*a*) Describe briefly how you would obtain pure ethanoyl chloride from ethanoic acid, giving equations, necessary experimental conditions and the purification procedure.

(*b*) Give a balanced equation for the reaction of ethanoyl chloride with each of the four compounds below, and state essential conditions for each reaction:

 (i) sodium ethanoate,

 (ii) phenylamine,

 (iii) ethanol,

 (iv) benzene.

(*c*) Reaction of benzoyl chloride with phenol in ice-cold aqueous sodium hydroxide gives phenyl benzoate in good yield, but the corresponding reaction of ethanoyl chloride under the same conditions does *not* give a good yield of phenyl ethanoate. Comment briefly on these observations.

(O 82)

33.16 POSTSCRIPT: ETHANOIC ANHYDRIDE

Ethanoic anhydride is used for the manufacture of aspirin...

Ethanoic anhydride is used in the manufacture of aspirin. The method of synthesis is

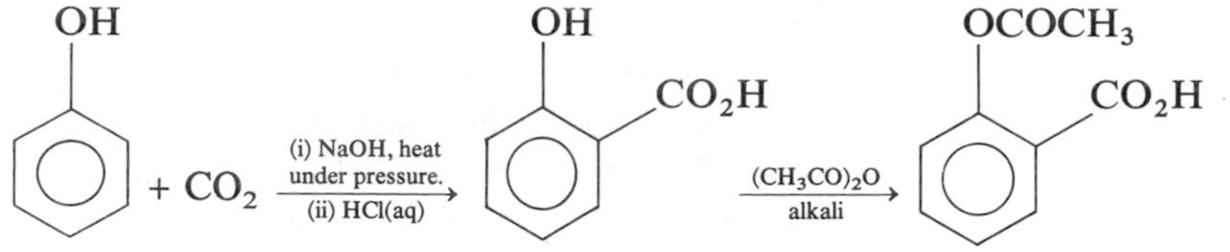

Phenol $+ CO_2$ $\xrightarrow[\text{(ii) HCl(aq)}]{\text{(i) NaOH, heat under pressure.}}$ 2-Hydroxybenzoic acid (*Salicylic acid*) $\xrightarrow[\text{alkali}]{(CH_3CO)_2O}$ 2-Ethanoyloxybenzoic acid (*Aspirin*)

...which is an analgesic...

The analgesic (pain-killing) effects of the esters of 'salicylic acid' have been known for centuries. Physicians in ancient times made up remedies for fever and pain from leaves which are now known to contain these compounds. 2-Hydroxybenzoic acid (salicylic acid) is the active analgesic. While it is a sufficiently strong acid to irritate the stomach, its ester, aspirin, is less acidic and less irritating. Aspirin passes through the acidic stomach contents unchanged, to be hydrolysed in the alkaline conditions of the intestines to sodium 2-hydroxybenzoate (sodium salicylate), which dissolves in the blood stream.

Ethanoic anhydride is also used for converting wood cellulose into cellulose ethanoate...

...which can be made into threads...

...to be used in the manufacture of acetate rayon

The major use of ethanoic anhydride is in the ethanoylation of cellulose. Cellulose is a polymer of 2000–3000 glucose units. In cotton and linen, the cellulose molecules are arranged into long fibres which are suitable for making into thread and fabric. Wood cellulose has the same chemical composition, but the molecules form a tangle which cannot be made into thread. Since wood cellulose is plentiful, chemists have tackled the problem of converting it into threadlike form. The problem is to get the cellulose into solution so that it can be forced through tiny holes into a medium that will precipitate cellulose as fine threads. The difficulty is that the large number of hydroxyl groups make cellulose insoluble in organic solvents, but it is not sufficiently polar to dissolve in water. Treatment with ethanoic anhydride converts the —OH groups into —OCOCH$_3$ groups. The reaction takes place quickly at room temperature to give a high yield of cellulose ethanoate. If ethanoic acid is used, prolonged heating and the presence of an acid catalyst are necessary, and degradation of the cellulose molecules results. With the hydroxyl groups esterified, cellulose ethanoate is soluble in organic solvents. From solution, it can be precipitated as fine threads of *acetate rayon*, which are made into fabric [see Figure 33.19].

FIGURE 33.19
Cellulose Ethanoate
(Acetate Rayon)
Ac is CH$_3$C—
‖
O
The monomer is shown

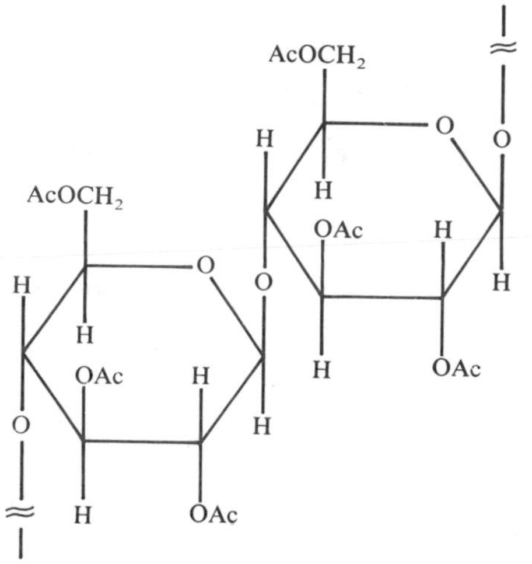

Problem (a)

Methylcellulose is used in the paper industry for imparting a resistance to oil and grease. It is made by converting the —OH groups in cellulose to —OCH$_3$ groups. Suggest how this could be done, and draw a structural formula for methyl cellulose.

Problem (b)

'Oil of wintergreen' has been used for centuries as a soothing liniment. Its IUPAC name is methyl 2-hydroxybenzoate. Devise a synthesis for this compound from phenol.

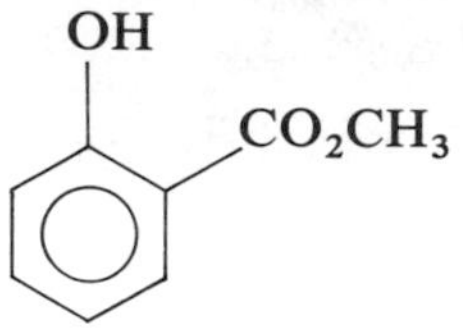

OH

CO$_2$CH$_3$

Methyl 2-hydroxybenzoate

34

THE IDENTIFICATION OF ORGANIC COMPOUNDS

Before a compound can be identified, it must be obtained in a pure state.

34.1 METHODS OF PURIFYING ORGANIC COMPOUNDS

Recrystallisation... **1.** Recrystallisation and fractional recrystallisation are used for solids [p. 177].

...sublimation... **2.** Sublimation can be used for a few substances, e.g., benzoic acid [see Figure 34.1].

...filtration... **3.** Filtration will remove suspended matter from liquids. Figure 9.2, p. 177 shows an apparatus which can be used for filtration under reduced pressure, in order to speed up the process.

FIGURE 34.1
Sublimation

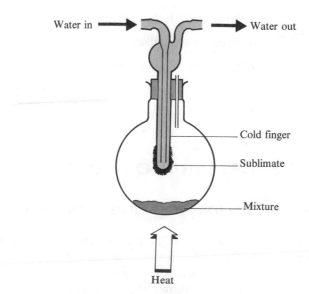

Water in → → Water out

Cold finger

Sublimate

Mixture

Heat

...and centrifugation... **4.** Centrifuging is used to remove solid particles from liquids [see Figure 34.2].
When a liquid is spun round in a centrifuge, the force of gravity acting on suspended

...are all methods used for purifying organic compounds particles is increased by a centrifugal force. The particles settle to the bottom of the liquid, and the liquid can be poured off, i.e., **decanted**.

FIGURE 34.2
An Analytical Chemist
at Work

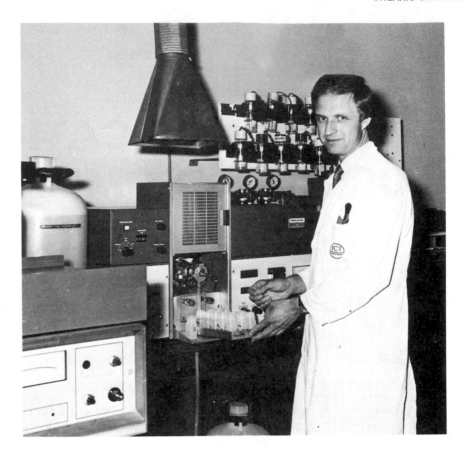

FIGURE 34.2
An Analytical Chemist
at Work

Other methods are distillation: simple, fractional and in steam...

5. Distillation is used for liquids [p. 155].

6. Fractional distillation is used for mixtures of liquids [p. 160].

7. Steam distillation is used for mixtures of liquids [p. 163].

...solvent extraction and chromatography

8. Solvent extraction is widely used for liquids and solids [p. 166].

9. Various types of chromatography are widely used [p. 170].

34.2 CRITERIA OF PURITY

34.2.1 SOLIDS: MELTING TEMPERATURE

A pure solid has a sharp melting temperature

A pure solid melts at a certain temperature, at which it changes sharply from solid to liquid. The presence of impurity lowers the melting temperature, and also results in a more gradual melting over a range of temperature [see Figure 9.5, p. 181]. Figure 34.3 shows an apparatus which can be used for finding a melting temperature.

A mixed melting temperature with an authentic specimen can be used to identify a compound

A still better test of purity is a mixed melting temperature determination. Imagine you have a compound **C** which melts at 93 °C. It could be impure **A**, of T_m 94 °C or impure **B** of T_m 97 °C. A mixed melting temperature determination on a mixture of **A** and **C** gives 85 °C, while a mixture of **B** and **C** melts at 95 °C. The compound you have made is a slightly impure specimen of **B**. [See freezing temperature depression, p. 181.]

FIGURE 34.3 Apparatus
for Finding a Melting
Temperature

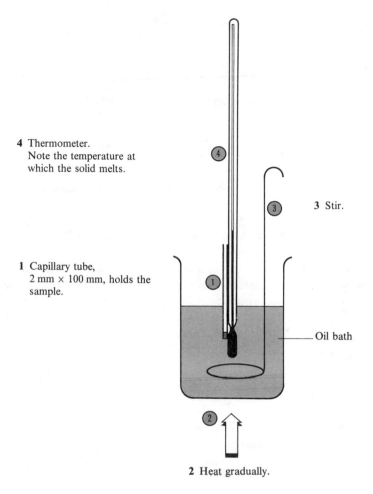

FIGURE 34.3 Apparatus
for Finding a Melting
Temperature

4 Thermometer.
Note the temperature at
which the solid melts.

3 Stir.

1 Capillary tube,
2 mm × 100 mm, holds the
sample.

Oil bath

2 Heat gradually.

34.2.2 LIQUIDS: BOILING TEMPERATURE

*Boiling temperatures are
less reliable indicators of
purity than melting
temperatures*

A pure liquid can be identified by its boiling temperature. An apparatus such as
that in Figure 8.5, p. 155 can be used. The thermometer measures the temperature of
the vapour in equilibrium with the boiling liquid. It should not be immersed in the
liquid because **superheating**, i.e., heating above the boiling temperature of the liquid,
can occur if the liquid is heated rapidly. Boiling temperatures tend to be less reliable
as a test of purity than melting temperatures because of the danger of superheating.
[See boiling temperature elevation, p. 178.]

34.3 ANALYSIS TO FIND OUT WHICH ELEMENTS ARE PRESENT

*The Lassaigne test for
sulphur, nitrogen and the
halogens*

Once a pure sample of the compound has been obtained, the next step is to find
out which elements it contains. It is assumed that carbon and hydrogen are present
in an organic compound. The **Lassaigne test** is done to test for the presence of nitrogen,
sulphur and the halogens. A small quantity of the substance is heated in an ignition
tube with sodium and then plunged into distilled water. This treatment converts nitrogen
into sodium cyanide, sulphur into sodium sulphide and halogens into halides.

The test for a cyanide is the formation of Prussian blue with a solution of iron(II)
sulphate and iron(III) sulphate.

The test for a sulphide is the formation of a purple colour with sodium pentacyano-nitrosylferrate(II), $Na_3Fe(CN)_5(NO)$.

Before being tested for halides, the solution is acidified and boiled to expel hydrogen cyanide and hydrogen sulphide. Then, silver nitrate solution is added.

Mass spectrometry [p. 9] and [p. 731] will detect the elements present.

34.4 QUANTITATIVE ANALYSIS

The percentage by mass of each element present can be found by quantitative analysis.

34.4.1 CARBON AND HYDROGEN

The masses of C and H in a compound can be found by weighing CO_2 and H_2O formed on combustion...

A known mass of the compound is heated in a stream of pure, dry oxygen in the presence of copper(II) oxide. Hydrogen is oxidised to steam, which is absorbed in weighed calcium chloride tubes. Carbon is oxidised to carbon dioxide, which is absorbed in weighed bulbs of concentrated potassium hydroxide solution. From the increases in mass, the masses of carbon and hydrogen in the sample can be found [see Question 1, p. 727].

34.4.2 NITROGEN

...N can be estimated after conversion to NH_4^+ ...

Nitrogen can be converted into ammonium sulphate by boiling a known mass of the compound with concentrated sulphuric acid and sodium sulphate. The ammonium salt is estimated by a titrimetric method [see Question 3, p. 727].

34.4.3 HALOGENS

...Halogens can be determined as AgX...

Heating a known mass of the compound in a sealed tube with fuming nitric acid and solid silver nitrate converts the halogens into a precipitate of silver halide. This can be filtered off, dried and weighed [see Question 2, p. 727].

34.4.4 SULPHUR

...S can be determined by precipitation as $BaSO_4(s)$...

The sulphur content of a known mass of the compound can be converted into sulphate ions by heating the sample with fuming nitric acid. The sulphate can be precipitated as barium sulphate, filtered off, dried and weighed [see Question 2, p. 727].

34.4.5 OXYGEN

...Oxygen is found by difference

The percentage by mass of oxygen is difficult to find by experiment, and is obtained by difference, once the other elements have been assayed.

34.5 EMPIRICAL FORMULA

The empirical formula is calculated from the percentage composition by mass, as outlined on p. 48.

34.6 MOLECULAR FORMULA

To convert the empirical formula into a molecular formula, the molar mass must be known. Molar mass determinations have been described for gases [p. 135, p. 139]; for volatile liquids [p. 156]; for solutes [p. 178, p. 180, p. 182 and p. 183] and the mass spectrometric method is covered on p. 9 and p. 731.

34.7 QUALITATIVE ANALYSIS

Chemical tests reveal the presence of hydroxyl groups, carbonyl groups, etc. You will be familiar with one of the schemes of qualitative analysis which arrange the tests for functional groups in a logical order.

34.8 STRUCTURAL FORMULA

The molecular formula and the results of chemical tests often enable a compound to be identified

The structural formula can often be worked out from the results of chemical tests and the molecular formula. There are, in addition, a number of physical methods available for identifying the groups present.

Once an opinion on the identity of the compound has been reached, it can be confirmed by matching the melting or boiling temperature with that of a known compound. Boiling temperatures are less reliable than melting temperatures. It is better to convert a liquid into a solid derivative, and find the melting temperature of the derivative. For example, ketones can be converted into solid 2,4-dinitrophenylhydrazones, and amines into solid benzoyl derivatives.

CHECKPOINT 34A: ANALYSIS

1. (*a*) When 0.200 g of a compound **A**, which contains only carbon and hydrogen, was burned completely in a stream of dry oxygen, 0.629 g of carbon dioxide and 0.257 g of water were formed. Find the empirical formula of the compound.

(*b*) When 0.200 g of **A** is vaporised, the volume which it occupies (corrected to stp) is 53.3 cm³.
The GMV = 22.4 dm³ at stp. Find the molar mass of **A**.

(*c*) Use the molar mass of **A** to convert the empirical formula, from (a), into a molecular formula.

2. A Lassaigne test showed that a compound **X** contained sulphur and chlorine. From 0.1000 g of **X** was obtained 0.1322 g of $BaSO_4$. Another 0.1000 g sample of **X** gave 0.0813 g of AgCl. The combustion of 0.2000 g of **X** gave 0.2992 g of CO_2 and 0.0510 g of H_2O. Find the empirical formula of **X**. (Remember that, if the percentages of C, H, S and Cl do not total 100%, the difference is due to oxygen.)

3. The compound **Y** contains nitrogen. After conversion of 0.1850 g of **Y** into an ammonium salt, ammonia was expelled by warming the salt with 75.0 cm³ of sodium hydroxide solution of concentration 0.100 mol dm⁻³. Titration showed that 50.0 cm³ of sodium hydroxide remained at the end of the reaction. Find the percentage by mass of nitrogen in **Y**.

When 0.146 g of **Y** was burned completely in dry oxygen, 0.264 g of CO_2 and 0.126 g of H_2O were formed. Find the percentage by mass of C and H in the compound.

Find the empirical formula of the compound.

4. 10 cm³ of a hydrocarbon C_aH_b is exploded with an excess of oxygen. A contraction of 35 cm³ occurs, all volumes being measured at room temperature and pressure. On treatment of the products with sodium hydroxide solution, a contraction of 40 cm³ occurs. Obtain the formula of the compound.
[If you need to, refer to p. 137.]

34.9 PHYSICAL METHODS OF STRUCTURE DETERMINATION

34.9.1 MOLECULAR SPECTROSCOPY

The electromagnetic spectrum is shown in Figure 34.4.

FIGURE 34.4
The Electromagnetic
Spectrum

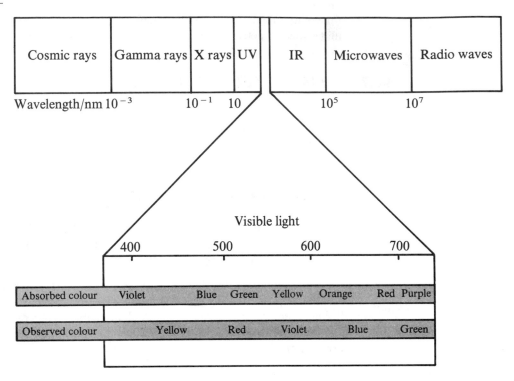

When electromagnetic radiation falls on to a compound, radiation of certain wavelengths is absorbed. In absorbing radiation, the molecules are excited to a higher energy level. The relationship expressed by Bohr is

$$hv = E_{\text{upper level}} - E_{\text{lower level}}$$

where v = frequency of radiation absorbed and h = Planck's constant [p. 29]. In atomic spectra, the transitions arise from changes in the distribution of electrons. In molecules, in addition to the excitation of electrons, the absorption of energy can increase the extent of vibration in the bonds and the speed of rotation of the molecule. The energy changes associated with changes in vibration and rotation are smaller than those that result in the excitation of electrons, and the frequency of radiation needed is less [see Figure 34.8, p. 730, for frequencies associated with vibration]. Electron transitions require radiation in the visible or ultraviolet region, vibrational changes require infrared light, and rotational changes require microwaves.

Molecular spectra arise from electron transitions...

...from changes in the extent of bond vibration and in the frequency of rotation

34.9.2 VISIBLE—ULTRAVIOLET SPECTROPHOTOMETRY

The promotion of electrons gives rise to UV spectra

Since all matter contains electrons, all substances absorb light at some wavelength in the visible–ultraviolet spectrum. Compounds with π electrons, since these are less strongly held than σ electrons, absorb light intensely. Benzene absorbs in the UV region, and appears colourless, naphthacene absorbs in the blue-violet region of the spectrum, and appears yellow.

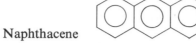

Naphthacene

Transmitted light is complementary to absorbed light.*

If the UV 'fingerprint' of a compound is on file, the compound can be identified

The ultraviolet spectrum of a compound is like a fingerprint. If it can be matched up with the spectrum of a known compound, it can be used as a means of identification. The spectrum is obtained by an instrument like that illustrated in Figure 34.5 and shown in Figure 34.6. The spectrum of ethyl 3-oxobutanoate in a number of solvents is shown in Figure 34.7 [see Question 6, p. 733].

FIGURE 34.5
A Visible–Ultraviolet
Spectrophotometer

2 Monochromator, M, selects wavelength.

4 One beam travels through a quartz cell containing the sample.

7 Electric circuitry compares the two currents. The difference depends on the absorption of light by the sample.

1 Light source, S

3 Quartz mirror splits light beam into a double beam.

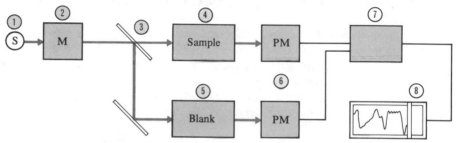

5 Second beam passes through a quartz cell containing the solvent.

6 Photomultiplier, PM, converts light into electric current.

8 Recorder. A pen traces the absorption spectrum.

FIGURE 34.6 A UV Spectrophotometer

*See R Muncaster, *A-Level Physics* (Stanley Thornes)

FIGURE 34.7 Ultraviolet Spectrum of Ethyl 3-oxobutanoate in A, Hexane; B, Ethoxyethane; C, Ethanol; D, Water

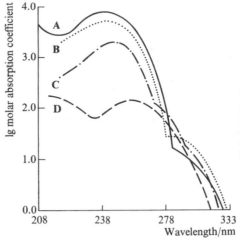

Molar absorption coefficient = $\lg(I_0/I)/cl$
I_0 and I are the intensities of the incident and transmitted light respectively
c = concentration of solution/mol dm^{-3}
l = length of cell/cm

34.9.3 INFRARED SPECTROSCOPY

All organic compounds absorb in the infrared

All organic molecules absorb strongly in the infrared [see Figure 34.8]. The operation of an infrared spectrophotometer is similar to that of a visible–UV spectrophotometer. The source of radiation is a hot wire. The detector is a **thermocouple** (instead of a photomultiplier), which converts heat into an electric current. Since glass and quartz absorb IR radiation, the cells, mirrors and prisms are cut from large sodium chloride crystals.

There are many peaks in an IR spectrum, and it is a good 'fingerprint'

Different bonds give rise to different absorption lines

There are many more peaks in an IR spectrum: it serves better than the UV spectrum to identify an organic compound. Books of recorded spectra are available for comparison. If the unknown compound is a new substance, its spectrum will not match up with any recorded IR spectrum; yet it is still possible to infer a great deal about the structure of the compound. The C=O bond and the C—OH bond and others have characteristic absorption frequencies, and can be identified. Figures 34.9 and 34.10 show the IR spectra of butanal and propanoic acid. It is customary to record IR spectra as percentage transmittance, so that the peaks extend from the top of the trace downwards.

FIGURE 34.8 Some Vibrations and Their Wavelengths

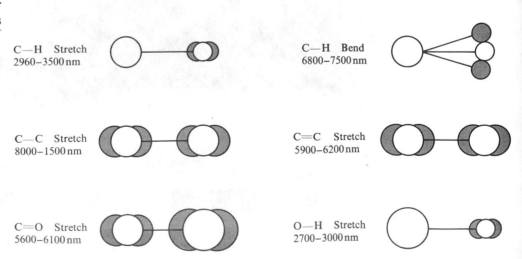

C—H Stretch
2960–3500 nm

C—H Bend
6800–7500 nm

C—C Stretch
8000–1500 nm

C=C Stretch
5900–6200 nm

C=O Stretch
5600–6100 nm

O—H Stretch
2700–3000 nm

FIGURE 34.9
IR Spectrum of
$CH_3CH_2CH_2CHO$

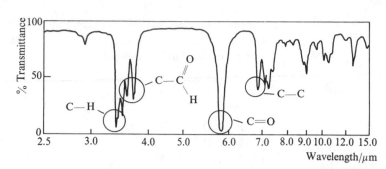

FIGURE 34.9
IR Spectrum of
$CH_3CH_2CH_2CHO$

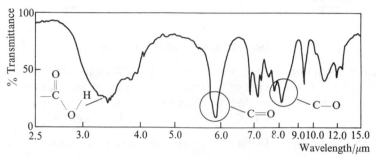

FIGURE 34.10
IR Spectrum of
$CH_3CH_2CO_2H$

34.9.4 NUCLEAR MAGNETIC RESONANCE SPECTROSCOPY

In a magnetic field, H atoms absorb radio waves...

Radio frequency spectroscopy becomes possible when substances are positioned in a strong magnetic field. In order to be detected by nuclear magnetic resonance spectroscopy, NMR, atoms must have a magnetic moment. Then the absorption of radio waves changes the orientation of the nuclei in the magnetic field. Any compound which has atoms of ^1_1H, which are magnetic because they have a nuclear spin, will absorb radio waves. The frequency of the radio waves absorbed depends on the magnetic moment of the nucleus and the size of the applied magnetic field.

...The frequency of radio waves absorbed gives information about the bonding of H atoms

The value of NMR as a tool in organic chemistry is that ^1_1H atoms in different chemical environments, e.g., in OH and in CH_3, absorb radio waves of different frequencies. In the NMR spectrum, the wavelength of the absorption reveals the environment of the H atoms, and the area of the peak reveals the number of H atoms in this environment. In the NMR spectrum for $C_6H_{12}O_2$ [see Figure 34.11], there are (a) 9H in CH_3—C groups and (b) 3H in —$COCH_3$ groups. The structural formula must be $(CH_3)_3COCOCH_3$.

FIGURE 34.11 The Shifts in Frequency for Hydrogen Nuclei in Different Environments. The zero point is the position of the peak for $(CH_3)_4Si$

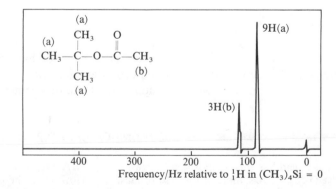

34.9.5 MASS SPECTROMETRY

The mass spectrum depends on the ratio m/e for the particles formed by the compound

Mass spectrometry [p. 9] is used to find the masses of atoms from the m/e (mass/charge) ratios of ionised atoms. Molecules also give mass spectra. The ionising beam of electrons is able to remove an electron from a molecule M to form the ion M^+. The ratio of m/e for this ion gives the molecular mass of the compound. The accuracy of the determination is such that molecular masses can be used to distinguish between compounds with the same mass number (nucleon number). The example in Table 34.1 will clarify the point.

Compound	Mass number/m_u	Molecular mass/m_u
$C_{10}H_{20}NO$	170	170.1540
$C_9H_{18}N_2O$	170	170.1415
$C_9H_{16}NO_2$	170	170.1177

The accurate value of the molecular mass will enable a distinction to be made between the compounds.

Molecular masses are found

The ionising beam of electrons has enough energy to split molecules into fragments. One of the fragments formed in each dissociation is ionised. The mass spectrum of methanol shows peaks at 32, 31, 29 and 28. These are produced by the ions CH_3OH^+ (the molecular ion), and CH_3O^+, CHO^+ and CO^+ [see Figure 34.12].

Fragmentation patterns can be used for fingerprinting

Isomers have the same molecular mass, but since the bonds are different, they often break up differently to give different ions. The **fragmentation patterns** of the isomeric pentanols shown in Figure 34.12 differ.

FIGURE 34.12
Fragmentation Patterns
for the Pentanols

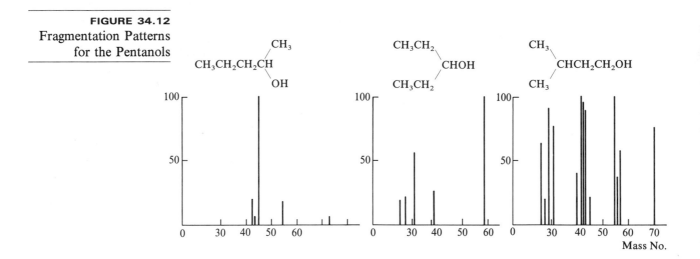

Fragmentation patterns of compounds have been recorded and catalogued. An unknown compound can often be identified by a comparison of its fragmentation pattern with a computerised bank of fragmentation patterns.

QUESTIONS ON CHAPTER 34

1. An organic dibasic acid has the composition by mass: C 41.4%, H 3.4%, O 55.2%. A sample of the acid of mass 0.189 g neutralised 32.5 cm³ of 0.100 mol dm⁻³ potassium hydroxide solution.

(a) Calculate the empirical formula and molecular formula of the acid.

(b) Write the names and structural formulae for the isomers with this molecular formula. State how you could distinguish between them.

2. A compound **B** contains bromine but no other halogen. How could the percentage by mass of bromine in the compound be found? Be careful to state the weighings that must be done.

3. At room temperature, 50 cm³ of an unsaturated hydrocarbon weigh 0.1167 g and react with 50 cm³ of hydrogen. Given the GMV at room temperature = 24.0 dm³, find the molar mass of the hydrocarbon. Suggest a formula for this compound.

4. A chemist is oxidising cyclohexanol. How can he find out when the conversion into cyclohexanone is complete?

5. Explain the reasons for the use of the following techniques in organic chemistry:

(a) distillation under reduced pressure,

(b) paper chromatography,

(c) steam distillation,

(d) recrystallisation,

(e) mixed melting temperature determinations,

(f) ethoxyethane extraction.

6. In Figure 34.7, p. 730, the peak at 238 nm is due to absorption by the $\diagdown C{=}O$ group. Can you explain why the absorption of light by ethyl 3-oxobutanoate, $CH_3COCH_2CO_2C_2H_5$, differs in different solvents, and is less in non-polar solvents than in ethanol and water? [See p. 646 if you need help.]

7. Synthesis is only the initial stage in the preparation of a pure organic compound and must be followed by separation, purification and characterisation. Discuss general methods for carrying out these steps subsequent to synthesis, illustrating your answer by reference to appropriate examples.

(JMB 81, S)

8. Suppose you were given an impure sample of an unknown organic compound. Describe the steps that may be taken to deduce the structure of the compound.

You should recognise the need to obtain first: a pure sample, the relative molecular mass, and at least one physical constant of the compound.

The emphasis in your answer may be on modern instrumental methods or on qualitative chemistry, but neither should be discussed to the total exclusion of the other.

(L 83)

9. Figure 34.13 shows the reactions of some propane derivatives and Figure 34.14 illustrates the infra-red spectra of propane and some of the same derivatives.

(a) Compound **Q** gives a positive result with Fehling's solution.

(i) Describe how tests with Fehling's solution are carried out and describe the appearance of a positive result.

(ii) What is the functional group in **Q**?

(iii) Draw a diagram of the structural formula of **Q**.

(b) Compound **P** is readily converted to **Q** in a single stage.

(i) Draw a diagram of the structural formula of **P**.

(ii) Give the name(s) of the reagent(s) needed to convert **P** to **Q**.

(iii) What experimental conditions would you use?

(c) Compound **R** is weakly acidic.

(i) Draw a diagram of the structural formula of **R**.

(ii) What is the formula of the product of the reaction of **R** with phosphorus pentachloride?

(d) (i) Suggest, as far as possible, which bonds may be responsible for the peaks in the infra-red absorption spectra at about

$3\,\mu m$:

$3.5\,\mu m$:

$5.9\,\mu m$:

$6.6{-}7.2\,\mu m$:

(ii) On a copy of the fourth diagram, draw the infra-red spectrum you would predict for compound **R**.

(e) What *type* of reaction occurs in the change of **Q** to **Z**?

(f) (i) What is the formula of the product **X**?

(ii) What is the formula of the product **Y**?

(iii) What reagent is used for identification of compounds with functional groups similar to **Y**?

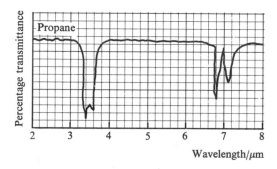

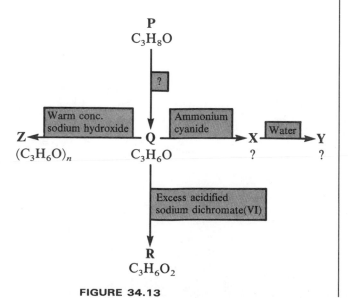

FIGURE 34.13

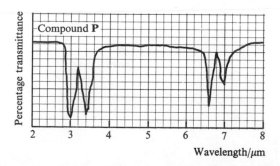

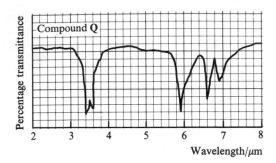

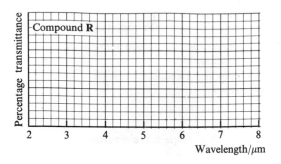

FIGURE 34.14

(L 80, N)

35

SOME GENERAL TOPICS

This chapter deals with a number of topics which have cropped up in different chapters. Polymerisation has been encountered in a number of chapters, and the pieces are here fitted together. Students complain of difficulty in getting from compound **A** to compound **D** via **B** and **C**; the pages on synthetic routes are an attempt to tackle this difficulty. The section 'What are these reagents used for?' is a summary of the plethora of different reagents and conditions which appear over the top of arrows, and which students find onerous to remember if tackled separately. There is a selection of questions on topics which span chapters, such as methods of distinguishing between compounds, reaction mechanisms, comparisons between functional groups and the identification of 'mystery' compounds from descriptions of their reactions.

35.1 PLASTICS

35.1.1 USES OF PLASTICS

The properties of plastics find them many uses

Plastics are polymers of carbon compounds. They are macromolecular substances with molar masses of some thousands. Plastics are light and strong, and soften when heated so that they can be moulded into a variety of shapes. They are resistant to chemical attack and therefore to corrosion. They are electrical insulators. A multitude of uses have been found for plastics, and the list grows daily.

Thermosoftening plastics are used when ease of moulding is important . . .

. . . Thermosetting plastics are used for articles which must not soften when heated

When plastics are heated, they soften, and in some cases melt. When they are cooled again, they behave in one of two ways. Some plastic materials harden when they are cooled, but soften again if reheated. Plastics which do this are called **thermoplastics**. Examples are polyethene, polystyrene and poly(chloroethene). Other plastic materials can be heated only once: on cooling, they harden and will not soften again when reheated. These are **thermosetting plastics** or **thermosets**. Examples are: polyurethane; phenolic resins, including Bakelite®; and melamine. Both types of plastics have their advantages. A manufacturer can buy a thermosoftening plastic in the form of granules, soften it, and mould it into articles by a mass production technique. Thermosetting plastics are used for articles which must not soften when heated, for example, electric light fittings, saucepan handles and bench tops.

35.1.2 THE STRUCTURE OF PLASTICS

There are differences in structure between thermoplastics and thermosetting plastics. Various types of structure are illustrated below:

There are linear and
cross-linked polymers

Linear polymer
(e.g., polyethene)

$$—A—A—A—A—A—A—A—A—A$$

Linear copolymer
(e.g., nylon)

$$—A—B—A—B—A—B—A—B—A$$

Minor cross-linked polymer
(e.g., vulcanised rubber)

$$—A—B—A—B—A—B—A—B—$$
$$\qquad\quad |\qquad\qquad\qquad |$$
$$\qquad\quad X\qquad\qquad\qquad X$$
$$\qquad\quad |\qquad\qquad\qquad |$$
$$—A—B—A—B—A—B—A—B—$$

Massively cross-linked polymer
(e.g., Bakelite® and melamine)

$$—A—B—A—B—A—B—A—B—$$
$$\;\;|\;\;|\;\;|\;\;|\;\;|\;\;|\;\;|\;\;|$$
$$\;\;X\;\;X\;\;X\;\;X\;\;X\;\;X\;\;X\;\;X$$
$$\;\;|\;\;|\;\;|\;\;|\;\;|\;\;|\;\;|\;\;|$$
$$—A—B—A—B—A—B—A—B—$$

Linear polymers have no cross-links between chains. On heating, the distance between chains increases and the polymer softens, becoming more flexible. On cooling, the process is reversed. Polymers of this type are thermoplastics. Cross-linked polymers are not softened easily. When sufficient heat has been supplied to break the cross-linkages, the whole polymer has decomposed and cannot be reformed on cooling. Polymers of this type are thermosetting, e.g., urea-methanal and phenol-methanal plastics.

35.1.3 METHODS OF POLYMERISATION

Addition polymerisation of alkenes, such as ethene, chloroethene and many others, has been described on p. 564.

References to material in
other chapters

Condensation polymerisation to give polyamides, such as nylon, has been described on p. 710. The formation of polyesters, such as Terylene® or Dacron®, by condensation polymerisation has been covered on p. 706. The important thermosetting plastics derived from methanal by polymerisation (Delrin®) and by condensation with phenol (Bakelite®) and with urea (melamine and Formica®) have been covered on p. 654 and p. 656.

CHECKPOINT 35A: PLASTICS

Some of the material covered is in the chapters referred to above.

1. Polymerisation reactions may be classified as **addition** or **condensation** reactions. Explain the meanings of these terms, and give two examples of each.

2. Give an example of, and explain the structure of (a) a polyalkene, (b) a polyester and (c) a polypeptide. Give examples of the everyday uses of all three.

3. Explain the terms *crosslinking* and *thermosetting* with reference to condensation polymers. For what purposes are thermosetting polymers suitable?

4. How does the chemical inertness of polyethene arise? How does it increase the usefulness of the material? How does it affect the disposal of waste polyethene?

5. (a) What type of functional group joins the repeating units in nylon?

(b) In what way does the structure of nylon resemble that of a polypeptide?

(c) What type of interaction takes place between polymer molecules which contain the functional group present in polypeptides?

6.

$$A = —(CH—CH_2)—_n$$

$$B = —(O_2C—\underset{\bigcirc}{}—CO_2CH_2CH_2)—_n$$

Outline the preparation of the polymers **A** and **B** from $C_6H_5COCH_3$ (which must first be converted into phenylethene) and $H_3CC_6H_4COCH_3$ (which must first be converted into benzene-1,4-dicarboxylic acid). If you wanted to manufacture plastics bottles, which would you choose, **A** or **B**, for making bottles to contain (a) concentrated sodium hydroxide, (b) concentrated hydrochloric acid and (c) concentrated sulphuric acid?

35.2 SYNTHETIC ROUTES

FIGURE 35.1 Some Methods of Increasing or Decreasing the Length of a Carbon Chain

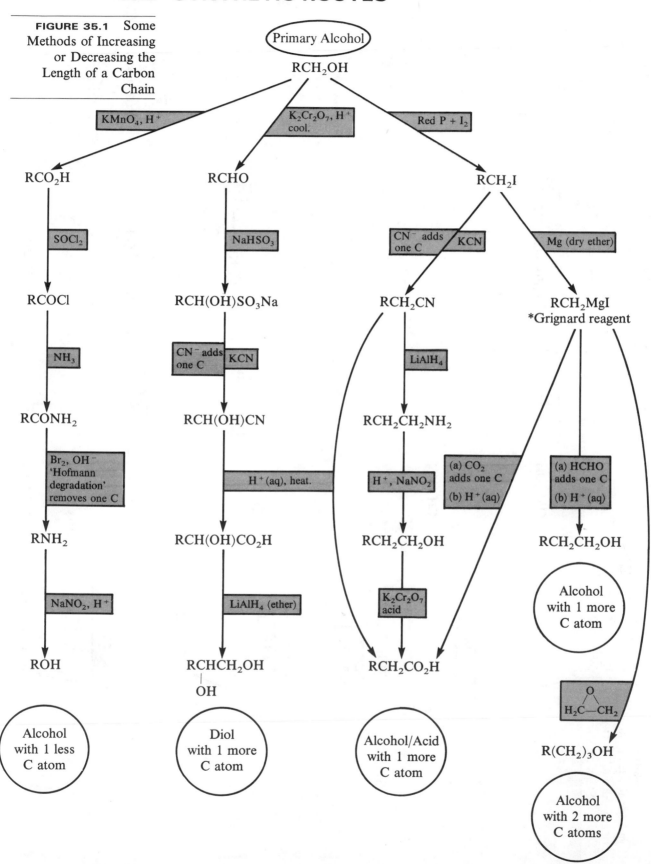

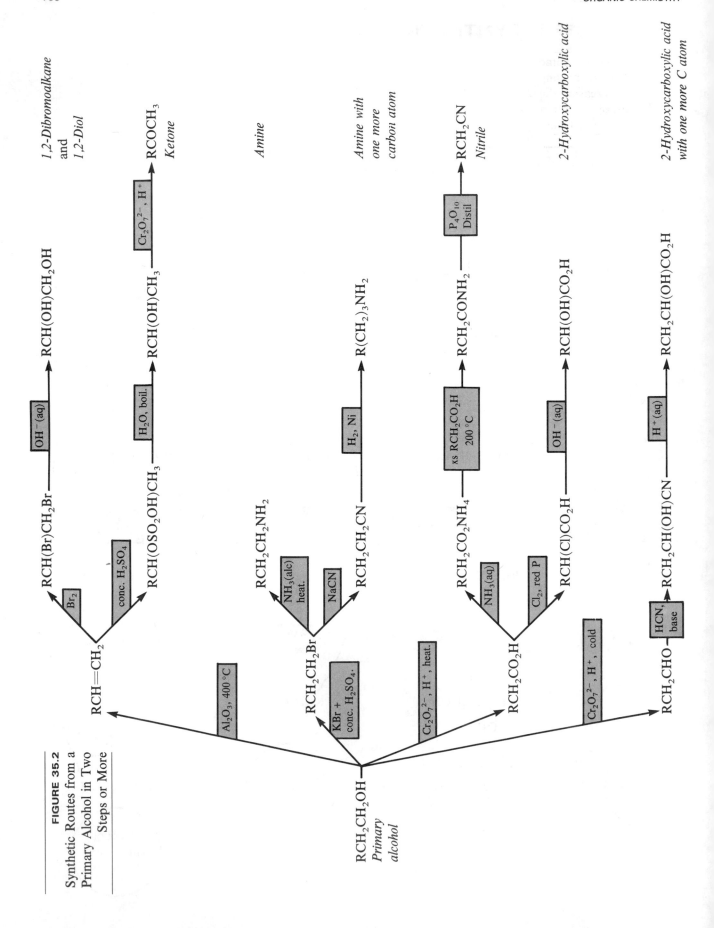

FIGURE 35.2
Synthetic Routes from a
Primary Alcohol in Two
Steps or More

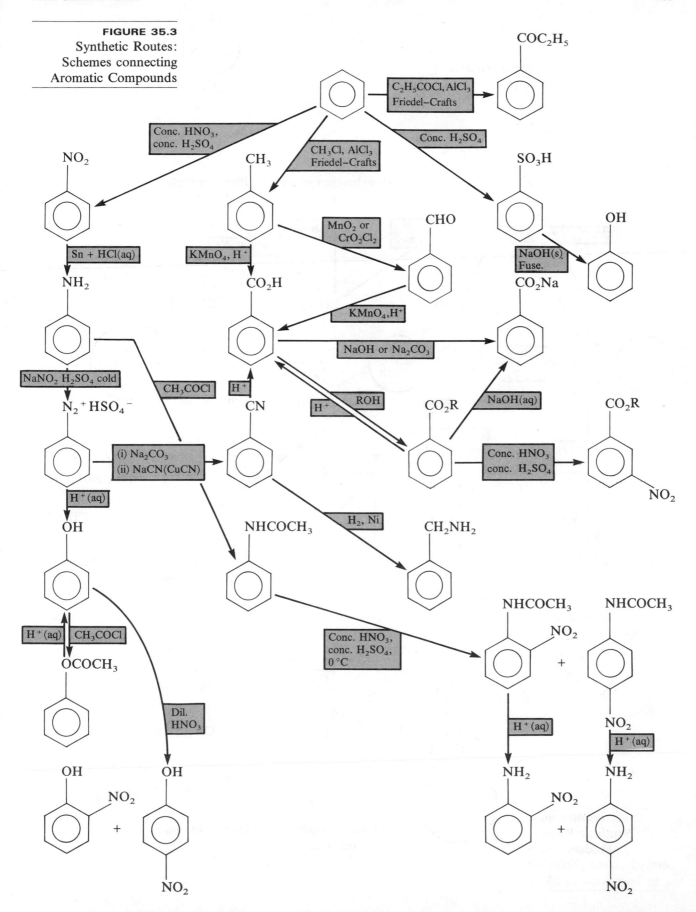

FIGURE 35.3
Synthetic Routes:
Schemes connecting
Aromatic Compounds

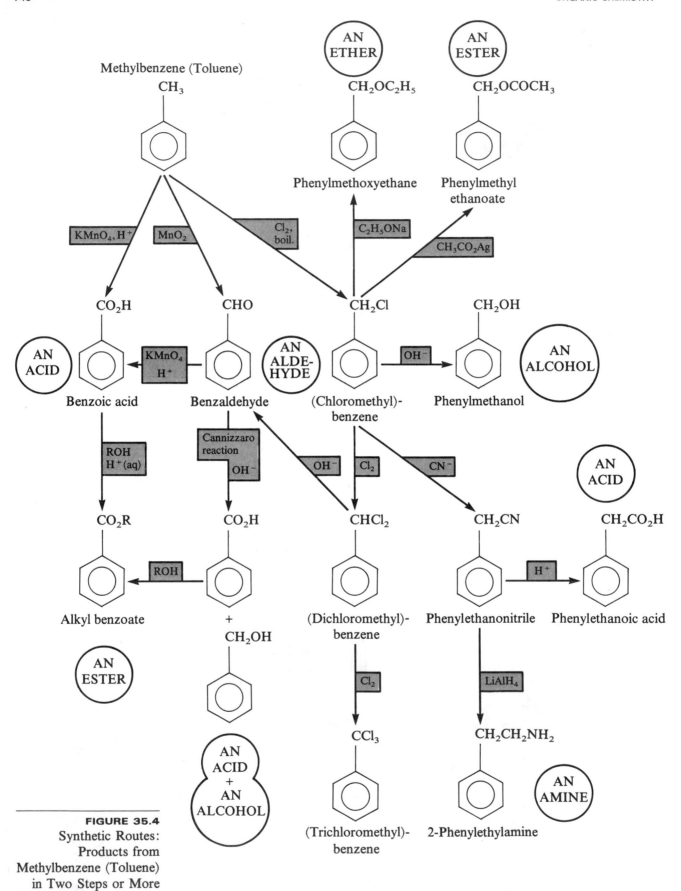

FIGURE 35.4
Synthetic Routes:
Products from
Methylbenzene (Toluene)
in Two Steps or More

CHECKPOINT 35B: SYNTHETIC ROUTES

1. Describe how you would carry out the following conversions:

(a) $CH_3CHO \rightarrow CH_3CH{=}CHCHO$

(b) $CH_3CHO \rightarrow HOCH_2CHO$

(c) $CH_3COCH_3 \rightarrow CH_3CO_2H$

(d) $CH_3CHO \rightarrow CH_3CH(OH)CO_2H$

(e) $CH_3CH_2CHO \rightarrow CH_3CH_2CO_2CH_2CH_2CH_3$

(f) $CH_3CH(OH)CH_3 \rightarrow (CH_3)_2C{=}NOH$

(g) $CH_3CH(OH)CH_3 \rightarrow (CH_3)_2C(OH)(CN)$

2. How could you carry out the conversions listed below?

(a) $CH_3COCH_3 \rightarrow CH_3CHClCH_2Cl$

(b) $(CH_3)_2CO \rightarrow (CH_3)_2CHBr$

(c) $CH_3C{\equiv}CH \rightarrow CH_3C(OH)(CN)CH_3$

(d) $CH_3CH_2CH_2OH \rightarrow CH_3C{\equiv}CH$

(e) $C_2H_6 \rightarrow C_2H_5OH$

(f) $C_2H_6 \rightarrow C_2H_5NH_2$

(g) $C_6H_6 \rightarrow C_6H_5COCH_3$

3. With C_2H_5OH as your only organic starting-material, explain how you could make: (a) $(C_2H_5)_2O$, (b) $C_2H_5NH_2$, (c) $CH_3CH_2CH_2NH_2$, (d) CH_3CONH_2, (e) CH_3COCl, (f) $C_2H_5CO_2C_2H_5$.

4. Starting from benzene or methylbenzene, say how you could make: (a) $C_6H_5CO_2H$, (b) $C_6H_5CH_2CO_2H$, (c) $C_6H_5CH(OH)CH_3$, (d) $C_6H_5CH_2CH_2NH_2$.

5. How can the following compounds be made from phenylamine? (a) C_6H_5OH, (b) $C_6H_5CO_2H$, (c) C_6H_5COCl, (d) $C_6H_5CO_2C_6H_5$, (e) C_6H_5I.

6. How could you make from propene: (a) CH_3COCH_3, (b) $(CH_3)_2CHCO_2H$, (c) $(CH_3)_2CHCH_2NH_2$ and (d) $(CH_3)_2CHOCOCH_3$?

7. How would you convert (a) a named hydrocarbon into methanol and (b) methanol into ethanol?

8. Explain how you could make from chloroethane and any inorganic materials you need (a) CH_3CHO, (b) $(CH_3CO)_2O$, (c) $CH_3CO_2C_2H_5$, (d) $CH_3CH_2CO_2C_2H_5$, (e) $CH_3CH_2CONH_2$ and (f) $(CH_3)_2C{=}NOH$.

9. Outline a synthesis from benzene of **A** and the conversion of **A** into **B**:

$$A = O_2N\!-\!\bigcirc\!-\!NHCOCH_3$$

$$B = HO\!-\!\bigcirc\!-\!OH$$

10. Devise syntheses for the following compounds, using no organic compounds other than those stated. State the reagents and conditions needed for each step in the synthesis:

(a) *N*-phenylbenzamide, $C_6H_5CONHC_6H_5$ from benzene and methylbenzene,

(b) phenylhydrazine from benzene,

(c) benzenecarboxylic acid (benzoic acid) from benzene and methane.

35.3 WHAT ARE THESE REAGENTS USED FOR?

TABLE 35.1
Reagents and their Uses

Reagent	Use	Example
$KMnO_4$, H^+(aq)	Oxidation	$C_6H_5CH_3 \rightarrow C_6H_5CO_2H$
MnO_2	Oxidation	$C_6H_5CH_3 \rightarrow C_6H_5CHO$
CrO_3 or CrO_2Cl_2	Oxidation	$C_6H_5CH_3 \rightarrow C_6H_5CHO$
$K_2Cr_2O_7$, H^+(aq) $Na_2Cr_2O_7$, H^+(aq)	Oxidation	$RCH_2OH \rightarrow RCHO$ (Add oxidant slowly to hot alcohol, and distil off aldehyde as fast as it is formed.) $RCH_2OH \rightarrow RCO_2H$ (Reflux the mixture until oxidation is complete.)
$KMnO_4$ (neutral)	Oxidation	Reflux: $C_6H_5{-}\overset{\mid}{\underset{\mid}{C}}{-}X \rightarrow C_6H_5CO_2H$ (where $X = {-}CH_3$, ${-}CH_2OH$ etc.)
H_2, Pt	Reduction	$C_6H_5NO_2 \rightarrow C_6H_5NH_2$
$NaBH_4$(aq)	Reduction of $C{=}O$, not $C{=}C$	$RCHO \rightarrow RCH_2OH$ $R_2C{=}O \rightarrow R_2CHOH$
Fe + HCl(aq)	Reduction	$C_6H_5NO_2 \rightarrow C_6H_5NH_2$

Reagent	Use	Example
Sn + HCl(aq)	Reduction	$C_6H_5NO_2 \rightarrow C_6H_5NH_2$
Zn·Hg + conc. HCl(aq)	Reduction	$RR'C{=}O \rightarrow RR'CH_2$
H_2, Ni	Reduction of $C{=}C$, not $C{=}O$	 $-C{\equiv}C- \rightarrow -CH_2CH_2-$ $RCN \rightarrow RCH_2NH_2$ Hydrogenation of oils to form fats
$LiAlH_4$ (ether)	Reduction of $C{=}O$, not $C{=}C$	$RCO_2H \rightarrow RCH_2OH$ $RCO_2R' \rightarrow RCH_2OH$ $RCl \rightarrow RH$ $RCONH_2 \rightarrow RCH_2NH_2$
O_3 followed by $Zn + CH_3CO_2H$(aq)	Ozonolysis to determine position of $C{=}C$ bond	$R{-}CH{=}CH{-}R' \rightarrow RCHO + R'CHO$ $RCH{=}CH_2 \rightarrow RCHO + HCHO$
$Br_2(CCl_4)$	Test for unsaturation	$RCH{=}CHR' \rightarrow RCH{-}CHR'$ (No HBr evolved) with Br Br
$KMnO_4$, OH^-(aq) dilute, 0 °C	Test for unsaturation	$RCH{=}CHR' \rightarrow RCH{-}CHR'$ with OH OH
$NaOH$, I_2	Iodoform test for $CH_3CO{-}$ or $CH_3CH(OH){-}$	$CH_3CHO + I_2 + NaOH \rightarrow CHI_3$
Cu^+(aq), Ag^+(aq)	Test for a terminal triple bond	$HC{\equiv}C{-} \rightarrow CuC{\equiv}C{-}$ or $AgC{\equiv}C{-}$
$NaNO_2$, H^+(aq)	Diazotisation	$C_6H_5NH_2 \rightarrow C_6H_5\overset{+}{N}{\equiv}N$
Al_2O_3	Dehydration	$C_2H_5OH \rightarrow CH_2{=}CH_2$
P_4O_{10}	Dehydration	$RCONH_2 \rightarrow RCN$
Hot conc. H_2SO_4	Dehydration	$RCH(OH)CH_3 \rightarrow RCH{=}CH_2$
Conc. H_2SO_4	Esterification	$ROH + R'CO_2H \rightarrow R'CO_2R + H_2O$
Cold conc. H_2SO_4 then H_2O	Hydration	$RCH{=}CHR' \rightarrow RCH{-}CHR'$ with H OH
H_3PO_4, then H_2O	Hydration	$RCH{=}CHR' \rightarrow RCH{-}CHR'$ with H OH
H_2SO_4(aq), $HgSO_4$	Hydration of an alkyne $\rightarrow$ ketone	$R{-}C{\equiv}CH \rightarrow RCOCH_3$
$FeCl_3$, $FeBr_3$	Halogen carriers	$C_6H_6 + Cl_2 \xrightarrow{FeCl_3} C_6H_5Cl + HCl$
$AlCl_3$	Friedel–Crafts catalyst	$C_6H_6 + RCl \xrightarrow{AlCl_3} C_6H_5R + HCl$ $C_6H_6 + RCOCl \xrightarrow{AlCl_3} C_6H_5COR + HCl$

Reagent	Use	Example
Conc. HNO_3 + conc. H_2SO_4	Nitration	$C_6H_6 + NO_2^+ \rightarrow C_6H_5NO_2$
SO_3 in conc. H_2SO_4	Sulphonation	$C_6H_6 \rightarrow C_6H_5SO_3H$
Conc. H_2SO_4	Sulphonation of phenols and aromatic amines	$C_6H_5OH \rightarrow HOC_6H_4SO_3H$ $C_6H_5NH_2 \rightarrow H_2NC_6H_4SO_3H$
'Soda lime'	Decarboxylation	$RCO_2H \rightarrow RH$
OH^- (ethanol)	Elimination of hydrogen halide	$RCH_2CH_2X \rightarrow RCH{=}CH_2$
OH^- (aq)	Hydrolysis	$RCO_2R' \rightarrow RCO_2^-$ (aq) + $R'OH$ $RCN \rightarrow RCO_2^-$ (aq) $RCONH_2 \rightarrow RCO_2^-$ (aq) + NH_3(g)
$(CH_3O)_2SO_2$, alkali (Dimethyl sulphate)	Methylation	$C_6H_5OH \rightarrow C_6H_5OCH_3$
C_6H_5COCl, NaOH	Schotten–Baumann benzoylation	$C_6H_5OH \rightarrow C_6H_5OCOC_6H_5$
$C_6H_5COOCC_6H_5$ (Dibenzoyl peroxide)	Initiating chain reactions	Chlorination of alkanes
$SOCl_2$ (Sulphur dichloride oxide)	Replacement of —OH by —Cl	$RCH_2OH \rightarrow RCH_2Cl$
H^+ (aq)	Hydrolysis	RCO_2R', $RCONH_2$, $RCN \rightarrow RCO_2H$
$*C_2H_5MgBr$ (Ethylmagnesium bromide)	Grignard reactions	HCHO → Primary alcohol $R'CHO$ → Secondary alcohol $\begin{matrix} R' \\ \diagdown \\ C{=}O \rightarrow \text{Tertiary alcohol} \\ \diagup \\ R'' \end{matrix}$
2,4-Dinitrophenyl hydrazine (DNP)	Test for a C=O group	Gives an orange ppt. with an aldehyde or ketone
Fehling's solution (Cu^{2+} complex ions)	Test for an aliphatic aldehyde	Gives a reddish ppt. of Cu_2O with aliphatic aldehydes, but not with aromatic aldehydes or with ketones
Tollens' reagent ($Ag(NH_3)_2^+$ (aq))	Test for an aldehyde	Reduced to a silver mirror by aliphatic aldehydes and, when warmed, by aromatic aldehydes, but not by ketones

CHECKPOINT 35C: REAGENTS

1. Compare the reactions of bromine with the following compounds. State the products of the reactions and the conditions necessary for reaction.
(*a*) methane, (*b*) benzene, (*c*) phenol, (*d*) ethene, (*e*) ethanamide.

2. Explain the following terms. Give an example of each type of reaction, indicating the reactants and the conditions needed for reaction.
(*a*) reduction, (*b*) ozonolysis, (*c*) alkylation, (*d*) acylation, (*e*) decarboxylation.

3. Give an example of each of the following types of reaction. Indicate the necessary conditions.
(*a*) nitration, (*b*) sulphonation, (*c*) oxidation, (*d*) cracking, (*e*) the halogenation of an aromatic ring.

4. Give examples of the reduction of organic compounds by (*a*) hydrogen and a catalyst, (*b*) zinc and hydrochloric acid, (*c*) lithium tetrahydridoaluminate, (*d*) sodium tetrahydridoborate.

5. Illustrate the use of the following reagents in organic chemistry. State the conditions necessary for reaction, and give equations.
(*a*) bromine, (*b*) aluminium chloride, (*c*) sodium nitrite, (*d*) hydrogen cyanide, (*e*) sodium hydrogensulphite, (*f*) nickel, (*g*) alkaline potassium manganate(VII), (*h*) ozone.

6. State the products of the reaction between sodium hydroxide and each of the following compounds. For each reaction, state the necessary conditions, and write the equation.
(*a*) ethanal, (*b*) benzenecarbaldehyde (benzaldehyde), (*c*) ethyl ethanoate, (*d*) 1-bromobutane, (*e*) 1-iodo-1,1-dimethylethane, $(CH_3)_3CI$, (*f*) 1,1,1-trichloropropanone, CCl_3COCH_3.

7. The following are well-known reagents. State what use each finds in organic chemistry. Describe the bonding in each compound. Explain how the type of bonding enables the reagent to function as it does.
(*a*) $AlCl_3$, (*b*) HCN, (*c*) $NaHSO_3$, (*d*) KOH(aq), (*e*) $LiAlH_4$.

8. (*a*) In Figure 35.5, name the reagents represented by the letters **A** to **F** inclusive and state the conditions under which each of the reactions take place.

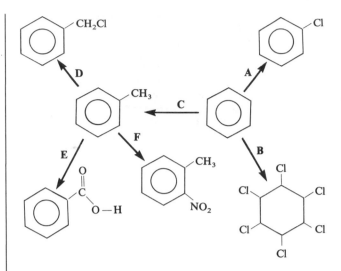

FIGURE 35.5

(*b*) Give the reaction conditions and the name or structural formula of the organic product formed when (chloromethyl)benzene (benzyl chloride) reacts with each of the following:
(i) aqueous sodium hydroxide;
(ii) potassium cyanide;
(iii) sodium;
(iv) ammonia.

(AEB 81)

35.4 HOW WOULD YOU DISTINGUISH BETWEEN THE MEMBERS OF THE FOLLOWING PAIRS OF COMPOUNDS?

CHECKPOINT 35D: PAIRS OF COMPOUNDS I

1.
A CH_3COCH_3
B C_2H_5OH

2.
A CH_3COCH_3
B C_2H_5CHO

3.
A $C_2H_5COCH_3$
B $C_2H_5CO_2H$

4.
COCH₃ (A)   CH₃...CHO (B)

5.
A CH_3CHO
B $CH_3CO_2CH_3$

6.
A $CH_3CH_2CH_2OH$
B $CH_3CH(OH)CH_3$

7.
A $(CH_3)_3COH$
B $(CH_3)_2CHCH_2OH$

8.
OH...CH₃ (A)   CH₂OH (B)

9.

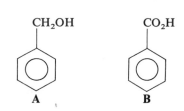

A   B

10.

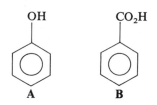

A   B

11.

A   CH₃CHO

B   CO₂H
     |
     CO₂H

12.

A   CH₃COCH₃

B   C₂H₅CO₂H

13.

A   CH₃CO₂H

B   HCO₂H

14.

A   CO₂Na
     |
     CO₂Na

B   CH₃CO₂Na

15.

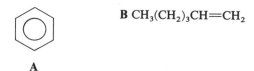

A

B CH₃(CH₂)₃CH=CH₂

16.

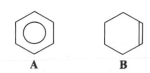

A   B

17.

A   CH₃C≡CCH₃

B   CH₃CH₂C≡CH

18.

A   CH₄

B   CH₃Cl

19.

A   CH₃COCl

B   C₆H₅COCl

20.

A   ClCH₂CO₂H

B   CH₃COCl

21.

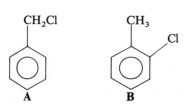

A   B

22.

A   CH₃CH₂I

B   CH₃CH₂Br

23.

A   HCO₂C₂H₅

B   CH₃CO₂CH₃

24.

A   CH₃CO₂C₆H₅

B   C₆H₅CO₂CH₃

25.

A   (CH₃)₃CCHO

B   (CH₃)₂CHCH₂CHO

26.

A   CH₃CH₂NH₂

B   CH₃CN

27.

A   CH₃CH₂NH₂

B   (CH₃)₂NH

28.

A   CH₃CONH₂

B   CH₃CH₂NH₂

29.

A   CH₃CONH₂

B   CO(NH₂)₂

30.

    A $C_2H_5CONH_2$

    B $C_6H_5NH_2$

31.

    A $C_3H_7NH_3{}^+Cl^-$

    B $C_3H_7CO_2{}^-Na^+$

32.

    A $CH_3CO_2NH_4$

    B CH_3CONH_2

33.

    A   **B**

CHECKPOINT 35E: PAIRS OF COMPOUNDS II

1. A, a ketone, reacts with 2,4-dinitrophenylhydrazine, DNP, to give an orange-coloured precipitate of the 2,4-dinitrophenylhydrazone.

B, an alcohol, reacts with $CH_3CO_2H(l)$ + conc. H_2SO_4 to give a fruity, ester smell of $CH_3CO_2C_2H_5$.

2. Both react with DNP. **A**, a ketone, is not easily oxidised. **B**, an aldehyde, reduces Tollens' reagent to silver and Fehling's solution to red Cu_2O.

3. A is a ketone and reacts with DNP. **B** is an acid, and liberates CO_2 from $NaHCO_3$.

4. A has no reducing action; **B** reduces Tollens' reagent but not Fehling's solution.

5. A + NaOH(aq, conc.) forms a brown resin, $(CH_3CHO)_n$, with a characteristic smell.

B + NaOH(aq) when warmed give $CH_3CO_2Na + CH_3OH$. On addition of HCl(aq), CH_3CO_2H, with a characteristic smell, is formed.

6. A, a primary alcohol, is oxidised by acidified $Na_2Cr_2O_7$ in the cold to an aldehyde. (Test as in 2.)

B, a secondary alcohol, is oxidised to a ketone. (Test as in 2.)

7. A, a tertiary alcohol, is oxidised by acid $K_2Cr_2O_7$ to a mixture of products, including an acid.

B, a primary alcohol, is oxidised to an aldehyde. (Test with DNP.)

8. A is a phenol. Its solution is acidic, reacting with NaOH(aq) to give the phenoxide, $H_3CC_6H_4O^-Na^+$. **A** gives a purple colour with $FeCl_3$(aq) and a white precipitate with Br_2(aq).

B is an alcohol, neutral in solution. It does not react with NaOH(aq), but does react with Na(s) to give H_2(g) and the alkoxide, $C_6H_5CH_2O^-Na^+$.

9. A is an alcohol, neutral in solution.

B is an acid: its solution is acidic, and it reacts with Mg to give H_2(g) and with $NaHCO_3$ to give CO_2(g).

10. A is weakly acidic: it dissolves in NaOH(aq), but does not give CO_2(g) with $NaHCO_3$. **B** is an acid, and liberates CO_2(g) from $NaHCO_3$.
A gives a purple colour with $FeCl_3$(aq) and a white ppt. with Br_2(aq).

11. A + $AgNO_3$(aq) give a silver mirror when warmed.
B + $AgNO_3$(aq) give a white ppt. of AgO_2CCO_2Ag.

12. A with conc. $NaHSO_3$(aq) gives a colourless, crystalline ppt.
B + C_2H_5OH + conc. H_2SO_4 gives, when warmed, an ester with a characteristic, fruity smell. **B** liberates CO_2(g) from $NaHCO_3$.

13. A smells of vinegar, **B** of formalin.
When warmed with $AgNO_3$, NH_3(aq), **B** gives a silver mirror. Conc. H_2SO_4 + **B** give CO(g), which burns with a blue flame.
B + Fehling's solution give a red ppt. of Cu_2O.

14. Conc. H_2SO_4 reacts with **A** to give CO_2 (limewater test) and CO (which burns with a blue flame), and with **B** to give CH_3CO_2H (which smells like vinegar).

15. A is benzene. It does not show unsaturation, as does **B**, which decolorises $Br_2(CCl_4)$ and alkaline $KMnO_4$(aq).

16. A burns with a smoky flame, as aromatic compounds do. **B**, cyclohexene, shows unsaturation (see above).

17. Both **A** and **B** are alkynes and decolorise $Br_2(CCl_4)$ and alkaline $KMnO_4$(aq). **B**, with a terminal triple bond, reacts with CuCl, NH_3(aq) to give a red ppt. of $CH_3CH_2C{\equiv}CCu$(s) and also with $AgNO_3$, NH_3(aq) to form a pale yellow ppt. of $CH_3CH_2C{\equiv}CAg$(s).

18. A is unreactive. **B** is hydrolysed by alkali to form methanol (Test for alcohol) and Cl^- ions (Test with $AgNO_3$, HNO_3(aq)).

19. In water, **A** immediately forms CH_3CO_2H and HCl. A strongly acidic solution is formed. (Test with $NaHCO_3$).
B is hydrolysed when warmed with NaOH(aq) to sodium benzoate. On acidification with HCl(aq), solid benzoic acid comes out of solution.

20. A, chloroethanoic acid, is a stronger acid than ethanoic acid. (Test: Mg; NaHCO$_3$).

B is not acidic, but it reacts with water to form a strongly acidic solution of $CH_3CO_2H + HCl$. Addition of $AgNO_3$, HNO_3(aq) will detect Cl^- ions, which are not present in a solution of **A**.

21. A —Cl in the side chain, as in **A**, is hydrolysed off by warming **A** with alkali, to give an alcohol (Test by oxidising it to an aldehyde), and Cl^- ions (Test with $AgNO_3$, HNO_3(aq)).

The —Cl in the ring in **B** cannot be removed in this way.

22. On hydrolysis with NaOH(aq), **A** gives I^- ions, and **B** gives Br^- ions. $Pb(NO_3)_2$(aq) gives a yellow ppt. of PbI_2 with the first, and a white ppt. of $PbBr_2$ with the second. With **A**, CH_3CO_2Ag(aq) gives a yellow ppt. of AgI; with **B** it gives a pale yellow ppt. of AgBr.

23. A is hydrolysed when warmed with OH^-(aq) to $HCO_2H + C_2H_5OH$. Ethanol gives the iodoform test: with $I_2 + OH^-$(aq), a ppt. of CHI_3 is formed.

B is hydrolysed to CH_3CO_2H (which smells of vinegar) and CH_3OH. On oxidation, CH_3OH gives methanal, HCHO, with the smell characteristic of formalin.

24. A is an ester which is hydrolysed to CH_3CO_2H and C_6H_5OH. Phenol in alkaline solution couples with a diazonium salt to give a dye.

B is hydrolysed to $C_6H_5CO_2H + CH_3OH$. Benzoic acid liberates CO_2 from NaHCO$_3$.

25. A has no H atoms on the C atom next to the CHO group. It therefore undergoes the Cannizzaro reaction: with OH^-(aq), **A** gives $(CH_3)_3CCO_2H + (CH_3)_3CCH_2OH$ (Test with NaHCO$_3$ for acid) **B** does not give a Cannizzaro reaction.

26. A is an amine, with a characteristic smell. Its aqueous solution is alkaline. With acids, it forms salts, which are crystalline solids. With nitrous acid, **A** gives $C_2H_5OH + N_2$(g).

B is a nitrile, hydrolysed by H^+(aq) to CH_3CO_2H (Test for carboxylic acid).

27. A is a primary amine. With nitrous acid, it gives $C_2H_5OH + N_2$(g). With CHCl$_3 + OH^-$(alc), it gives a foul-smelling isonitrile.

B is a secondary amine. With nitrous acid, it gives an oily nitrosoamine.

28. A is an amide. It is hydrolysed by NaOH(aq) to give NH_3(g) (which is basic) and CH_3CO_2Na(aq).

B is an amine, and reacts with nitrous acid to give N_2(g).

29. A, an amide, is hydrolysed by OH^-(aq) to NH_3(g).

B, urea, is hydrolysed by OH^-(aq) to NH_3(g) and by H^+(aq) to CO_2(g).

30. A + Br_2(aq) + OH^-(aq) give an amine, which is basic and has a 'fishy' smell.

B + Br_2(aq) gives a white ppt. of $C_6H_2Br_3NH_2$.

31. A solution of **A** in water gives a white ppt. of AgCl on treatment with $AgNO_3$, HNO_3(aq). Addition of alkali to **A** liberates the free amine, $C_3H_7NH_2$, which has the characteristic smell.

B gives a yellow flame test. Addition of HCl(aq) to **B** liberates butanoic acid, $C_3H_7CO_2H$, with a rancid smell.

32. With cold NaOH(aq), **A** gives NH_3(g).

When strongly heated with NaOH(aq), **B** gives NH_3(g).

33. A is an aryl amine, which can be diazotised with cold $NaNO_2 + HCl$(aq). The diazonium salt formed gives, when warmed, N_2(g) and a smell of phenol. It gives a dye when coupled with phenol in alkaline solution. **B** reacts with $NaNO_2 + HCl$(aq) to give N_2(g).

35.5 QUESTIONS ON SOME TOPICS WHICH SPAN CHAPTERS

QUESTIONS ON FUNCTIONAL GROUPS AND REACTION MECHANISMS

1. Explain these statements:

(a) Water and petrol do not mix.

(b) Halogenoalkanes are more reactive than alkanes towards nucleophilic reagents.

(c) Alkenes, unlike alkanes, react readily with bromine water.

(d) Phenol, unlike benzene, reacts readily with bromine water.

(e) Ethanoyl chloride is more reactive towards water than is chloroethane.

2. Compare and contrast the reactions of:

(a) sulphuric acid with ethanol and phenol,

(b) phosphorus(V) chloride with ethanol and phenol,

(c) sodium hydroxide with 1-chlorohexane and chlorobenzene,

(d) nitrous acid with ethylamine and phenylamine.

3. Describe the mechanisms of the reactions between the following substances:

(a) ethene and bromine,

(b) methane and chlorine in sunlight,

(c) benzene and chlorine in the presence of iron filings,

(d) benzene and ethanoyl chloride in the presence of aluminium chloride,

(e) 1-iodopropane and aqueous sodium hydroxide,

(f) propanal and hydrogen cyanide.

4. Explain, by means of examples, what is meant by the following terms:

(a) homolytic fission,

(b) heterolytic fission,

(c) nucleophilic substitution,

(d) an addition–elimination reaction,

(e) reductive ozonolysis,

(f) catalytic cracking.

5. Explain the following observations:

(a) Ethene does not react with water, but, when ethene is passed into concentrated sulphuric acid and water is added to the solution, ethanol is formed.

(b) Hydrogen cyanide adds to propanone in the presence of bases but not in the presence of acids.

*(c) Esterification of ethanoic acid is catalysed by acids but not by bases. Hydrolysis of ethyl ethanoate is catalysed by both acids and bases.

(d) Ethers, although they are relatively unreactive, dissolve in concentrated sulphuric acid.

(e) Both propanone and ethanol give yellow precipitates when treated with iodine and alkali. Neither pentan-3-one nor pentan-3-ol gives a positive result in this test.

6. Carboxylic acids, acid halides, amides, esters, aldehydes and ketones all contain the group

$$\diagdown C{=}O.$$

Point out the similarities in the behaviour of these groups of compounds which arise from the possession of a carbonyl group.

Why do carboxylic acids and their derivatives not show the addition reactions of aldehydes and ketones with nucleophiles?

7. Adrenalin has the formula shown below. It is a water-soluble hormone:

(a) Name the functional groups in the molecule.

(b) Describe how you could test for the presence of each of these groups.

(c) Where does the optical activity of adrenalin arise? Outline a method for resolving a racemic mixture of the two enantiomers.

8. Discuss the way in which the properties of the groups (a) —Cl and (b) —NH_2 are influenced by the remainder of the molecule in which they are present.

9.

Imagine that you have synthesised the compound which has the formula shown above. Describe how you could test for each of the functional groups, and say what the results of these tests would be if the compound you have made is the correct substance.

Say what further tests you could do to confirm that your product is indeed the substance you intended to make.

10. (a) Explain, with the aid of an example, what is meant by *homolytic fission*.

(b) Write equations to illustrate the mechanism of the chlorination of methylbenzene (*toluene*) to form (chloromethyl)benzene (*benzyl chloride*).

(c) How and under what conditions does (i) (chloromethyl)benzene and (ii) chloro-4-methylbenzene (p-*chlorotoluene*) react with aqueous sodium hydroxide?

(d) A hydrocarbon C_8H_{10} was heated with an excess of bromine in the presence of ultraviolet light. It was found that a maximum of six of the ten hydrogen atoms per molecule could be replaced by bromine. What can be deduced about the nature of the hydrocarbon? Suggest a structure for it.

(e) Why is the bromination of propane likely to be, in general, an unsatisfactory method for the preparation of 2-bromopropane? Suggest *one* better method for its preparation.

(JMB 82)

11. 'The properties of an organic functional group are influenced by the structure of the molecule of which it is a part.'

Discuss and explain this statement, illustrating your answer by reference to *three* reactions which may be taken from the following or from reactions of your own choice.

(a) Sodium hydroxide with (i) bromopropane and (ii) bromobenzene,

(b) Nitrous acid with (i) aminopropane and (ii) phenylamine (aniline),

(c) Sulphuric acid with (i) propan-1-ol and (ii) hydroxybenzene (phenol).

(L 81)

12. Give examples of the following types of reaction, indicating the conditions necessary and naming the organic product in each case:

(a) nucleophilic substitution of a primary alkyl halide with hydroxide (hydroxyl) ions;

(b) electrophilic substitution of benzene by the ion, $NO_2{}^+$, including the formation of this ion;

(c) electrophilic addition of bromine to an alkene;

(d) nucleophilic addition of hydrogen cyanide to a carbonyl compound.

Discuss in simple terms the mechanism of these reactions.

(NI 82)

CHECKPOINT 35F: DRAW YOUR OWN CONCLUSIONS

1.

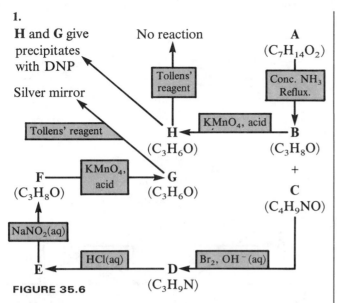

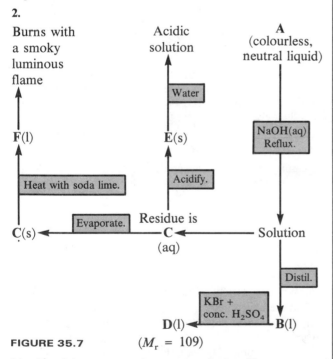

FIGURE 35.6

Identify the compounds **A** to **H**. Explain the reactions depicted.

2.

FIGURE 35.7

Identify, giving reasons, the compounds **A** to **E**.

3. The isomers **A** and **B** have the molecular formula $C_9H_8O_2$. They are insoluble in water but soluble in aqueous sodium carbonate. Both **A** and **B** are reduced by hydrogen in the presence of a platinum catalyst to **C**, $C_9H_{10}O_2$. Benzoic acid is formed when **A** or **B** or **C** is oxidised by alkaline potassium manganate(VII), and the solution is later acidified. **C** reacts with phosphorus(V) chloride to give **D**, and reacts with sodium carbonate solution to give carbon dioxide.

Identify the compounds **A** to **D**, giving reasons, and explaining the reactions mentioned.

4. **P** is a crystalline solid which melts at 200 °C. It dissolves in water to give a neutral solution. **P** forms salts with both acids and bases. When heated with soda lime, it gives **Q**, a pungent-smelling gas. **Q** burns in air and dissolves in water to give an alkaline solution. When treated with nitrous acid, **P** gives nitrogen and an acid, **R**, with a molar mass of $76\,\mathrm{g\,mol^{-1}}$. Identify, with reasons, **P**, **Q** and **R**.

5. Identify compounds **A** to **G**. Explain the reactions mentioned.

A is a crystalline solid which is very soluble in water and gives a yellow flame test. When **A** is heated with ethanoyl chloride, **B** distils. **B** is a neutral liquid with a relative molar mass of 102. **B** reacts with water to give an acidic solution.

C is a colourless liquid which is sparingly soluble in water but dissolves readily in hydrochloric acid. Evaporation of the solution yields **D**, a crystalline solid. **D** reacts with alkali to give **C**. A solution of **C** in hydrochloric acid reacts with a cold solution of sodium nitrite, followed by an alkaline solution of phenol to give an orange dye. When **C** is treated with bromine water, it gives a white precipitate of **E**, which has a relative molar mass of 330.

F is a colourless solid with a distinctive smell. It dissolves in water to give a weakly acidic solution, which does not liberate carbon dioxide from a solution of sodium hydrogencarbonate. When an alkaline solution of **F** is shaken with benzoyl chloride, a solid, **G**, of molar mass $198\,\mathrm{g\,mol^{-1}}$, is formed.

6. **W** is a colourless aliphatic liquid with a relative molar mass of 123. **W** is insoluble in water. When refluxed with aqueous sodium hydroxide, it gives a solution of **X** and **Y**.

A solution of **X** gives a creamy yellow precipitate with silver nitrate solution.

Y can be distilled from the solution. It gives a positive iodoform test and is oxidised by chromic acid to **Z**. **Z** gives a positive iodoform test, and reacts with 2,4-dinitrophenylhydrazine but not with ammoniacal silver nitrate.

Identify **W**, **X**, **Y** and **Z**. Write equations for the reactions involved.

7. **A** ($C_7H_7NO_2$) is reduced by tin and concentrated hydrochloric acid, followed by alkali, to **B** (C_7H_9N).
B is converted by sodium nitrite and dilute hydrochloric acid into **C** (C_7H_8O).
C is converted by (i) sodium, (ii) iodomethane into **D** ($C_8H_{10}O$).
D is oxidised by acidified dichromate to **E** ($C_8H_8O_3$).
On treatment with concentrated hydriodic acid, **E** gives **F** ($C_7H_6O_3$).
When heated with soda lime and acidified, **F** gives **G** (C_6H_6O).
B dissolves in acid; **E**, **F** and **G** dissolve in alkali; **C** and **G** give a violet colour with iron(III) chloride.

Give the names and structural formulae for **A** to **G**. Explain what happens in each of the reactions mentioned.

8.

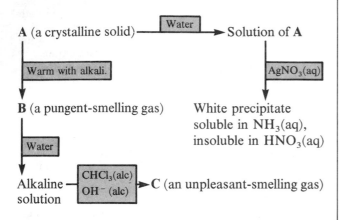

FIGURE 35.8

Identify the compounds, **A**, **B** and **C**. Explain the reactions depicted above.

9. The gas **P** burns with a smoky but luminous flame. It reacts with chlorine under carefully controlled conditions to give **Q**, a liquid of molar mass $168\,\mathrm{g\,mol^{-1}}$. **P** reacts with ammoniacal copper(I) chloride solution to form a reddish brown precipitate of **R**.

Identify **P**, **Q** and **R**. Write equations for the reactions mentioned.

10. **X** is a volatile liquid. Its aqueous solution is neutral, but becomes acidic on exposure to the air. **X** reduces ammoniacal silver nitrate, and reacts with aqueous sodium hydrogensulphite to form **Y**. When a drop of concentrated hydrochloric acid is added to **X**, an oily product, **Z**, is formed. **Z** is almost insoluble in water, and has a molar mass of $132\,\mathrm{g\,mol^{-1}}$.

Identify **X**, **Y** and **Z**, and explain the reactions described.

11. An organic compound **H** ($C_2H_4O_3$) is oxidised to **I** ($C_2H_2O_3$), which can be oxidised to **J** ($C_2H_2O_4$). **H**, **I** and **J** dissolve in water to give acidic solutions which decolorise potassium manganate(VII) on warming. On treatment with phosphorus(V) bromide, all three react, **H** and **J** giving 2 moles of hydrogen chloride per mole, and **I** giving 1 mole of hydrogen chloride per mole. **I** gives a precipitate with 2,4-dinitrophenylhydrazine.

Deduce the identity of **H**, **I** and **J**. Explain the reactions described.

12. Deduce the structural formulae of the aromatic compounds **B** to **J** in Figure 35.9. Give full reasons for your deductions.

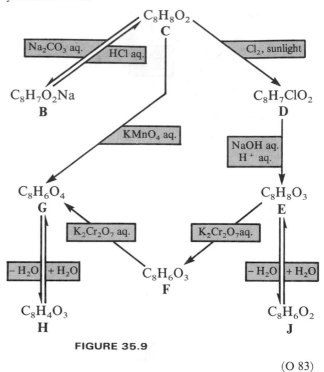

FIGURE 35.9

(O 83)

13. The organic compound **P**, C_8H_8O, gave the tri-iodomethane (iodoform) reaction and on vigorous oxidation yielded **Q**, $C_7H_6O_2$, which was sparingly soluble in cold water.

After neutralisation, **Q** gave a buff precipitate with aqueous iron(III) chloride.

Treatment of **Q** with phosphorus pentachloride gave **R** which reacted with concentrated aqueous ammonia to give a precipitate of **S**, C_7H_7ON.

Prolonged treatment of **S** with phosphorus(V) oxide gave **T** which could be reduced to **U**, C_7H_9N.

On reduction with sodium amalgam, **P** gave **V**, $C_8H_{10}O$. **V** gave the tri-iodomethane (iodoform) reaction, but did not give a coloration with aqueous iron(III) chloride.

(*a*) Write the structural formulae for substances **P** to **V**.

(*b*) Write equations for

 (i) the reaction of **P** with iodine and alkali (the iodoform reaction);

 (ii) the reaction of **Q** with phosphorus pentachloride;

(iii) the reaction of **R** with water.

(SUJB 82)

APPENDIX 3

A SELECTION OF QUESTIONS ON ORGANIC CHEMISTRY

1. (*a*) Write down structural formulae for **four** organic compounds each of which has a molecular formula $C_2H_xNO_y$, where x is not less than 3 and y can be 0 or 1. For each of your structural formulae give the name of the class of compounds to which it belongs. [Before making your choice of possible structures, read part (*b*) of this question.]

(*b*) You are given an unlabelled sample of each of your four compounds chosen in part (*a*). Using test-tube experiments, explain how you would assign the correct formula to each sample.

(O 82)

2. (*a*) Give a mechanism for the conversion of an alkane into corresponding halogeno-compounds.
Further give a mechanism for the conversion of a halogeno-compound into a hydroxy-compound.

(*b*) Describe how a typical hydroxy-compound can be formed directly from an alkene.

(*c*) Starting from the halogeno-alkane RCH_2X (where R = alkyl and X = halogen) suggest how the amines RCH_2NH_2 and $RCH_2CH_2NH_2$ and the carboxylic acids RCOOH and RCH_2COOH can each be synthesised.

(*d*) Discuss the relative reactivity of the chlorine atom in 1-chlorobutane, (chloromethyl)benzene and chlorobenzene and describe how you would establish, experimentally, the differences in reactivity.

(WJEC 82)

3. Comment on the following observations.

(*a*) Treatment of trichloromethane with D_2O and NaOD gives a compound of relative molecular mass 120.5.

(*b*) Treatment of propanone with D_2O and NaOD gives a compound of relative molecular mass 64.

(*c*) Treatment of pentan-3-one with D_2O and NaOD gives a compound of relative molecular mass 90.

(*d*) Treatment of ethanonitrile (methyl cyanide) with D_2/Pd gives a compound (**A**) of relative molecular mass 49. Shaking (**A**) with water gives a compound of relative molecular mass 47.

(*e*) Treatment of ethanal with D_2/Pd gives an equimolar mixture of two compounds of formula $C_2H_4D_2O$.

(O & C, 82, S)

4. It may be said that 'organic chemistry is the chemistry of functional groups'. Discuss this statement by comparing and contrasting the reactions of

(*a*) the OH groups in C_6H_5COOH and C_6H_5OH,

(*b*) the unsaturated carbon to carbon bonds in $CH_2{=}CHCH_3$ and C_6H_6,

(*c*) the Cl groups in CH_3COCl and CH_3CH_2Cl.

(JMB 83)

5. Write an account of the reactions of chlorine with organic compounds. Your answer should include the reactions of ethane, propanone, ethanoic acid, and ethylbenzene.

(O & C, 82, S)

6. 'The properties of an organic functional group are influenced by the structure of the molecule of which it is a part'.

Discuss and explain this statement, illustrating your answer by reference to *three* reactions which may be taken from the following or from reactions of your own choice.

(*a*) Sodium hydroxide with (i) bromopropane and (ii) bromobenzene.

(*b*) Nitrous acid with (i) aminopropane and (ii) phenylamine (aniline).

(*c*) Sulphuric acid with (i) propan-1-ol and (ii) hydroxybenzene (phenol).

(L 81)

7. (*a*) Give examples of reactions of different types of compounds containing a carbon–oxygen double bond which are initiated by attack by a nucleophilic reagent and which lead ultimately to the formation of compounds containing (i) a carbon–oxygen single bond, (ii) a carbon–oxygen double bond and (iii) a carbon–nitrogen double bond. Discuss the mechanisms of such reactions and explain how the chemical nature of both the initial carbonyl compound and the nucleophilic reagent affects the type of product which is formed.

(*b*) Condensation of the aldehyde **G** with the amine **H** gives a product **J**, $C_8H_{17}N$, containing a carbon–carbon double bond. What is the structure of **J** and how is it formed?

G   **H**

(O 83, S)

Periodic Table

Key:
Relative atomic mass — (top) ; Proton (atomic) number — (bottom)

H	Hydrogen
1 (mass)	1 (number)

TRANSITION ELEMENTS

1A	2A						TRANSITION ELEMENTS						3B	4B	5B	6B	7B	0
																		He Helium 4, 2
Li Lithium 7, 3	**Be** Beryllium 9, 4												**B** Boron 11, 5	**C** Carbon 12, 6	**N** Nitrogen 14, 7	**O** Oxygen 16, 8	**F** Fluorine 19, 9	**Ne** Neon 20, 10
Na Sodium 23, 11	**Mg** Magnesium 24, 12												**Al** Aluminium 27, 13	**Si** Silicon 28, 14	**P** Phosphorus 31, 15	**S** Sulphur 32, 16	**Cl** Chlorine 35.5, 17	**Ar** Argon 40, 18
K Potassium 39, 19	**Ca** Calcium 40, 20	**Sc** Scandium 45, 21	**Ti** Titanium 48, 22	**V** Vanadium 51, 23	**Cr** Chromium 52, 24	**Mn** Manganese 55, 25	**Fe** Iron 56, 26	**Co** Cobalt 59, 27	**Ni** Nickel 59, 28	**Cu** Copper 63.5, 29	**Zn** Zinc 65, 30		**Ga** Gallium 70, 31	**Ge** Germanium 73, 32	**As** Arsenic 75, 33	**Se** Selenium 79, 34	**Br** Bromine 80, 35	**Kr** Krypton 84, 36
Rb Rubidium 85, 37	**Sr** Strontium 88, 38	**Y** Yttrium 89, 39	**Zr** Zirconium 91, 40	**Nb** Niobium 93, 41	**Mo** Molybdenum 96, 42	**Tc** Technetium 99, 43	**Ru** Ruthenium 101, 44	**Rh** Rhodium 103, 45	**Pd** Palladium 106, 46	**Ag** Silver 108, 47	**Cd** Cadmium 112, 48		**In** Indium 115, 49	**Sn** Tin 119, 50	**Sb** Antimony 122, 51	**Te** Tellurium 128, 52	**I** Iodine 127, 53	**Xe** Xenon 131, 54
Cs Caesium 133, 55	**Ba** Barium 137, 56	**La** Lanthanum 139, 57 *	**Hf** Hafnium 178, 72	**Ta** Tantalum 181, 73	**W** Tungsten 184, 74	**Re** Rhenium 186, 75	**Os** Osmium 190, 76	**Ir** Iridium 192, 77	**Pt** Platinum 195, 78	**Au** Gold 197, 79	**Hg** Mercury 201, 80		**Tl** Thallium 204, 81	**Pb** Lead 207, 82	**Bi** Bismuth 209, 83	**Po** Polonium 210, 84	**At** Astatine 210, 85	**Rn** Radon 222, 86
Fr Francium 223, 87	**Ra** Radium 226, 88	**Ac** Actinium 227, 89 †	**Ku** Kurchatovium 260, 104	**Ha** Hahnium 260, 105														

*58–71 Lanthanum series

†90–103 Actinium series

Lanthanum series

Ce Cerium 140, 58	**Pr** Praseodymium 141, 59	**Nd** Neodymium 144, 60	**Pm** Promethium 147, 61	**Sm** Samarium 150, 62	**Eu** Europium 152, 63	**Gd** Gadolinium 157, 64	**Tb** Terbium 159, 65	**Dy** Dysprosium 162, 66	**Ho** Holmium 165, 67	**Er** Erbium 167, 68	**Tm** Thulium 169, 69	**Yb** Ytterbium 173, 70	**Lu** Lutetium 175, 71

Actinium series

Th Thorium 232, 90	**Pa** Protactinium 231, 91	**U** Uranium 238, 92	**Np** Neptunium 237, 93	**Pu** Plutonium 244, 94	**Am** Americium 243, 95	**Cm** Curium 247, 96	**Bk** Berkelium 247, 97	**Cf** Californium 251, 98	**Es** Einsteinium 254, 99	**Fm** Fermium 257, 100	**Md** Mendelevium 258, 101	**No** Nobelium 259, 102	**Lr** Lawrencium 260, 103

BASIC SI UNITS

Physical Quantity	Name of unit	Symbol
Length	metre	m
Mass	kilogram	kg
Time	second	s
Electric current	ampere	A
Temperature	kelvin	K
Amount of substance	mole	mol
Light intensity	candela	cd

DERIVED SI UNITS

Physical Quantity	Name of unit	Symbol	Definition
Energy	joule	J	$kg\,m^2\,s^{-2}$
Force	newton	N	$J\,m^{-1}$
Electric charge	coulomb	C	$A\,s$
Electric potential difference	volt	V	$J\,A^{-1}\,s^{-1}$
Electric resistance	ohm	Ω	$V\,A^{-1}$
Area	square metre		m^2
Volume	cubic metre		m^3
Density	kilogram per cubic metre		$kg\,m^{-3}$
Pressure	newton per square metre or pascal		$N\,m^{-2}$ or Pa
Molar mass	kilogram per mole		$kg\,mol^{-1}$

With all these units, the following prefixes (and others) may be used.

Prefix	Symbol	Meaning
deci	d	10^{-1}
centi	c	10^{-2}
milli	m	10^{-3}
micro	μ	10^{-6}
nano	n	10^{-9}
kilo	k	10^3
mega	M	10^6
giga	G	10^9
tera	T	10^{12}

ANSWERS TO NUMERICAL PROBLEMS AND SELECTED QUESTIONS

PART 1: THE FOUNDATION

CHAPTER 1: THE ATOM

Checkpoint 1B: Mass Spectrometry [p. 12]
2. 35.45
3. $CH_2{}^{35}Cl_2$, $CH_2{}^{35}Cl^{37}Cl$, $CH_2{}^{37}Cl_2$,
   ^{35}Cl is $3 \times$ as abundant as ^{37}Cl.
4. 6   5. 91.3 m_u and 207.2 m_u

Checkpoint 1C: Nuclear Reactions (I) [p. 19]
3. (a) $^{14}_{7}N$ (b) $^{19}_{10}Ne$ $^{19}_{9}F$
   (c) $^{226}_{88}Ra$ $^{222}_{86}Rn$ (d) $^{73}_{33}As$ $^{73}_{32}Ge$
   (e) $^{27}_{14}Si$ (f) $^{1}_{0}n$

Questions on Chapter 1 [p. 25]
2. $1 = \beta$ particle, $2 = \alpha$ particle, Pb in Gp 4, **X** in Gp 5,
   Y in Gp 3, **Z** in Gp 4
3. 1.88 and 1.94; $^{32}_{15}P \rightarrow {}^{4}_{2}He + {}^{28}_{13}Al$; 1.87
6. 64 and 73; $^{13}C_2{}^{1}H_5{}^{35}Cl$, $^{12}C_2{}^{1}H_5{}^{37}Cl$, $^{12}C_2{}^{1}H_3{}^{2}H_2{}^{35}Cl$,
   $^{12}C^{13}C^{1}H_4{}^{2}H^{35}Cl$
8. (a) $a = 35$, $b = 16$, **X** = S; (b) $c = 4$, $d = 2$, **Y** = He
11. $^{228}_{90}Z$
12. $^{12}C_5{}^{13}C^{1}H_5{}^{37}Cl$ (0.269%), $^{12}C_3{}^{13}C_3{}^{1}H_5{}^{35}Cl$,
   $^{12}C_6{}^{1}H^{1}H_4{}^{37}Cl$, $^{12}C_6{}^{2}H_3{}^{1}H_2{}^{35}Cl$

CHAPTER 2: THE ATOM: THE ARRANGEMENT OF ELECTRONS

Questions on Chapter 2 [p. 45]
3. BCl_3

4. (c) 17% and 19%
5. (c) 240 years

CHAPTER 3: EQUATIONS AND EQUILIBRIA

Checkpoint 3A: The Mole [p. 48]
1. (a) $482 \, g \, mol^{-1}$ (b) $342 \, g \, mol^{-1}$ (c) $368 \, g \, mol^{-1}$
2. (a) $2.50 \times 10^{-3} \, mol$ (b) $5.00 \times 10^{-2} \, mol$
   (c) $5.40 \times 10^{-2} \, mol$
3. (a) 3.0×10^{19} (b) 7.5×10^{16} (c) 3.0×10^{11}

Checkpoint 3B: Formulae and Percentage Composition [p. 50]
1. (a) 72%   (b) 39%   (c) 80%
2. (a) CO_2 (b) C_3O_2 (c) $Na_2S_2O_3$ (d) Na_2SO_4
3. (a) MgO (b) $CaCl_2$ (c) $FeCl_3$
4. $a = 7$, $b = 2$, $c = \frac{1}{2}$

Checkpoint 3C: Masses of Reacting Solids [p. 51]
1. 94 tonnes
2. 25.4%
3. 21.0 g, 93%
4. 2.12 g, 96%

Checkpoint 3D: Reacting Volumes of Gases [p. 52]
1. $2.80 \, dm^3$ at stp
2. 26.8 g
3. 26.1 g NaCl, 43.8 g H_2SO_4

Checkpoint 3E: Titrations [p. 54]
1. 98.9%   2. $n = 10$ 3. 15.0%
4. $4CuSO_4 + 6NaOH \rightarrow Cu_4(OH)_6SO_4 + 3Na_2SO_4$

Checkpoint 3G: Oxidation Numbers [p. 57]
1. (a) $+2$, $+4$, $+4$, $+1$, $+3$, $+5$, $+5$
   (b) $+2$, $+3$, $+4$, $+7$, $+6$
   (c) $+3$, $+3$, $+5$, $+3$
   (d) $+6$, $+6$, $+6$
   (e) -1, $+1$, $+5$, 0, $+3$, $+1$

Checkpoint 3H: Equations and Oxidation Numbers [p. 59]
2. $ICl_3 + 3KI \rightarrow 2I_2 + 3KCl$; $+3$
3. (a) -2, -1, -2, 0
   (b) 0, $+5$, $+2$, $+2$
   (c) $+2$, $+6$, $+3$, $+3$
   (d) $+2$, 0, $+2.5$, -1
   (e) $+5$, -1, -1, -1, $+1$, -1
   (f) $+6$, $+6$

Checkpoint 3I: Redox Titrations [p. 62]
1. 96%   2. $5.50 \times 10^{-2} \, mol \, dm^{-3}$, $30.8 \, cm^3 \, O_2$
3. 95.0%   4. $60.0 \, cm^3$
5. $6Hg + 2KMnO_4 + H_2O \rightarrow 2MnO_2 + 3Hg_2O + 2KOH$

Questions on Chapter 3 [p. 66]
3. $30.0 \, cm^3$ 5. 20.0%   6. 20.1%
7. $S_2O_3{}^{2-} + 4Cl_2 + 5H_2O \rightarrow 2SO_4{}^{2-} + 8Cl^- + 10H^+(aq)$
8. $4Fe^{3+} + 2NH_3OH \rightarrow 4Fe^{2+} + N_2O + H_2O + 6H^+(aq)$

PART 2: PHYSICAL CHEMISTRY

CHAPTER 7: GASES

Checkpoint 7A: Correcting Gas Volumes   [p. 134]
1. (a) $224\,cm^3$ (b) $70.2\,cm^3$ (c) $54.3\,dm^3$
   (d) $3.21\,dm^3$
2. (a) $13.7\,dm^3$ (b) $439\,cm^3$

Checkpoint 7B: Diffusion and Effusion   [p. 136]
1. $8.28\,cm^3$ 2. $1520\,s$ $(25.3\,min)$
3. $44\,g\,mol^{-1}$, C_3H_8

Checkpoint 7C: Reacting Volumes of Gases   [p. 138]
1. C_2H_4 2. NH_3 3. C_2N_2
4. $10\,cm^3\,CH_4 + 15\,cm^3\,C_2H_6$

Checkpoint 7D: Gas Molar Volume   [p. 140]
1. $64\,g\,mol^{-1}$ 2. $71\,g\,mol^{-1}$
3. $4.65 \times 10^{-2}\,mol$ 4. $27.6\,dm^3$

Checkpoint 7E: Partial Pressures   [p. 141]
1. $2.5 \times 10^4\,N\,m^{-2}$
2. (a) $3.03 \times 10^4\,N\,m^{-2}\,CO$, $5.05 \times 10^4\,N\,m^{-2}\,O_2$,
   $2.02 \times 10^4\,N\,m^{-2}\,CO_2$
(b) $3.03 \times 10^4\,N\,m^{-2}\,CO$, $5.05 \times 10^4\,N\,m^{-2}\,O_2$
3. $5.33 \times 10^5\,N\,m^{-2}$
4. (a) $p(NH_3) = 1.96 \times 10^4\,N\,m^{-2}$;

$p(H_2) = 5.39 \times 10^4\,N\,m^{-2}$; $p(N_2) = 2.40 \times 10^4\,N\,m^{-2}$
(b) No change

Checkpoint 7F: Solubility of Gases   [p. 142]
1. $32\%\,O_2$: $68\%\,N_2$ 2. $35\%\,H_2$, $49\%\,O_2$, $16\%\,N_2$

**Checkpoint 7G: The Kinetic Theory and the
Ideal Gas Equation**   [p. 146]
3. $c_{rms}(H_2)/c_{rms}(O_2) = 4.00$
4. $c_{rms}(^{235}UF_6)/c_{rms}(^{238}UF_6) = 1.004$
5. 1.34×10^{22} molecules

Questions on Chapter 7   [p. 150]
1. No, PV is not constant.   3. $780\,cm^3\,dm^{-3}$
6. (b) $3.40\,kJ\,mol^{-1}$
7. $65.3\,g\,mol^{-1}$; NO_2 molecules are dimerised;
   degree of dimerisation = 0.59
8. 1.22×10^{17}
9. $126\,g\,mol^{-1}$; PF_5
10. (a) $x = 8$ (b) NO
11. (c) (i) $M = 102$ (ii) ClO_3F
12. (c) (i) $2.50\,atm$ (ii) $p(A) = 0.25\,atm$, $p(B) = 2.25\,atm$
    (iii) $x_A = 0.1$, $x_B = 0.9$
13. (a) $p_i = 4.99 \times 10^6\,N\,m^{-2}$

CHAPTER 8: LIQUIDS

Checkpoint 8A: Vapour Pressure   [p. 156]
7. $0.013\,g$

Checkpoint 8B: Molar Mass of Volatile Liquids [p. 157]
1. $86\,g\,mol^{-1}$; $84\,g\,mol^{-1}$ 2. $90\,g\,mol^{-1}$; $99\,g\,mol^{-1}$
3. $53\,g\,mol^{-1}$ (experimental); $56\,g\,mol^{-1}$ (correct)
4. $119\,g\,mol^{-1}$

**Checkpoint 8C: Vapour Pressures of Solutions of
Two Liquids**   [p. 162]
1. (a) $42\,kPa$ (b) $31\,kPa$ 2. $36\,kN\,m^{-2}$
3. (a) $21\,kN\,m^{-2}$ (b) 0.29

Checkpoint 8D: Steam Distillation   [p. 165]
1. $156\,g\,mol^{-1}$ 2. 71%

Checkpoint 8E: Partition   [p. 170]
1. B   2. $1.88\,g$ 3. (a) $4.55\,g$ (b) $4.63\,g$
4. $Cu(NH_3)_4^{2+}$

Questions on Chapter 8   [p. 174]
7. $78\%\,CH_3OH$, $22\%\,C_2H_5OH$ 8. (a) $98.5\,°C$
   (b) 23% 10. $3.56\,g$ 11. $66.5\,mol\,dm^{-3}$
13. (b) $36\,mmHg$ (c) $78\,mmHg$

CHAPTER 9: SOLUTIONS

Checkpoint 9B: Vapour Pressure Lowering   [p. 178]
1. $9.19 \times 10^3\,N\,m^{-2}$ 2. $88\,g\,mol^{-1}$

Checkpoint 9C: Boiling Temperature Elevation [p. 180]
1. (a) $W = 60\,g\,mol^{-1}$ (b) $X = 150\,g\,mol^{-1}$
   (c) $Y = 180\,g\,mol^{-1}$ (d) $Z = 62\,g\,mol^{-1}$
2. $120\,g\,mol^{-1}$ 3. S_8

Checkpoint 9D: Freezing Temperature Depression [p. 182]
1. (a) $A = 62\,g\,mol^{-1}$ (b) $B = 58\,g\,mol^{-1}$
   (c) $C = 343\,g\,mol^{-1}$ (d) $D = 84\,g\,mol^{-1}$
2. 10.0% 3. $161\,g\,mol^{-1}$

Checkpoint 9E: Osmotic Pressure   [p. 185]
1. (a) $1.42 \times 10^5\,N\,m^{-2}$ (b) $100.109\,°C$
2. (a) $1.62 \times 10^4\,g\,mol^{-1}$ (b) $6 \times 10^{-5}\,K$
3. $1.60 \times 10^4\,g\,mol^{-1}$

Checkpoint 9F: Colligative Properties   [p. 187]
2. $74\,g\,mol^{-1}$ In benzene, it is almost completely dimerised.
3. $333\,g\,kg^{-1}$; $1.67\,kg$ 4. $2.51 \times 10^4\,g\,mol^{-1}$
6. NaCl; NaCl   7. C_6H_6

Questions on Chapter 9   [p. 190]
3. (a) 122 in water, 243 in benzene
   (b) 156 in water, 156 in benzene
   Benzoic acid dimerises in benzene by intermolecular
   hydrogen bonding; 1-chlorobenzoic acid does not
   because hydrogen bonding is intramolecular.
5. (b) $100.104\,°C$ (c) (ii) NaCl 0.030; H_2O 0.097
6. (c) NaCl   (d) $643\,kg\,mol^{-1}$
7. (b) $M_r = 212$

CHAPTER 10: THERMOCHEMISTRY

Checkpoint 10A: Combustion   **[p. 199]**
1. 803 MJ, 18.8 GJ   2. 983 kJ; $\Delta H_c^{\ominus}$ at 37 °C rather than at 25 °C   3. Propane   4. 12.0 g

Checkpoint 10B: Enthalpy Changes   **[p. 203]**
1. (a) −   (b) −   (c) −   (d) +   (e) −
3. (d) < (e) < (c) < (f) < (b) < (a)

Checkpoint 10C: Standard Enthalpy of Reaction and Average Standard Bond Enthalpies   **[p. 208]**
1. (a) $-76 \, kJ \, mol^{-1}$ (b) $+48 \, kJ \, mol^{-1}$
   (c) $-484 \, kJ \, mol^{-1}$ (d) $-246 \, kJ \, mol^{-1}$
   (e) $-202 \, kJ \, mol^{-1}$
2. $-474, -246, -484, -286$ and $-30 \, kJ \, mol^{-1}$
3. $-604 \, kJ \, mol^{-1}$
4. (a) $-1560 \, kJ \, mol^{-1}$ (b) $-1370 \, kJ \, mol^{-1}$
   (c) $-286 \, kJ \, mol^{-1}$ (d) $-5520 \, kJ \, mol^{-1}$
5. $-372 \, kJ \, mol^{-1}$

Checkpoint 10D: Entropy   **[p. 211]**
*9. $110 \, J \, mol^{-1} \, K^{-1}$ *10. $97 \, J \, mol^{-1} \, K^{-1}$

*Checkpoint 10E: Free Energy   **[p. 214]**
2. 2100 K   3. $\Delta G^{\ominus} = +54.6 \, kJ \, mol^{-1}$ at 300 K — no reaction; $\Delta G^{\ominus} = -14.4 \, kJ \, mol^{-1}$ at 800 K — reaction is feasible.
4. (a) $\Delta G^{\ominus}$ for the reaction
   $2Mg(s) + CO_2(g) \rightarrow 2MgO(s) + C(s)$ is negative.
   (b) It melts.   (c) 1900 K upwards

Questions on Chapter 10   **[p. 216]**
3. $-436 \, kJ \, mol^{-1}$ 4. $\Delta U^{\ominus} = -3267 \, kJ \, mol^{-1}$,
   $\Delta H^{\ominus} = -3271 \, kJ \, mol^{-1}$ 5. (a) $331 \, kJ \, mol^{-1}$
   (b) $3990 \, kJ \, mol^{-1}$ 6. $2346 \, kJ \, mol^{-1}$
7. (b) $-140 \, kJ \, mol^{-1}$
8. (a) $C_6H_6 \, 64 \, kJ \, mol^{-1}$; $C_6H_{12} \, -154 \, kJ \, mol^{-1}$
9. $87 \, kJ \, mol^{-1}$
10. (b) both $-240 \, kJ \, mol^{-1}$ 11. $-57 \, kJ \, mol^{-1}$
12. $103 \, kJ \, mol^{-1}$
*15. (b) Yes, $\Delta G^{\ominus} = -191 \, kJ \, mol^{-1}$
17. (e) $57 \, kJ \, mol^{-1}$ 20. (c) $-76 \, kJ \, mol^{-1}$
19. (b) $\Delta H^{\ominus} = +76 \, kJ \, mol^{-1}$,   (d) $T = 382 \, K$

CHAPTER 11: CHEMICAL EQUILIBRIUM

Checkpoint 11B: Equilibrium Constants   **[p. 227]**
3. 2.5   4. $0.510 \, dm^6 \, mol^{-2}$
5. (a) $0.316 \, atm^{-1/2}$ (b) $1.60 \times 10^4 \, atm^{-1}$ 6. 4.0

Checkpoint 11C: Association and Dissociation   **[p. 230]**
1. 0.43   2. $\alpha = 0.405$, $K_p = 1.98 \times 10^4 \, N \, m^{-2}$
3. $(2 + \alpha) \, mol$, $\alpha = 0.41, 17\%$ 4. 0.75
5. $9.99 \times 10^4 \, N \, m^{-2}$

Questions on Chapter 11   **[p. 234]**
1. (c) $\alpha = 0.30$, $K_p = 4.0 \times 10^4 \, N \, m^{-2}$
2. (a) $3.73 \, atm^{-1}$ 3. (b) $K_p = 4.0$

4. $2.89 \, mol^{-1} \, dm^3$ 5. $8.91 \, mol \, dm^{-3}$
6. (a) Concentration^{-1} (b) No, ratio = 0.33
   (c) $18.0 \, dm^3$
7. (a) No, ratio = 19.6   (b) L → R
9. $-196 \, kJ \, mol^{-1}$
10. $-65.0 \, kJ \, mol^{-1}$ 11. (a) (i) 67.8 kPa   (ii) 1.44
   (b) (ii) $58 \, kJ \, mol^{-1}$
12. (b) (i) $0.51 \, atm^{-1}$ (ii) $p(SO_2) = p(SO_3) = 4 \, atm$;
   $p(O_2) = 2 \, atm$ (c) (ii) $\Delta H^{\ominus} = -198 \, kJ \, mol^{-1}$
14. (a) (i) 7.59 kPa   (ii) 21.4 g   (b) (i) L → R   (ii) L → R
   (iii) R → L

CHAPTER 12: ELECTROCHEMISTRY

Checkpoint 12A: Electrolysis   **[p. 243]**
1. 1.8 g
2. (a) 2.40 g   (b) 7.10 g   (c) 6.35 g   (d) 20.7 g
   (e) 0.200 g
3. (a) Double   (b) No change   (c) No change
   (d) Double
4. $1.23 \times 10^{-3} \, mol$ metal, $2.46 \times 10^{-3} \, mol$ electrons, $+2$

Checkpoint 12B: Conductivity   **[p. 248]**
2. $0.177 \, \Omega^{-1} \, cm^{-1}$ 3. $133 \, \Omega^{-1} \, cm^2 \, mol^{-1}$
4. $3.91 \times 10^{-2} \, \Omega^{-1} \, m^2 \, mol^{-1}$ 5. $1.09 \times 10^{-5} \, mol \, dm^{-3}$

Checkpoint 12C: Acids and Bases   **[p. 253]**
5. (a) $pK_w = 15$ (b) pH = 13

Checkpoint 12D: pH and Dissociation Constants   **[p. 258]**
1. Less   2. CH_3CO_2H is incompletely dissociated
3. HA is the stronger acid.
5. (a) 3.0   (b) 1.6   (c) 4.4   (d) 12.4   (e) 11.8
6. (a) 3.38   (b) 2.38
7. (a) $3.97 \times 10^{-10} \, mol \, dm^{-3}$ (b) $3.02 \times 10^{-11} \, mol \, dm^{-3}$
8. (a) 4.6   (b) 3.0

Checkpoint 12E: Titration   **[p. 265]**
4. (a) 3.00   (b) 3.22   (c) 3.70   (d) 10.3
   (e) 11.0   5. 37% Na_2CO_3

Checkpoint 12F: Buffers   **[p. 268]**
4. $1.79 \times 10^{-5} \, mol \, dm^{-3}$ 5. (a) 4.76   (b) 5.06
6. pH = 4.76; pH = 4.67

Checkpoint 12G: Salt Hydrolysis   **[p. 269]**
2. (a) 8.7   (b) $0.167 \, mol \, dm^{-3}$

Checkpoint 12H Solubility Products   **[p. 272]**
2. $a^2 \, mol^2 \, dm^{-6}$, $4b^3 \, mol^3 \, dm^{-9}$, $27c^4 \, mol^4 \, dm^{-12}$
4. (a) $4.5 \times 10^{-3} \, mol \, dm^{-3}$ (b) $2.0 \times 10^{-4} \, mol \, dm^{-3}$
   (c) $1.0 \times 10^{-4} \, mol \, dm^{-3}$

Checkpoint 12I: Complex Ions   **[p. 273]**
*2. $3.9 \times 10^{-3} \, mol \, dm^{-3}$ 3. $1.1 \times 10^{-3} \, mol \, dm^{-3}$

Questions on Chapter 12   **[p. 273]**
4. $1.40 \times 10^{-2} \, \Omega^{-1} \, m^2 \, mol^{-1}$ 6. 287 mins
7. (c) $5.19 \times 10^{-2} \, \Omega^{-1} \, m^{-1}$ 11. $56 \, cm^3$
12. (a) 0.11 A   (b) $6.35 \times 10^{-2} \, g$ (c) $11.2 \, cm^3$
   (d) 0.0347 g
13. $292 \, cm^3$ 17. (a) $350 \, \Omega^{-1} \, cm^2 \, mol^{-1}$
   (c) $126.4 \, \Omega^{-1} \, cm^2 \, mol^{-1}$
18. $K_{sp}(CaSO_4) = 6.41 \times 10^{-5} \, mol^2 \, dm^{-6}$;
   $[CaSO_4] = 6.41 \times 10^{-4} \, mol \, dm^{-3}$
   $K_{sp}(AgCl) = 1.55 \times 10^{-10} \, mol^2 \, dm^{-6}$;
   $[AgCl] = 7.78 \times 10^{-10} \, mol \, dm^{-3}$
   pH = 3.02
19. (c) (i) 2.0   (ii) 3.4   (iii) 5.8
20. (c) $1.84 \times 10^{-5} \, mol \, dm^{-3}$
21. (c) (ii) $4.0 \times 10^{-11} \, mol \, dm^{-3}$
22. (b) $1.61 \times 10^{-5} \, mol^3 \, dm^{-9}$
   (c) (i) $4.03 \times 10^{-6} \, mol \, dm^{-3}$ (ii) $2.00 \times 10^{-3} \, mol \, dm^{-3}$

════════════ **CHAPTER 13: OXIDATION–REDUCTION EQUILIBRIA** ════════════

Checkpoint 13A: Electrode Potentials   [p. 282]
3. (a) $-0.46\,V$ 5. (b), (e) and (h)

Questions on Chapter 13   [p. 288]
5. $-0.27\,V$ 6. (a) $-0.10\,V$ (b) $+0.26\,V$
    (c) $-0.27\,V$ (d) $+0.94\,V$ (e) $-0.46\,V$

(f) $+0.78\,V$ (g) $+0.63\,V$ 7. (c) $+0.36\,V$
8. (c) $1.0 \times 10^{-5}\,mol\,dm^{-3}$
    (d) (i) $1.0 \times 10^{-10}\,mol^2\,dm^{-6}$
9. (d) (i) $-0.76\,V$ (g) $-1.10\,V$
10. (a) $+1.617\,V$
    (b) $2Zn(s) + 4OH^-(aq) + O_2(g) \rightarrow 2ZnO_2^{2-}(aq) + 2H_2O(l)$

════════════ **CHAPTER 14: REACTION KINETICS** ════════════

Checkpoint 14A: Initial Rates   [p. 299]
3. (a) Rate $= k[A]^2$ (b) $k = 5.0\,dm^3\,mol^{-1}\,s^{-1}$
    (c) $1.8\,mol\,dm^{-3}\,s^{-1}$
4. (a) 1 wrt **A**; 2 wrt **B**; 3
    (b) $k = 1.5 \times 10^{-3}\,mol^{-2}\,dm^6\,s^{-1}$
    (c) $8.7 \times 10^{-6}\,mol\,dm^{-3}\,s^{-1}$

Checkpoint 14B: First-Order Reactions   [p. 302]
1. $8047\,s$ $(134\,min)$ 2. 6.25% 3. (a) 1
    (b) $6.3 \times 10^{-4}\,mol\,dm^{-3}\,s^{-1}$ (c) $7.9 \times 10^{-4}\,s^{-1}$
4. $25.5\,min$ 5. (b) $14.6\,cm^3$ (c) $10.3\,min$ $(618\,s)$
    (d) $t_{1/2}$ is the same for different concentrations
    (e) $0.0673\,min^{-1}$ $(1.12 \times 10^{-3}\,s^{-1})$
    (f) $1.06 \times 10^{-6}\,mol\,dm^{-3}\,s^{-1}$ (g) $9.6 \times 10^{-4}\,s^{-1}$
    (h) $1.19 \times 10^{-3}\,s^{-1}$

Checkpoint 14C: Order of Reaction   [p. 305]
1. (a) $1.7 \times 10^{-4}\,mol\,dm^{-3}\,s^{-1}$
    (b) $5.1 \times 10^{-4}\,mol\,dm^{-3}\,s^{-1}$
    (c) $15.3 \times 10^{-4}\,mol\,dm^{-3}\,s^{-1}$
2. 1st order; $k = 1.75 \times 10^{-3}\,s^{-1}$
3. (a) $k = 0.125\,min^{-1}$ (b) $320\,min$ (c) $100\,min$
4. (a) $0.0128\,min^{-1}$ (b) $126\,min$ 5. Order $= 1$,
    $k = 2.44 \times 10^{-4}\,s^{-1}$

Checkpoint 14D: Reaction Kinetics   [p. 315]
6. Ratio $= 1.2$ *9. (a) $52.9\,kJ\,mol^{-1}$
    (b) $83.8\,kJ\,mol^{-1}$ *10. $166\,kJ\,mol^{-1}$

Questions on Chapter 14   [p. 317]
5. $50\,h$ 6. 3700 years
7. $\Delta H^{\ominus}_{Reaction} = -19\,kJ\,mol^{-1}$
8. (a) 1   (b) 1   (c) $6.0 \times 10^{-3}\,dm^3\,mol^{-1}\,s^{-1}$;
    $1.08 \times 10^{-5}\,mol\,dm^{-3}\,s^{-1}$
10. Reaction 1: 1st order, $t_{1/2} = 6.7\,min$, $k = 0.100\,min^{-1}$
    Reaction 2: zero order, $t_{1/2} = 5.0\,min$,
    $k = 0.100\,mol\,dm^{-3}\,min^{-1}$
    Reaction 3: 2nd order, $t_{1/2} = 10.0\,min$,
    $k = 0.100\,dm^3\,mol^{-1}\,min^{-1}$
18. $E = 107\,kJ\,mol^{-1}$, $A = 9.1 \times 10^{13}\,s^{-1}$
20. Order $= 1$, $k = 6.0 \times 10^{-4}\,s^{-1}$
21. Rate $= kp_{NO}^2 p_{H_2}$, $k = 3.10 \times 10^{-8}\,mm^{-2}\,s^{-1}$
23. (c) $45.9\,kJ\,mol^{-1}$

**Appendix 1: A Selection of Questions on
Physical Chemistry**   [p. 322]
1. **A** $= {}^2H_2$, **B** $=$ He   5. (a) 1.90×10^{11} atoms
7. (d) (ii) $4.33 \times 10^{-4}\,mol^3\,dm^{-9}$

PART 3: INORGANIC CHEMISTRY

════════════ **CHAPTER 17: HYDROGEN** ════════════

Questions on Chapter 17   [p. 351]
7. (b) (i) $436\,kJ\,mol^{-1}$ (ii) 0   (iii) $592\,kJ\,mol^{-1}$

════════════ **CHAPTER 19: GROUP 3B** ════════════

Questions on Chapter 19   [p. 384]   **Checkpoint 19C: Aluminium**   [p. 386]
7. $KAlSO_4 \cdot 12H_2O$ 8. $M_r = 267$ 4. $10.7\,MC$

════════════ **CHAPTER 21: GROUP 6B** ════════════

Checkpoint 21B: Bonding   [p. 415]   **Questions on Chapter 21**   [p. 429]
2. (a) $+950\,kJ\,mol^{-1}$ (b) $+555\,kJ\,mol^{-1}$ 8. (a) (i) SF_4
3. (a) $+388\,kJ\,mol^{-1}$ (b) $+856\,kJ\,mol^{-1}$

════════════ **CHAPTER 22: GROUP 5B** ════════════

Questions on Chapter 22   [p. 452]
14. Urea

━━━━━━━━━━━━━━ CHAPTER 23: GROUP 4B ━━━━━━━━━━━━━━

Checkpoint 23B: Carbon [p. 462]
2. (a) $-216\,kJ\,mol^{-1}$ (b) $+46\,kJ\,mol^{-1}$

━━━━━━━━━━━━━━ CHAPTER 24: THE TRANSITION METALS ━━━━━━━━━━━━━━

Questions on Chapter 24 [p. 514]
14. $0.036\,mol\,dm^{-3}$ 15. Cr^{2+}
25. $[Co(NH_3)_4Cl_2]^+ Cl^-$
27. $[Cr_2O_7{}^{2-}] = 0.417\,mol\,dm^{-3}$; $[Cr^{3+}] = 0.667\,mol\,dm^{-3}$
28. (c) (i) $102\,g\,dm^{-3}$ (ii) $336\,cm^3$

Appendix 2: Topics which Span Groups of the Periodic Table
A2.1: Some Detective Work [p. 519]
1. (a) $A = CO$ (b) $B = O_3$ (c) $C = PH_3$
2. (a) $A = Cr$ (b) $B = Pb_3O_4$ (c) $C = P$
   (d) $D = FeCl_2$
3. $E = HF$, $F = H_2O$, $G = H_2O_2$, $H = CrOCl_2$
4. $J = K_2CrO_4$, $L = KI$, $M = SnCl_2$, $K = MnO_2$

5. $P = (NH_4)_2Cr_2O_7$, $Q = NaNO_2$, $R = K_2SO_3$,
   $S = HCO_2Na$, $T = KCN$, $U = CrCl_2 \cdot nH_2O$
6. $A = MnO_2$, $B = K_2MnO_4$, $C = KMnO_4$,
   $D = Mn^{2+}(aq)$, $E = MnS$
7. $P = FeCl_4{}^-(aq)$, $Q = Fe(OH)_3$, $R = Fe(NO_3)_2$,
   $S = Fe(OH)_2$, $T = FeCO_3$
8. $K = a\ Co^{2+}$ salt, $L = a\ Cu^{2+}$ salt, $M = NaNO_2$,
   $N = NO_2$, $O = N_2$, $P = C_2H_5OH$, $Q = Al$, $R = H_2$
9. $A = Fe(NO_3)_3$, $B = Fe(NO_3)_3(aq)$, $C = FeCl_4{}^-(aq)$,
   $D = Fe(OH)_3$, $E = Fe(OH)_2$, $F = FeSO_4$,
   $G = K_4Fe(CN)_6$, $H = $ Turnbull's blue, Red

PART 4: ORGANIC CHEMISTRY

━━━━━━━━━━━━━━ CHAPTER 26: THE ALKANES ━━━━━━━━━━━━━━

Questions on Chapter 26 [p. 550]
10. $6.4\,kg$
11. (a) (i) [1] $+242$ [2] -19 [3] $-107 = 116\,kJ\,mol^{-1}$

(ii) [1] $+193$ [2] $+46$ [3] $-100 = 139\,kJ\,mol^{-1}$

━━━━━━━━━━━━━━ CHAPTER 27: ALKENES AND ALKYNES ━━━━━━━━━━━━━━

Checkpoint 27A: Alkenes [p. 565]
9. 64

Questions on Chapter 27 [p. 572]
5. 56, C_4H_8
6. (a) (i) C_5H_{10}

━━━━━━━━━━━━━━ CHAPTER 31: ALDEHYDES AND KETONES ━━━━━━━━━━━━━━

Questions on Chapter 31 [p. 667]
7. $C_6H_5COCH_3$, $C_6H_5CH_2CHO$, $CH_3C_6H_4CHO$
8. $P = $ cyclohexene, $Q = $ hexan-1,6-dial
9. $A = C_6H_5CHO$
11. $A = (CH_3)_2CHCH(OH)CH_3$
12. $P = CH_3COCH_2CH_2CHO$

13. $Q = CH_3CH(OH)CH_2CH_2CH_2OH$
17. $X = (CH_3)_3CCHO$ $Y = (CH_3)_3CCO_2H$
   $Z = (CH_3)_3CCH_2OH$ 19. (d) $+23\,kJ\,mol^{-1}$
   (e) $\Delta H^{\ominus}_{Vaporisation}$ for CH_3CHO, H_2O, $CH_3CH(OH)_2$
   and $\Delta H^{\ominus}_{Dissolution}$ for $CH_3CH(OH)_2$

━━━━━━━━━━━━━━ CHAPTER 32: AMINES ━━━━━━━━━━━━━━

Questions on Chapter 32 [p. 686]
8. $A = (CH_3)_2CHCN$, $B = (CH_3)_2CHCH_2NH_2$,
   $C = (CH_3)_2CHCH_2NHCOCH_3$,

$D = (CH_3)_2CHCH_2NHCH_3$,
$E = (CH_3)_2CHCH_2\overset{+}{N}(CH_3)_3I^-$, $F = (CH_3)_2CHCO_2H$,
$G = (CH_3)_2CHCH_2OH$, $H = (CH_3)_2C{=}CH_2$

CHAPTER 33: ORGANIC ACIDS AND THEIR DERIVATIVES

Checkpoint 33A: Acids   [p. 696]
8. $A = C_6H_5CHO$, $B = C_6H_5CH_2OH$, $C = C_6H_5CO_2H$

Checkpoint 33C: Acid Chlorides and Anhydrides   [p. 704]
4. 0.196 g

Checkpoint 33D: Esters   [p. 709]
4. 10.6 g   5. 88 g mol^{-1}, $C_4H_8O_2$

Checkpoint 33E: Amides and Nitriles   [p. 713]
5. $A = C_2H_5COCl$ 6. $A = C_6H_5CHO$
   $B = C_2H_5CONH_2$ $B = C_6H_5CH_2OH$
   $C = C_2H_5OH$ $C = C_6H_5COCl$
   $D = CH_3CO_2C_2H_5$ $D = C_6H_5COCH_2C_6H_5$
   $E = C_2H_5CO_2CH_3$ $E = C_6H_5CONH_2$
   $x = Br_2 + CHCl_3 + KOH$ $F = C_6H_5NH_2$
   $y = HNO_2$ $G = C_6H_5NHCOC_6H_5$
   $z = CH_3OH$

Checkpoint 33F: Amino Acids and Proteins   [p. 717]
5. 166 g mol^{-1} 6. 0.69 g

Questions on Chapter 33   [p. 719]
2. $A = CH_3CH_2X$, $B = CH_3CH_2OH$,
   $C = CH_3CH(OH)CN$, $D = CH_3CH_2CN$,
   $E = ClCH_2CO_2H$
5. $C = C_6H_5COCl$
9. A

          $B = 1,2\text{-}HO_2CC_6H_4CONH_2$
          $C = 1,2\text{-}HO_2CC_6H_4NH_2$
          $D = 1,2\text{-}HO_2CC_6H_4N_2Cl^-$
          $E = 1,2\text{-}HO_2CC_6H_4CN$
          $F = 1,2\text{-}HO_2CC_6H_4CO_2H$
13. $A = CH_3CO_2CH(CH_3)C_2H_5$,
    $B = C_2H_5CH(CH_3)CO_2CH_3$
15. $C_2H_5CH(CH_3)CO_2CH(CH_3)_2$
20. $C_6H_5CO_2C_2H_5$

CHAPTER 34: THE IDENTIFICATION OF ORGANIC COMPOUNDS

Checkpoint 34A: Analysis   [p. 727]
1. (a) CH_2 (b) 84 g mol^{-1} (c) C_6H_{12}
2. $C_6H_5SO_2Cl$ 3. C_3H_7NO 4. C_4H_{10}

Questions on Chapter 34   [p. 732]
1. (a) CHO (b) $C_4H_4O_4$, $HO_2CCH{=}CHCO_2H$ (cis and
   trans isomers)   3. 56 g mol^{-1}, C_4H_8

CHAPTER 35: SOME GENERAL TOPICS

Checkpoint 35F: Draw Your Own Conclusions   [p. 749]
1. $A = CH_3CH_2CH_2CO_2CH_2CH_2CH_3$,
   $B = CH_3CH(OH)CH_3$, $C = CH_3CH_2CH_2CONH_2$,
   $D = CH_3CH_2CH_2NH_2$, $E = CH_3CH_2CH_2\overset{+}{N}H_3Cl^-$,
   $F = CH_3CH_2CH_2OH$, $G = CH_3CH_2CHO$,
   $H = CH_3COCH_3$
2. $A = C_6H_5CO_2C_2H_5$, $B = C_2H_5OH$, $C = C_6H_5CO_2Na$,
   $D = C_2H_5Br$, $E = C_6H_5CO_2H$, $F = C_6H_6$
3. A and B are cis and trans $C_6H_5CH{=}CHCO_2H$,
   $C = C_6H_5CH_2CH_2CO_2H$, $D = C_6H_5CH_2CH_2COCl$
4. $P = H_2NCH_2CO_2H$, $Q = CH_3NH_2$,
   $R = HOCH_2CO_2H$
5. $A = CH_3CO_2Na$, $B = (CH_3CO)_2O$, $C = C_6H_5NH_2$,
   $D = C_6H_5\overset{+}{N}H_3Cl^-$, $E = 2,4,6\text{-}Br_3C_6H_2NH_2$,
   $F = C_6H_5OH$
6. $W = (CH_3)_2CHBr$, $X = NaBr$, $Y = (CH_3)_2CHOH$,
   $Z = (CH_3)_2CO$
7. $A = CH_3C_6H_4NO_2$, $B = CH_3C_6H_4NH_2$,
   $C = CH_3C_6H_4OH$, $D = CH_3C_6H_4OCH_3$,
   $E = HO_2CC_6H_4OCH_3$, $F = HO_2CC_6H_4OH$,
   $G = C_6H_5OH$

8. $A = C_2H_5\overset{+}{N}H_3Cl^-$, $B = C_2H_5NH_2$, $C = C_2H_5NC$
9. $P = CH{\equiv}CH$, $Q = CHCl{=}CHCl$, $R = Cu_2C_2$
10. $X = CH_3CHO$, $Y = CH_3CH(OH)SO_3Na$,
   $Z = (CH_3CHO)_3$
11. $H = CH_2OH$ $I = CHO$ $J = CO_2H$
        | | |
       CO_2H CO_2H CO_2H
12. $B = 1,2\text{-}CH_3C_6H_4CO_2Na$, $C = 1,2\text{-}CH_3C_6H_4CO_2H$,
   $D = 1,2\text{-}ClCH_2C_6H_4CO_2H$,
   $E = 1,2\text{-}HOCH_2C_6H_4CO_2H$,
   $F = 1,2\text{-}OHCC_6H_4CO_2H$, $G = 1,2\text{-}HO_2CC_6H_4CO_2H$
   $H =$ $J =$

13. (a) $P = C_6H_5COCH_3$, $Q = C_6H_5CO_2H$,
   $R = C_6H_5COCl$, $S = C_6H_5CONH_2$, $T = C_6H_5CN$,
   $U = C_6H_5CH_2NH_2$, $V = C_6H_5CH(OH)CH_3$

Index

INDEX OF SYMBOLS AND ABBREVIATIONS

The page listed is that on which the symbol or abbreviation is first used or explained.

Greek letters:
α = degree of association 186
α = degree of dissociation 186
α = degree of ionisation 254
α = position of group 658
α = change in volume 149
κ = electrolytic conductivity 244

λ = wavelength 7
Λ = molar conductivity 245
Λ_o = molar conductivity at infinite dilution 246
v = frequency 28
π = osmotic pressure 183
ρ = density 134
ρ = resistivity 244